1st International Conference on Petroleum Technology and Petrochemicals 2019

IOP Conference Series: Materials Science and Engineering
Volume 579

Baghdad, Iraq
10 – 11 April 2019

ISBN: 978-1-5108-9450-1
ISSN: 1757-8981

Printed from e-media with permission by:

Curran Associates, Inc.
57 Morehouse Lane
Red Hook, NY 12571

Some format issues inherent in the e-media version may also appear in this print version.

Institute of Physics
Dirac House, Temple Back
Bristol BS1 6BE UK

Phone: 44 1 17 929 7481
Fax: 44 1 17 920 0979

techtracking@iop.org

Additional copies of this publication are available from:

Curran Associates, Inc.
57 Morehouse Lane
Red Hook, NY 12571 USA
Phone: 845-758-0400
Fax: 845-758-2633
Email: curran@proceedings.com
Web: www.proceedings.com

Table of contents

Volume 579

1st International Conference on Petroleum Technology and Petrochemicals

10–11 April 2019, University of Technology Campus, Baghdad, Republic of Iraq

Accepted papers received: 26 June 2019
Published online: 16 August 2019

Preface

Papers

IOP Conf. Series: Materials Science and Engineering **579** (2019) 011001 doi:10.1088/1757-899X/579/1/011001

PTP 2019

1st International Conference on Petroleum Technology and Petrochemicals -2019

10-11 April 2019

Organized by

Petroleum Technology Department

University of Technology

Baghdad - Republic of Iraq

IOP Conf. Series: Materials Science and Engineering **579** (2019) 011001 doi:10.1088/1757-899X/579/1/011001

1st International Conference on Petroleum Technology and Petrochemicals -2019

10-11 April 2019

Petroleum Technology Department /University of Technology

Preface:

Petroleum and Petrochemical Engineering is a field of engineering concerned with the activities related to the production of hydrocarbons, which can be either crude oil or natural gas. Exploration and Production are deemed to fall within the upstream sector of the oil and gas industry. Exploration, by earth scientists, and petroleum engineering are the oil and gas industry's two main subsurface disciplines, which focus on maximizing economic recovery of hydrocarbons from subsurface reservoirs. Petroleum geology and geophysics focus on provision of a static description of the hydrocarbon reservoir rock, while petroleum engineering focuses on estimation of the recoverable volume of this resource using a detailed understanding of the physical behavior of oil, water and gas within porous rock at very high pressure.

Introduction:

The Petroleum Technology Department at University of Technology, Baghdad invite all the researchers to attend the "1st International Conference on Petroleum Technology and Petrochemical" (PTP2019) scheduled on 10-11 April, 2019, at University of Technology, Baghdad, Iraq. This conference directs towards addressing main issues as well as future strategies of global energy industry. This is going to be the most promising international conference where petroleum and petrochemical engineering (e.g. oil and gas engineering) professionals as well as decision makers will come to discuss and debate on various aspects of the challenges, risks and investment opportunities throughout the complete integrated energy and utilities supply chain.

IOP Conf. Series: Materials Science and Engineering **579** (2019) 011001 doi:10.1088/1757-899X/579/1/011001

1st International Conference on Petroleum Technology and Petrochemicals -2019
10-11 April 2019

Petroleum Technology Department /University of Technology

Target Participants
- Petroleum Engineers.
- Scientists, Researchers & Technical Staff from petroleum engineering field.
- Professors, Students & Delegates from petroleum engineering Department.
- The people who are specialized with geologists, geophysicists, reservoir engineers, production engineers, Chemical Engineers.
- Production engineers are the Potential Exhibitors for the Respective Conferences.
- Local and International Oil Companies.

Conference Topics:

1. Reservoirs Engineering.
2. Drilling Engineering.
3. Petroleum Production Engineering.
4. New Techniques of Petroleum Exploration & Investigation.
5. Oil transportation&Storage Management
6. Engineering Structures and Foundations in Oil Projects
7. Petroleum Geology.
8. Petrochemicals and Natural Gas.
9. Environmental Pollution and Treatment Technologies in Petroleum Processing.
10. Artificial Intelligence and Neural networks.
11. Sustainable Developments in Petroleum Industry.

IOP Conf. Series: Materials Science and Engineering **579** (2019) 011001 doi:10.1088/1757-899X/579/1/011001

1st International Conference on Petroleum Technology and Petrochemicals -2019

10-11 April 2019

Petroleum Technology Department /University of Technology

Conference logo:

Sponsors logos:

IOP Conf. Series: Materials Science and Engineering **579** (2019) 011001 doi:10.1088/1757-899X/579/1/011001

1st International Conference on Petroleum Technology and Petrochemicals -2019

10-11 April 2019

Petroleum Technology Department /University of Technology

Editorial Board

IOP Conf. Series: Materials Science and Engineering **579** (2019) 011001 doi:10.1088/1757-899X/579/1/011001

1st International Conference on Petroleum Technology and Petrochemicals -2019
10-11 April 2019

Petroleum Technology Department /University of Technology

IOP Conf. Series: Materials Science and Engineering **579** (2019) 011001 doi:10.1088/1757-899X/579/1/011001

1st International Conference on Petroleum Technology and Petrochemicals -2019
10-11 April 2019

PTP 2019

Petroleum Technology Department /University of Technology

Photographs:

IOP Conf. Series: Materials Science and Engineering **579** (2019) 011001 doi:10.1088/1757-899X/579/1/011001

1st International Conference on Petroleum Technology and Petrochemicals -2019

10-11 April 2019

Petroleum Technology Department /University of Technology

IOP Conf. Series: Materials Science and Engineering **579** (2019) 011001 doi:10.1088/1757-899X/579/1/011001

1st International Conference on Petroleum Technology and Petrochemicals -2019

10-11 April 2019

PTP 2019

Petroleum Technology Department /University of Technology

IOP Conf. Series: Materials Science and Engineering **579** (2019) 011001 doi:10.1088/1757-899X/579/1/011001

1st International Conference on Petroleum Technology and Petrochemicals -2019

10-11 April 2019

Petroleum Technology Department /University of Technology

IOP Conf. Series: Materials Science and Engineering **579** (2019) 011001 doi:10.1088/1757-899X/579/1/011001

1st International Conference on Petroleum Technology and Petrochemicals -2019

10-11 April 2019

Petroleum Technology Department /University of Technology

Peer review statement

All papers published in this volume of *IOP Conference Series: Materials Science and Engineering* have been peer reviewed through processes administered by the proceedings Editors. Reviews were conducted by expert referees to the professional and scientific standards expected of a proceedings journal published by IOP Publishing.

IOP Conf. Series: Materials Science and Engineering **579** (2019) 012001 doi:10.1088/1757-899X/579/1/012001

Performance Simulation for Gas Flow through a Porous Media in Packed bed Columns using CFD

Asaad Salim Bded

Petroleum Engineering Department, Kerbala University, Karbala, Iraq

Corresponding email: asaadsalim78@gmail.com

Abstract. Two dimensions of Computational Fluid Dynamics simulation for gas flow in the column with filling material as bed at low particle Reynolds numbers and high particle Reynolds numbers have been presented. The simulation test has been done with the column length 0.9 m. Porous media model was used to simulate the gas flow inside porous granular media. Model of standard k-e for the flow turbulent range was used; the difference on the mesh size used to study effect it on the performance of gas flow through porous media. The result of flow rate of the gas was an important variable influencing bed and there are good results between the CFD simulation of pressure drop and the predicted values from Ergun model, and Computational Fluid Dynamics gave us simulation with high details for porous flow in packed bed equipment.

1. Introduction

The flow through column filled with granular particles is of frequent occurrence in petroleum industry and chemical process, therefore an expression are needed to predict the pressure drop within material of beds due to the resistance caused by the filling material inside of column [1]. The flow at low velocity and high velocity through the spaces between the particles have been retained by porous medium to allow us obtaining the influence of some of parameters on the performance of flow inside bed columns. Therefore this study carried out a computational fluid dynamics (CFD) based on the standard k-e model investigation of flow through porous media ranging from laminar to turbulent flow. CFD is relatively lower in cost than the physical testing, ability to simulate real conditions, many flows and transfer processes cannot be easily tested [2]. By using a Computational Fluid Dynamics (CFD) the studies were extended to treat various geometries with different boundary conditions, more applications for flow studies in porous media have produced extensive basis information in published literature, of which R.S.Maier & R.S. Bernard.[3]used the lattice Boltzmann method to the velocity distribution under the spatial resolution in packing without gives the details of the packing to simulate viscous fluid flow through a column of glass beads.

J. Gunnar I. Hellstr, & P. Jonsson [4] analysis of a quadratic array of cylinders shown that a number of different Reynolds number values can be defined for porous media, CFD simulations for flow through a quadratic array of cylinders at three solid fractions with a turbulent, found that turbulence flow needs be considered when Rep is above 300, since the laminar setup of equations failed to predict the experimental results over this value F.Augier&J.Y.Delenne[5] verified the transport and transfer properties inside packed beds of particles by CFD simulations at low to moderate Reynolds numbers. They found a problem of contacting points between particles, which is inherent to finite-volume numerical methods and solved by applying a shrinking of 2% along of material of the bed when transfer properties been estimated of increasing in arranging packing complexity. Kamyar M. & Ali A.&

Eckehard S.[6] used the porous media model as a method for simulation to ability the prediction of CFD using as a tools design for such packed columns to flow in the reaction media, The results of computational fluid dynamics simulation was excepted with results of the experimental in all cases studies.
The present research uses the software ANSYS-Fluent to develop the porous media in column with filling materials as bed to simulate of performance the flow of gas from the laminar to turbulent gas flow through the filling materials (porous media) within the range of Rep (7.87 – 1115.87) respectively, throughout change in axial temperature distributions along materials column.

2. Material Properties

The details of system are a gas phase without change in its density and uniform of Glass balls size. The properties, both of the (air) and material (Glass balls), are tabulated in Tables (1) to (3). Based on an experimental setup [5] and the thermal properties of glass balls are taken from Peter, S.T. [8], and the gas properties based on the ANSYS database.

Table 1. Properties of gas phase at the temp. 293 K

Parameter / unit	value
(air) density (kg/ m^3)	1.225
(air) viscosity (kg/ m. s)	1.7894
Heat capacity (j/kg. k)	1006.43
Coefficient of thermal conductivity (w/m. k)	0.0242

Table 2. the Physical properties of column (Steel) from ANSYS database

Parameter	value
density (kg/ m^3)	8030
specific heat (j/ kg. k)	502.48
heat conductivity coefficient (w/ m. k)	16.27

Table 3. Material properties of simulated sphere [7]

Parameter	value
Particle diameter (m)	0.005
Density (kg/ m^3)	2590
Heat capacity (j/ k_g.k)	780
Heat conductivity coefficient (W/m.k)	0.9656

3. Models of Porous Media

Forchheimer suggested the flow with in porous medium and described his suggestion when the flow and the pressure gradient have a linear relationship (linear Darcy's law) as shown in below form [9]:

$$\frac{\Delta p}{L} = \frac{\mu u}{k} + \rho\, k_F u^2 = \alpha u + \beta u^2 \qquad (1)$$

Where k_F is the inertial resistance (Forchheimer coefficient) at low Reynolds number, the equation (1) considered (Darcy law) that the viscous forces be controlled in porous media and Δp is proportional to the Darcy velocity (inertial effects can be neglected), while as the Reynolds number increases, the inertial forces are considered and equation (1) gives non-linear relationship between Δp and Darcy velocity, and at high Reynolds number, the inertial forces became strong, Δp is proportional to the squared velocity [10]

In the ANSYS-FLUENT the superficial phase velocities can be calculated depending on volumetric flow rate in a porous media of the regime [11, 16]. Porous Media Model for homogeneous medium consists of two portions when the momentum term is added; inertial loss and viscous resistances portions. Will be mixed with other factors of Darcy's equation to predict the pressure drop through the porous zone inside the gas cell zone (fluid volume) in the axial direction, this is called a Porous Media Model by cell zone conditions [11].

The concept of permeability was developed by Kozeny and Carman the following equation (2). It depends on the porosity and glass sphere diameter and medium's permeability also depends on physical and geometric model characteristics.

$$K = \frac{\rho g}{5\mu} \frac{\varepsilon^3}{(1-\varepsilon)^2} \left(\frac{d}{6}\right)^2 \qquad (2)$$

Where ε is the porosity of medium's, It is calculated from equation (3) Furnas [12] is proposed an equation for the porosity in bed column as function of particle and bed diameter, as in the following equation: $\varepsilon = 0.375 + 0.34\,\frac{d}{D}$ (3)

The coefficient β is the inertial resistance in the porous media along the flow direction per unit length at high velocities [10]. Values of equation (1) are obtained from physical regime by solution of Navier-Stokes equation relating to velocity and pressure subject to corresponding boundary condition in CFD simulation. In this paper, air was passed through the bed to simulate the flow and to estimate the pressure drop depending on properties of the fluid, and porosity of medium. Ergun equation [13], is defined as follows:

$$\frac{\Delta p}{L} = 150\,\mu\,\frac{(1-\varepsilon)^2}{\varepsilon^3\ d^2}\,u + 1.75\rho\,\frac{(1-\varepsilon)}{\varepsilon^3\ d\phi}\,u^2 \qquad (4)$$

The pressure drop predicted from CFD simulation of bed columns has been compared with values from equation (4).

4. Modelling Fluid Flow

The basis for all CFD fluid flow modeling is the Navier-Stokes equations. The term Navier-Stokes equations are used to describe three equations; the momentum equation, the continuity equation, and the energy equation.

5. Navier- Stokes Equation

The Navier Stokes equation is the equation of motion for a Newtonian fluid with constant viscosity and density. The equation is greatly simplified when applied to 2D flow with the assumption that velocity is only in the axial direction [11 14 17].

6. Standard k-e Turbulence Model

IOP Conf. Series: Materials Science and Engineering **579** (2019) 012001 doi:10.1088/1757-899X/579/1/012001

At high Reynolds numbers, the viscous dissipation of kinetic energy (ε) will rise because of high inertial forces, multiplied by the fluctuating vorticity and from Navier-Stokes equation it can be derived the viscous dissipation at the exact transport fluctuating vorticity equation [10 14 17].

7. Preparation of media and design of column

The first step in CFD simulation of bed columns is Pre-processing, which has been done by ANSYS Workbench 12.0 tool to form the geometric model with (hexahedral grids) of numerical density: dividing 20 interval counts and 40000 quadrilateral cells (40k), 1600 2D wall faces, 40851 nodes, (40k) computational cells as shown in Fig.(2).The geometric model was designed as two dimension (2D geometry) on the (x, y) coordinates, with 0.041m in a diameter of column, 0.9 m length with four surface boundaries: inlet (for inflow, t, u), outlet (for outflow, t, P,.etc), wall and porous medium. The bed materials in this model of glass balls are packed into the column randomly as porous media. The model assumption is shown in Fig. (1).

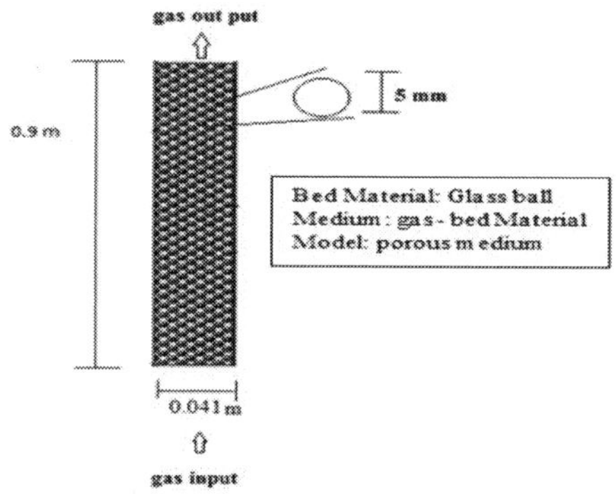

Figure 1. 2D schematic of the Bed column

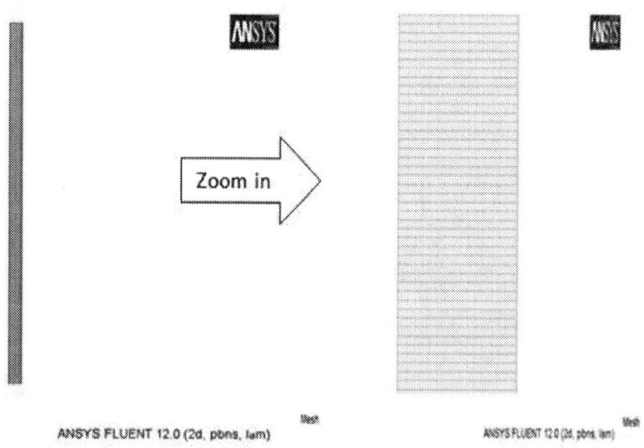

Figure 2. 2D hexahedral meshing bed column.

8. Computational Fluid Dynamics (CFD) model using (ANSYS- FLUENT)

The method of (CFD) used to simulate gas flow in the assumed bed column by using the ANSYS-FLUENT, the pre-processor, solver and post-processor are the basic parts of the ANSYS-FLUENT software to analyse the process. The pre-processor step is usually carried out through a graphical interface and describe in section 3. The velocity of gas, and the data of the porous cell condition, density of the gas and all coefficient values are entered to the solver software (user input values), to solve the numerical equations based on the entered values and the results are obtained to provide the post-processor step software [15]. In this work the single phase flow is initially in static condition inside the bed column and the gas passes through bed from bottom of the column by velocity inlet at the central of porous cell zone that is specified for the porous medium conditions, there are options related to inertial loss and viscous resistances and axial direction vectors to be solved , void fraction has been calculated from Equation (3), and the permeability has been calculated from Equation (2) for each of the space coordinates. Flow Analysis by FLUENT through bed column using 2D model is solved in steady state laminar and turbulent flow simulation. all models that described previously are available in FLUENT software [15]. Select physical velocity under porous formulation, will take the volume blockage into consideration and provides a more accurate representation of flow within bed column. The boundary condition for laminar model and turbulent model flow will range from 0.023 to 3.23 as inlet velocity and is used for inlet temperature equals 293°k (isothermal boundary condition), column wall temperature be constant and equals to 373°k (heating wall), boundary condition for porous zone are estimated in section 2. After the simulation is iterated, the final steps are post-processing and displaying the output in contours or in vectors, the data that is obtained by the solver can be visualized and displayed such as static pressure, velocity magnitude, static temperature as shown in next section. The procedure analysis by CFD using model porous flow is shown in the Flowchart below:

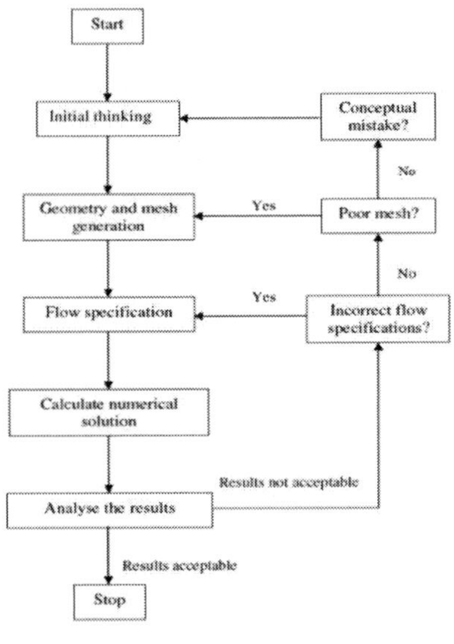

Figure 3. Flowchart of (CFD) analysis process [15]

9. Results and Discussion

9.1. Effect of porous flow properties on pressure drop

The different velocity flow (porous flow) ranges have effected on pressure drop through the packed column, the pressure difference between the bottom and top of the bed is (ΔP), the profile of pressure has been simulated over the Reynolds numbers range of (7.8 – 1115.876) as shown in Fig (4) CFD contours simulation of static pressure, and as function to superficial gas velocity and length of bed column (0.9) m, the pressure drop CFD results are compared with Ergun correlation as shown in Table (4). The Fig (5) shows the results from both models and gives reasonable data with error (0.6% – 0.63%) between data results from (CFD) analysis and values from Ergun equation, this error comes from the equation (3) for the porosity of medium. for bed height with (0.9 m), maximum pressure drop is (33381.57) (Pascal) under Reynolds numbers (turbulent flow), the reason is the effect of changes in porous flow regime to turbulent flow (strong inertial forces) when the bed length increases and (ΔP) along column increases

Table.4. CFD results compared with Ergun equation

Re_p	(ΔP) (Pascal) CFD model (0.9) m length column	(ΔP) (Pascal) Ergun model (0.9) m length column
7.872750643	2.7494	10.52869951
188.2614284	972.0385	946.9017
290.94948	2299.466	2261.608
328.60177	2927.186	2884.842
376.522857	3835.525	3787.607
492.90265	6551.711	6490.893
554.51548	8282.316	8215.037
835.196155	18729.81	18636.28
1115.8768	33381.57	33267.08

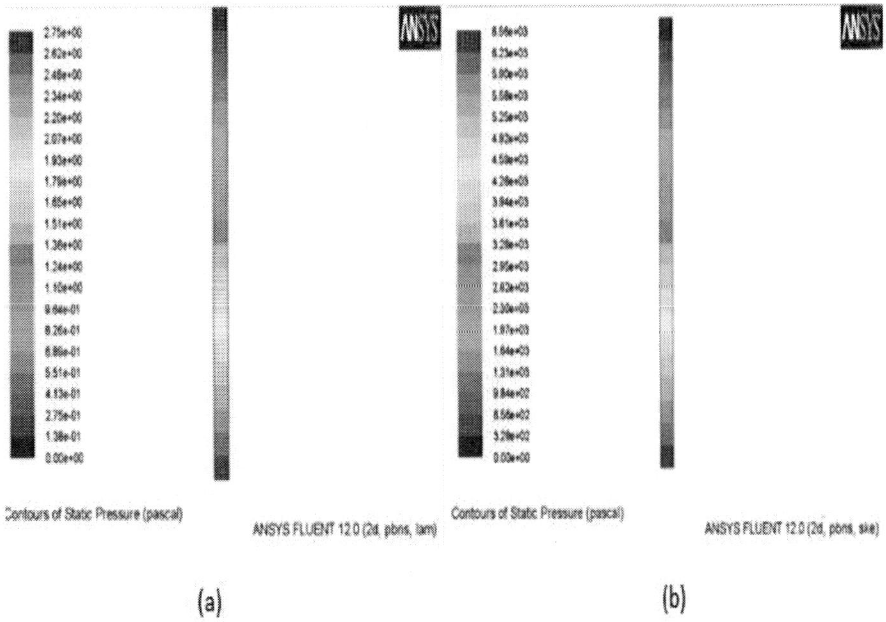

Figure 4. CFD contours simulation of static pressure (Pascal) (a) laminar flow (b) Turbulent flow.

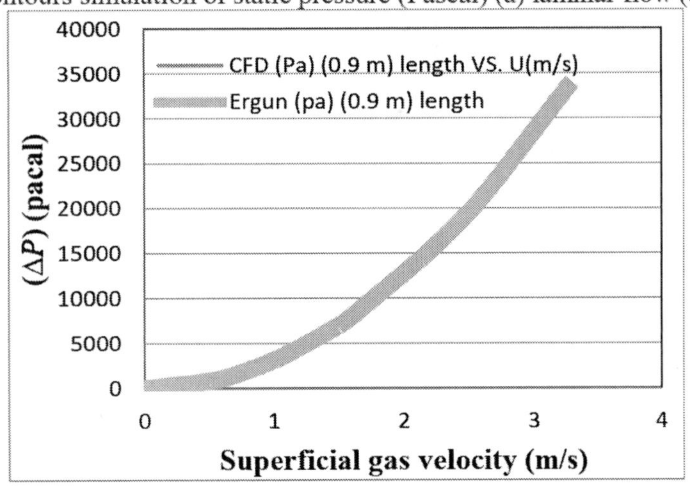

Figure 5. (ΔP) (Pascal), CFD results compared with Ergun equation variations with superficial gas velocity (m/s) along length of column 0.9 m.

9.2. Effect of Mesh size:

the concept of mesh density can be examined at different mesh sizes by temperature distribution in the model as shown in Fig.(6).The comparison between steady-state temperature distributions is the plot of temperature (K) vs. radial position (m) for a velocity of 0.023 (m/s) for four mesh sizes (12k, 30k, 40k, 64k) at high that equals to 0.594 m from bottom of packed bed model, the profiles of change in temperature inside column model comes from changes in mesh sizes on the flow gas for 0.9 m model with the appearance of very little difference in temperature between the mesh (40k and 30k) in the middle of the column and a great convergence in temperature at the wall of the column temperature. The mesh 64k has a clear difference in temperature, indicated to mesh size and refinement created a major little difference in temperature. So the variation in the mesh density should be smooth to reduce numerical diffusion.

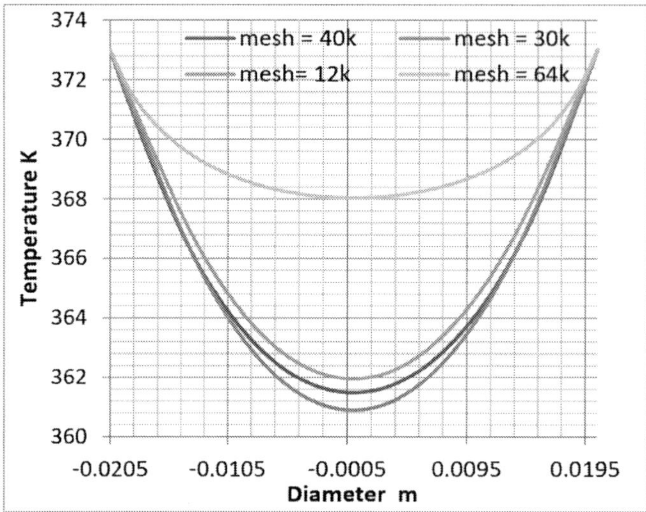

Figure 6. Temperature distributions at four different mesh sizes from bed model.

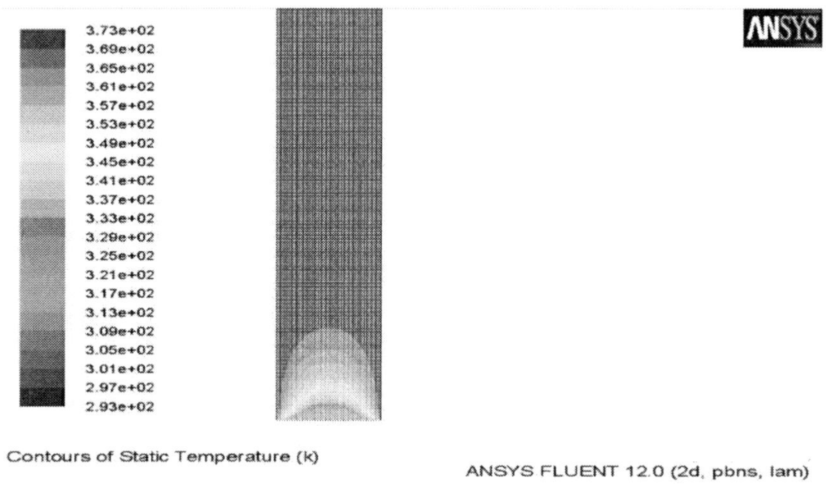

Figure 7. Static temperature in (K) for axial position (m) at a velocity = 0.023 (m/ s) with 40K mesh size.

9.3. Effect of gas flow rate:

The gas flow rates have an important variable in bed column system and have direct influence on the process model. Fig.8 shows simulation Ansys-Fluent for Contour of static temperature values along bed model and affects it by gas flow inside system model at laminar and turbulent flow. from Fig.8 the total difference in temperature for laminar flow (Rep=7.8) is (79K- 80K) to reach steady – state condition, but for turbulent flow Rep = 1115.87 the total difference in temperature is (59K) along bed column model, and this difference is ranging by an increase in the gas velocity with (x/L) for model column, The increase in temperature with increasing Reynolds number has the effect on amount of heat into the bed at wall temperature is constant.(CFD) simulations give suitable prediction in amount of heat inside geometry model column at change in gas flow regime. Table.(5) illustrates CFD results temperature (K) profiles inside bed column model at Reynolds number ranged from laminar to turbulent flow for different positions. Fig.9 illustrates porous medium model distribution of radial temperature by CFD simulation at different positions with high Reynolds number that equals to 1115.87 variations with radial

(r) m. so the prediction from CFD simulation gives high details for porous flow in packed bed equipment.

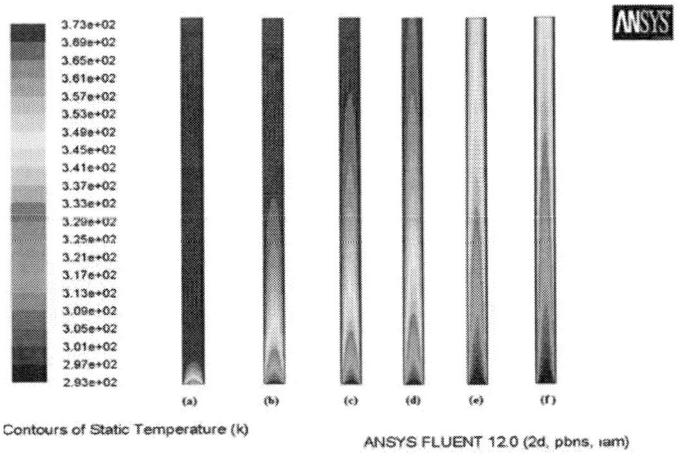

Fig.8. Static temperature in (K) along geometry model (a) Re_p=7.872, (b, c, d, e, f) Re_p=turbulent flow

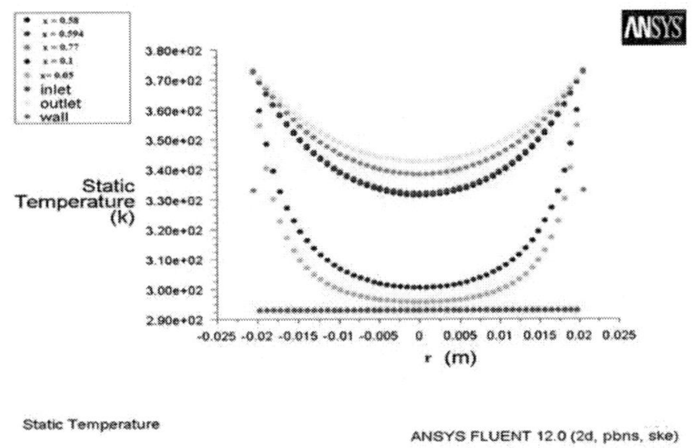

Figure 9. Porous media model illustration radial temperature distributions by CFD simulation at different positions.

Table.5.CFD results temperature (K) profiles inside bed column model at laminar and turbulent flow for different position

Reynolds number	Different positions along bed material column (m)					
	0.05	**0.1**	**0.4**	**0.6**	**0.8**	**0.9, outlet**
7.872	370.00	372.82	372.89	373.00	373.00	373.00
188.26	335.55	352.34	368.61	371.04	372	372.6
290.94	326.98	341.20	363.28	371.15	371.06	371.21
328.60	324.81	338.14	361.37	369.28	370.49	370.78
376.52	322.51	334.87	359.05	367.32	368.94	369.49

492.90	318.28	328.98	353.94	362.40	365.58	366.97
554.51	316.56	326.65	351.54	360.03	362.92	364.76
835.19	311.2	319.38	342.76	350.24	355.89	358.15
1115.8	308.02	314.96	336.48	344.55	348.64	352

10. Conclusion

The pressure drop simulation has been compared with the values calculated from Ergun equation, the relative error is (0.6% – 0.63%) for our model and It was found that there is a good agreement between CFD and Ergun model, and maximum (ΔP) for model bed with (0.9 m) height is (33381.57) (Pascal) under different in flow regimes and It illustrated the effect of changes in porous flow regime to turbulent flow, It becomes (strong inertial forces), and gives increases in the (ΔP) along column.

The profiles of change in temperature inside column model comes from changes in mesh size on the flow gas, there is a very little difference in temperature between the meshes (40k and 30k) in the middle of the column it is and very close in the wall.

influence on the process model. The total difference in temperature for laminar flow (Rep=7.8) is (79K- 80K) to reach steady – state condition, but for turbulent flow Rep = 1115.87 the total difference in temperature is (59K) along bed column model, and this difference is ranging by an increase in the gas velocity with (x/L) for model column. The prediction from Computational Fluid Dynamics gives us simulation with high details for porous flow in packed bed equipment, design its and performance flow of gas through porous media in packed bed column.

References

[1] Burke S P and Plummer W P 1928 Gas flow through packed column *Ind. Eng. Chem.* **20** 1196
[2] Bark H 2009 *CFD Advantages and Practical Applications* Glumac Inc.
[3] Maier R S, Kroll D M, Kutsovsky Y E, Davis H T and Bernard R S 1998 *Phy. Fluid*, **10** 60
[4] Gunnar J, Hellstr I, Jonsson P and Staffan Lundstr 2010 *Journal of Porous Media* **13** 1073
[5] Kamyar M, Alkhalaf A and Specht E 2018 *Computational Particle Mechanics* **6** 157
[6] Augier F, Idoux F and Delenne J Y 2010 *Chemical Eng. Science* **65** 1055
[7] Wen D and Ding Y 2006 *Chem. Eng. Sci.* **61** 3532
[8] Peters S T 1998 *Handbook of composites* Springer-Verlag (Houston: Springer-Verlag) ed 2 p 1118
[9] G K Batchelor 1967 *An Introduction to Fluid Dynamics* (Cambridge: University Press Cambridge) ed 1 p 615
[10] Hlushkou D and Tallarek U 2006 *Journal of Chromatography* **1126** 70
[11] ANSYS Fluent Documentation ANSYS Inc. 2010 ANSYS Fluent 12.0/12.1 Documentation
[12] Furnas C C 1931 Ind. *Eng. Chem* **23** 1052
[13] Stewart B R B 1960 *Transport Phenomena* (New York: John Wiley and Sons Inc.) ed 1 p 780
[14] Launder B E and Spalding D 1972 *Lectures in Mathematical Models of Turbulence* (London: Academic Press) ed 1 p 24
[15] ANSYS Fluent Documentation ANSYS Inc. 2009 Release 12.0
[16] Gunnar J, Hellstrom I, Jonas P and Jonsson P 2010 *Journal of Porous Media* **13** 1073
[17] Festuccia A and Bella G 2018 Journal *of Porous Media* **21** 5

IOP Conf. Series: Materials Science and Engineering **579** (2019) 012002 doi:10.1088/1757-899X/579/1/012002

Experimental Study on Tanuma Shale Stability Using Drilling Fluids with Different Additives

Khalid Mohammed, Asawer A. Alwassiti and Mayssaa Ali AL-Bidry

Petroleum Technology Department, University of Technology-Iraq, Baghdad, Iraq

Corresponding author's e-mail address: kaled_mohamed54@yahoo.com

Abstract: Shale formation is represented as one of the challenge formations during drilling wells because it is a strong potential for wellbore instability. Tanuma formation in Iraqi oil fields (North Rumila) is located at a depth from 2102 m MD – 2126 m MD. It is considered one of the most problematic formations through drilling wells in North Rumila. Most problems of Tanuma shale are sloughing, swelling, caving, cementing problem and casing landing problem caused by the interaction of drilling fluid with the formation. This paper focuses on studying the chemical and minerals properties of Tanuma shale. In addition, study the effect of drilling fluids and additive salts by using different techniques; (XRD, XRF, reflected and transmitted microscope, scanning electron microscope and EDS)Also, the study includes experiments to observe and enhance the shale stability by using two types of drilling fluids, API and polymer type, with different additives. Four types of additives (KCl, NaCl, $CaCl_2$ and Sodium silicate) in different concentrations (0.5, 1, 5 and 10) wt % and different immersion period (1, 24 and 72 hours) were used. The results show that adding 10 wt. % of sodium silicate to API drilling fluid results in a high percentage of shale recovery (77.99%). While, the maximum shale recovery was (79.76%) in polymer drilling fluid type gained by adding 10 wt % of sodium silicate.

1. Introduction

Shale is a fine-grained, clastic sedimentary rock collected of mud that is a mix of flakes of clay minerals and tiny fragments of other minerals. It is characterized by parallel layering or thin laminae or bedding less than one centimeter in thickness, called fissility[1]. Shales contain clay minerals and non-clay minerals, clay minerals can be grouped as Smectite, Vermiculite and Kaolin, Illite, and chlorite and Non-Clay minerals are quartz, feldspars [2]. Shale is recognized as one of the most suspicious rock types in petroleum engineering applications, it makes up over 75% of the drilled formations. Shale behavior is critical because its transition to an unstable situation may occur rather easily and quickly. The chemical properties of shale rock were playing a substantial effect on its behavior like the toughness and strength properties. It can be affected strongly by drilling fluid type. For example, the interaction of shale with drilling fluid, or movement of drilling mud into the shale matrix, may occur within a few hours, leading to tremendous practical problems, more than 70% of the wellbore problems are related to shale instability. The swelling is the measure of the reactivity of the fluids with shale samples [3]. Immersion (fracture development) tests were used in the last few years to evaluate and understand the contact between the drilling fluids and shale formations. Immersion tests were intended to directly evaluate and observe the rock-fluid interactions and fracture development when a piece of shale samples was immersed in the drilling fluids [4]. Tanuma formation in Iraqi oil fields (North Rumila) is located at a depth from (2102 m MD – 2126 m MD). It considered as one of the most problematic formations through drilling wells in North Rumila. Most problems of

IOP Conf. Series: Materials Science and Engineering **579** (2019) 012002 doi:10.1088/1757-899X/579/1/012002

Tanuma shale formation are sloughing, swelling, caving, cementing problem and casing landing problem caused by the interaction of drilling fluid with the formation. The description of Tanuma shale and the drilling fluids additive are mentioned in (table 1) and (table 2) respectively. This paper focuses on studying the properties of Tanuma shale as well as studying the effect of chemical interaction between the shale and salts additives in drilling fluid. The problems of shale formation are improved by using additives that have the ability to react with shale and reduce the reaction between water and shale formation.

Table 1. Description of Tanuma shale.

Depth	Description of Tanuma formation
2102-2104m	Shale (50%): Medium dark grey, olive grey, slightly hard, commonly fissile and slightly calcareous in place, occasionally thinly laminated.
	Limestone (50%): Mudstone to Wackestone, very light grey, yellowish grey, light olive grey, soft, firm in place, fine crystalline, earthy luster, argillaceous, no visible porosity, no oil show
2104-2110m	Shale (70%)
	Limestone (30%)
2110-2120m	Shale (100%)
2120-2126m	Shale (60%)
	Limestone (40%)

Table 2. Drilling Fluid Additives in Tanuma shale.

Products	Product Description	Product Function
Soda Ash	Sodium carbonate (Na_2CO_3)	Precipitate calcium in the water
Caustic Soda	Sodium Hydroxide (NaOH)	Alkalinity Source (pH)
BARAZAN D	Dispersion-enhanced xanthan biopolymer	Viscosifier / Hi-Vis
PAC L	Polyanionic Cellulose	Filtration Controller
DEXTRID LTE	Modified potato starch	Filtration Controller
NaCl Salt	Sodium Chloride	Shale inhibitor/weighting Material
KCl	Potassium Chloride	Shale/Clay inhibitor
EZ MUD (PHPA)	Polyacrylamide/Polyacrylate (PHPA)	Shale stabilizer/Encapsulator
BAROTROL-PLUS	Fine-grind blended hydrocarbon powder	Shale stabilizer
GEM GP or GEM CP	Poly-Alkaline glycol	Shale stabilizer, lubricity
BAROLUBEGOLDSEAL	Blend of surfactants and lubricants	Lubricant, reduce torque and drag and sticking shale problems.
Barite API	Ground Barium Sulphate	Weighting agent for slug
$CaCO_3$ Fine	Calcium Carbonate	Weighting & Bridging agent

2. Methodology (ANALYTICAL METHODS)

A. Samples Preparation

Samples have been taken from Basra (North Rumila) at 2120 m depth. Two methods were used to clean the samples from hydrocarbons and drilling additives. First, soxhlet method to remove hydrocarbons from shale by using toluene, methanol, and benzene then heated in soxhlet device for three hours. The second method was using wet sieving to remove additive salts.

B. Shale test

Forty-five samples of Tanuma shale have been studied by using different techniques, to investigate petrography and the effect of drilling fluid additives after and before the clean sample. Forty-five thin sections and polish thin sections were prepared to be examined by transmitted and reflected light microscope type BX51M /Olympus. Ten samples were examined using a scanning electron microscope and Energy dispersive X-ray spectroscopy (EDS). Mineral and chemical compositions of shale were examined by X-Ray Diffraction and XRF.

3. Experimental Procedure

In this work, 128 shale samples were immersed in two types of drilling fluids with different additives. Four types of salts (KCl, NaCl, $CaCl_2$ and Sodium silicate) with different concentration (0.5, 1, 5 and 10) wt % and different immersion period (1, 24 and 72) hours were used. Shale samples are prepared to soak with different drilling fluids [5]. The samples were studied by the microscope before and after immersion to determine the effect of the drilling fluids on the stability of the shale.

Drilling fluid Preparation

Two types of drilling fluids were used in this paper WBM (API Manual) and PM. The drilling fluid was prepared then add additive with mixing for 2 minutes. Water-based mud was prepared by mixing water with the bentonite by using Hamilton Beach mixer for 20 min then the suspension is aged in a sealed container for 24 hours to ensure good hydration of bentonite after that the salts with concentrations of 0.5, 1, 5 and 10 wt % were added to WBM then mixed for 10 min. and Polymer based mud was prepared by mixing water with the bentonite by using Hamilton Beach Mixer for 20 min then the suspension is aged in a sealed container for 24 hours to ensure good hydration of bentonite then KCl, KOH, PAC Polymer, XC polymer were added. Each material was let to mix with the suspension for 2 minutes to ensure dispersion of particles into the drilling fluid matrix; finally, the whole mixture was mixed for 10 minutes after that the salts with concentrations of 0.5, 1, 5 and 10 wt % were added to PM then mixing for 10 min.

4. Results and Discussion

Various methods such as X-Ray Diffraction analysis, Reflected microscope and Scanning electron microscope and EDS used to detect the clay minerals in shale. Mineralogy analysis detected the quantities of compounds present in the shale. Clay and non-clay minerals were found in the shale samples. The clay minerals found in the shale indicates the degree of its hydration.

The mineralogy of Tanuma shale

1. Analysis of XRD and XRF

The mineralogical compositions of the samples were analyzed using X-ray diffraction. This is the method of analysis provides the mineralogy of shale; clay minerals composed mainly of kaolin and little Illite also found non-clay minerals quartz and calcite as shown in (figure 1). The chemical composition was measured before and after cleaning the samples, (table 3). The concentration of CaO,

Na_2O, and K_2O are changed after cleaning sample so those were referred to additive salts. The SiO_2, Fe_2O_3, Al_2O_3, and MgO referred to the main composition of shale.

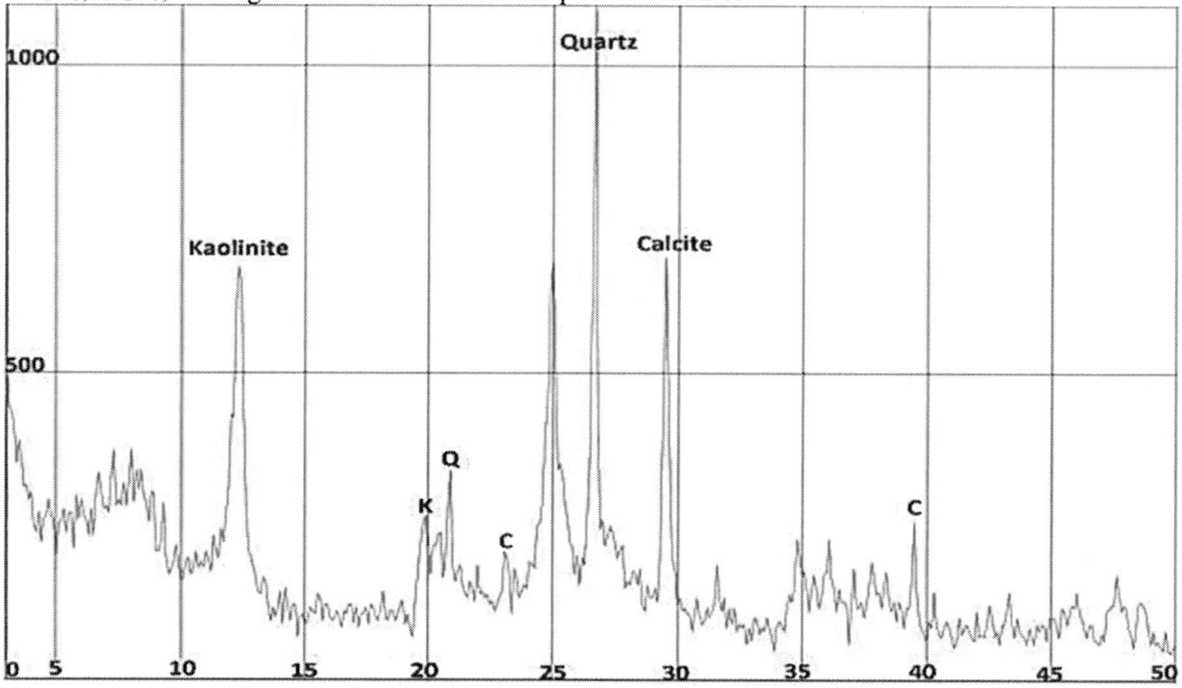

Figure 1. X-Ray Diffraction.

Table 3. X-Ray fluorescence.

Specimen	cleaning	Sio_2 %	Fe_2O_3 %	Al_2O_3 %	CaO %	MgO %	So_3 %	Na_2O %	K_2O %	Lol %
1	Before	40.67	10.33	16.85	10.2	0.8	0.510	0.91	3.42	14.64
2	After	47.96	10.22	17.9	5.93	0.86	0.440	0.25	2.98	11.79

2. Analysis of Reflected, transmitted microscope, SEM and EDS

Before cleaning the shale, samples was found a lot of salt covered the surface and accumulation through caves, micro-fracture, and between sheets. The maximum accumulation at the edge and between sheets is illustrated in (Figure 2). The weakness of kaolin minerals is represented on the edge of the sheets so the cation exchange of ions occurs on the edge of the sheet [6].

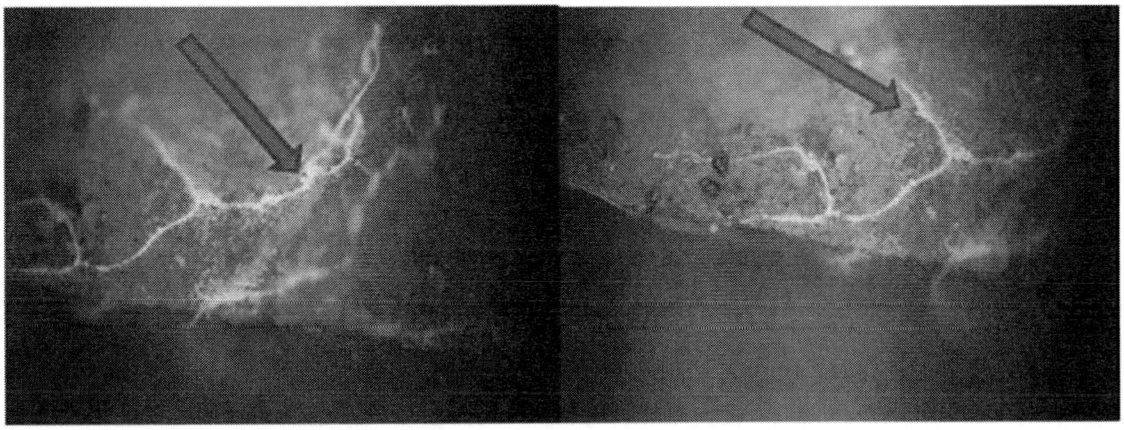

Figure 2. Distribution Salts in the edge of the sheet.

Salts accumulation was at the edge due to sorption on shale surface, as well as cations exchange between salt in drilling fluids and shale as shown in (Figure 3 D).

Gypsum veinlet scattered on the surface of samples each veinlet twisted with another veinlet like a network [7] as illustrated in (figure 4). After cleaning shale samples, could recognize a small grains of magnetite distributed in shale and also, Biotite with red-brown color (figure 3 A), (figure 3 B) and (figure 7). Some of origin Biotite alteration due to an accumulation of salts on the surface of minerals (figures 3 C).

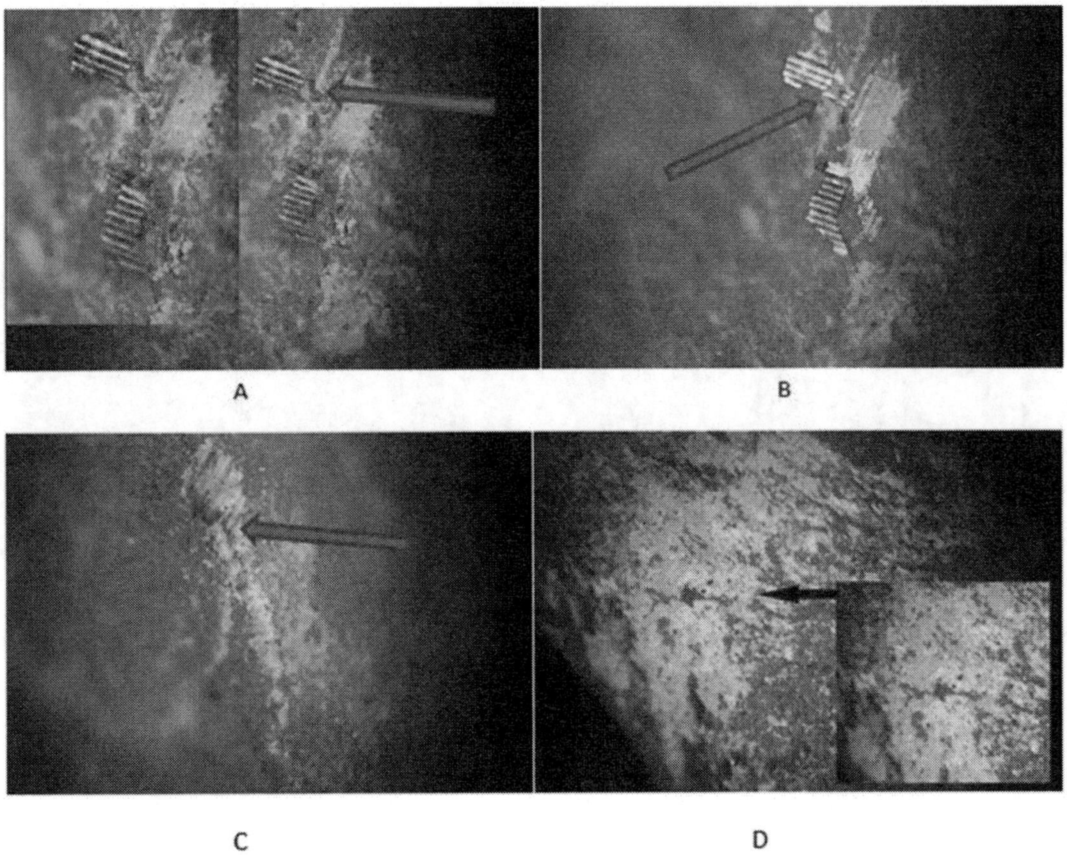

Figure 3. A: Mica, B: Mica by (Xn), C: Alteration Mica, D: coated salts.

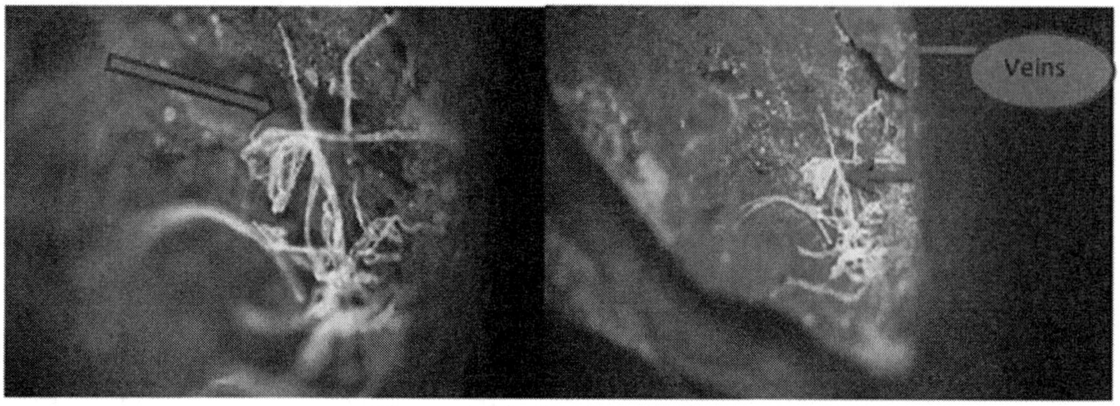

Figure 4. Distribution of veins.

Salts in drilling fluids had been affected on the parts of Mica [8]. The Scanning electron microscope and EDS were used to detect the major elements in shale and chemical element that distribution on the surface due to sorption and cation exchange. (Figure 5) shows a high concentration of Si^{+4} and Al^{+3}.

Clay minerals are hydrous aluminum phyllosilicates. The drilling fluids were used for Tanuma shale led to a high concentration of Ca^{+2} and K^{+}. As we know the attractive of kaolin with water is little but the cation concentric on the edge of the sheets often cation exchanges take place in the weak edge of the sheet.

(Figure 3 D) shows, samples before cleaning there are a lot of salts distributed on samples, as salt grains aggregated between sheets and on edge of the sheet, and covering clay grains completely.

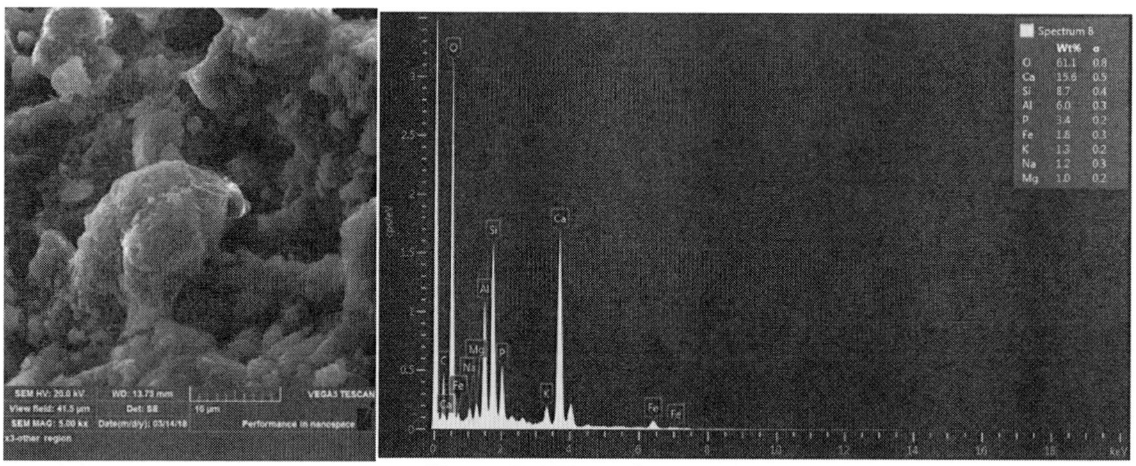

Figure 5. SEM and EDS.

The grains of clay were covered by salts because of the reaction between the salts and clay minerals. Clay grains coated with salts and salt aggregate caused sheets separation and fractures. These causes reduce the stability of the shale. (Figure 6) shows the shale samples after cleaning. Note decreased the concentration of Ca^{+2}, Na^{+}, and K^{+}.

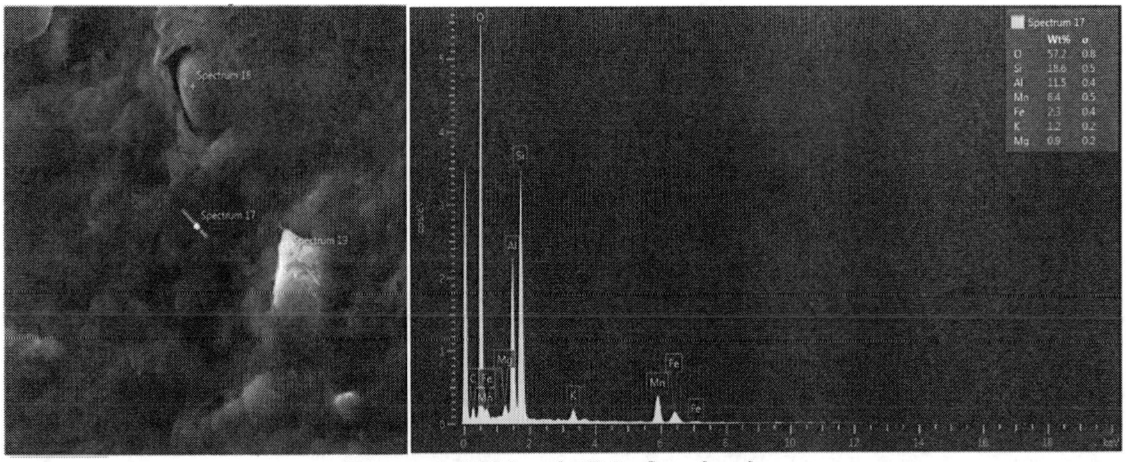

Figure 6. SEM and EDS after cleaning.

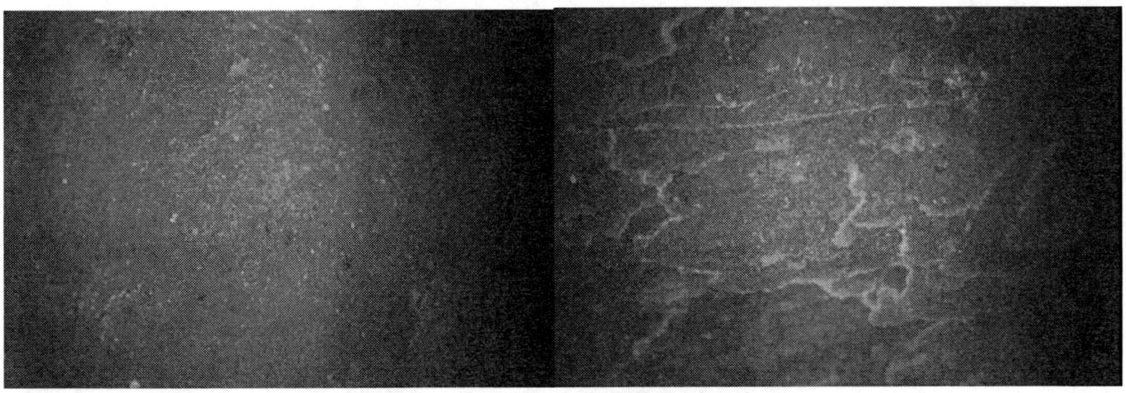

Figure 7. Shale samples after cleaning.

Effect of drilling fluids on the Tanuma shale

1. Immersion test

The recovery percentage values were calculated after immersion the samples in drilling fluid. These values were varied depending on the activity between the additive and shale samples. (Figure 8) shows the effect of 0.5% addition of different salt with water-based mud on the shale samples it can be seen that KCl salt gives the best recovery after 1 hour with a percentage of 70.1 %. After 72 hours the KCl recovery decreases to 60.2. In this point, the water got enough time to interact with shale samples and the KCl concentration is not enough to reduce the effect of water on shale samples. Increasing the concentration to (1, 5 and 10) wt % of KCl salt causes higher percentage recovery after one hour (73.2, 78.4, and 80.7) wt % in comparison with other salts. Using the same additives at the same concentration in the polymer mud, the recovery percentage increased as shown in (Figure 8). At 1 wt % concentration, the KCl salt gives the high recovery percentage in the water-based mud and polymer mud as shown in (figure 9).

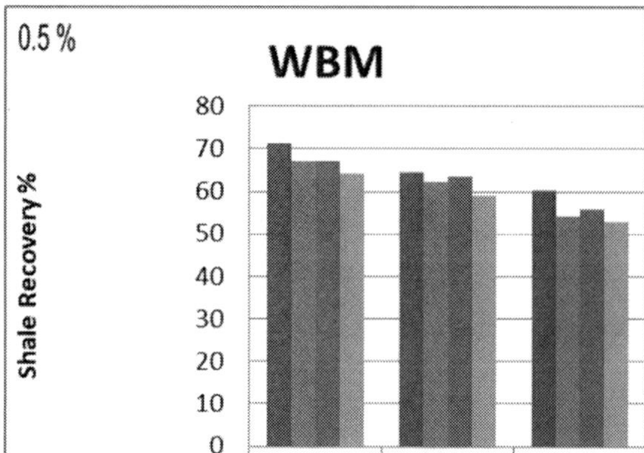

	1 hr	24 hr	72 hr
KCl+WBM	71.3	64.6	60.2
CaCl2 + WBM	67.2	62.3	54.3
NaCl + WBM	67.1	63.4	55.7
Sodium Silicate + WBM	64.28	59.17	52.86

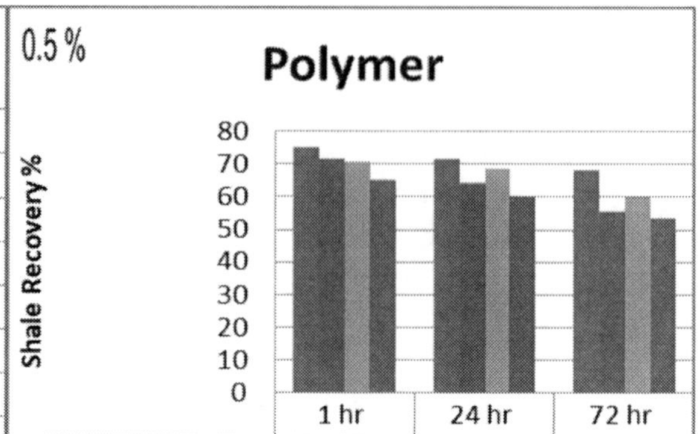

	1 hr	24 hr	72 hr
KCl + Polymer	75.12	71.52	68.41
CaCl2 + Polymer	71.64	64.29	55.81
NaCl + Polymer	70.6	68.8	60.1
Sodium silicate + Polymer	65.37	60.1	53.43

Figure 8. Recovery for WBM and Polymer Mud with concentration 0.5% Wt.

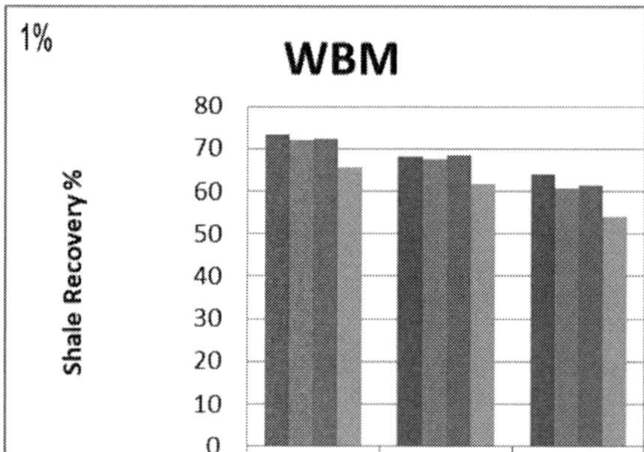

	1 hr	24 hr	72 hr
KCl + WBM	73.2	68.1	63.9
CaCl2 + WBM	72	67.4	60.8
NaCl + WBM	72.4	68.4	61.3
Sodium silicate + WBM	65.55	61.68	54.13

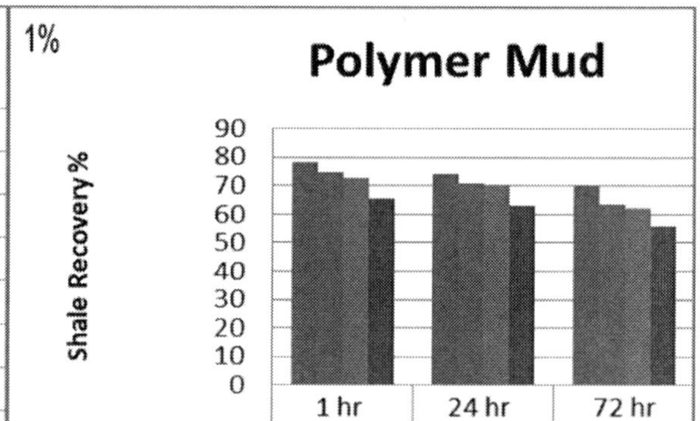

	1 hr	24 hr	72 hr
KCl + Polymer	78.31	74.47	69.72
CaCl2 + Polymer	74.94	70.72	63.49
NaCl + Polymer	72.8	70.1	62.3
Sodium silicate + Polyemr	65.73	63.24	55.67

Figure 9. Recovery for WBM and Polymer Mud with concentration 1% Wt.

The effect of sodium silicate noticed at the concentration 5-10 wt %. The recovery percentage is high for sodium silicate at 5 and 10 wt % as shown in (figures 10) and (figure 11). Increasing the concentration of the salts will give a good recovery as noted in the tests.

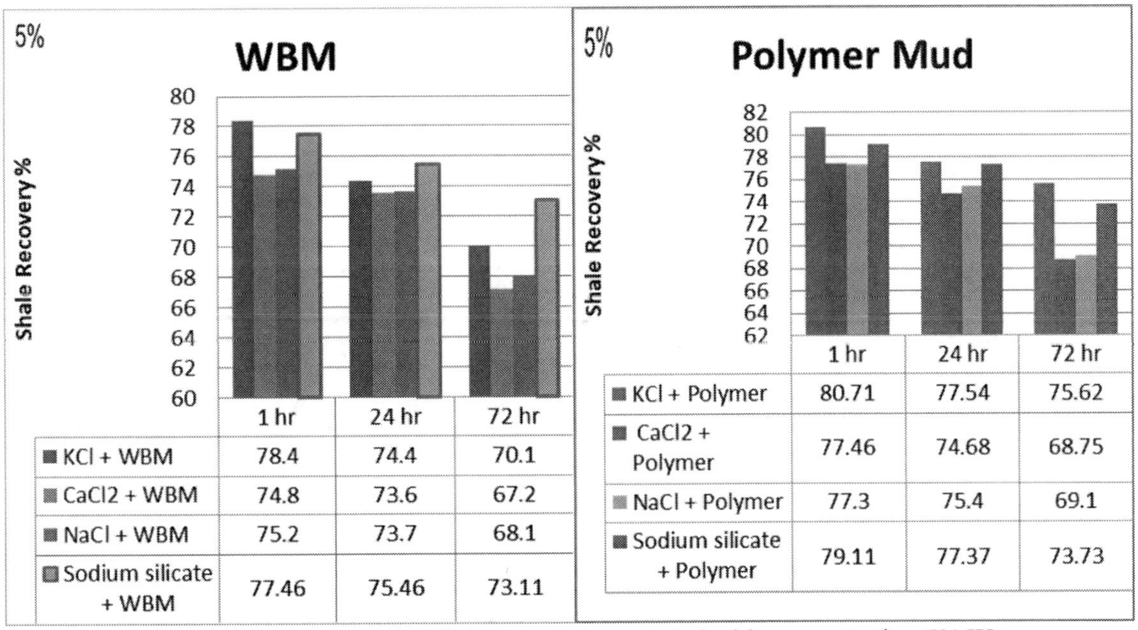

5% WBM	1 hr	24 hr	72 hr
KCl + WBM	78.4	74.4	70.1
CaCl2 + WBM	74.8	73.6	67.2
NaCl + WBM	75.2	73.7	68.1
Sodium silicate + WBM	77.46	75.46	73.11

5% Polymer Mud	1 hr	24 hr	72 hr
KCl + Polymer	80.71	77.54	75.62
CaCl2 + Polymer	77.46	74.68	68.75
NaCl + Polymer	77.3	75.4	69.1
Sodium silicate + Polymer	79.11	77.37	73.73

Figure 10. Recovery for WBM and Polymer Mud with concentration 5% Wt.

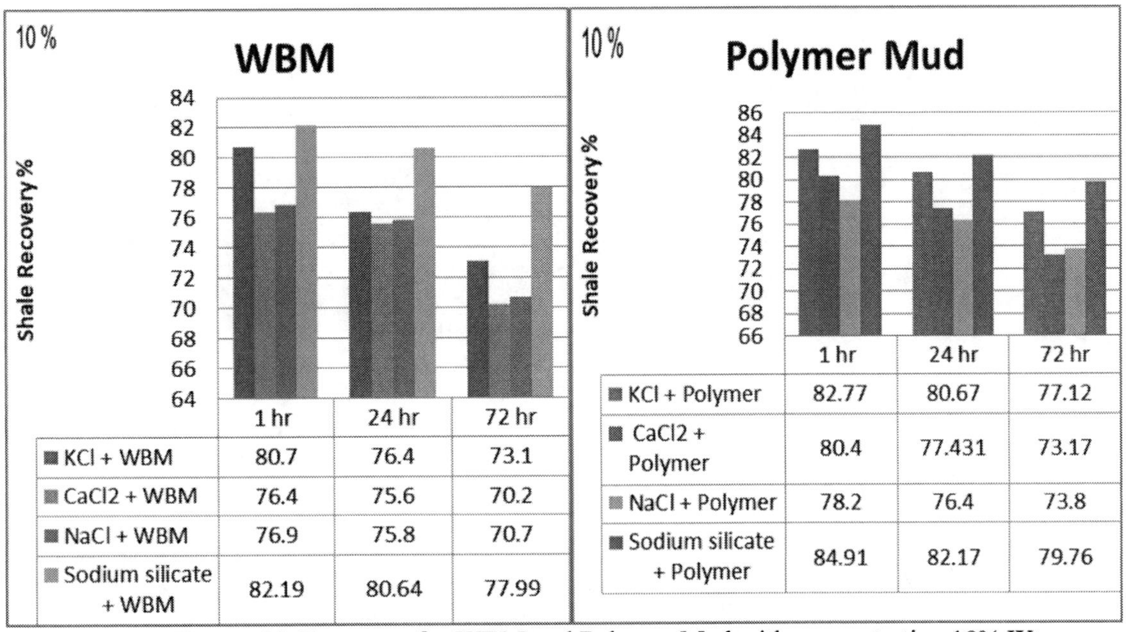

10% WBM	1 hr	24 hr	72 hr
KCl + WBM	80.7	76.4	73.1
CaCl2 + WBM	76.4	75.6	70.2
NaCl + WBM	76.9	75.8	70.7
Sodium silicate + WBM	82.19	80.64	77.99

10% Polymer Mud	1 hr	24 hr	72 hr
KCl + Polymer	82.77	80.67	77.12
CaCl2 + Polymer	80.4	77.431	73.17
NaCl + Polymer	78.2	76.4	73.8
Sodium silicate + Polymer	84.91	82.17	79.76

Figure 11. Recovery for WBM and Polymer Mud with concentration 10% Wt.

KCl is considered as one of the best inhibitors for shale problems as noticed in results, the effect of KCl was very clear to increase the recovery percentage of shale samples. K^+ has the ability to adsorb on the surface of shale and react with the ions on the surface of shale sheets. Due to replace of Si^{+4} with Al^{+3} in octahedral sheets and Al^{+3} changed with Mg^{+2} in tetrahedral sheets this will increase the negative charge and causes the attractive of the positive charge on the surface of shale. Therefore, the positive ions like K^+, Ca^{+2}, Na^+, and Si^{+4} were adsorbed on the surface of sheets. NaCl has a higher preliminary viscosity relative to KCl in the mud and it has lower water activity which gives rise to higher osmotic pressures. NaCl is represented a better equipped for reducing filtrate invasion for

IOP Conf. Series: Materials Science and Engineering **579** (2019) 012002 doi:10.1088/1757-899X/579/1/012002

drilling fluids to the shale [1]. The microscope shows the effect of samples immersion with KCl additives and other salts. (Figure 12) shows an accumulation of salts on the surface and edge of shale that's happened due to adsorb of the sheet of shale for K+. Na+ and Ca+2 ions are accumulated and the crystals of salts consisted of the samples and coated it as shown in (figure 13).

Figure 12. KCl accumulations on the surface.

Figure 13. Crystals of the salts on the surface of the shale.

Silicate additive was used with drilling fluids. Soluble sodium silicate has the ability to invade the shale and react with available ions in the shale to consist insoluble precipitation. (Figure 14) shows the sodium silicate accumulation on the shale samples and sodium silicate invaded into the fracture and built a bridge between two sides then covered the shales to prevent any interaction with other fluids.

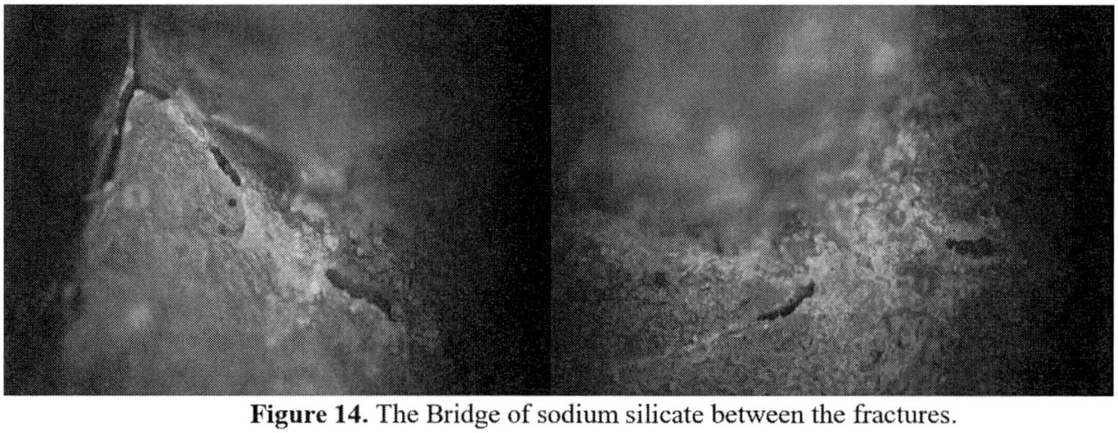

Figure 14. The Bridge of sodium silicate between the fractures.

2. Swelling Test

The samples were placed in the different drilling fluids to calculate the swelling percentage for Tanuma shale. As shown in (Figure 15), (figure 16), (figure 17) and (figure 18) the swelling affected by the salts concentration. Its percentage decreased by increasing the salts concentration. Sodium silicate with concentration 0.5 wt % gives a high swelling percentage while at 10 wt % concentration the swelling reduced to 3 percentages as shown in (figure 18).

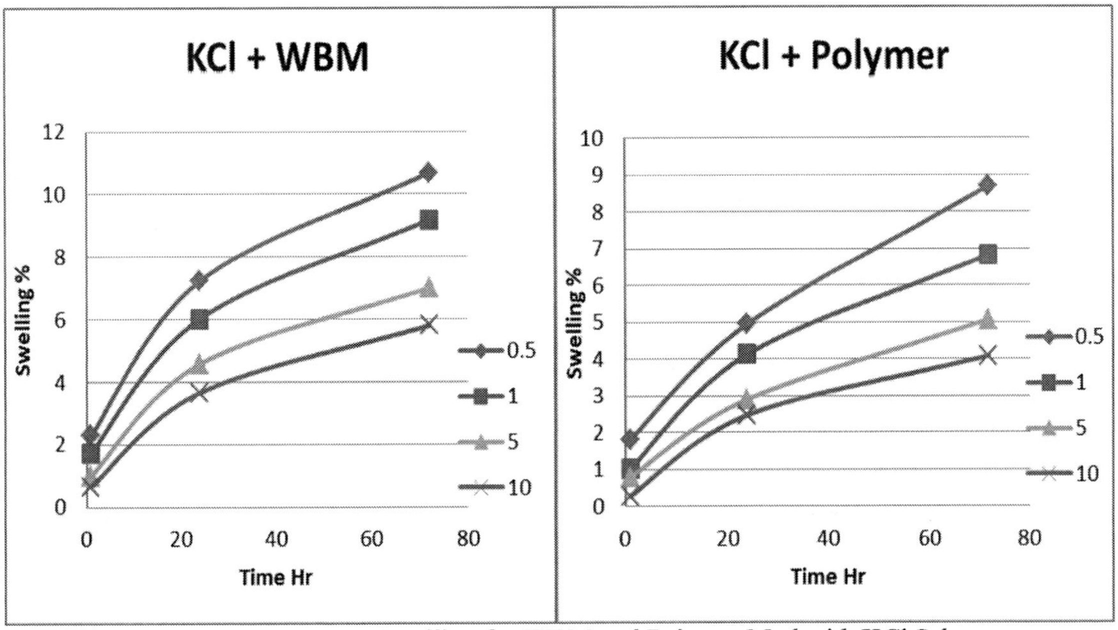

Figure 15. Swelling for WBM and Polymer Mud with KCl Salt.

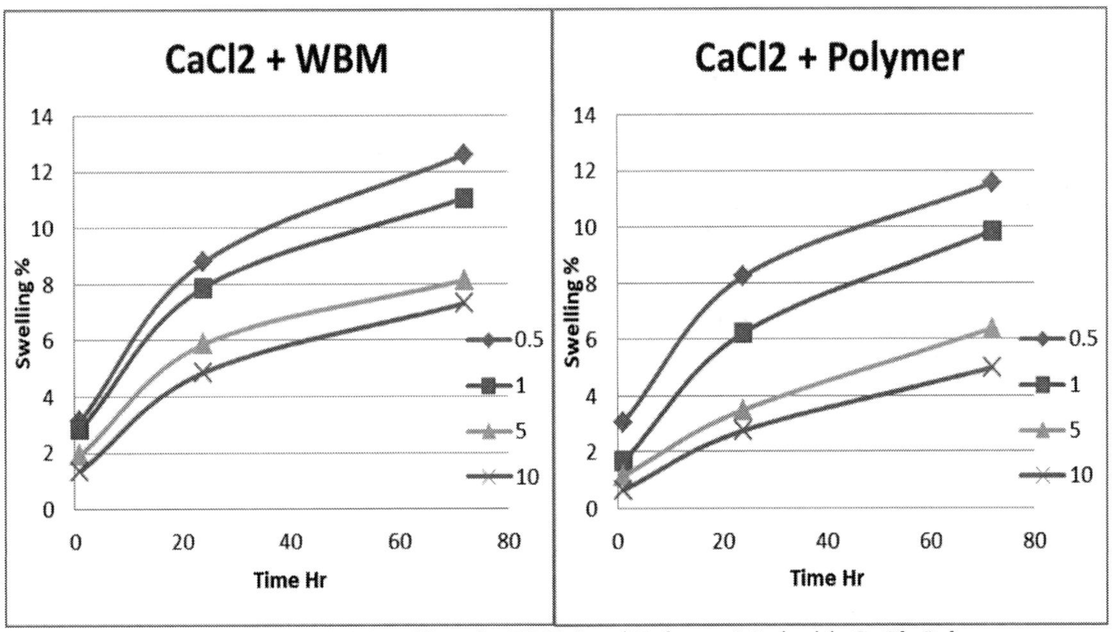

Figure 16. Swelling for WBM and Polymer Mud with $CaCl_2$ Salt.

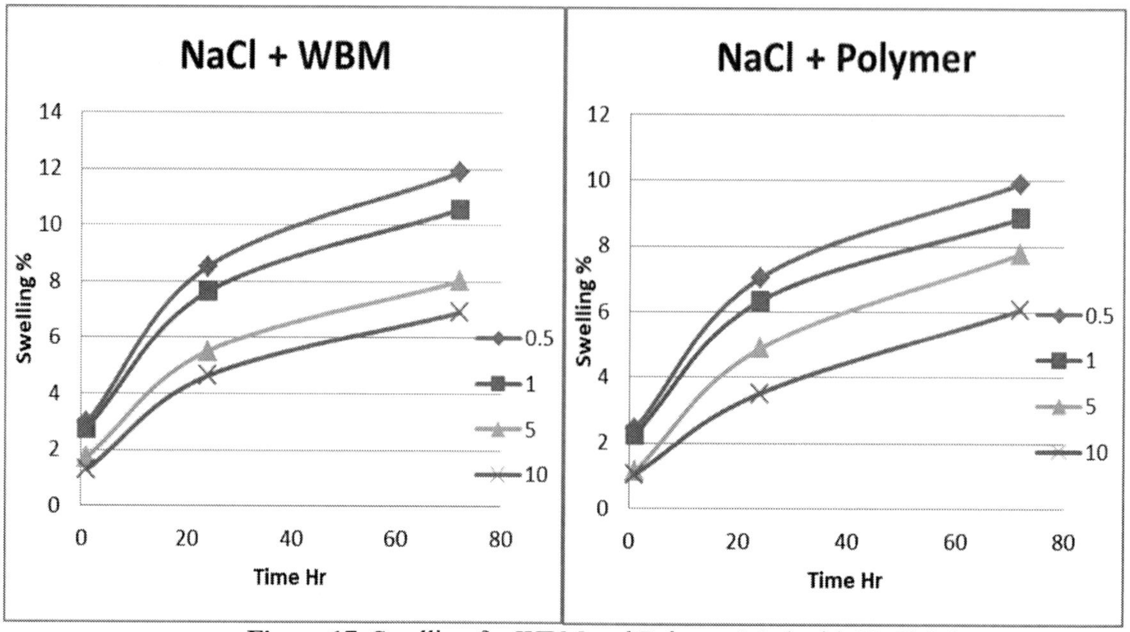

Figure 17. Swelling for WBM and Polymer Mud with NaCl Salt.

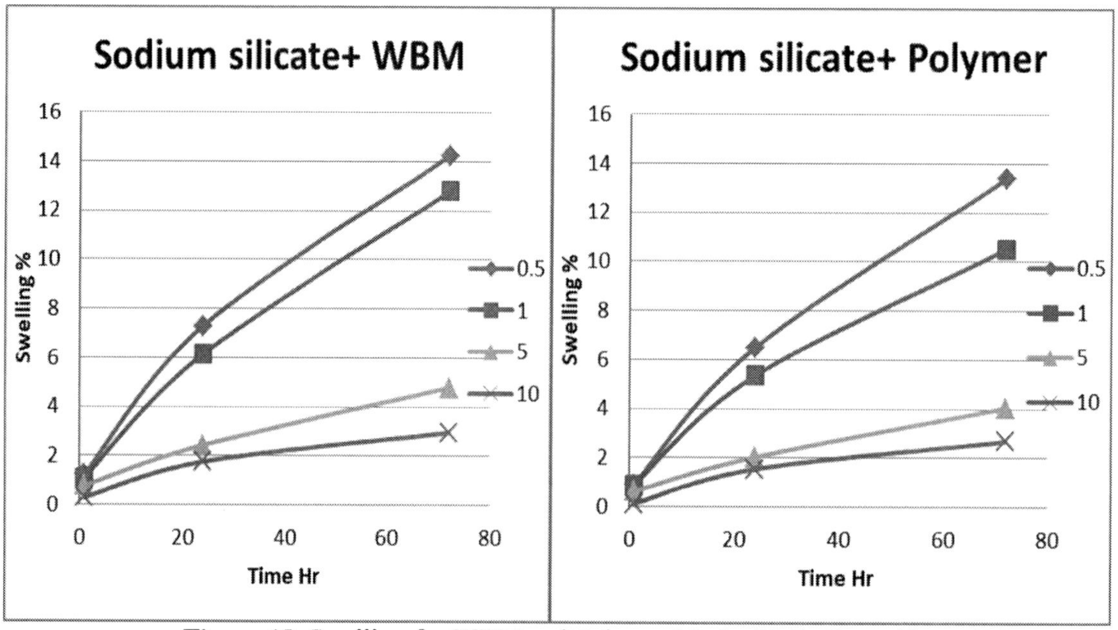

Figure 18. Swelling for WBM and Polymer Mud with Sodium silicate Salt.

As noticed in the swelling test of the Tanuma shale the swelling percentage is very low compared to other shale formations because of the swelling reactivity of the kaolin minerals very little. The quantity of swelling for Kaolin is very little compared with another clay mineral like Smectite. KCl can decrease the swelling in the shale formations. The results show the increasing of KCl concentration in the drilling fluids that lead to a decrease in the values of swelling as shown in (figure 15), indicating that KCl plays a role in inhibitor shale expansion. The results show the K^+ is more effective comparing with Ca^{+2} and Na^+ in decreasing the swelling of shale samples. Sodium silicate is normally preferred to use for formations those mainly containing swelling clay minerals such as shale.

It has the ability to attract the water and makes colloidal molecules then creating a large aggregation for suspending negative charge particles in the water. The swelling is reduced when using 5-10 wt % for Sodium silicate due to seal the fracture of samples and prevents further interaction with water as illustrated in (figure 14).

5. Conclusion

Tanuma shale composes mainly kaolin with a little amount of Illite. The concentrations of salts were accumulated on the surface of the shale sheet due to adsorb of salts on the surface of shale. The salts had differently affected on the shale samples depending on the ability of salts to adsorb on the surface of sheets. The appearance of veins on the shale surface of samples and twisted like a network represents gypsum veins. The microscope showed the distribution of salts on the surface of samples especially on the edge of shale that appeared the distribution of cations on the edge of kaolin mineral. Increasing the concentration of the salts will give a good recovery. The swelling was affected by the salts concentration, its percentage decreased by increasing the salts concentration. The high percentage recovery was found in the sodium silicate and KCL additives at concentration 5-10 % with polymer mud. One of the reasons causes of shale instability is related to the transportation of water/ions into or out of shale.

References

[1] Farrokhrouz M and Asef M R 2013 *Shale Engineering* (Leiden: CRC Press) ed 1 p 304
[2] Charles E Weaver 1989 *Clays, Muds, and Shales* (United States and Canada: Elsevier Science Publishers B.V.) ed 1 p 818
[3] Stephens, M. S. 2009 *AADE National Technical Conference and Exhibition* (New Orleans) (Houston: AADE) p 7
[4] Gomez, S. A. 2006 *AADE Drilling Fluids Technical Conference* (Houston: AADE)
[5] Sandra Gomez and Wenwu He. 2012 *IADC/SPE Asia Pacific Drilling Technology Conference and Exhibition* (Tianjin: China/ SPE) p 4
[6] Varga G 2007 *The structure of kaolinite and metakaolinite* Constantine the Philosopher University p 6
[7] Hawkins B 2015 *13th ISRM International Congress of Rock Mechanics* (Montreal: Canada/ International Society for Rock Mechanics ISRM) p 3
[8] Stoch L and Sikora W 1975 *Clays and Clay Minerals* **24** 156

IOP Conf. Series: Materials Science and Engineering **579** (2019) 012003 doi:10.1088/1757-899X/579/1/012003

Optimization of Drilling Parameters with Aid of Real Time Data for Buzargan Oil Field

Khalid Hazem Hassan[#] and Hassan A. H. Abdul Hussien
University of Baghdad, Baghdad, Iraq

[#]Corresponding auther e-mail address: eng.khalid.1994@gmail.com

Abstract: This paper showed how real-time data that obtained during drilling operation can be used for optimizing drilling parameter such as weight on bit (WOB), and rotary speed (RPM) in term of drilling cost per foot. The real-time data that obtained from mud logging unit included the rate of penetration (ROP), WOB, RPM, depth ..... etc.

In this work, the real-time data obtained from Buzurgan oil well (BU-50) were used for drilling optimization. Rock strength for each drilling foot of section obtained from a sonic log, using SPSS (Statistics) software to calculate it correction factors a,b, and c for Mishrif formation and apply nonlinear regression method after that make drill behind for validation purpose. Finally, the optimal drilling parameters were estimated with the cost model.

Based on the results obtained, a good agreement achieved between results of rate of penetration that obtained from drilling model and actual penetration rate along the drilling section. Also, cost analysis shows that the optimal weight on bit and rotary speed for drilling that section are 18 klb and 116 rpm respectively which achieves a minimum cost per foot.

1. Introduction

The drilling is one of the most expensive operations in exploration well and development well for oil and gas field. More than (25 – 35 %) of total development costs, is the cost of drilling operation[1]. In addition, exploration in extreme environment, complex well designs, deeper well, and environmental pressure lead to more increase in drilling costs. In a most drilling operation, rotating on time has less than half of the total find spent on drilling well. Thus, optimizing of drilling velocity leads to more reduction in the total drilling component cost like daily rig rate and fuel consumption [2].

Generally, drilling optimization is the application of technology that yields a decrease of drilling prices related to drill the well, also there are a different definition of drilling optimization: "is Optimized drilling techniques have considerably reduced drilling price. Results indicate that higher knowledge, additional expertise, and confidence can end in great savings within the future [3]. The process involves the post-appraisal of offset well record to calculate the costs effectiveness of choice control variables.

To understand optimization drilling operations should know all drilling variable which effects on drilling cost.The drilling parameters such as rotary speed (RPM), weight on bit, pump parameters, the rate of penetration (ROP), fluid properties, inclination and azimuth, depth and torque[4]

2. Real Time Data

The data which obtained from the previous well are very necessary for optimization operation where all drilling model work on data acquisition from old drilling well. Advanced computer technologies are working on storing a lot of data, and calculate difficult problems. From the beginning, all companies seek to reduce the cost of drilling by increasing weight on bit and rotary speed[5].

3. PDC Bit Model

The PDC bit model is created depending on design comprehensive for the cutter of PDC bit in (1994). The equation of rate of penetration for PDC bit some times utilized to determine confined compressive strength for rock[6].

Also, there is simple form derived from the same model and modified in (2008) for predicting the rate of penetration (ROP) for the polycrystalline diamond compact bit (PDC) that take in consider dealings between the rock strength and cutters, wear of bit, and lithology. New parameters are presented for the equivalent bit radius, lithology coefficients, dynamic cutter action, and cutter wear[7] as illustrated below:

$$ROP = W_f \frac{14.14 * WOB * RPM * \cos(\alpha)}{S * d_b * Tan(\theta) * f_c(P_e)} * \frac{a}{RPM^b * WOB^c} \quad (1)$$

Where:
S: Confined compressive strength,psi
WOB:Weight on bit,lbf
d_b: bit diameter, in.
θ : PDC cutter backrake angle, degrees
α : PDC siderake angle
RPM: rotary speed
W_f:Wear function,dimensionless
$F_c(p_e)$:Chip hold down function,dimensionless.
a,b,c:correction factors for cutter geometry

The values of correction factors(a,b, and c) were determined in laboratory or information apply nonlinear regression on drill-off tests data for each formation. fc(Pe) function is an estimate of the resultant forces on-chip when it is generated by the bit. The correlation for estimating this fc(Pe) function as follows[8]:

$$f_c(P_e) = C_c + a_c(P_e - 120)^{bc} \quad (2)$$

Where P_e is the difference between hydrostaic mud pressure and pore pressure (in psi) as follows;

$$P_e = P_{hy} - P_p \quad (3)$$

Where P_p is the pore pressure in psi.

Bit wear function (Wf) is the effect of bit wear on the rate of penetration. Eqs. (4) and (5) are used for determining wear of bit in term of RPM, WOB, confined rock strength and abrasiveness of rock as follows:

$$\Delta BG = W_c \sum_{i=1}^{n} WOB_i \,.\, RPM_i \,.\, A_{ABRI} \cdot S_i \quad (4)$$

$$W_f = 1 - \frac{\Delta BG}{8} \quad (5)$$

Where
ΔBG: the change in the bit tooth wear, dimensionless.

WOB_i: weight on bit.
S_i: confined compressive rock strength.
A_{ABRi}: the relative abrasiveness.

4. Rock Strength

Rock strength value along the drilled wellbore is very imported. This value is important when detection the safe drilling fluid window for determining the stability of the well, choosing mud weights and designing casing program, and for completion and stimulation purposes. Assessment of rock strength along the wellbore could be achieved by well logs, mechanical rock tests, and from drilling data.

In drilling operations, the rock strength of drilled formation has a major impact on the velocity of drilling. Its well known from field practice that as the compressive strength of rocks increases with depth, the rate of penetration decreased which leads to an increase in cost per foot[9].

In normal drilling situations, the drill bit is operated at a condition where the mud hydrostatic pressure is greater than formation pressure(overbalance drilling). This means that there is confined pressure on the drilled formations. Thus, the compressive strength derived from drilling models is confined compressive strength(CCS). Unconfined compressive strength (USS) is another measurement of rock strength but without applying confining pressure. In order to relate these two rock strength parameters, an empirical expression was formulated as a function of effective differential pressure as follows[10].

$$S = S_o\left(1 + a_s P_e^{b_s}\right) \tag{6}$$

So : (CCS) in MPa ,S_o: (UCS) in MPa ,P_e different between hydrostatic and pore pressure,(a_s, bs)are fitting constants. The permeability coefficients for chip hold down are show in table (1).

Table 1. Permeability Coefficients for chip hold down[8]

Coefficients	formations	
	permeable	Impermeable
P_e	P_{hy}-P_p	P_{hy}
a_c	0.00497	0.0141
b_c	0.757	0.470
c_c	0.103	0.569
a_s	0.0133	0.00432
b_s	0.577	0.782

4.1. Calculation of rock strength from sonic log

The usage of sonic velocity log for calculates rock elastic properties are good established. At present, there are many correlations between sonic travel time and rock strength or a grouping of different logs[11]. The rock strength depends mainly on lithology, so the rock strength was high for low porosity rock or low traveling time. The equation used in this study is shown below[12].

$$S_{os} = \left(\frac{1}{K_1(\Delta t_c - K_2)^2} + 2\right) * \frac{1}{145.08} \tag{7}$$

Where
Δtc : time of traveling,msec/ft.
S_{OS} : unconfined compressive strength (UCS),mpa.
k_1, k_2 are constants obtaind from table(2)

Table 2. values of k_1, k_2 constant

K_1	$5.15*10^{-8}$
K_2	23

5. Calculation of Drilled cost per foot

The calculation of cost per foot is done by the conventional cost equation as follows:[13]

$$COST = \frac{cost_{Rig}(T_{Rotating}+T_{Triping})+cost_{bit}}{footage\ drilling} \quad (8)$$

Where
$t_{Rotating}$:Rotating time ,hr
$t_{Tripping}$: Tripping time,hr

$$T_{triping} = (D_1 + D_2)\left(\frac{0.75hr}{1000ft}\right) \quad (9)$$

$$T_{Rotating} = \sum_{i=1}^{n} \frac{\Delta D_i}{ROP_i} \quad (10)$$

D1: depth in, D2: depth out, ΔDi : drilling depth.

6. Workflow of study

The work planner summary that will be used during this study is shown in the figure (1). This figure is described all the steps that will take place during this research.

6.1 Filtaration of sonic and drilling parameter Data

This is the first step of this process. The data of transition from the travel time of references well formation are filtered and averaged for every ten meters because of there high quality. Also, the measured drilling parameters such as rate of penetration, weight on bit and rotary speed that recorded in mud log were subjected to the same proceeding procedure in order to apply the model. The implementation of this procedure was an aid with the Microsoft Excel program.

6.2 Calculation of formation strength for refrence well(BU-50)

The second step of this process is to calculate the formation strength of drilled formations in the reference well (BU-50). The unconfined compressive rock strength (UCS) was generated from sonic traveling time measurements. While confined compressive rock strength (CCS) was created from calculated (UCS) by depending on over balance pressure value (100-300 psi) that was used during drilling this formation. The rock strength values were used in drilling model calculation while drilling the measured formation is the confined strength. These strengths represented resistant to the bit penetration during drilling and can be used in drilling simulation and bit selection. This lead to produced unconfined and confined formation strength for reference well (BU-50) along formations for each meter. Figure (2) presents the calculated UCS & CCS for Mshrif formation of well (BU-50).

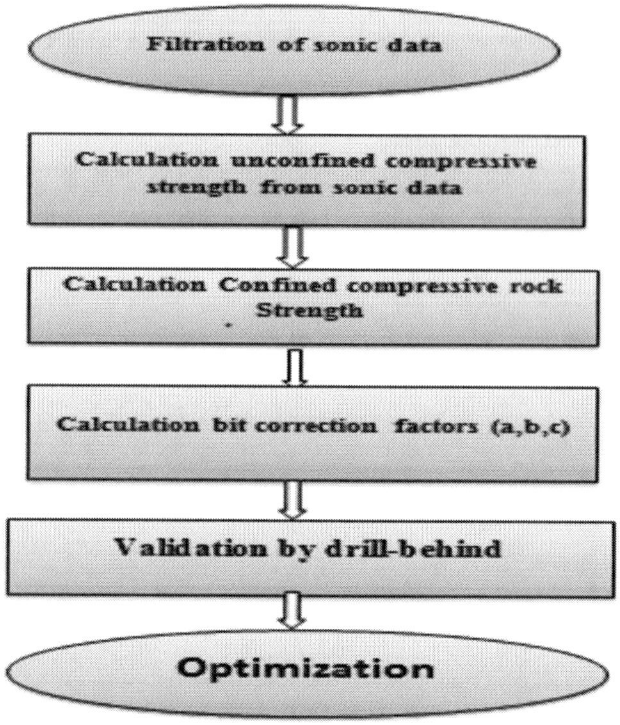

Figure 1. Workflow of study

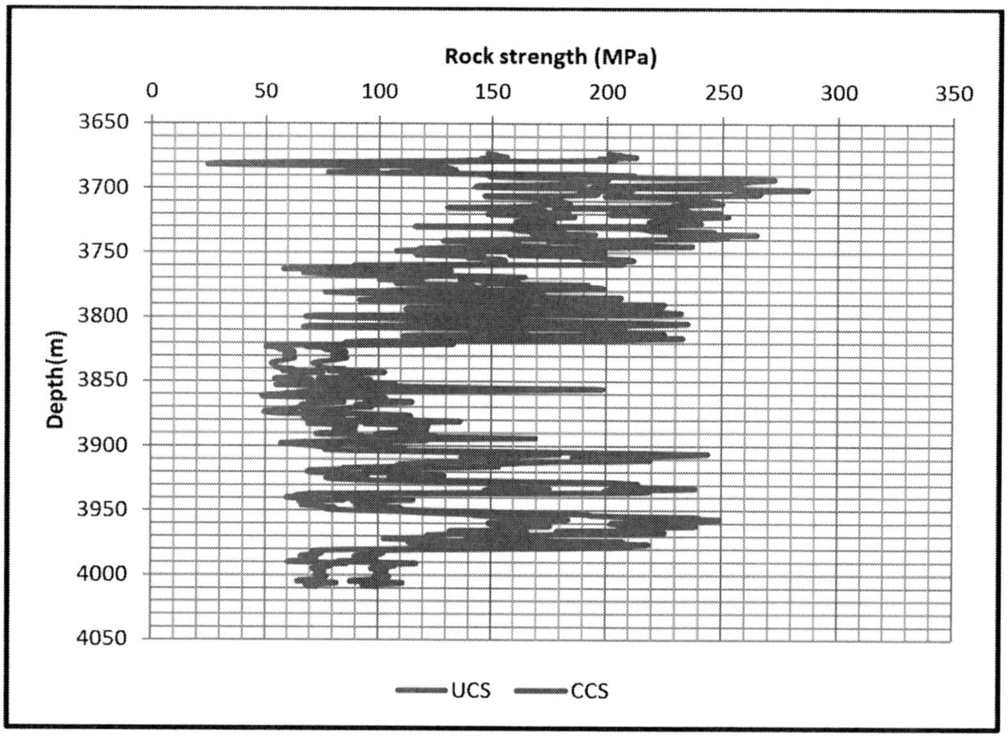

Figure 2. Unconfined and confined rock strength from sonic logs for mishrif formation of (BU-50)well.

6.3 Calculation of bit correction factors (a,b,c)

The third step of this process is to calculate bit correction factors a,b and c. As mentioned above, this correction factor was calculated for each drilled formations in the reference well in order to accounts for many effects that are not considered in the theoretical modeling of the penetration rate model. In addition, these correction factors will be the same for particular PDC bit, with a particular geometry, and specific type lithology.

Table (3) shows the calculated bit correction factors values by applying the nonlinear regression method. Real-time data of mud log (ROP, weight on bit, rotary speed) for mishrif formations are fed into SPSS (Statistics) software and apply nonlinear regression method. This bit correction factor values will be used for the same PDC in the future drilling for the same formation. In addition, the value of bit wear coefficient is calculated by using drilling parameter and rock strength.

Table 3. (PDC) bit correction factors for Mishrif formation.

Formation	*a*	*b*	*C*	*Wc*
Mishrif	317	0.04	0.46	3.2E-11

6.4 Validation of drilling simulation (Drill-behind procedure)

The objective of the drill-behind procedure is to check the accuracy of the results calculated from the sonic log model and bit correction factors used in this study. This process uses the unconfined compressive rock strength which is generated from reference well (BU-50) to calculate the rate of penetration for another well(in this study we use well BU-47) for Mishrif formation, and comparing with actual ROP of this well.

First, the unconfined compressive rock strength for well (BU-50) for Mishrif formation which is calculated from necessary data in past step converted to confined compressive strength according to hydrostatic and pore pressure for (BU-47)well, where the top of Mishrif formation in well(BU-50) matched to the top of the same formation in well(BU-47). The results of this step for Mishrif formation are shown in figure 3.

As illustrated in figure (3), there is very good match between the rate of penetration which calculated from the model in well (BU-47) using rock strength that generated from reference well (BU-50) and the actual value of penetration rate which acquisition from drilling recorder for (BU-47) , the value of matching is 94%.

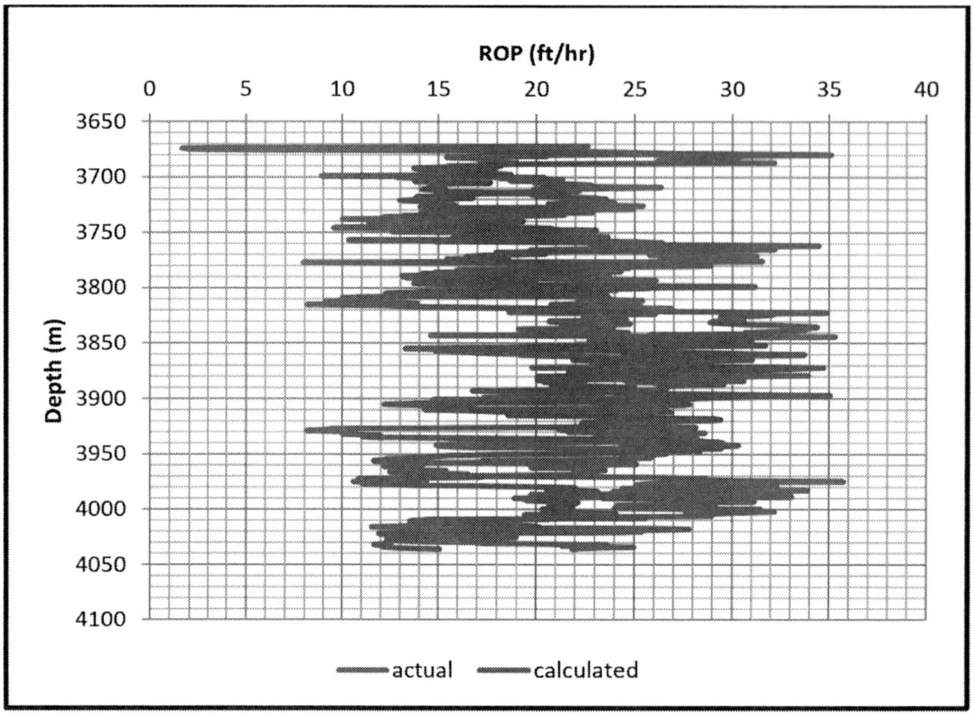

Figure 3. Comparison ROP simulation to the actual for Mishrif formation, Well BU-47.

6.5 Optimization of weight on bit and rotary speed

Optimization is the procedure to find the optimum drilling parameter which gives a maximum rate of penetration or minimum cost. Now we will find the optimum weight on bit and rotary speed. The first step will express a relationship between weight on bit and rotary speed with cost when all drilling parameters are constant. Figure (4) shows the relationship between WOB (klb) , RPM, and cost ($/ft) for Mishrif formation in Buzargan oil field.

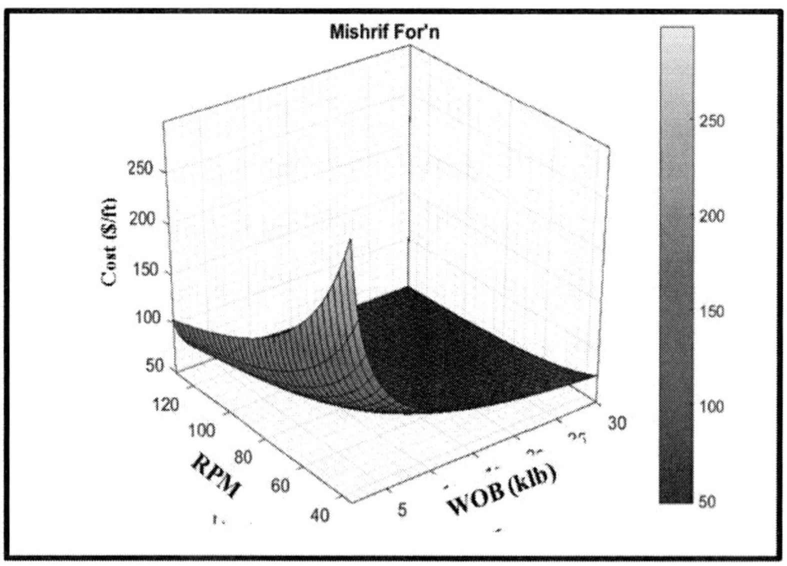

Figure 4. Relationship between WOB (klb) , RPM, and cost ($/ft) for Mishrif formation .

It is noticed that the cost of drilling decreases as the weight on bit and rotary speed increase and the optimum drilling when the WOB and RPM are in maximum value through the parameter bounds for bit manufacture. The rate of penetration as discussed above decreases when WOB and RPM decrease, then the cost per foot increases.

Finally, In this study, it can be trying to reduce the cost of drilling by increasing the rate of penetration during drilling formation. After modelling the equations on the real-time data by using Microsoft Excel program, the aim of the work is to increase the rate of penetration by making all drilling parameters constant and change the values of weight on bit(WOB) and rotary speed(RPM) until high rate of penetration and less cost of drilling ($/ft)are achieved. These values of Weight on bit(WOB) and rotary speed(RPM) which give good results will assume the optimum drilling parameters. The results in a table (4) show the optimum values of weight on bit and rotary speed which causes higher ROP and lower cost per foot of mishrif formation in comparison with actual data.

Table 4. Result optimization of drilling parameter for Mishrif formation in well(BU-50)

Formation	Parameters	WOB (klb)	RPM	ROP (ft/h)	Cost ($/ft)
Mishrif	Actual	14	104	23	77.7
	Optimization	18	116	27	70
		Saving			7.7

7. Conclusion:

This study presented a methodical approach to optimize drilling parameters for Mishrif formations from Buzargan oil field utilizing real-time data for two wells. The following conclusions of result include:

1- The results of unconfined and confined compressive strength of reference that generated from the sonic log for Mishrif formations increase with depth.

2- Applying the rate of penetration model on data using rock strength, which obtained from reference well for each formation in order to find optimum operational parameters during the wells in Buzargan oil field.

3- The results which obtained from optimization operational parameter, rotary speed and weight on bit for Mishrif formation, could potentially reduce the cost drilling by 7.7$/ft.

4 - Analysis of the drilling parameters showed that an increase in WOB and RPM value led to a decrease in cost per foot.

References

[1] Barker P 2004 *Problem based learning approach to Drilling engineeering* (Higher Education Academy Engineering Subject Centre, Loughborough Universit)

[2] Oscar P U, Mohammed J, Muhammad N, Mohamed A, Badar A, Rashid A and Michael B 2012 Drilling Optimization In Deep Horizontal Wells *Int. Pet. Tec. Conf*. (Thailand: Bangkok)

[3] Chen D K 2004 New Drilling Optimization Technologies Make drilling More Efficient *Canadian International Petroleum Conference* (Calgary, Alberta)

4] Tuna E and Mehmet E 2010 Real-Time Optimization of Drilling Parameters During Drilling Operations *Middle East Tech SPE Oil and Gas India Con and Ex.* (Mumbai : India) **2** 129126

[5] Wallace, S.P., Hegde, C.M. and Gray 2015 A system for real-time drilling performance optimization and automation based on statistical learning methods. In *SPE Middle East Intelligent Oil and Gas Conference and Exhibition*. Society of Petroleum Engineers

[6] Hareland G, Rampersad PR 1994 Drag-bit model including wear *SPE Lat Am P Eng Con* (Argentina) p 657–667
[7] Hareland G, Motahhari HR, Hayes JP, Qureshi A 2008 Increased Drilling Efficiency of Gas Storage Wells Proven Using Drilling Simulator. *InCIPC/SPE Gas Technology Symposium 2008 Joint Conference 2008 Jan* 1. SPE
[8] Rampersad PR, Hareland G, Boonyapaluk P 1994 Drilling optimization using drilling data and available technology. *InSPE Latin America/Caribbean Petroleum Engineering Conference 1994* Jan 1.SPE
[9] Adel A 2017 Application of Artificial Neural Networks in Prediction of Uniaxial Compressive Strength of Rocks Using Well Logs and Drilling Data (in Iran) *Procedia Eng* vol 191 (Tehran: Islamic Azad University) p. 279–286
[10] Rastegar M, Hareland G, Nygaard R, Bashari A. 2008 Optimization of multiple bit runs based on ROP models and cost equation: a new methodology applied for one of the Persian Gulf carbonate fields *InIADC/SPE Asia Pacific Drilling Tech Conf and Ex*
[11] Onyia EC.1988 Relationships Between Formation Strength, Drilling Strength, and Electric Log Properties *InSPE An. Tech. Conf. and Exh*
[12] Kerkar PB, Hareland G, Fonseca ER, Hackbarth CJ. 2014 Estimation of rock compressive strength using downhole weight-on-bit and drilling models *InIPTC: In. Pet. Tech. Conf*
[13] Self R, Atashnezhad A, Hareland G. 2016 Reducing Drilling Cost by finding Optimal Operational Parameters using Particle Swarm Algorithm. *InSPE Deep. Dr. and Com. Con* SPE

IOP Conf. Series: Materials Science and Engineering **579** (2019) 012004 doi:10.1088/1757-899X/579/1/012004

Improving Drilling Fluid Properties at High Pressure Conditions Using Selected Nanomaterials

Nada Sabah[1], Asawer A. Alswasiti[2#], Massara Salam

[1] Petroleum Department, College of Engineer, University of Baghdad, Baghdad-Iraq
[2] Petroleum Technology Department, University of Technology, Baghdad-Iraq

[#]Corresponding author e-mail address: 150003@uotechnology.edu.iq

Abstract. Drilling fluids with conventional lubricants (including solid and liquid lubricants) will produce lubricant film between drill string and wellbore/casing, but this film will not be strong enough to protect the bit bearing surfaces subjected to high load conditions during the drilling operations, resulting in shortening the life of drill bit. In this research, the possibility of using nanomaterials as extreme pressure lubricants was investigated by adding different concentrations (0.02, 0.05 and 0.1) wt. %of TiO_2nanoparticles (NPs) and Cu NP into water- based mud. The lubricating film strength of water based mud samples under extreme pressure conditions were measured using Extreme Pressure (EP) and Lubricity Tester. With the purpose of getting the best performance from extreme pressure lubricants, the effect of nanomaterials on rheological (including plastic viscosity, yield point, apparent viscosity, and gel strength) and filtration (including filtrate loss volume and mud cake thickness) properties were measured. TiO_2 NPs and Cu NP showed a good enhancement in lubricating film strength under extreme pressure conditions. The maximum value of lubricating film strength increasing percentage (69.6%) was obtained using TiO_2 NPs with concentration of 0.1wt.%.

1. Introduction

One of the significant functions of drilling fluid is lubricating the bits and drill string. They fail more rapidly if it was not lubricating effect of drilling fluid, where the bits are subjected to heavy weight load which increase the bearing and tooth wear of roller cone bits. Also during drilling extended and horizontal wells the torque and drag between the drill string and wellbore increase due to high friction which leads to reduction in the equipment life and decrease in the well length [1].Oil and synthetic base muds lubricate better than most water base muds, but lubricants can be added to water base muds to improve them.Two main properties are important to choose the lubricantmaterials; they are coefficient of friction(COF) and extreme pressure(EP). The extreme pressure property of lubricant concerns with the strength of the film produced by the lubricant where the metal surfaces are in contact with each other under high pressure. Most of lubricant losses their hydrodynamic lubricant at high pressure load. The extreme pressure lubricant provides a strong protection film of metals that are subjected to high pressure load. It is not necessary that lubricant materials that have low coefficient of friction can have a strong lubrication film[2].

In recent years nonmaterials have been used in all fields of petroleum industry and drilling process was one of them. Adding nanomaterials to drilling fluid in order to improve lubricity, improve rheological properties,improve fluid loss, improve BHA's life, and improve thermal stability of drilling fluid. Some researchers [3],[4][5]and [6]used nano-additive to enhance the rheological,filtration and coefficient of friction reduction properties of drilling fluid. Jahns [7] used alumina, titania, and silica nanoparticles to improve water base mud, he found that titania, and silica nanoparticles reduced friction factor effectively, while alumina nanoparticles had a limited friction reduction. Taha [8] added (1-5) vol% of nanographene to water base mud, he gained a 50% torque reduction in high temperature-high pressure (HTHP) water-based mud by using 4% - 5% by volume of the nonmaterial. In Krishna work [9], a boron-based a nonmaterial was added to water based mud, a 80% torque reduction in 10 lb/gal water-based mud and 52% torque reduction in 13.5 lb/gal water-based mud were obtained by using 5% by volume of the product. [10]and[11]used chemically and mechanically generated barite nanoparticles at different concentrations to enhance the lubrication properties of water base mud. Al_2O_3, Fe_3O_4, and SiO_2nanoparticles with KH550 coating effect on water based mudlubricity have been studied by [12]. He found that lubricity was improved with the addition of SiO_2nanoparticle with and without coating, performing of SiO_2nanoparticle with KH550 coating was better than SiO_2nanoparticle without coating. [13]used Cu, XC polymer, barite, and lignite nanoparticles. She gained good reduction in COF for both water-basedmud and KCl polymer mud. Themaximum value of COF reduction percentage (more than 60%) wasobtained by adding 4 gm XC Polymer toKCl polymer mud.[1]investigated the usage of adding multi-walled carbon nanotubes (MWCNT) and graphene-nanplatelets (GNP) to water base mud in order to reduce coefficient of friction. Their results showed that the addition of 0.01 ppb of MWCNT and 0.2 ppb of GNP gave torque lubricityreduction between 35 to 59 %.

In spite of the above workers, studying the combined effect of nanomaterilas on both of lubricity coefficient and extreme pressure has not been done yet. Hence, the main objective of this research is to investigate the possibility of using nanomaterial as lubricant materials and study the stability of the thin film formed between the two metal surfaces under extreme pressure conditions.

2. Experimental Work

2 . 1 Materials

Commercial bentonite (supplied by Oren Hydrocarbons Middle East Inc.)wasusedin the preparation of water-based mud. The nanomaterials that had been used in this study were Copper NP* (supplied by Nanjing Nano Technology Co., Ltd.)with (10-30)nmparticle size and purityof 99.99% and Titanium dioxide NP of 25nm particle sizeand 99.98% purity (supplied by Cheng Du Micxy Chemical Co., Ltd.)

2.2 Procedure of Experiments

2.2.1 ClayCharacterization Analysis:

Bentonite was quantitatively and qualitatively analyzed using XRF-1800 Sequential X-ray Fluorescence Spectrometerand XRD-6100/7000 X-ray Diffractometerrespectively. While Particle-size distributions of bentonite was measured using a Malvern Mastersize 2000 particle size analyzer, these analyses were carried out in Geological Survey/Central Laboratories Department.

2.2.2Drilling Fluid Preparation:

To prepare the blank sample of water-based mud (according to API Specification 13A, 2010), the bentonite is mixed with fresh water using Hamilton Beach Mixerfor 10 minutes and then the suspension is aged in a sealed container for 16 hours to ensure good hydration of bentonite. It is important to mention that the blank sample of water-based mud didn't have any other chemicals, just waterandbentonite, in its formulation.This is in order to investigate the effect of nanomaterials on the drilling fluid properties without the side effect of chemicals. Thenanomaterials with concentrations of 0.02, 0.05, 0.1wt% were added to water-based mud and mixed for 10 minutes using Hamilton

Beach Mixer then the samples were continued to mix using Ultrasonic Bathfor 10 minutes to ensure a good dispersion of nanomaterials.

2.2.3 Measurements and Instruments:

The lubricating quality of water-based mud with and without nanomaterials had been measured by using OFITE EP and Lubricity Tester. The standard lubricity coefficient test is run at 60 rpm with 150 in-lb of force (the equivalent of approximately 600 psi pressure of the intermediate fluid) is applied to two hardened steelsurfaces, a rotating ring and a stationary block. Before any test, the apparatus was calibrated with distilled water so as to calculate the value of correction factor (CF) using the following equation:

$$CF = 34 / \text{Meter Reading} \quad (1)$$

The lubricity coefficient (LC) of water based mud was calculated manually using the data obtained from apparatus as follows:

$$LC = CF * \text{Meter Reading} / 100 \quad (2)$$

Where:
CF = Correction Factor
LC = Lubricity Coefficient

The film strength of water-based mudunder extreme pressure conditions (extreme pressure testwas run at a high shear rate, 1,000 RPM, and torque ranging from 150 to 270 inch-pounds)was calculated manually using the data obtained from apparatus as follows:

$$P = T / 1.5 * L * W \quad (3)$$

Where:
P = Film Strength (psi)
T = Torque Meter Dial Reading (in-lb)
W = Scar Width (inches)
L = Scar Length (inches)

The rheological properties of water based mud with and without nanomaterials had been measured using OFITE Hand Crank Rheometeras follows:

The rheological properties including plastic viscosity (PV),yield point (YP), and apparent viscosity (AV) were calculated manually using the data obtained from the rheometer as follows:

$$PV = \phi600 - \phi300 \quad (4)$$
$$YP = \phi300 - PV \quad (5)$$
$$AV = \phi600 / 2 \quad (6)$$

Where:
PV = Plastic Viscosity (cP)
Ø600 = Dial Reading at 600 RPM
Ø300 = Dial Reading at 300 RPM
YP = Yield Point (lb/100 ft2)
AV = Apparent Viscosity (cP)

Also the gel strength was measured from the rheometer.Filtrate volumeand mud cake thickness were measured using OFITE Low-Pressure Filter Press with Dead Weight Hydraulic Assembly.

IOP Conf. Series: Materials Science and Engineering **579** (2019) 012004 doi:10.1088/1757-899X/579/1/012004

3. Results and Discussion

3.1 Clay Characterization

The X-Ray fluorescence (XRF) and the X-Ray diffraction (XRD analyses of bentonite that was used in the preparation of water-based mud are shown in Tables 1 and 2 and figure 1. The analyses show that the major component of bentonite is montmorillonite and a high percentage of Silicon Oxide (SiO2) along with Aluminum Oxide (Al2O3), as well as a low percentage of Potassium Oxide (K2O).

Table 1. XRF Analysis of Bentonite.

Component	Quantity	Component	Quantity
SiO2 %	49.98	SO3 %	0.16
Fe2O3 %	10	LOI %	8.49
Al2O3 %	17.5	P2O5 %	0.62
CaO %	4.4	Na2O %	2.24
MgO %	4.9	K2O %	0.12

LOI: Loss On Ignition

Table 2. XRD Analysis of Bentonite.

Minerals	
Major	Montmorillonite, Quartz
Minor	Palygorskite, Calcite

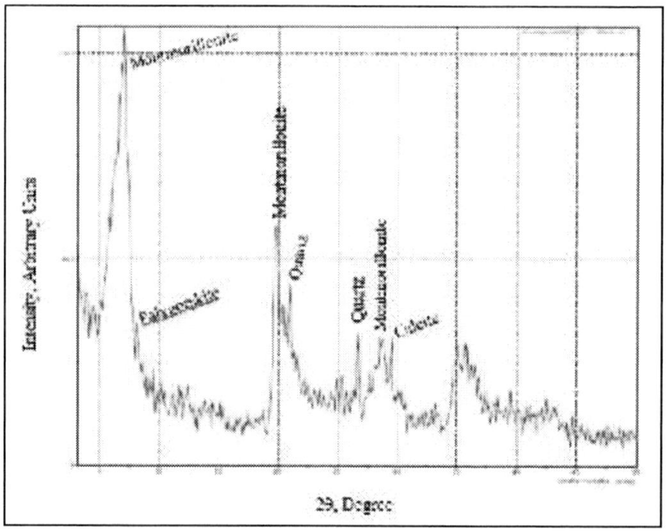

Figure 1. XRD Analysis of Bentonite.

The particle size distribution of bentonite is within average size of 238 nm, as shown in Figure 2.

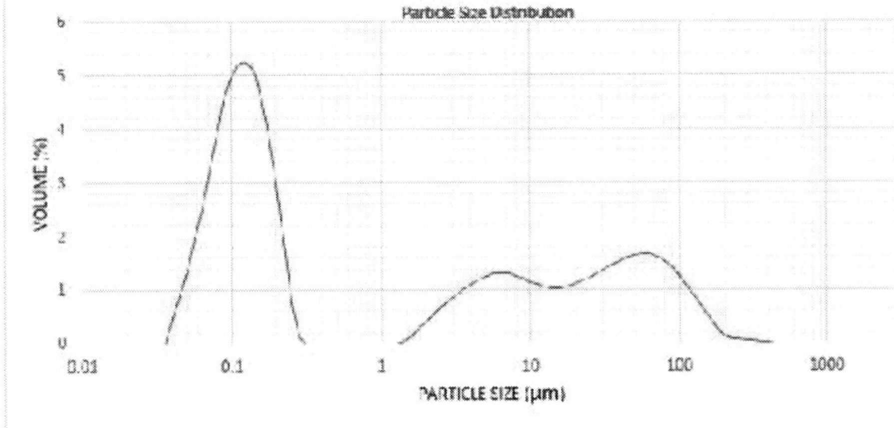

Figure 2. Particle Size Distribution of Bentonite.

3.2 EPLubricating Quality Measurement:

Effect of TiO2 NP and Cu NP addition to water-based mud with different concentrations on EP film strength is illustrated in Table 3 and shown inFigures 3&4.

Table 3. EPFilm Strength and EPFilm StrengthIncreasing % of Water Based Mud with Nanomaterials.

weight %	Lubricity coefficient	Lubricity Reduction %	Film Strength	Film Strength Increasing %
Blank	0.532	/	2373.6	/
TiO2 NPs				
0.02	0.483	9.21	3371.6	42
0.05	0.513	6	3491.13	47
0.1	0.491	7.7	4025.7	69.6
Cu NP				
0.02	0.5	5.4	3326.94	40.16
0.05	0.503	5.45	3306.87	39.3
0.1	0.482	9.39	3748.43	57.92

The addition of both TiO2 NP and Cu NP to water-based mud caused a significant increase in lubricating film strength at extreme pressure conditions, the greatest increase (69.6%) was observed by adding TiO2 NP at concentration of 0.1 wt%. However, the effect of these additives on the reduction of coefficient values is so slight and insignificant. Also, they increase slowlywith increasing the concentrations of nanomaterials.

According to the layered shape of montmonolite of bentonite, the nanoparticles move towards the surface of these layers due to some physicalforces (vanderwalls and coulombic), and this will make a change in the surface of bentonite. Indeed, due to the high temperature produced from the friction at the point of rubbing contact, a chemical reaction occurs leading to form a film on the surface of metal. However, this film is thin and this can be concluded from the low effect of nanoparticles on lubricity coefficient. Increasing the weight of nanomaterials showed a regular effect on these values. In spite of that, the strength of this film at extreme pressure increased with the addition of nanomatierals addition and its weight values.

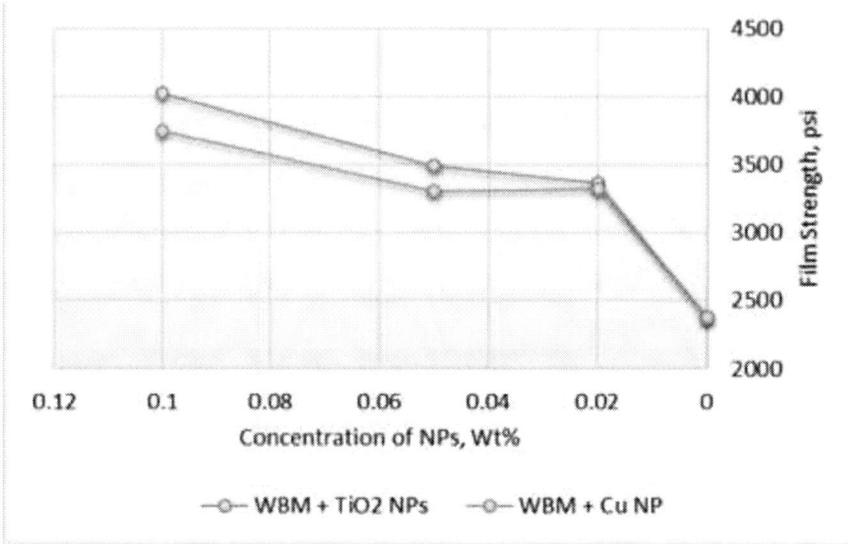

Figure3. Effect of Nanomaterial's Concentration on EP Film Strength of Water Based Mud.

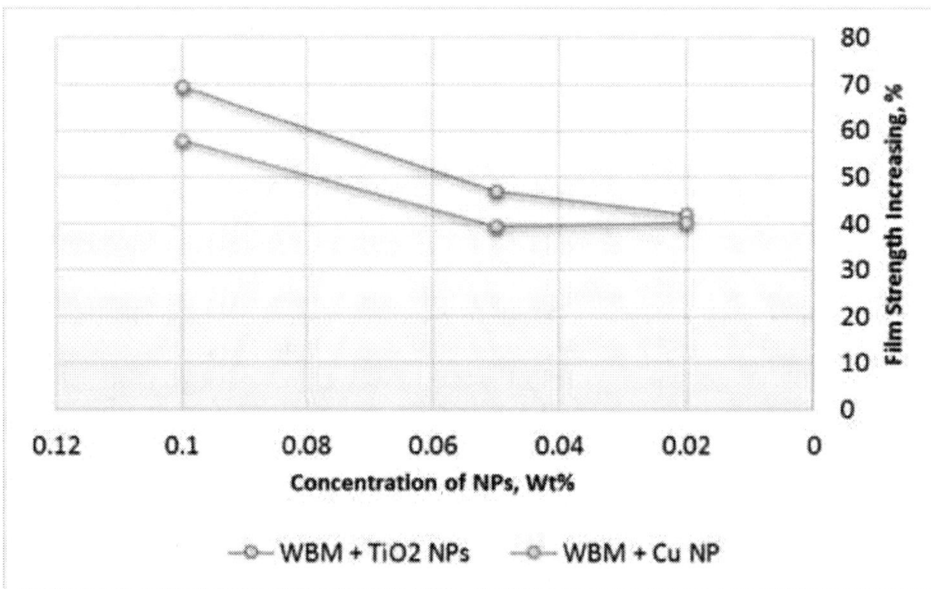

Figure 4. Effect of Nanomaterial'sConcentration on EP Film Strength Increasing Percentage of Water Based Mud.

3.3 Rheological Properties:

The rheological properties of water-based mud with different concentrations of both TiO2 NP and Cu NP are illustrated in Table 4 and shown in Figures 5 to 8.

Table 4. Rheological Properties of Water Based Mud with Different Concentrations of Nanomaterials.

weight %	PV, cP	YP, lb/100 ft^2	AV, cP	Gel Strength, lb/100 ft^2	
				10 sec	10 min
Blank	9	27	22.5	12	15
NPs					
0.02	11	28	25	13	19
0.05	12	28	26	20	22
0.1	10	30	25	20	20
Cu NP					
0.02	10	28	24	15	20
0.05	11	27	24.5	17	20
0.1	12	30	27	20	20

As mentioned above, adding nanomaterials can affect on the surface of bentonite that lead to change in the drilling fluid properties. Indeed, these nanomaterials have large surface area per volume causing an increase in the interaction between the particles. A slight increasein plastic viscosity, yield point and apparent viscosityare shown in figures 5 to 7. Also, a slight increase ingel strength values with an increase in nanoparticle concentration as a result of the increasing inattraction forces and electrostatic force[14][15] and [16].

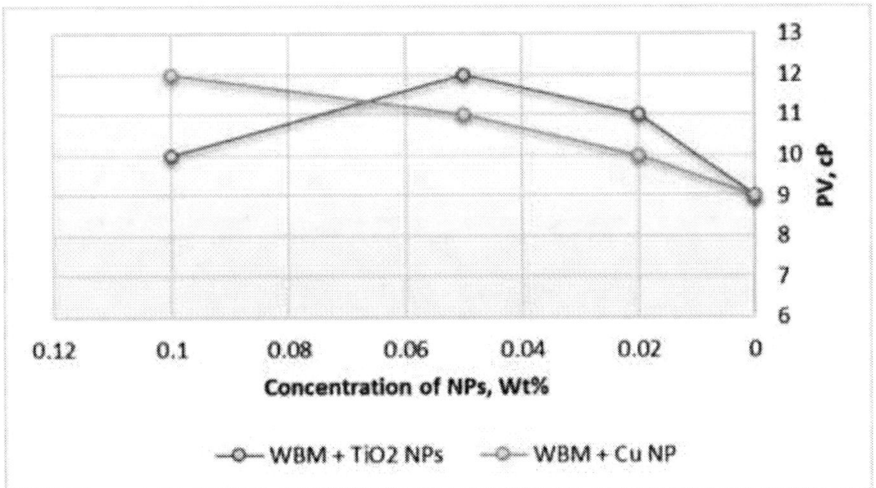

Figure 5. Effect of Nanomaterial's Concentration on Plastic Viscosity of Water Based Mud.

IOP Conf. Series: Materials Science and Engineering **579** (2019) 012004 doi:10.1088/1757-899X/579/1/012004

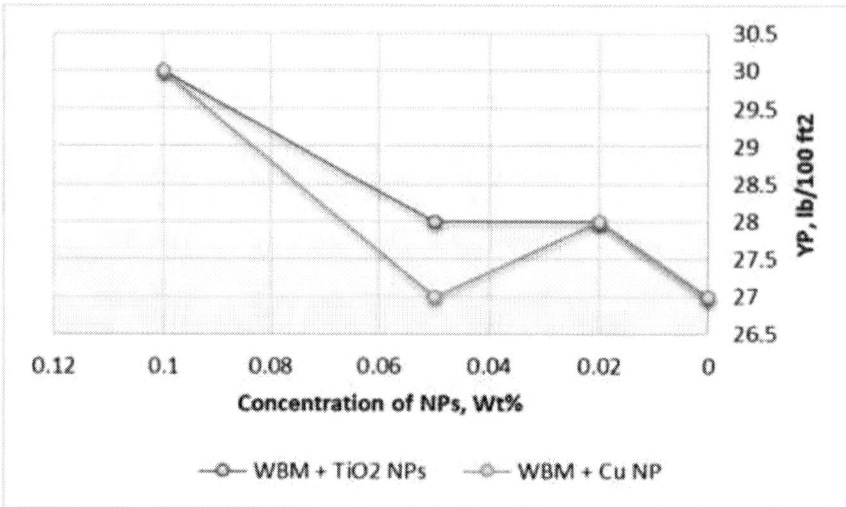

Figure 6. Effect of Nanomaterial's Concentration on Yield Point of Water Based Mud.

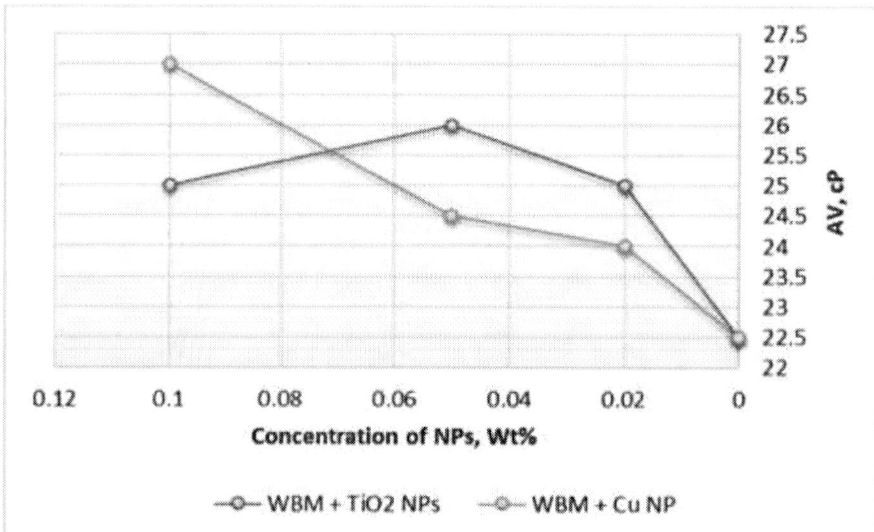

Figure 7. Effect of Nanomaterial's Concentration on Apparent Viscosity of Water Based Mud.

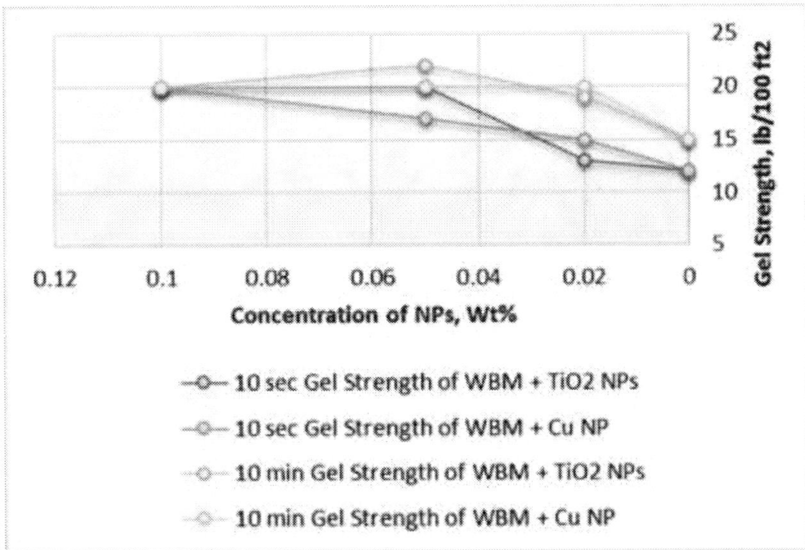

Figure 8. Effect of Nanomaterial's Concentration on Gel Strength of Water Based Mud.

3.4 Filtration Properties:

The filtration properties of water-based mud with different concentrations of TiO2 NP and Cu NP are illustrated in Table 5 and shown in Figures 9 and 10.

Adding TiO2 NP to water-based mud caused a slight decrease in filtrate volume, and approximately had no effect on mud cake thickness. While the addition of Cu NP caused an increase in mud cake thickness at concentrations of 0.05 and 0.1 wt%, and had no effect on filtrate volume.

Nanomaterials may cause a decrease in filtrate loss volume, this may be explained due to the ability of nanomaterials to seal the nanopore throats of the wellbore formation and prevent water infiltration [17] and [18].

Table 5. Filtration Properties of Water Based Mud with Different Concentrations of Nanomaterials.

Weight, %	Filtrate Volume, ml	Mud Cake Thickness
Blank	10	0.2
TiO2 NPs		
0.02	8.8	0.23
0.05	9.6	0.28
0.1	8.8	0.25
Cu NP		
0.02	10	0.17
0.05	10	0.35
0.1	10	0.3

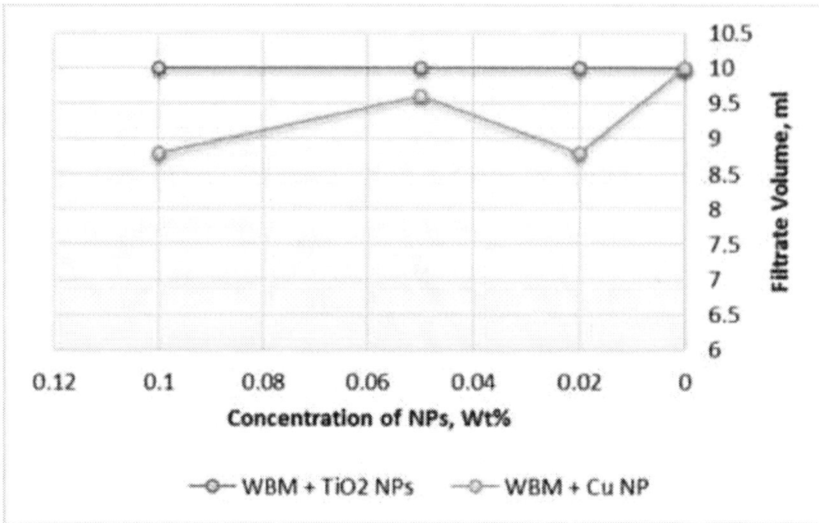

Figure 9. Effect of Nanomaterial's Concentration on Filtrate Volume Loss of Water Based Mud.

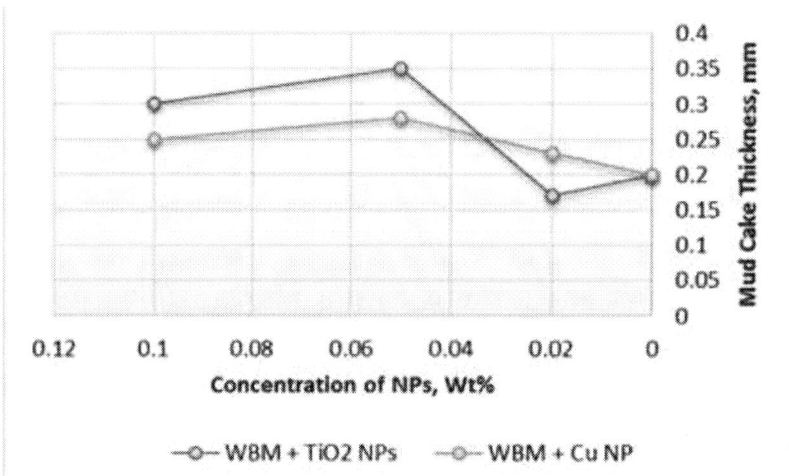

Figure 10. Effect of Nanomaterial's Concentration on Mud Cake Thickness of Water Based Mud.

4. Conclusions

1. Both TiO2 NP and Cu NP caused a significant increase in lubricating film strength at extreme pressure conditions.
2. TiO2 NP provide better effect on lubricating film strength
3. Slight decrement of lubricity coefficient upon adding various concentrations of TiO2 NP and Cu NP
4. Adding TiO2 NP and Cu NP causes a slight change in rheological and filtration properties.

Acknowledgements
This study was carried out on equipment of the petroleum technology research center at Petroleum Technology Department.

References

[1] A. R. Ismail, M.S.A. Rashid,and B.Thameem 2018, *IOP Conf. Series: Materials Science and Engineering* vol 380
[2] M. Rosenberg and R. Tailleur 1959 *Increased Drill Bit Life Through Use of Extreme Pressure Lubricant Drilling Fluids* (Society of Petroleum Engineers)
[3] E. Jabrayilov 2014, Friction Reduction by Using Nanoparticles in Oil-Based Mud, Trondheim: (Norwegian University of Science and Technology / Department of Petroleum Engineering and Applied Geophysics)
[4] F. H. A. Al-Ogaili and N. Suripis 2016 *International Journal of Scientific &Engineering Research* **7** 1489-1497
[5] B. Vegard and M. Belayneh 2017 *International Journal of NanoScience and Nanotechnology*, **8** 25-35
[6] S. Wrobel and M. Belayneh 2017 *International Journal of NanoScience and Nanotechnology* **8** 59-71
[7] C. Jahns 2014 Friction Reduction by Using Nano-Fluids in Drilling, Trondheim: (Norwegian University of Science and Technology / Department of Petroleum Engineering and Applied Geophysics)
[8] N. M. Taha and S. Lee 2015 Nano Graphene Application Improving Drilling Fluids Performance *International Petroleum Technology Conference* (Doha)
[9] S. Krishnan, Z. Abyat and C. Chock 2016 Characterization of Boron-Based Nanomaterial Enhanced Additive in Water-Based Drilling Fluids: A study on Lubricity, Drag, and Fluid Loss Improvement the *SPE/IADC Middle East Drilling Technology Conference and Exhibition* (Abu Dhabi)
[10] V. Caldarola, S. Akhtarmanesh, A. Cedola, R. Qader and G. Hareland 2016 Potential Directional Drilling Benefits of Barite Nanoparticles in Weighted Water Based Drilling Fluids *The 50th US Rock Mechanics / Geomechanics Symposium* (Houston)
[11] G. Alshubbar, T. Coryell, Atashnezhad, A. Akhtarmanesh, S. Hareland and G. Hareland 2017 The Effect of Barite Nanoparticles on the Friction Coefficient and Rheology of Water Based Mud *The 51st US Rock Mechanics / Geomechanics Symposium* (California)
[12] P. Dhiman 2016 Experimental Investigation of the Imfluence of Various Nanoparticles on Water-Based Mud Thesis (Fairbanks: University o f Alaska Fairbanks/College of Engineering and Mines/Department o f Petroleum Engineering)
[13] M. Salam 2018, Improving Drilling Fluid Lubricity Using Nanomaterials (Petroleum Technology Department/University of Technology, Baghdad)
[14] A. R. Ismail, T. C. Seong, N. A. Buang and W. R. W. Sulaiman 2014 Improve Performance of Water-Based Drilling Fluids Using Nanoparticles *The 5th Sriwijaya International Seminar on Energy and Environmental Science & Technology* (Palembang)
[15] A. H. Salih, T. A. Elshehabi and H. I. Bilgesu 2016 Impact of Nanomaterials on the Rheology and Filtration Properties of Water-Based Drilling Fluids *The SPE Eastem Regional Meeting* (Canton)
[16] A. E. M. and A. M. 2015 *Journal of Physical Science and Application* **5** 415-422
[17] A. R. Ismail, N. M. Rashid, M. Z. Jaafar, W. R. W. Sulaiman and N. A. Buang 2014 *Journal of Applied Sciences* **14** 1192-1197
[18] A. Aftab, A. Ismail and Z. Ibupoto 2017 *Egyptian Journal of Petroleum* **26** 291–299

IOP Conf. Series: Materials Science and Engineering **579** (2019) 012005 doi:10.1088/1757-899X/579/1/012005

Applying Selected Lost Circulation Materials to Study Lost Circulation of Azkand Formation in Khabaz Oil Field

Mazin J. Shibeeb[1#], Asawer A. Alwasiti[2], Nada S. Al_Zubaidi[3]

[1, 2] Department of Petroleum Technology, University of Technology, Baghdad-Iraq
[3] Department of Petroleum Engineering, University of Baghdad, Baghdad-Iraq

[#]Corresponding author e-mail address: mazin.shibeeb@gmail.com

Abstract. Managing lost circulation during drilling is a great challenge issue in naturally fractured or vugy formations. This situation becomes further complicated if there are gas or oil returns at the same formation or at a formation above it. The lost circulation problem happens in some of the northern Iraqi oil fields such as Khabaz oil field. Particulate lost circulation materials (LCM) have been used for many years to control lost circulation. However, these traditional LCMs are not efficiently applicable to cure sever to total losses, such as highly fractured formations and formation with large vugs. In this work, an experimental attempt was done to solve such problem in Khabaz oil field. The experiments were carried out using a closed loop circulation system that simulates the drilling fluid loss into formations. Five core plugs from Azkand formation of Khabaz oil field from different depths were used. The experiments were grouped into three, the first was done with the fresh-water-based drilling fluid to measure the severity of losses. The second, fresh-water-based drilling fluid was treated with traditional LCM (cotton seeds) to investigate how the severity of losses is affected by these materials. Adding polymer lost circulation material were the last sets to modify the drilling fluid rheological properties which prepare a medium within the lost circulation zone that increases the friction inside the losses paths to increase the LCMs (the polymer itself) sealing efficiency.

1. Introduction

Lost circulation can be defined as the reduced or total absence of drilling fluids flow up the annulus when drilling fluids are pumped through the drill string. Or it can be defined as a lack of mud returning to the surface after being pumped down a well [1] . Lost circulation occurs when the drill bit encounters natural fissures, fractures or caverns, and mud flows into the newly available space. Also, it may be caused by applying more mud pressure (that is, drilling overbalanced) on the formation than it is strong enough to withstand, thereby opening up a fracture into which mudflows [2] . Even though lost circulation is a well-known problem for years, lost circulation still considered one of the main providers to non- productive time (NPT) [3] .

During drilling mud circulate back to the surface, the fluid comes into contact directly with the wellbore inner walls. In traditional drilling practices, the wellbore pressure exceeds that of the formation to allow drilling to be continued without allowing the formation fluids from entering the wellbore. As was mentioned previously, in order for a fluid to be lost from the wellbore into the formation, the wellbore pressure has to exceed the formation pore pressure. In one method of fluid loss from the wellbore into the permeable rock, a filtration process takes place, whereby the solid particulates and the emulsion droplets accumulate on the wellbore wall and form a filter cake after the liquid components have left them and moved into the rock. The low permeability of this cake works as a barrier to keeps the volume of the fluid lost by leak-off relatively low. Knowing that this is not considered a lost circulation event. But when the rock is fractured, vugular, highly porous or induced fractures when the wellbore pressure is higher than the rock's tensile strength, lost circulation occurs with large volumes of drilling fluid into the thief zone [4] .

According to[4] a comprehensive lost circulation management program consists of four tiers. The bottom three tiers focus on lost circulation prevention measurements, while the top one represents remediation measures by using lost circulation materials where LCMs can be pumped with the drilling mud and can start curing losses directly as soon as they have reached the thief zone [5] . [3] had found that when a system controls a certain severity level it will always control the less severe conditions, and vise versa, when it fails to control a certain severity level it will never control more severe conditions.

The best method to govern lost circulation is to make an evaluation of the severity of a loss section (seeping, partial, severe or complete) and match the remedial material and technique to it in terms of both the size of the material and its purpose [6] .

In general, when classifying LCMs into sizes, three classes may appear and they are[7] :

1. Fine materials: for most cases, they will pass through the shaker screens and remain in the system.
2. Medium size materials: where they lean towards screening out of the flowing system, but in most cases will not plug the jets or MWD tools.
3. Course materials: they can plug off most tools except open-ended drill pipe.

Moreover, there are LCMs that are chemically reactive such as using an internally activated silicate solution which permits to gain a solid free Newtonian solution with a very low early viscosity. After a certain period, which depends on the fluid temperature and design, the viscosity of the solution rises rapidly to form a gel. This gel is strong, coherent, and does not develop free water with time. The very low initial viscosity of this solid free fluid permits it to be pumped right through the bottom hole drilling assembly, therefore saving time and provides deep penetration of the final gel [8] .

Conventional LCM was used instead of cement or hydratable LCM pill to seal relatively large fractures. Their results showed that using LCMs that have irregularity in particles shapes and degree of deformability can seal the wide fracture when used individually or in combination with other LCM

Some researchers concluded that using large multi-model particle size distribution (PSD) and acid – soluble solution can allow for curing severe –to-total loss circulation in naturally fractured reservoir formation [10] .

Using environmentally friendly LCM was the study of [11] . They used two new eco- friendly LCMs, one of them was from a special type of grass while the other was made from a flowering plant. They concluded that the performance of these new additives was superior to widely used LCM.

The aim of this study includes using traditional lost circulation material LCM as cotton seeds and comparing it to new LCM which is a cross-linked copolymer. These results were conducted using a laboratory model test.

2. Data Gathering

Five Dolomite core plugs were used in this study. These core Plugs were collected from Khabaz oil field, Azkand formation well KZ-35. Core plugs 2, 4 and 6 are got from C1 sub-formation, while plugs 1 and 3 from the sub-formation C2 that is located below C1. All the core plugs characteristics are shown in Table 1. While the drilling mud data, depths of drilled formations and used LCMs were gathered from wells KZ-30 and KZ-35.

The Khabaz oil field represents a small subsurface asymmetrical anticline, and its northeast limb dipper than the southwest limb. The Khabaz structure is located between Jambur and Bai Hassan oil fields. This oil field located 23 km to northwest of Kirkuk city northeast Iraq as shown in Figure 1 [12] .

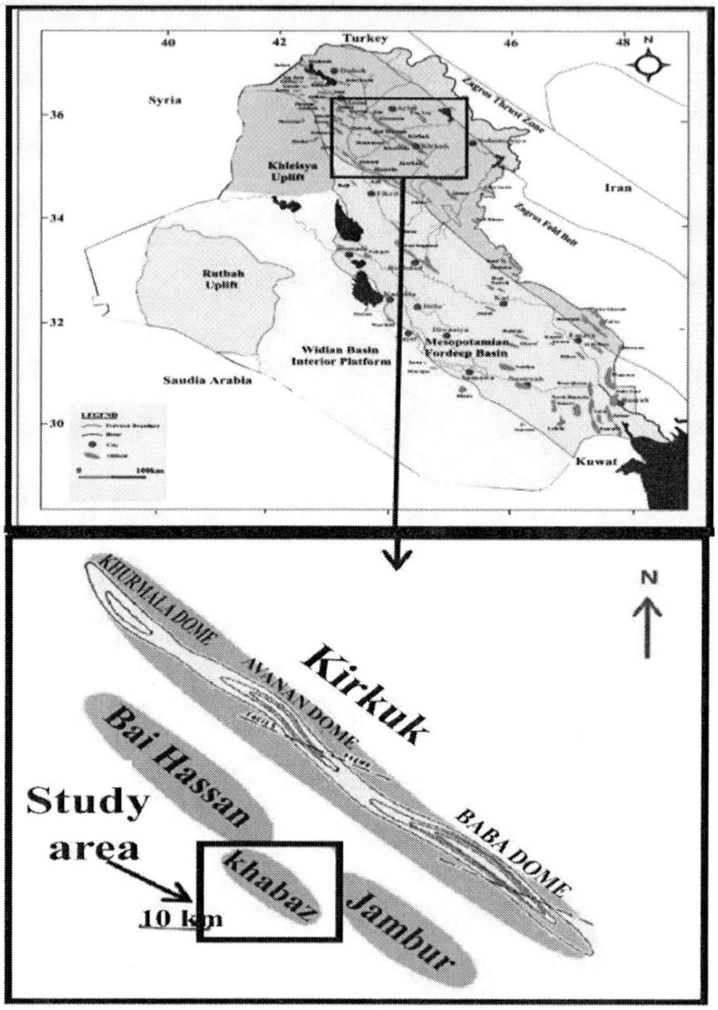

Figure 1. Major petroleum fields of Iraq shows the location of the study area, modified from[12] .

3. Experimental Work

3.1 Materials

This study compares the efficiency of the reference mud and the effect of the addition of two different types of lost circulation materials (LCM) to it. The reference mud system is composed of 60 gm bentonite, 5gm XC polymer, 5gm starch, and 1gm soda ash where all these materials mixed in one litter of fresh water. The cotton seeds and Cross-Linked Copolymer of Acrylamide and Potassium Acrylate (FLOSORB CE 300 S) were used as LCMs'. The materials are shown in Table 2. And the mud system rheological properties are shown in Table 3.

3.2 Rheological and filtration properties

The rheological and filtration properties were measured for the reference mud and the five types of mud systems with LCMs as shown in Table 2. These properties are very important due to their high impact on their performance in sealing the pores, vugs, and fractures as well as controlling the volume of mud losses.

Adding 100 gm of cotton seeds lead to a decrease in all properties. The large size of the seeds (5 to 8 mm) leads to separating the bentonite plates apart that cause a decrease in yield point, plastic viscosity, and gel strength. While a decrease in the volume of filtration is observed due to the existence of cotton seeds which close the pores of mud cake. The wide range of mud cake thickness can be directed to the irregular gathering of cotton seeds.

When Flosorb CE 300 S in four concentrations (0.2, 0.5, 2, and 5 gm/l) were added to the reference mud leads to an increase in yield point and plastic viscosity this is due to the flocculation of the bentonite plates. An improvement in filtration properties is obtained due to the effect of polymer addition.

Table 1. Core plugs characteristics

Core No	Depth (ft)	L (cm)	D (cm)	K (md)	Appearance
2	7226.5	4.37	3.77	30.92	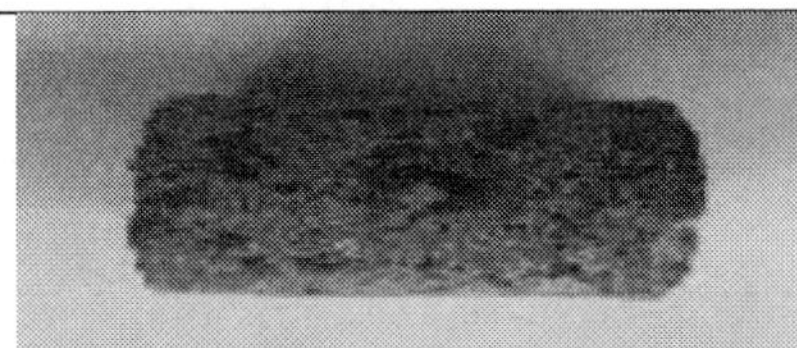
4	7236.4	4.38	3.79	59.28	
6	7246.2	4.4	3.77	39.44	
1	7251.1	4.37	3.79	319.93	
3	7261	4.37	3.79	14.43	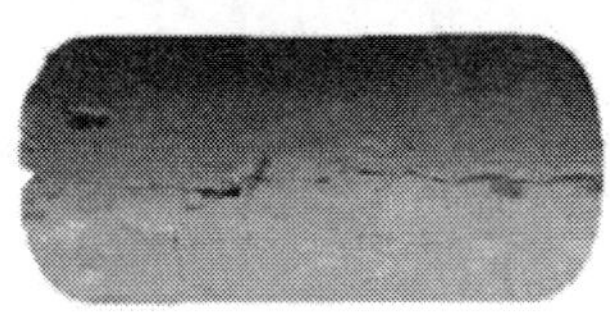

Table 2. Specifications of the Materials in Mud System

Materials	Concentration (gm/litre)	Primary Function	Solubility in Water
Bentonite	60	Viscosity and fluid- loss control	Insoluble
Soda Ash , Sodium Carbonate, Na_2CO_3	1	pH controller	Soluble
XC Polymer	5	Viscosity and fluid-loss control	Soluble
Starch	5	Fluid- loss control and viscosity	Insoluble
Flosorb CE 300 S	2 or 5	LCM	Insoluble
Cotton Seeds (5-8 mm size)	100	LCM	Insoluble

Table 3. Rheological filtration properties of drilling mud with different concentrations of LCMs

Properties	Rf. Mud	Rf. Mud+100 gm/l Cottonseed	Rf. Mud+0.25 gm/l FLOSORB CE 300S	Rf. Mud+0.5gm/l FLOSORB CE 300S	Rf. Mud+2.0 gm/l FLOSORB CE 300S	Rf. Mud+5.0 gm/l FLOSORB CE 300S
SP.GR	1.02	0.97	1.01	1	1.02	0.995
10 sec gel, lb_f /100 ft^2	78	67	72	85	132	124
10 min gel, lb_f/100 ft^2	117	88	110	119	154	125
Filter Volume, ml	7.4	6.6	7.4	7.8	5.8	5.6
Mud Cake Thickness, mm	1.1	7	0.9	0.9	1	0.8
pH	10.66	10.66	10.4	10.3	10.3	10.25
YP, lb_f/100 ft^2	116	117	108	116	178	165
PV, cP	13	1.4	15.6	16.7	59	34

*depends on whether there is a cotton seed or not

3.3 Closed loop Circulation System

A closed loop circulation system was constructed to simulate the event of lost circulation into a formation. Figure 2 shows the closed loop circulation system and its schematic is demonstrated in Figure 3. The details for the preparation and developing of the model are described in the following sections.

3.4 Dynamic Mud Loss Cell

The dynamic mud loss cell is a core holder similar to standard Hassler type core holder. It is made of stainless steel that can house a core plug of 4.4 cm diameter.

The core plug was positioned inside a rubber sleeve and the two sealed together by applied overburden pressure by an air compressor. The pressure was applied to prevent bypassing the injected fluids to the annulus between the rubber sleeve and the core samples, as shown in Figure 4.

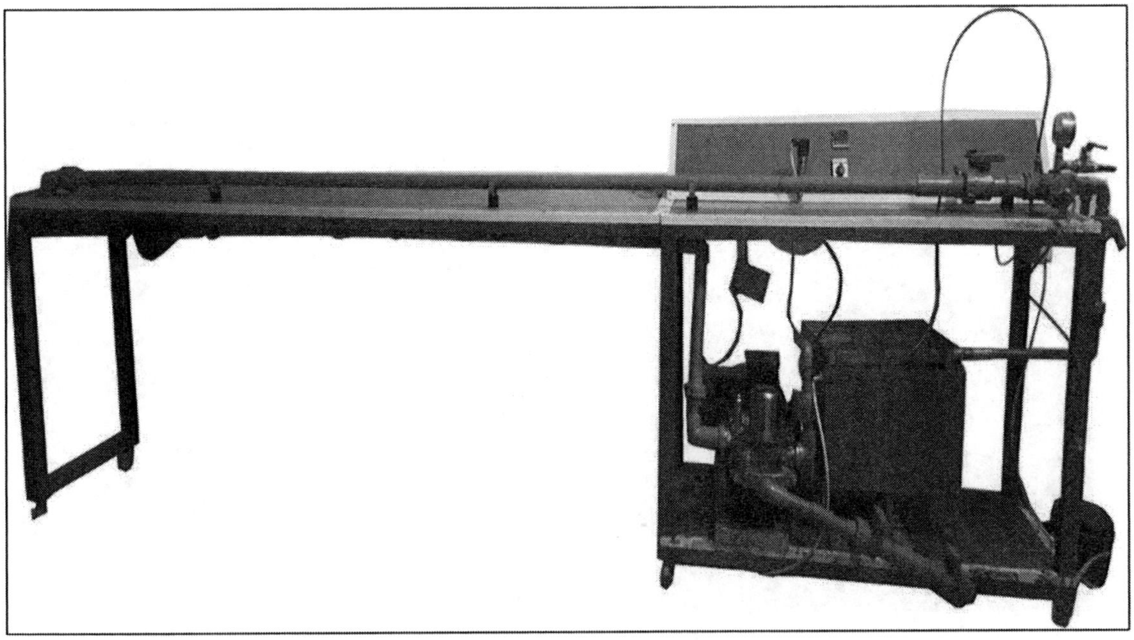

Figure 2. Main components of the closed loop circulation system.

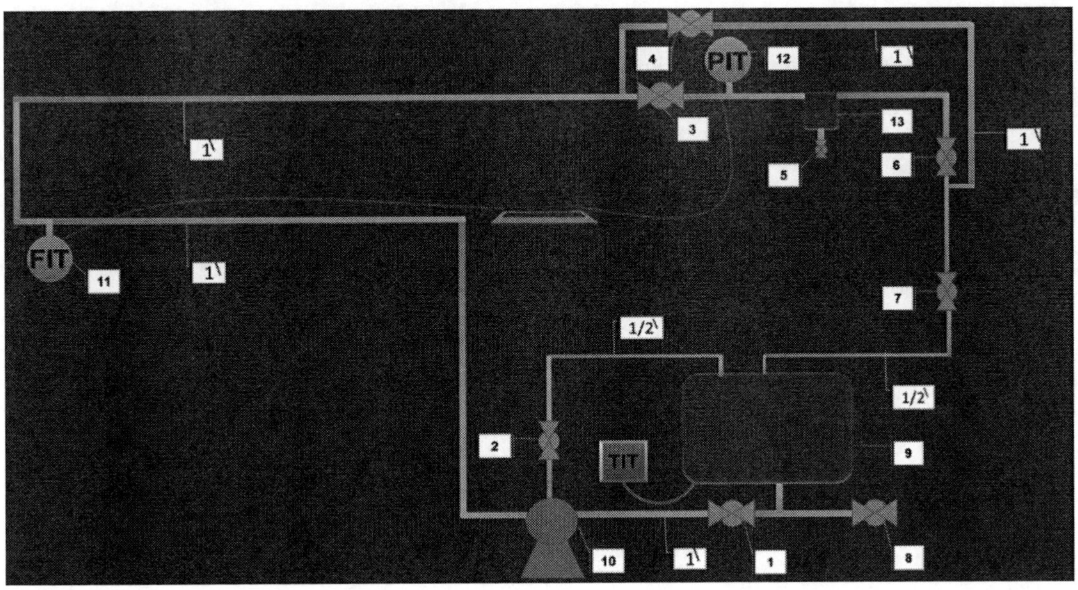

Figure 3. Schematic Closed Loop Circulating System. Where 1-8 isolation valves, 9- mud tank, 10- pump, 11- flow rate indicator transmitter,12- pressure indicator transmitter, and 13- core holder.

3.5 Test Procedure

Prior to starting the test; eight to ten litres of mud were prepared. The procedure of the test is explained as follows, where the core plugs are fixed inside the rubber sleeve of the core holder and valve 4 is opened, as shown in Figure 3. After applying the desired pressure by an air compressor, valve 4 is closed. The test starts with opening valves 1 and 2, as shown in Figure 3. The mud is circulated by the pump, while the mud temperature is observed until it reaches 60 °C. After that, valves 3, 6 and 7 is opened and valve 2 is closed, as shown in Figure 3. The pressure is controlled by closing valve 7 gradually until the pressure reaches 120 psi. Start recording lost mud through the core plug with time until the test is finished. After the test is finished, valve 7 is opened to reduce the pressure before the pump is shut down.

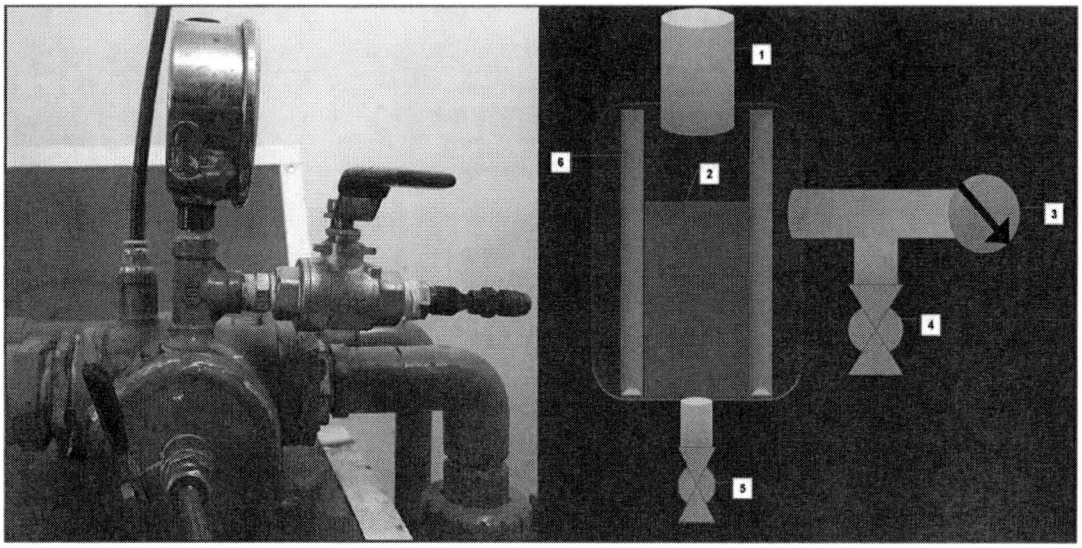

Figure 4. The core holder. Where 1- inlet line, 2- core sample, 3- pressure gauge, 4- gas inlet valve, vent valve, 5- lost mud outlet valve, 6- rubber

4. Results and Discussion

The performance of the materials added to the reference mud as LCMs is presented in the Figures 5 to 8 and demonstrate how efficient each additive preformed in other words how much mud is lost in a unit time.

The cotton seeds and the fibre attached to them gathered on the core plug face to reducing lost circulation. Cotton seeds have reduced the volume of lost mud when it was added at a concentration of 100 g/l about 7.7 % - 20.7 % of lost volume of the reference mud for core plugs 6 and 5 respectively. The same previously mentioned figures also demonstrate the effect of FLOSORB CE 300 S that have reduced the mud losses to 37.5 % - 60 % for the core plugs 1C2 and 6 respectively when they were add at 0.25 g/l to the reference mud. This reduction in drilling mud losses was improved with increasing FLOSORB CE 300S concentration to 0.5 g/l where the efficiency of curing the losses reached 80% with core plug 3 where the crossed-liked polymer involved participate on the plug face to bridge and seal permeable formations.

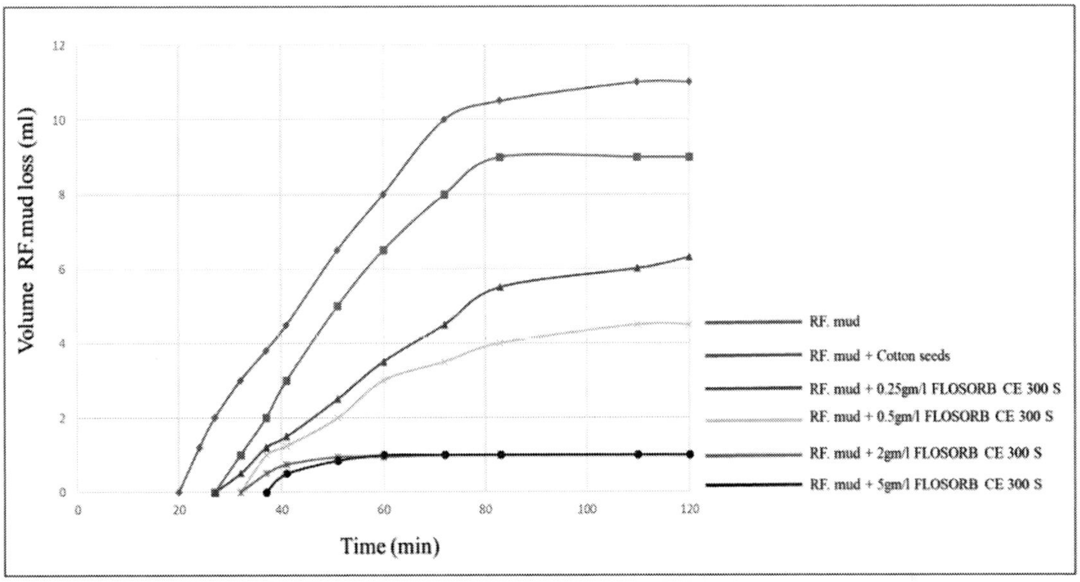

Figure 5. Effect of cotton seeds and polymer LCM on lost volume for core No 2.

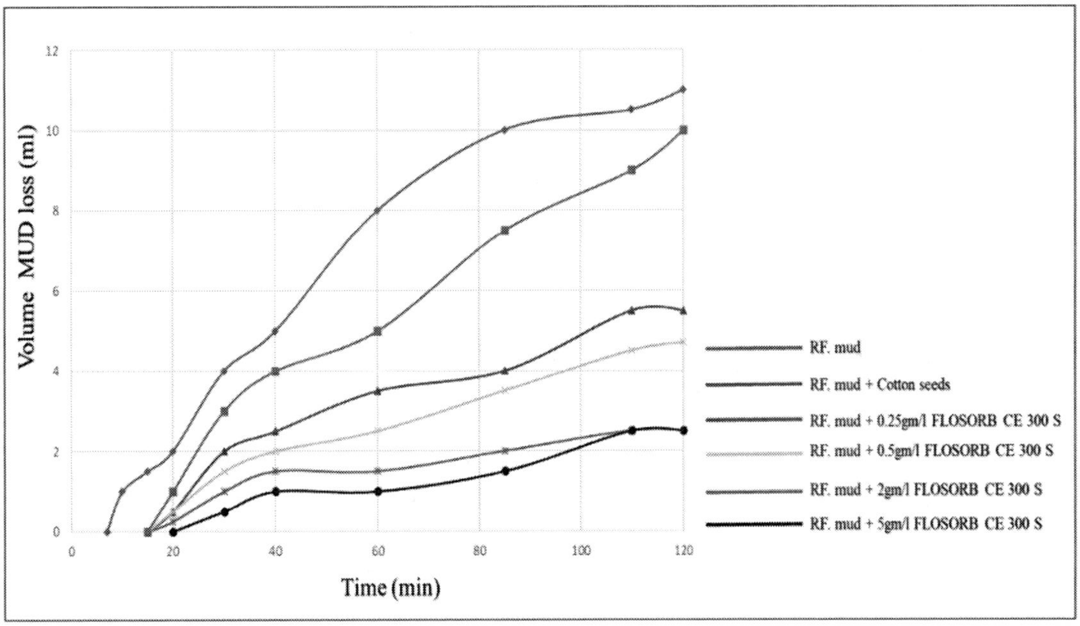

Figure 6 . Effect of cotton seeds and polymer LCM on lost volume for core No 4.

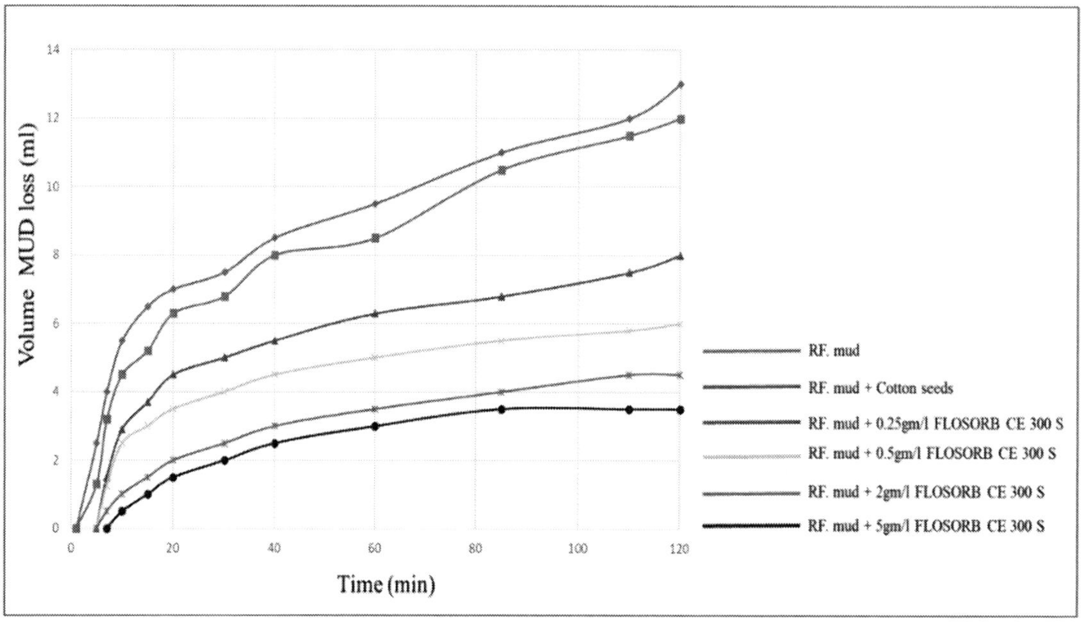

Figure 7. Effect of cotton seeds and polymer LCM on lost volume for core No 6.

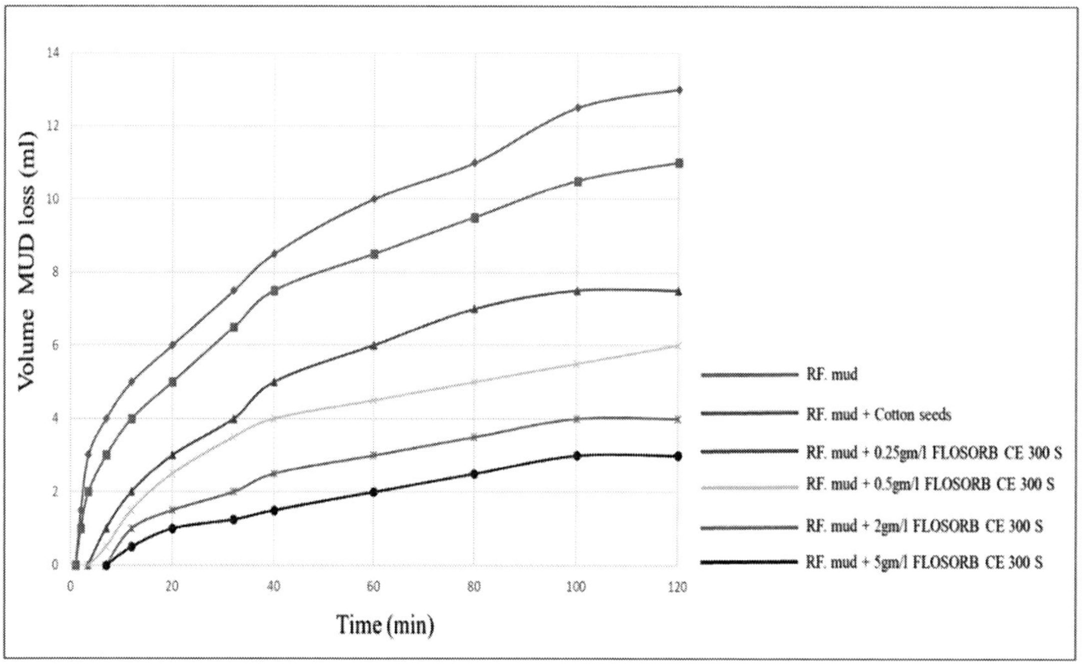

Figure 8. Effect of cotton seeds and polymer LCM on lost volume for core No 1.

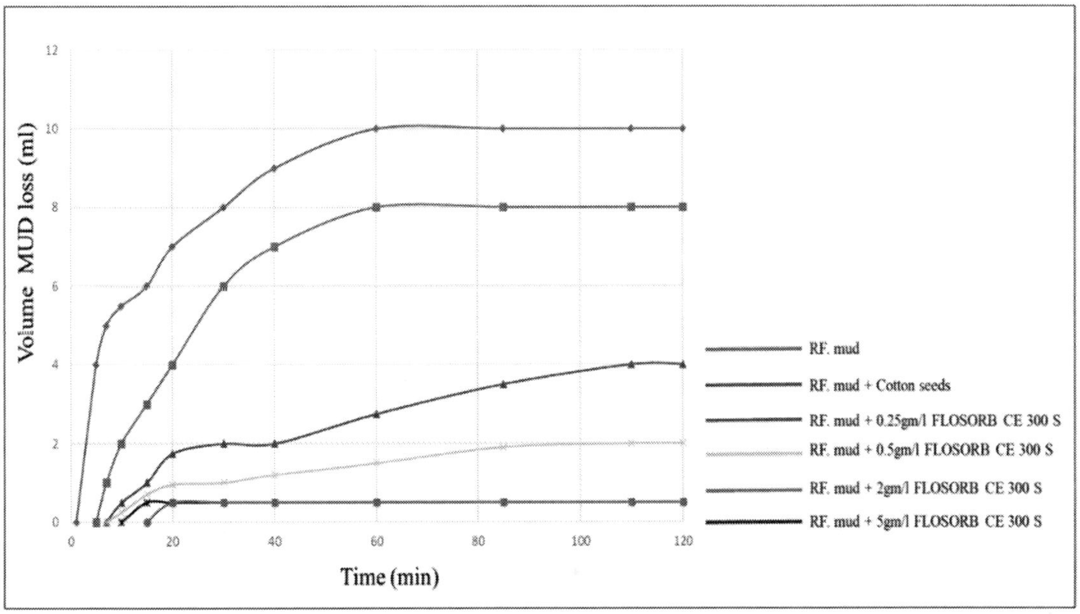

Figure 9. Effect of cotton seeds and polymer LCM on lost volume for core No 3.

Adding 5gm/l FLOSORB CE 300S leads to 70% reduction in lost volume for core No. 1 and 90% for core No.2 and core No.3 that have 30.92 and 14.43 md permeability respectively. Cores No.4 and No.6 with 59.28 and 39.44 md permeability have 80% and 20% reduction in lost volume respectively. As the permeability increases it becomes more challenging to cure the losses especially when fractures are presented in the core plug, and this is illustrated in Figure 8. Where core No.1 has 319.93 md permeability and it is clear that the plugging efficiency of the cotton seeds is poor in comparison with the FLOSORB CE 300S. This is due to the wide range of sizes that FLOSORB CE 300 S provide after it absorbs water giving the crossed-link polymer the ability to bridge and block the permeable sections Adding relatively high concentrations of FLOSORB CE 300S such as 2 or 5 gm/l will increase the plugging and the loss control capability to 95%, but it will affect the rheological properties dramatically.

5. Conclusion

1. Drilling mud loss into fracture are more severe than that into a vuggy formation.
2. For all of the cases, FLOSORB CE 300S acted better than the cotton seeds where it managed to cure the losses in plugs 2 and 3of the lower permeability.
3. For the core plugs, 4, 6, and 1, FLOSORB CE 300S did not cure the losses 100% but it gave a better result.
4. Adding only (0.2, 0.5, 2, to 5 gm/L) of FLOSORB CE 300S gave better results than adding 100 gm/L of cotton seeds, which gives a great advantage when it comes to the needed storage area.
5. Adding (2 to 5 gm/L) of FLOSORB CE 300S dose give better sealing properties but it increases the rheological properties too much.
6. FLOSORB CE 300S can be pumped through the relatively small bit nozzles.
7. The LCMs used in the presented study for the purpose of reducing or preventing lost circulation have to be applied on other formations suffering lost circulation in the adjacent wells and presumably in other fields.

References

[1] Sclumberger Glossary Oilfield, [Online]. Available: http://www.glossary.oilfield.slb.co,/terms/l/lost_circulation.aspx.

[2] Y. J. J. F. G. K. Feng 2016 A review on fracture-initiation and propagation pressure for lost circulation and wellbore strengthening SPE *Drilling & Completion* **31** 134-144
[3] C. R. Mirand, J. L. Oliveira, G. M. S. Cavalcante, A. R. d'Almeida, R. F. L. Pereira, R. L. L. Santos and R. Surmas 2017 Materials for controlling severe lost circulation – laboratory evaluation *in SPE Latin America and Caribbean petroleum engineering conference* (Buenos Aires)
[4] J. Cock 2012 A New Approach to Prevent Lost Circulation While Drilling *Schlumberger*
[5] A. Lavrov, Alexandre Lavrov 2016 Lost circulation mechanisms and solutions (Gulf Professional Publishing, Elsevier)
[6] M-I Manual, M-I SWACO 1998
[7] R. I. McNeil and S. C. Zeilinger 2001 Designing an LCM Program for Drilling Depleted Zone *AADE National Drilling Conference Drilling Technology- The Next 100 years* (Omni in Houston, Texas)
[8] B. Vidick, J. Yearwood and H. Perthuis 1988 How to Solve Lost Circulation Problems *International Meeting on Petroleum Engineering* (Tianjin)
[9] M. Alsaba and R. Nygaard 2014 Lost Circulation Materials capability of sealing wide fractures *SPE deep-water drilling and completion conference* SPE-170285-MS
[10] S. Savari, D. L. Whitfill and J. Walker 2016 Lost circulation management in naturally fractured reservoirs *SPE/IADC Middle east drilling technology conference and exhibition* (Abu Dhabi)
[11] A. Nasiri, M. A. Shahrabi and M. k. Moraveji 2018 Application of new eco-friendly LCMs for combating the lost circulation in heavy-weight and oil-based mud *RSC Adv* **8** 9685-9696
[12] R. K. AlAtroshi, F. Yagmurlu and A. R. Al-Khatabi 2018 Petrographical and Petrophysical Features of the Jeribe Formation at the Khabaz Oil Field, Kirkuk, *Northern Iraq Journal of Natural and Applied Sciences* **22** 1283-1295

Activation Iraqi bentonite for using as drilling mud

A S Ibrahim[#] **and M A Al-Bidry**

Petroleum Technology Department, University of Technology-Iraq, Baghdad, Iraq

[#]Corresponding author's e-mail address: ahmedsattarmaster@gmail.com

Abstract. Drilling fluid is very important in drilling a well because it plays a vital role in all drilling operations. Drilling fluid should have popular properties that simplify safe and easy completion of the well. Because of the big cost of transporting and imported clays that used as drilling fluid, Iraqi benronite must be enhanced to use as drilling fluids due to its poor rheological and filtration properties, and stability. Iraqi bentonite has a limited usage in oil field because it is Ca- bentonite. The research has studied the usage of bentonite and its activation by NaoH in drilling fluid. The results show that Iraqi Ca-bentonite reaches to API specifications while the indirect activation includes activated Ca-bentonite by NaOH with different concentrations, HCL then NaOH without temperature and HCL then NaOH with temperature. Then activated bentonite is used as drilling mud. The results show that the indirect activated bentonite does not reach API specification, where XRF analysis show that an ion exchange between Ca and Na is not enough to convert Ca- bentonite to Na- bentonite.

1. Introduction

Drilling fluids are important for the oil, gas and geothermal drilling industry because they perform many functions like transporting rock cuttings to surface, lubricating the drill bit, applying hydrostatic pressure in the well bore to insure well safety and minimizing fluid loss across permeable formations by forming a filter cake on the well of the well bore [1]. Bentonite is the main constituent of water- based mud due to high viscosity, good swelling and lower filter loss [2]. Bentonite is a smictite clay [3], because montmorillonite clay has the highest swelling capacity (which is responsible for viscosity build up and formation of low permeability filter cake), the presence of other materials will have an adverse effect on bentonite quality. Bentonite often contains clay minerals such as montmorillonite, illites, kaolinites, chlorites and non-clay components such as quartz and feldspar in appreciate amounts. A good quality bentonite should contain mainly montmorillonite [4].There are two types of bentonites: sodium bentonite (Na-bentonite), which has a high swelling capacity, and calcium bentonite, which is non-swelling clay and bentonite spontaneously forms colloidal dispersion in water. Bentonite is mainly composed of layers including silica and alumina sheets linked together and arranged on top of each other. Bentonite lamellar platelets are packed together by electrochemical forces and form aggregates that contain interposition water. The platelets are composed of a central octahedral alumina (Al_2O_3) layers and two tetrahedral silica (SiO_2) layers [5].

Iraqi bentonite exhibits high filtration loss and does not develop sufficient viscosity so, It cannot meet the API (15 cp minimum viscosity at 600rpm, 15cm^3 filtration loses) standards. By alkali activation or by introducing some polymers additives, it is possible to upgrade raw bentonite to meet the above standard and thus require appropriate activation formulation. This application typically employs various additives such as inorganic salts, MgO salts; the in organic salts improve the swelling or viscosity [2]. The chemical compositions which determine properties of the clays i.e. layer charge, cation exchange capacity,

adsorption capacity and morphology, can be varied depending upon their origins. These factors play a significant role in the modification of the natural clays. A number of methods for modifying clay minerals have been studied with a view to enhancing their adsorption capacities, including acid activation, treatment with cationic surfactants, clay-rubber composite, thermal activation, polymer addition by interparticle polymerization, binding of inorganic and organic anions, and grafting of organic compounds. Among these, acid activationis one of the most commonly used modification techniques because it is a simple and low-cost process [6].

The aims of this work are to enhance the quality of Iraqi bentonite by decreasing non-clay minerals, and activate it by Na ions to enhance the rheological and filtration properties to be used as drilling fluid.

2. Iraqi Bentonite Charactarization

2.1 Mineral Composition Analysis

Mineral composition of Iraqi bentonite was measured by X-Ray diffraction (XRD) using SHIMADZU XRD-7000.

2.2 Chemical Composition Analysis

Chemical composition for blank bentonite and activated bentonite were determined by (XRF) Spectrometry using SHIMADZU XRF-1800, the analyses for Iraqi bentonite are shown in Table 3, Table 6 and Table 12 respectively

2.3 Specific Surface Area (SSA) and Particle Size Distribution (PSD)

The technique selected for establishing the correlations in the determination of the specific surface area is wet laser diffraction technique. Wet laser diffraction allows the particle size distribution of the test sample to be determined as direct parameter. The distribution was determined using MALVERN MASTERSIZER 2000 laser diffraction instrument. In this instrument, the particles cross a laser light beam and the scattered light is collected at 52 detectors [7].

3. Experimental Works

3.1 Preparing Iraqi bentonite

Iraqi Bentonite taken from Trefawi formaion/AL-Anbar region/ Iraq, was grinded to powder and sieved into size <75μm, after that it was dried in oven at 180 ^{0}C for 12 hr. Then, 22.5 g of Iraqi bentonite was mixed with 350 ml of distilled water for 20 minutes and left for 24 hr to hydrate, then the mud properties were measured.

3.2 Activation Iraqi Bentonite

3.2.1 Direct Activation

Iraqi bentonite was activated by adding 0.8, 1.6, 2.4, 3.2 and 4 g of NaoH directly to 22.5 g of Iraqi bentonite and 350 ml of distilled water.

3.2.2 Indirect Activation

Many processes (heating at different heating temperature, mixing at different hours and addition of inorganic materials) were made for Iraqi bentonite before API standards mud (22.5 g bentonite with 350ml distilled water for 20 minutes) to make a good bentonite that can be used directly by mixing with distilled water without any addition and give it to oil companies for using it instead of commercial bentonite.

3.2.2.1 Alkali Activation

10 g of prepared Iraqi bentonite was mixed with 1, 3, 5, 7 and 9 g of NaoH dissolved in 100 ml distilled water with mixing time 2, 4 & 6 hours. The mixture was dried in at 105 ^{0}c. Then, 22.5 g of the dried mixture was mixed with 350 ml distilled water for 20 minutes and left for 24 hours in a closed container to hydrate. After that, the mud properties were measured.

IOP Conf. Series: Materials Science and Engineering **579** (2019) 012006 doi:10.1088/1757-899X/579/1/012006

3.2.2.2 Alkali Activation and Thermal Activation

10 g of prepared Iraqi bentonite was mixed with 1, 3, 5, 7 and 9 g of NaoH dissolved in 100 ml distilled water with mixing time 2, 4 & 6 hours and mixing temperature 70 & 100 ^{0}C. The mixture was dried in Electrical Oven at 105 ^{0}C, 22.5 g of the dried mixture was mixed with 350 ml distilled water for 20 minutes and left for 24 hours in a closed container to hydrate. Then, the mud properties of the mixture were measured.

3.2.2.3 Alkali Activation and Acid Activation

10 g of prepared Iraqi bentonite was mixed with 2 ml (0.5 M) of HCl solution for 30 minutes. The acid to bentonite ratio was 2:10 (ml/g). Then, the mixture was washed several times until removing cl^- ions and pH approximately reached 7. After that, the mixture was mixed with 1, 3, 5, 7 and 9 g of NaoH dissolved in 100 ml distilled water for 2 hours at room temperature (35 ^{0}c).The mixture was dried for 105 ^{0}c. 22.5 g of the dried mixture was mixed with 350 ml distilled water for 20 minutes and left 24 hours in a closed container to age [8]. The mud properties of the mixture were measured.

4. Mud Testing

4.1 Rheological Properties

Rheological properties were measurd using OFFITE Viscometer (Model 900). Rheological properties were used to indicate solids buildups flocculation or de flocculation of solids, lifting and suspension capabilities, and to calculate hydraulics of a drilling fluid [9]. Rheology is the study of how matter deforms and flows, it is primary concerned with the relationship of shear stress and shear rate and the impact these have on flow characteristics inside pipes and annular spaces. Viscosity is a resistance to flow. The apparent viscosity is as half of the viscometer reading at 600RPM [10]. Plastic Viscosity (PV) is defined as the 600RPMminus the 300RPM shear stress reading. The plastic viscosity depends mainly on the friction between particles. Yield Point (YP) is defined as the 300RPM shear stress reading minus the plastic viscosity. Yield point is a measure of the attractive forces between clay particles in the mud under flowing conditions. It is also a measure of hole cleaning capabilities of the mud[11]. Gel Strength denoted the thixotropic properties of the mud and is a measurement of the attractive forces of the mud while at rest or under static conditions.

4.2 Mud Density

Density is defined as weight per unit volume. It is expressed in pounds per gallon (lb/gal), pounds per cubic foot (lb/ft^3), and kilograms per cubic meter (kg/m^3), or compared to the weight of an equal volume of water, as specific gravity (SG). The pressure exerted by a static mud column depends on both the density and the depth; therefore it is convenient to express density in terms of pounds per square inch per foot (psi/ft). To prevent the inflow of formation fluids in the wellbore, the pressure of the mud column must exceed the pore pressure, which is pressure exerted by the fluids in the pores of the formation. Mud density was measured using mud balance.

4.3 pH

pH is a value representing the hydrogen ion concentration in a liquid. The pH indicates acidity or alkalinity of a drilling fluid. The pH of a drilling fluid can affect its perfomance and is important for corrosion control. Low pH (Acidic) fluid is undesirable because it can corrode the drill string. Normal drilling fluid range between 9.5 to 10.5, higher values are not common [12].

4.4 Filtration properties

There are two types of filtration: Static and dynamic. Static filtration test is used to indicate filter cake quality and filtrate volume loss for a drilling mud under specific test conditions. Filtration characteristics are affected by the types and quantities of solids and their physical and chemical interactions. Temperature and pressure further affect these solids and their interaction [13].

API fluid loss test was carried out using a static filtration apparatus (API Filter press offite apparatus) by placing the mud into a stainless steel chamber with an opening at the bottom. A filter paper was placed on the bottom. The mud was exposed to 100 psi for 7.5 min. and the amount of filtrate volume was measured.

4.5 Stability
Stability of drilling mud was measured by placing drilling mud in a 250 ml glass cylinder and checked the settlement after 2hrs, 4hrs and 24hrs.

5. Results and Discussions

5.1 Mineral Composition of Iraqi Bentonite
XRD shows that Iraqi bentonite contains montmorillonite and palygorskite as main minerals and quartz and calcite as impurities. By using special process of grinding, quartz was removed and montmorillonite was increased.

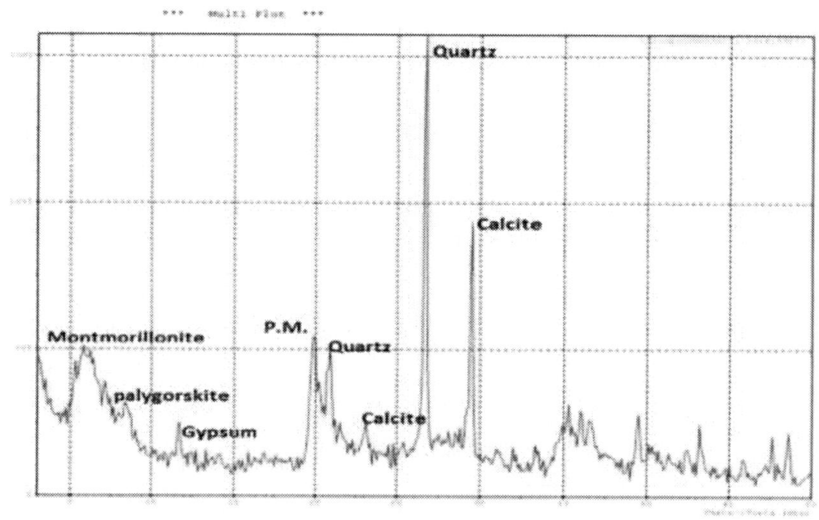

Figure 1. XRD of Iraqi bentonite

5.2 Chemical Composition of Iraqi Bentonite
XRF analysis is shown in Table 1, the ratio of Al_2O3/SiO_2 was 1/3 as expected for montmorillonite which is the main component of bentonite. The ratio of $[(Na_2O+K_2O)/ (CaO+MgO)]$ for Iraqi bentonite was 0.143 confirming that Iraqi bentonite was Ca-bentonite.

Table 1. XRF Analysis of Iraqi bentonite.

SiO_2 %	Fe_2O_3 %	Al_2O_3 %	CaO %	MgO %	Na_2O %	K_2O %
52.1	5.71	15.91	7.06	1.79	0.60	0.67

5.3 Particle Size Distribution and Specific Surface Area
Particle size distribution (PSD) analysis is shown in figure 2, 10% of the particle size was smaller than 5.354μm, 50% of the particle size was smaller than 24.512μm and 90% of the particle size was smaller than 69.96μm. the average particle size of Iraqi bentonite inside the powder was 36.0825μm.

The specific surface area (SSA) of Iraqi bentonite diffraction was 0.449m²/g. The technique that the wet laser diffraction was used to measure the specific surface area depends mainly on particle size.

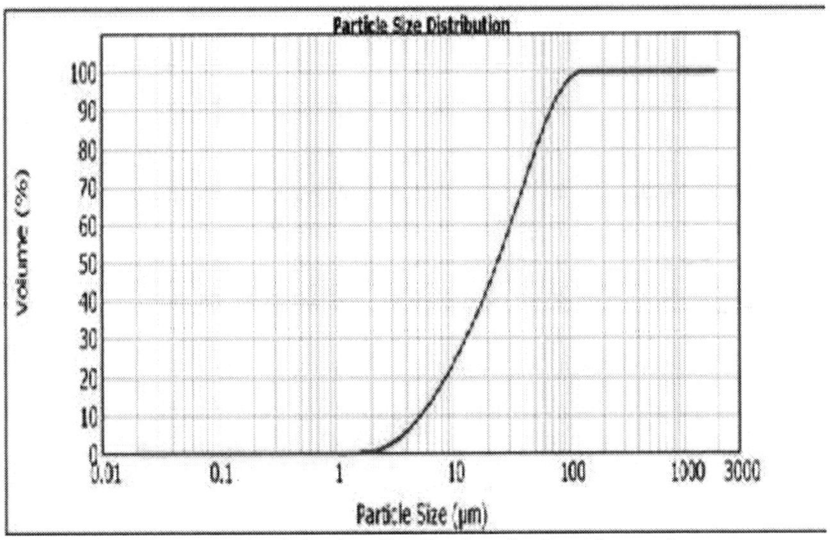

Figure 2. Particle size distribution for Iraqi bentonite

5.4 Properties of Iraqi Bentonite

22.5g of prepared Iraqi bentonite was mixed with 350ml distilled water for 20 minutes and left for 24hr to to hydrate. Then, Iraqi bentonite properties was measured. The results were shown in Table 2.

The rheological properties of Iraqi bentonite and gel strength were very bad in compared with API specifications. The filtration loss of Iraqi bentonite was very high and the stability is very low. The bad result that Iraqi bentonite has is because of the Ca in the composition of Ib. Iraqi bentonite was undesired clay because it minimized the adsorption and minimized the swelling of Iraqi bentonite. Therefore, many processes of activation were carried out to enhance its properties.

Table2. Properties of Iraqi bentonite

AV,cp	2.27
PV,cp	2.35
YP,(lb/100ft^2)	0.73
YP/PV,(lb/100ft^2/cp)	0.31
10s gel,(lb/100ft^2)	0.8
10m gel,(lb/100ft^2)	1.9
P,ppg	8.6
pH	7.1
$V_{7.5}$,ml	27
V30m,ml	54
Tmc,mm	2.13
Stability,2hr,%	76
Stability,4hr,%	67.2
Stability,24hr,%	50.8

5.5 Activation Iraqi Bentonite

Two Processes of activation were used; direct activation and indirect activation

5.5.1 Direct Activation

Iraqi bentonite was activated with NaoH. 0.8, 1.6, 2.4, 3.2 and 4 g of NaoH added to the API mixture of 22.5g Iraqi bentonite and 350ml distilled water in order to enhance the rheological properties.

Table 3. Direct Activation Iraqi bentonite by NaoH

W,g	AV,cp	PV,cp	YP, (lb/100ft^2)	YP/PV, (lb/100ft^2/cp)	Gel Strength, (lb/100ft^2)	
					10,sec	10,min
0.8	8.84	3.94	9.79	2.48	5.9	6.2
1.6	11.6	4.68	13.85	2.96	7.7	6.6
2.4	11.93	4.05	15.76	3.89	7.8	6.9
3.2	14.5	5.67	17.65	3.11	8.6	7.9
4	13.84	5.85	15.98	2.73	8.3	7.3

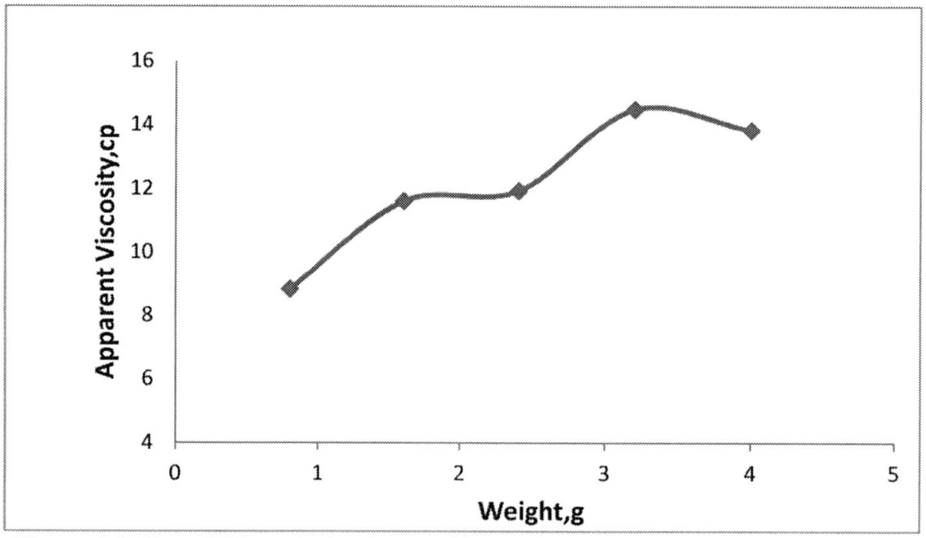

Figure 3. Effect of Direct Activation Iraqi bentonite by NaoH on Apparent Viscosity

By adding NaoH, the viscosity of mud increased and reached the API specifications at 3.2g NaoH. From Figure 3, the viscosity increased when the weight of NaoH increased until adding 4g of NaoH, the viscosity becomes decreasing because there was foam in the mixture. The foam contains many bubbles which disappear causing the mud loss and decreasing in viscosity. other rheological properties such as plastic viscosity, yield point, yield point to plastic viscosity ratio and gel strength were enhanced and approximately accepted compared with API specification.

Table 4. Stability Direct Activation Iraqi Bentonite

W,g	Stability %		
	1hr	2hr	24hr
0.8	78.4	73.6	59.2
1.6	98.8	98	88
2.4	100	100	100
3.2	100	100	100
4	100	100	100

When adding 2.4, 3.2 and 4 g of NaoH, the stability has become 100%. The effect of NaoH on stability was very good and very big enhancement in comparison with Iraqi bentonie before activation, which Na ion is interacted with chemical composition of bentonite and Na inters to the composition and becomes Na-bentonite, and Ca was interacted with OH^{-1} ion and became $Ca(OH)_2$. There is cation exchange capacity between bentonite and NaoH.

IOP Conf. Series: Materials Science and Engineering **579** (2019) 012006 doi:10.1088/1757-899X/579/1/012006

Table 5. Density, pH and Filtration of direct activation Iraqi Bentonite

W,g	P,ppg	pH	$V_{7.5}$,ml	V_{θ} ,ml	Tmc,mm
0.8	8.6	12.5	65.5	131	5.13
1.6	8.58	12.7	57	114	5.66
2.4	^8.55	12.8	50.2	100.4	6.28
3.2	8.53	13	43	86	7.46
4	8.5	13.4	35.4	70.8	8.4

The density of activated Iraqi bentonite by NaoH is accepted while the pH value is approximately high, the filtration loss is very high and must use filtration treatment if activated bentonite could be used as drilling fluid. CMC-LV or polyacrylamide or starch could be added in a small weight to minimize the filter loss.

XRF, particle size distribution and specific surface area were carried out for 3.2 g NaoH add to Iraqi bentonite as the best result of activation as shown in Table 6 and Figure 4 respectively.

Table 6. XRF Analysis of activated Iraqi bentonite by NaoH

SiO_2 %	Fe_2O_3 %	Al_2O_3 %	CaO %	MgO %	Na_2O %	K_2O %
38..3	4.56	8.34	8.3	1.89	5.41	0.42

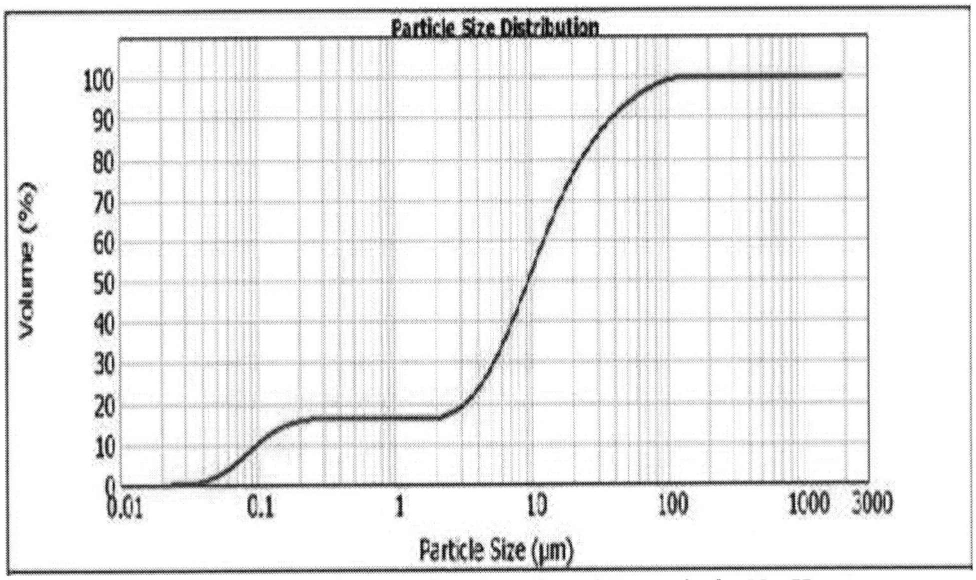

Figure 4. PSD of Activated Iraqi Bentonite by NaoH

XRF of direct activation Iraqi bentonite by 3.2g NaoH shows a big increase in the Na ion compared with Iraqi bentonite before activation, and the ratio of $[(Na_2O+K_2O)/(CaO+MgO)]$ was found to be 0.57 confirming that Iraqi bentonite, composition. By XRF, The value of Na_2O before activation was 0.6 and the value after adding 3.2g NaoH directly was 5.41, indicating that the NaoH interacted with ca-bentonite.

Ca-bentonite+NaoH→Na-bentonite+$Ca(OH)_2$

Most of times, the composite $Ca(OH)_2$ was undesired in the drilling mud. The activated bentonite can be used in formations that have rich calcium or formations that was not affected by calcium.

By PSD analysis, it is shown that the particle size of activated bentonite decreased, where 10% volume of particle size inside bentonite was smaller than 0.103µm, 50%volume of particle size inside bentonite was smaller than 0.103µm, 50%volume of particle size inside bentonite was smaller than 9.433µm, and 90%volume of particle size inside bentonite was smaller than 39.255µm.

IOP Conf. Series: Materials Science and Engineering **579** (2019) 012006 doi:10.1088/1757-899X/579/1/012006

The specific surface area of activated bentonite was 13.5m²/g. Because the particle size decreased, the specific surface area increased. The increasing of specific surface area gives good indication of high adsorption and high swelling index of activated bentonite.

5.5.2 Indirect Activation

The aim of this activation is to make a good bentonite by enhancing its rheological properties. This process was made before the standard API mud (22.5g bentonite and 350ml of distilled water). There were three work activation processes as shown below.

5.5.2.1 Alkai Activation

The process includes adding 1,3, 5, 7 and 9 g of NaoH dissolved in 100ml distilled water to 10g Ib and mixing for 2, 4 and 6hr to maintain good interaction between molecules. Then, the mixture was washed several times to remove $Ca(OH)_2$ and decrease high PH resulting from add NaoH. After the pH had become less 12, the mixture dried in oven at 105^0C and then 22.5g of dried mixture was mixed with 350 of distilled water to test mud properties.

The previous research shown that there is no effect on bentonite properties after 6 hr mixing time[14]. Rheological properties of activated Iraqi bentonite by alkali activation are shown in Table 6.below.

Table 7. Rheological Properties of Alkali Activation Iraqi bentonite

M	W,g	AV,cp	YP,(lb/100ft²)	PV,cp	YP/PV, (lb/100ft²/cp)	Gel Strength, (lb/100ft²)	
						10 sec	10 min
	1	1.92	0.85	1.49	0.57	0.2	0.5
	3	1.9	0.9	1.45	0.62	0.8	0.4
2	5	2.24	1.27	1.6	0.79	0.5	06
	7	2.8	1.2	2.2	0.56	0.7	1.1
	9	3.14	0.74	2.77	0.44	0.5	0.4
	1	1.56	0.31	1.5	0.21	0.3	0.5
	3	1.93	0.66	1.6	0.41	0.6	0.9
4	5	1.76	0.53	1.49	0.35	1.3	0.4
	7	2	0.4	1.8	0.22	1.1	1.4
	9	2.29	0.32	2.13	0.15	1.4	0.9
	1	1.38	0.84	0.96	0.88	0.1	0.2
	3	1.77	0.9	1.32	0.68	0.8	1.3
6	5	1.38	0.21	1.28	0.16	0.4	0.6
	7	2.4	1.4	1.7	0.82	0.9	0.9
	9	2.29	1.6	1.49	1.07	1	0.7

This process failed to activate Iraqi bentonite, the rheological properties were very bad and did not reach the API specification. From the results above, Plastic Viscosity (PV) was very low that mean there was no resistance to flow caused by mechanical friction. From the results above, Yield Point (YP) was very low because there were no attractive forces in muds under flow conditions.

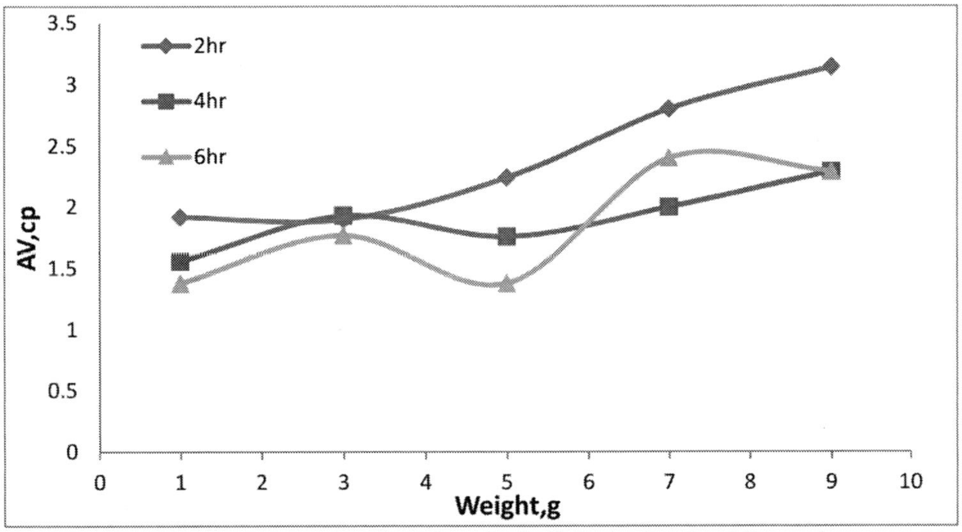

Figure 5. Apparent Viscosity of Alkali Activation Iraqi Bentonite

The apparent viscosity has compound effect between plastic viscosity (PV) and yield point (YP). Because of the small values of plastic viscosity and yield point, the apparent viscosity became very low small values. But from figure above, the apparent viscosity was decreased when mixing time was increased, that led to fix the mixing time for 2 hr in alkali and thermal activation, and alkali and acid activation.

Table 8. Density, pH and Filtration of Alkali Activation Iraqi Bentonite

S	W,g	P,ppg	pH	$V_{7.5}$,ml	V_{θ} ,ml	Tmc,mm
	1	8.55	11.5	19.9	39.8	0.79
	3	8.6	10.7	22.1	44.2	1.1
2	5	8.55	11.4	28.6	57.2	1.24
	7	8.5	10.8	22.9	45.8	1.4
	9	8.6	11.66	27.3	54.6	1.17
	1	8.6	11.34	17.4	34.8	1.22
	3	8.6	11.2	25.2	50.4	0.8
4	5	8.55	10.61	22.9	45.8	0.82
	7	8.5	10.9	29.5	٥٩	0.99
	9	8.55	10.93	34.8	69.6	1.3
	1	8.55	11.22	17.6	35.2	1.29
	3	8.6	10.7	20.3	40.6	1.22
6	5	8.55	10.79	25.3	50.6	1.3
	7	8.5	10.9	27.4	54.8	1.17
	9	8.55	10.54	32.7	65.4	1.08

The filtration loss was very high because the NaoH has no any effect to filtration loss.

5.5.2.2 Alkali Activation and Thermal Activation

By repeating the alkali activation processes and fixing the mixing time for 2hr, in addition to use the heating in mixing. The heat that was used in mixing was 60^{0}C and 90^{0}C and brought the first results that it was for 2hr without heating in mixing (at room temperature 30^{0}C) just for comparison. The rheological properties of activation Iraqi bentonite are shown in Table 8.

Table 9. Rheological Properties of Alkali and Thermal Activation Iraqi Bentonite

T	W,g	AV,cp	PV,vp	YP, (lb/100ft²)	YP/PV, (lb/100ft²/cp)	Gel Strength, (lb/100ft²)	
						10 sec	10 min
	1	1.92	0.85	1.49	0.57	0.2	0.5
	3	1.9	0.9	1.45	0.62	0.8	0.4
30	5	2.24	1.27	1.6	0.79	0.5	0.4
	7	2.8	1.2	2.2	0.56	0.7	1.1
	9	3.14	0.74	2.77	0.44	0.5	0.4
	1	1.81	1.7	0.22	0.13	0.3	0.2
	3	2.1	1.8	0.6	0.33	0.4	0.7
60	5	2	1.7	0.6	0.35	0.4	0.3
	7	2.3	1.8	1	0.55	1.1	0.9
	9	2.3	2.4	1	0.43	0.6	1.1
	1	1.7	1.5	0.2	0.34	0.3	0.4
	3	1.9	1.58	0.64	0.41	1.2	0.8
90	5	1.91	1.7	0.43	0.25	1.3	0.4
	7	2.4	1.8	1.2	0.67	1.2	1.1
	9	2.29	1.49	1.6	1.07	1	0.7

The benefit of heating the mud is to increase the velocity by inceasing the interaction between molecules and capacity of ion exchange. Because the rheological properties was not enhanced and reached the API standards, the activation failed. The Plastic viscosity values and yield point values were very low. The adsorption was very low because there was no attractive between the molecules of bentonite.

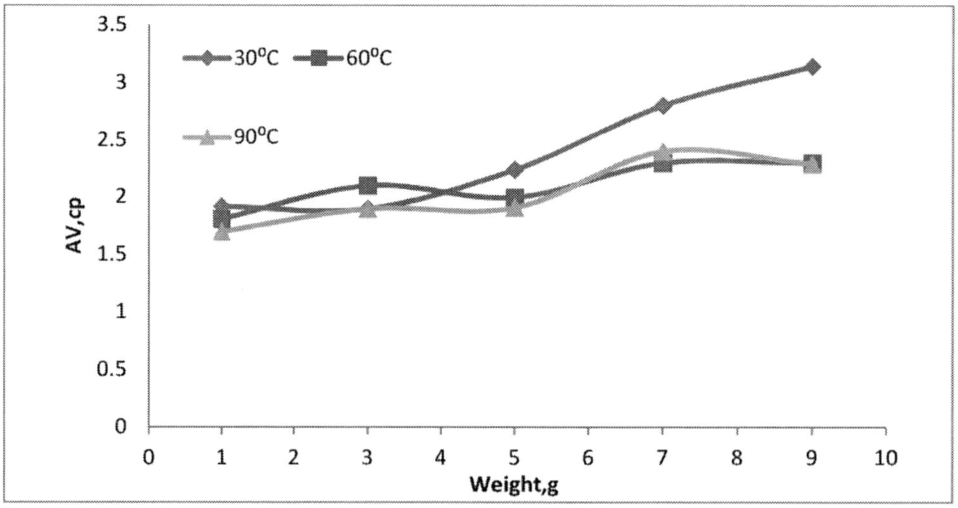

Figure 6. Apparent viscosity of alkali and thermal activation Iraqi bentonite

The Apparent Viscosity was not enhanced by alkali and thermal activation and did not reach the API specification. Apparent Viscosity was depended on plastic viscosity and yield point. Therefore the plastic viscosity and yield point had very small values, the apparent viscosity had low values. Figure 6 shows that the apparent viscosity decreased when the heating was increased.

Table 10. Density, pH and Filtration of Alkali and Thermal Activation Iraqi Bentonite

T	W,g	ρ,ppg	pH	$V_{7.5}$,ml	V_{θ} ,ml	Tmc,mm
	1	8.55	11.5	19.9	39.8	0.79
	3	8.6	10.7	22.1	44.2	1.1
30	5	8.55	11.4	28.6	57.2	1.24
	7	8.5	10.8	22.9	45.8	1.41
	9	8.6	11.66	27.3	54.6	1.17
	1	8.55	10.76	18.4	36.8	0.82
	3	8.6	10.8	22.8	45.6	1.42
60	5	8.55	11.61	26.3	52.6	1.71
	7	8.5	11.11	28.7	57.4	1.53
	9	8.5	11.1	33.8	67.4	1.03
	1	8.55	11.22	31.5	63	1.45
	3	8.5	10.72	25.1	50.2	1.2
90	5	8.6	10.86	31.2	62.4	1.32
	7	8.6	10.82	33.3	66.6	1.3
	9	8.55	11.4	45.8	90.6	1.65

Filtration losses are very high because Iraqi bentonite did not make an adsorption with NaoH molecules. Table 9 shows the thermal activation effect on density, pH and filtration properties of Iraqi bentonite.

5.5.2.3 Alkali and Acid Activation

HCl is very important in activation processes. By HCl, impurities such as calcite can be removed and by washing the solution of bentonite and HCl, the Cl^{-1} is removed and increased the value of PH [15]. Iraqi bentonite was treated with HCl first (Acid Activation) and then treated with NaoH (Alkali Activation). Rheological properties of alkali and acid activation Iraqi bentonite are shown in Table 10.

Table 11. Rheological Properties of Alkali and Acid Activation Iraqi Bentonite

W,g	AV,cp	PV,cp	YP, (lb/100ft²)	YP/PV, (lb/100ft²/cp)	Gel Strength,(lb/100ft²)	
					10 sec	10 min
1	1.65	1.49	0.32	0.21	0.2	0.3
3	1.6	1.4	0.4	0.29	0.8	0.9
5	1.76	1.6	0.32	0.2	0.4	0.6
7	2.5	1.94	1.12	0.58	0.5	1.1
9	2.93	2.13	0.23	0.11	0.5	1.2

The rheological properties of alkali and thermal activation for Iraqi bentonite failed to be enhanced and unable to reach API specifications. Because of acid activation by HCl, there was an enhancement in mineral composition of Iraqi bentonite by removing calcite and increasing the percentage of montmorillonite. . In spite of the mineral enhancement, it could not be used as drilling fluids because of the bad values of rheological properties.

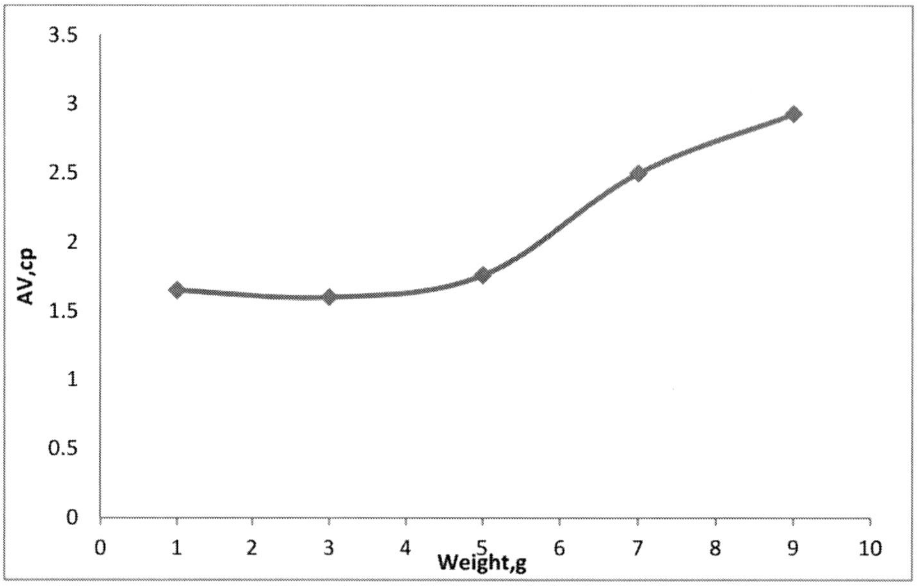

Figure 7. Apparent Viscosity of Alkali and Acid Activation Iraqi Bentonite

Apparent Viscosity had very low values and did not reach the API specifications. Figure 7, shown an increase in weight of NaoH, may increase the apparent viscosity.

Table 12. Density, pH and Filtration of Alkali and Acid Activation Iraqi Bentonite

W,g	P,ppg	pH	$V_{7.5}$,ml	V_{θ} ,ml	Tmc,mm
١	8.55	11.4	14.1	28.2	0.96
٣	8.5	11.23	28.3	58.6	1.43
5	855	11.6	37.3	74.6	1.86
٧	8.6	11.81	31.3	62.6	2.12
9	8.6	11.7	26.4	52.8	2.4

The filtration loss was very high and it was not fulfill API standards.

By direct and indirect activation process, results improve that NaoH could not work on filtration enhancement.

The XRF, SSA and PSD were carried out to the addition of 2 ml of HCl (0.5M) and 1g of NaoH (sample of alkali and acid actvation Iraqi bentonite) and the results are shown in Table 12 and Figure 8.

Table 13.XRF of Alkali and Acid Activation Sample Iraqi Bentonite (2ml Hcl and 1g NaoH)

SiO_2 %	Fe_2O_3 %	Al_2O_3 %	CaO %	MgO %	Na_2O %	K_2O %
53.39	6.27	14.50	3.26	3.18	1.48	0.49

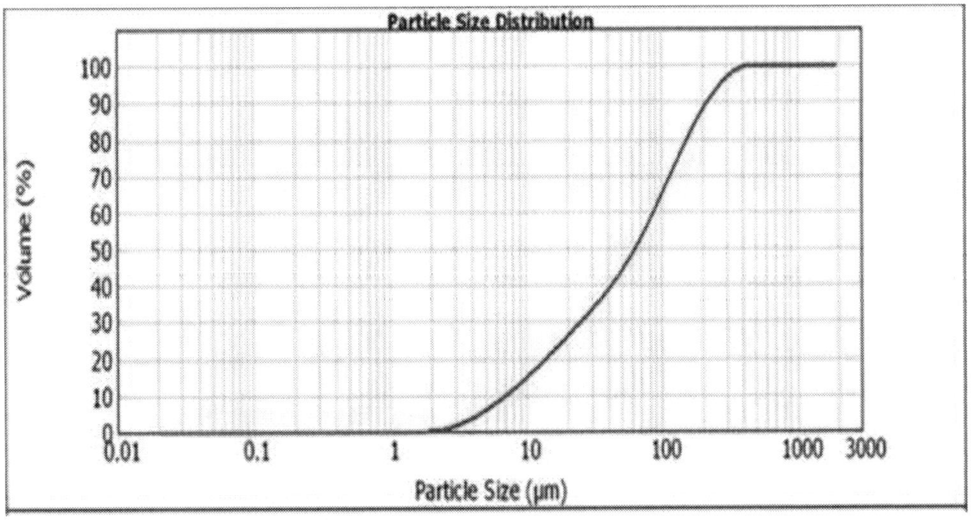

Figure 8. PSD of Alkali and Acid Activation Iraqi Bentonite Sample (2ml HCl and 1g NaoH)

The XRF analysis shows that the ratio of [(Na_2O+K_2O)/(CaO+MgO)] of activated Iraqi bentonite was found 0.31, confirming that bentonite was Ca-bentonite and there was no difference in comparison with Iraqi bentonite before activation. The particle size distribution of activated Iraqi bentonite shown that there was 10% volume of particle sizes under 7.198µm, 50% volume of particle sizes under 63.245µm, and 90% volume of particle size 217.771µm. The value of specific surface area was 0.289(m^2/g). This value was smaller than the value of Iraqi bentonite before activation. Because the particle size was increased, the value of specific surface area was decreased.

To search about why the indirect activation Iraqi bentonite failed, the first reason was that the solution of the mixture after mixing time before first wash out, was taken and made XRF analysis in Varian AA240FS Atomic Absorption. The analysis shows that the concentration of Ca in the solution is 2.49 ppm and the concentration of Na in the solution is 118709 ppm. The Na did not interact with Iraqi bentonite but it was removed in washing. The washing process was incorrect and failed the activation.

Another reason was that when taking picture to the three samples of three indirect activation processes and analyzing them in OLYMPUSVBX51M Microscope, the microscope picture shown that there is $Ca(OH)_2$ in the composition of activated bentonite. The microscope pictures of three indirect activation processes are shown in Figure 9, 10 and 11.

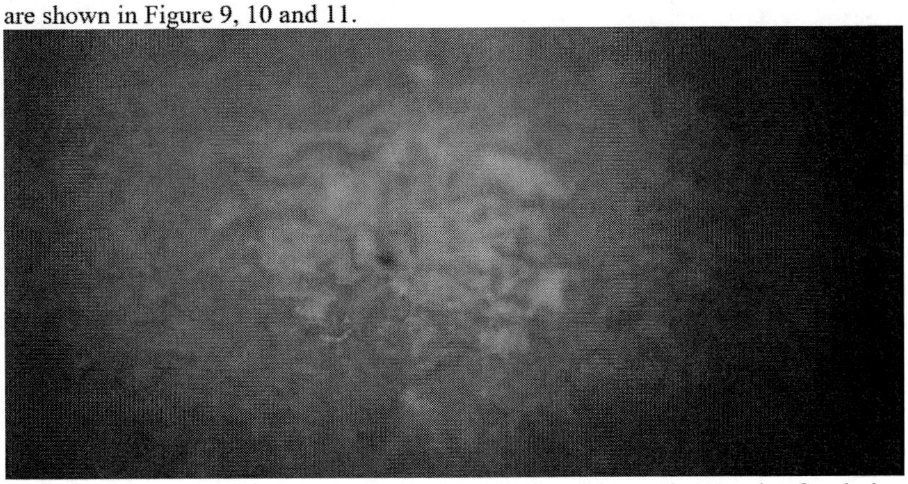

Figure 9.Microscope Picture of Alkali Activation Iraqi Bentonite sample after drying

Figure 10. Microscope Picture of Alkali and Thermal Activation Iraqi Bentonite after drying

Figure 11.Microscope Picture of Alkali and Acid Activation Iraqi Bentonite Sample after drying

6. Conclusions

Iraqi bentonite obtained from Trefawi Formation/ Al-Anbar Region/ Iraq was characterized using XRF, PSD and SSA, and measured its properties. The results shown that Iraqi bentonite is calcium type and cannot use as drilling fluid because of its bad rheological properties and has highly percentage of impurities.

Iraqi bentonite contains quartz and calcite as non-clay minerals or impurities. Quartz was removed by using special method of grinding.

By direct Activation for Iraqi bentonite when 3.2 g NaoH was added, Ib was activated and enhanced its properties except the filtration. The Filtration can be enhanced by adding CMC-LV or PAC or starch and then activating directly Iraqi bentonite by 3.2 g NaoH can be used as drilling fluids.

In direct activation processes failed to activate Iraqi Ca-bentonite, so It cannot be used as drilling fluid.

Nomenclature

Symbol	Description	Unit
W	Weight of sample	g
AV	Apparent viscosity	cp
PV	Plastic viscosity	cp
YP	Yield point	$lb/100ft^2$
YP/PV	Yield point to plastic	$lb/100ft^2/cp$

	viscosity	
ρ	Density of the mud	ppg
$V_{7.5}$	Filtrate volume at 7.5min	ml
V_{θ}	Filtrate volume at 30min	ml
Tmc	Thickness of mud cake	mm
H	Temperature	^0C
M	Mixing Time	hours

References

[1] Abdou M I, Al-sabagh A M and Dardir M I 2013 Evaluation of Egyptian bentonite and nano-bentonite as drilling mud *Egyptian Journal of Petroleum* **22** 53-59

[2] Dardir M M, Awadallah I A, Abd El-Wahab S I and Al-Shefey H I 2016 Improvement of the rheological properties of local bentonite clay using new blending polymers for water-based drilling fluids *Egyptian Journal of Chem*istry **59** 1 59-78

[3] Rasin F A and Hamad E A 2013 Study on the activation of Qara Tappah Iraqi bentonite and its test as gelling agent *International Journal of Application or Innovation in Engineering & Management (IJAIEM)* ISSN2319-4847

[4] Erdogan B and Demirci S 1996 Activation of some Turkish bentonite to improve their drilling fluid properties *Applied Clay Science* **10** 401-10

[5] Magzoup M I, Nasser M S, Hussein I A, Benamour A Onaizi S A, Sultan A S and Mahmoud M A 2017 Effects of sodium carbonate addition, heat and agitation on swelling and rheological behavior of Ca-bentonite colloidal dispersions *Applied Clay Science* **147**167-183

[6] Toor M, Jin B, Dai S, and Vimonses V 2014 Activation natural bentonite as a cost-effective adsorbent for removal of Congo-red in wastewater *Journal of Industrial Engineering and Chem*istry 1-9

[7] Gomez-Tena M P Gilabert J, Toledo J, Zumaquero E and Machi C 2014 Relationship between the specfic surface area parameters determined using different alalytical techniques *Qualicer'14* 1-10

[8] Agwu O E, Okon A N and Akpanika O I 2016 Activation of local bentonite clays for use as viscofiers in water-based drilling fluids *Journal of Scientific Researchs & Reports (JSRR)* **12** 2320-0227

[9] Amoco Production Company 2001*Drilling Fluid Manual* p82

[10] M-I SWACO 1998 M-I Manual

[11] Soony W M 2000 *IADC drilling manual*

[12] Chemwotei S C 2011*Geothermal Drilling Fluids* **10** 149-77

[13] Annis M R and Smith M V1996 *Drilling Fluids Technology* (Revised Edition, EXXON COMPANY, U.S.A) p 79

[14] Boussen S, Sghaier D, Chaabani F, Jamoussi B and Bennour A 2015 The rheological, mineralogical and chemical characteristic of the original and the Na_2CO_3-activated Tunisian swelling clay (Aleg Formation) and their utilization as drilling mud *Applied Clay Science* **10** 17

[15] Aramco S, Samy M, Garcia M, Frenier W and Samuel M 2004 A single-stage acid treatment to remove and mitigate clacium carbonate scale in sandstone and carbonate reservoirs *SPE* 87454

IOP Conf. Series: Materials Science and Engineering **579** (2019) 012007 doi:10.1088/1757-899X/579/1/012007

Better Hole Cleaning In Highly Deviated Wellbores

Raed H. Allawi[#1], Faleh H. M. Almahdawi[2]
[1]Ministry Of Oil, Thi-Qar Oil Company, Thi-Qar,Iraq
[2]University of Baghdad, Petroleum Engineering Department, Baghdad, Iraq.

[#]Corresponding author's e-mail address: Raedallawi80@gmail.com

Abstract. Many drilling problems (lost circulation, hurdle in running casing, poor cement jobs, high torque and drag and the necessity of re-drilling and, mechanical stuck caused by cutting accumulation in the wellbore) can be avoided by good hole cleaning during drilling operation especially in directional wells.
In this experimental work, the cutting transport ratio (CTR) in hole inclined (60 degree) from vertical was estimated by using three values of drilling mud viscosities, three annular velocities, two types of cuttings size and two rotation speed of inner pipe (RPM). Note that the rig consisted from PVC outer casing 4 1/2 inch as OD diameter and inner drill pipe 2 inch as OD diameter.
The experimental results showed that the Cuttings Transport Ratio (CTR) increased with increasing the annular velocity and better hole cleaning which obtained from the third velocity (3.247 ft/sec) for all cases. Moreover, increasing the mud yield point (YP) helped to improve the cuttings transport ratio; however increasing the mud yield point slightly affected with increased the annular velocity. Furthermore, it was noticed that the rotation of drill pipe (RPM) is one of the most important parameter for cuttings removal especially in laminar flow and transition flow. The cuttings sizes 1.7 mm are also easier to transport than 3.35 mm cuttings sizes for all cases except low yield point (YP).

Keywords: Cuttings removal efficiency; cuttings transport ratio; drill cuttings; drilling fluid; hole cleaning.

1. Introduction

In drilling of direction a wells, the gravity force has impact on motion drilling cuttings. It causes the settling cuttings along the bottom side of the wellbore. When accumulative of drill cuttings to form bed that it can lead to indicate of drilling problems such as: stuck pipe, lose circulating, hardly running casing, bad cement, slow rate of penetration, excessive over-pull on trips, high torque and drag [1]. For example, one stuck pipe leads to waste more time and money. Therefore, the main aim of study is hole cleaning and cutting transport ware to minimum the cost of drilling operation.

The most significant parameters effecting on the carrying capacity of drilling mud can be brief as annular velocity of muds, rotation of drill pipe, inclined of hole, hole geometry, penetration rate, pipe/hole eccentricity, mud properties and cuttings characteristics [2,3]. Therefore, annular mud velocity is the most factor that impact on hole cleaning because it is overcome on gravity forces of drill cuttings [4]. So, high annular mud velocity gives better hole cleaning but cannot be used because of hydraulic constraint.

Also, fluid yield point improves cuttings transport ration and then good hole cleaning. Moreover, rotation of drill pipe assists to remove cuttings transport because it is improved the flow regime profile and it improves annular velocities lowers than critical annular velocities to remove has drilling cuttings[5].

The type of drilling formation is effect on hole cleaning especially soft formation because it accumulated more cuttings comparison with hard information. Moreover, the small size of cuttings is easier to remove but the larger size and heavier cutting made hole cleaning more difficult and require higher pump rates for high- yield point fluids [6, 7].

In general, altering cutting size alone does not have significant role in enhancing hole-cleaning actually because in drilling operations we can't control the cuttings size while drilling.

In this study, most factors that effect in hole cleaning are taken in the one hole inclined angle (60 degree). Some researchers are considered the hole inclined (from 40 degrees to 60 degrees) as worst angles comparison with another hole angles.

The problem of cuttings transport in directional wells was the object of research in petroleum engineering. Initially, the researchers concentrated on the impacts of drilling factors on the removal of cuttings for oil and water based drilling muds. Thus, more experimental and theoretical researches, that are wanted in order to realize the mechanism of motion of cuttings and to determine the significant of drilling muds properties inside annulus.

The aim of this research study is knowing as the effect of most parameter that related with cuttings transport performance such as: annular fluid velocity, mud yield point, cuttings size, hole inclination and drill pipe rotation. The fluid velocity is very important because it is more contributed in hole cleaning. In this study all flow regimes laminar, transition and turbulent flow are investigated. Also the fluid yield point that has been tested to estimate the effect of fluid yield point on hole cleaning and knowing the best value of fluid yield point for better hole cleaning. The drill pipe has more impact especially in high hole inclination. Also, the high hole inclination has impact in drilling cuttings transport because the increase in hole inclination is assisted the rolling force of forward.

2. Experimental work

The experimental data are obtained from loop system that was built to investigate the cuttings transport. Elaborations of each part was summarized and discussed in this section:

2.1 Cuttings transport flow loop

The rig has the possibility to run all set of experiments in a normal pressure and normal temperature system (Figure 1). This rig was simulation for field condition by assumed drilling bit 8.5 inch (215.9 mm) and the drill pipe 5 inch (127 mm). Therefore, the rig designed to be 14 m (46 ft) as long from PVC which fits the test section that long of 5.5 m (18 ft) consists of two pipes outer pipe: outer diameter was 4½ in (110 mm) (inner diameter 4 inch) and inner pipe: 2 in (50.8 mm) outer diameter. The inner pipe designed to be centralized and has the ability to rotate.

The solid-liquid mixture is provided from a 1.5 m3 (1500 liters) container (Feed basin) where the liquid pumped and combined with the drilling cuttings. The mixer's impeller is grebes pump at the bottom depth inside the basin. A 10 H.P. centrifugal pump used to pump the muds and meet with the cuttings inoculated into the test section. This pump is convenient for situations where dissolution of the solid particles has to be avoided. The pump has been chosen to overlay the prospective range of operational conditions, which supplies 150 m3/hr (660 gpm).

The coarse sands taken from Nassiriya field from depth 1650 m to 1700 m were selected as solid particles. After cuttings, they were sift and washed and separated for two size 1.7 mm and 3.35 mm then weight of 204 g (0.449743 lb) as test sample. The total amount of sift cuttings which were ready prior to each run exceeded 15 kg (33.07 lb).

IOP Conf. Series: Materials Science and Engineering **579** (2019) 012007 doi:10.1088/1757-899X/579/1/012007

2.2 Drilling Mud

The drilling muds consisted of bentonite, caustic soda (Sodium hydroxide) and water to be used in this experimental work. Adding bentonite to water and mixing at a high speed for five hours and finally, drilling mud has been mixing at a middle speed.

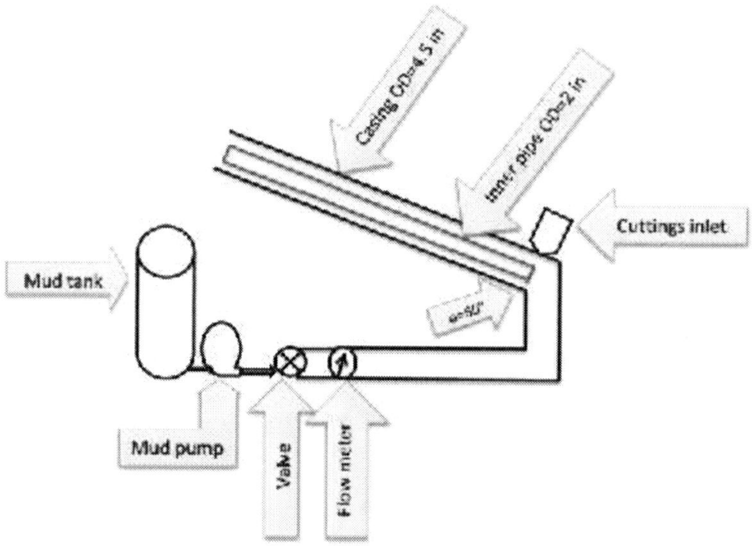

Figure 1. Cutting Transport Flow Loop

2.3 Experimental procedure

After sieving and cleaning the drill cutting and then weight them to 204 g (0.449743 lb). Each run was consisted of 6 experiments and was set up for high inclination angle (60 degree). All of these experiments were run with one rotating speed of inner pipe (80 rpm) and without rotating speed (0 rpm) for three flow rates (44, 66, and 92.46 gpm). For the next experiment, the same parameters were followed except using a new size of cuttings. After all experiments the termination of first run ware completed, the second run was done with a new mud yield point (higher yield point). To do so, caustic soda was used as another yield point control agent.

Percentage of recovery is calculated using the following equation:

$$\text{Recovery (\%)} = (\text{Final dried weight} \times 100) / (\text{Initial dried weight}) \quad (1)$$

3. Results and discussion

3.1 Effect of yield point (YP)

According to this study, two types of cutting 3.35 mm and 1.7 mm, cuttings transport ratio (CTR) improved to increase mud yield point (YP) as showing in Figures (2 a – b).

Both figure (2a) for 3.35 mm cuttings size and figure 2b for 1.7 mm cutting size, the CTR is drew with multiple mud viscosities (YP) for three annular velocities. Each carve in the diagram describes a constant muds velocity for the three characteristic of drilling mud with three various viscosities YP (8, 21, and 28 lbf/100ft^2). It is visible from Figures (2a - b) that primary increasing yield point from (YP=8 lbs/100ft^2) to (YP=21 lbs/100ft^2) results in better CTR for all the three different flow rates. Also the secondary increasing yield point from (YP=21 lbs/100ft^2) to (YP=28 lbs/100ft^2) results in better CTR for all the three different flow but has less effect in the primary increasing yield point. Therefore, the increase in YP is the main reason for increase they the amount of cuttings removal for both cuttings size, Also the effect of yield point is decreased with the increase in annular velocity.

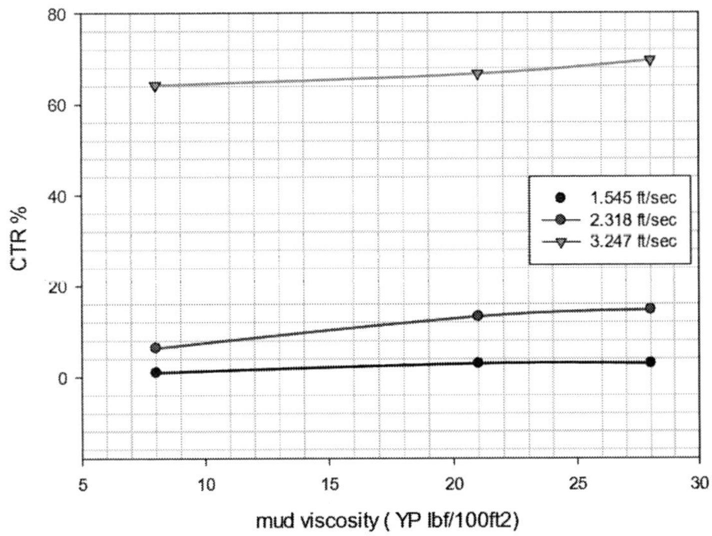

Figure 2a. cutting size 3.35 mm with RPM=0

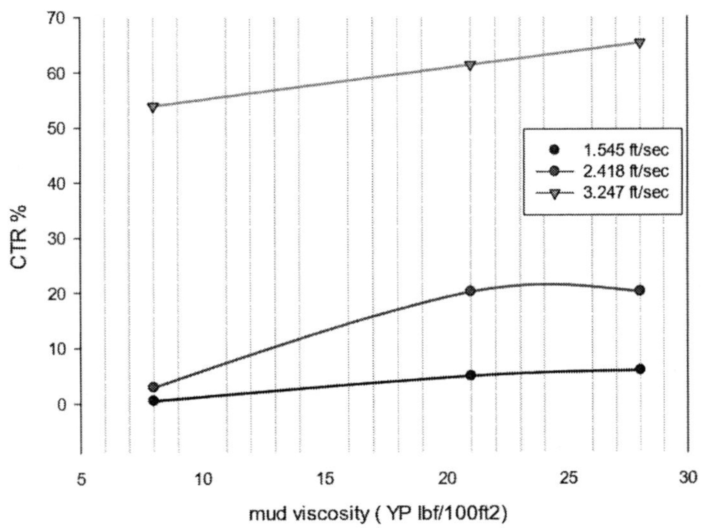

Figure 2b. cutting size 1.7 mm with RPM=0

3.2 Effect of annular fluid velocity

Annular fluid velocity has significant impact on cutting transport ratio as it showed in Figures (3 a – b). Figure (3a) for 3.35 mm cuttings size and Figure (3b) for 1.7 mm cuttings size in which, the CTR is drew with multiple annular velocities for three mud viscosities (YP).

As it can be seen in figures, any small increase in annular velocity will have more increased in CTR for the same yield point. Moreover, the increase in CTP, when the laminar flow converted to transient and then to turbulent flow regime.

IOP Conf. Series: Materials Science and Engineering **579** (2019) 012007 doi:10.1088/1757-899X/579/1/012007

The CTR was increased with the increased of annular velocity and raise maximum value in the third annular velocity (3.247 ft/sec Therefore, to get good cuttings transport (better hole cleaning), the annular velocity must be enough to prevent settling of cuttings.

The Most of injected cuttings were recovered by highest velocity of the drilling fluid, with more viscous mud and smaller cutting size. This fact also supports previous discussion about mud yield point in which increasing yield point.

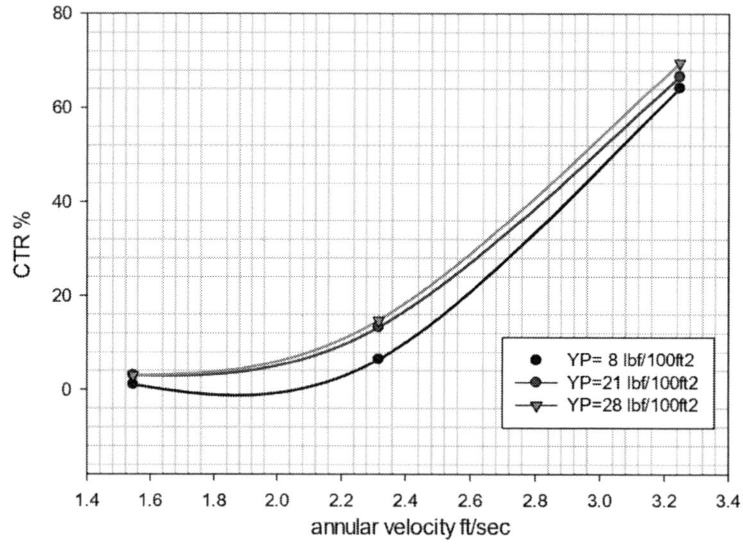

Figure 3a. Cutting size 3.35 mm with RPM=0

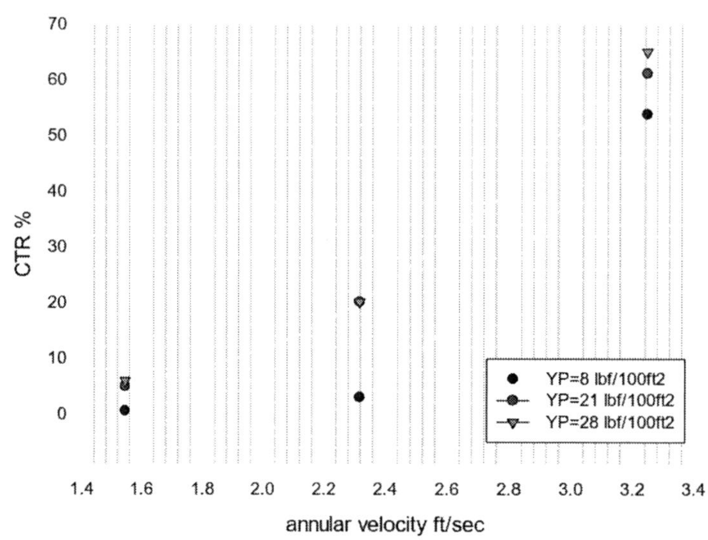

Figure 3b. Cutting size 1.7 mm with RPM=0

IOP Conf. Series: Materials Science and Engineering **579** (2019) 012007 doi:10.1088/1757-899X/579/1/012007

3.3 Effect rotation of drill pipe

Drill pip rotation (RPM) observingly has an important role in the removal of cuttings especially in laminar and transition flow as seen in figures (4 a – b). In figures (4a) for 3.35 mm cuttings size and (4b) for 1.7 mm cutting size, the CTR is drew with multiple annular velocities, three value of yield point (YP) and two rotation of drill pipe (RPM). From fingers, the drilling cutting is become faster with rotation of drill pipe. Attributed the effect of drill pipe rotation to several reasons as follows:

1- Rotation of drill pipe drives the drilling cuttings to the stream of fluid flow.
2- Rotation of drill pipe assist to improve the flatting of flow regime.
3- Rotation of drill pipe irritate of bed cuttings.

Therefore, the effect of drill pipe rotation is slight in turbulent flow and decreased with increasing in fluid yield point. In low yield point (8 lbf/100ft^2) and turbulent flow (3.247 ft/sec), the effect of drill pipe rotation is very slight for both cutting size 3.35 mm and 1.7 mm. The rotation of drill pipe effect on cuttings 1.7 mm more than on cutting 3.35 mm.

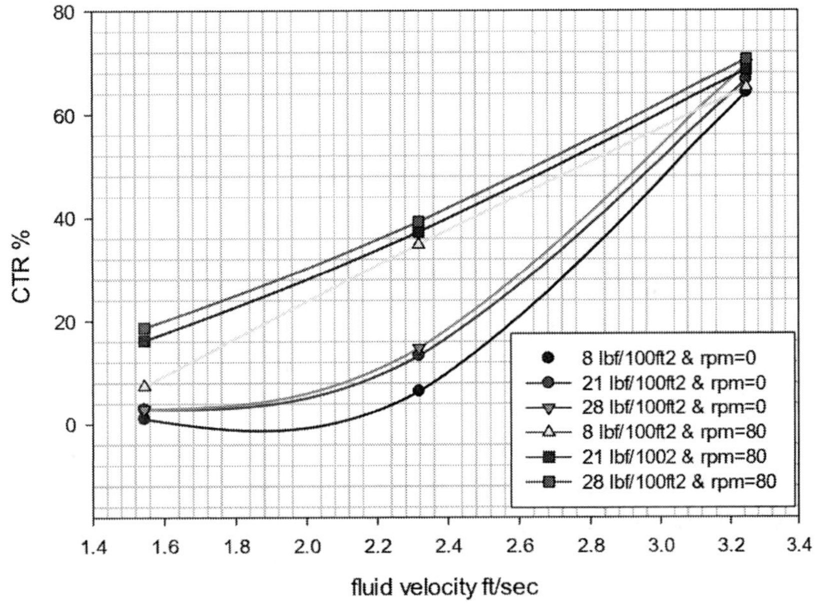

Figure 4a. Cutting size 3.35mm

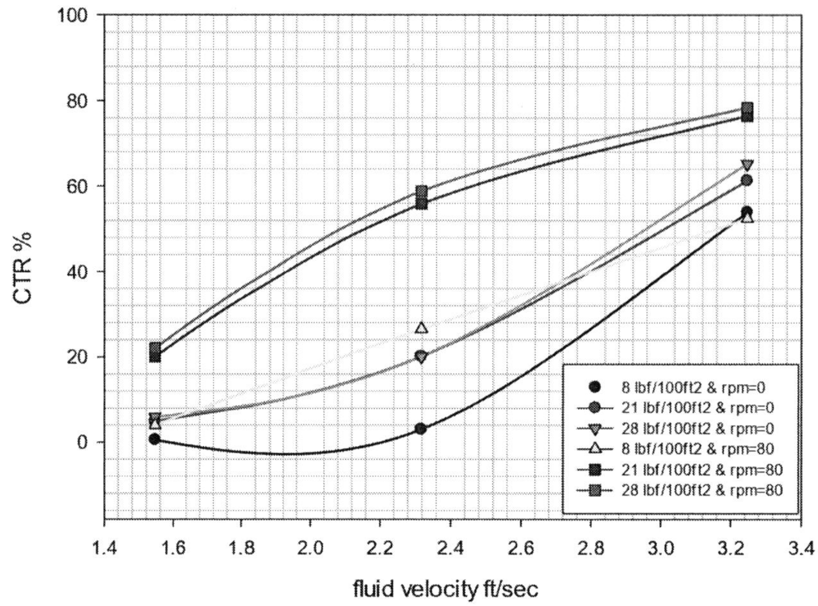

Figure 4b. Cutting size 1.7mm

3.4 The Effect size of cutting

The size of cuttings is one of the parameter that has an influences on the hole cleaning. For this study the small and medium sized cuttings have been considered for the analysis as showing in figures (5 a – d).

Figure 5a for cuttings size 3.35 mm and 1.7 mm, the CTR is drew with multiple annular velocities for three mud viscosities (YP) with RPM=0. Also figure 5b for cuttings size 3.35 mm and 1.7 mm, the CTR is drew with multiple mud viscosities (YP) for three annular velocities with RPM=0.

From figures noticed that the cuttings with 1.7 mm transported faster than cutting with 3.35 mm but in low yield point the effect became inversely phenomena. As shown in figures, the drilling cutting become slowly with decries in annular velocity for the same size.

The effect of drill pipe rotation is big on small cutting size 1.7 mm because of slip velocity as show:

$$V = \frac{(\rho c - \rho m)^{0.6} * 1 * d}{M^{0.3} * \mu^{0.3}} \tag{2}$$

Therefore, in laminar flow with rotation of drill pipe and high yield point, the cuttings 1.7 mm are transported faster than 3.35 mm comparison with laminar flow without rotation of drill pipe as mentioned previously. Also in this case, in low yield point the cuttings 1.7 mm transport less than cuttings 3.35 mm. In general, altering cutting size alone does not have significant role in enhancing hole-cleaning actually because of drilling operation that cannot control the cuttings size diameter.

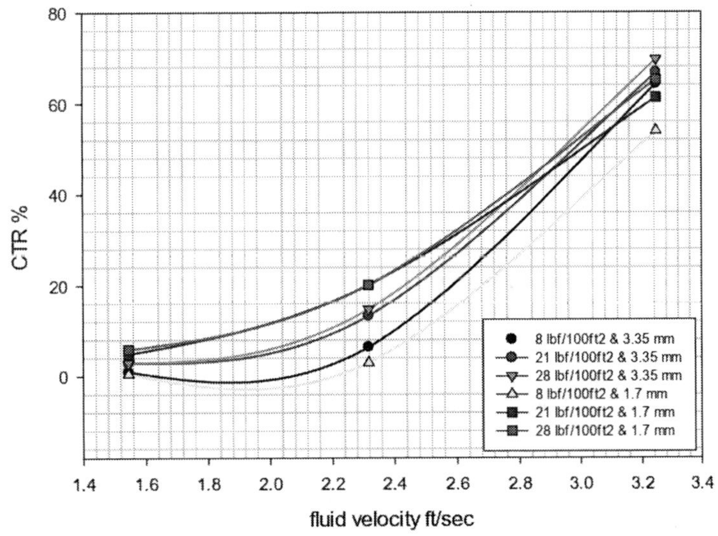

Figure 5a. Cutting size 3.35mm and 1.7 mm with RPM=0

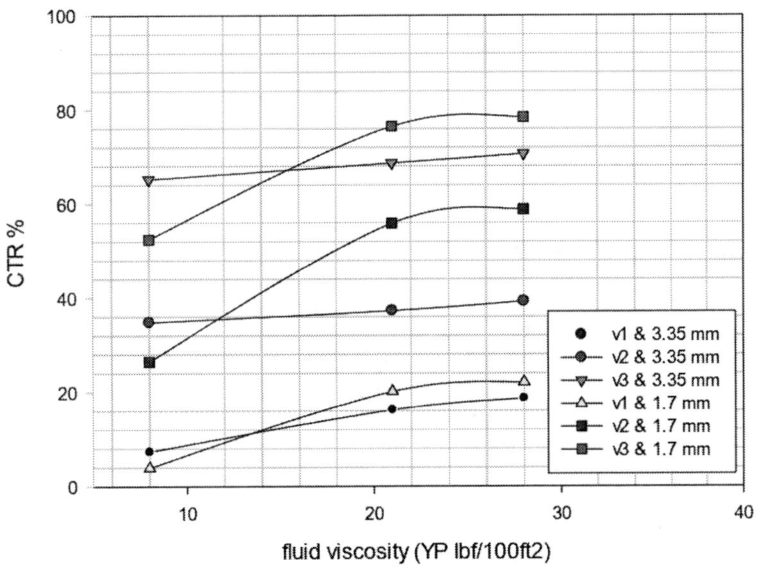

Figure 5b. Cutting size 3.35mm and 1.7 mm with RPM=0

4 Conclusions

Only four effective factors were chosen to the rotation of drill pipe, fluid yield point and velocity as the topical of this research. The most points noticed summered as:

1- when mud yield point (YP) increase, the CTR is increased. The increase in YP is the main reason for the increase in the amount of removed cutting. Increasing of fluid yield point at the same flow rate,

the turbulent flow tends to shift to laminar flow. However, it should be observed that converting a turbulent flow to laminar flow requires crossing through a large range of Reynolds number including transition zone. If velocity and hole inclination are kept constant, the CTR has been greater by upper fluid yield point until the flow is still turbulent. However, when muds yield point are increased that the CTR has been slightly increased gradually until the end of transition zone. Also the CTP has been more increased after transition zone. Finally, for better hole cleaning must use high mud yield point yield point with laminar flow and low yield point with turbulent flow.

2- Fluid velocity has a significant impact and more contribution in removal of drilling cutting. Any small increase in annular velocity well be creates a great increase in cuttings transport.

3- Rotation of drill pipe was noticed an impact part in the removal of cuttings especially in laminar and transition flow. Its effect leads to increase the ability of cutting removal and the maximum effect was in transition flow. Also this effect was slightly in turbulent flow. Through the previous point, increase impact of drill pipe rotation with increase fluid yield point.

4- In high yield point 1.7 mm cuttings are transport better than 3.35 mm cuttings but in low yield point going on the contrary. Moreover, actually in drilling can't control the cuttings size while drilling.

Nomenclatures:

CTR: Cutting transport ratio
YP: Yield point
RPM: Rotation of drill pipe
V: Annular velocity
Vs is slip velocity in ft/min.
ρc is cutting density in ppg
dc is cutting diameter in inch

References

[1] Bourgoyne, A.T., Millheim, K.K., Chenevert, M.E. and Young, F.S. 1986 *SPE Textbook Series*, Vol. 2 (Richardson, TX)
[2] Ettahadi R.O., Sorgun M. Energy Sources Part B: Economics, Planning, and Policy *Esb-08-046 (accepted).*
[3] Francis Effiong 2013 Thesis (Norwegian University of Science and Technology, Department of Petroleum Engineering and Applied Geophysics)
[4] Sorgun M., Aydin I., Ozbayoglu M.E. *Journal of Petroleum Science and Engineering* (submitted).
[5] Sun Xiaofeng, Wang Kelin, Yan Tie, Zhang Yang, Shao Shuai and Luan Shizhu 2013 *The Open Petroleum Engineering Journal* **6** 25-32
[6] J.J. Azar, G. Robello Samuel 2007 *Drilling Engineering* (PennWell Corporation1421 South Sheridan Road Tulsa, Oklahoma USA)
[7] Neal J. Adams 1985 *Drilling Engineering A Complete Well Planning Approach* (PennWell Publishing Company 1421 South Sheridan Road Thlsa, Oklahoma)

IOP Conf. Series: Materials Science and Engineering **579** (2019) 012008 doi:10.1088/1757-899X/579/1/012008

Smart Cement Modified with Nano Fly Ash and Micro Fiber for Oil Well Applications.

A A Abdul-hamead, F M Othman, D M Abdullah

University of Technology, Materials Engineering Department, Iraq, Baghdad

Corresponding author's e-mail address: adr.alaaa@gmail.com

Abstract. Adequate slurry design is critical to the success of a cementing work. In this study, a ball mill procedure was used to prepare a nano powder from a fly ash material supplied by Al-kufa Cement Factory for re-inforcing the oil well cement through using it as a partial replacement of oil well cement class G by a weight percentage of(0.25%,0.5%, 0.75% and1%)and micro fiber polypropylene at percentage (0.25%,0.5%,0.75%,and 1%).The mix with a water to cement ratio was made at 0.44.Characterizations of the prepared samples were conducted by X-Ray Diffraction (XRD), Atomic Force Microscope (AFM), density and compressive strength .The results manifested that the structural properties were improved with the development of calcium silicate hydration (C-S-H)phase ,the density as well as the enhancement in compressive strength .That makes it proper for the oil well applications.

1. Introduction

Nanotechnology is not a new science and it is not a new technology .It is rather an extension of the science and technologies that have already been in development for many years and it is the logical progression of the work that has been done to examine the nature of our world at an ever smaller scale. Nearly every aspect of the construction process will be touched by the foregoing innovations of nano scale technologies [1].

The nano scale size of particles can result in dramatically improved properties from conventional grain size materials of the same chemical composition [2].

Processing of the nano materials and their applications, normally if the particle sizes are in the ranges of (1-100) nm ranges, they are generally called nano-particles or materials [3].

Materials at Nano scale have a relatively larger surface area when compared to the same mass of material produced in a larger form that's make it more chemically reactive (in some cases materials that are inert in their larger form are reactive when produced in their nanoscale form), and affect their strength or electrical properties [4].

The cement used in oil wells is a classify as class G cement, that variation of the Portland cement with low content of celite (C3A) and with bigger grains than the other common Portland cements [5]. Oil well cement (OWC) slurries is generally more complicated than that of conventional cement paste, In order to contend with bottom hole conditions (wide range of pressure and temperature), a number of additives are usually used in the OWC slurries, which exhibit different characteristics depending on the combination of admixture used incorporating various chemical and mineral admixtures used [5].

Cementing job is the process of placing cement slurry to the desired location in the annular between casings and well bore with general function to bond the casing and formation, to protect producing formations, to prevent migration of formation fluids between zones and to control lost circulation. The slurry properties which are generally considered to be important include; density, fluidity, setting time or pumping time, strength, stable suspension of particles , free fluid or free water; ability to resist dehydration against permeability – fluid loss; permeability, shrinkage and durability. The cement must last a long time – longer than the well – and prevent fluids moving between the subsurface and surface [6].

Several studies were conducted for many years on the using of nano-powders and fiber in well industry and their effects on the properties, performance, and application of the cementious materials. In 2015 J. D. Mangadlao et al [7] made a review of summarized the importance and approaches of smart cement materials. For a variety of industries including the oil and gas production sector, the use of smart cements can lead to cost-effective solutions that may well result in on-time drilling operations and new monitoring protocols in asset integrity management. The review has emphasized the features and the stages of the cement materials and current efforts underway to fabricate smart cements for a variety of applications. The application of nanotechnology and stimuli-responsive systems will result in true advances in materials and technologies for the oil and gas production in upstream operations. Have study the use of a polycarboxylate superplasticizer to improve the dispersion of graphine oxides(GO) in cement. A 48% increase in tensile strength was observed for graphene-cement nanocomposites containing 1.5wt% GO and 0.5 wt% plasticizer. They have demonstrated that the GO nanoplatelets were well dispersed and no GO agglomerates were seen in the matrix. Babak et al., 2014[8] studied the aforementioned advantages that include higher compressive and tensile strength; excellent durability and corrosion resistance make this new class of material highly attractive for applications in construction industry. Particularly, it is highly applicable for well cementing and construction on offshore and other structures in marine environments.Wu et al., year [9].

GO as an additive to concrete do not only enhance mechanical property but may also prevent Microbial Induced Corrosion (MIC) of concrete, which is a very common problem in oil wells. Graphene oxide has antibacterial property that can inhibit the growth of a wide range of bacteria such as Escherichia coli (E. coli) and Bacillus subtilis (B. subtilis) and Cupriavidus metallidurans (C. metallidurans) [10].

In 2017 by S.M. Ragehet al [11], prepared the cement prepared for testing at a temperature of less than 25 0c. In this research, cement class G is used, in addition to accelerators and retardants, was used at certain percentages after completion. They measurement of thickening time, free water content, viscosity, rheological properties, fluid loss and various other properties and compressive strength after two daythey note that the addition of the Accelater.fcOur results do support my hypothesis. The test we did went on very smoothly only place where we struggled was in finding the best equipment to perform the experiment, since the rate of temperature has to be constant for the mixing. Thickening time and compressive strength correlations for Portland – class "G" cement slurries have been estimated .The test further went on ahead to come to conclusion that the additives effect the compressive strength and the thickening time of class G of cementing oil well. In 2015 C. Vipulanandan et al[12], study the objectives was to quantify the effect of different w/c ratio on the electrical resistivity and piezoresistive behavior of smart oil well cement. In this study, Class H cement with water-to-cement of 0.38, 0.44 and 0.54 was used. The samples were prepared according to the API standards (API 1997, 2002). To improve the sensing properties and piezoresistive behavior of the cement modified with 0.1% of conductive fillers (CF) by the weight of cement was mixed with all the samples. After mixing, cement slurries used for rheological, curing and piezoresistivity studies. Two conductive wires were placed in all of the molds to measure the changing in electrical resistivity. At least three spe- cimens were used for each type of test for all the water-to-cement ratio mixes investigated in this study. experimental and analytical modeling of the rheological and piezoresistivity behavior on smart cement with w/ c ratio of 0.38, 0.44 and 0.54, show that The rheological test showed that class H oil well cement had shear-thinning behavior and a hyperbolic model was proposed to predict shear stress–shear strain rate relationshipThe hy-perbolic rheological model predicted the test results very well compared to Herschel Bulkley model. The yield stress and maximum shear stress limit reduced with increase in water-to-cement ratio while increasing the tem- perature increased these properties and The initial resistivity (ρo) of the smart cement decreased from1.03 Ω-m to 1 Ω-m and 0.9 Ω-m, a 3% and 13% reduction with increasing the water-to-cement ratio from 0.38 to 0.44 and 0.54 respectively. Hence the electrical resistivity can be used for quality control. Water-to-cement ratio affected the curing characteristics (ρmin and tmin) of the cement.The smart cement showed enhanced piezoresistive behavior compared to unmodified cement. With 0.1% conductive filler (CF) modification the piezoresistivity at peak stress was en- hanced by over 700 times the unmodified cement. The piezo- resistivity enhancement was depended on the water-to-cement ratio and curing time. The new piezoresistive constitutive model predicted the compressive stress—changes in resistivity relationship very well. Additional of 0.1% CF also increased the28 day compressive strength by over 10%.and The resistivity index (RI24 h) of the smart cement with lower w/c ratio was higher than that of the cement with higher w/c ratio. Linear correlations were determined between resistivity index and compressive strength of smart cement for different curing times.

There were two local researches on the effect of adding Nano material to the cement slab in terms of compressive resistance ,thickening time ,density where, In2017 H. A. Hadi et al[13], study experimental investigation of nano alumina and nano silica on strength and consistency of oil well cement. In this study The cement powder used Glass G (HSR) cement according to API standards, was provided by Missan Oil Company and Nano Alumina and Nano Silica are used as additives to cement slurryat perecentage (1%, 2%) of Nano silica by weight and (1% ,2% ,3%)of Nano Alumina by weightThey placed cement slurry to be tested in the prepared molds and placed the specimens in the curing bath at atmospheric pressure and heating to test temperature 38 °C for 8 hr and to the second test temperature 60 °C for 8 hr.after tested it they show Compressive strength of oil well cement at 38°C is increased when amount of NS&NAL is increased. The reason of this strength improvement is due to the small size of nano particles in pores between cement particles; therefore, it increased the efficiency of the packing. The second reason is due to the effect of (NS&NAL) on promoting the pozzolanic interaction. That means it reacts chemically with Ca(OH)2 (formed from dehydration interaction) at ordinary temperature in the presence of water and produce compounds having cementitious properties. At 600C compressive strength is decreased when the amount of NS exceeds 2% bwoc. This means that raising the quantity of NS is good in increasing compressive strength to a specific cutoff after which further increase in the NS quantity leads to a reducing in the compressive strength. The reason for that truth is the additional rate of NS could prompt agglomeration of its particles under the present dispersion circumstance. Nano materials (NS and NAL) accelerated the thickening time of cement slurry class G. and in 2014 by A. A. Abdulrahman et al[14] was studied. , Characterization of Nanosilica from Iraqi Rice Husk and its Application in oil well cement. The objectives of this research are Preparation of nanosilica from Iraqi rice husk by two chemical precipitation methods. And Studying the compressive strength of oil well cement class G by using different percentage of nano silica.

Cement class (G) used in this study was from South Oil Company and used Nano silica as additive

For cement slurry the affect of adding nanosilica to oil well cement class G for 8h at temperature 380C and 60 0C as specified in API Specification (10A)for cements and materials for well cementing is at 8h Compressive strength In psi , 38 0c without Nano silica is879 andwith Nano silica is (953 , 963 , 919) and at8h Compressive strength ,In psi , 60 0c with out Nano silica is 2090 and with Nano silica is (2094 , 2306 , 2262) . It is obviously noted that compressive strength increased by increasing the percentage of nanosilica additive, but increasing in amount of this additive decreased the compressive strength as it happens when 2.5% percent of nanosilica added that is attributed to nanosilica is considered as a lightener agent , therefore it was showed this specification due to the fineness of the particles and agglomeration caused by difficulties to disperse the particles during mix so increasing the amount of nanosilica reduced slurry density which ultimately leads to reduction of compressive strength. Aim of this paper is Preparation of nano powder from fly ash material resulted from kufa cement factory

IOP Conf. Series: Materials Science and Engineering **579** (2019) 012008 doi:10.1088/1757-899X/579/1/012008

for supporting OWC using produced nano powder as apartial replacement OWC and Characterization the prepared sample usinScanning Electron Microscope (SEM) observation, X-Ray Diffractometer (XRD) and Atomic Force Microscope (AFM) •Studying the compressive strength8hr,density test of oil well cement class G slurry at different percentage by - Nano fly ash and Polypropylene fiber in order to improve cement properties such as cement improvement and mechanical properties improvement to make it suitable for well properties.

2. Experiment

The experimental step of this section of study was designed to investigate the effect of adding nanoparticles on the mechanical,physical,structural properties of cement glass g.cement glass mixtures were prepared using two different groups of nano materials and one micro fiber which are(Nano flyash, , polypropylene fiber) in five concentrations of(0.25%, 0.5% ,0.75% ,1%) per weight for nano materials but by volume for micro fiber .In order to achieve these steps,number of sample cement are designed For oil well cement class G must apply 792gm weight of cement and 349gm water, according to API (American Petroleum Institute, 1997) in nine classes (from A to J) .the peresentage of nano materials and micro fiber was(0.25%,0.5%,0.75%,1%)by weight of the cement material for nano materials but by volume for micro fiber .Cement glass G with nano materials (flyash)and micro rod(polypropylene fiber) were compared with refrence mixture to evaluate the effect of the dosage of nano materials and micro fiber.

Chemical composition test for cement and fly ash was performed in the Iraqi geolological survey department ministry of Industry Baghdad Iraq. Density test was done according to (ASTM C-373) standard[15], the true density(pt) is prepared by setting model in the water (Archimedes rule) via following connection [16-17]:

$$pt= Wd/(Ws-Wn)*D \quad ...(1)$$

where ρt is true density (gm/cm3), D is density of distilled water(1gm/ cm3), Wd is dry weight of sample (gm), Ws is weight of sample after saturation in water (gm), and Wn of sample when submerged in water (gm). Atomic force Microscopy (AFM) was taken with digital instruments; typical data have been taken from height images including root mean squre (RMS) and roughness. made in USA model AA300220V.Compressive strength test for cubes samples was determined by using auniversal measurement machine (EVERY DENISON) of 2000KN capacity. The average result of three specimens was reported for each sample, The following relationship was used:

$$\sigma = P/A \quad .....(2)$$

Where P is ultimate compressive load, (N) Ais asample area, (mm2) and σ is compressive strength (MPA).

The X-ray diffraction was taken by diffractometer with radiation CuKa (λ=1.54060 A) this inspection was carried out in Nano technology and Advance Material Research center at the university of technology . particle size test for powder fly ash was performed in nanotechnology and advanced research center (model:Brookhaven Nano Brook 90 plus USA), it was 100nm.

TABLE 1. Chemical composition of raw materials(Cement Glass G)

Material	SiO_2	Fe_2O_3	Al_2O_3	TiO_2	CaO	MgO	SO_3	LOI
weight%	30.38	2.61	5.95	0.33	42.64	3.80	1.840	7.97

TABLE 2. Chemical composition of raw materials (Fly Ash)

Mate rial	SiO 2	Fe2 O3	Al2 O3	Ti O2	Ca O	Mg O	SO 3	L.O .I	Na2 O	K2 O	CI	P2O 5	Cd O
weight %	13.8 8	3.22	4.21	0.2 3	39.6 5	3.1	4	24.6 5	2.41	2.4	0.00 56	0.0 2	2.2 3

3. Results and Discussion

Figure1 shows the Density of the prepared samples. Nano powder plays a filler role in order to enhance the density of the samples, which can decrease the porosity of mixture; on the one hand powder can accelerate cement hydration by virtue of their high free energy. Also by adding the PF improve the density of the samples as it can shows in the fig.1.

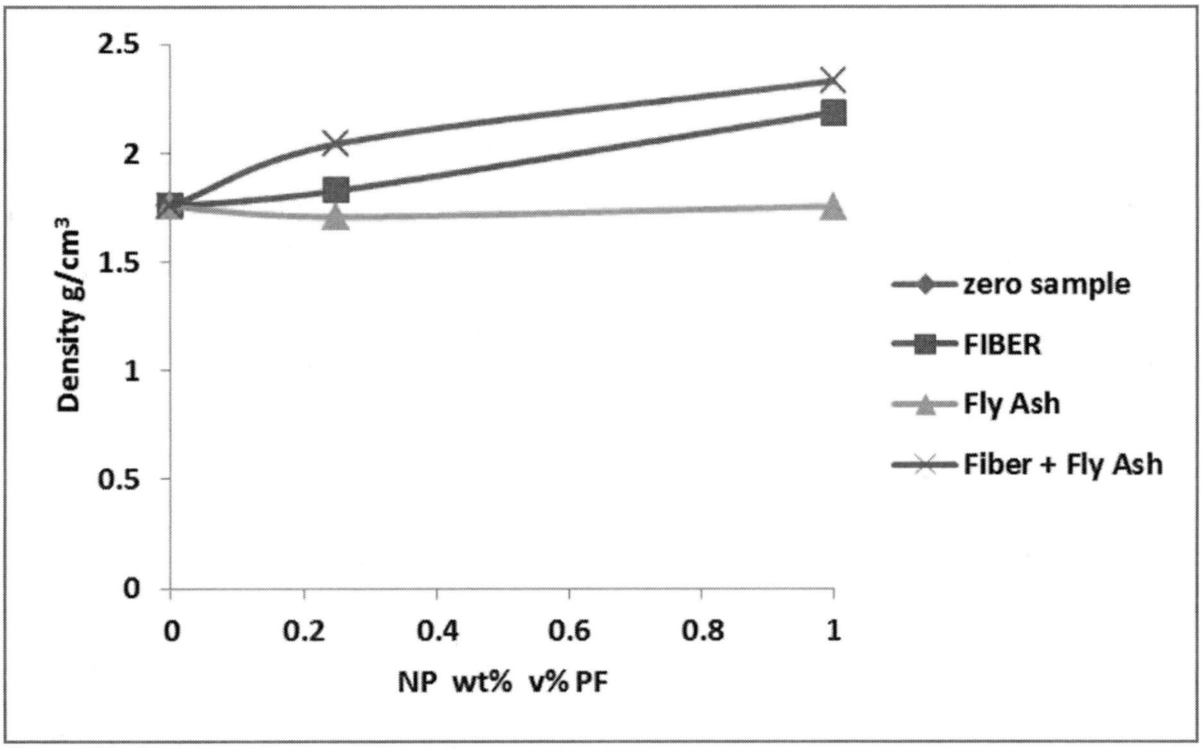

Figure; The Density of the prepared samples.

Material	SiO_2	Fe_2O_3	Al_2O_3	TiO_2	CaO	MgO	SO_3	L.O.I	Na_2O	K_2O	CI	P_2O_5	CdO

Material	SiO_2	Fe_2O_3	Al_2O_3	TiO_2	CaO	MgO	SO_3	L.O.I	Na_2O	K_2O	CI	P_2O_5	CdO
Weight%	13.88	3.22	4.21	0.23	39.65	3.1	4	24.65	2.41	2.4	0.0056	0.02	2.23

Compressive Strength Result

Figure2 shows the compressive strength: The compressive strength and capability of mortar with "nano fly ash" were the outcomes linked with G class Cement Mortar GC, mortar and "nano fly ash" were produced to be long-lasting more than Normal Cement Mortar "NCM" and the measuring size will be growing by the strong point of mortar with "nano - fly ash" with different to GC that founded to be in the diversity between 17 % to 50 % for a range of symbols of mortar, the capability of mortar with "nano - fly ash" has been created to be very much more than GC, It is recognized that the porosity and pore volume distributing are critical combination of "microstructure hydrated cement dentifrice", the effect strength orderly to gain high strength reducing the permeability and a toughing form of mortar is urgent to reduce the porosity of cement dentifrice it is very familiar that the mixture of "pozzolanic (CSH)" materials of cement smooth the porosity and pore size distributing of the dentifrice. Fly ash is identified to be a high-quality "pozzolanic" material to be used with mortar, many researchers investigator have documented these effects on the physical characteristics and pore structure, researching proof the effect of "fly ash" purity at the compressive strength pore size distributing and porosity of reinforcement cement dentifrice. category C fly ash "high - calcium" with nano particle dimension about (100 nm) also it has been using with partially replaced with Portland cement on (0 , 0 . 25 , 0 . 5 , 0 . 75 , and 1 wt%) by loading the water to cement proportion (w / c) of (0.44) was intended for all the blended cement dentifrice mixture, the totality porosity and capillary pores were decreasing whilst the gel pore has better than previous situation because adding "nano fly ash" at all substitute levels [18 , 19 , 20]. The examination sequel submit that the blended cement dentifrice at high (1 %),"nano fly ash" has been shaped a dentifrice with better compressive strength than (0.25) nano fly ash, the porosity and pore size of blended cement dentifrice were severely affect by the adding more" nano fly ash", the substitute of Portland cement by "nano fly ash" has bigger than previous to the porosity other than reduce the average of pore size of dentifrice the calculated gel porosity (5.7-10 nanometer) had improve by increasing the "fly ash content".

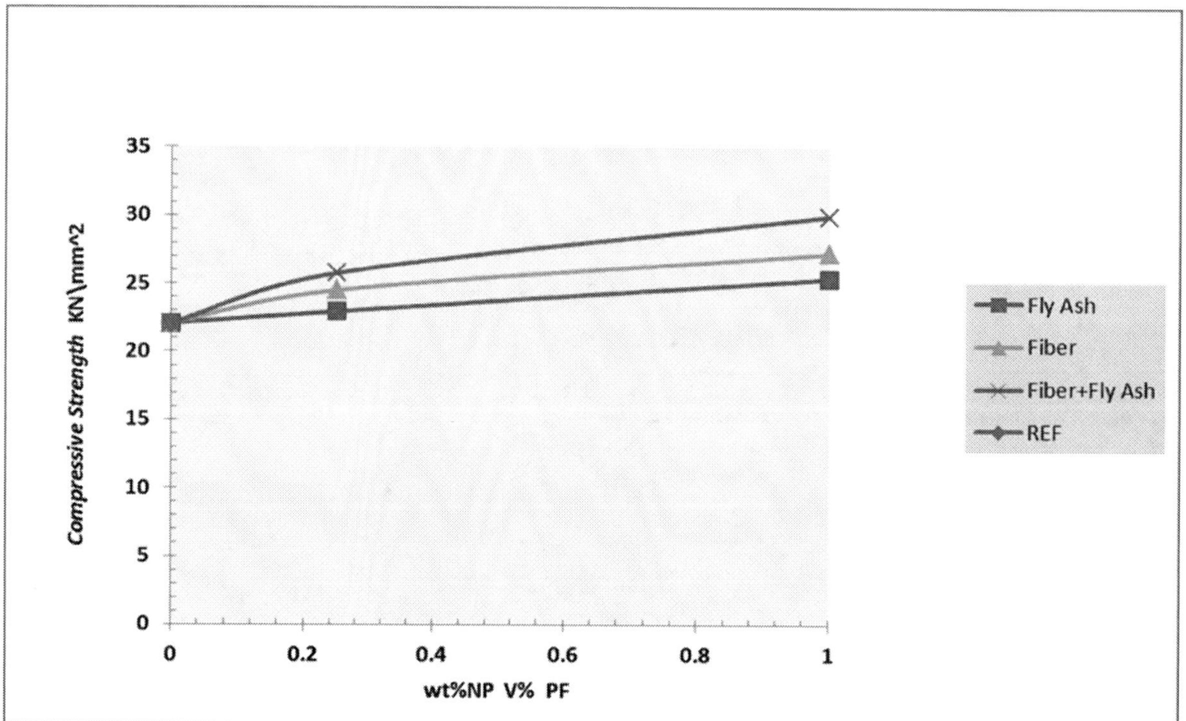

Figure2: The compressive strength of the prepared samples.

Atomic Force Microscopy Result

AFM is a powerful technique to investigate the surface morphology at nano to microscale. The surface of mortar is shown in fig. (3) 3D of the surface, the roughness of mortar surface without the addition of

fly ash was 18.79 nm, while in the presence of 0.25 Wt% partial replacement of cement by nano fly ash was15.38 nm and in percentage of 1% fly ash surface was20.17nm .The difference depends on the rate of replacement of cement with nano powder. From AFM test, also it can be seen that mortar without adding Average Diameter103.01nm in 0.25 FA was. Diameter184.82 nm and in 1%FA was Avg. Diameter: 158.30. nm..
Table 3. shows the AFM result of Glass G only and with additives.

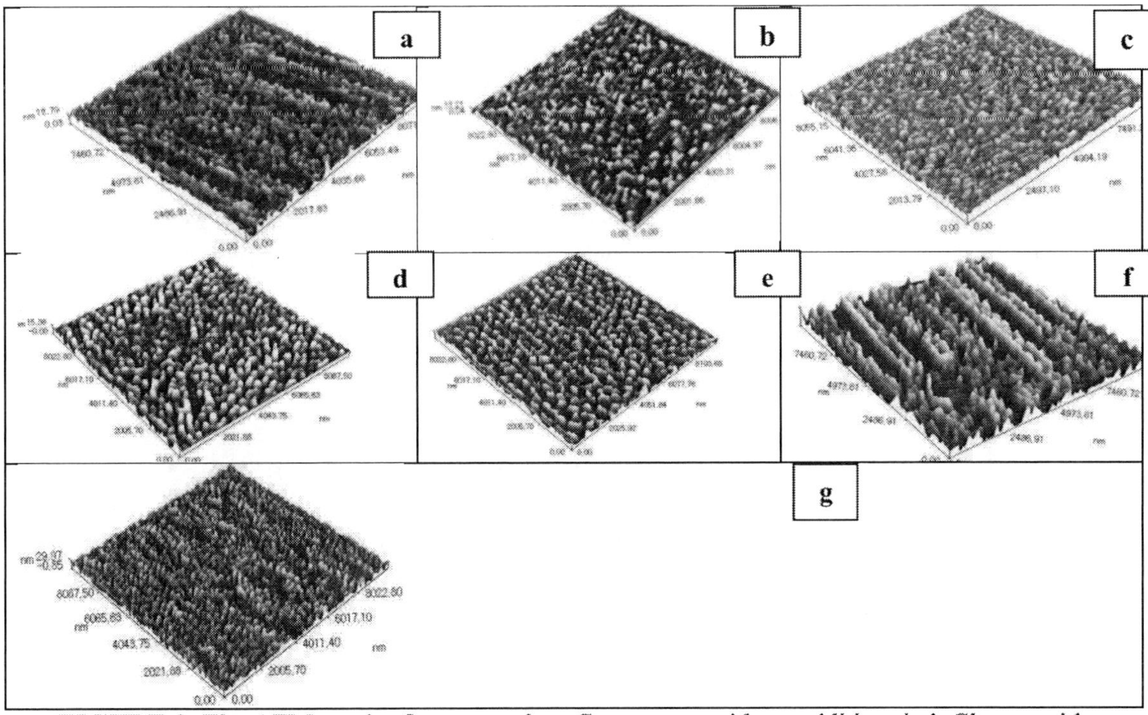

FIGURE 3 :The AFM result of cement glass G a:cement without addition ,in b:Glass g with 0.25%fiber, in C: Glass g with1%fiber,in d: Glass g with0.25%Fly Ash, in e: Glass g with 1%fly ash, in f: Glass g with 0.25% Fiber + Fly Ash, in g: Glass g with 1% Fiber+ Fly Ash.

Table 3. The AFM result of Glass G only and with additives.

Percentage%	Roughness nm	Diameter nm
Zero sample	18.79	103.01
0.25% PF	37.21	232.64
1% PF	73.76	124.39
0.25% FA	15.38	184.82
1% FA	20.17	158.30
0.25%FA + PF	49.29	206.84
1% FA + PF	29.27	133.20

X-XRay Diffraction Result
Figure 4 shows the XRD pattern of the cement mortar without nano powders, in cement mortar (after curing time) of the (reference and mixed specimens), the components of phases are shown in the Figs from 4-b to 4-c, Were: Portlandite: Ca(OH)2, hexagonal crystallized, JCPDS (04-0733)/(CH), Tobermorite:Ca5Si6(O,OH,F)18.5H2O,orthorhombic crystallized, JCPDS (45-1480) (Tob), Ettringite:Ca6Al2(SO4)3(OH)12.26H2O,hexagonal crystallized, JCPDS (41-1451) (CASH).

IOP Conf. Series: Materials Science and Engineering **579** (2019) 012008 doi:10.1088/1757-899X/579/1/012008

Furthermore, the phases shown were: - Calcium Silicate Hydrate:CaO.SiO2.H2O, poor crystallized, JCPDS (34-0002) (CSH), Wollastonite: CaSiO3 , monoclinic crystallized, JCPDS (43-1460) (CS). It was observed that the ettringite and portlandite are involved in all stages of the hydration. In addition, the changes that happened in mineralogical components during the hydration processes were observed, where hydro-silicates and hydro- aluminates were present (portlandite, tobermorite and ettringite), the highest peaks correspond to portlandite (CH) and calcium silicate hydrated (C-S-H) gells [21], XRD characterization were done in order to investigate the activity of the incorporated nano powders with cement mortar specimens after hydration, from the XRD patterns it is clear that the (CH) peaks are nearly decreased with mixing by (fly ash) nano powders, it is therefore inferred from the figures below that nanopowders can react with the produced CH through the hydration. Hence, the reactivity of nanopowders is significantly high and has developed the microstructure of cementitious materials, thereby it has improved the mechanical properties of cementitious materials, and CH compound has a sharp peak in the cement mortar acting as the pure hydration product, which is released from the hydration of cement. Obviously, the peak/intensity of CH is reduced due to the nanopowders as a cement replacement as shown in the Figs. from (4-b) to (4-c), which show the consumption of CH via the pozzolanic reaction [22].

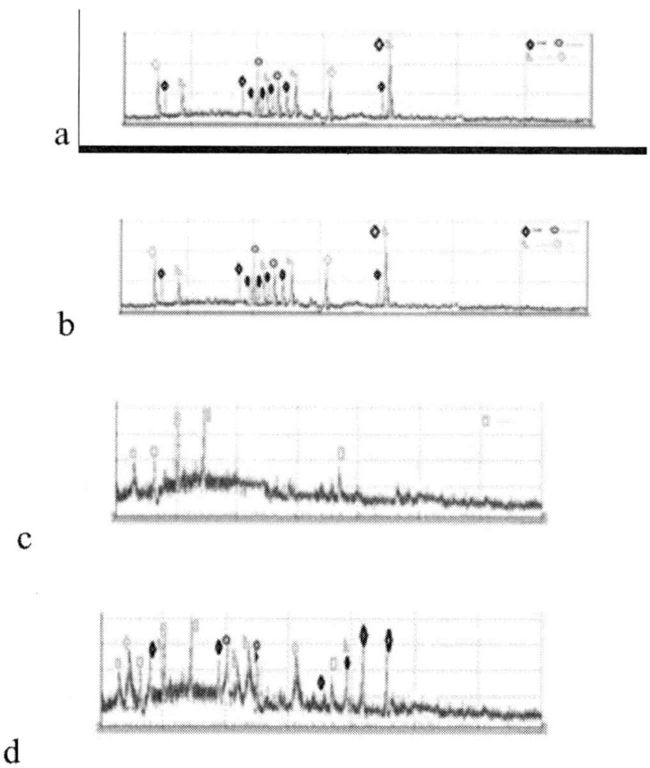

FIGURE 4. XRD pattern of the G cement mortar; a: without nanopowders (b): 0.75% FA, (c): 1% PF (d):1%FA+PF

4. CONCLUSION

When nano powder as partial replacement of Portland cement in mortar result increase in the density and a rise in the compressive resistance, increased pozzolan interaction this leads, as well as roughness of the surface decrease with increase fly ash. The Polymer fiber also act as binder material, mainly used to enhance the shrinkage cracking resistance and toughness Polypropylene fibers are com• mercially utilized at relatively low volume fractions to control plastic shrinkage cracking of concrete. Polypropylene fibers are not expected to increase the strength of concrete, but to improve its ductility and toughness, and impact.

References

[1] Mann S 2006 Nanotechnology and Construction, *Nano Forum Report* 5.
[2] Li G 2004 *Cement and Concrete Research* **34** 1043
[3] Bandyopadhyay A K 2008 *Nano Materials* (New Delhi: New Age International) ed 1 p 324
[4] The Royal Society 2004 Nanoscience and Nanotechnologies: Opportunities and Uncertainties *Royal Society and Royal Academy of Engineering's report*
[5] Jadid K M 2011 *Chemical alteration of oil well cement with basalt additive during carbon storage application* MSc Thesis Middle East Technical University
[6] American Petroleum Institute 2010 Specification for Cements and Materials for Well Cementing *API report*
[7] Vipulandan C and Mohammed A 2015 *Smart Mater. Struct.* **24** 125020
[8] Fakhim B, Abolfazl H, Alimorad R and Parviz G 2014 *The Scientific World Journal* **2014** 276323
[9] Pan Z, Duan W, Li D and Collins F 2012 *Graphene oxide reinforced cement and concrete* Patent WO2013096990 A1
[10] Rageh S M, Nezami M, Dhanalakshmi K and Ali Basha Sh 2017 *International Journal of Engineering Science Invention* **6** 1
[11] Vipulandan C and Mohammed A 2015 *Journal of Petroleum Science and Engineering* **135** 50
[12] Abdul Hadi H and Abdul Ameer H 2017 *Journal of Engineering* **23** 51
[13] Salah N, A. Abdulrahman and Yousif Z 2015 University OF Technology **22** 233
[14] ASTM 2006 *Annul book of ASTM Standads* ASTM
[15] Kondraivendhan B and Bhattacharjee B 2015 *Int .j. Sust .Built Environ* **4** 270
[16] Phoo-Ngernkham T, Sata V and Hatanaka S 2016 *Arab. j .Sci. ENG* **41** 1263
[17] Gopalan M K 1993 *ACI Mater J.* **90** 117
[18] Isaia G, Gastaldini A and Moraes R 2003 *Cement and Concrete Composites* **25** 69
[19] Arulraj P and Carmichael J 2011 *international journal of civil and structural engineering* **2** 475
[20] J.Elenal, and M.Lucia(2011), JAES,1,79 86.
[21] Aly M, Hashmi M, Olabi A, Messeiry M, Abadir E and Hussain A 2012 *Materials and Design,* **33** 127

IOP Conf. Series: Materials Science and Engineering **579** (2019) 012009 doi:10.1088/1757-899X/579/1/012009

The effect of some variables on the removal of synthetic bentonite suspension in water by electrocoagulation using turbidity measurements

N A Al-Rubaiey[#] and M G Albrazanjy

Petroleum Technology Department, University of Technology-Iraq, Baghdad, Iraq

[#]Corresponding author's e-mail address: 100108@uotechnology.edu.iq

Abstract. Electrocoagulation can be considered as a simple and an efficient method for treating both wastewater and water. In this work, a synthetic bentonite suspension water was treated using this technique to reduce turbidity caused by solid particles pollutants. This process was carried out in a batch glass reactor of 1000 ml operated with aluminium electrodes with and without agitation. The removal efficiency as a turbidity removal procedure has been examined using bentonite clay as a turbidity source. The effects of some parameters such as contact time, and electrolyte concentration on the removal efficiency were investigated and compared. The mixing effect over the range (0-800 rpm) on the removal efficiency was also evaluated. The process was found to achieve excellent turbidity removal with mild mixing (550 rpm). In addition, the effect of using Ultrasonic as a mixing source has also been studied. The lowest residual turbidity was found to be 1.0 NTU for the samples with initial turbidities of 500 NTU. Similar final turbidity of 1.0 NTU was obtained after 25 min with initial turbidity of 200 NTU. This was obtained at a current of 750 mA, a contact time of 60 min, and a calculated dissolved aluminium concentration of 10 mg/l.

1. Introduction

Electrochemical treatment of turbid water is an effective method of increasing the clarity and purity of water. The electrocoagulation (EC) process has been recently employed in separating suspended solids materials in water and wastewaters. This method functions in a series of steps which start with breaking down forces that stabilize charged particles found in water or wastewater. This allows inter-particle collision to occur to generate what is known as flocs [1].

The chemistry of this process has been utilized from the subject of electricity. The electricity explains the performance of charged particles given their attraction/repulsion affinities. The suspended solids have negative charge in water and they inclined to stabilize and repel each other when they get closer to each other that are because of their charge surface similarity [2].

The EC process intends to destabilize these negative charged particles. Appropriate application of the process takes into consideration satisfactory empathetic of certain interaction parameters. These comprise of the source of the charge, composition of the charge, particle size, particle shape, and density of the suspended particles. EC starts usually with the dissolution of Al or Fe as hydroxides in the wastewater. Thus these metal hydroxides are used as coagulants material. Electro-addition of coagulants material with opposite charges is the first approach to destabilize the particles' charge. They are added to water and wastewater to neutralize the negative charge of suspended particles. Upon neutralization, the suspended solid particles try to stick together in order to form slightly larger

IOP Conf. Series: Materials Science and Engineering **579** (2019) 012009 doi:10.1088/1757-899X/579/1/012009

particles. Furthermore, mixing is preferred here to scatter the coagulant and encourage particle collision. It is later followed by the "flocculation process" where gentle mixing increases the particle size from micro flocs to visible suspended solids. Particles are thus bounded together to produce larger macro flocs. Finally, a floating process is proceeding due to the formation of bubbles of hydrogen and oxygen gas on the electrodes. This process moves up most of the flocs to the top of the water and separates them [3].

Bentonite was designated to represent clays which are often found in the oil industry drilling mud. Bentonite behaves as hydrophobic colloids in water. These compounds involve flat sheets of alternating layers of silicon oxides and aluminum oxides, held together by ionic attraction for cations sandwiched between the sheets. In water solutions, aluminum (+3) or silicon (+4) can be replaced with sodium (+1), potassium (+1), or ammonium (+1) ions resulting in an overall negative charge of the particle. This charge is responsible for the electric repulsion of bentonite particles and, thus, of the stability of the colloidal suspension [4].

Most previous studies show that mixing could cause an increase in the removal efficiency of the pollutants. However, it has not been clear on this subject. Studies of Sitorus et al. [5] showed a proportional relation in the range (10-200 rpm) whereas Eskibalci et al. [6] revealed a reduction in turbidity with increasing mixing speed up to 360 rpm, yet an increase happened after that speed up to 700 rpm. On the other hand, Ilhan et al. [7] has proposed that mixing is not good for the EC process where they showed a decrease in COD removal efficiency with increasing mixing. Derayat et al. [8] investigated the EC treatment of surface water showing that turbidity rate increased with increasing mixing rate and they preferred low mixing rate.

In this paper, the mixing parameter besides basic other parameters such as initial concentration, pH and adding NaCl salt have been studied to elucidate their effect on the EC process using turbidity measurements.

2. Materials and Methods

Detail description of the experimental system can be found elsewhere [9]. A photograph for the experimental set up is given in Figure 1. In this study, the EC reactor (Pyrex glass beaker) with active volume of 1000 ml was used. Two home-made circular metal electrodes were Ø95mm×2 mm, and operative area for coagulation was 235 cm2. Commercially available power supply with 220v AC input voltage and 0~16v output voltage was used. The anode and cathode electrodes were fixed horizontally paralleled (9 mm distance) in the bottom of the reactor, and then 800 ml of wastewater was poured into the reactor. Samples of treated water-clay were synthetically prepared in the lab using mixer blender. The bentonite suspension is prepared using fresh tap water. This suspension is then succumbed to a strong mixing to obtain a well homogenization. Later, bentonite suspension samples after EC were submitted to sedimentation for 5 min before aspiration for their analyses. Hanna Multimeter (model HI-9828)- Multisensor probes was used to monitor changes in conductivity, TDS, pH and temperature. The electrolytic cell was equipped with a magnetic stirrer (BOECO MSH-300N, 50-1250RPM) in order to keep the electrolyte well mixed. Ultrasonic bath (frequency of 40 kHz and power of 50W SilverCrest model, China) was used in some experiments. Turbidity was determined by using Microprocessor Turbidity Meter, Model HI93703, HANNA instruments where samples of the liquid were taken periodically and analyzed using standard procedure.

Figure 1. The EC experimental setup.

3. Results and Discussion

This study was part of series of experiments carried out in our labs to investigate the treatment of various water and wastewater pollutants using EC technique [9]. In the current work, pollutant was chosen as suspended bentonite water. Thus, the purpose of this investigation was to evaluate the effect of various factors including mixing with magnetic stirrer/ or ultrasonic waves on the treatment efficiency of EC process. Turbidity removal kinetics was investigated by varying the values of time, initial concentration, added NaCl salt, pH, applied voltage and mixing parameters. In this research, turbidity measurements were utilized as a measure of the bentonite concentration in the tap water.

The effect of time was studied at constant applied voltage of 10v, mixing speed of 400rpm and initial concentration of 100NTU. Figure 2 illustrates the turbidity as a function of operating time. It is clearly seen that the operating time has an important effect on the turbidity removal.

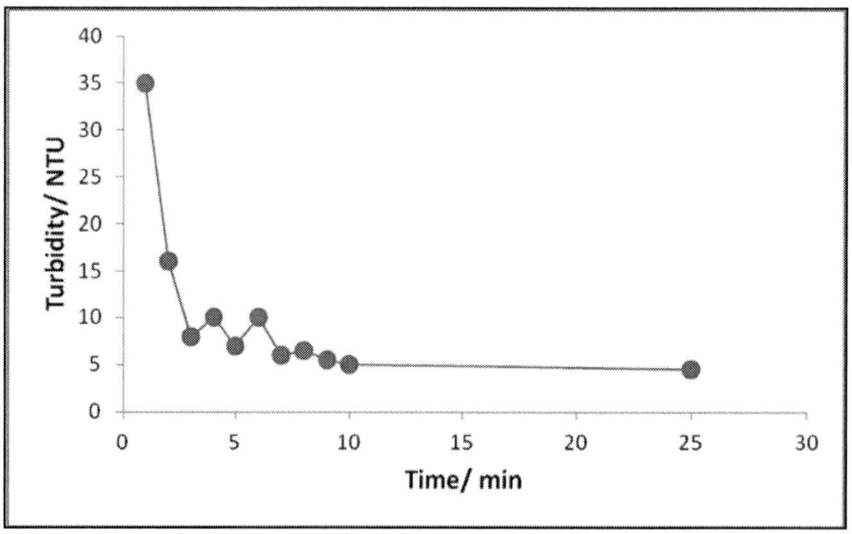

Figure 2. The change in turbidity with operating time during the EC process.

During the period of 10 min time, the turbidity has been dropped from 100 NTU to 5 NTU with an efficiency of 95%. Figure 2 also indicates that the EC process encompasses of two stages that are destabilization and aggregation, the first stage is fast, whereas the second stage is relatively slow. To explain that, Faraday's law stated that the current time of passing raises the applied electricity to the electro-cell. This may lead to more ions of aluminum to be liberated at the anode electrode. These positive charges of these ions would neutralize the negatively charged particle droplets. This effect would improve the split-up of the particle droplets and thus they float on the water surface. Alternatively, more bubbles (conceivably H_2) would be created at the cathode by increasing the operating time. These bubbles would drive the produced flocs upwards by "bouncy force", thus increasing the removal efficiency and decreasing turbidity of the water (improving separation).

On the other hand, and in order to explore the effect of the operating time and pH value, the initial pH value was taken at 9. Figure 3 shows the reduction of turbidity with the operating time at initial pH of 9. pH parameter is a vital factor influencing the treatment performance of EC process. The pH of the solution varies little during electrolysis, as shown in Figure 4. It is the result of several reactions occurring in the synthetic solution : hydrolysis reaction of Al^{3+}, formation of hypochlorous acid from chlorine (existing in municipal water) reduced to the anode, etching of the electrodes by the Cl^- and OH^- ions and formation of the OH^- ions to the cathode. It may be the reason behind the decrease in pH was due to the hydrolysis and the formation of the hypochlorous acid.

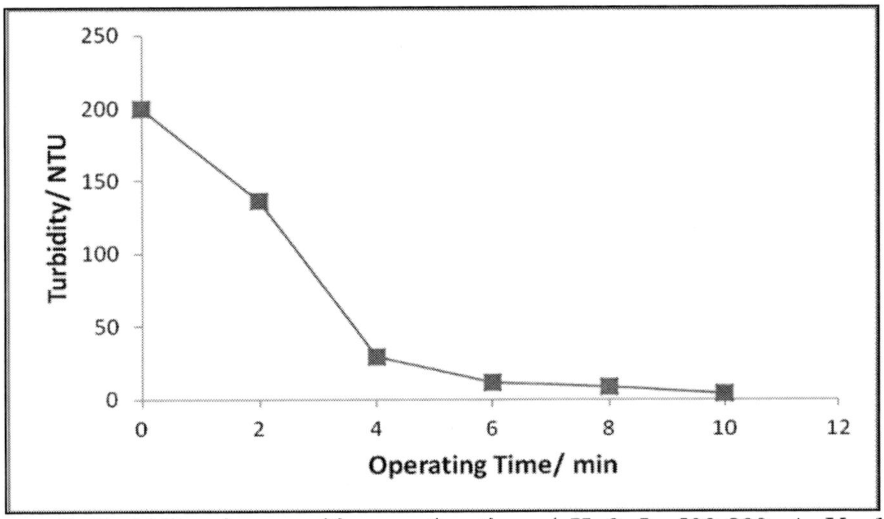

Figure 3. Turbidity change with operating time. (pH=9, I= 500-800mA, V= 10v, Ci=200NTU)

Figure 4 shows the change trend of pH with time for the EC process where a clear small reduction in pH was found from 8.2 to approximate 7.5 over a period of 10 min.

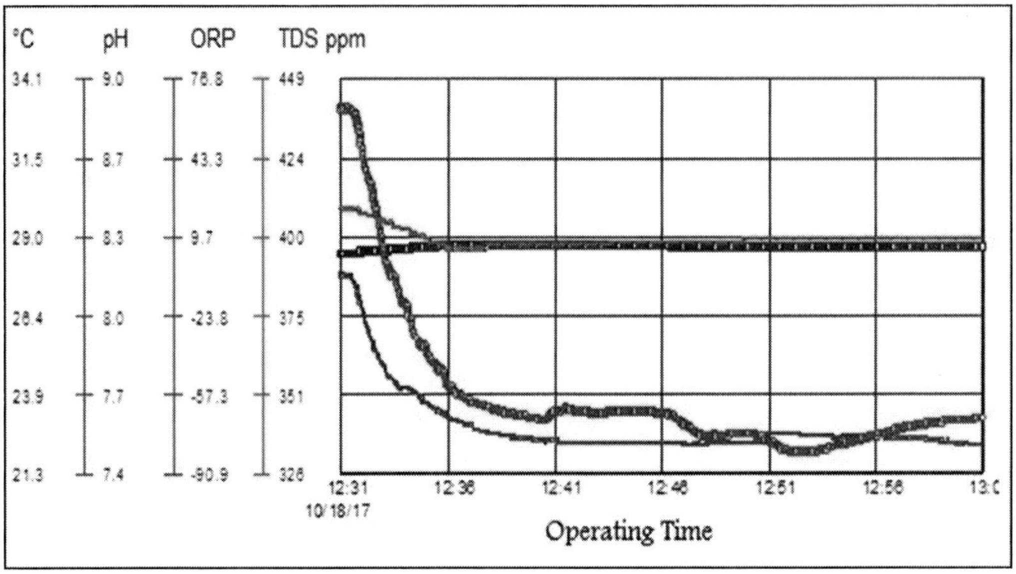

Figure 4. The change of pH with a long period of operating time of EC process.

In figure 5, other experiments were carried out to evaluate the effect of increasing initial concentration on final turbidity. Increasing turbidity of more than 400 NTU would reduce separation efficiency as expected. Since the formation amounts of the coagulant are insufficient, and the voltage to achieve the same removal efficiency as concentration is lowered.

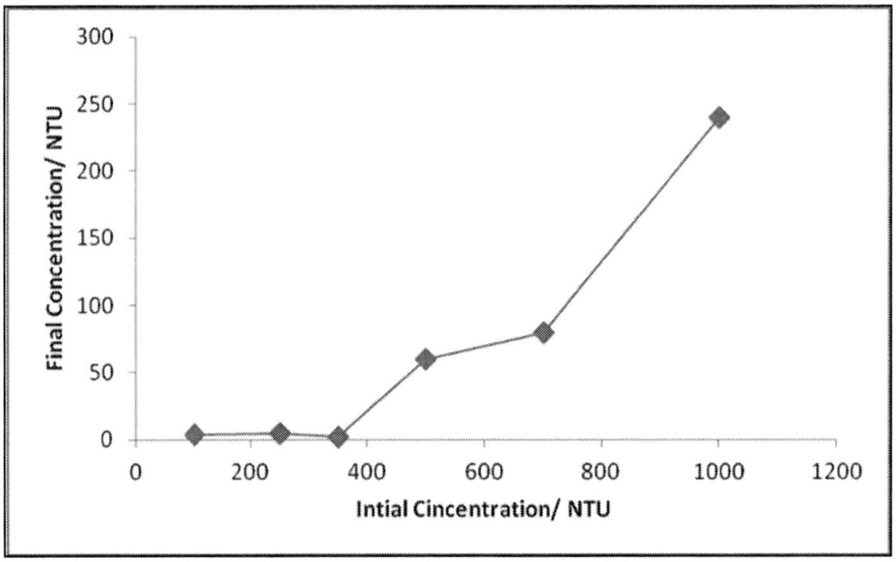

Figure 5. variation of initial concentration with final concentration. (Time= 5min, V= 10v, Speed= 400rpm,I= 400-1000mA)

The outcome of NaCl concentration on the EC separation of clays from water was plotted in Figure 6.

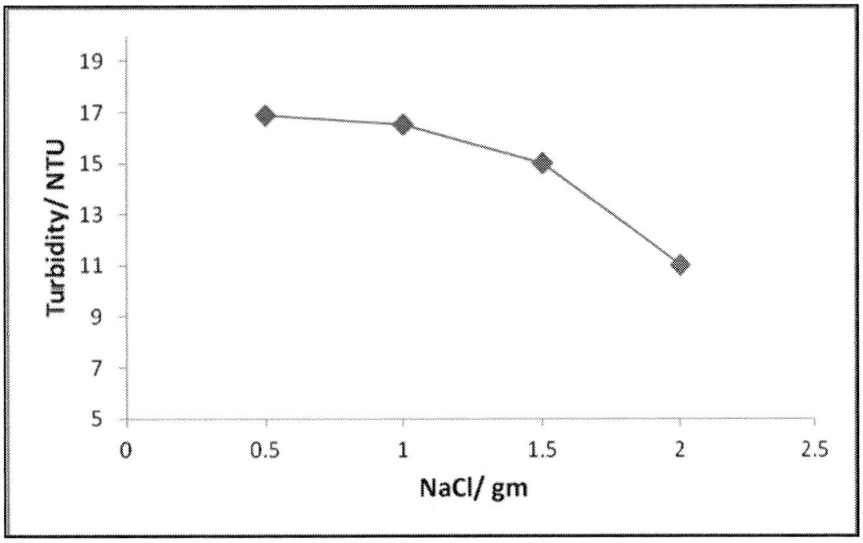

Figure 6. The effect of adding NaCl on the removal efficiency of bentonite suspension waste. (Ci= 100 NTU, V= 4v, Time= 5min, mixing speed= 400rpm)

From this figure it can be seen that increasing NaCl concentration leads to reduce turbidity with decreasing in treatment time that is required to achieve the highest removal percent. The presence of NaCl causes the formation of Na^+ ions and Cl^- ions. Accordingly, presenting NaCl salt to the water would increase the conductivity of the solution and decrease the portion of the applied potential that is spent to overwhelm the resistance of the solution. Consequently, it raises the fraction of the applied potential engaged to the EC process. The same results of sodium chloride effect were reported by [10].

Figure 7 also shows the change of turbidity removal with NaCl added at moderate mixing. Results show higher efficiency removal at shorter time compared to the study without adding any salt. However, the difference was reduced at longer operating time.

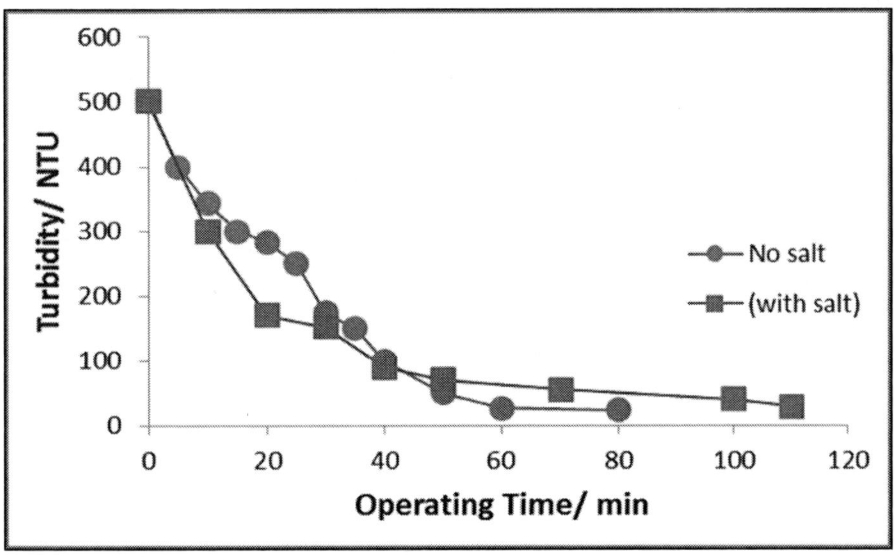

Figure 7. Comparison results of the EC process with and without adding NaCl.

The mixing speed has also been studied in this work to determine its effect on the interaction of ions and molecules in suspension with particles. Mixed solution would produce a distressed of ions placement and some inadequately electricity consumption. The most effective interaction rate may produce a suspension and may also increase coagulation efficiency considerably. When the mixing speed and operating time exceed certain restrictions, the produced flocs might be fragmented. This also reduces the efficiency of coagulation. In order to conclude the optimum stirring speed (0-800 rpm) for the EC experiments in the absence of NaCl, the other parameters namely current density, operating time, and pH were all retained constant.

The result of agitating speed on the EC is revealed in Figure 8. The turbidity of the suspension at a stirring speed of 550 rpm dropped from 100 NTU to 10 NTU level with a 90% removal efficiency. With mixing speeds over 600 rpm, it was determined that turbidity values started to increase again in parallel with the disintegration of formed flocs. Therefore, 500 rpm was chosen to be the optimum mixing speed for this study.

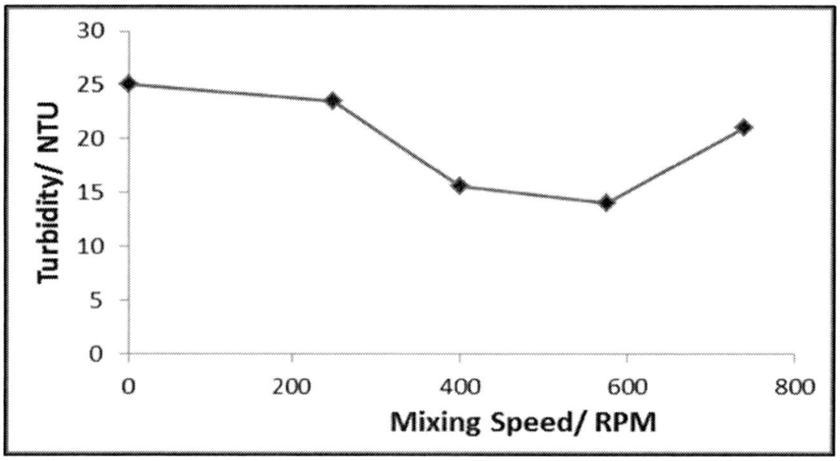

Figure 8. The change in turbidity with the mixing speed of the magnetic stirrer (voltage=15v, Ci= 110NTU, time= 5min, Ø=10cm,electrodes distance= 9mm).

Figure 9 has shown that changing applied voltage (10v and 16v) has not changed Turbidity considerably. However, it seems lower voltage was preferred.

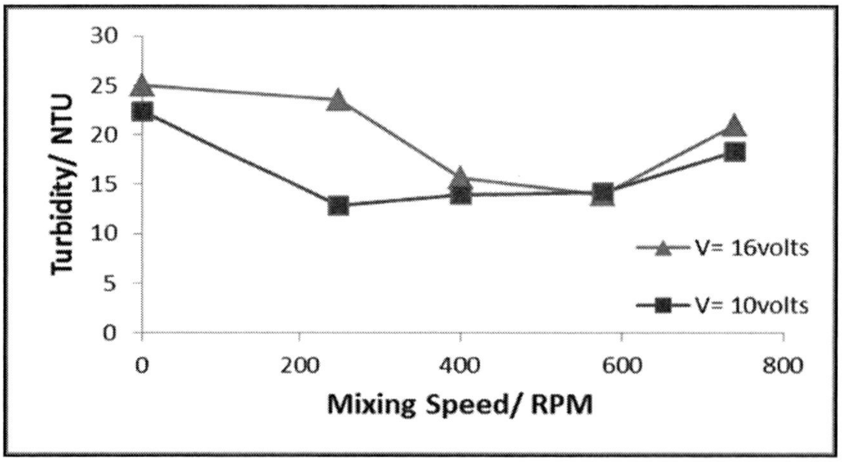

Figure 9. The change in turbidity with mixing speed using magnetic stirrer at two different applied voltages.(Ci=110NTU, Mixing speed=250NTU, 0.5gm NaCl)

The key cause behind this result is the fact that mixing operation influences the mobility of ions in water (negatively charged ions towards the anode and positively charged ions towards the cathode). Considering the fact that mixing may cause the break-up of flocks, mixing is not always recommended to be a good choice for EC treatment, especially at higher speed (above 600 rpm). Taking into consideration these points, it was established that low to medium mixing should be included in EC treatment process to obtain the best removal efficiencies and to maintain the system working. Thus, another experiment was carried out with adding NaCl and using the optimum mixing condition of 500 rpm. This case has provided approximately 98% efficiency.

With increasing mixing rate, flocs in solution are formed and sedimentation gets easier. The results display that the removal efficiency of study turbidity rate parameter increased with increasing mixing rate. For the prevention of flocs break and the release of metal is better using moderate mixing rate. Similar results of investigating the effect of mixing rate on removal efficiency was found by Modirshahla *et al.* [11].

Figure 10 represents the comparison of current consumption with and without mixing.

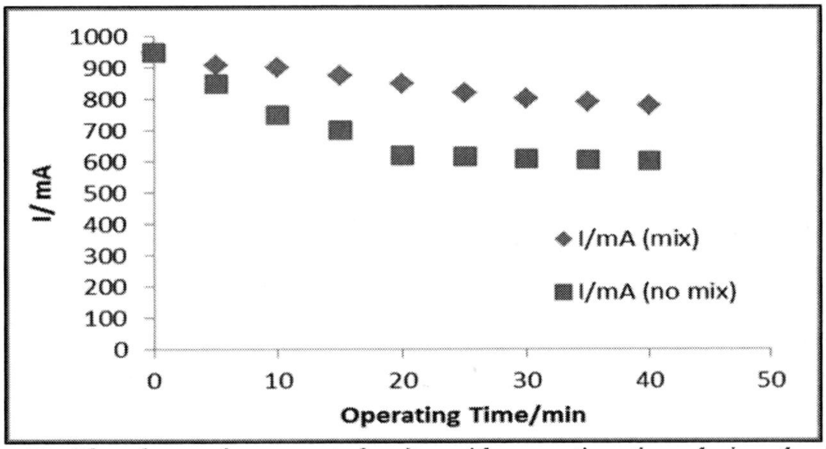

Figure 10. The change in current density with operating time during the EC process for both mixing and no mixing process.

It can be seen clearly that the current required with mixing is higher compared with no mixing. The reason behind that is lying in the process of making the solution more conductive in mixing process thus the amount of energy consumed is higher.

Furthermore, comparison results of using agitating mixing by magnetic stirrer and when the process carried out with no mixing (the mixer turned off) were done. Data demonstrate an affected reduction in removal efficiency without mixing, this reached to about 50% at about 15 min operating time. Mixing is important due to the fact that this process improved conductivity and thus the current density was increased [8]. With lower to moderate mixing, flocs were formed which led to facilitate sedimentation as stated previously.

To evaluate the effect of Ultrasonic Waves on the EC processes, a number of experiments were carried out using bentonite suspended water and zinc wastewater [15] (for comparison). The results are presented in figures 11 and 12.

The results in figure 11 illustrate a serious deterioration in the wastewater turbidity. Similar results have been reported from other researchers [5]. On the other hand, other researchers proposed a positive impact using combined ultrasonic-electrocoagulation technique [9,12]. Ultrasonic-EC technique was proposed by Chu Jinyn *et al.* [13] to treat the car-washing wastewater. They reported the COD removal efficiency by ultrasound-EC technique to be better than the conventional EC. They also proposed a mechanism to clarify the reason behind the enhancement in the efficiency. In addition,

He *et al.* [14] concluded that ultrasonic tool may improve the EC work by eliminating the passive film of Reactive Blue 19 that produced on the surface of the electrode during EC process. They proposed a mechanism showing that the passive film could be detached from the sacrificial electrode under the effect of ultrasound causing to generate more coagulant metals and thus reducing the size of the metal hydroxide which in turn adsorbs more contamination in wastewater.

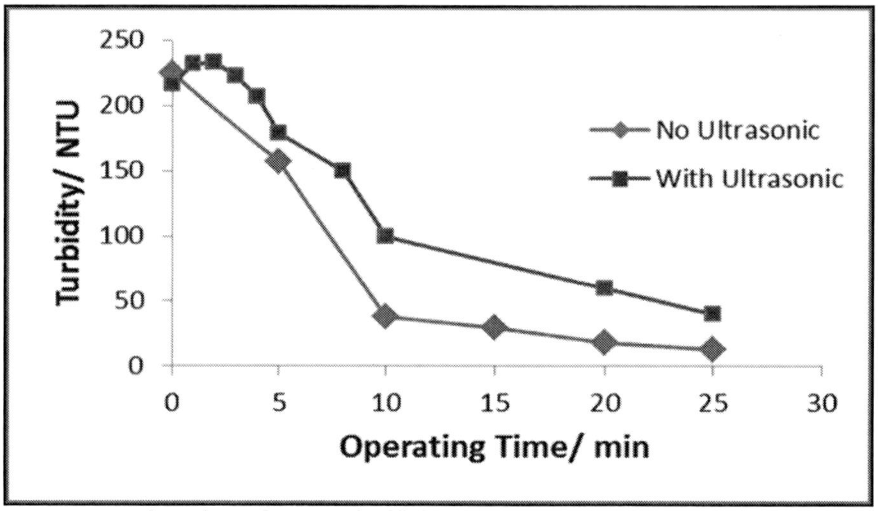

Figure 11. Changing of removal efficiency with operating time for EC of bentonite wastewater (comparison Results).

On the other hand, the situation was different for the zinc pollutant where the removal efficiency was found to improve. A comparison was made between the treatment with and without Ultrasonic in figure 12 [15]. The same results of positive ultrasonic effects were reported elsewhere [16].

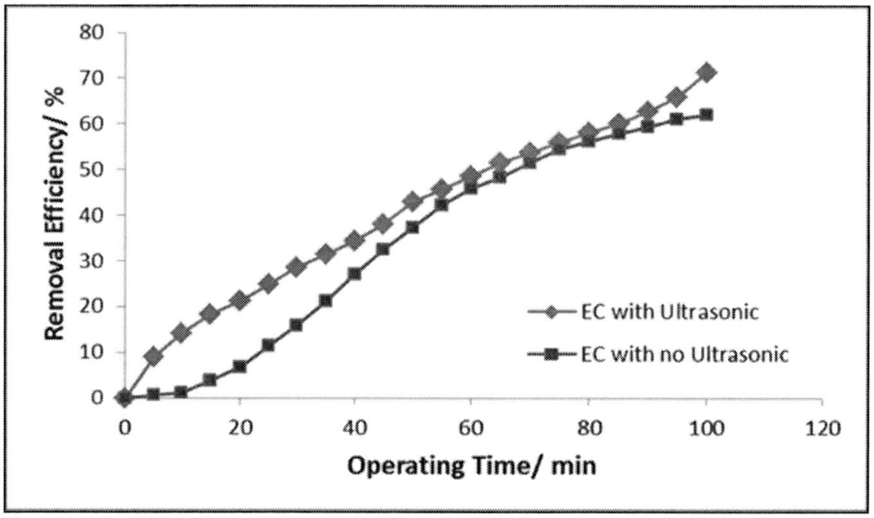

Figure 12. Changing of removal efficiency with operating time for EC of Zn wastewater (comparison Results) [9]

4 Conclusions

In this study, the efficiency of the EC process applied to the treatment of bentonite suspended water was investigated. It was observed that the EC treatment achieved a fast and effective removal of solid particles turbidity. During the operating time, two steps are eminent, the most important of which is related to the removal of the turbidity. During this process, there is the creation of flocs, a quantity of which is moved to the free surface by the bubbles generated in the surrounding area of the surface of the electrodes. The turbidity is removed by adsorption on the $Al(OH)_3$ flocs. In this step, the electrochemical kinetics is fast and increases when the initial electrical conductivity of the solution increases. The treatment efficiency was found to be a function of the NaCl electrolyte dose, mixing process using magnetic stirrer and operating time under the optimal values of the process parameters. The results showed removal efficiencies of 95% turbidity at moderate mixing speed of 500 rpm This technique thus seems to be a promising alternative for the treatment of clay particles suspensions wastewater under low to moderate mixing.

Acknowledgment
The authors wish to thank the Petroleum Technology Department at the University of Technology - Iraq for facilitating this work. A special gratitude is given to Engineer: Haithem Shihab Ahmed.

References

[1] Sczukarczuk J, Kan J, Hassan T and Donini J 1994 Electrochemical coagulation of clay suspensions *Clays and Clay Minerals* **42** 6 667-673
[2] Ghernaout D, Ghernaout B and Boucherit A 2018 Effect of pH on Electrocoagulation of bentonite suspensions in batch using iron electrodes *Journal of Dispersion Science and Technology* 29, 1272–1275
[3] Abuzaid N, Bukhari A and Al-Hamouz Z 1998 Removal of bentonite causing turbidity by electro-coagulation *Journal of Environmental Science and Health, Part A, Toxic/Hazardous Substances and Environmental Engineering* **33** 7 1341-1358
[4] Cañizares P, Martínez F, Carmona M, Lobato J and Rodrigo M 2005 Continuous electrocoagulation of synthetic colloid-polluted wastes *Ind. Eng. Chem. Res* **44** 22 8171–8177
[5] Sitorus I, Astono W and Iswanto B 2018 Batch leachate treatment using stirred electrocoagulation reactor with variation of residence time and stirring rate *The 4th International Seminar on Sustainable Urban Development, IOP Publishing, IOP Conf. Series: Earth and Environmental Science*, 106 012095
[6] Eskibalci M and Ozkan M 2018 An investigation of the effect of NaCl concentration on the electrocoagulation of coal preparation plant tailings *Physicochem. Probl. Miner. Process* **54** 3 934-943
[7] Ilhan F, Kurt U, Apaydin O and Gonullu M 2008 Treatment of leachate by electrocoagulation using aluminum and iron electrodes *Journal of Hazardous Materials* 154 381–389
[8] Derayat J, Motlagh Z, Pirsaheb M, Sharafi K and Motlagh R 2015 Survey of electrocoagulation process Performance (with iron-rod electrode) in treatment of drinking water-case study: Kermanshah Soleymanshah dam *Intl. Res. J. Appl. Basic. Sci.* **9** 8 1453-1459
[9] Al-Rubaiey N and Al-Barazanjy M 2017 Study the Efficiency of Electrocoagulation System Using Conductivity Measurements for the Removal of Zinc Heavy Metal *EIOGI 2017 conference*, Koya, Kurdistan Region, Iraq, 17th-19th April
[10] Bejjany B, Lekhlif B, Eddaqaq F, Dani A, Mellouk H and Digua K 2017 Treatment of the surface water by electrocoagulation-electroflotation process in internal loop airlift reactor: conductivity effect on turbidity removal and energy consumption *JMES* **8** 8 2757-2768
[11] Modirshahla N, Behnajady M and Mohammadi-Aghdam S 2008 Investigation of the effect of different electrodes and their connections on the removal efficiency of 4-nitrophenol from aqueous solution by electrocoagulation *Journal of Hazardous Materials* 154 778–786

[12] Sadeddin K, Naser A and Firas A 2011 Removal of turbidity and suspended solids by electro-coagulation to improve feed water quality of reverse osmosis plant *Desalination* 268 204–207

[13] Jinyu C, Yurong L, Ning L and Weihong H 2012 Treatment of car-washing wastewater by electrocoagulation-ultrasound technique for reuse *Advanced Materials Research.* **433-440**, 227-232

[14] He C, Hu C and Lo S 2016 Evaluation of sono-electrocoagulation for the removal of **reactive b**lue 19 passive film removal by ultrasound *Separation and Purification Technology* 165, March 2016

[15] Al-Rubaiey N and Al-Barazanjy M 2018 Ultrasonic technique in treating wastewater by electrocoagulation *Engineering and Technology Journal* **36** C1 54-62

[16] Doosti M, Kargar R and Sayadi M 2012 Water treatment using ultrasonic assistance: a review *Proceedings of the International Academy of Ecology and Environmental Sciences* **2** 2 96-110

Zn/HY-Zeolite as a Catalyst for Upgrading Iraqi Heavy Crude Oil Using Aquathermolysis Method

Amani Q. Saeed[1], Adel S. Hamadi[2#], Bashir Y. Sherhan[3]
[1, 3]Chemical Engineering Department/ University of Technology-Baghdad-Iraq
[2]petroleum Technology Department/ University of Technology-Baghdad-Iraq

#Corresponding author's e-mail address: adel_al_obaidi@yahoo.com

Abstract. Upgrading heavy crude oil from East of Baghdad oilfield using a high pressure reactor with steam supplied and toluene as hydrogen donor has been investigated through aquathermolysis process. Synthesized Zn/HY Zeolite, which is characterized by XRD, SEM, EDX, FT-IR, BET-surface area and pore volume has been employed as a bi-functional catalyst. Statistical design of experiment (DOE) is also applying in determine the optimum operating conditions of aquathermolysis reaction under different temperatures, time and weight percentage of catalyst with the targeted response being the viscosity after upgrading. The extent of upgrading might be established through various analyses of the initial unconventional oil and final produced crude oil. These were included the measurements of oil density, API gravity, oil viscosity, flash point, pour point, sulfur content and yield of light and intermediate fractions. The results show that the optimum operating conditions are found at 0.25 wt% of catalyst, 231.25°C and 30 h reaction time. The experimental data under these conditions has confirmed that the viscosity and sulfur are decreased by 67.25% and 55.58% respectively, whilst the API has increased up to 24.5. The response surface regression model equation in coded unit for viscosity reduction percentage has been established.

1. Introduction

Recently, with the depletion of conventional oil reserves in the world, heavy oil becomes a worldwide interest to meet future demand for petroleum products [1], in which 70% of remaining oil reserves is bitumen and heavy oil [2]. But the exploitation, transportation and utilization of heavy crude oil is still limited [2, 3], because of its high density and viscosity and high contents of asphaltene and resin in addition to abundant amount of heteroatom such as O, S and N [4]. Therefore, various methods are used to the recovery of heavy oil (i.e. Cyclic Steam Stimulation (CSS), Steam Flooding, Steam-Assisted Gravity Drainage (SAGD), In Situ Combustion (ISC) etc.) [5, 6]. Aquathermolysis is one of the popular and effective technologies to retrieve the heavy oil [7], in which cracking, hydrogenation, and desulfurization can take place by using steam under a high pressure in order to improve the crude oil properties [8].

Hyne et al. in 1982 had simplified the reaction of aquathermolysis according to the following chemical equation [9]:

$RCH_2CH_2SCH_3 + 2H_2O = RCH_3 + CO_2 + H_2 + H_2S + CH_4$

During this reaction, the free radicals may be created with the polymerization reactions and composing bigger molecules .This guides to the creation of more viscous oil in place of decreasing the viscosity. Thus, various catalysts should be used to prevent free radical formation and increase the viscosity diminution [5, 8]. Catalysts development is very important which can enable the extent of oil conversion to be increased and the aquathermolysis process to become more economically attractive [9]. Catalysts can be classified according to the stage within which they are active. As a rule, catalyst might be dissolved in the oil, water, or suspended as particles [10]. The dispersed and oil soluble catalysts are mentioned to be partially more active than dissolved catalyst in water [8, 11]. This study is focusing on Zn/HY-Zeolite as a bi-functional catalyst for aquathermolysis reaction because it has a highly surface area and larger-pore structures in combination with both acidic and metallic sites. Zeolites are hydrated aluminosilicate minerals produced by using interlinked tetrahedral of silica (SiO_4) and alumina (AlO_4). Y zeolite has 3D network of cavities and inter channels, consequently it can be separated to molecular of various sizes and therefore it can be utilized in various industrial application [12].

2.Experiments

2.1. Materials

Ludox AS-40, Sodium Aluminate, Sodium Hydroxide (96%) and Ammonium Chloride (99.8%) all have purchased from Sigma Aldrich Germany, Zinc Chloride (99%) from Geekee Biotech China. Additionally, Toluene (99%) as a hydrogen donor is supplied from Alpha Chemika India. The heavy crude oil is used as a feed has been obtained from East of Baghdad oilfield with the properties scheduled in (Table 1).

Table 1. The physical properties of used heavy oil.

DENSITY, SP.GR	0.9224 G/CM^3,(20°C)	
VISCOSITY	47.87 g/cm.s or cP, (50°C)	
API	21.9, (20°C)	
POUR POINT	-25°C	
FLASH POINT	80°C	
SULFUR CONTAIN	4.52 wt.%	
YIELD OF LIGHT AND INTERMEDIATE FRACTIONS	Temperature for fraction (C°)	%Yield of fraction
	Light Naphtha (IBP 45-75)	1.2
	Heavy Naphtha (75-175)	13.2
	Kerosene (175-225)	6.4
	gas oil (225-340)	20
	Total	40.8

2.2 Synthesis and impregnation processes

2.2.1. Synthesis of Na-Y zeolite

The NaY-Zeolite is prepared by making two types of solutions depending on a chemical recipe for the FAU [13]. The first one is called the "Seed gel" is prepared by blending: distilled water, ludox "SiO_2", sodium aluminate "$NaAl_2O_3$" and sodium hydroxide "NaOH" in a plastic bottle and then mixing for at least 10 min. The product is left for aging at 25°C for 24 h. The other called "Feedstock gel" is prepared by blending: distilled water, ludox "SiO_2", sodium aluminate "$NaAl_2O_3$" and sodium hydroxide "NaOH". All these components are stirred for 10 min. A certain amount of the seed gel has been added slowly to the feedstock gel. The mixture is also stirred for 20 min then being kept for crystallization purpose in the dryer at 100°C for 18 h. The sample is taken out of the dryer and centrifuged for 10 min and washed with distilled water until the pH is decreased to about 8. Finally, the NaY-Zeolite sample is dried overnight at 110°C as shown in block diagram (Figure 1).

2.2.2. Preparation of NH_4Y and HY zeolite types

Ion exchange method is used to exchange Na^+ ions within NaY zeolite framework with NH_4^+ ions within NH_4Cl solution by preparing 100 ml of NH_4Cl solution (0.5M) and then adding 1gm of synthesized NaY-Zeolite to this solution. The mixture is heated at 80°C beneath reflux condition with regular mixing for 1h. After that, the NH_4Y-Zeolite form is filtered off, washed several time with distilled water, is centrifuged and then dried at 100°C. The ion exchange process is repeated several times under the same conditions until the level of Na^+ within Y-framework becomes less than 0.5%. Heat treatment (i.e. calcination process) is required after multi-stage ion exchange procedure in order to prepare an anhydrous crystalline HY-Zeolite (i.e. acidic form of Y-Zeolite) and decomposes the NH_4^+ cations into ammonia and protons according to the following equations [14]. This process is performed in a furnace for 2 h. at 450°C under dry airflow.

$$NaZ_{(s)} + NH^+_{4\,(aq)} \longleftrightarrow Na^+_{(aq)} \downarrow + NH_4Z_{(S)} \quad (1)$$

$$NH_4Z_{(S)} \xrightarrow{\Delta} NH_{3(g)} \uparrow + HZ_{(S)} \quad (2)$$

2.2.3. Preparation of Zn/HY-Zeolite

Basically, 1 gm of Y-Zeolite catalyst is placed in a 100 ml solution of zinc chloride in order to the impregnation of 1% from Zn metals on the framework structure of HY-Zeolite. Under reflux condition the mixture is heated at 80 °C with constant mixing for 4 h. After that, the impregnated Y-Zeolite is separated off, washed several times with distilled water, and centrifuged for 15 min. and then is dried at 110 °C overnight to obtain the product of bi-functional zeolite catalyst. After that, the furnace is used to reduce metal at 350°C for 2 h under gas flow rate of 10 °C/min.

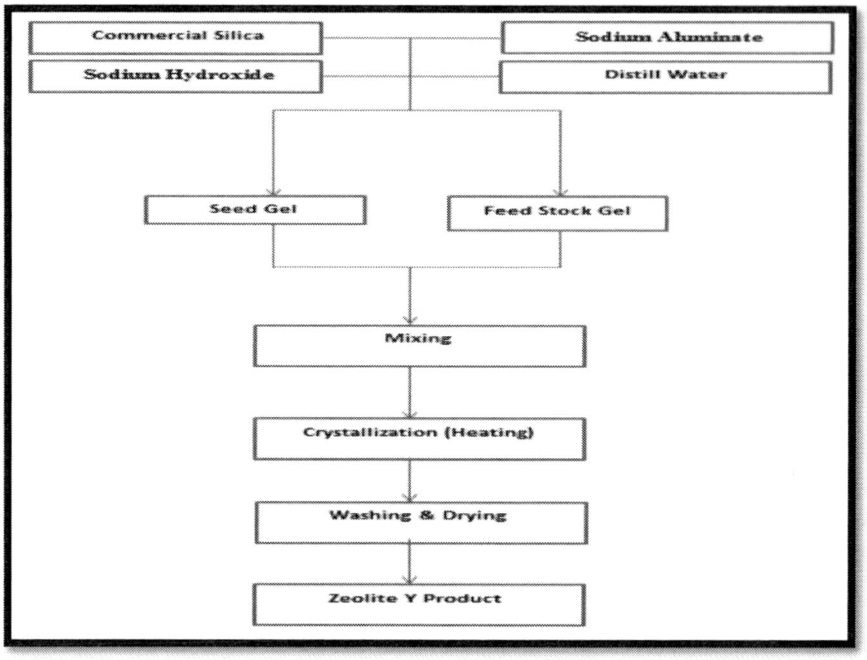

Figure 1. Manufacturing procedures for NaY zeolite

2.3. Catalytic aquathermolysis reaction of the heavy crude oil

Catalytic aquathermolysis reaction is carried out in a 300 ml stainless steel autoclave (high pressure reactor) as shown in (Figure 2). Briefly, heavy oil, water and hydrogen donor (Toluene) are added to the reactor at 75%, 15%, 10 vol% ratio, respectively with certain weight percentage of Zn/HY-Zeolite. Then the reactor is closed and heated at a rate of 4°C/min until reached the reaction temperature under a high pressure (e.g. pressure range of 30 to 45 bar according to the steam temperature). When the aquathmolysis reaction is completed, the reaction mixture is cooled to room temperature and afterward transferred to a 500 mL flask, after that utilizing a condenser to get rid of residual water, catalysts and affords the heavy oil sample ready for further analysis. In this study, Central Composite design (CCD) has been chosen for the aquathermolysis process to discover the optimum conditions. CCD is used in this work with three independent variables: reaction time, temperature, and catalyst weight percent. The CCD matrices consist of 20 runs for three variables covering the full design and working up models of the second order by using regression analysis program (Design-Expert 11 Software).

IOP Conf. Series: Materials Science and Engineering **579** (2019) 012010 doi:10.1088/1757-899X/579/1/012010

Figure 2 Photograph of the Aquathermolysis reactor

3.Results and Discussion

3.1 Characterization of the catalyst

3.1.1 X-ray Diffraction (XRD) analysis

The XRD patterns for the zeolite samples are shown in Fig. 3 in which the major diffractions of the samples are occurred at a 2-theta range of 5 to 40 degrees. It can be seen that a structure of synthesized zeolite-NaY sample has approximately the same pattern in comparison with the standard zeolite Y [15]. The percentage degree of crystallinity of each zeolite Y sample is calculated from the intensities of the main peaks given from the XRD machine according to the below equation by dividing the sum of main peak intensities of the synthesized zeolite Y on the sum of main peak intensities of the reference zeolite Y [14, 15]. (Table 2) shows the degree of crystallinity of synthesized samples, which almost high crystallinity. The results show that the crystallinty of catalysts follows the trend: Y-Zeolite > Zn/Y-Zeolite.

$$\text{Deg. of cryst. \%} = \frac{\sum \text{intensitie s of the peaks of zeolite sample}}{\sum \text{intensitie s of the peaks of reference sample}} \quad (3)$$

Zinc metals within Zn/Y-Zeolite pattern in (Figure 3) have no diffraction peaks because the quantity of zinc loaded over zeolite is too small; and can also be attributed to interference within zeolite-Y structure. Another possible explanation is that the zinc dispersed well during metal impregnation, and therefore cannot be detected when using XRD. Moreover, the intensities of some characteristic peaks decrease within Zn/Y-Zeolite pattern due to the loaded transition metals into zeolite framework which can slightly decrease the crystallinity and that agree with Kristiani et. at. [16], and also indicates that the impregnation process does not obviously influence the zeolite Y structure.

Table 2. XRD-data of the synthesized zeolite-Y samples.

Standard Faujasite		Samples		
2θ °	Intensity (I)	2θ °	Intensity (I) Zeolite	Intensity (I) Zn/Zeolite
6.233	100	6.2459	100	100
15.640	53	15.5844	60	57
20.360	38	20.2703	37	40
23.635	59	23.5348	92	81
27.042	39	26.9170	63	62
30.730	22	30.6023	29	34
31.378	49	31.2487	72	74
32.432	16	32.3059	26	30
34.053	15	33.9239	29	27
Total	∑391	Total	∑508	∑505
% Degree of crystallinity using the standard faujasite as a reference			130%	129%

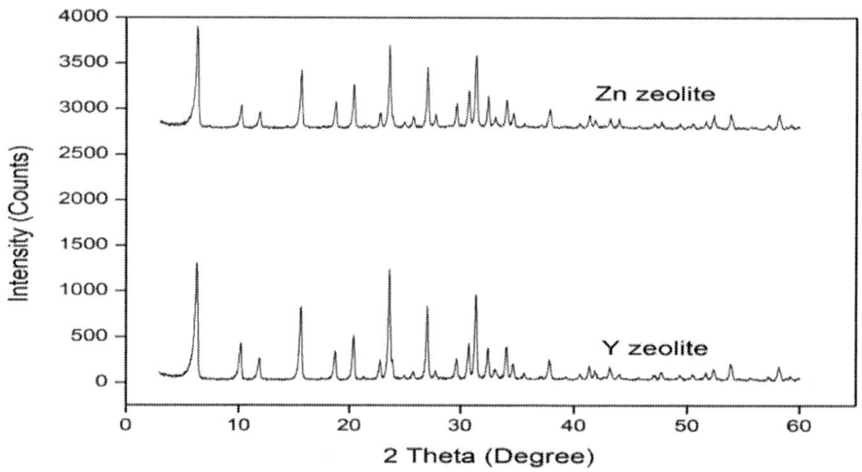

Figure 3 XRD-pattern of synthesized Y Zeolite, Zn Zeolite

3.1.2 Fourier Transform Infrared Spectroscopy (FTIR) Analysis

The FTIR spectra for the zeolite samples are given in (Figure 4) generally, every infrared spectrum of prepared zeolite seems to be comparable to the structural units for the reference commercial Y-zeolite, and is appropriate with XRD results. Absorption bands in the 453−490 cm^{-1} region can be referred to the T-O bending mode (i.e. Si-O/Al-O). The band watched in the 715−816 cm^{-1} region is owing to T-O-T symmetric stretching. This band is seen to be minimal intense with reference to the asymmetric stretching vibration of (Si/Al)O4 units, which may be observed as a large peak with high intensity near on 1014−1173 cm^{-1}. An absorption band at 1400 cm^{-1} is attributed to the NH_4. Strong IR absorption in the spectral region of 1618−1647 cm^{-1} is ascribed to the vibrational bands of framework water molecules. Likewise, the broad band in the 3159−3583 cm^{-1} region is almost because of the surface hydroxyl group (i.e. OH: hydrogen linked to the oxygen ions of the lattice structure).Generally all metals and its oxide, give the FT-IR peaks at lower wave number ranging from 424-459cm^{-1}.

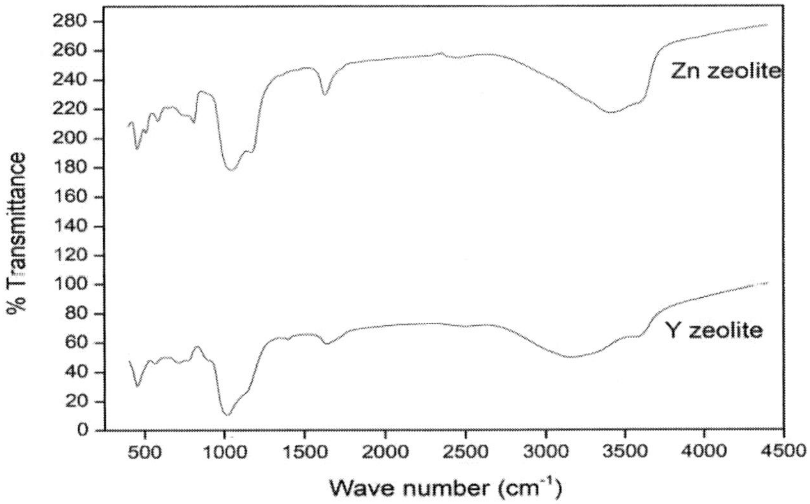

Figure 4 The infrared spectrum of Y Zeolite, Zn/Y Zeolite

3.1.3 Energy dispersive X-Ray (EDX) Analysis

It is important for zeolite catalysts to have a small amount of sodium, since sodium reduces the acidity of the zeolite as it reacts with the zeolite acid sites and by these decreases the catalytic activity of catalyst. Sodium is usually reported as the Wt. % on catalyst. The synthesized zeolite is modified by multi-stage ion exchange procedure to produce the acidic form of Y-Zeolite with less than 0.5% of Na^{+}. The Energy dispersive X-Ray (EDX) analysis has been employed to analyze the content of sodium within the zeolite structure before and after ion exchange. Shown in (Table 3). The extent of exchange is determined by the equation below:

$$\%\text{Exchange} = \frac{\mathit{Orgin\ quantaty\ of\ Na-Removal\ of\ Na}}{\mathit{Orgin\ quantaty\ of\ Na}} * 100\% \quad (4)$$

Table 3. Sodium Wt. % exchange after four stages of ion exchange.

Stages	% exchange	Wt.% Na^{+} Content within the zeolite
Before ion exchange	0.0	11.8
First ion Exchange	64.4	4.2
Second ion Exchange	86.4	1.6
Third ion Exchange	88.1	1.4
Fourth ion Exchange	98.3	0.2

The overall (Si/Al) ratio in the bulk structure of prepared zeolite can be determined by EDX analysis as shown in (Figure 5-a), Bulk (Si/Al) ratio was found to be (2.06) and this value is matching the outcomes of Matti et. al. [17] and Marcelo et. al. [18] researches.

Figure 5-a. EDX-spectrum of NaY zeolite catalyst

Furthermore, EDX gives an indication about the weight percent of transition metal that impregnated within the zeolite structure (i.e. approximately 0.7 wt.%) as shown in (Figure 5-b).

Figure 5-b. EDX-spectrum of Zn/HY zeolite catalyst

3.1.4 Scanning electron microscopy (SEM) Analysis

The surface morphologies of Y-zeolite samples are obtained by utilizing scanning electron microscopy. (Figure 6-a) with magnification up to 10 μm for prepared NaY zeolite appears the segregated Y-crystals with sharp edges and without clear agglomeration and that agrees with Nsaif et. al. [19]. SEM also has demonstrated that no geometric change can be noticed in Zn/HY-zeolite crystals after multi-stage ion exchange and metal loading as shown in (Figure 6-b). It indicates that no crystalline transformation has occurred within Y- structure during the modifications.

Figure 6-a. SEM image of Zeolite-NaY catalyst

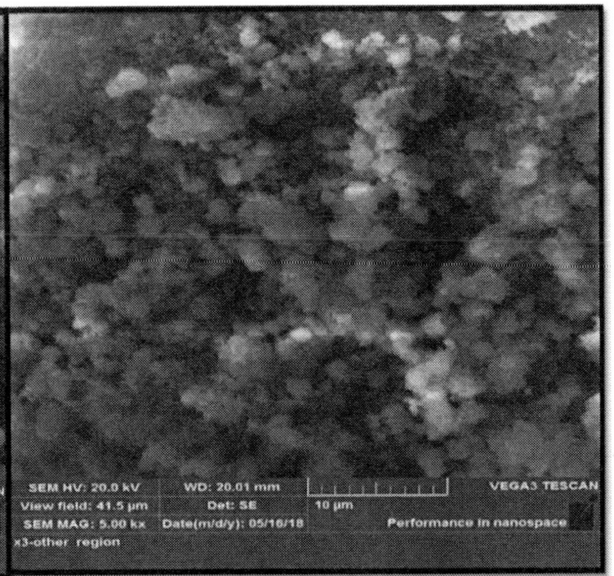

Figure 6-b.SEM image of Zn/HY-Zeolite catalyst

3.1.5 Surface Area and Pore Volume Analysis

Surface area and pore volume are determined based on Brunauer Emmett and Teller (BET) method and are summarized in (Table 4). The surface area of HY-Zeolite presents a good agreement with Scott et al. [20]. BET surface area of Zn/HY-Zeolite appears to be less than HY-Zeolite. These results may be caused by several factors. Metal ions can fill part of the channels and/or pores within the catalyst structure and cause some of the opining pores to be blocked, thereby reducing their volume and decreasing the overall BET-surface area of zeoile-Y this agrees with the results obtained by Ramli et. al. [21] and Lopes et. al. [22].

Table 4. Surface area and pore volume texture of catalysts.

Catalysts	Surface Area (m^2/g)	Pore Volume (cm^3/g)
HY-Zeolite	428.12	0.22
Zn/HY-Zeolite	416.46	0.20

3.2 Central composite design for aquathermolysis process by used Zn/HY-Zeolite

The proposed data from CCD model is used as a reaction conditions to carry out the experimental work inside the pressurized rig. Twenty experiments have been done under pressure ranging from 30 to 45 bar according to the steam temperature with the viscosity of products measured by means of BROOKFIELD DV-II Viscometer. Summarized in (Table 5). The results indicate that the increase in the percentage of viscosity reduction due to the increasing of: Y-catalyst amount, temperature, and time within the reaction. Yet, increasing the temperature more than 235° C leads to a slow increase of the viscosity reduction. Also, the viscosity reduction has increased by increasing the time of reaction, but when the time becames more than 30 h, the viscosity reduction has slightly decreased because the deactivation of the catalyst along the reaction time.

Table 5. Experiments experienced by a design expert.

#	Catalyst (wt%)	Temperature (°C)	Time (h)	% Viscosity reduction
1	0.2	212.5	24	57.83
2	0.2	212.5	12	33.38
3	0.25	193.75	18	43.17
4	0.2	212.5	24	57.16
5	0.2	212.5	36	55.75
6	0.2	250	24	60.71
7	0.2	212.5	24	56.39
8	0.25	231.25	30	70.73
9	0.25	193.75	30	67.92
10	0.15	193.75	30	61.35
11	0.25	231.25	18	48.25
12	0.1	212.5	24	54.56
13	0.2	212.5	24	62.71
14	0.2	175	24	43.64
15	0.2	212.5	24	58.83
16	0.15	231.25	18	45.71
17	0.15	193.75	18	41.38
18	0.15	231.25	30	64.48
19	0.2	212.5	24	55.97
20	0.3	212.5	24	61.47

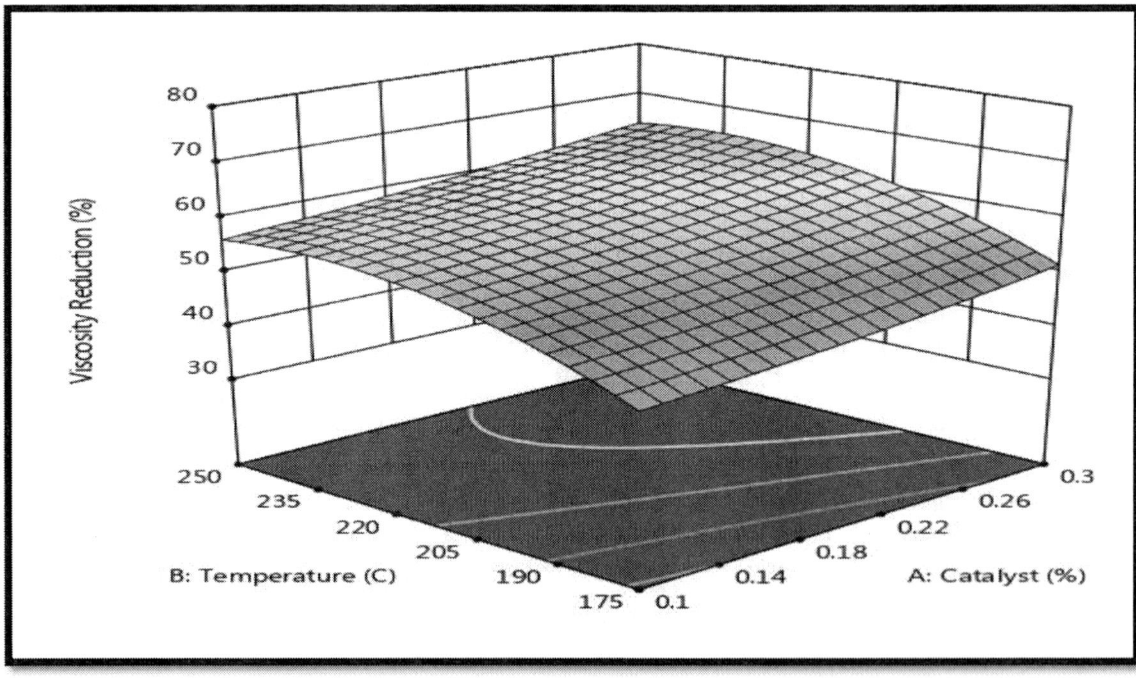

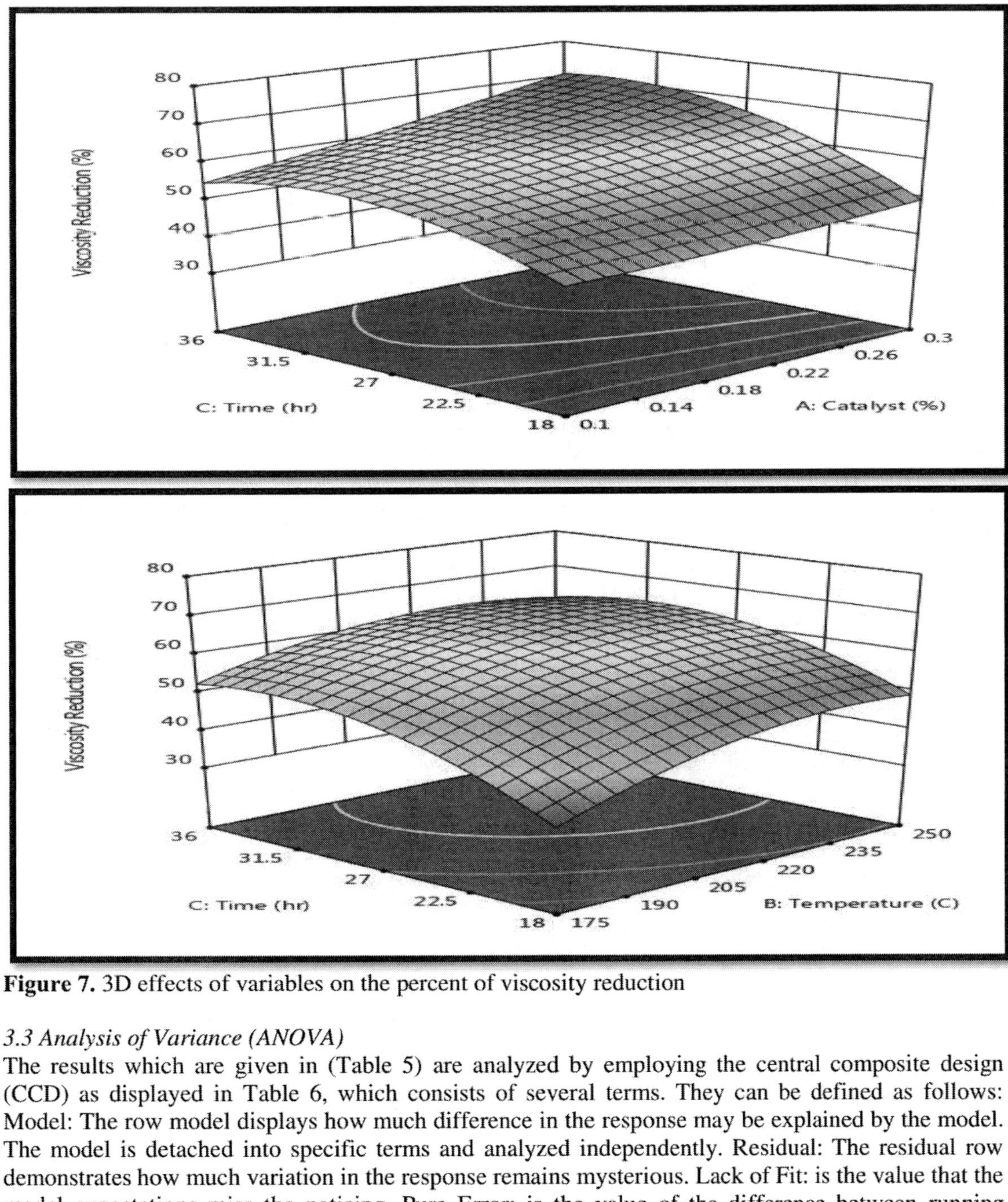

Figure 7. 3D effects of variables on the percent of viscosity reduction

3.3 Analysis of Variance (ANOVA)

The results which are given in (Table 5) are analyzed by employing the central composite design (CCD) as displayed in Table 6, which consists of several terms. They can be defined as follows: Model: The row model displays how much difference in the response may be explained by the model. The model is detached into specific terms and analyzed independently. Residual: The residual row demonstrates how much variation in the response remains mysterious. Lack of Fit: is the value that the model expectations miss the noticing. Pure Error: is the value of the difference between running replication. Sum of Squares: is the sum of the squared variations between the global mean and the amount of variance shown in this row source. df: Degrees of Freedom; number of parameters evaluated to calculate the source's sum of squares. Mean Square: also called variance, the sum of squares divided by df. F Value: investigation for comparing the source's mean square to the residual variance. P Value: also called Prob > F, which is the Probability, equals the integral under the curve of the F-distribution that lies beyond the observed F-value.

The ANOVA analysis shows that the F value is 9.80. The P value is 0.0007 (i.e. lower than 0.05). This indicates that the quadratic model chosen is significant due to magnitudes being higher than 0.05 are not significant. The suitability of the models is actually regulated in with the coefficient of determination R^2. The models give high $R^2 = 0.8982$. Besides, a reasonable agreement with the adjusted determination coefficient is found. In this work, the Adj-R^2 value is 0.8066. The values of R^2 and Adj-R^2 are higher than 0.8 which is high enough to support a high connection between the predicted and observed values. This indicates that the regression model provides a great explication of the interaction between the factors and the response (Viscosity Reduction %). Equation of the surface regression response model in coded unit for the viscosity is illustrated by means of the following equation:

$$R1 = 58.54 + 1.94A + 3.09B + 8.17C + 0.0538AB + 1.06AC - 0.4337BC + 0.1619A^2 - 1.30B^2 - 3.20C^2 \quad (5)$$

Where: R1 = viscosity reduction (%), A = catalyst weight percent (%), B = temperature (°C), C = time (h)

According to the above equation, the coefficient values of variables (i.e. catalyst amount, temperature and time) are + 1.94, +3.09, and + 8.17, respectively. The highest coefficient value is of the time and this means that the time being the most effective and significant variable then temperature and catalyst amount. The positive values of coefficients mean that the increase of all variables lead to increase the percentage of viscosity reduction. Also, these results and (Figure 7) show the interaction effect of the three variables, where the interaction coefficient values of amount of catalyst/temperature, amount of catalyst/time and temperature/time are + 0.0538, + 1.06, - 0.433, respectively. This means that a good interaction effect is occurred between amount of the catalyst and time than other interaction variables and the negative charge of (-0.433) means a negative relation between temperature and time. But in general, the interaction effect of catalyst amount /temperature and temperature/time are not significant in the process.

Table 6. ANOVA regression model of process; analysis of variance.

Source	Sum of Squares	df	Mean Square	F-value	p-value	
Model	1583.66	9	175.96	9.80	0.0007	significant
A-A	59.95	1	59.95	3.34	0.0976	
B-B	153.08	1	153.08	8.53	0.0153	
C-C	1067.82	1	1067.82	59.50	< 0.0001	
AB	0.0231	1	0.0231	0.0013	0.9721	
AC	9.01	1	9.01	0.5020	0.4948	
BC	1.51	1	1.51	0.0839	0.7780	
A^2	0.6593	1	0.6593	0.0367	0.8518	
B^2	42.37	1	42.37	2.36	0.1554	
C^2	257.55	1	257.55	14.35	0.0036	
Residual	179.47	10	17.95			
Lack of Fit	149.28	5	29.86	4.94	0.0521	not significant
Pure Error	30.19	5	6.04			

The outline of analysis of variance (ANOVA) is displayed in (Figure 8) and (Table 7). The obtained results prove that the quality of the model is appropriate with the assumption of ANOVA. (Figure 8-a) describes the plotting of normal probability of residuals from the least square fit and once plotted, the points where located almost over a horizontal line and the normality assumption seems to be convincing. (Figure 10-a) reveals the normal distribution indicating that the assumptions can be accepted and there is no visible problem with normality. In addition, the input factors have had an impact on the responses. (Figure 10-b) describes the actual and predicted values which are closed to each other. (Figure 10-C) illustrates the random scatter plot of residuals against the predicted values. The anticipated results were displayed as haphazardly dispersed about the zero line, signifying that the difference being constant and nonlinear connection can be existed.

Table 7. Report for the result from ANOVA.

Run Order	% Actual value of viscosity reduction in Table 5	% Predicted value of viscosity reduction	Residual
1	57.83	58.54	-0.7089
2	33.38	29.40	3.98
3	43.17	43.33	-0.1565
4	57.16	58.54	-1.38
5	55.75	62.08	-6.33
6	60.71	59.53	1.18
7	56.39	58.54	-2.15
8	70.73	68.08	2.65
9	67.92	62.66	5.26
10	61.35	56.77	4.58
11	48.25	50.49	-2.24
12	54.56	55.32	-0.7553
13	62.71	58.54	4.17
14	43.64	47.16	-3.52
15	58.83	58.54	0.2911
16	45.71	48.63	-2.92
17	41.38	41.69	-0.3053
18	64.48	61.98	2.50
19	55.97	58.54	-2.57
20	61.47	63.06	-1.59

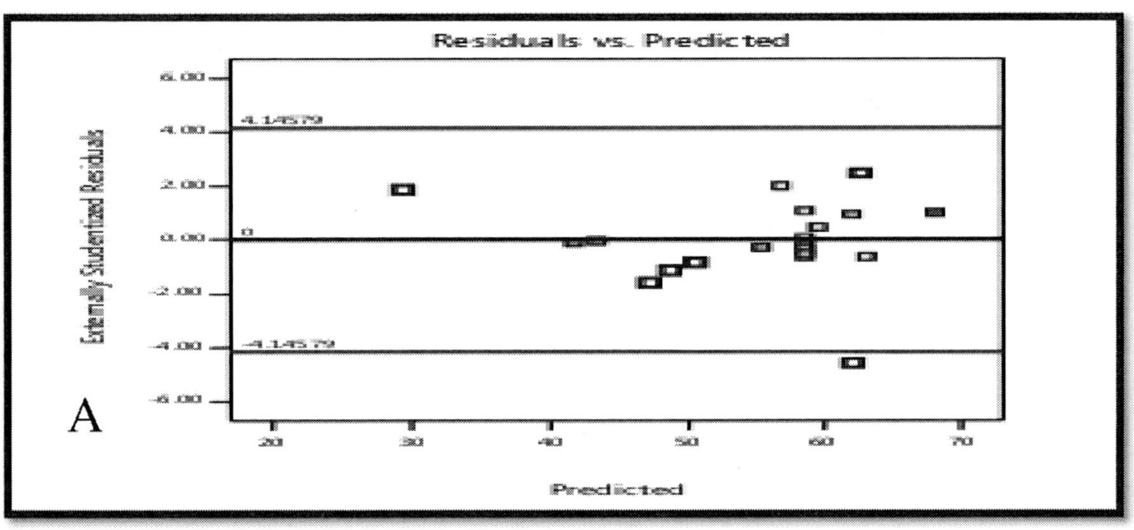

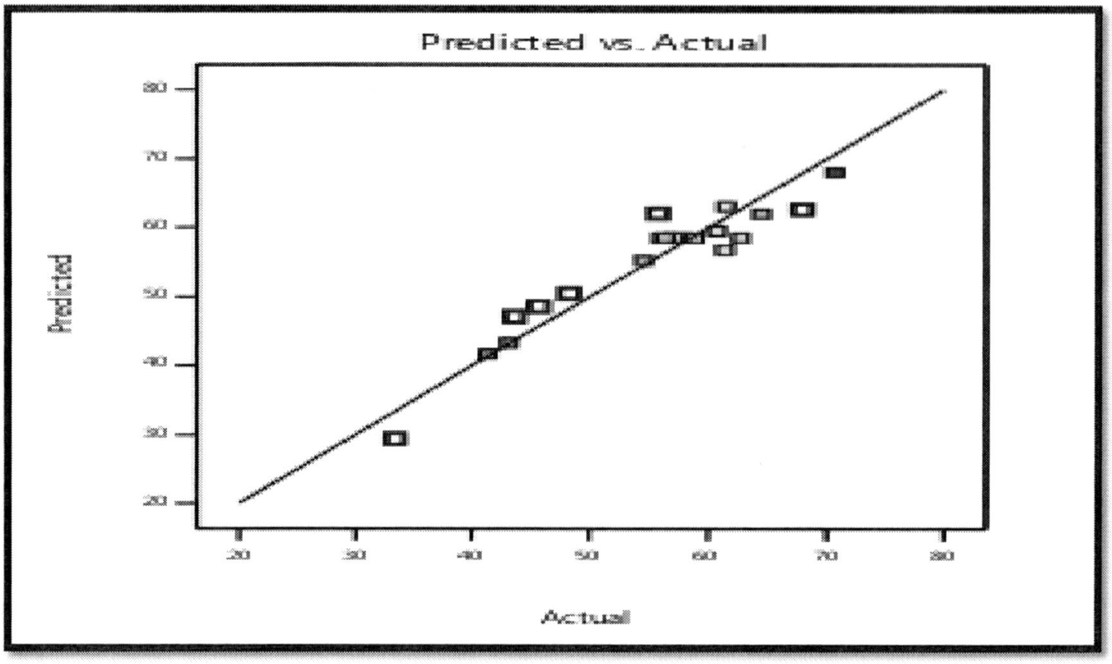

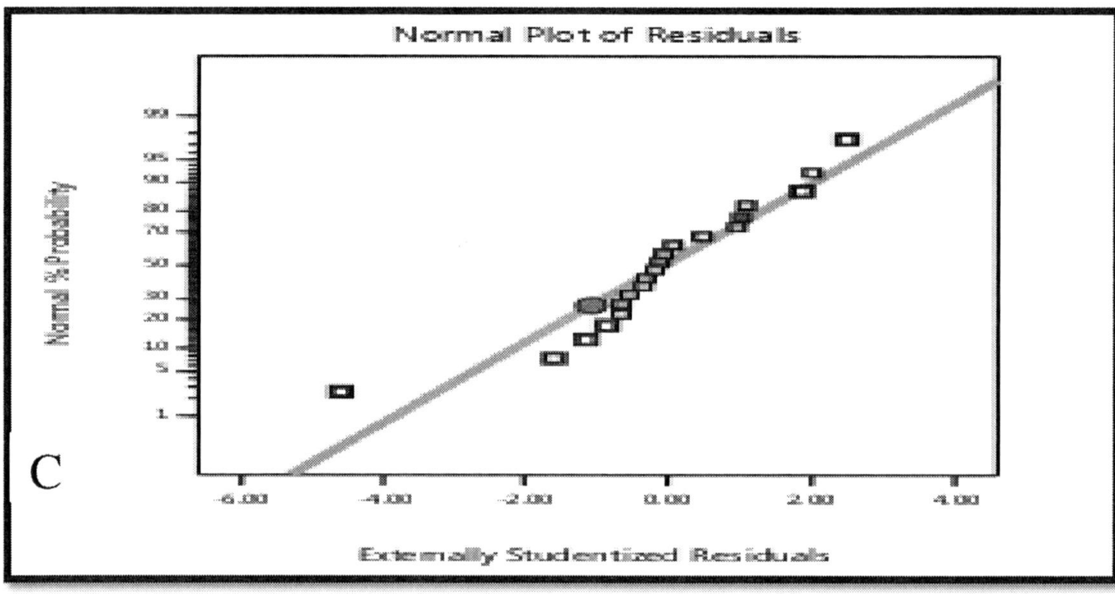

Figure 8 (a-c). Residual plots for viscosity reduction

3.4 The evaluation of the optimum conditions

The essential objective of this study is to discover the best operating conditions (time, temperature and catalyst weight percent) which give the highest viscosity reduction by means of employing a response surface method. The desirability function has utilized to discover the optimal operating conditions as shown in (Figure 9) in which 0.25 wt% of catalyst, 231.25°C and 30 hour are the best conditions of the reaction in case of using Zn/HY-Zeolite catalyst and these values are considered economically good when compared with the other researchers.

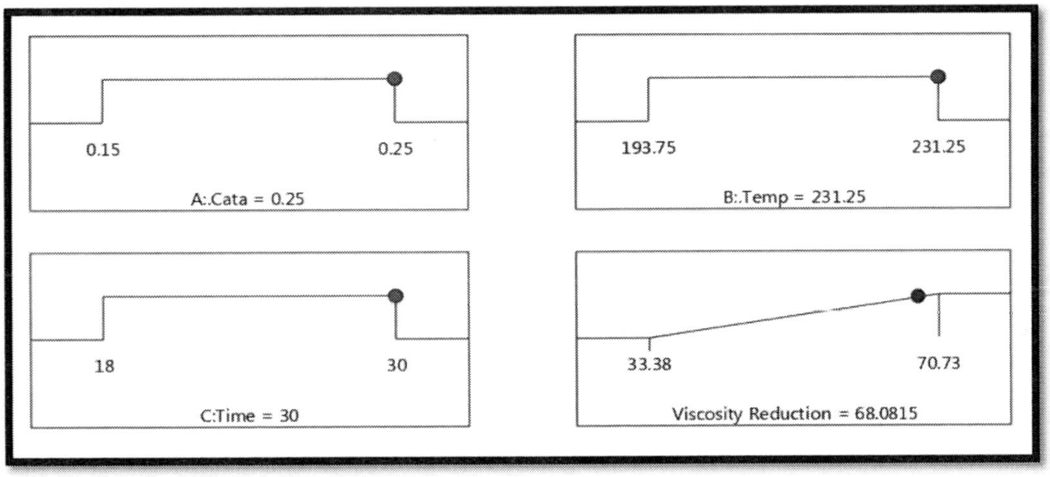

Figure 9. Optimum conditions for Zn/HY-Zeolite

The analysis of the produced crude oil from the aquathermolysis reaction at these optimum conditions is shown in (Tables 8 and 9). According to Tables 1 & 8, it can be seen that the viscosity is decreased from 47.87 to about 15.68 g/cm.s and sulfur contain is decreased from 4.520 to about 2.007 wt.%, while the API is increased from 21.9 to 24.5 (i.e. by 2.6). Both flash point and pour point are decreased by 10 and 5 Celsius respectively, which indicates an increase in the volatile hydrocarbon and the ability of crude oil to flow. In addition, (Table 9) indicates an increase in the percentage of yield fractions from 40.8 to 45.6. Increasing the yield after upgrading reaction can be obviously observed in the case of kerosene and gas oil.

Table 8. The characterization of crude oil sample at the optimum conditions.

Viscosity Reduction (%)	67.25%
Sulfur contain Reduction (wt%)	55.58%
Density (g/cm^3)	0.9070
API	24.5
Flash point (°C)	70
Pour point (°C)	-30

Table 9. Yield of light and intermediate component of crude oil after upgrading at the optimum conditions.

Temperature for fraction (°C)	%Yield of fraction
Light Naphtha (IBP 37-75)	1.6
Heavy Naphtha (75-175)	14
Kerosene (175-225)	7.2
gas oil (225-340)	22.8
Total	45.6

4. Conclusions

Conclusions can be summarized as follows:

1. Zeolite-Y in various formats (i.e. NaY, NH_4Y, and HY) has successfully prepared and the acidic form is loaded with Zn metal to produce a bi-functional zeolite catalyst (Zn/HY). It seems to be an efficient and useful catalyst for aquathermolysis reaction because it has a high ability to the reduction of viscosity and sulfur contain of heavy oil.
2. Improvement of the physical properties of the crude oil can be achieved by aquathermolysis process under solid catalyst and low reaction temperature, which is a good way to increase both the API and the yield of the light hydrocarbons' cuts (i.e. Reduce the crude oil density).
3. Response surface methodology (RSM) has been used to enhance aquathermolysis process variables, and the results obtained are confirmed that the reaction time is the most important factor affecting aquathermolysis reaction in comparison with temperature and the amount of catalyst used.
4. According to the statistical method (i.e. Design Expert) and the experimental results, the optimum operating conditions for aquathermolysis reaction are 0.25 wt. % of catalyst, 231.25°C and 30-hour reaction time to obtain the highest reduction in the viscosity and sulfur contain within the heavy oil by 67.25% and 55.58%, respectively.
5. The response of surface regression model equation in coded unit for viscosity reduction percentage is established to be: $R1 = 58.54 + 1.94A + 3.09B + 8.17C + 0.0538AB + 1.06AC - 0.4337BC + 0.1619A^2 - 1.30B^2 - 3.20C^2$
6. The enhancement of the transportation characteristics of the crude oil can be accomplished through the catalytic aquathermolysis process, both flash and pour point can be decreased and that means an increase in the volatile hydrocarbon and the ability of crude oil to flow.

References

[1] Wu, C., Lei, G., Yao, C., and Jia, X 2010 In situ upgrading extra-heavy oil by catalytic aquathermolysis treatment using a new catalyst based anamphiphilic molybdenum chelate. *International Oil and Gas Conference and Exhibition* (China)

[2] Noorlaily, P., Nugraha, M. I., Abdullah, M., and Iskandar, F 2013 Ethylene glycol route synthesis of nickel oxide nanoparticles as a catalyst in aquathermolysis. *Materials Science Forum* **737** 93-97

[3] Xu, H., Pu, C., and Wu, F 2012 Low frequency vibration assisted catalytic aquathermolysis of heavy crude oil. *Energy & Fuels* **26** 5655-5662

[4] Zhao, F., Liu, Y., Fu, Z., and Zhao, X 2014 Using hydrogen donor with oil-soluble catalysts for upgrading heavy oil. *Russian Journal of Applied Chemistry* **87** 1498-1506

[5] Desouky, S., Betiha, M., Badawi, A., Ghanem, A., and Khalil, S 2013 Catalytic aquathermolysis of Egyptian heavy crude oil. *Proceedings of World Academy of Science, Engineering and Technology* **80** 887

[6] Guo, K., Li, H., and Yu, Z 2016 In-situ heavy and extra-heavy oil recovery: A review. *Fuel* **185** 886-902

[7] Zhao, F. J., Liu, Y. J., Zhang, B., Dong, L., and Yu, H. T 2011 Hydrogen Donor Catalytic Aquathermolysis Upgrading for Heavy Oil. *A Laboratory Study. In Key Engineering Materials* **480** 142-147

[8] Maity, S. K., J. Ancheyta, and G. Marroquín 2010 Catalytic aquathermolysis used for viscosity reduction of heavy crude oils: a review. *Energy & Fuels* **24** 2809-2816

[9] Galukhin, A. V., A. A. Erokhin, and D. K. Nurgaliev 2015 Effect of catalytic aquathermolysis on high-molecular-weight components of heavy oil in the Ashal'cha field. *Chemistry and Technology of Fuels and Oils* **50** 555-560

[10] Wu, Y., Li, Y., Chen, G., Zhao, W., and Zhang, J. 2016 Urea enhanced aquathermolysis of heavy oil catalyzed by hydroxamic acid-co (II) complex at low temperature. *MATEC Web of Conferences* **67** 06038

[11] X. Yang; I.D. Gates 2009 Design of hybrid steam in-situ combustion bitumen recovery. *Process Nat. Resour. Res.* **18** 213

[12] Jha, Bhagwanjee, and Devendra Narain Singh 2016 Basics of Zeolites. *Fly Ash Zeolites. Springer, Singapore* **5** 31

[13] Bashir Y Al-Zaidi, Richard J. Holmes, and Arthur A. Garforth 2012. Study of the

IOP Conf. Series: Materials Science and Engineering **579** (2019) 012010 doi:10.1088/1757-899X/579/1/012010

relationship between framework cation levels of Y-zeolites and behavior during calcination, steaming, and n-heptane cracking processes. *Industrial & Engineering Chemistry Research (I&EC)* **51** 6648-6657
[14] B. Y. Al-Zaidi 2010 The effect of modification techniques on the performance of zeolite-Y catalysts in hydrocarbon cracking reactions. PhD Thesis (School of chemical Engineering and analytical science – The University of Manchester – United Kingdom)
[15] Treacy M.M.J. and Higgins J.B 2001 Collection of Simulated XRD Powder Patterns for Zeolites 4th edition (Amsterdam)
[16] Kristiani, A., Sudiyarmanto, S., Aulia, F., Hidayati, L. N., and Abimanyu, H 2017 Metal supported on natural zeolite as catalysts for conversion of ethanol to gasoline. *MATEC Web of Conferences* **101**
[17] Matti, Aveen H. and Surchi, Kafia M 2014 Comparison the Properties of Zeolite Nay Synthesized by Different Procedures. *International Journal of Innovative Research in Science* **3** 6
[18] Marcelo J., and Antonio S 2003 Catalytic Cracking of C5+ Gasoline over HY Zeolite. *React. Catal. Lett* **79** 257-262
[19] Mohammed Nsaif, Ahmed Abdulhaq, Ali Frhan, Safa Neamat 2012 Catalytic Cracking of Heptan Using Prepared Zeolite. *Journal of Asian Scientific Research* **2** 927-948
[20] Scott M.A., Kathleen A.C., Brabir K.D 2003 *Handbook of zeolite science and Technology* (Marcel Dekker Inc. New York)
[21] Ramli, Nur Aainaa Syahirah, and Nor Aishah Saidina Amin 2015 Fe/HY zeolite as an effective catalyst for levulinic acid production from glucose: characterization and catalytic performance. *Applied Catalysis B: Environmental* **163** 487-498
[22] Lopes, J. H., Nogueira, F. G., Gonçalves, M., and Oliveira, L. C 2015 Modified Zeolite with Transition Metals Cu and Fe for Removal of Methylene Blue from Aqueous Medium: Mass Spectrometry Study. *Bulletin of Chemical Reaction Engineering & Catalysis* **10** 237-248

The use of Mixed Bacterial Culture to improve the Biodegradation of Diesel Pollution.

N M Jabbar[1#], A K Mohammed[1], S M Jabber[2], E H Kadhim[3]

[1]Al-Khwarizmi Engineering College, Biochemical Engineering Department, University of Baghdad, Baghdad, Iraq

[2]Nanotechnology research center and advanced materials / University of Technology.

[3]Treatments and Disposal of Hazardous Waste Directorate/ Ministry of Higher Education and Scientific Research

[#]Corresponding author's e-mail address: noor.uot252@kecbu.uobaghdad.edu.iq

Abstract. This study was conducted to evaluate the hydrocarbon biodegradation abilities of Enterobacter cloacae, Staphylococcus aureus, Sphingomonas paucimobilis, and Pentoae species which were isolated from different diesel-contaminated soil samples. The isolates were identified by the Vitek 2 system. Fourier-transform spectroscopy (FT-IR) tested the potential of these isolates to biodegrade the diesel according to the peak areas, a significant decrease in the area of the peaks at 2856-2928 cm-^{1} corresponds to aliphatic hydrocarbons. The appearance of small peaks at 900-1032 cm-^{1} refers to substituted benzene derivative compounds. An appearance of some new peaks at 3010- 3030 cm-^{1} which indicate the presence of alcohol (-OH) and ketones (RC=O). A sharp peak appeared at 1712 cm-^{1} refers to the carbonyl group (C=O). The potential of biological surfactant production was tested using the Sigma 703D stand-alone tensiometer showed that these isolates were biological surfactant producers. The better results of the surface tension reduction test were obtained when using the mixed bacterial culture which reduced the surface tension of the medium from 66 mN/m to 35.15 mN/m. Single isolates and mixed bacterial culture have investigated their ability to degrade 3.0 % (v/v) of diesel as sole source of carbon and energy in Bushnell- Haas medium. The results demonstrated that the bacterial isolates could be effective in biodegradation of diesel spills individually and showed good biodegradation abilities when they are used together in the mixed bacterial culture.

Keywords: Biodegradation, Diesel, Mixed-culture, biological surfactant, Pollution.

1. Introduction

Petroleum hydrocarbons are the most common environmental pollutants in the world, and oil spills show a great hazard to terrestrial and aquatic ecosystems. Oil pollution may arise either accidentally or operationally whenever oil is produced, transported, stored and treated, or when it is used on land or at sea. Oil spills are a major threat to the environment because they severely damage surrounding ecosystems [1]. Soil pollution with diesel fuel may be toxic to microorganisms and plants and acts as a

source of groundwater pollution [2]. Diesel fuels are complex mixtures of saturated hydrocarbons (primarily paraffins including n, iso, and cycloparaffins), and aromatic hydrocarbons (including naphthalenes and alkylbenzenes) obtained from the middle-distillate, gas-oil fraction during petroleum separation [3]. Remediation of petroleum contaminated sites could be achieved by either physicochemical or biological methods. Apart from the physicochemical treatment techniques, bioremediation is a clean and environmentally friendly treatment that is widely used to control environmental pollution [4]. Bioremediation is the use of microorganisms, mainly bacteria and fungi, or plants to utilize and break down environmental contaminants such as petroleum into less harmful substances [5]. It involves breakdown organic compounds through a biological transformation to less complex metabolites such as carbon dioxide and water. These techniques have a number of key advantages over traditional technologies including the fact that they are simple to implement, applicable over large areas, cost-effective and can lead to the complete removal of different contaminants [6; 7]. Due to the complexity of petroleum products, biodegradation caused by mixed cultures is more effective than that caused by pure culture since a broader enzymatic capability can be achieved [8]. Bacteria are considered to be efficient biodegrades due to the wide variety of species, their abundance, and their catabolic and anabolic versatility, as well as their ability to adjust to adverse environmental conditions. Therefore, mixed bacterial culture of microbial community is required to complete biodegradation of petroleum pollutants [1].
The aim of this research was: studying the biodegradation of diesel by mixed bacterial cultures isolated from diesel-polluted soil for bioremediation process applications.

2. Materials and methods
2.1 Collection of samples

The soil samples were collected from different locations, including the local generators and Al Dura refinery in Baghdad. All samples were kept in sterile plastic bags and transported to laboratory for analysis [8].

2.2 Isolation and identification of Diesel degrading bacteria

Four grams of contaminated soil sample were suspended in 20 mL of sterile Bushnell-Haas medium (BHM) containing (g/l): $MgSO_4.7H_2O$, 1.0; K_2HPO_4, 1.0; KH_2PO_4, 1.0; $FeCl_3$, 0.05; NH_4NO_3, 1.0; $CaCl_2$, 0.02; pH to 7±0.2 which was sterilized at 121°C for 15 min. Bacteria were grown in 250 ml Erlenmeyer flasks for one week in a cool incubator shaker at 30 °C and 180 rpm. Isolation of pure cultures was carried out by streaking method on nutrient agar plates. The plates are placed in an incubator at 30 °C for 48 hours. Specific colonies were sub-culture on nutrient agar plates and incubated for another 48 hours. The pure cultures were preserved in the refrigerator at 4 °C for further study and identification. The isolates were classified based on their shape, color and biochemical properties than identified using VITEK 2 device [1].

2.3 Inoculum preparation

Bacterial isolates were cultured on nutrient agar at 30 °C for 48 h. Aqueous culture was then suspended, standardizing to the Macfarland scale tube number 8 [9]. Tube 8 represents turbidity of bacteria in the concentration (2.1 x 10^9 CFU/mL). The volume of inoculum was 4 ml.

2.4 Measuring Biodegradation of Diesel
Measuring the biodegradation of diesel by isolated bacteria was done with the following ways:

2.4.1 The use of FT-IR Spectrum technique
The four bacterial isolates individually were inoculated in of 93 ml BHM in Erlenmeyer flasks and 3.0 % (v/v) of diesel was added as the sole source of carbon. Flasks were incubated at 30°C in a cool incubator shaker at 180 rpm. Samples were withdrawn after 7 days during the experiment period.

Then 15 ml of medium was transferred into separating funnel containing chloroform (1sample: 1chloroform) with vigorous shaking. The chloroform layer which contains the remainder of diesel was tested using FT-IR spectrum [10].

2.4.2 The reduction of surface tension

The bacterial isolates were inoculated with 93 ml of BHM containing 3.0 % (v/v) of diesel in 250 ml Erlenmeyer flask and incubated at 30 °C in a cool shaker incubator at 180 rpm for 8 days. A medium was prepared by adding 10 ml of culture to 20 ml of carbon tetrachloride (CCL_4) to separate hydrocarbons from the liquid culture. Then the medium was poured in centrifuge tubes and centrifuged at 10000 rpm for 15 minutes at -5 °C to precipitate the cells [11]. The upper free cells layer that containing the diesel was used for surface tension test using a Sigma 703D Du-Nouy-Ring tensiometer [12].

2.4.3 The quantitative Loss of Diesel

Each of individual and mixed bacterial culture was suspended in 250 mL Erlenmeyer flasks which containing 93 mL of BHM. Diesel was added at concentration of 3.0 % (v/v) as a sole carbon source. Flasks were incubated at 30 °C in a cool incubator shaker at 180 rpm for 28 day. The experiment conducted in duplicate [13]. Then, to determine the remaining concentration of diesel at the end of incubation period, 15 ml of medium from each flask was taken. The acidity of the sample culture was adjusted to (pH <= 2) by adding HCl. The extraction was carried out by liquid-liquid extraction method by mixing equal volumes of sample and solvent (1sample: 1CCL_4). The amount of residual diesel was measured in the medium by using (Horiba oil analyzer ocma-350) device [14]. The degradation efficiency was calculated according to the following formula:

$$\text{Degradation rate \%} = \frac{\text{Intial conc. of diesel} - \text{Final conc. of diesel}}{\text{Intial conc. of diesel}} \times 100 \qquad (1)$$

3. Results and Discussion

3.1 FT-IR analysis of Diesel degradation

The biodegradation of diesel fractions was observed using Fourier-transform spectroscopy to identify the absorption of beams before and after exposure to bacterial isolates. In figure 1, the control sample showed the absorption bands locations of aliphatic compounds (C-H), (C-C) at 2856-2928 cm^{-1} which appears in large and broad peaks, while a small peak at 720 cm^{-1} refers to aromatic compounds (benzene ring) [15; 16]. Absorption bands at 1376- 1464 cm^{-1} observed the symmetrical -CH3, -CH2 of aliphatic compounds [17].

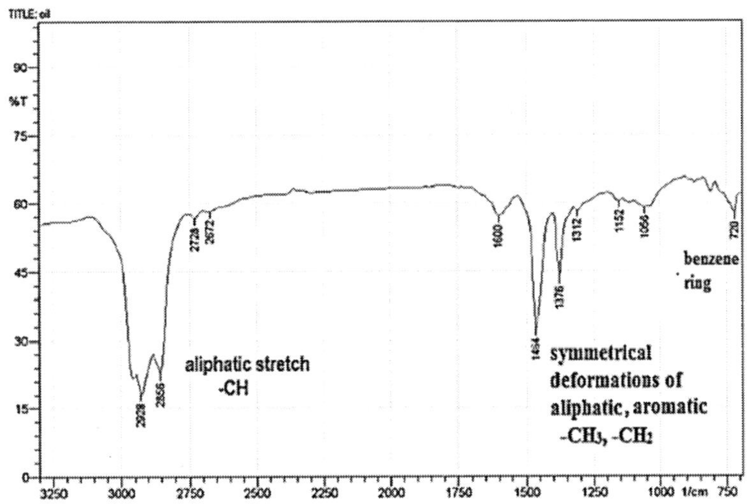

Figure1. The spectrum of control diesel by FT-IR

After 7 days of incubation, figures from 2 to 5 revealed that the spectrum reflected more absorption bands than the control sample under the effecting of bacterial isolates at concentration of 3 % (v/v) of diesel. Appearance of some new peaks at 3010- 3030 cm-[1] indicates the presence of alcohol (-OH) and ketones (RC=O). A sharp peak appeared at 1712 cm-[1] refers to the carbonyl group (C=O). Moreover, the peak 1216 cm-[1] in the wave range 1032-1320 cm-[1] showed the presence of primary alcohol (C-OH) and carboxylic acid. A significant decrease in the area of the peaks at 2856-2928 cm-[1] was corresponding to aliphatic hydrocarbons. The appearance of small peaks at 900-1032 cm-[1] refers to substituted benzene derivative compounds [18]. A change in the absorption band at wave range 1360-1496 cm-[1] refers to the transformation and consuming of hydrocarbons (C-H bond) in the diesel.

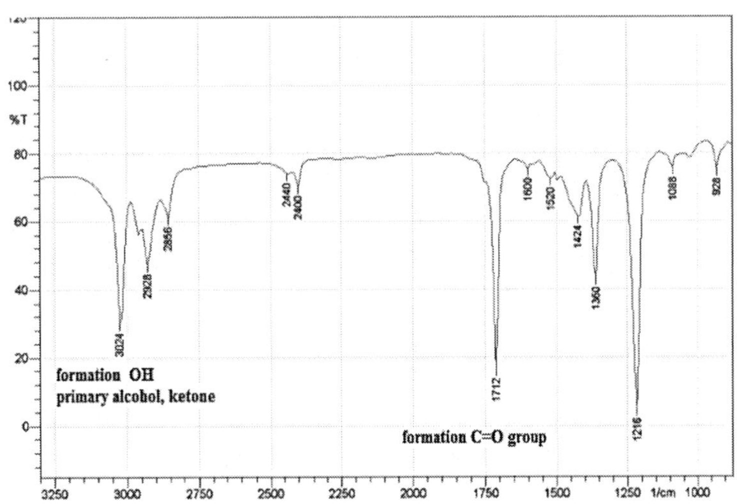

Figure2. FT-IR spectrum of diesel at concentration 3.0 % treated by <u>*Enterobacter*</u> <u>*cloacae*</u> after 7 days, (at 30 °C, pH to 7±0.2)

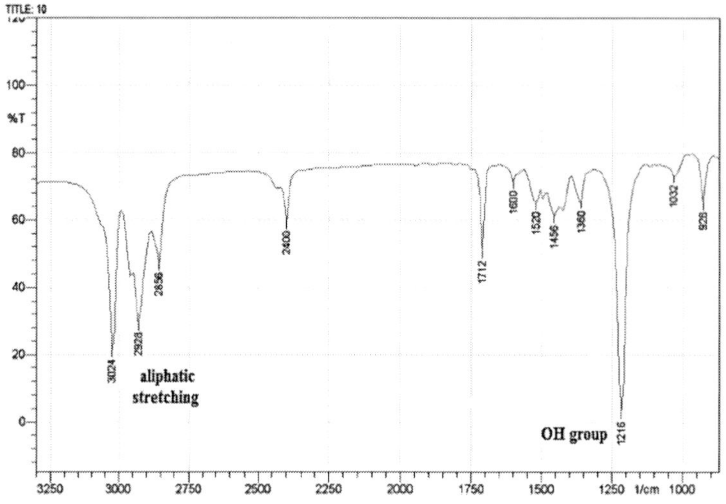

Figure3. FT-IR spectrum of diesel at concentration 3.0% treated by *Staphylococcus aureus* after 7 days, (at 30 ˚C, pH to 7±0.2)

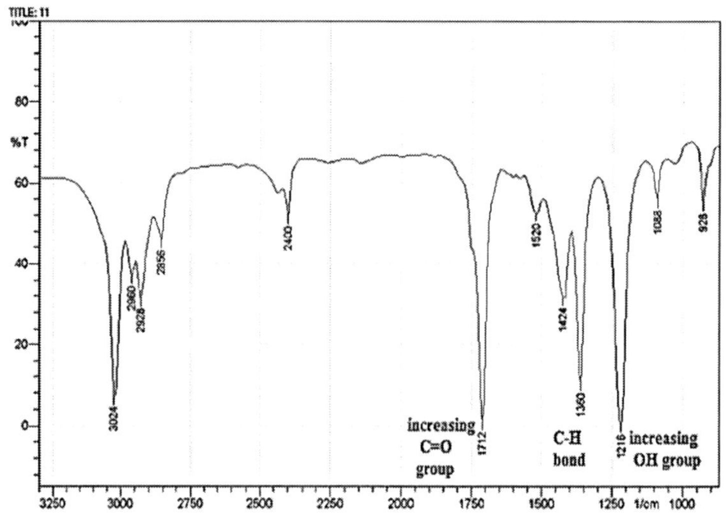

Figure4. FT-IR spectrum of diesel at concentration 3.0 % treated by *Sphingomonas paucimobilis* after 7 days, (at 30 ˚C, pH to 7±0.2)

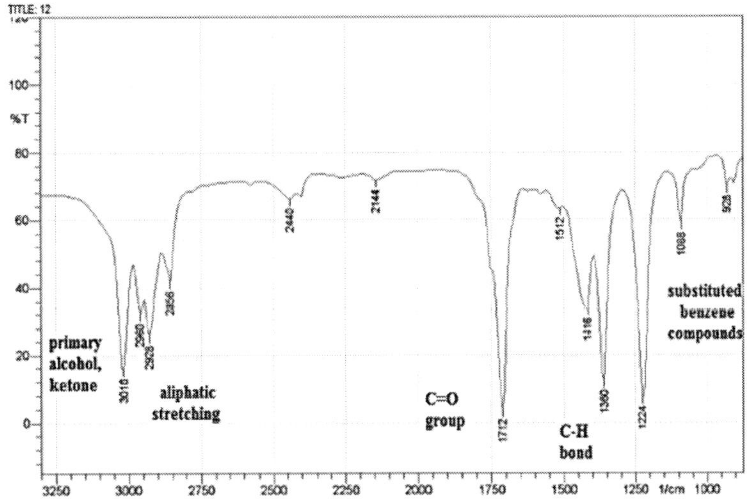

Figure5. FT-IR spectrum of diesel at concentration 3.0 % treated by <u>Pentoae species</u> after 7days, (at 30 °C pH to 7±0.2)

The presence of alcohol and ketone groups during biodegradation indicates the breaking down of aliphatic and aromatic compounds of diesel by bacterial isolates studied. Also, the changes in bands locations and area indicate the role of bacterial isolates in the degradation of hydrocarbons, which attributed to the fact that these compounds are naturally hydrophobic and have high molecular weights which have been converted by bacteria into low molecular weights and more hydrophilic compounds [19].

3.2 Determination of Surface Tension Reduction

The test of surface tension reduction by the individual isolates and by their mixed culture has been measured after 8 days of incubation. The results demonstrated that all isolates and the mixed culture were able to produce biological surfactants in BHM containing 3.0 % (v/v) of diesel as the only carbon source. The greatest surface tension reduction of the medium was achieved with the mixed culture which reached a value 35.15 mN/m compared to the control 66 mN/m. While the other data of the surface tension obtained from individual isolates still high and ranged 43.15 mN/m to59.01 mN/m, as shown in Table 1. These results correspond to the findings of Cerqueira et al. (2011) [20] who reported that a microbial consortium that included (*B. cereus, P. aeruginosa, B. cibi, S. acidaminiphila and B. megaterium*) was reduced the surface tension of the medium containing oily sludge from 60.4 mN/m to 36.6 mN/m. However, the other surface tension values as a result of the singular action for the isolates ranged from 41mN/m to 42.6mN/m.

The reduction in surface tension of the medium depends on the type of biological surfactants produced by bacterial isolates as well as the critical micelle concentration (CMC). The micelles are large molecules which when aggregated the surface tension begins to decrease [21]. The biodegradation process for petroleum hydrocarbons depends strongly on the hydrocarbons emulsifying. The biological surfactants which produced by bacterial species seems to be more potent than chemical surfactant, since they are environmentally friendly and non-toxic biodegradable compounds [12].

Table1. Surface tension reduction by hydrocarbon-degrading bacterial isolates

Bacterial Isolates	Surface tension (m N/m) at concentration 3.0 % of diesel
E. cloacae	59.01
S. aureus	43.15
S. paucimobilis	46.55
Pentoae species	54.33
Mixed bacterial culture	35.15
Surface tension of the control = 66 mN/m	

3.3 Decomposition of Diesel

The biodegradation percentage of diesel at concentration of 3.0 % (v/v) by the individual isolates and their mixed culture after incubation at 30 °C for 28 days was demonstrated in figure 6. The efficiency of biodegradation varies between the isolates and ranged from 44 % to 70 % according to their ability to degrade diesel. At concentration 3.0 % (v/v) of diesel the mixed bacterial culture performed a maximum biodegradation up to (75.31 %) followed by *S. paucimobilis, Pentoae sp., E. cloacae,* and *S. aureus* (70 %, 67.6 %, 56 %, and 44 % respectively) as shown in figure 6 below. The individual isolates metabolize a limited range of hydrocarbon substrates and diesel is made of a blend of compounds, so the biodegradation of it needs a mixture of different bacterial groups with broad enzymatic capabilities competent to degrade a wider range of hydrocarbons.

In fact, a mixed culture of the bacterial community is required to complete the biodegradation of petroleum contaminants because the hydrocarbon mixture varies markedly in the volatility, solubility, tendency to biodegradable and the certain enzymes cannot be gained in a single organism [1]. These results are consistent with Shahaby (2014) [22] who studied the biodegradation of complex mineral oil at concentrations 3 and 5% (v/v) by a mixed culture. The highest biodegradation rate occurred at 3 % (v/v) and reached 41.8%. Al-Wasify and Hamed (2014) [13] proved that after 28 days of incubation the mixed bacterial was degraded up to 88.5% of crude oil.

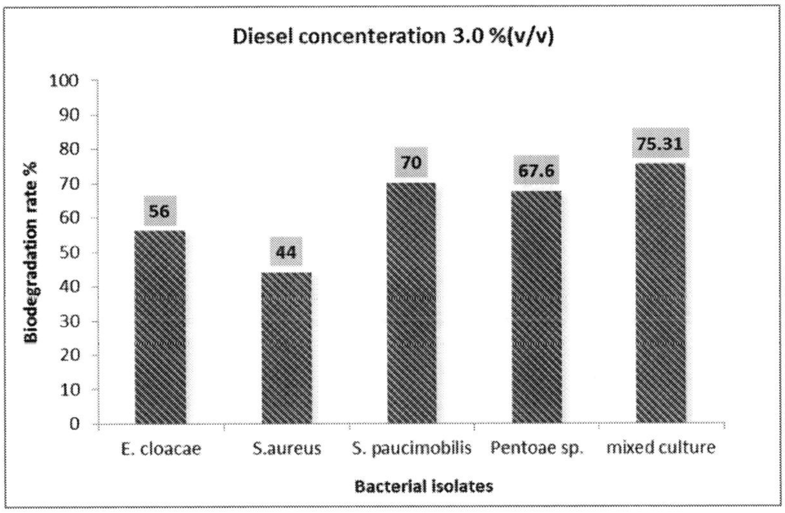

Figure6. Biodegradation rate at concentration 3.0 % of diesel by individual bacterial isolates and mixed culture after 28 days, (at 30 °C, pH to 7±0.2)

4. Conclusions

The present study showed that the bioremediation of diesel with a mixed culture of *Enterobacter cloacae, Staphylococcus aureus, Sphingomonas paucimobilis,* and *Pentoae species* were more efficient than individual isolates after 7 days of incubation, Because of the ability of the mixed culture to produce the biological surfactants and reduced the surface tension of the medium from 66 mN /m to 35.15 mN/ m. Also, FT-IR spectrum analysis confirmed the biodegradation of diesel by the four bacterial isolates when used individually. Biodegradation of diesel means the transformation of its complex, harmful components like polycyclic aromatic hydrocarbons, and long chain alkanes into simpler fractions. Accordingly, these bacteria could be useful in treating diesel polluted environment.

References

[1] Jabbar N M, Kadhim E H and Mohammed A K 2017 *ALKEJ* **13** 80
[2] Serrano A, Tejada M, Gallego M and Gonzalez J L 2009 *Sci Total Environ* **7** 56
[3] Zanaroli G, Toro S and D Todaro 2010 *Microb Cell Fact* **9** 1
[4] Jasmine J and Mukherji S 2014 *Environ Sci Process Impacts* **16** 1889
[5] Koshlaf E and Ball A S 2017 *AIMS Microbiol* **3** 25
[6] Haritash A and Kaushik C 2009 *J Hazard Mater* **169** 1
[7] Jabbar N M, Kadhim E H and Mohammed A K 2018 *ALKEJ* **14** 48
[8] Darsa K, Thatheyus A J and Ramya D 2014 *Science International* **2** 20
[9] Morais E and Tornisielo S 2009 *Braz Arch Biol Techno* **52** 1571
[10] Thamer M, Al-Kubaisi A, Zahraw Z, Abdullah H, Hindy I and Abd Khadium A 2013 *N S journal* **5** 865
[11] Yasin A R *Biodegradation of polluted water with crude oil discharged from south Baghdad power plant* 2013 M.Sc. Thesis. College of Science, University of Baghdad: pp-136.

[12] Jabbar N. M. *Design of bio-pile reactor for bioremediation of petroleum hydrocarbon contaminated soil* 2018 M.Sc. Thesis. Al-Khwarizi Engineering College, University of Baghdad: pp-90.
[13] Al-Wasify R and Hamed S R 2014 *Int J Microbiol* 1
[14] Yasin A R, Salman S A and Al-mayaly I 2014 *IJS* **55** 113
[15] Ali H R, El-Gendy N S and Hashem A I 2012 *ISRN Soil Science* 1
[16] Neif R A and Al-mayaly I K 2015 *IJS* **56 (3C)**: 2574-84.
[17] Abbas O, Rebufa C and Kister J 2008 *Talanta* **75** 857
[18] Sharma A, Kumar P and Rehman M B 2014 *Int J Environ Bioremediat Biodegrad* **2** 202
[19] Owaied H Y 2008 *Raf. J. Sci* **19** 101
[20] Cerqueira V S, Hollenbach E B, Maboni F, Vainstein M. H, Camargo F, Peralba M and Bento F 2011 *Bioresource Technology* **102** 11003
[21] Jabbar N M, Mohammed A K and Kadhim E H, 2019 *Baghdad Sci. J* **16** 185
[22] Shahaby A F 2014 *Int. J. Curr. Microbial App. Sci.* **3** 401

IOP Conf. Series: Materials Science and Engineering **579** (2019) 012012 doi:10.1088/1757-899X/579/1/012012

Decreasing the Sulfur Content of Crude Oil by Ultra-Sound and Activated Carbon Assisted Oxidative Method

Tahseen Hameed Khlaif[1#], Asaad Salim Bded[1]

[1]Petroleum Engineering Department, Kerbala University, Karbala, Iraq

[#]Corresponding author email: dr.alyasari@gmail.com

Abstract. There is a large amount of sulfur compound in crude oil, sulfur gases such us H2S for example are emitted and can affect the environment and finally can cause problems during the processing operation of crude oil. Therefore, in this study, a method for the removal of sulfur from crude oil using an ultrasound device has been proposed. This method is a combination of oxidative method and applying ultrasonic waves to achieve the allowable sulfur content in crude oil. Samples of crude oil have been taken that we got from an al-Najaf refinery to study the effect of three parameters (activated carbon weight, hydrogen peroxide volume, and acetic acid volume) at 50oC on the desulfurization process. the optimum result of desulfurization has been found to be (81.325%) at activated carbon amount of (9gm) with (H2O2/acetic acid)(40/30)ml.

I. Introduction

The development of technology and the growing desire to crude oil derivatives have led to the increasing amount of sulfur content in the crude oil resulting in poor oil quality [1]. Poor oil quality is because of the presence of some dissolved sulfide, for instance hydrogen sulfide, sulfoether, mercaptan, disulfide, and thiophene [2]. Sulfur components that exist in crude oil are unfavorable in refining process because they lead to first suppression of some of the catalysts used in crude oil processing and second cause erosion problems in pipeline, pumping, and refining equipment [3]. By nature, sulfur compounds that are maintained in fuels cause to release acid gases like sulfur oxides. water Reacting with these gases to form acid rain which harms the buildings and glaze finishes as well [4]. The percent of sulfur content in crude oil is varying between 0.04% w/w for crude oil with low density to about 5% w/w for crude oil with high density [5].

The sulfur compound that exist in crude oil is produced by the process of putrefaction of organic substance, sulfur compound can be segregated from the oil in a shape of hydrogen sulfide which emerges in the correlating gas, an amount of it remains in the liquid [6].Sulfur compounds in crude oil are found in many types such as: Free sulfur is scarcely present in crude oil [7] .The appearance of unleashed sulfur is linked to the denudation process of extra complicated sulfur compound [8], H2S is a harmful, colorless, blazing gas which smells as rotten eggs [9], Thiols are organosulfur compounds which consist of a sulfhydryl group [10],The sulfides with linear structure are considered a compounds of organosulfur. They are chemically neutral. sulfides boiling point is above than mercaptans with regard to molecules of the same carbon number [11], The disulfide (R – S – S – R') exists in small amount in petroleum fractions which boiling point reach to 300°C [12].The accounts prove for 0.07 to 0.15 of the total amount of sulfur, Thiophene and derivatives [13]. Thus, to minimize the effect of the sulfur compounds that are exist in crude oil and petroleum fractions, several desulfurization methods have been used around the world in the last and current century, such us :

1- Hydrodesulfurization (HDS) method which needs rigorous conditions of operation [14].

2- Adsorptive desulfurization the effectiveness of this method depends on the quality of the sorbent material [14].

3- Oxidative desulfurization is considered to be an auspicious one for its profound desulfurization technology so it can be accomplished under moderate situations [14]:

Otsuki et (2000) has used a blend of H2O2 and HCOOH to oxidize a model of diesel (dibenzothiophene derivatives, benzothiophene derivatives, and thiophene derivatives), gas oil (S: 2.17 wt %) Come from vacuum process and light gas oil come from atmospheric process (S: 1.35 wt %), he also has tested the relationship between the electron density of the sulfur atoms and the reactivity of the different sulfur-containing compounds and thiophene are usually oxidized with hardness because of lower electron densities while BT as well as other sulfur compounds that have higher electron densities are oxidized to erect their conformable sulfones [12]. Flores et. (2004). Have studied oxidative desulfurization to heavy fuel oil and diesel using Fenton-like catalysts and hydrogen peroxide followed by extraction with acetonitrile [13]. Oxidative desulfurization of a commercial diesel fuel has been studied by Yu et al. (2005). (The content of sulfur, 800 wt. ppm) using hydrogen peroxide, activated carbon, and formic acid system [14]. Oxidative desulfurization has been used by Reda (2006) and that 35 kHz ultrasonic irradiation supports this process to lower the sulfur proportion of the crude oil of Baghdad. The empirical results show the highest rate of the elimination efficiency is was about 76% at 60 oC, 80W/cm2 sonication energy density that is applicated for 12 minutes [15].

Chan et. (2008). expound potassium superoxide as a substitutional oxidizing agent for the desulfurization process. Sulfur removal has reached the high is percent that is about 99% For epitome compounds of benzothiophene, dibenzothiophene, as well as a number of selected diesel oil patterns [16]. A study about the process of Sulfur removal of Azeri crude oil has accomplished by Lin et al (2010) using electrical field. This process is achieved in the electrician desalination unit. The results revealed that density and viscidity of crude oil are to be decreased with the extraction of sulfur content whereas some other basic qualities of crude oil tend to be less affected by the process of desulfurization [17]. Hammad et al. (2012) deals with the feed stream of high sulfur content in crude oil by combining solvent plucking out, in addition to hydrotreating and presented a processed crude oil product that is substantially reduced from sulfur content and without any considerable volume waste [18]. Wang et al. (2013) inspected the process of synchronous elimination of sulfur and demetallization of crude oil by electrician desalination based on the corrosion of alkyl thiophene and nickel compounds with the component of Chitosan Schiff under the status of microwave irradiation. By the use of suitable conditions of the microwave irradiation, reaching up to 56 % of sulfur and 82 % of nickel, the removal proportions are accomplished for ideal compounds in crude oil patterns [19].

In the recent study, the main target of the research is to consider the desulfurization process of crude oil by the ultrasound and AC assisted oxidative desulfurization of crude oil.

2. Experimental Part:

2.1. Material and Instruments

Ten specimens of crude oil (100 ml of each sample) have been provided from al-Najaf refinery and have been mixed with activated carbon, hydrogen peroxide (H_2O_2) and acetic acid to shape ten mixtures as listed in Table (1)

IOP Conf. Series: Materials Science and Engineering **579** (2019) 012012 doi:10.1088/1757-899X/579/1/012012

Table1. The used samples in this project

Sample	H_2O_2 content (50%)	Acetic acid (ml)	AC (g)	Crude oil (ml)
1	40	30	3	100
2	40	30	7	100
3	40	30	9	100
4	40	30	5	100
5	40	50	5	100
6	40	40	5	100
7	40	20	5	100
8	30	20	5	100
9	50	20	5	100
10	60	20	5	100

Ultrasound Device from Hans-Ulrich Petermann GmbH & Co. KG, German with a power of 400w and frequency of 24 kHz has been used. The used crude oil forms originally comprised of about 1.95 wt. % sulfur content as shown in Table (2).

Table 2. Crude oil sample properties.

Item	Measurement	Measurement unit	Test Result
1	Water Cut	%	Trace
2	Salt	PPM	13.27
3	Sulfur	%	1.95
4	API@60F	--	27.61
5	Density@15C	Gm/cm3	0.8941

2.2 Procedure

All the samples (10 once) are entered into an ultrasound device (2 hrs for each sample) and then to be exposed to ultrasound waves to detach the sulfur from the crude oil. After the ultrasound, the shaker device is to be used to mix the components (2 hrs for each sample at a temperature of 50 C) to get excellent mixing state. By using Separation pipe, the crude oil is separated from the hydro solution, add (200 ml) of filtered water(for each sample) to the secession funnel and shake the funnel. Then we separate the aquatic solution again (washing process), by adding (100 ml) of Methanol (for each sample) to the funnel and shake it, after that, the second separation process stats (second washing process). Finally, a filtration paper is used the residual sample has lixiviate after them separation process (for each sample), the specimen has been entered to a sulfur analyzer device and the left sulfur content is measured after the removal process.

3. Results and Discussion

3.1. Effect of activated carbon changing

Through the present study four samples have been used of crude oil which contain steady amount of H_2O_2 and acetic acid contents with a change in activated carbon content. Thus, the results according to this study reveal that the sulfur content in crude oil samples will be decrease if the activated carbon content has increase. this is because of the growing surface area, therefore offering more dynamic spots for the bound of sulfur types. Furthermore, activated carbon stimulates the putrefaction of H_2O_2 so that hydroxyl radicals can be produced which work as strong oxidants like per hydroxyl ions (OH-2) which in turn will oxidize the uncompromising sulfur compounds with much ease into sulfones [20]. The outcomes of this study are shown in Table (2) and figure (1).

Table 3. The result of changing A.C content in samples

Original sulfur content (Weight)	H2O2 (ml)	Acetic acid (ml)	AC (g)	Remaining sulfur content	Desulfurization %
	40	30	9	0.3642	**81.325**
	40	30	7	0.3974	**79.622**
1.95	40	30	5	0.4387	**77.5**
	40	30	3	0.79279	**59.344**

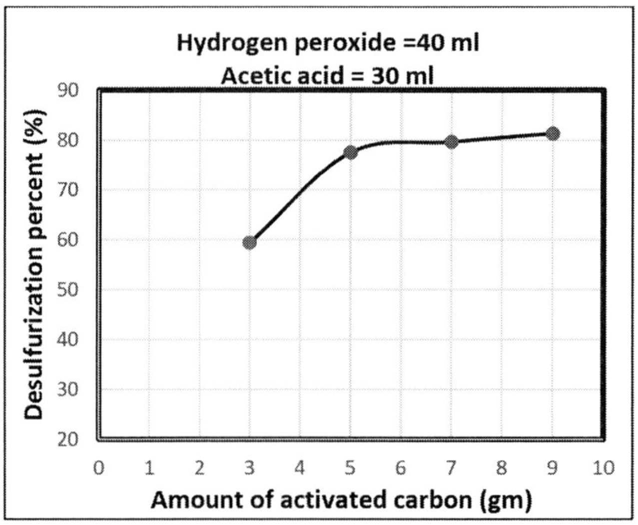

Figure.1. The relation between the percent of desulfurization and AC content in the used samples.

3.2. The effect of hydrogen peroxide

Here three samples of crude oil are used which contain a constant amount of Acetic acid and AC besides changing the amount of H_2O_2. As a result, this study shows the same behavior of AC that the sulfur content is reduced by increasing the used volume of H_2O_2. This is because of the reaction between hydrogen peroxide and hydrogen sulfide as well as Mercaptan and Dialkyl Sulfides in addition to another sulfur compounds that exist in crude oil under acid, neutral conditions to form sulfones or sulfoxides [9].

$H_2S + H_2O_2 \rightarrow S + {}_2H_2O$

From the above reaction, it will be absorbed by AC that presents in the mixture. The results of this study are shown in Table (4) and figure (2).

Table 4. The result of changing hydrogen peroxide in the used sample

Original sulfur content (Weight)	H_2O_2 (ml)	Acetic acid (ml)	AC (g)	Remaining sulfur content	Desulfurization %
	30	20	5	0.6978	**64.215**
	40	20	5	0.630	**67.684**
1.95	50	20	5	0.5854	**69.978**
	60	20	5	0.5692	**70.806**

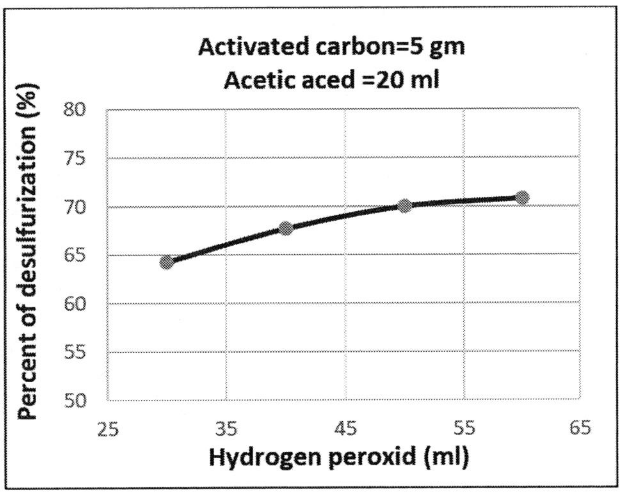

Figure 2. The relation between the percent of desulfurization and H_2O_2 content in the used samples.

3.3. Effect of acetic acid

In The last three samples through this study there are H_2O_2 and AC amount which is fixed whereas acetic acid volume is changed. Increasing the amount of acetic acid in the mixture leads to a rising in the percent of desulfurization because the acetic acid 1s found to have better oxidation-desulfurization effect than other acids, Also a large amount of acetic acid leads to increase in peracetic acid (CH_3COOOH) amount that results from the reaction between acetic acid and (H_2O_2) [20]. The (CH_3COOOH) will react with a large number of sulfides and will oxidize it into sulfones or sulfoxides. The results of this study are shown in Table (5) and figure (3).

Table 5. The result of changing acetic acid content in the used sample

Original sulfur content (Weight)	H_2O_2 (ml)	Acetic acid (ml)	AC (g)	Remaining sulfur content	Desulfurization %
1.95	40	20	5	0.6301	**67.684**
	40	30	5	0.43875	**77.5**
	40	40	5	0.40540	**79.21**
	40	50	5	0.3871	**80.15**

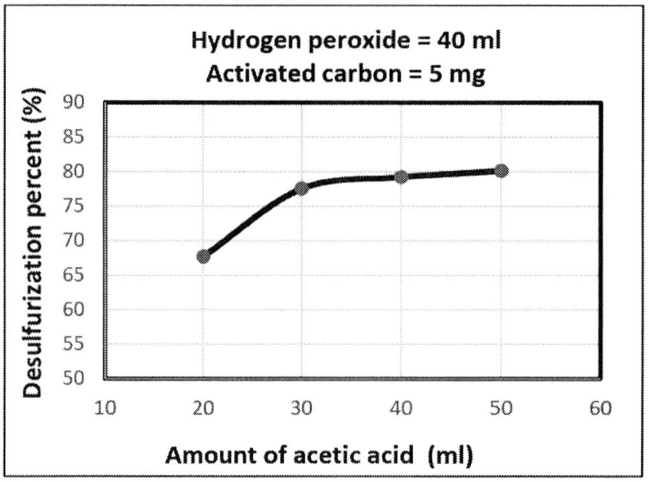

Figure 3. The relation between the percent of desulfurization and H_2O_2 content in the used samples.

4. Conclusions

Depending on this experiment and the results of sulfur removal of crude oil by ultra-sound assisted oxidative desulfurization method, it can be concluded that:

Ultrasonic irradiation enhanced desulfurization reaction and this leads to better desulfurization efficiency.

The highest Desulfurization efficiency obtained is about 81 % of activated carbon amount of (9 gm) with (H_2O_2/acetic acid) (40/30) ml.

The desulfurization efficiency has increased by increasing the used amount of (AC).

The desulfurization efficiency has increased with the increased amount of oxidant acids (H_2O_2 and acetic acid) which has been used in the mixture.

References

[1] Qingtang Y 2008 *China Petroleum Processing and Petrochemical Technology* **1** 1
[2] Shafi R and Hutchings G J 2000 Catalysis today **59** 423
[3] Zhao D, Liu R, Wang J, and Liu B 2008 *Energy Fuels* **22** 1100
[4] USEPA. 2019 Washington, D.C. United States Environmental Protection Agency
[5] Speight J G 2014 *The Chemistry and Technology of Petroleum* (Florida: CRC Press) ed 5 p 953
[6] Wauquier J P 1995 *Petroleum Refining: Crude Oil, Petroleum Products* (Paris: Edition Technip) ed 1 p 470
[7] Eccleston C, Williams A C and Morley S 2009 *Cochrane Database Syst Rev.* 1465
[8] United States Environmental Protection Agency Hydrogen Sulfide and Sulfur Bacteria in Well Water
[9] Nehlsen J P 2006 *Developing Clean Fuels Novel Techniques for Desulfurization* Ph.D. Thesis Princeton University Princeton NJ USA
[10] Shiraishi Y, Hirai T and Komasawa I 1998 *Ind. Eng. Chem. Res* **37** 203
[11] Gray J M M 2010 *Fundamentals of oil sands Upgrading Custom Courseware* Calgary Alberta Department of Chemical & Materials Engineering
[12] Otsuki S, Nonaka T and Takashima N 2000 *Energy Fuels* **14** 1232
[13] Flores *et al.* 2008 *Journal Petroleum Science and Technology*
[14] Bailey A E and Shahidi F 2005 *Bailey's Industrial Oil and Fat Products* (Hoboken: John Wiley & Sons) ed 6 p 3616
[15] Reda A N 2006 *Desulfurization of Crude Oil Using Ultrasound Technique* MSc Thesis University of Baghdad Iraq
[16] Chan Y, Yoon J, Wu J T, Kim H J, Pan K T and Yim J Chien 2008 *Potassium Superoxide as a Substitutional Oxidizing agent for the Desulfurization process*
[17] Liu L, Lu H, Meng F, Wang N and He Y *2011 China Petroleum Processing and Petrochemical Technology* 59
[18] Wang G Zhang J and Liu Y *Korean Journal of Chemical Engineering* **30** 1559
[19] Hosseini H and Hamidi A 2014 *Int. Conf. on Biological Civil and Environmental Engineering* Dubai UAE/ International Institute of Chemical, Biological & Environmental Engineering IICBE) p 10
[20] Afzlinia A, Mirzaie A, Nikseresht A and Musabeygi Tahereh 2017 *Ultrasonics Sonochemistry* **34** 713

Environmental pollution control emitted from diesel engines by Al_2O_3 nanoparticles.

Adel Sharif Hamadi[1], Ali Ghufran Khidhir[2]

[1]Petroluem Technology Department, University of Technology, Baghdad, Iraq

[2]North Gas Company, Ministry of Oil, Kirkuk, Iraq

Abstract. One of the most important problems for diesel engines is the emissions of nitrogen oxides (NOx) and smoke, the use of emulsification technology is one of the most influential ways to control this emissions. In this present work Aluminum oxide (Al_2O_3) are blended by rotary-stator homogenizer at three different levels (50,100,150 ppm) with prepared Nano-emulsion at optimum composition and circumstances that are analyzed experimentally by Design Expert software. Alumina nanoparticle with its high thermal conductivity (30^{-1} Wm K^{-1} compare to diesel and high calorific value which helps to promote more internal combustion due to making higher thermal efficiency. Direct injection (DI), water cooled four cylinders, in-line, natural aspirated Fiat diesel engine was used and run at a constant speed (1500 rpm) and constant fuel injection pressure (400 bars) with varying the operation load. Multigas analyzer model 4880 was used to measure the concentration of the emission gases such as NOx, unburned total hydrocarbon HC, CO_2 and CO. The engine exhaust smoke emissions were measured using the AVL-415 smoke meter. Based on these observations, the rate of CO and NOx would be reduced significantly up to 24 and 5.8 %, at 150ppm to net diesel respectively, noting that the net smoke of diesel would undergo the highest change (up to 26.5 %). The results also indicate a 6.97 % fuel consumption reduction accompanied with 6.2 % improvement in the thermal efficiency, utilizing Al_2O_3 nanoparticles in blended in Nano-emulsion. Enhancement in break thermal efficiency (BTE), reduction in (NOx) and smoke are noticed after mixing Al_2O_3 into prepared nano-emulsion.

Keywords: Nanoemulsion, Aluminum oxide , Diesel engine, nanoparticle, emissions.

IOP Conf. Series: Materials Science and Engineering **579** (2019) 012013 doi:10.1088/1757-899X/579/1/012013

1. Introduction

In last past years scientists and engineers have applied nanotechnologies to human lives in a wide variety of subjects such as biomedical, material, computer, and fuel engineering fields[1]. Diesel fuel (CnHm) is a stable organic molecule. Its electronic structure, its lack of polarity and in the absence of any functional group makes it difficult to decompose into its constituent parts [2]. Many researchers found that nanoparticles can be used as an additive along with the conventional fossil fuel to enhance the combustion and reduces the emission significantly [3].

Adding metallic powder to ordinary fossil fuel usually increases the combustion efficiency and improves combustion stability[4]. adding 0.5 % aluminum nanopowder to a rocket's solid fuel can improve combustion efficiency by 10 to 25 % and increase combustion speed by an order of magnitude [1].

Aluminum nanopowder has a very high activity and can react with water at temperatures from 400 to 660°C to generate hydrogen and improve fuel combustion [5].

Some studies show that the hydrogen is a high quality fuel that burns rapidly and cleanly [6]. Accordingly, hydrogen can be obtained in the proposed manner from water to burn cleanly and efficiently in diesel engines [1].

Research on the topic of water/diesel mixtures in compression ignition engines is aimed at fuel conservation and reduction of undesirable emissions and has been presented in several papers [7–8].This was explained by the presence of nanoparticles in the emulsified fuel which made the hydrogen atom present in the water molecule to be broken easily which helped in improving the combustion process [3]. Hydrogen is clearly a promising alternative to hydrocarbon fossil fuels since it has higher energy efficiencies and is associated with lower emissions [2].The reduction in smoke emission was explained by the micro explosion of the emulsified fuel[9,10].

Aluminum nanopowder can react with water at high temperature and generate hydrogen, promoting the combustion of the fuel [11, 12]. Because the aluminum is nanometer size, it has more surface area and higher activity to decompose the hydrogen from water and increase combustion heat.[23]

BTE was enhanced and CO, HC emissions reduced by adding aluminum, iron, and boron nanoparticles with diesel fuel [13]. This enhancement is due to complete combustion by rapid oxidation, decreased in ignition delay and burning period. Enhancement in surface area to volume ratio and catalytic activity is increased by Nano-additive. [14].

Catalytic effects of high energy density nanoparticles have benefits to improve fuel oxidation and shorten the ignition delay time by catalytic effects. The ignition chance for the fuel blends that having nanoparticles was considerably more than that of pure diesel [15].

Peak cycle temperature is reduced by presence of water in emulsified fuel which had high latent heat of vaporization causing in reduction of NOx emission [16]. The carbon monoxide emissions were also found to be lower with water – diesel - nano fluid emulsion as compared to conventional diesel at all power outs[17]. This trend was explained by the presence of nano particles in the emulsified fuel which helped in breaking the water molecule to separate the oxygen atom which resulted in complete oxidation of the carbon atoms in the fuel. It is concluded that water diesel emulsion can be used as a promising fuel for simultaneous reduction of smoke and NOx emissions in diesel engines. In addition the hydrocarbon and carbon monoxide reduction with improved engine performance can be achieved by adding nanofluids along with the emulsified fuel.

2. Materials and methods

First, for preparation of suitable Nano-emulsion with optimum circumstances, response surface method (RSM) is used for analyzing the factors effecting on prepared fuel according to water droplet size in Nano emulsion.

Pure diesel fuel provided by (Doura Refinery) had (320 ppm wt) Total Nitrogen content. Two surfactants are blended to obtain certain required HLB for the optimum Nanoemulsion composition condition. Low HLB surfactant 4.3 had Mwt 428.61 with 0.99 density (g/ml) at 20 °C is lipophilic brown viscous, while high HLB emulsifier 15 had Mwt 1310 and 1.08 density (g/mol) at 20°C ,both surfactants are provided by CDH group compay.

After preparation the surfactant with HLB 6 it was added (5 wt %) to net diesel fuel, by rotor-stator homogenizer (Heidolph homogenizer DIAX 900 homogenizer) it is dispersed inside the diesel for 30 min at 1500 rpm. Water with10.8 pH will add with 12wt% form the total weight of solution and stirring for 30 min more at the same speed.

(Al_2O_3) nanoparticles with purity 99% had a diameter around 80nm and specific surface area >15 (Supplied by Hongwu international Group Ltd. company, China). (Al_2O_3) nanoparticles was added directly to a prepared nanoemulsion at different levels (50 ,100,150 ppm)and blended for extra 30 min more for achieving complete distribution in all part on the fuel equally before ignition inside the engine.

The prepared fuel from obvious step will examine in diesel engine type 4cycle; 4 strokes at compression ratio 17.The engine are cooled by water because of with direct injection of fuel into the compression vessel which is directly faced high pressure and temperature by piston lead to explosive of Nano size of water droplets. Hydraulic dynamometer was coupled to the engine through which the load was applied by increasing the torque.

Automotive emission analyzer (Air rex-HG-550) by EGMA Company is used for measure the concentration of HC, CO, NOx and CO_2 for all samples at different loads. All sample are implemented twice to ensure the results and its repeatability, the tests were repeated twice.

AVL – 415 smoke meters is used for the engine exhaust smoke emissions. Its measurement principle are depend on filter paper blackening with the measurement range 0 to 10 FSN and detection limit from 0.002 FSN. The apparatus taking sample 10 sec intake time with resolution from 0.01 mg/m^3.

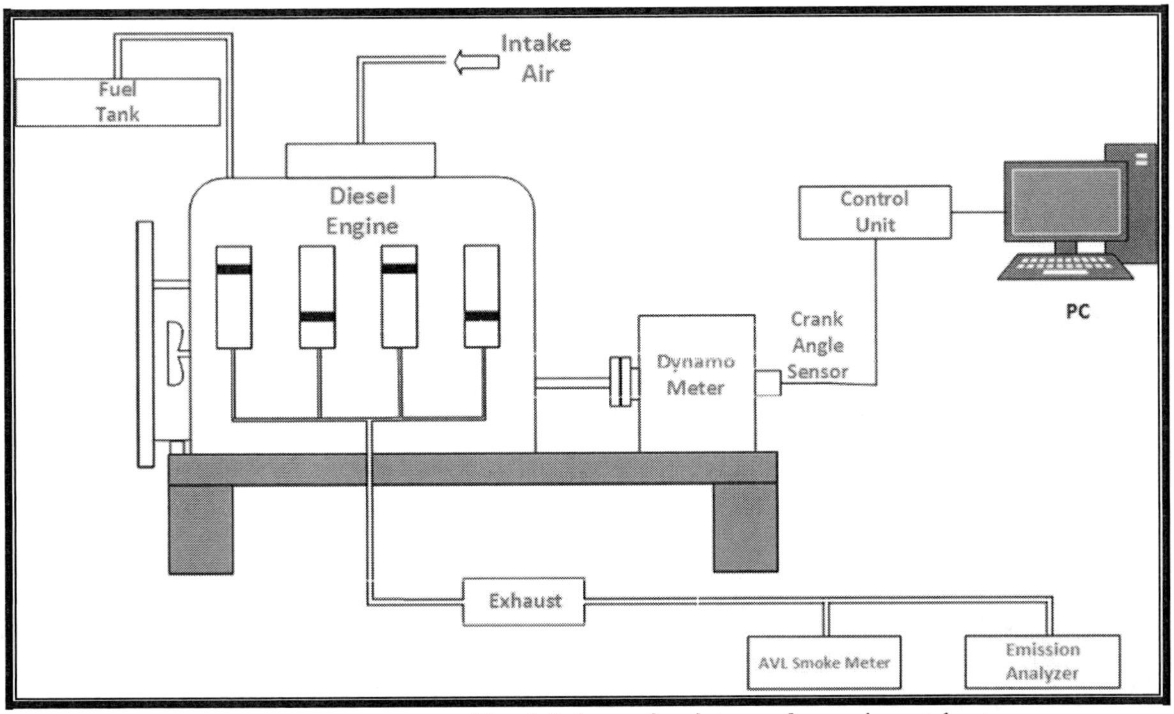

Fig. 1 Schematic picture of experimental set-up

3. RESULTS AND DISCUSSION

Performance Characteristics

For analyzing BSFC experiments diesel engine was conducted with pure diesel, WiDNE and WiDNE+AL_2O_3 in three different levels, (50,100,150 ppm) at constant compression ratio 17.The load of the diesel engine was varied from 5 to 15 kg with 2.5 kg step size at constant speed 1500 rpm during the of each run of experiments.

From figure (2) it is noticed that the BSFC decreased when the load was increased for all processes of diesel and Nanoemulsions blends due to the consumption of more fuel [7]. It is observed that WiDNE shows higher BSFC as compared to conventional diesel and nanoparticle blended water in diesel Nanoemulsions. This can be due to the higher viscosity and lower heating value of water diesel emulsions as compared with diesel [8, 9] .It was also observed that BSFC decreases when nanoparticle volume fraction in the Nanoemulsions blend was increased for any given load. This reduction in fuel consumption can be due to increase in calorific value with the addition of the nanoparticle [10, 11].

Potential advantage of the water in diesel Nanoemulsions fuels is the integrated influence of secondary atomization and microexplosion phenomenon, resulting in strong vaporization and mixing better of air, turns in improving the combustion reactions inside the engine cylinders. Influence of Al_2O_3 on BSFC was investigated and it is illustrated through figure 2.There was an apparent effect on reduction of BSFC with increasing of Al_2O_3concentration in WiDNE. With increasing of $Al_2O_3$from 50 ppm to 150 ppm, BSFC reduced. This is due to the shortening of ignition delay features of nanoparticle by virtue of improved surface area to volume ratio. Al_2O_3 nanoparticles with 80 nm with specific surface area > 15 m^2/g, makes it require more energy required for atomization.

From figure (3) it was perceived that (BTE) increased when the load was increased for all operations of diesel and nanoemulsion blends as obvious in below figure 3. It was observed that the conventional diesel shows lower Brake thermal efficiency as compared to WiDNE and nanoparticle blended water in diesel Nanoemulsions [20, 21].Table (1) show the effect of water for preparation of nanoemulsion with the pure diesel, in spite of addition of water contributes in reduction of emissions but it is also cause reduces in fuel properties as increase flash point and viscosity.

Table 1: Comparison of diesel and Nanoemulsion

Properties (measured)	**Diesel**	**Nanoemilsion**
1. Calorific value (kj/kg)	44800	38850
2. Flash point (°C)	54	61
3. Viscosity at 40 °C (cSt)	3.268	4.56
4. Pour point (°C)	+9	+7
5. Density (g/cm³)	0.87	0.882

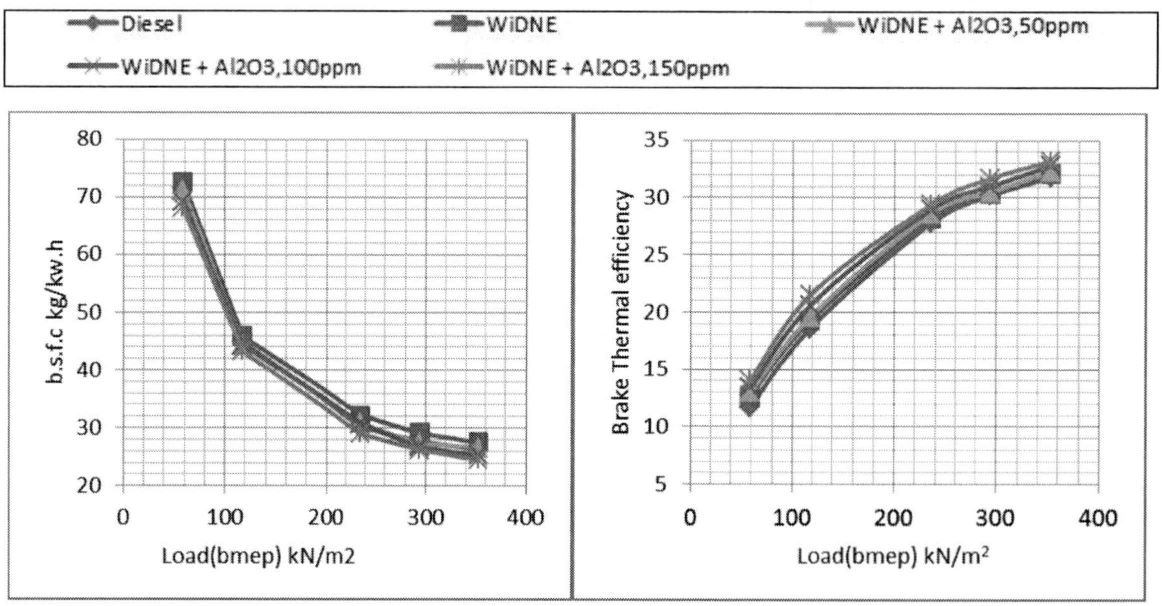

Figure 2: Variation of BSFC with Load (bmep) kN/m^2 Figure 3: Variation of BTE with Load (bmep) kN/m^2

As concentration of (Al_2O_3) was increased from 50 ppm to 150 ppm, the thermal efficiency increased from 32.3 to 33.2 at higher load.This is due to the Al_2O_3 nanoparticles encapsulation in water molecule, which offers the secondary atomization directly after the primary micro explosion phenomenon taking place due to the Nanoemulsion.

Exhaust Emission Analysis

It was noticed that the NOx emission increases when the load was increasing with the all operations of diesel and Nanoemulsion blends as shown in below figure 4. It is observed that WiDNE reduces NOx emission as compared to conventional diesel. Water molecules presents in emulsified fuel possess very high latent heat of vaporization, which tends to reduce the peak cycle temperature, hence oxides of nitrogen emission of water diesel nanoemulsion were found on the lower side as compared to diesel fuel for all the power output.It is found that the water in diesel emulsion with 15% water content produces the least amount of NOx emission as compared to other fuels [22]. The variation of NOx emission was observed with Al_2O_3 nanoparticle blended with nanoemulsion in their respective concentration, as in figure 4. As concentration was increased from 50 ppm to 150 ppm the NOx emission barely changes could be noticed. This is due to the Al_2O_3 nanoparticles size 80 nm.

Diesel fuel showed the higher magnitude of Smoke opacity over the entire load range. It was noticed that the Smoke opacity increased when the load was increased for all operations of diesel and Nanoemulsion mixers, as shown in below figure 5. It is observed that WiDNE reduces Smoke opacity as compared to conventional diesel. Micro explosions of emulsified fuel result in secondary atomization of injected fuel which ends up with reduced smoke intensity as compared to diesel fuel. The presence of nanoparticle in the emulsion tends to separate the oxygen atom in the water molecule which rises up the availability of oxygen. Excess oxygen present in the combustion chamber tends to oxidize the partially burned hydrocarbon; thereby reduced smoke intensity is witnessed for water diesel nanoemulsion as compared to diesel for all power outputs. It was also observed that Smoke opacity reduced when nanoparticle volume fraction in the nanoemulsion blend was increased. The enhancement of Smoke opacity was observed with Al_2O_3 nanoparticle blended with nanoemulsion in their respective concentration, as in figure 6. As concentration was increased from 50 ppm to 150 ppm, Smoke opacity decreased from 5 to 4.7 at higher load.

Carbon dioxides are coming as exhaust resulting from combustion of CO inside cylinder and complete combustion of carbon particles of the fuel. Carbon monoxide is extremely depending upon the air-to-fuel ratio comparing with stoichiometric ratio. In most cases, CI engine functions with lean mixture and thus results in lower CO emission. It was noticed that the CO emission decreased when the load was increased for all operations of diesel and Nanoemulsion blends, as shown in below figure 6.

IOP Conf. Series: Materials Science and Engineering **579** (2019) 012013 doi:10.1088/1757-899X/579/1/012013

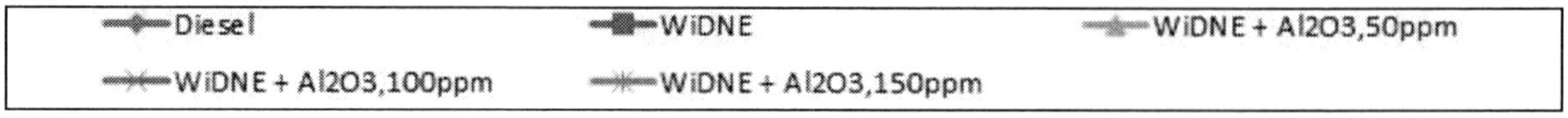

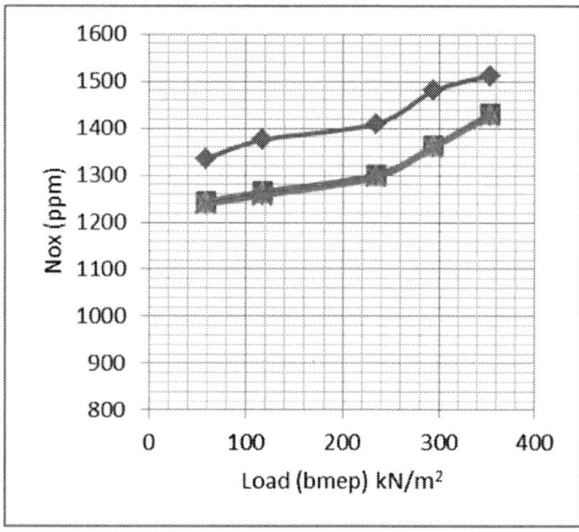

Figure 4: Variation of NOx to Load (bmep) kN/m^2

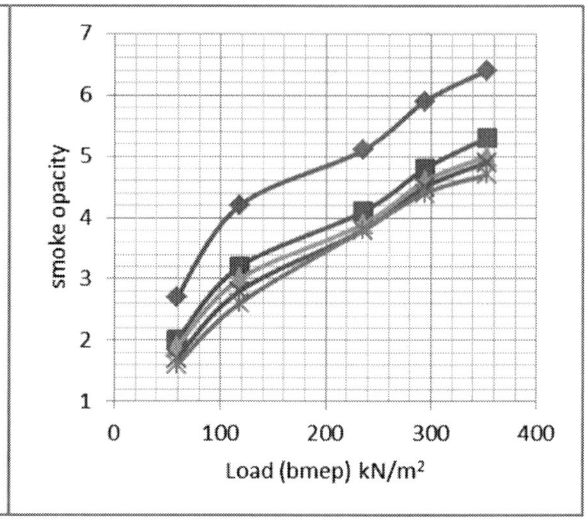

Figure 5: Variation of Smoke to Load (bmep) kN/m2

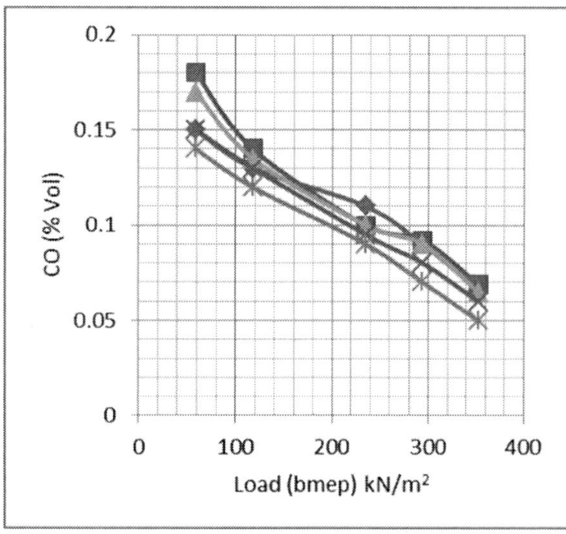

Figure 6: Variation of CO to Load (bmep) kN/m^2

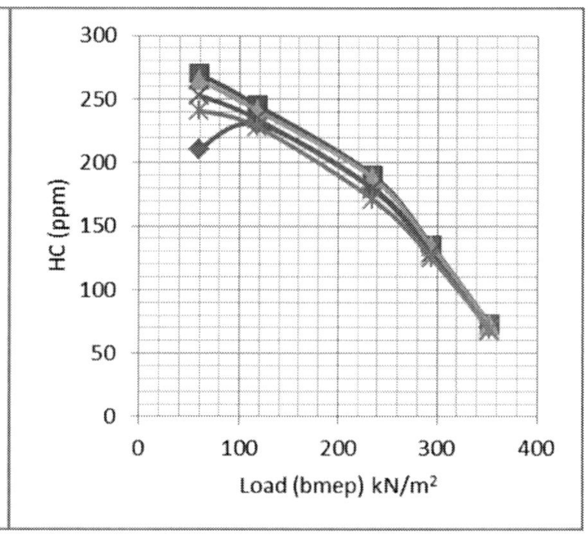

Figure 7: Variation of HC to Load (bmep) kN/m^2

WiDNE fuel showed the higher magnitude of CO emission over the entire load range. It was also observed that the CO emission decreased when nanoparticle volume fraction in the Nanoemulsions blend was decreased for any given load. This reduction in CO emission because of nanoparticles will easily separate the oxygen atom from the water molecule which ensures complete oxidation of fuel droplets. In addition to it better air and fuel mixing rate was achieved inside the combustion chamber due to finer atomization of emulsified fuel droplets, which ends up with complete combustion. There are some fluctuations in curves with increasing the load, in 250 kN/m^2 CO is less than net diesel, in continued increasing load, the CO emission again backs to rise. The variation of CO emission was observed with

Al_2O_3 nanoparticle blended with nanoemulsion in their respective concentration, as in figure6. As concentration was increased from 50 ppm to 150 ppm, CO emission reduced from 0.065 to 0.05. This is due to the Al_2O_3 nanoparticles association in atomization process.

HC emissions are producing due to the incompletion combustion and they tend to decline with load increasing. The total HC emissions are commonly made as a consequence of flame quenching. The formation of higher HC and CO emissions is extremely relied on the viscosity of fuel. More viscosity is also leading to longer spray penetration rate. The formation of higher HC emissions is due to the wetting cylinder of engine by incomplete combustion .It was observed that the HC emission decreased when the load was increased for diesel and Nanoemulsions blends as shown in figure 8 due to the increase of combustion temperature. It is observed that WiDNE showed higher HC emission as compared to conventional diesel and nanoparticle blended water in diesel Nanoemulsions.

It was also noticed that the HC emission decreased as nanoparticle volume fraction in the Nanoemulsions blend was increased for all loads. At peak load condition, the HC emission of water diesel Nanoemulsions was quite higher than diesel fuel.

From figure 7 shows that at low load situation, the HC emission of diesel was less than WiDNE. With increasing the load, the HC emission will reduce with increasing the load. At high load the emissions by WiDNE will be less.

When using diesel fuel, there is a small increase of HC emission with increasing the load, but the curve will back down to become less than the net diesel in HC emission.

The improvement of HC emission was observed with Al_2O_3 nanoparticle. As concentration was increased from 50 ppm to 150 ppm, the HC emission decreased from 72 ppm to 69 ppm at higher load.

4. CONCLUSIONS

According to the obtained results of experiments from the diagram, it was observed decrease in brake specific fuel consumption with the addition of Al_2O_3nanoparticles. Brake thermal efficiency improved in blending of 150 ppm Al_2O_3 to nanoemulsion in comparison with other fuels. Addition of nanoparticles improve combustion process due to fuel mixing with air which Contribution of nanoparticle in reduction of NOx ,HC,CO and smoke emission.

References

[1] Mu-Jung K, Chen-Ching T, Bai-Fu L, and Tsing-Tshih T 2007 *Journal of Testing and Evaluation* **36** 186

[2] Grégoire-Padró C E and Lau F 2000 *Advances in Hydrogen Energy* (New York: Kluwer Academic/Plenum Publishers) ed 1 p 192

[3] Surendra S and Haritha M 2017 *Int. J. of Sci. Deve. and Res.* **2** 22

[4] Beloni E, Hoffmann V and Dreizin E L 2008 *J. of propulsion and power.* **24** 1403

[5] Barnaud F, Schmelzle P and Schulz P 2000 *Society of Automotive Engineerers.* **1** 95

[6] Saetre T 1997 *Conf. on the HYPOTHESIS II Symposium* (Boston: Kluwer Academic Publishers) p18

[7] Abu-Zaid M 2004 *Energy Convers. Manage* **45** 697

[8] Harbach J and Agosta V 1991 *J. Ship Res.* **35** 356

[9] Nitish K 2012 *International Journal of Mechanical Engineering* **2** 33

[10] Yang W M H, Chou S K, Chua K J and Mohan B Sivasankaralingam V Raman V Maghbouli A and Li J 2013 *Applied Energy* **112** 1206

[11] Harish V and Venkataramanan M 2017 *Journal of Mechanical Science and Technology* **31** 409

[12] Mench M M Yeh C L and Kuo K K 1998 *Proceedings of the 29th International Annual Conference of ICT* (Karlsruhe: Federal Republic of Germany) p 30

[13] Rakhi N Mousumi C Parimal A 2014 *Fuel.* **120** 91

[14] Basha J Anand R 2014 *Alexandria Engineering Journal* **53** 259

[15] Beloni, E Hoffmann V K and Dreizin E L 2008 *Journal of propulsion and power* **24** 1403

[16] Sasi K and Senthil k 2015 *J. of Chemical and Pharmaceutical Sciences* 37

[17] Bidita B, Suraya A, Shazed M, Salleh M and Idris A 2014 *Energy Fuel* **28** 4149

[18] Nithin S and Muhammed S 2015 *International J. of science and research* **4** 673

[19] Arul V, Anand R and Udayakumar M 2009 *ARPN J. of engineering and applied science* **4** 1

[20] Singh N 2012 *International J. of Mechanical Engineering* **2** 33

[21] Yang W An H Chou K Chua J Mohan B Sivasankaralingam V Raman V Maghbouli A and Li J 2013 *Applied Energy* **112** 1206

[22] Lin C Y and Li W 2008 *Fuel* **87** 2154

[23] Soukht H, Jafarmadar Taghavifar H and Ashrafi S 2015 *Int. J. Environ. Sci. Technol.* **12** 2245

IOP Conf. Series: Materials Science and Engineering **579** (2019) 012013 doi:10.1088/1757-899X/579/1/012013

Abbreviations

Symbols	Defenition
APS	Average particle size
bmep	Brake mean effective pressure
BP	Brake power
BSFC	Brake specific fuel consumption
BTE	Break thermal efficiency
CI	Compression ignition
CO	Carbon monoxide
CO_2	Carbon dioxide
FSN	Filter smoke number
HC	Unburned hydrocarbon
HLB	Hydrophilic lipophilic balance
N_2	Nitrogen
NO	Nitrogen oxide
NOx	Nitrogen oxides
PM	Particulate matter
rpm	Round per minute
SSA	Specific surface area
WiDNE	Water in diesel nanoemulsion
Al_2O_3	Zirconium oxide

Specifications of Heavy Diesel Lubricating Oil Improved by MWCNTs and CuO as Nano-additives

Thulfiqar N Jaber[1#], Khalid A Sukkar[1], and Abullhassan A Karamalluh[2]
[1]Chemical Engineering Department/ University of Technology-Baghdad-Iraq
[2]Mechnical Engineering Department/ University of Technology-Baghdad-Iraq

[#]Corresponding author's e-mail address: 080503@student.uotechnology.edu.iq

Abstract. In this investigation, the properties of heavy diesel lubricants were enhanced by adding the nano-additives. The purpose of this research is verifying the enhancement of the rheological and thermal characteristics of the lubricating oils by adding the nano-additives, in order to upgrade to the highest level for performance lubricant in the system. The SAE 15W40 (Oil parent) was mixed with two types of nanomaterials (MWCNTs and CuO) at four various concentrations by (0.1, 0.2, 0.5, and 1 % wt.) to prepare the nano-lubricants, studying and comparing their properties with parent lubricants. The results indicated that the thermal conductivity values of (Oil_{MWCNTs}) and ($Oil_{CuO\ NPs}$) at 1 wt. % concentration of both additives increased by 23%, 15.3% respectively. Also, the flash point of both nano-lubricants comparison with parent oil increased by (19.54%, 15.9%) respectively. It was concluded that (Oil_{MWCNTs}) showed higher lubricant's specification than ($Oil_{CuO\ NPs}$) due to the high surface area of MWCNTs. Hence, the operating properties of lubricating oil much enhanced when scattered the carbon nanotubes and copper oxide NPs.

Keywords: *Heavy Lubricating Oil; MWCNTs and CuO Nano-additives; Thermal & Rheological Properties.*

1. Introduction

Nowadays, petroleum refineries produce many types of lubricating oils with different properties and applications [1]. The main purpose of lubricating oil is to lubricate the moving parts, prevent corrosion, improve performance, and cool the engine. These lubricants must be qualified the requirements engines operation of high performance and long lifetime. Nano-lubricants (or Nanofluids) are the most important application of nanotechnology. Lubricant oils based nanofluids are also called nano-lubricants. Many nanoparticle additives have been investigated in order to improve the tribological performance of the base lubricating oils for diesel engine. Most designers and authors have focused on nanotribology in internal combustion engines as the key strategy for minimizing frictional power losses, excessive heat generation, exhaust emission, combustion, saving fuel, enhance heat transfer rate, reduction of consumed fuel and cost of maintenance [2]. The addition of MWCNTs to lubricating oils can effectively improve the tribological properties through the formation of a protective film on surfaces and creating a rolling effect between friction surfaces [3, 4]. Many authors prepared nano-lubricants containing different nanoparticles and nanomaterials such as BN, TiO_2, SiO_2,

CuO, ZrO_2, Al_2O_3, nano-diamond, and carbon nanotubes [5–8]. In fact, nanoparticles dispersed in base fluids have a significant impact on the rate of increase in the thermal conductivity of nano-fluids, making these liquids highly efficient in heat transfer [9, 10]. For example, carbon nanotubes, due to their high thermal conductivity coefficient, are known as an ideal material for making nanofluids [11, 12]. Hwang et al. studied the heat transfer properties of nanofluids in water base and ethylene glycol base which contain CuO nanoparticles and multi-walled carbon nanotubes [13]. Also, the MWCNTs thermal conductivity nano-lubricants, which contain different nanostructures such as copper and copper oxide nanoparticles with spherical structures and single walled and multi-walled carbon nanotubes with tubular structures, have been studied [13-18]. However, only a few of these published articles have been on other oils' properties including viscosity, flash point and pour point of nano-lubricants from theoretical and experimental approaches [19, 20, 21]. Indeed, it has been proved that carbon nanotubes have a high thermal conductivity, high electrical conductivity and efficient mechanical properties [22, 23, 24]. The effects of base fluid and concentration on the rheological behavior of MWCNTs were investigated by Chen et al. [25]. Thus, by dispersing CNTs into a liquid phase such as engine oil, water, and ethylene glycol, can improve its thermal properties and transport. Choi et al. stated apparently that nanoparticles in lubricating oils can pad the scar and channels of the friction, and wear surfaces [26] as shown in Fig. 1.

It is important to mention here, that the investigated present work focus on the enhanced of the high-grade lubricating oil such as SAE 15W40 engine oil. Therefore, the main aim of the present work is to study the effect of adding MWCNTs and CuO nanoparticles on the properties of diesel engine oils and will look at their potential in terms of their viscosity, stability, flash point, pour point of nano-lubricants, and also thermal conductivity in heat transfer applications.

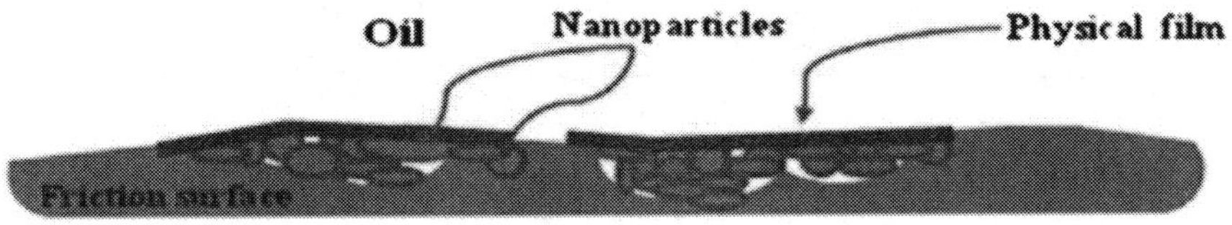

Fig. 1: Mechanism appearing how to reduce friction between oil surfaces when mixed with nanoparticles [26].

2. Experimental work

2.1 Materials

The SAE 15W40 engine oil and diesel fuel were attended from (Al-Dura Refinery, Middle Petroleum Company, Baghdad, Iraq). The SAE 15W40 lubricating oil is used as a parent fluid. **Table 1** shows the specifications of lubricating oil. In the present work the CuO NPs was bought from (Nanjing XFNANO Material Tech Co., Ltd). As for the multi-walled carbon nanotubes (MWCNTs) were manufactured by a new production process. The production process of MWCNTs included the burning of hydrocarbons source fuel in a special combustion chamber unit, under controlled operating parameters of temperature and fuel to oxygen ratio. Tests tribological were conducted on heavy-duty multi-grade diesel engine oils (SAE 15W40). **Table 2** shows main properties of MWCNTs and CuO NPs.

2.2 Measuring and analysing apparatus

A bath ultrasonic (P120 Elmasonic, Germany) was used for homogenizing, dispersing and mixing of CuO and TiO_2 NPs and making so-called nanofluids. The thermal conductivity was measured by KD2-Pro (Decagon Devices, USA) of parent oil and nano-lubricant at different temperatures and concentration. Also, field emission scanning electron microscope (FESEM, Japan, JEOL 7610F), X-ray Diffractmeter (XRD, Japan, Shimadzu-6000), and Fourier transform infrared spectroscopy (FTIR-8400S, Shimadzu) was used for characterizing the properties of copper oxide and titanium oxides.

Table 1: Properties of SAE 15W40 engine oil.

Property	Value
Viscosity at 100 °C (cSt)	14.5
Viscosity at 40 °C (cSt)	109.84
Viscosity index	135
Density at 15 °C (kg/m^3)	904
Pour Point (°C)	−27
Flash Point (°C)	220

Table 2: The main properties of MWCNTs and

Specification	CuO	MWCNTs
Color	black	Black
Purity (%)	>99.9	99.1
surface area (m^2/g)	263	433
Diameter (nm)	25	11.4
Volume density (g/cm^3)	0.20-0.35	0.271
Density (g/cm^3)	6.4	3.9
Crystal form	sphere	Tubes

2.3 Preparation of nano-lubricant

The nano-lubricants were prepared by adding MWCNTs, copper oxide and titanium oxides at different concentrations of 0.1 wt. %, 0.2 wt. %, 0.5 wt. %, and 1 wt. %, to parent lubricating oil (SAE 15W40). The required amount of multi-walled carbon nanotubes was carefully weighed using an accurate electronic balance and mixed with the parent oil. In the nanofluids preparation process, the (Two-step method) was used, due to more common, easier and more economical than the (One-step method) for production nanofluids. Due to the high viscosity of the base fluid, the mixing process was achieved by using a mechanical mixer and bath ultrasonic (type, P120 Elmasonic, Germany) method for dispersing of MWCNTs, another nanoparticles CuO inside the base fluid to obtain high desperation for nano-lubricants. The operating conditions parameters of the bath ultrasonic process are shown in **Table 3**.

Table 3: Specifications of Bath Ultrasonic.

The property	Value
Power	**100%**
Time	**2-3 h**
Temperature	**40 °C**
Frequency	**20 kHz**

2.4 Rheological and Thermal Measurements of Nano-Lubricant

The thermal conductivity of nano-lubricants has been measured, for this purpose, KD2-pro, a laboratory and portable device, was used to increase thermal properties. The device has a probe length of about 60 mm and diameter of 1.3 mm, and the device measures the thermal conductivity based on hotwire technology. The parent oil and Nano-lubricants thermal conductivity that was measured at various concentrations and temperatures. The viscosity of parent oil and nano-lubricants with concentrations from 0.1 wt. % to 1 wt. % concentrations were measured at a temperature of 40 °C and 100 °C. Kinematic viscosity, flash and pour point of the parent oil and nano-lubricants were measured based on ASTM D-445, ASTM D-97, and ASTM D-92 respectively.

3. Results and Discussion

3.1 XRD Pattern and FE-SEM

The XRD pattern in the current investigation indicates the MWCNTs as shown in **Fig. (2. a)**. The result indicated that there were no carbon oxides or other crystalline materials and only pure carbon nanotubes were obtained under applied experimental conditions. The peaks around 26.1° indicated the presence of $(MWCNTs)_1$, the peaks around 42.9°, indicated the existence of oxygen. The (002) peak indicated the graphitic structure of the MWCNTs presented. More over a comparison between the JCPDS card No. (PDF - C - 00-026-1076) with XRD patterns that the same file data is also confirm, the results that obtained from multi-walled carbon nanotubes are in good agreement with many authors [27, 28] who found that MWCNTs observed at 26.1° with high purity. Additionally, from the XRD pattern in **Fig. (3. a)**, it can be found that the CuO nanoparticles in the current investigation have the same major characterization peaks of pure copper oxide card No. (PDF Card - Cu O - 00-048-1548) at 2θ values of 32.2°, 35.4°, 38.6°, 48.5°, 52.9°, 58°, 61.3°, 65.5°, 67.8°, 72.3°, and 74.5°. Otherwise, in the XRD pattern, no other impurity peak was noticed, observing the formation of the single sample phase of CuO nanoparticles.

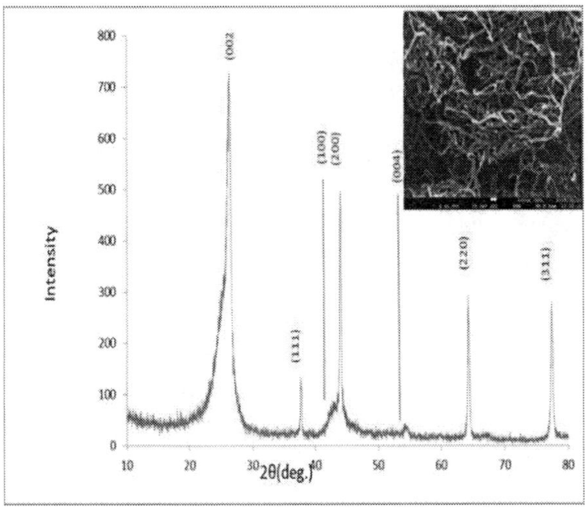

Fig.2: XRD pattern (a) and FE-SEM (b) image of MWCNTs

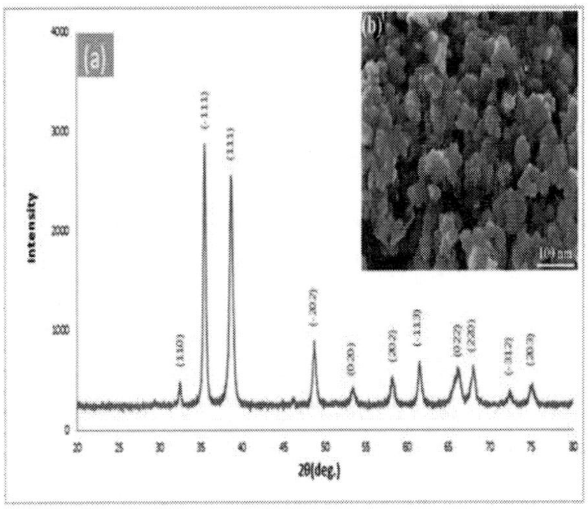

Fig. 3: XRD pattern (a) and FE-SEM (b) image

The FE-SEM pictures in **Fig. (2.b and 3.b)** showed the morphology of the nanoparticles of MWCNTs and CuO respectively. The shape of MWCNTs is generally cylindrical and has helical forms because the inter-molecular force of MWCNTs has various dimensions and sizes are presented in their structure. From the results of FE–SEM, a few impurities were observed on the surface of the tube for the prepared materials which can be removed by the purification process of MWCNTs. The morphology of copper oxide nanoparticles was spherical to some extent, which provided a good rolling mediator inside lubricating oil. The FE-SEM results show that the average particle size of MWCNTs and CuO is about 11.4 and 25 nm respectively.

3.2 Properties of Prepared Nano-lubricants

In fact, the estimating of the suitable concentration is a well important issue to be realized best characteristics [12, 16]. For dispersing MWCNTs and CuO NPs inside the base oil, mechanical methods of bath ultrasonic and magnetic stirrer were used. In the present investigation, it has been observed that samples which prepared by the bath ultrasonic method the NPs are better distribution, not agglomerated, and more uniform dispersed compared with previous samples that prepared from other methods.. All the inside samples were kept in transparent glass containers in a completely stagnant state for about 60 days to assess their own stability conditions as shown in **Fig. 4**.

Fig. 4: The stability condition of nano lubricants after 60 days (a) CuO/oil, (b) MWCNTs/oil, (c) base oil.

During this interval, the stability is recorded for all the samples that checked visually. On the other hand, it has been observed that the dispersion of MWCNTs and CuO nano-additives within the base oil was very high through the period of checking; this gives an impression of the high stability of these materials within the lubricants, because of the high dispersion caused by the bath ultrasonic device.

3.3 Viscosity

Fig. (5 and 6) show that the viscosity amount in total has a rising tendency, with increased concentration of (MWCNTs and CuO) NPs at 40 °C. At higher temperature, the viscosity decreased of all the samples significantly. And also it seems, by increasing the concentration of (MWCNTs and CuO) NPs, for each 40°C and 100°C the viscosity increase.

The viscosity relative to the base lubricant increased by 5.38%, 9.97%, 8.93% for each concentration for MWCNTs. And for copper oxide the viscosity increased by 2.42%, 7.1%, 10.88%, which is linked to the nano-lubricant with concentration 0.2 wt.%, 0.5 wt.%, and 1 wt.% respectively and at 40 °C temperature, especially the concentration at 1 wt.% gave the highest ratio of viscosity improvement. The point interestingly with respect to the viscosity of nano-lubricants for both additives nanomaterials with concentration 0.1% at both temperatures is that the amount of viscosity has a little reduction with respect to the base lubricants, but the viscosity gets it an upward direction with increased concentration of multi-walled carbon nanotubes and copper oxide. Nanoparticles take an intermediate position between the oil layers when added, thus facilitating movement between the oil layers and sliding each other.

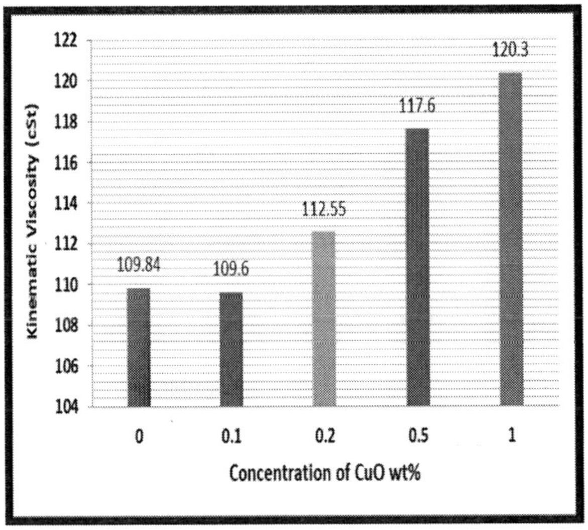

Fig. 5: The kinematic viscosity of nano-lubricants based on MWCNTs at 40 °C and different concentration.

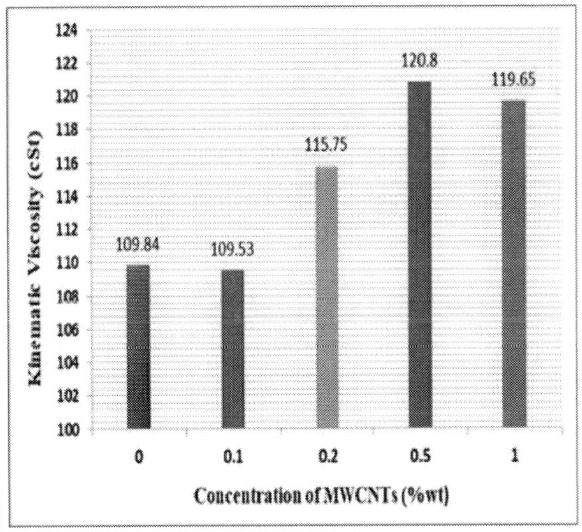

Fig. 6: The kinematic viscosity of nano-lubricants based on CuO at 40 °C and different concentration.

Fig. (7 and 8) show the viscosity of nano-lubricants at 100 °C. In the current work, the lubricating oils viscosity was decreased after adding MWCNTs and CuO NPs at the level by 0.28%, 0.3% at temperature 40 °C at 0.1 wt.%. On the other hand, it can be concluded that the viscosity of the SAE 15W40 base lubricating oil did not undergo any significant changes when adding nanomaterials to a few concentrations. Therefore, the engine oil for the Considered is a function of the concentration of the added nanoparticles.

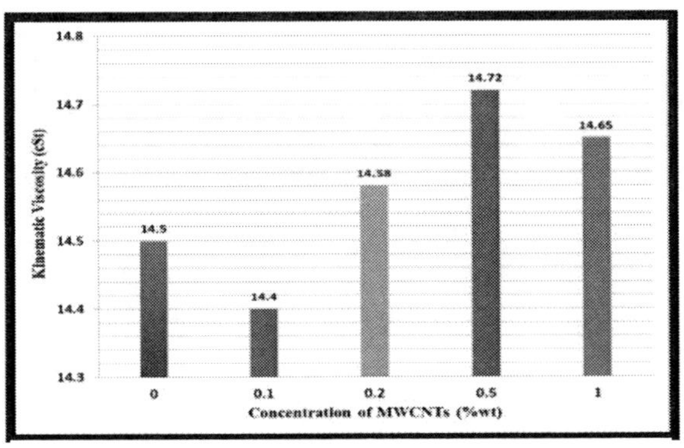

Fig. 7: The kinematic viscosity of nano-lubricant at 100 °C with different concentration of CuO.

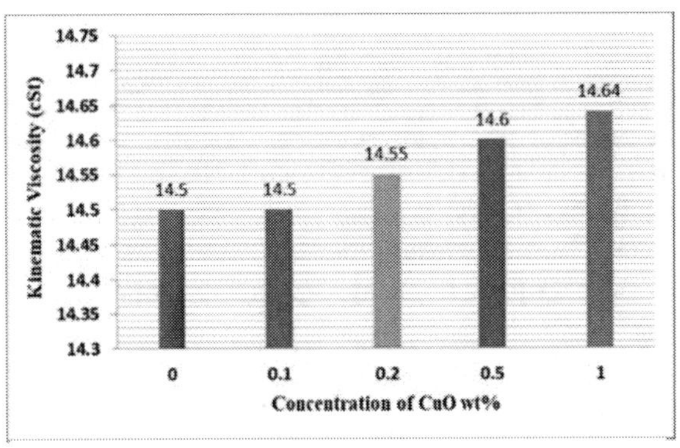

Fig. 8: The kinematic viscosity of nano-lubricant at 100 °C with different concentration of MWCNTs.

3.4 The thermal conductivity measurement

Fig. (9 and 10) show the results for thermal conductivity determined of nano-lubricants after adding nanoparticles to parent oil and in different concentrations and also with different temperatures. It has also been seen, that the results obtained from engine parent oil by the addition of MWCNTs and CuO NPs at specific rates, was the thermal conductivity coefficient of the nano-lubricants much greater than the parent oil without additives.

Based on the results obtained, adding nanoparticles to SAE 15W40 engine parent oil, based on the method which has been done in current work, the amount of thermal conductivity at 0.1 wt%, 0.2 wt%, 0.5 wt%, and 1 wt% concentrations of MWCNTs and CuO NPs had increased 11.4%, 16.76%, 21.48%, and 23.72% respectively with respect to MWCNTs, while CuO had increased 9.85%, 15.2%, 19.73%, and 21.84% respectively. During this research, it can be concluded that MWCNTs has a high impact on the thermal conductivity coefficient of the engine oil compared to copper oxide which have a positive effect of the major, but less than when the addition of MWCNTs to lubricants.

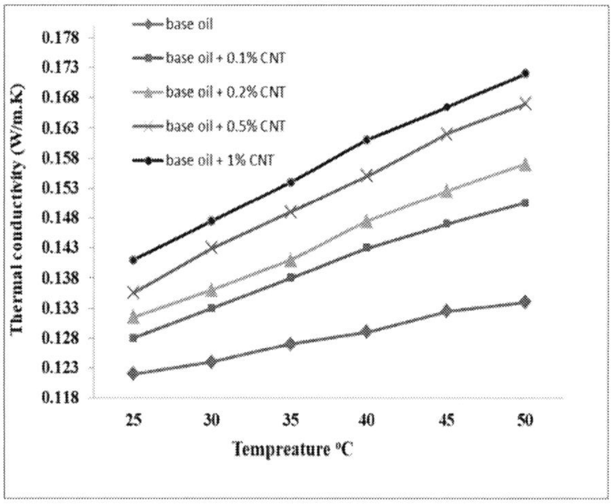

Fig. 9: Thermal conductivity of Nano lubricant (Oil_{MWCNTs}) and base lubricant.

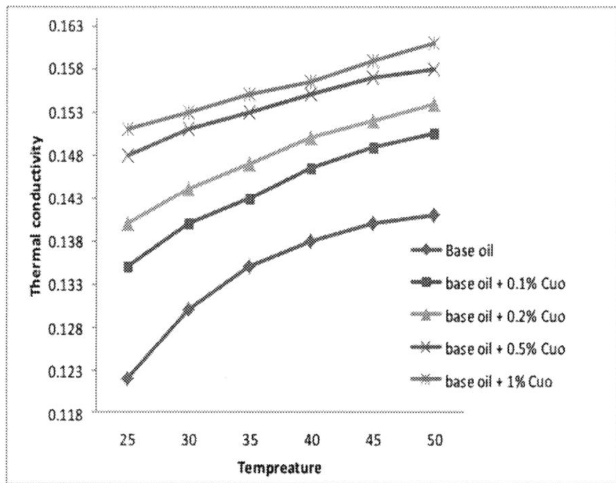

Fig. 10: Thermal conductivity of nanolubricant (Oil_{CuO}) and base lubricant.

As well, it can be observed that, when the temperature increased, the thermal conductivity of the nano-oil increased, this also applies to the concentration of nanoparticles, as the concentration increases, the quantity of thermal conductivity coefficient of the engine oil increased. However, it should be noted that, in the case of different type of basic fluids, the rate of concentration the nanoparticle, kind of nano-additives, the rate of dispersion of nanoparticles within the base fluid, and the method used for dispersion of nanoparticles inside the base fluid. The different parameters resulting from the previous reasons are the effect of the resulting changes in properties.

MWCNTs and copper oxide have greater potential than other nanoparticles to improve thermal conductivity coefficient of engine oils, due to the high thermal conductivity of these materials, especially the MWCNTs. By taking a look at this property, that one of the functions of engine oil reducing the erosion of moving parts, removing some heat generated by friction and combustion between the parts to outside environment, so that helps to cool the engine. Oil works as a cleaner and removes dust and carbon particles between moving parts. Therefore, the using carbon nanotubes structures can be considered as a method to improve the thermal conductivity of diesel engine oils.

3.5 Flash point and Pour point

The flash point is the lowest temperature at which the vapour of the lubricating oil is exposed to the air to ignite; the oil takes the ignition curve in an instant and then turns off quickly in a moment. Therefore, it can be said that the flash point in the lubricating oils of heavy engines is in fact, the main determinant of the maximum temperature limits in the functions of that oil.

Fig. 11 shows the results of the flash point of the nano-lubricant that prepared from MWCNTs and CuO respectively. The increasing of flash point as a function of increasing the concentration of MWCNTs. The high thermal conductivity of carbon nanotubes made the flash point for prepared lubricants significantly higher than base lubricating oil and thus improved remarkably.

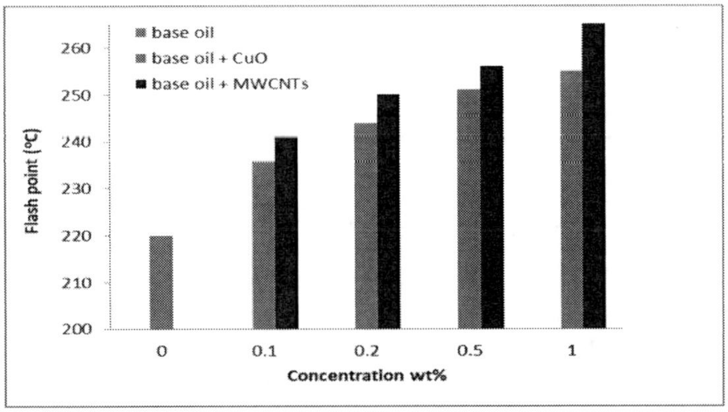

Fig. 11: The flash point values of parent oil and nano-lubricant.

From **Fig. 11**, it has been observed that the flash point of base oil increased by 4.09%, 8.18%, 16.36%, and 24.09% with the value of MWCNTs concentration 0.1 wt%, 0.2 wt%, 0.5 wt%, and 1 wt% respectively. But shows other values of the flashpoint for CuO when the flash point of base oil increased by 3.18%, 6.81%, 11.8%, and 15.45% respectively. It was concluded that the flash point was increased for both types of nano-lubricants concentration, and the highest values of the flash point were 273°C and 254°C for MWCNTs and CuO respectively.

The pour point this property of lubricating oil is appraised as the marginal point of temperature, in which liquid cannot flow at this stage. The moment in which the engine starts working is the greatest rate of wear up to the engine machine. The lubricating oil does not reach all parts of the engine quickly when it is pumped, and to avoid and prevent such a problem and reduce its impact the lubricating oil must be sufficiently viscous stirred to flow easily and fast to reach all parts of the engine. The resulting changes in the rate of the parent oil point were studied under the effect of the addition of nanoparticles MWCNTs and CuO at different concentrations as shown in **Fig. 12**.

Trends in changes in the pour point of oil as a function of the concentration of nanoparticles in oil lubrication with MWCNTs and CuO nanoparticles with a concentration of 0.1% by weight, the spill point was reduced slightly, At 0.2 wt. %, high rate values of pour point were observed at about 7.4% and 3.7% with the addition of MWCNTs and CuO respectively. Yet, when the concentration of nanoparticles increased to 0.5% by weight and 1% by weight, the pour point was reduced again.

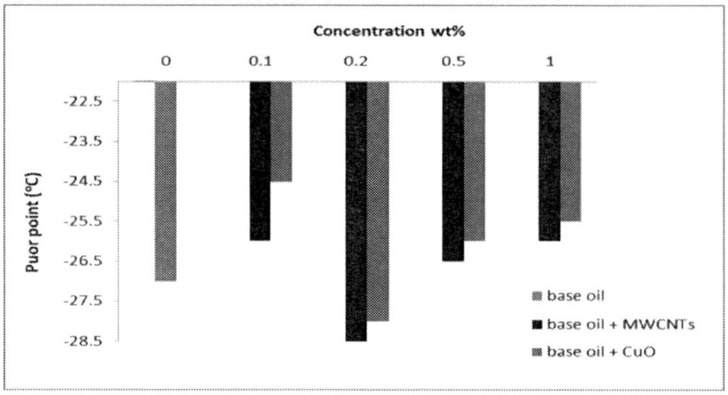

Fig. 12: The pour point values of parent oil and nano-lubricant.

It can be concluded that the increasing thermal conductivity coefficient by adding copper oxide and titanium oxides nanoparticles is due to increased oil resistance against ignition, It can be seen, by estimate the produced change in the rate of pour point and flash point of nano-oil after adding nanoparticles, which were dependence as a function of concentration, the best amount noticed of both parameters have 0.2% by weight concentration.

4. Conclusions

It has been concluded that the dispersing of NPs in lubricating oil to form the nano-lubricant is a very complicated process in high viscosity oil. It has been observed that MWCNTs and CuO have been distributed in different concentrations by using the bath ultrasonic method within the lubricating oil to obtain the best stability, due to the fact that they provide high energy and prevent their aggregation and precipitation again.

The results observed that thermal conductivity and flash point by the effected addition NPs may change for better at a high rate, but the oil properties such as viscosity and pour point have no perceivable concrete changes in the least nanoparticles concentration 0.1% by weight. The thermal conductivity increased MWCNTs/oil and CuO/oil about 11.4% and 9.85% respectively at 0.1 % wt. concentration. Also, a flashpoint for both nano-lubricants prepared increased by about 7.27% and 4.45% respectively.

It can be concluded that the nano-lubricant prepared with 0.2 wt. % concentration of MWCNTs/oil & CuO/oil, it seems that is the better sample for both because in this sample viscosity had unchanged much, this is desirable, thermal conductivity and other oil specifications have been improved. On the other hand, a comparison that observed results from assay the effects of MWCNTs and CuO on the characterizations of engine oil shows that, MWCNTs have a better functionality for improving the characteristics of (SAE15W40) engine oil than CuO NPs.

References

[1] Ferguson C R, Kirkpatrick A T 2001 (New York John Wiley & Sons) p134
[2] Choi Y, Lee C, Hwang Y, Park M, Lee J, Choi C, Jung M 2009 *Current Applied Physics* **9** 124
[3] Dai W, Kheireddin B, Gao H and Liang H 2016 *Tribol Int* **102** 88
[4] Ali M, Xianjun H, Turkson R, Peng Z and Chen X 2016 *RSC Adv.*
[5] Wu Y, Tsuia W and Liub T 2007 *Wear* 819
[6] Liu L, Fang Z, Gu A and Guo Z 2011 *Tribology Letters* **42** 59
[7] Peng Y, Hu Y and Wang H 2007 *Tribology Letters* **25** 247

[8] Mingwu S, Jianbin L, Shizhu W and Junbin Y 2001 *Chinese Science Bulletin* **46** 1227
[9] Ji X, Chen Y, Zhao G, Wang X and Liu W 2011 *Tribology Letters* **41** 113
[10] Hwang Y, Lee C, Choi Y, Cheong S, Kim D, Lee K, Lee J and Kim S H 2011 *Journal of Mechanical Science and Technology* **25** 2853
[11] Ma S, Zheng S, Cao D, Guo H 2010 *Particuology* **8** 468
[12] Chen L, Xie H, Yu W, Li Y 2011 *J. Dispersion Sci. Technol.* **32** 550
[13] Hwang YJ, Ahn YC, Shin HS, Lee CG, Kim GT, Park HS and Lee J.K 2006 *Current Applied Physics* **6** 1068
[13] Hwang Y J, Ahn Y C, Shin H S, Lee C G, Kim G T, Park H S, Lee J.K 2006 *Current Applied Physics* **6** 1068
[14] Kim D, Kwon Y, Cho Y, Li C, Cheong S, Hwang Y, Lee J, Hong D and Moon S
2009 *Current Applied Physics* **9** 119
[15] Choi C, Yoo H S, Oh J M 2008 *Current Applied Physics* **8** 710
[16] Hwang Y, Park H S, Lee J K, Jung W H 2006 Current Applied Physics **6S1** 67
[17] Saeedinia M, Akhavan-Behabadi M A, Razi P 2012 *International Communications in Heat and Mass Transfer* **39** 152
[18] Ruan B, Jacobi A M *Nanoscale Research Letters* 2012 **7** 127 ().
[19] Vakili-Nezhaad G R, Dorany A 2009 *Chemical Engineering Communications* **196** 997
[20] Chen L, Xie H, Yu W and Li Y 2011 *Journal of Dispersion Science and Technology* **32** 550
[21] Vasheghani M H, Marzbanrad E, Zamani C, Aminy M, Raissi B, Ebadzadeh T and Barzegar-Bafrooei H 2011 *Heat and Mass Transfer* **47** 1401
[22] Harris P J, *Carbon nanotubes and related structures: new materials for the 21st century*, 1999 (Cambridge: University Press) p 459
[23] Otsubo Y, Fujiwara M, Kouno M and Edamura K 2007 *Rheol. Acta* **46** 905
[24] Yu M F, Files B S, Arepalli S, Ruoff R S 2000 *Physical Rev Lett* **84** 5552
[25] Chen L, Xie H, Yu W and Li Y 2011 *J. Disp. Sci. Tech.* **32** 550
[26]] Choi Y, Lee C, Hwang Y, Park M, Lee J, Choi C, *et al.* 2009 Curr Appl Phys **124** 7
[27] Mohamed I M, Saleh N J, Mohmood L H 2012 *Journal of Engineering* **18** 905
[28] Anyuan, C, Cailu, X, Ji, L, Dehai, W, Bingqing, W 2000 *chemical physics letters* **344** 13

IOP Conf. Series: Materials Science and Engineering **579** (2019) 012015 doi:10.1088/1757-899X/579/1/012015

Performance enhancement of trickle bed reactor for the removal of organic compounds from petroleum industrial wastewater.

S A Al-Naimi, F T Al-Sudani, G D A Al-Rawi

Chemical Engineering Department, University of Technology, Baghdad, Iraq

Corresponding author's e-mail address: 80028@uotechnology.edu.iq

Abstract. Catalyst Wet Air Oxidation of phenol Promoted with Hydrogen Peroxide (PP-CWAO) was studied experimentally in this work using a trickle bed reactor operating in both steady state and periodic operation (liquid flow modulation). The effect of cyclic parameters (cycle period and split) are investigated under the same conditions chosen on the basis of the steady state results for PP-CWAO. The results showed that the PP-CWAO of phenol are kinetically controlled for steady state operation. For periodic operation, the time average conversion enhancement is indeed quite considerable in liquid flow modulation as split is lowered from steady state cycle split (σ =1) to a value of (σ =0.5) and reached as much as 6.66 % over the steady state at cycle period (τ_p) = 15 sec.

1. Introduction

Wastewater from a petroleum refinery usually contains hazardous chemicals, as hydrocarbons and phenols. The chemical composition and concentrations of phenolic compounds, even if small, have an undesirable effect on the environment as well as dangerous to human health. Effluents with medium to high phenolic loadings must be subjected to a pre-treatment process that may involves a conventional biological treatment. Catalytic wet air oxidation (CWAO) is an important process for the treatment or pre-treatment of wastewaters with phenolic content because of its high removal [1-5]. Moreover, a set of strategies, such as chemical oxidation, coagulation, biological treatment physical adsorption, and electro-catalytic oxidation technology, have been used in the removal of phenol from liquid effluent [4,6]. All stated processing can be applied individually or combined with other processes to develop the treatment efficiency.

There are considerations that must be taken into account when treating wastewater in any selected method, namely toxicity, concentration and loading of pollutants in the energy requirements of the waste stream and / or the economic aspects and the level of mineralization required [7-12].Of between it, Wet Air Oxidation (WAO) is a beneficial technology because it can be applied to treat wastewaters with organic loads that are as high as (10–100 gm/Liter COD); in addition, at high flow rates toxic contaminants can be treated with this method if high pressures (0.5–20 MPa) and temperatures(125–320 °C) are used , despite that this method does not require high operation costs and does not result in any type of pollution in the air, its commercial use is not possible due to the high capital investments required for such process and the requirement of high pressure and temperature for high removal of organic materials [4-7]. The previously mentioned requirements all have driven the attention away from the (WAO) to the more efficient method which is the catalytic wet air oxidation (CWAO) which is a method that was distinguished by operating conditions that are lower than those required in the (WAO) due to the use of a catalyst and this also resulted in reducing the residence time required to achieve an acceptable percentage of organic materials removal from wastewaters which in result reduced the operating cost by half; in addition, if the CWAO method was conducted in a FBR, the separation stage can be excluded[4, 13-14]. From literature review, we can extract some conclusions that are relevant to the work on the CWAO of phenol concern on the obtained durable and efficient catalysts and determine the optimal process conditions at which the process can be done effectively for treating phenolic wastewaters [15]. Also, studied the kinetic of mineralization rate of CWO of phenol [16-17]. Singh et al.,[18], Guo et al.,[19] and Fadel et al.,[20-21] demonstrated the interaction of mass transfer with reactor hydrodynamics and kinetics for CWAO of phenol.

The present H2O2 increased catalytic Wet Oxidation (PP-CWAO) process and biodegradability of phenolic aqueous solution [22-28]. Since 1989, there are numerous studies on the liquid flow interruption(periodic) operation to unveil the transient behavior of cyclic trickle bed reactor (TBR). Tukac et al., [29] found that if the period length of CWAO of phenol was comparable with an effective liquid residence time and thus phenol conversion in wet oxidation was about 10% more than that aquired at steady state operation. Massa et al., [30], observed that the ON-OFF mode liquid flow modulation (slow-mode cycling) had just a small impact on CWAO of phenol conversion but more effect on selectivity to total mineralization, particularly at longer cycle period. The effect of gas feed flow and/or gas composition modulation on activated carbon performance in phenol wet air oxidation also studied [31-32].

Only a few works have been dedicated in the literature for periodic operation of gas and/or liquid flow rate and composition utilized for the removal of phenol in TBRs by CWAO. Austere experimental effort is necessary and needed, to recognize the phenomena of periodic operation in order to organize a direction for commercial achievement. The present study continued to study further points utilizing the advantages associated with periodic operation of catalytic wet air oxidation promoted with H2O2 (PP-CWAO) to improve the mass transfer characteristics of the limiting reactant and catalyst utilization which are the most attractive features of trickle bed reactors and in PP-CWAO of phenol.

2. Experimental work

The continuous promoted CWAO of phenol with hydrogen peroxide was conducted in a TBR as shown in Figure.1. The reactor is a 0.018 m inside diameter stainless steel tube that is 0.76 m length and thickness of 2 mm enough to operate at a pressure up to 2.5 MPa (25 bar), packed with 70 gm of (0.5% Pt/Al2O3) catalyst particles. The reactor is constructed of two parts with stainless steel flange welded on each end. Characteristics of the bed and catalyst specification, material used through the experiments and operating conditions are given in Tables 1 and 2. The operating conditions used during the experiments cover trickle to pulse flow regimes, as shown in Figure.2. a stainless steel screen with appropriate mesh openings was put at the bottom of the reactor to keep the packing and prevent catalyst loss and bed flooding. Uniform radial liquid distribution was ensured by using a 0.2 m of pre-packing of 2 mm glass beads that was put ahead of the catalyst pack ,

Then a 0.35 m catalyst bed which is comprising of 0.5%Pt/Al2O3 particles and finally a 0.21 m of glass beads as post-packing to support the catalyst bed and this will result in the overall pack which was 0.76 m. An electric heater was used as an external heat source to heat the reactor while the bed temperature was controlled within ±3°C by (DT109A-230VAC-R) through and on-off type regulator that adjusted the heat supply to the external heater. Three (type T) thermocouples were placed at different positions along the length of the reactor. The pressure inside the reactor was measured by pressure taps at the head of the reactor.Two (Dose pump, DDE6-10B-PVCLVLC-X311001FG, Italy) were used to supply the phenol and hydrogen peroxide solutions into the pre-heater where the two solutions were sufficiently heated and mixed. The periodic flow modulation was obtained by using two solenoid valves (EVR 15, Danimark) connected through liquid delivery line which were used to switch between steady flow and flow modulation required under change ful condition.

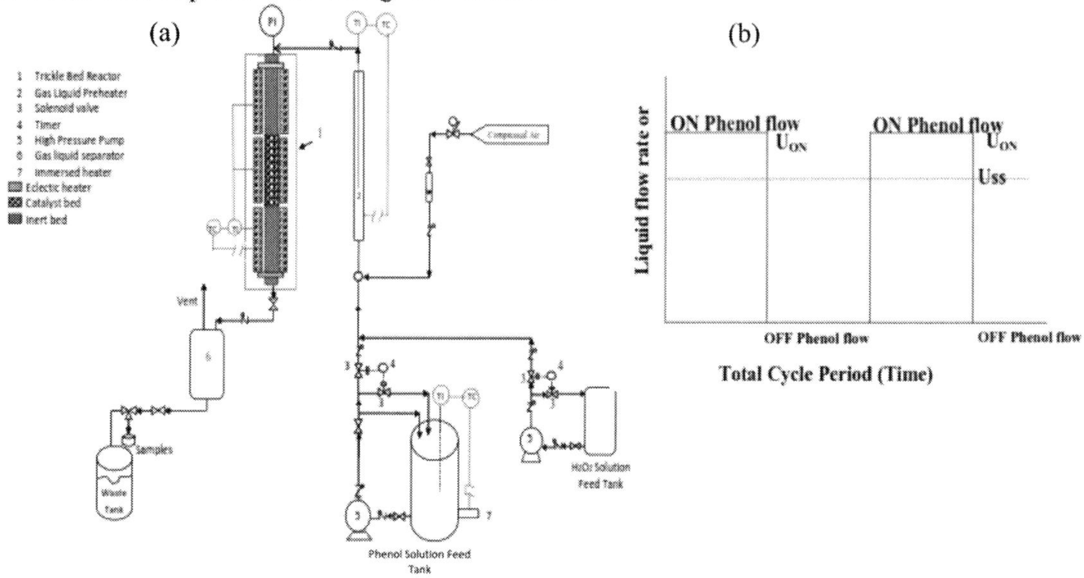

Figure 1. Experimental Setup for steady state and unsteady state experiment
a) experimental setup, b) ON-OFF liquid flow modulation arrangement

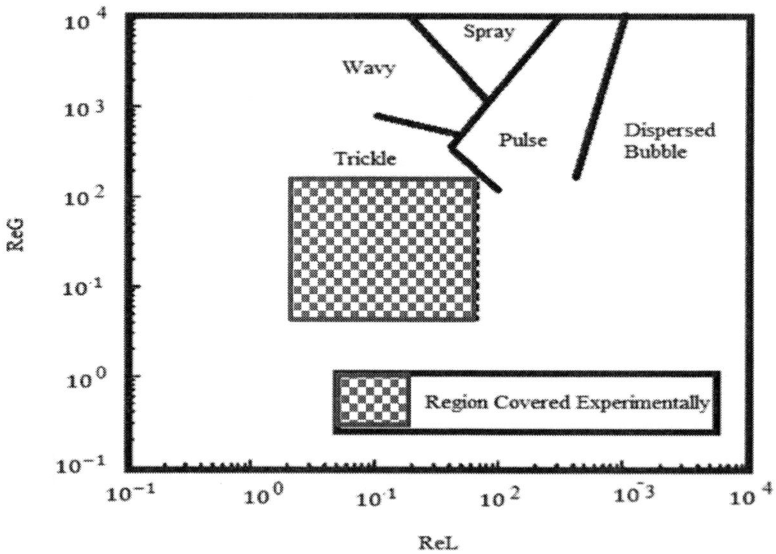

Figure 2. Flow region map [33-34].

The solenoid valves were dominance by an electronic timer (2TIL GARANTILIDIR) which supplied the cycle period and split data were.The exodus solutions from the reactor go through the gas-liquid line separator and then split into a sampling stream and a waste stream. The sampling stream was linked to a needle valve for withdrawal of samples to be tested. To prevent pressure build in the gas and liquid delivery and exit stream pressure indicator and safety valve were installed, also to assure the flow in one direction, one-way valve was located in gas and liquid line.During the pulse time (ON), the liquid

IOP Conf. Series: Materials Science and Engineering **579** (2019) 012015 doi:10.1088/1757-899X/579/1/012015

was allowed to reach the reactor through the first solenoid valve. While during the base time (OFF), it was recycled to the feed tank through the second solenoid valve, so that the pump operation remained unaffected. The performance enhancement conversion was evaluated to identify the degree of improvement over the steady state operation [29-30]: -.

$$\text{Enhancement Conversion} = \frac{\text{Average Conversion(Un} - \text{steady state)}}{\text{Conversion(Steady state)}} \quad (1)$$

Table 1. Catalyst, reactor and material characteristics.

Reactor characteristics		Phenol characteristics	
Reactor diameter m (I.d)	0.018	Color	White
Wall Thickness m	0.002	Molecular weight, (gm/gmol)	94.11
Total length m	0.76	Chemical structure	C_6H_6O
Pre packing depth m	0.2	Purity, (%)	99.5
Post packing depth m	0.21	Freezing point, °C	40-41
Type of inert bed(Glass bead)	0.2m.		White
Catalyst characteristics		Hydrogen peroxide characteristics	
Active metal	0.5%Pt	Colorless in solution	
Catalyst support	γ-Al_2O_3	Molecular weight, (gm/gmol)	34.01
Particle shape	Sphere	Chemical structure	H_2O_2
Particle diameter (cm)	0.16	Purity, (%)	50
Surface area($\mathbf{m^2}$/gm)	250	Boiling point (°C)	150.2
Pellet density(gm/$\mathbf{cm^3}$)	0.56		34.01
Boiling point (°C)	150.2		

Table 2. Range of operating condition for steady and unsteady state experiments.

Operating Conditions	Liquid flow modulation
Pressure, P (bar)	1
Oxygen partial pressure, P (bar)	0.021
Inlet phenol concentration, Cph,o (g/L)	1.35
Bed temperature, T (°C)	80
Inlet hydrogen peroxide concentration %	35%
Liquid flow rate (phenol solution), FL ,ph(L/h)	0.224
Superficial liquid velocity(phenol solution); uL (m/s)	0.000245
Gas flow rate, FG(NL/h)	150
Superficial gas velocity, uG (m/s)	0.163
hydrogen peroxide flow rate, FL , H2O2 (L/h)	0.099
Superficial liquid velocity (H2O2 solution); uL (m/s)	0.00011
LHSV, (h-1)	1.16
Space time, (h) τ	0.86163

Cycle time (period), (Sec) τ_p	5 - 180
Cycle split (ON flow fraction) σ	0.2, 0.5, 0.7

3. Theoretical aspect

3.1. Reactant limitation

The overall reaction for the complete oxidation (mineralization) of phenol to inorganic product proceeds as: -

$$C_6H_6O + 7O_2 \rightarrow 6CO_2 + 3H_2O \quad (2)$$

For the study of CWO of phenol in TBR, it is an important information to determine in which phase-gas or liquid-the limiting reacting is in under the operating conditions in order to choose the appropriate reactor configuration (steady state and/or periodic operation). According to Beaudry et al., [35], the criterion was particularly useful to diagnose the gas and liquid reactant limitation.

$$\gamma = \frac{D_{e_B} C_B}{v\, D_{e_A} C_A^*} \quad (3)$$

If $\gamma > 1$ the reaction is gas limited reactant.
$\gamma < 1$ the reaction is liquid limited reactant.

Where v stoichiometric coefficients of oxygen, De_A , De_B effective diffusivity of species A and B component in the catalyst m^2/s, C_B concentration of liquid reactant kmol /m^3, C_A^* equilibrium concentration of gaseous reactant in liquid phase kmol /m^3 ,

3.2 External mass transfer (interface)

The possibility of external mass transfer limitations of oxygen has to be addressed prior to any kinetic analysis of the obtained results. For this purpose, α_{gL} *and* α_{LS} ,and Mears' criterion (C_m) were used to diagnose the external diffusion limitation from the bulk gas phase to the bulk liquid phase, from the bulk liquid phase to the catalyst surface , and from the bulk gas phase to the catalyst surface [17,36] :-

$$\alpha_{gL} = \frac{(-\acute{r}_A)_{obs}\rho_p}{(k_{gL}a)_{O_2} C_A^*} \leq 0.05 - 0.1 \quad (4)$$

$$\alpha_{LS} = \frac{(-\acute{r}_A)_{obs}\rho_p}{(k_{Ls}a)_{O_2} C_A^*} \leq 0.05 - 0.15 \quad (5)$$

$$C_m = \frac{(-\acute{r}_A)_{obs}(d_p/2)\,\rho_p\; n}{(k_{gs})_{O_2} C_{Ab}} < 0.15 \quad (6)$$

According to these criteria the external mass transfer can be neglected when α_{gL} *and* $\alpha_{LS} \leq 0.05 - 0.1$, *and* $C_m < 0.15$. Where a interfacial area m^2/m^3, $\acute{r}_{A_{obs}}$ observed reaction rate kmol/kg$_{cat}$.h, C_{Ab} bulk reactant concentration kmol/m^3, k_{gL} gas-Liquid mass transfer coefficient m/s , k_{Ls} liquid-solid mass transfer coefficient m/s , k_{gs} gas-solid mass transfer coefficient m/s, n reaction order, ρ_p particle density kg/m^3.

3.3. Intra particle diffusion (internal diffusion)

The internal diffusion limitation was estimated by using modified Weisz-Prater criterion for nth order reaction [17,36,38]:

$$\varphi = \frac{(-\acute{r}_A)_{obs}\, dp^2\, \rho_p}{36\, D_{e_A}} \ll \frac{2}{n+1} \quad (7)$$

If $\varphi \ll \frac{2}{\beta+1}$ it can be considered that the internal diffusion limitations can be neglected. The present work will be considered as nth order reaction [37-38]. Where dp particle diameter m.

3.4. Axial dispersion

Liquid misdistribution (channeling, wall flow) and Axial dispersion can reduce the obtained conversion. According to the Young and Finlayson Criterion [42], the axial dispersion /diffusion can be neglected when: -

$$\left|\frac{(-r_A')_{obs}\, dp_p\, \rho_p}{U_0\, C_{A_b}}\right| \ll \left|\frac{U_0\, dp_p}{D_{a_x}}\right| \tag{8}$$

Where D_{ax} Dispersion Coefficient m^2/s , Uo initial Liquid velocity m/s

4. Results and discussions

4.1. Kinetic regime and reactor performance

The performance of the TBRs depend upon the type of reactant limitation used, the results of the diffusion fluxes (γ) of both reactants from Eq.(2) are shown in Table 3. It can be noticed that the results of the diffusion fluxes (γ), confirms that the oxygen is the limiting reactant (near - liquid-limited reactant) in our operating conditions. Furthermore, the liquid phase oxygen concentration which has a value of about (1.71) $*10^{-4}$ kmol/m^3 was 29 times lower than the initial phenol concentration (8.92-16)$*10^{-3}$ kmol/m^3.

Table 3. Values of diffusion fluxes (γ)

C_{phenol}^0		γ
ppm	(kmol/m^3)	
1350	0.0144	3.742

Table 4 shows the results of α_{gL} and α_{LS} and Mears' criterion for the external mass transfer limitations of oxygen , it can be noticed that the gas liquid mass transfer limitation slightly influences in the performance of TBR. However, as the reactor operates under partial wetting conditions, the direct gas - solid mass transfer will further diminish the weak influence of gas – liquid mass transfer limitations.

Table 4. External mass transfer limitations

LHSV h^{-1}	$-r_{Phenol}'$ $*10^5$ kmol/kg .h	k_{gL} a (s^{-1})	k_{Ls} a (s^{-1})	k_{gs} (m/s)
1.16	1.233	0.041	1.613	0.00275
		α_{gL}	α_{LS}	C_m
		0.117	0.0029	0.00072

As shown in table 5, the overall rate of reaction was found to be free from the effects of internal diffusion as the results falls within the limits of the most appropriate criteria [REF].

Table 5. Intra particle diffusion limitations (φ)

C_{phenol}^0 ppm	LHSV h^{-1}	$-r_{Phenol}'$ $*10^5$ kmol/kg .h	$\phi<2/\beta+1$ ($\phi<1.538$)
1350	**1.16**	**1.177**	0.000704

Such effects of the axial dispersion and liquid misdistribution were minimized due to the appropriate geometry of both the reactor and the particle (L_R= 35 cm, Dr=1.8 cm and d_p= 1.6 mm). The value of the criteria L_R/dp was found to be 218.75 while Dr /dp took a value of 11.25, which are similar to the critical values of 50 and 10, respectively established for safe operation [39-41]. In addition, from Young and Finlayson Criterion Eq. (6), the results ensure that our process was free from the axial dispersion effects as can be seen in Table 6.

Table 6. Axial dispersion /diffusion (young and finlayson criterion) for liquid and gas phase.

LHSV h^{-1}	$-r'_{Phenol}$ kmol/kg .h $*10^5$	Axial Dispersion of Liquid Phase	
		$\left\lvert\frac{(-r'_A)_{obs}\, dp_p\, \rho_p}{U_0\, C_{A_b}}\right\rvert$	$\left\lvert\frac{U_0\, dp_p}{D_{a_x}}\right\rvert$
1.16	1.4349	0.013387	0.092789
		Axial Dispersion of Gas Phase	
		$\frac{(-r'_A)_{obs}\, dp_p\, \rho_p}{U_0\, C_{A_b}}$	$\left\lvert\frac{U_0\, dp_p}{D_{a_x}}\right\rvert$
		2.012E-05	59.4465

Several conclusions can be withdrawn from the TBR results, which are as follows : -

- The wet catalytic oxidation of phenol was controlled by chemical reaction in this case of study.
- The effects of the hydrodynamics were at least with respect to the wetting efficiency, external mass transfer, intra particle diffusion and axial dispersion/diffusion, which are the key parameters for the TBRs performances

4.2 Effect of cyclic parameters for PP-CWAO of phenol.

A set of cyclic experiments (cycle split and cycle period) were performed at a varied cycle period and split for ON-OFF mode of operation for promoted CWAO of phenol with hydrogen peroxide at reaction temperature =80°C, LHSV = 1.16 h-1, initial phenol concentration =1350 ppm, gas flow rate =2.5 liter/min, 35%hydrogen peroxide and U_{L,H_2O_2}=1.08933*10-4 m/sec were shown in Figure 3 and 4. Increasing gaseous reactant supply to the catalyst; varying the ratio of the gas to liquid access times at a constant average liquid flow rate (i.e., varied cycle split (σ) at a given cycle period).They were experimented to investigate the possibility of an additional enhancement in the time average conversion as compared to the conventional conversion obtained from operating the reactor at the steady state mode.Figure 3 and 4 show the results of time average conversion and enhancement conversion.It can be seen that at a given cycle period, the time average conversion enhancement is in fact considerable as split is less than steady state operation (cycle split σ =1) to a value of cycle split σ =0.5 and reached as much as 6.66 % over the steady state for CWAO at τp = 15 sec. These results are similar to those of Lange et al [43]; Banchero et al., [44]; Liu et al., [45] and [46] and Al-Sudani, [47].

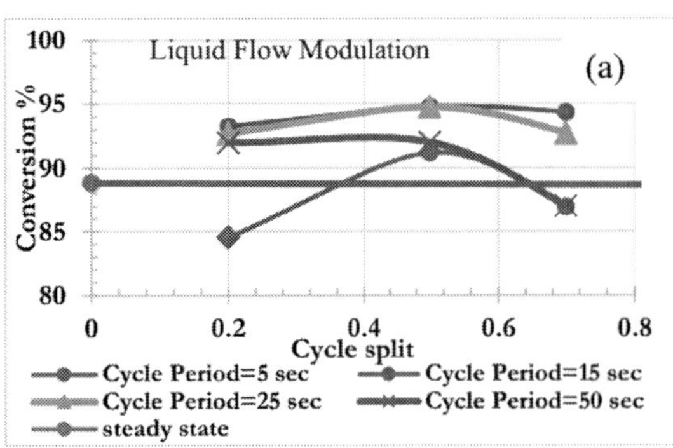

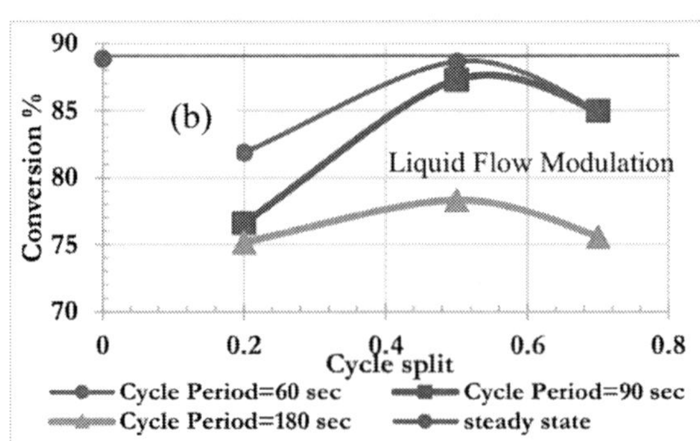

Figure 3. Conversion vs. cycle split at H_2O_2 %=35, LHSV=1.16 h-1, $Cphenol_0$=1350 ppm, superficial gas velocity 0.163 m/s and reaction temperature = 80 °C, (a) at Cycle Period =5-50 sec, (b) at cycle period =60-90 sec.

When the liquid flow stops, reaction proceeds between the flowing gas reactant and the liquid reactant retained in the reactor. Performance is ascribed to the direct on coming of access of oxygen to the catalyst that tremendously reduction of liquid films during OFF period as compared with steady-state operation, and that high accumulated oxygen concentration on the catalyst surface and increased the rate of reaction during ON period.At higher cycle period > 25 sec, it can be noticed from Figure 3 and 4 that the time average conversion and performance enhancement conversion drops even below the steady-state value. Similar behavior was showed by Haure et al., [49]; Lange et al., [43]; Khadilkar et al., [50]; Massa et al., [30]; Liu et al., [46] and Al-Sudani, [47]. However, the results are not as expected, where all the scientists in this field agree with that. There are different results on the impact of split and there is an optimal period whatever its length.

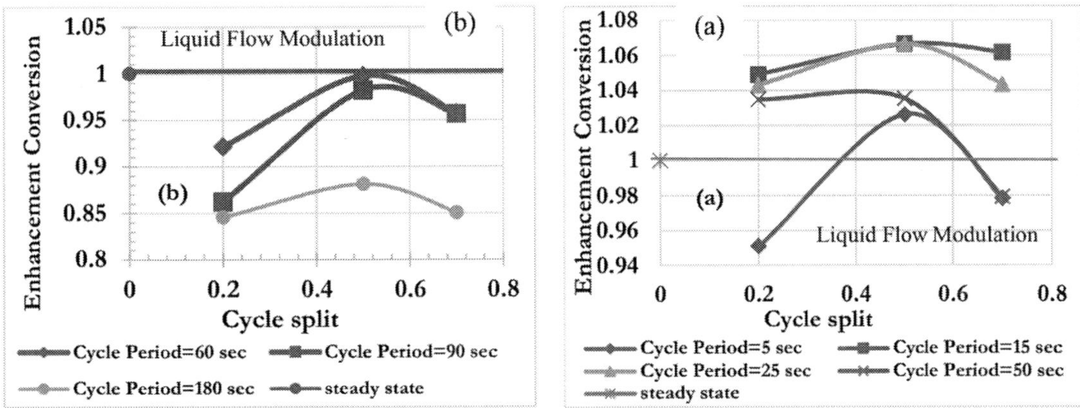

Figure 4. Enhancement conversion vs. cycle split at H_2O_2 %=35, LHSV=1.16 h-1, $Cphenol_0$=1350 ppm, superficial gas velocity 0.163 m/s and reaction temperature = 80 °C, (a) at Cycle Period =5-50 sec, (b) at Cycle Period =60-90 sec.

The optimal period can only be explained by the following qualitative reasons: time periods are shorter than necessary for their consumption produce to exacerbation of oxygen (i.e. gas reactant starvation). Adverse behavior is shown at longer cycle periods (for the same cycle split). Several factors can negatively affect the attained conversion for long cycle period, since the accumulation of oxygen (longer gas access time) is over the limit of the liquid required for the ON period which leads to the starvation of the liquid reactant (the liquid limitation). Also, for low cycle splits, there is lower chance to reload the liquid through the ON period [43], [48-49].

5. Conclusions

The essential points concluded from the present study are:

1. The PP-CWAO of phenol are kinetically controlled. Where external mass transfer and the internal diffusion limitations can be neglected.
2. The performance of TBR is enhanced under periodic operation to a maximum value at 0.5 cycle split and τp = 15 sec for liquid flow, while it drops to lower steady state value at higher cycle period and split under identical operating conditions with steady-state operation.

Acknowledgment

The authors would like to express their gratitude to chemical engineering department- university of technology,Baghdad, Iraq for financially supporting this research.

Appendix A : Transport Parameters

Axial Liquid and Gas Dispersion Coefficients[50] :-

$$D_{a,G} = \frac{u_G\, d_p}{0.042 R_{eL}^{0.5}} \qquad D_{a,L} = \frac{u_L d_p}{1.8 R'^{-0.7}_{eG} * 10^{-0.005 R'_{eL}}}$$

Wetting efficiency f_w [51]:- $f_w = 1 - \exp[-1.35 Re_L^{0.333}\ Fr_L^{0.235}\ We_L^{-0.17}\ (\frac{a_s\, d_p}{(1-a_s)^2})^{-0.0425}\ ]$

Diffusion in Liquid Phase ,(Wilke and Chang correlation [52]):- $D_{AB,L} = 7.4 * 10^{-8} \left(\frac{M_{wt,B}^{0.5} T}{\mu_B \widetilde{V}_A^{0.6}} \right)$

Were A =Ph , B=O_2

Diffusion in gas Phase [53]:- $D_{AB,G} = \frac{1.013 * 10^{-2} T^{1.75} \left(1/M_{wt,A} + 1/M_{wt,B} \right)^{1/2}}{P\left(\Sigma \vartheta_A^{1/3} + \Sigma \vartheta_B^{1/3}\right)^2}$

Effective Diffusivity :- $D_{ei} = \frac{D_{AB} \varepsilon_p}{\tau''}$

Mass Transfer Coefficients:- $\frac{K_{gl} d_P}{D_{O2L}} = 25.1\ Re_L{}^{0.45} Sc_{O_2L}{}^{0.5}$ [54]

$\frac{K_{ls} d_P}{D_{PhL}} = 3.77\ Re_L{}^{0.8} Sc_{Phl}{}^{1/3}$ [54] , $K_{O2GS} = 0.0138\ R_{eG}^{0.79}\, Sc_{Co2G}^{0.5}\, \frac{D_{O2G}}{d_p}$ [36]

Henry Law constant [55]: - $H(T) = 761.1 - 108.9 \ln(T) - 40785.5/T$

Where D_{AB} molecular diffusivity m^2/s, D_r reactor diameter m, u_G and u_G superficial gas and liquid velocity m/s, $\widetilde{V}_A$ molar volume $m^3/kmol$, ε_L dynamic liquid holdup, ρ_L *and* ρ_g density of liquid and gas kg/m^3, τ'' catalyst pellet tortuosity, F_r froude number $\frac{U_L^2}{g d_e}$, R_e reynolds number $\frac{\rho_L u_L d_e}{\mu_L}$, Sc schmidt number $\frac{\mu}{D_{AB}\rho}$, Sh'' sherwood number $\frac{k_{Ls} d_e}{D_{AB}}$, W_{e_L} weber number $\frac{\rho_L d_p u_L^2}{\sigma_L}$.

References

[1] Mohammadi S, Kargari A, Sanaeepur H, Abbassian K, Najafi A and Mofarrah E 2015 *Desalin. Water Treat* **53** 2215
[2] Aneggi E, Trovarelli A, Goi D 2017 J. *Environ. Chem. Eng.* **5** 1159
[3] Hancock F E , 1999 *Catal. Today* **53** 3
[4] Kolaczkowski S T, Plucinski P, Beltran F J, Rivas F J and McLurgh D B 1999 *Chem. Eng. J.* **73** 143
[5] Fadel M A, Jasim F T and Ahmed L S 2014 *Chem. Biochem. Eng. Q.* **28** 319
[6] Shah MP 2016 *Adv Recycling Waste Manag* **2** 1
[7] Sassi H, Lafaye G, Amor H B, Gannouni A, Jeday M R and Barbier-Jr J 2018 *Front.Environ. Sci. Eng.* **12** 1

[8] Ayral C, Lebigue C J, Stuber F, Wilhelm A M and Delmas H 2010 Ind. *Eng. Chem. Res.* **49** 107
[9] Fadel M A, Alwan G M and Abdul-Ridha L A 2016 *Arab. J. Sci. Eng.* **41** 2659
[10] Liotta L F, Gruttadauria M, Carlo G D, Perrini G and Librando V 2009 *J. Hazard. Mater* **162** 588
[11] Guo J and Aldhhan M 2004 *Chem. Eng. Sci* **59** 2023
[12] Gutierrez M , Pin P , Torres M , Cauqui M A and Herguido J 2010 *Catal. Today* **149** 326
[13] Levec J and Pinter A 2007 *Catal.Today* **124** 172
[14] Guo J and Al-Dahhan M 2005 *Chem. Eng. Sci* **60** 735
[15] Miro C, Alejanre A, Fortuny A, Bengoa C, Font J and Fabregate A 1999 *Wat. Res.* **33** 1005
[16] Santos A , Yustos P, Durbon B, and Ochoa F G 2001 *Eng. Chem.* **40** 2773
[17] Eftaxias A, Font J, Fortuny A and Fabregat A 2005 *Journal of Chemical Technology and Biotechnology* **80** 677
[18] Singh A, Pant K K, and Nigam K D P 2004 *Chem.Eng.J.* **103** 31
[19] Guo J and Al-Dahhan M 2005 *Chem. Eng. Sci.* **60** 735
[20] Fadel M A, Jasim F T and Ahmed L S 2013 *Eng. &Tech. Journal* **31** 147
[21] Fadel M A, Jasim F T and Ahmed L S 2014 *Environ. Eng. & Man. J.* **13** 991
[22] Rubalcaba A, Suarez-Ojeeda M E, Carrera J, Font G, Stuber F, Bengoa C, Fortuny A, and Fabregate A 2007 *Catal.Today* **124** 191
[23] Quintanilla A, Fraile A F, Cass J A, and Rodriguez J J 2007 *Journal of Hazardous Materials* **146** 582
[24] Wittine O, Keav S,Valkaj K M, Barbien J, Zrncevic S 2011 *12th Int. Conf. on Enviromental Scinece And Technology* (Rodes: Greece) p 8
[25] Inchaurrondo N, Cechini J, Font J, Haure P 2012 *Appl. Catal. B: Environ.* **111** 641
[26] Rui S R, Raquel O R,Adrián M T S, Pedro B T, Ana M C, José L F, Joaquim L F and Helder T G 2017 *Applied Catalysis B: Environmental* **219** 645
[27] Yang G, Chen H, Qin H, Zhang X and Feng Y 2017 *Can. J. Chem. Eng.* **95** 1518
[28] Songshan J, Huiping Z and Ying Y 2018 *Separation and Purification Technology* **190** 243
[29] Tukac V, Hania J and Chyba V 2003 *Catal. Today* **79-80** 427
[30] Massa P, Ayude M A, Ivorra F, Fenoglio R and Haure P 2005 *Catal. Today* **107** 630
[31] Larruy B, Ayude A, Font J, Fortuny A, Bengoa C, Fabregat A and Stuber F 2007 *Chem. Eng. Sci.* **62** 5560
[32] Ayude A, Rodriguez T, Font J, Fortuny A, Bengoa C, Fabregat A, and Stuber F 2007 *Chem. Eng. Sci.* **62** 7348
[33] Fukushima S and Kusaka K 1977 *J. Chem. Eng. Japan* **10** 461
[34] Beaudry E G, Dudukovic M P and Mills P L 1987 *AlChE. J* **33** 1435
[35] Khadilkar M R, Wu Y X, Al-Dahhan M H, Dudukovic M P and Colakyan M 1996 *Chem. Eng.Sci.* **51** 2125
[36] Fogler S H 2006 *Element of chemical reaction engineering* (Indiana: Pearson Education Inc) ed 4 p 734
[37] Ahmed L S 2012 *Hydrodynamic and Kinetic of Industrial Wastewater Treatment in Trickle bed Reactor* MSc. Thesis University of Technology-Iraq
[38] Al-Rawi G D A, 2014 *Performance Enhancement in Unsteady State Operated Trickle Bed Reactor of Organic Compounds* MSc. Thesis University of Technology-Iraq
[39] Hurtado P, Ordóñez S, Sastre H and D´ıez F V 2004 *Applied Catalysis B: Environmental* **51** 229
[40] Levenspiel O 1999 *Chemical reaction engineering* (New York: John Wiley and Sons) ed 3 p 376
[41] Mears D E 1971 *Chem. Proc. Des. Dev* **10** 541
[42] Young L C and Finlayson B A 1973 *Ind. Eng. Chem. Fund.* **12** 409
[43] Lange R, Hanika J, Stradiotto D, Hudgins, R R and Silveston P L 1994 *Chem. Eng Sci.* **49** 5615
[44] Banchero M, Manna M, Sicardi S and Ferri A 2004 *Chem. Eng. Sci.* **59** 4149
[45] Liu G, Duany Y, Wang Y and Wang L 2005 *Chem. Eng. Sci.* **60** 6270
[46] Liu G, Zhang X, Wang L, Zhang S and Mia Z 2008 *Chem. Eng. Sci.* **63** 499

IOP Conf. Series: Materials Science and Engineering **579** (2019) 012015 doi:10.1088/1757-899X/579/1/012015

[47] Al-Sudani F J 2007 *Enhancing performance of trickle bed reactor by periodic operation: experimental study and modeling* PhD Thesis, University of Technology-Iraq
[48] Haure P M, Hudgins R R and Silveston P L 1989 *AIChE. J.* **35** 1437
[49] Khadilkar M R, Al-Dahhan M H and Dudukovic M P 1999 *Chem. Eng. Sci.* **54** 2585
[50] Jack M, Hochman and Edward E 1969 *AIChE. J.* **8** 63
[51] Mills P L and Dudukovic, M P 1981 *AIChE. J* **27** 803
[52] Wilke C R and Chang P 1955 *AIChE. J.* **1** 2643
[53] Reid R C and Sherwood T K 1977 *The properties of gases and liquids* (New York: McGraw-Hill) ed 2 p100
[54] Gosset J M, Cameron C E, Eckstrom B P, Goodman C and Lincoff A H 1985 *Mass Transfer Coefficients and Henry's Constants for Packed-Tower Air Stripping of Volatile Organics: Measurements and Correlation* (Florida: Defense Technical Information Center) p 302
[55] Wu Q, Hu X and Yue P 2003 Chem.Eng.Sci 58 923

IOP Conf. Series: Materials Science and Engineering **579** (2019) 012016 doi:10.1088/1757-899X/579/1/012016

Heavy metals concentration in the recent sediments along the sides of the road between Bangazi and Sirt Cities – Libya

M A Basi[1], K F Abbas[1#], M I Al-Sharqawi[2]

[1]Al Farabi University College, Baghdad-Iraq
[2]College of Science and Technology, Brack Al Shati - Libya

[#]Corresponding author's email: kafaa_fadhil@yahoo.com

Abstract. Heavy metals concentrations of Pb ,Cd ,Zn ,Mn ,Cu , Co, Ni , and Fe , have been studied in the recent sediments along the sides of the road between Bangazi and Sirt cities. The results indicate normal range of the concentrations, which reflect no pollution. Moreover, the study indicates the increase in concentrations of Pb, Cd, Zn, Mn, Cu, Co, and Ni, on the edges of both sides of the road. The increase in concentration of most heavy metals in the sediments at the edges of the road is most probably related to hydrocarbon's exhaust of various vehicle.

1. Introduction

Environmental pollution is a problem of the present age, not only in the third world, but also at the global level. The pollution coincided with the emergence of the industrial revolution at the beginning of the eighteenth century. This revolution coincided with the increase in population in the world. The pollution includes the disposal of liquid waste, and factory waste beyond reasonable limits. Even the seas and oceans have not been spared from pollution through the quantities of oil and petroleum products that are heavily affected by marine life, especially fish wealth. This leads to the spread of diseases whose emergence is associated with pollutants and leads to thousands of victims on the other hand.

In order to be able to deal with the effects of pollution, individuals must have both their environmental conscious and environmental protection. The archaeological elements involved in the evolution of human life, such as lead, cadmium and zinc, lead to environmental disasters not in the short term only, but in the long term as it affects our future and the future of our generations [1,2].

This study involves the concentration of Pb, Cd, Zn, Mn, Cu, Co, Ni, and Fe in the sediments along the edges of the road between Banghazi and Sirt cities, Libya. Human activities along the road involve the effect of transport vehicles only. The effect of hydrocarbons emitted from vehicles on the sediments and plants have been indicated by various authors [3- 9].

2- Location of study area

The study area is located on both sides of the coastal road connecting the cities of Benghazi and Sirt in Libya (Figure 1). The coastal road is about 2 km east of the Mediterranean Sea, where recent sand deposits surround both sides of the road with various desert plants.Samples from the sediments were taken along vertical paths along the studied road (Figure 2).

IOP Conf. Series: Materials Science and Engineering **579** (2019) 012016 doi:10.1088/1757-899X/579/1/012016

1-12

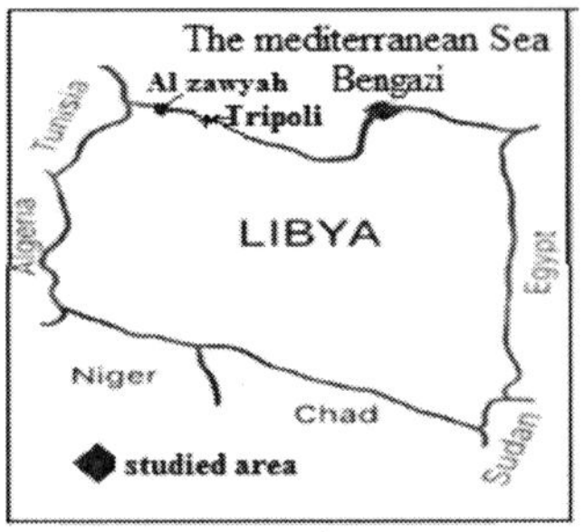

Figure 1. The location of the studied area
scale 18000000:1

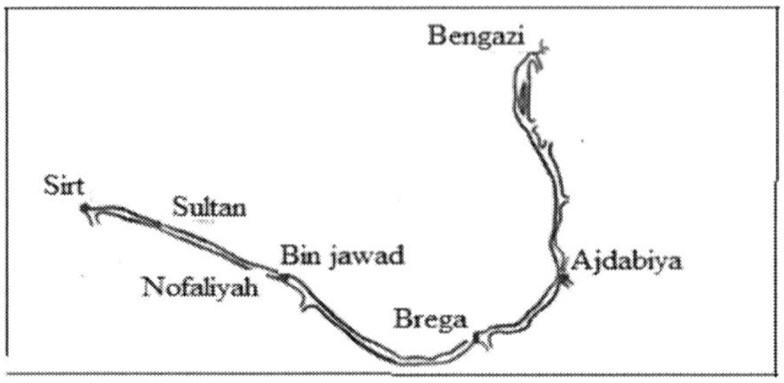

Figure 2. Location of the verticle paths in the studied area
scale 3500000:1

3- Experimental work

40 samples of recent sediment were collected from both sides of the public road linking the cities of Ajdabiya and Sirt. The samples were collected on May1997, in both sides of the road at Ajdabiya, Brega, Bin Jawad, Noufliya, and Sultan, as follow:

1- Deeply picked up 0-2 cm samples from the Ajdabiya, Brega, Bin Jawad, Nofaliya and Sultan paths at 5m, 20 m and 50 m distance from the edges of the road.
2- In addition to a depth 0-2cm, two additional samples were picked up at a depth of 2-5 cm, 5 -10 cm at 5m distance from the edges of the road for some locations at Bin Jawad, Nofaliya and Sultan paths.

All the sediment samples were dried at laboratory temperature, disintegrated and sieved to the size of less than 2 mm for chemical analysis. The samples were treated by concentrated nitric and hydrochloric acids, then analyzed by atomic absorption (Perkin- Elmer model 2380 AAS).The organic materials were analyzed by Page method [10].

Field study in the studied area (between Banghazi and Sirt) showed that the sediments are recent and consist mainly of fine to medium sand and have a reddish-brown color.

4. The results

The results of the chemical analysis for the studied elements (mostly at depth 0-2 cm) and the organic materials are listed in tables 1 and 2.

The analyzed elements are:

4.1 Lead

The concentration of lead ranged between 0.00-57.78 ppm with an average of 16.37ppm. The highest value was in samples **B7 and S7** (57.78 ppm), at Brega and Sultan paths, which are 5 meters from the edge of the road and the lowest value was in sample **N10** and **F14** (0.00 ppm) at Bin Jawad and Nofiliya paths, 50 meters from the edge of the road (Table-1).

4.2 Cadmium

The concentration of cadmium ranged between 0.11 - 3.47-ppm, with an average of 1.28 ppm. The highest value was in sample **J3** (3.47 ppm) at Ajdabya path, which is 5 meters from the edge of the road. The lowest value was in sample **F3** (0.11 ppm) at Nofiliya path, 5 meters from the edge of the road and at a depth 5-10 cm.

4.3 Zinc

Zinc concentration ranged between 3.81 - 10.61 ppm, with an average of 6.44 ppm. The highest value was in sample **J2** (10.61 ppm) at Ajdabiya path, which is at 5 meters from the edge of the road. The lowest value was in sample **N3** (3.81 ppm) at Bin Jawad path, 5 meters from the edge of the road and at a depth 5-10 cm.

4.4 Manganese

The concentration of manganese in the sediments ranged between 18.33 - 97.78 ppm, with an average of 53.76 ppm. The highest value was in sample **B1** (97.78 ppm) at Brega path, which is 5 meters from the edge of the road and the lowest value was in sample **N10** (18.33 ppm) at Bin Jawad path, 50 meters from the edge of the road.

4.5 Copper

The concentration of copper ranged between 1.08-6.61 ppm, with an average of 3.44 ppm. The highest value was in sample **J1** (6.61 ppm) at Ajdabiya path, which is 5 meters from the edge of the road. The lowest was in sample **N10** (1.08 ppm) at Bin Jawad path, 50 meters from the edge of the road.

4.6 Cobalt

Cobalt concentration ranged between 0.89 - 11.34 ppm, with an average of 4.81 ppm. The highest value was in sample **J3** (11.34 ppm) at Ajdabiya path, which is 5 meters from the edge of the road. The lowest value was in sample **N10** (0.89 ppm) at Bin Jawad path, 50 meters from the edge of the road.

Table 1: Heavy metals concentrations (ppm) in the studied sediments

Ser. No.	The dept. cm	The distance, m	Samp. No.	Pb	Cd	Zn	Cu	Mn	Co	Ni	Fe
1	0-2		J1	41.67 ± 0.00	3.39 ±0.04	9.86 ± 0.16	6.61 ± 0.57	82.5 2.45 ±	10 ±0.00	14.55 0.32 ±	6250 490 ±
2	2-5	5	J2	38.61 ± 3.93	3.03 ± 0.04	10.61 ± 0.63	5.94 ± 0.63	76.6 ±2.36	9.39 1.02 ±	13.59 0.47 ±	5861 628 ±
3	5-10		J3	28.33 ± 0.00	3.47 ± 0.04	7.42 ± 0.07	5.28 ± 0.16	67.7 2.08 ±	11.3 0.24 ±	14.56 ±0.16	4138 218 ±
4	0-2	20	J4	21.67 ± 0.00	2.78 ± 0.08	8.56 ± 0.10	5.17 ± 0.00	91.1 2.08 ±	9.67 0.24 ±	13.61 ±0.32	5833 ±360.
5		50	J5	19.44 ± 1.57	2.94 ± 0.08	7.97 ± 0.19	5.17 ± 0.00	89.4 2.08 ±	10.0 0.39 ±	13.75 +0.00	5611 437 ±
6	0-2	5	B1	31.67 ± 0.00	1.67 ± 0.00	7.75 ± 0.07	4.44 ± 0.08	97.7 3.14 ±	6.78 0.16 ±	10.28 ±0.16	6333 ±360.
7		20	B2	13.33 ± 0.00	2.36 ± 0.08	7.08 ± 0.20	3.89 ± 0.16	79.4 0.79 ±	6.34 0.24 ±	10.72 ±0.16	5000 180 ±
8		50	B3	16.67 ± 0.00	1.83 ± 0.00	6.83 ± 0.00	3.78 ± 0.16	77.8 3.14 ±	7.33 0.24 ±	11.28 ±0.16	4777 78. ±
9	0-2	5	B4	22.78 ± 0.78	2.42 ± 0.00	8.53 ± 0.87	5.2 ± 0.24	77.2 2.75 ±	7.83 0.00 ±	10.72 +0.42	4138 463. ±
10		20	B5	17.22 ± 0.78	2.42 ± 0.00	5.08 ± 0.25	3.56 ± 0.16	63.8 4.16 ±	7.61 0.16 ±	9.39 0.16 ±	2916 556 ±
11		50	B6	17.22 ± 0.78	2.2 ± 0.04	6.17 ± 0.00	3.67 ± 0.00	80.5 2.83 ±	8.0 0.24 ±	12.1 ±0.00	4611 578 ±
12		5	B7	57.78 ± 6.33	2.0 ± 0.14	8.64 ± 0.28	4.33 ± 0.00	82.7 ± 5.11	7.44 ± 0.34	10.1 ± 0.47	5111 ± 207
13	0-2	5	N1	7.22 0.78	1.39 ± 0.12	5.5 ± 0.38	2.22 ± 0.16	31.1 ± 2.08	2.11 ± 0.31	6.06 ± 0.55	3944 ± 257
14	2-5		N2	2.78 ± 0.78	0.41 ± 0.12	4.3 ± 0.24	1.75 ± 0.00	24.4 ± 0.79	1.67 ± 0.00	4.89 ± 0.34	3944 ± 171
15	5-10		N3	4.44 ± 1.57	0.33 ± 0.00	3.81 ± 0.08	2.89 ± 0.31	23.8 ± 0.79	2.22 ± 0.39	4.28 ± 0.16	3750 ± 117
16	0-2	20	N4	1.67 ± 0.00	0.31 ± 0.08	4.81 ± 0.16	2 ± 0.00	32.2 ± 0.78	1.95 ± 0.39	5.0 ± 0.24	4083 ± 245
17		50	N5	5.56 ± 1.57	0.42 ± 0.10	4.43 ± 0.25	2.45 ± 0.32	27.78 2.08	2.89 ± 0.23	5.5 ± 2.5	4972 ± 141
18	0-2	5	N6	5.56 ± 1.57	0.42 ± 0.13	4.78 ± 0.16	2.33 ± 0.47	28.8 ± 2.08	2.33 ± 0.49	5.16 ± 0.24	3583 ± 180
19	2-5		N7	7.22 ± 0.78	0.33 ± 0.00	4.92 ± 0.45	2.11 ± 0.16	28.3 ± 2.36	2.5 ± 0.00	5.44 ±0.55	4472 ±257.
20	5-10		N8	8.33 ± 0.00	0.5 ± 0.24	4.03 ± 0.16	1.92 ± 0.12	26.1 ± 1.57	2.5 ± 0.00	4.28 ±0.16	363 ±463.
21	0-2	20	N9	3.33 ± 0.00	0.25 ± 0.00	5.03 ± 0.91	1.75 ± 0.00	23.8 ± 2.08	1.95 ± 0.39	4.06 ±0.16	3694 ± 78.
22		50	N10	0 ± 0.00	0.42 ± 0.12	4.53 ± 0.93	1.08 ± 0.00	18.3 ± 1.36	0.89 ± 0.04	3.61 ± 0.16	372 ± 903

4.7 Nickel

The concentration of nickel ranged between 3.61 - 14.56 ppm with an average of 7.88 ppm. The highest value was in sample **J3** (14.56 ppm) at Ajdabiya path, which is 5 meters from the edge of the road and the lowest value was in samples **N10 and F3** (3.61 ppm) at Bin Jawad and Nofiliya paths, 50 meters from the edge of the road.

4.8 Iron

The concentration of iron ranged between 2916.67 – 8000 ppm, with an average of 5152.25 ppm. The highest value was in sample S8 (8000 ppm) at Brega path, which is 5 meters from the edge of the road and the lowest value was in sample B5 (2916.67 ppm) at Bin Jawad path, 20 meters from the edge of the road.

IOP Conf. Series: Materials Science and Engineering **579** (2019) 012016 doi:10.1088/1757-899X/579/1/012016

23	0-2	5	F1	10.83 ±2.04	0.53 ±0.08	5.42 ±0.13	1.92 ±0.12	37.5 ±2.45	3.33 ±0.00	5.05 ±0.32	4527 ±141
24	2-5		F2	10 ±2.36	0.33 ±0.00	5.53 ±0.19	1.75 ±0.00	35 ±1.36	2.78 ±0.39	5.5 ±0.00	4889 ±157
25	5-10		F3	2.78 ±0.78	0.11 ±0.04	3.97 ±0.08	1.08 ±0.00	27.22 ±0.78	1.17 ±0.35	3.61 ±0.16	3667 ±180
26	0-2	20	F4	6.67 ±0.00	0.61 ±0.04	5.61 ±0.17	2.67 ±0.27	49.44 1.96	3.33 ±0.00	6.09 ±0.12	6056 ±79
27		50	F5	6.11 ±2.08	0.89 ±0.56	4.67 ±0.18	1.28 ±0.28	41.11 ±3.14	3.0 ±0.24	4.83 ±0.00	6278 ±463.
28		20	F13	6.11 ±2.08	0.39 ±0.04	4.61 ±0.24	1.83 ±0.12	34.17 ±0.68	2.94 ±0.28	4.39 ±0.31	4229 ±341
29		50	F14	0 ±0..00	0.61 ±0.04	4.97 ±0.25	3.22 ±0.16	51.11 ±18.07	2.61 ±0.08	5.89 ±0.39	4972 ±142
30	0-2	5	S1	13.33 ±4.08	0.78 ±0.08	7.08 ±0.27	2.11 ±0.16	47.78 ±0.78	4 ±0.00	7.11 ±0.16	5778 ±695
31	2-5		S2	27.78 ±4.78	1.11 ±0.08	9.75 ±0.48	1.83 ±0.12	62.78 ±2.08	4.56 ±0.34	9 ± 0.7	5944 ±671
32	5-10		S3	22.78 ±0.78	1.0 ±0.00	9.5 ±0.07	2.44 ±0.16	64.44 ±1.57	4.67 ±0.00	8.95 ±0.16	6500 ±118
33	0-2	20	S4	10.83 ±0.00	1.03 ±0.26	7.64 ±0.63	3.67 ±0.00	66.67 ±7.58	4.33 ±0.24	8.33 ±0.95	7556 ±681
34		50	S5	10 ±1.18	1.0 ±0.00	6.72 ±0.27	5.2 ±0.24	65 ±4.08	3.78 ±0.32	8.0 ±0.00	7694 ±258
35	0-2	5	S6	10 ±1.18	1.36 ±0.08	5.89 ±0.10	4.78 ±0.19	56.11 ±3.14	5.44 ±0.55	8.11 ±0.16	5806 ±453
36	2-5		S7	57.78 ±6.32	1.36 ±0.08	9.72 ±0.39	4.28 ±0.21	62.22 ±2.08	6.06 ±0.16	8.78 ±0.34	7472 ±528
37	5-10		S8	54.17 ±12.42	1.36 ±0.08	6.53 ±0.32	3.78 ±0.16	59.72 ±6.95	5.0 ±0.00	8.39 ±0.34	8000 ±1047
38	0-2	20	S9	23.33 ±4.08	1.25 ±0.00	8.11 ±0.04	3.67 ±0.00	58.33 ±0.00	4.67 ±0.24	7.89 ±0.16	6278 ±393
39		45	S10	3.33 ±0.00	0.67 ±0.00	5.11 ±0.16	5.28 ±0.16	41.67 ±2.72	3.33 ±0.00	4.64 ±0.19	4528 ±142
40		50	S11	6.67 ±0.00	1.36 ±0.08	6.46 ±0.31	4.64 ±0.19	58.33 ±1.36	4.56 0.08	7.56 ±0.16	5527 ±157
The average				16.37	1.28	6.44	3.44	53.76	4.81	7.88	5152

Table 2. The concentrations of organic substance, % in the studied

Serial No	Sample No.	% of organic Subs.	Serial .No	Sample No.	% of organic subs.
1	J1	1.037	21	N9	0.027
2	J2	0.451	22	N10	0.113
3	J3	0.323	23	F1	0.054
4	J4	0.369	24	F2	0.137
5	J5	0.301	25	F3	0.002
6	B1	0.181	26	F4	0.131
7	B2	0.188	27	F5	0.158
8	B3	0.195	28	F13	0.182
9	B4	0.219	29	F14	0.029
10	B5	0.057	30	S1	0.369
11	B6	0.147	31	S2	0.301
12	B7	0.874	32	S3	0.183
13	N1	0.040	33	S4	0.279
14	N2	0.027	34	S5	0.321
15	N3	0.099	35	S6	0.306
16	N4	0.048	36	S7	1.146
17	N5	0.069	37	S8	1.093
18	N6	0.212	38	S9	0.209
19	N7	0.346	39	S10	0.347
20	N8	0.029	40	S11	0.349

5. Discussion

The average concentrations of elements in the sediments of the studied area are: Lead 16.37 ppm, Cadmium 1.28 ppm, Zinc 6.44 ppm, Manganese 53.76 ppm, Copper 3.44 ppm, Cobalt 4.81 ppm, Nickel 7.88 ppm, and Iron 5152.25 ppm.

The average concentrations of the elements reveal no pollution in the studied area. The concentrations are within the limits of non-polluted terrestrial and marine sediments of Africa, Saudi Arabia, Kuwait and Bahrain in which the concentrations are ranged between 8-60 ppm lead, 0.1-2.5 ppm cadmium, 95-ppm zinc, and 770-ppm magnesium [11]. Aslam et al, [3] reported the concentrations of Pb (259.66-2784.45ppm) and Fe (325.64-5136.37ppm) in the United Arab Emirates road of traffic signals more than two. They [3]
considered the iron is within the range of background level and lead is high in concentration due to high traffic density. The average concentrations are very low also as compared with the concentrations of cobalt 9-29 ppm and nickel 13-90 ppm in the sediments of Greece and Egypt respectively [6].

Except the iron, the study also indicates the decrease in the concentrations of the elements away from the edges of the road, but there is no decrease in the concentration with depth. However, few samples did show an increase with depth. The wide range of concentrations of the elements and the decrease in concentration away from the road might be related to the size of minerals components, organic materials, wind direction and the effect of vehicles exhaust. [1, 2, 4].

Field and laboratory study shows no difference in the size of minerals composition of the sediments, and the sediments are fine to medium sands. The winds were not effective as they were variable in directions and some paths were taken along the opposite side of the road and the results show no difference in concentration. For organic materials (Table 2) the sediments have low concentration and therefore don't play a role in concentration of elements as the correlation coefficient between the organic materials and the elements concentrations is low. These are Pb 0.5908, Cd 0.163, Zn 0.337, Mn 0.162, Cu 0.397, Co 0.191, Ni 0.2, Fe 0.402.The concentrations for the elements except the iron shows clear trend of decrease in elements concentrations away from the edges of the road. The decreases were obvious in lead as indicated in table-3 and Figure 3 and for other elements (Table 3 and Figures 4-11).

Therefore, the variations in the concentration of the most elements and the decreases in concentration away from the edges of the road most properly related to the effect of transport vehicles exhaust, particularly from hydrocarbon emission. Moreover, electrical and engines of the vehicles, dyes, paints, tires, fats, formed additional sources for Zn, Cu, and Cr. These sources were reported for the high concentration of Zn, Cu, and Cr in the other sediments, [11,12]. Lead is the highest value of correlation coefficient with organic materials than other elements r = 0.5908 indicating that there are slight contribution by organic materials for lead and for other elements also.

The results of the study also showed that in exception of iron, the concentration at the distance of 20 meters, 50 meters for all elements were very close, and the concentrations are high at 5 meters only, and convergence in concentration at 20 and 50 meters, (Figures 3-11).

Table 3. Average concentration of elements at the surface sediments, at the depths 0-2 cm and at 5, 20 and 50 meters distance from the edges of the road.

Serial no.	The metals	5 meter	20 meter	50 meter
1	Pb	22.32	10.10	8.50
2	Cd	1.55	1.26	1.23
3	Zn	7.05	6.28	5.78
4	Mn	60.19	55.46	55.11
5	Cu	3.90	3.28	3.00
6	Co	5.47	4.77	4.64
7	Ni	8.58	7.72	7.72
8	Fe	5052.5	5071.7	5169.4

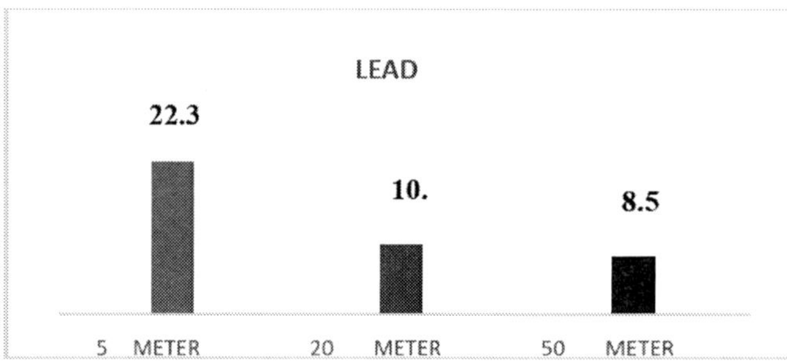

Figure 3. Average concentration of lead (ppm) in the surface sediments (0-2 cm) depth at different distances from the edge of the road.

IOP Conf. Series: Materials Science and Engineering **579** (2019) 012016 doi:10.1088/1757-899X/579/1/012016

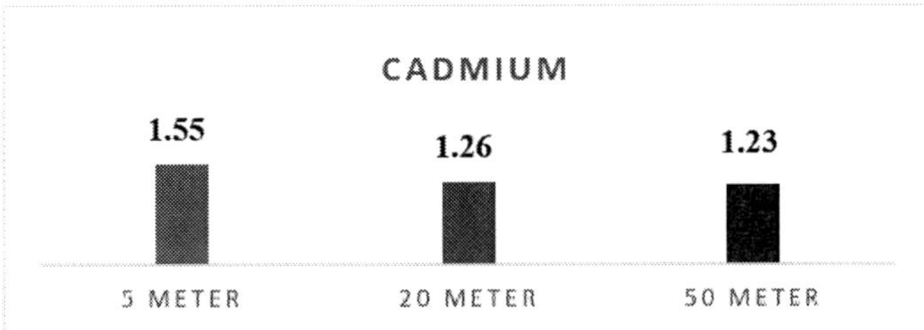

Figure 4. Average concentration of Cadmium (ppm) in the surface sediments (0-2 cm) depth at different distances from the edge of the road.

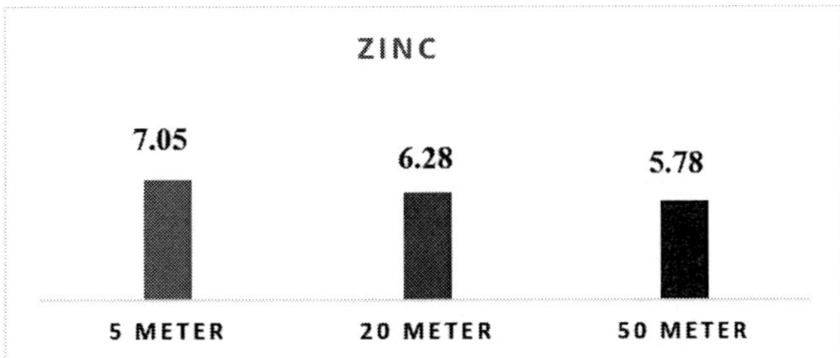

Figure 5. Average concentration of Zinc (ppm) in the surface Sediments (0-2 cm) depth at different distance from the edge of the road.

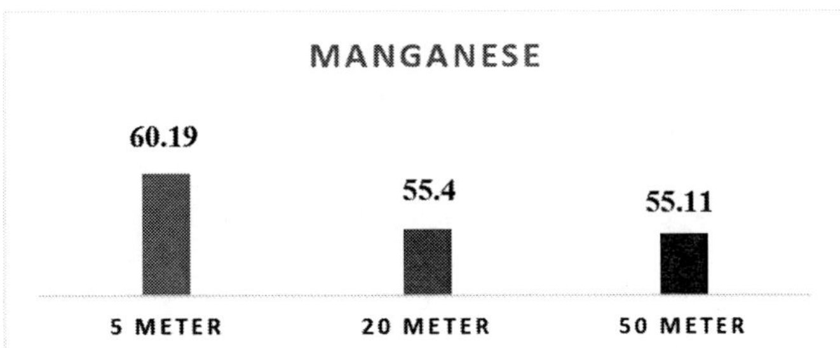

Figure 6. Average concentration of Manganese (ppm) in the surface sediments (0-2 cm) depth at different distances from the edge of the road.

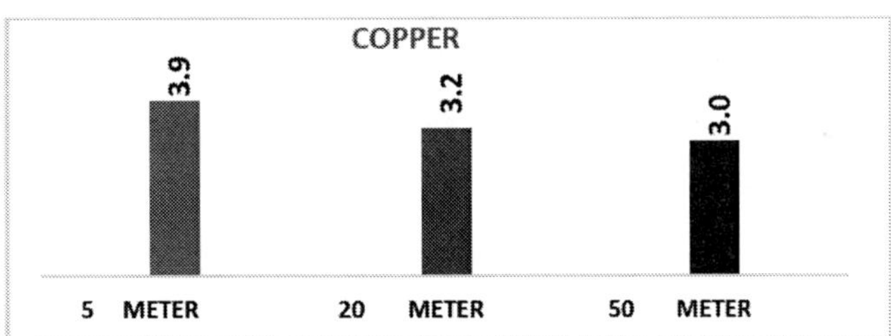

Figure 7. Average concentration of Copper (ppm) in the surface sediments (0-2 cm) depth at different distances from the edge of the road.

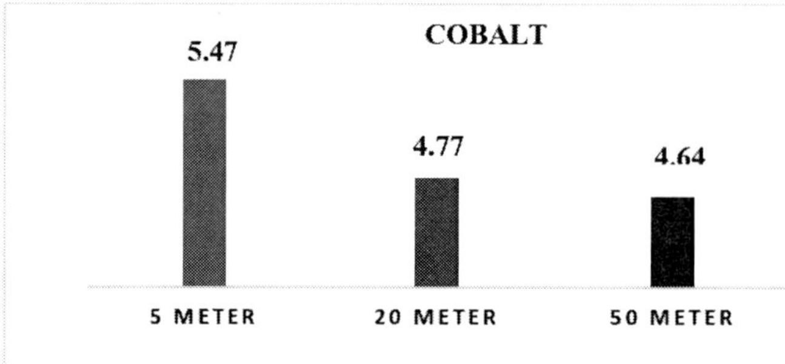

Figure 8. Average concentration of Cobalt (ppm) in the (0-2cm) depth at different distances from the edge of the road.

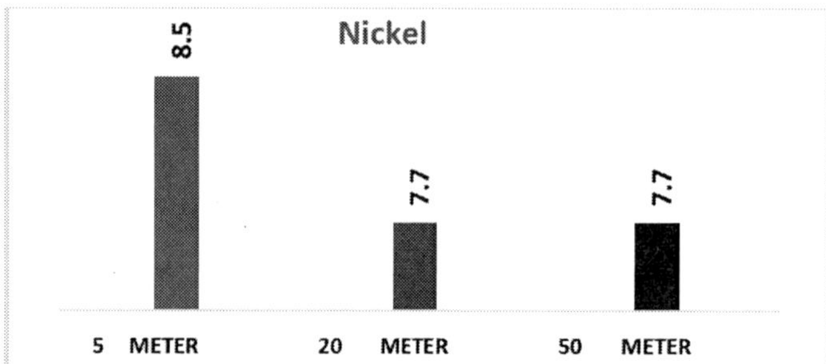

Figure 9. Average concentration of Nickel (ppm) in the surface sediments (0-2cm) depth at different distances from the edge of the road.

The high concentration at 5 meter and the convergence in concentration at 20 meter and 50 meter for most of the elements is most probably attributed to the less effect of vehicles exhaust at a distance more than 20 meter. Carson [13] pointed that the concentration of lead can reach 15 meter from the edge of the road while Abu Dhahi and Yunis [14] indicated that the concentration of lead might be reached up to 100 meter from the edge of the roads. Swaileh et al [4] indicated that the road side contamination by traffic Pb, Cu and Zn did not extend 20 m distance from the road edges. Moreover, the slight variation in the concentration of these elements might be related to the local differences in percentage of the minerals composition, organic material and chemistry at 20 and 50 meters from the edges of the road.

The study also showed that in the exception of iron, the concentration of the elements at Ajdabiya , Brega, and Sultan paths was higher than those in Bin Jawad and Nofaliya paths (Table 4). This increase is due to the high traffic density and low speed at Ajdabiya , Brega and Sultan, therefore more hydrocarbon emission and effect of vehicles at Ajdabiya, Brega , and Sultan paths in comparison with the other paths

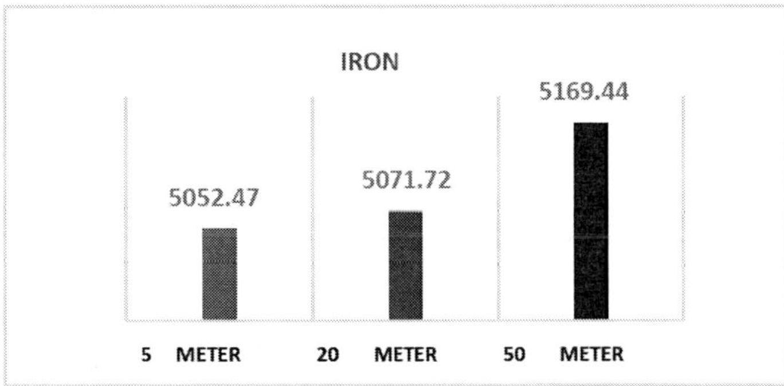

Figure (10) . Average concentration of iron (ppm) in the surface sediment at depth of (0 – 2 cm) depth at different distances from the edge of the road.

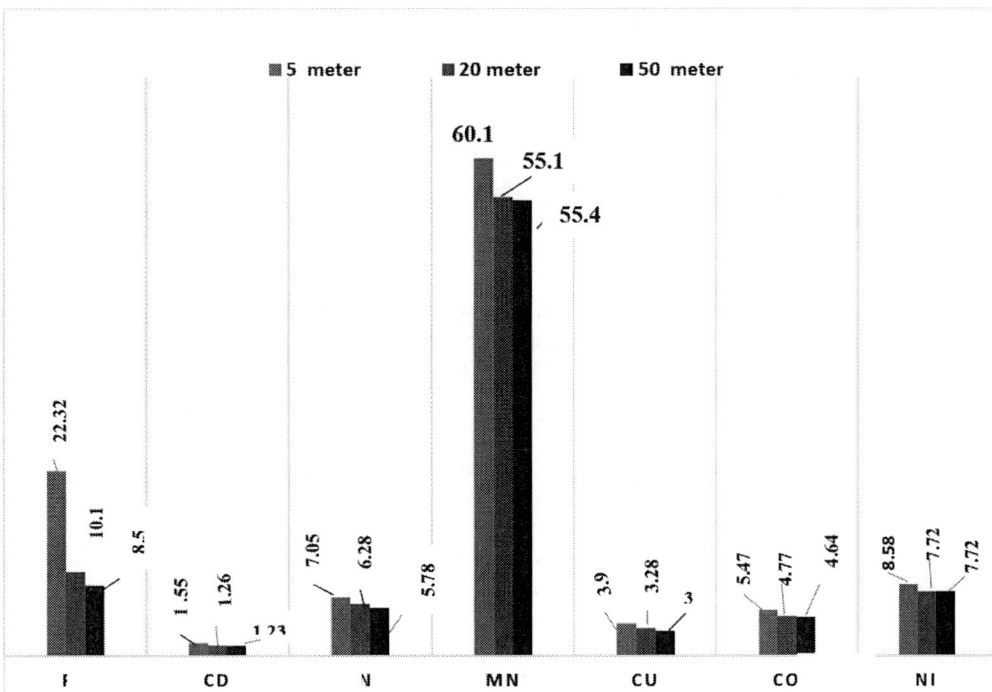

Figure (11): Average concentration of elements (ppm) in the surface sediments (0 – 2 cm) depth at different distances from the edge of the road.

The absence of a specific trend and the high concentration of the iron in the sediments of the studied area, which is arid to semiarid climate, might be related to the local difference in the rate of oxidation of iron. The oxidation of iron is the characteristic features of arid desert areas forming desert varnish [15]. However, organic materials do not exclude for the source of iron in this study as r =0.402 between iron concentration and organic materials.

Table 4. Average concentration of elements (ppm) in the surface sediments at depth of (0-2 cm). depth at different distances from the edge of the road.

Serial No.	The path	Pb	Fe	Ni	Co	Mn	Cu	Zn	Cd
1	Ajdabyah	27.59	5898.15	13.97	9.91	87.69	5.65	8.8	3.04
2	Brega	25.24	4698.1	10.68	7.33	79.92	4.12	7.15	2.13
3	Nofiliya	5.94	5212.45	4.9	2.02	27.04	1.97	4.85	0.53
4	Bin Jawad	3.89	3999.99	4.9	2.02	27.04	1.97	4.85	0.53
5	Sultan	11.07	6166.66	7.37	4.30	56.27	3.55	6.71	1,11

6. Conclusions

Heavy metal concentrations in the recent sediments along the sides of the road between Bangazi and Sirt cities in Libya indicate the following conclusions:

1- Normal range of concentrations of the studied elements reflect no pollution.
2- Except the iron, the study indicates an increase in concentrations of the elements at 3m distance from the edges of the road.
3- The increase in concentrations of the elements at the edges of the road might be related to the effect of transport vehicles exhaust from hydrocarbon emission and vehicles effect.
4- Except the iron, the convergence in concentration of the elements at 20 m and 50 m distances is most probably indicated that the contamination caused by traffic was less than 20 m distance from the road edges.
5- The absence of the specific trend in the concentrations of the iron might be related to local differences in the rate of oxidation of the iron in the studied sediments.

Acknowledgements
The authors would like to acknowledge Mr. B. B., Fakbulo for chemical analysis, and Mr.Z.M. Albasher and Miss. K.M. Al Juhimy for their help in the samples preparation.

References

[1] Wang M and Zhang H 2018 *Int. J. Environ. Res. Public Health* **15** 1064
[2] Kheir R B, Shomar B, Greve M B and Greve M.H 2014 *J. Geochem Explor* **147** 250
[3] Aslam J, Khan A S and Khan H S 2011 *J. Saudi Chem. Soc.* **17** 315
[4] Swaileh K M, Hussein R M and Abu Elhaj S 2004 *Arch. Environ. Contam. Toxicol.* **47** 30
[5] Fergusson J E 1989 *heavy metals; chemistry (Pergamon:* Press Oxford) p 614
[6] Varnavas S P, Panagos A G and Laios R 1987 *Environ Geol Water Sci.* **10** 159
[7] Keller E A 1981 *Environmental Geology* (Charles E. Merri Publishing Co. Ohio) p 488
[8] Ward N I, Reeves R D and Brooks R R 1975 *Environmental Pollution* **9** 243
[9] Dedolph R, Tehaar G, Holtzman R and Lucas H. *Environ. Sci. Techno.* **4** 271
[10] Page A L, Miller R H and Keeney D R 1982 *Methods of soil analysis* (Madison: wiley) ed 2 577
[11] FAO, *FAO fisheries Report*, **No.471** 43
[12] Ndiokwere C L 1984 *Poll. (Ser. B)* **7** 35
[13] Carson E W 1974 *Univ.Press of Virginia, Charlottesville*, 691

[14] Abu Dhahi G Y M, Yunis M A 1988 *Key to the nutrition of plants (in Arabic)* (Ministry of Higher Education and Scientific Research) 382
[15] Pettijohn E J 1975 *Sedimentary rocks* (New York: Harper and row publishing), 628

IOP Conf. Series: Materials Science and Engineering **579** (2019) 012017 doi:10.1088/1757-899X/579/1/012017

Effect of Ce-promotion on iron catalysts activity through the synthesis of liquid fuels by the Fischer-Tropsch process

F K AL-Zuhairi and W A Kadhim
Petroleum Technology Department , University of Technology,Iraq

Corresponding author's e-mail address: 150009@uotechnology.edu.iq

Abstract. The main objective of this work is studying the effect of cerium promotion and reaction temperature on the catalysts activity and product selectivity, the iron-based catalysts utilized in this study were prepared by impregnation methods and characterized via temperature programmed reduction (TPR) and N_2 adsorption-desorption isotherms. Reaction experiments were accomplished in a stainless steel fixed bed reactor. The reaction experiments for unpromoted and promoted catalysts were done at pressure 20 bar, space velocity 1400 h^{-1}, $H_2/CO=2$ and different reaction temperature (250, 275, 300 and 325 °C). The results have shown the addition of promoter to iron-based catalyst enhanced the reducibility of Fe_2O_3 through a shift down in reduction temperature, furthermore, for both catalysts, conversion of carbon monoxide (CO) and selectivities of undesired products (CO_2, CH_4, and C_2-C_4) were found to be increased whereas the selectivity of desired products (C_{5+}) was decreased with increasing in a reaction temperature from 55.87 to 35.65% and from 73.03 to 61.59 % for unpromoted and Ce-promoted catalysts respectively. A higher selectivity for high molecules weight hydrocarbons was about 73.03% detected at lower reaction temperature (250 °C) when a promotion catalyst was used.

Keywords: GTL, Iron catalyst, promotion, FTR, Liquid fuels.

1. Introduction

Natural gas was deliberated as one of the cleanest and abundant fossil fuels, that can be converted to liquid fuels form to be more economical, have lower environmental harm through transportation and uses [1], 40 to 60 % of the world's confirmed natural gas reservoirs are distant or stranded, thus transporting of natural gas from the reservoir to the users by pipeline is costly and difficult [2]. Liquefied of natural gas (LNG) is a present solution for transporting a natural gas, but rigorous transportation demand, energy safety of supply and environmental interest create a requirement search to alternate ways to profit natural gas resource [3].

Gas-to-Liquids (GTL) technology has been confirmed to be a promising method for the exploitation of a natural gas [4]. In the GTL process, the methane from a natural gas firstly convert to synthesis gas (syngas) by many methods such as partial oxidation [5], reforming [6] or auto-thermal reforming [7] which is then converted catalytically by Fischer–Tropsch synthesis (FTS) technology to synthetic crude that contained a different fraction of beneficial hydrocarbon which can be upgraded and separated into suitable liquid transportation fuels such as (gasoline and diesel) [8].

IOP Conf. Series: Materials Science and Engineering **579** (2019) 012017 doi:10.1088/1757-899X/579/1/012017

Increasing the high molecular weight hydrocarbons products are the main objective of FTS reaction [9]. So, selecting an appropriate catalyst and reaction conditions are actually significant in the FTS [10].

Iron-based catalysts were utilized as commercial catalysts for the production of a high molecular weight hydrocarbon at different reaction temperature [11, 12]. On the way to increase the activity and selectivity of iron- catalysts towards C_{5+} contents, recently various studies have recognized by the addition of different promoters, Mangaloğlu et al.,2018[13] found the addition of alkali promoters namely potassium, copper, and manganese to iron catalysts increases the amount of total produced liquid fuel from 48% to 64-79%, Beasley et al.,2018[14], showed that gallium addition to iron catalysts affected the reduction of Fe_3O_4 to metallic iron and the reactant conversions were reduced by an increase of gallium content in Fe catalysts, Park et al., 2018[15], presented the high catalytic activity and product selectivity for (C_{5+}) under the optimum FTS reaction, based on the well-dispersed of iron-carbide surface with potassium promoter, the CO conversion was about 94.1% and the products selectivities were (41.5,7.6,15.6 and 35.3%) for CO_2 , CH_4, C_2–C_4 and C_{5+} respectively they were noted at GHSV = 14 NL•gcat^{-1}•h^{-1}, 15 bar, 340 °C, and H_2/CO =1 ,Li et al., 2019[16],observed that the addition of manganese promoter into the iron based catalyst has a great effect on the products distribution and a higher selectivity for long-chain hydrocarbons (C_{5+}) was about 63.9 % they were obtained when 10 wt% of Mg was used.

There were limited studies in the literature that focus on the low loaded iron-based catalysts promoted with cerium to enhance the FTS product selectivity towards the high molecular hydrocarbon. Accordingly, the main objective of the present work is to find the effects of cerium addition as the promoter on the structure and activity of iron-based catalysts besides the range of reaction temperature.

2. Experimental Work

2.1. Catalysts preparation

γ-Al_2O_3 (Axens) was used as a support after thermal treatment under air flow at 500 for 4 h, the iron-based catalyst was prepared by incipient wetness impregnation methods through the following steps, firstly, the treated γ-Al_2O_3 was impregnated in calculated amount of aqueous solutions of iron nitrate $Fe(NO_3)_3.9H_2O$ (Sigma-Aldrich, ≥ 99.95% purity), with continuous mixing by a magnetic stir bar at room temperature until 20 wt.% of Fe was loaded, followed by drying overnight at 110 °C, subsequent the dried catalyst was calcined under air flow at 550 °C for 4 h with heating ramp 0.5 °C/min [17], to achieve 20Fe/ γ-Al_2O_3 catalyst after reduction.

In the cases of promotion, some of the dried catalyst before the calcination was promoted with 5wt. % of Ce by the co-impregnation of dried 20Fe/ γ-Al_2O_3 in an aqueous solution of $Ce(NO_3)_2.6H_2O$ (Merck) , then the promoted catalyst was dried and calcined by following the same above procedure.

2.2. Catalyst characterization

Temperature programmed reduction (TPR) of the calcined promoted and unpromoted catalysts were recorded by using TP-5000 analyzer equipped with a quartz tube reactor and thermal conductivity detector, in the first 100 mg of each sample was charged into quartz tube and heated under an argon flow to 400 °C for 2 h to eliminate the water, followed by cooling to room temperature, then the TPR program test was started by flowing 5 vol% H_2/Ar gas mixture at a flow rate of 30 mL/min with heating from room temperature to about 850 °C with heating ramp 10 °C/min for 12 h.

BET surface area and porosity characteristics of calcined promoted and unpromoted catalysts were measured by N_2 adsorption-desorption isotherms at -196 °C by using the Micrometrics ASAP- 2020 system. Before accomplishment the test, the temperature was slowly ramped to 200 °C and the sample was evacuated at this temperature for 10 hours to about 0.07 bar, then the BET surface area, pore volume, and an average pore radius were obtained for each sample.

2.3. Fischer-Tropsch reaction

The catalyst tests for converting a syngas to liquid fuels (high molecular weight hydrocarbons) were carried out in a single stainless steel fixed bed reactor (ID: 10mm and L= 900 mm) as shown a schematic diagram in (Figure1). Firstly 5 grams of calcined catalysts were loaded in a middle of the reactor and surrounded by two quartz beds, the reactor was heated by electrical furnace.

The flow rates of the reactant (H_2 > 99.5% purity, CO > 99.3% purity, and inert gas (Ar) ≥ 99.999% purity) and the reduction/reaction temperatures were controlled by a separate electronic mass flow controller and a proportional integral derivative controller respectively. Prior to Fischer–Tropsch reaction, the promoted and unpromoted iron catalysts were reduced in-suites by 5 % H_2/Ar. at flow rate 30 mL/min for 12 hours, Later on in-suites reduction the temperature was down gradually to the ambient temperature in the same gas mixture and with the same under reduction gas flow, after that the reactor was heated and pressured to the reaction temperature and pressure under inert gas (Ar) flow with heating ramp 5 °C/min, then followed by syngas mixture flow. The FTR was carried out at optimum reaction conditions pressure 20 bar, space velocity 1400 h^{-1}, H_2/CO=2 [18], and four different temperatures (250, 275, 300 and 325 °C). Every temperature was keep for 24 hours and then changed using the same ramp rate.

The products stream was removed continuously and reduced to atmospheric pressure by a back pressure regulator, then passed through hot and cold traps at 100 and 0 °C respectively, the compositions of gas has were analyzed on-line gas chromatograph (GC) Shimadzu-2014 fitted with a GS-GASPRO capillary column connected with flame-ionization detector (FID) for analysis of light hydrocarbons and Carboxen-1000 packed column connected with thermal conductivity detector (TCD) for the analysis of CO_2, H_2, and CH_4, however the compositions of liquid phase were collected in the two traps and analyzed off-line by using Varian CP 3800 Gas Chromatograph fitted with (FID).

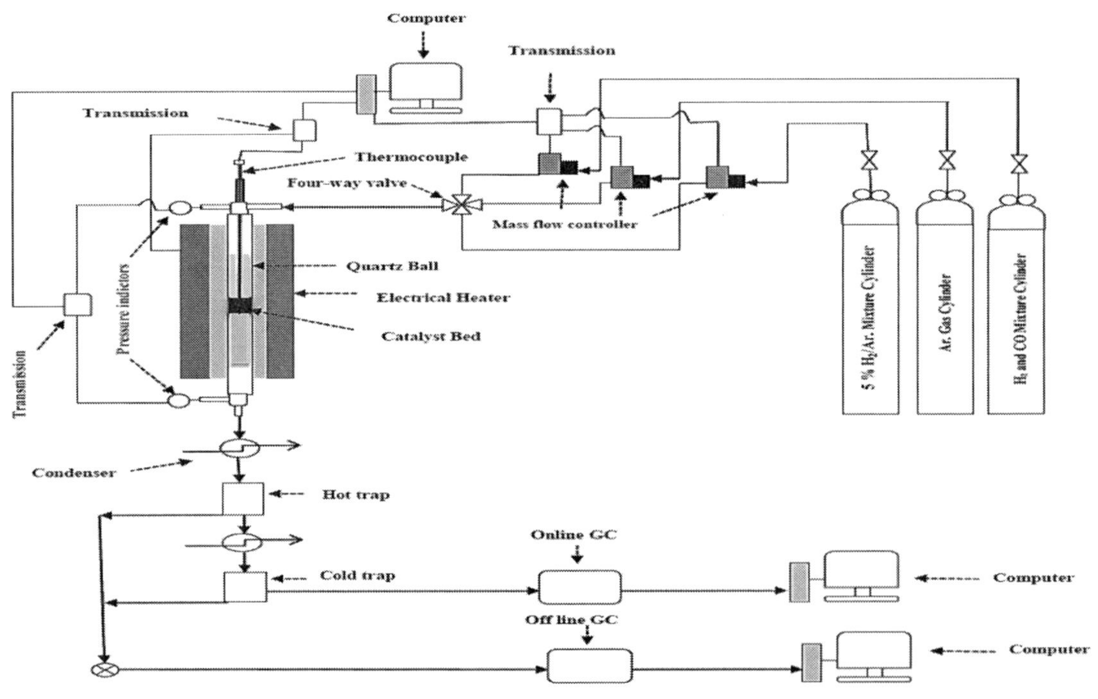

Figure 1. Schematic diagram of Fischer – Tropsch reaction system on single stainless steel fixed bed reactor.

3. Results and Discussion

3.1. Catalyst characterization

H_2-TPR profiles of the calcined (20% Fe/γ-Al_2O_3) and (5%Ce-20% Fe/γ-Al_2O_3) catalysts are presented in (Figure 2). Two hydrogen reduction peaks were observed for reduction of iron oxides. In the first peak, Fe_2O_3 was reduced to Fe_3O_4 and the second peak assigned to the reduction of Fe_3O_4 to FeO which was subsequently reduced to Fe. For the (20% Fe/γ-Al_2O_3) catalyst, reduction of Fe_2O_3 to Fe_3O_4 occurred at 450°C, and then consequently followed by the reduction of Fe_3O_4 to Fe at 625°C. The promoting of iron-catalyst with 5% Ce has strongly affected on the TPR profile (demonstrated in Figure 2), shifted to lower reduction temperature where the first reduction peak shifted to 310°C although the second peak reduced to 460°C. The similar trend has also been stated in the literature by Beasley et al., 2018 [14]. Moreover, the TPR results assured that the chosen parameters (5 % H_2/Ar gas mixture with the flow rate of 30 mL/min for 12 h) were suitable for reducing the Fe_2O_3 to Fe in-suitc reactor before starting the reaction.

BET surface area, single point pore volume and single point average pore radius measurements for calcined support (γ-Al_2O_3), iron-based catalyst (20% Fe/γ-Al_2O_3) and cerium promoted iron-based catalysts (5%Ce-20% Fe/γ-Al_2O_3) are listed in (Table 1). The surface area of support was 145 m^2/g which plunged to 98.8 m^2/g for the unpromoted catalyst, and then little decreases was observed when the catalyst was promoted with 5wt. % of Ce.

The single point pore volume and single point average pore radius for the support were 0.542 cm^3/g and 6.8 nm respectively that decreases to 0.312 cm^3/g and 5.2 nm after the support was loaded with 20 wt. % of Fe, whereas the addition 5 % of promoting resulting smaller reductions in a single point pore volume and single point average pore radius of the catalyst, the reduction of pore volume and average pore radius values as the metal loading, indicates that Fe and Ce were effectively introduced into the catalysts porous structure [19], the similar tendency was observed by Peña et al.,2018 [20].

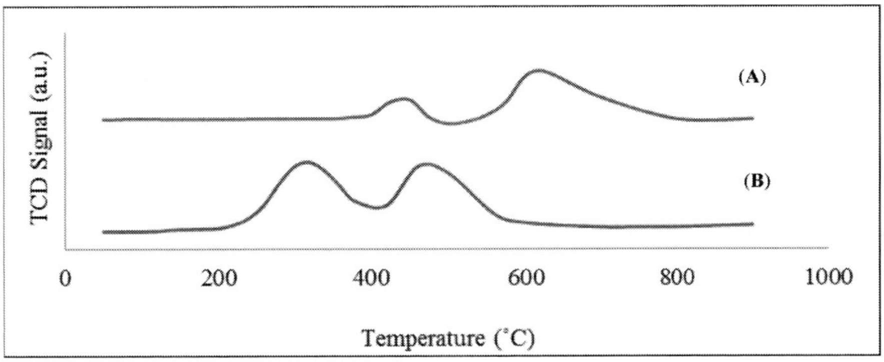

Figure 2.TPR profiles for unpromoted (20%Fe/γ-Al_2O_3) and promoted (5%Ce-20Fe /γ-Al_2O_3) calcined catalysts, where (A) and (B) for unpromoted and promoted catalysts respectively.

Table 1. BET surface area and pore volume.

Catalyst	BET SA (m^2/g)	Single point pore volume (cm^3/g)	Single point average pore radius (nm)
γ-Al_2O_3	145	0.542	6.8
Fe	98.8	0.312	5.2
Ce-Fe	95.3	0.297	4.9

3.2. Effect of reaction temperature on catalyst activity and products selectivity

To investigate the effect of reaction temperature on the activity of catalysts, and the selectivity towards liquid fuels production, several experiments were done at pressure 20 bar, space velocity 1400 h^{-1},

H_2/CO=2 and different reaction temperature.

The results obtained for the catalysts activity and products selectivity were considered when steady-state conditions were reached, presented in (Figure 3) and (Figures 4a, b, c, and d). The results shown an increase reaction temperature from 250 to 325 °C the conversion of CO for both catalysts were gradually increased, this indicated that the reaction temperature is affecting an activity of iron catalysts [18], Moreover, it can be seen, at increases reaction temperature the selectivity of light hydrocarbons (CH_4 and C_2-C_4), and CO_2 was increased. Consequently, lead to a decreases in the selectivity of desired products (C_{5+}), since the higher reaction temperature is favoured for the chain termination essential to produce low molecular weight hydrocarbons, whereas a low reaction temperature was favoured for chain growth and production of high molecular weight hydrocarbons [21]. From results, it can be noted that the promoted catalyst was found to be affected by reaction temperature at a lowered level than unprompted one, because of the promoter addition increase the dispersion of the active FTS metal according to the metal-support interaction [22], for that reason, the addition of cerium to the iron-based catalyst increased the forming of high molecular weight hydrocarbons, the similar trends were obtained in other studies using iron-based catalysts promoted with another promoter Peña et al., 2018[20], Feyzi et al., 2013 [23], Chernavskii et al., 2017[24], Comazzi et al., 2017 [25], Nikbakht et al.,2018[26].

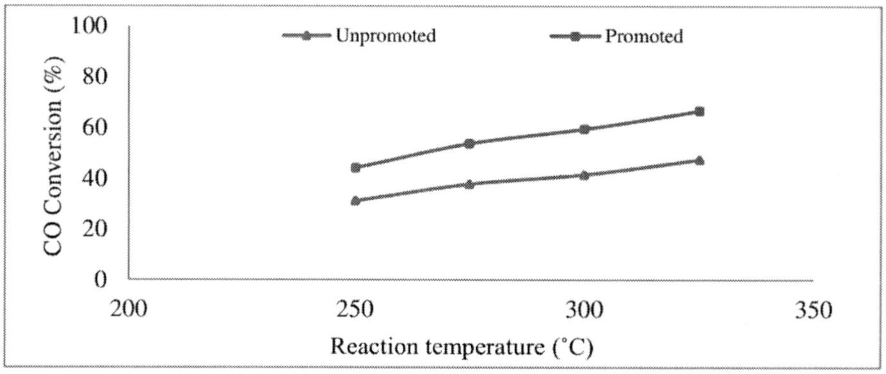

Figure3. Effect of reaction temperatures on carbon monoxide conversion for unpromoted and Ce-promoted iron based catalysts at P = 20 bar, SV= 1400 h^{-1} and H_2/CO= 2.

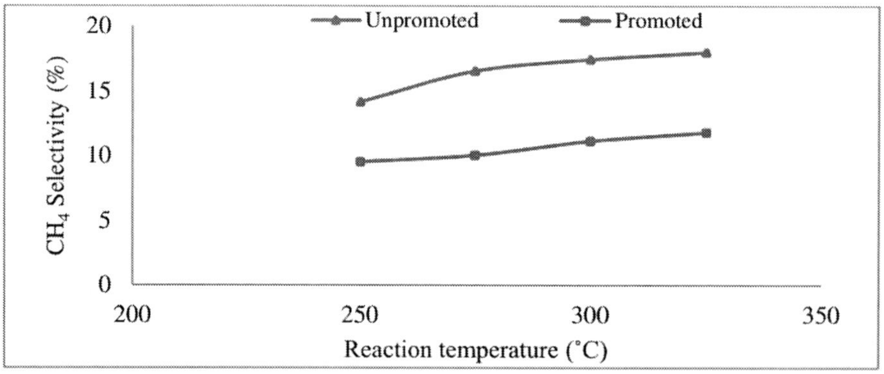

Figure 4a. Effect of reaction temperatures on CH_4 selectivity for unpromoted and Ce-promoted iron based catalysts at P = 20 bar, SV= 1400 h^{-1} and H_2/CO= 2.

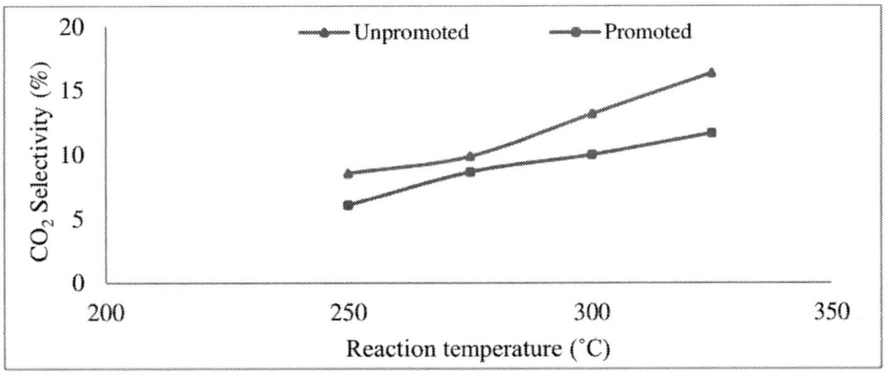

Figure 4b. Effect of reaction temperatures on CO_2 selectivity for unpromoted and Ce-promoted iron based catalysts at P = 20 bar, SV= 1400 h^{-1} and H_2/CO= 2.

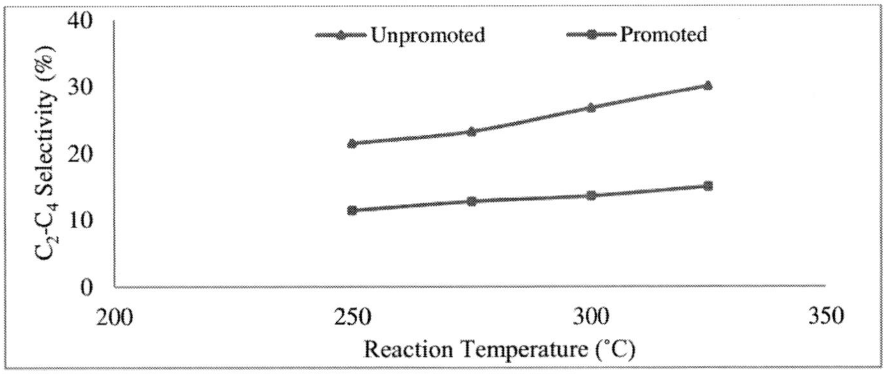

Figure 4c. Effect of reaction temperatures on C_2-C_4 selectivity for unpromoted and Ce-promoted iron based catalysts at P = 20 bar, SV= 1400 h^{-1} and H_2/CO= 2.

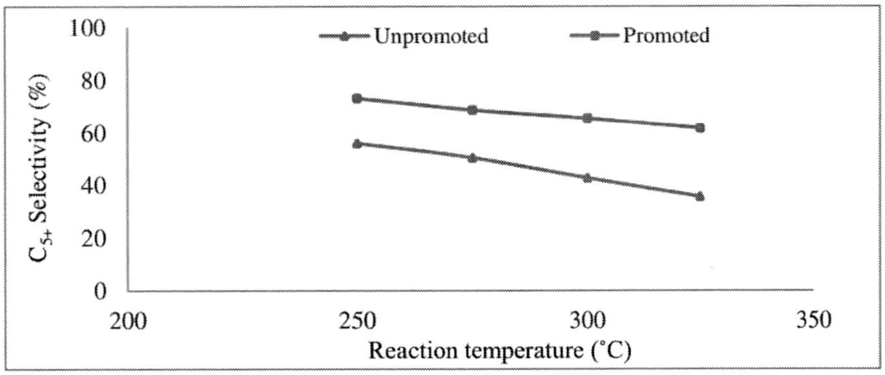

Figure 4 d. Effect of reaction temperatures on C_{5+} selectivity for unpromoted and Ce-promoted iron based catalysts at P = 20 bar, SV= 1400 h^{-1} and H_2/CO= 2.

4. Conclusions

The FTS of syngas derived from a natural gas was studied using an iron-based catalyst prepared by incipient wetness impregnation methods and promoted with cerium in order to synthetic a liquid fuels, The prepared catalysts were characterized by TPR and N_2 adsorption-desorption isotherms, The results shown the addition of cerium to iron-based catalyst enhanced the reducibility of Fe_2O_3 through a shift down in reduction temperature from 450 to 310°C and from 625 to 460°C for reduction of Fe_2O_3 to Fe_3O_4 and Fe_3O_4 to FeO respectively. The FTS reaction was carried out in a fixed bed reactor under range of temperatures (250- 325 °C), pressure 20 bar, space velocity 1400 h^{-1} and $H_2/CO=2$, for both catalysts, the results show that the conversion of CO and selectivity of light hydrocarbon products (C_1-C_4) and CO_2 increased but, the selectivity of desired products (C_{5+}) was decreased with an increase in a reaction temperature, the higher selectivity for the liquid product was about 73 % obtained at lowered reaction temperature (250 °C) when promotion catalyst was used.

References

[1] Hall K R 2005 *Catal. Today* **106** 243
[2] Thomas S and Dawe R A 2003 *Energy* **28** 1461
[3] Ao M, Pham G H , Sunarso J, Tade M O and Liu S *ACS Catal.* **8** 7025
[4] Mills G A 1994 *Fuel* **73** 1243
[5] Kleinert A, Feldhoff A, Schiestel T and Caro J 2006 *Catal. Today* **118** 44
[6] Rostrup-Nielsen J R 1984 *J. Catal.* **85** 31
[7] Gao J, Guo J, Liang D, Hou Z, Fei J and Zheng X 2008 *Int. J. Hydrog. Energy* **33** 5493
[8] Bao B, El-Halwagi M M and Elbashir N O 2010 *Fuel Process. Technol.* **91** 703
[9] Osa A R, Lucas A D, Valverde J L, Romero A, Monteagudo I, Coca P and Sáncheza P 2011 *Catal. Today* **167** 96
[10] Dry M E 1996 *Appl. Catal. A Gen.* **138** 319
[11] Li S, Meitzner G D and Iglesia E 2001 *J. Phys. Chem. B* **105** 5743
[12] Steynberg A P, Dry M E, Davis B H and Breman B B 2004 *Stud. Surf. Sci. Catal.* **152** 64
[13] Mangaloğlu D U, Baranak M, Ataç Ö and Atakül H 2018 *J. Ind. Eng. Chem.* **66** 298
[14] Beasley C, Gnanamani M K, Hamdeh H H, Martinelli M and Davis B H 2018 *Catal. Lett.* **148** 1920
[15] Park J C, Jang S, Rhim G B, Lee J H, Choi H, Jeong H D, Youn M H, Lee D W, Koo K Y, Kang S W, Yang J, Lee H-T, Jung H, Kim C S and Chun D H 2018 *Appl. Catal. A Gen.* **564** 190
[16] Li J, Wu L, Zhang S, Wen J, Liu M , Wang C and Li X 2019 *Sustain Energy Fuels* **3** 219
[17] Loiland J A, Wulfers M J, Marinkovic N S and Lobo R F 2016 *Catal. Sci. Technol.* **6** 5267
[18] Feyzi M, Irandoust M and Mirzaei A A 2011 *Fuel Process. Technol.* **92** 1136
[19] Das S K, Majhi S, Mohanty P and Pant K K 2014 *Fuel Process. Technol.* **118** 82
[20] Peña D, Jensen L, Cognigni A, Myrstad R, Neumayer T, van Beek W and Rønning M 2018 *ChemCatChem* **10** 1300
[21] Liu Y, Teng BT, Guo X H, Li Y, Chang J, Tian L, Hao X, Wang Y, Xiang HW, Xu Y Y and Li Y W 2007 *J. Mol. Catal. A: Chem.* **272**182
[22] Gnanamani M K, Jacobs G, Shafer W D, Ribeiro M C, Pendyala V R R, W. Ma and Davis B H 2012 *Catal. Commun.* **25** 12
[23] Feyzi M, Nadri S and Joshaghani M 2013 *J. Chem.* **973160** 1
[24] Chernavskii P A, Kazak V O, Pankina G V, Y D Perfiliev G V, Li T, Virginie M and Khodakov A Y 2017 *Catal. Sci. Technol.* **7** 2325
[25] Comazzi A, Pirola C, Longhi M, Bianchi C L M and Suslick K S 2017 *Ultrason. Sonochem.* **34** 774
[26] Nikbakht N, Mirzaeia A A and Atashi H 2018 *Fuel* **229** 209

IOP Conf. Series: Materials Science and Engineering **579** (2019) 012018 doi:10.1088/1757-899X/579/1/012018

Thermal Pyrolysis of Municipal Solid Waste (MSW)

S I Hammoodi[1#] and R S Almukhtar

[1]Chemical Engineering Department, University of Technology-Iraq, Baghdad, Iraq

Corresponding author's e-mail address: sally.hammoodi@visitor.uoz.edu.krd

Abstract. Plastics are very desired components, used in a lot of fields, from domestic packaging, agrarian, automotive, to building structure. The frequent use of plastic products related to their lightweight, high chemical stability and low cost.
The present work aims to illustrating the thermal pyrplysis of two types of plastic: Polyethylene terephthalate (PET) and polystyrene (PS) in a batch tubular reactor. The reaction indicated that PS and PET were converted more meaningfully into useful liquid products out of thermally route compared to PET pyrolysis.The resulted liquid could be used as fuels or raw materials in chemical industries. Formation of oxygenates and aromatics were noticed.The maximum liquid produced at optimum temperature and optimum heating time (90 min) is 77% , 67.56%, 54% and 37.48% for PS, 10%PET + PS, 20%PET + PS and 30%PET + PS respectively. The liquid product is analysed by GC-MS at optimum condition. The compositions of the liquid produced are styrene monomer, toluene, alpha-methylstyrene, 1,3 Diphenylpropane, 1-Phenylnaphthalene, 4-phenyl-1 butene. The main product in the pyrolysis of PS is styrene monomer, 10%PET+PS, 20%PET+PS and 30%PET+PS.

1. Introduction

Polymers are considered natural or synthetic molecules that consist of a large number of smaller moieties, monomers, which form a long chain by reacting with each other's. Polymers can be classified depending on two main criteria: according to their thermal behavior and polymerization mechanism. The classification of polymer is significant because the most suitable way of degradation for a given polymer is related to the properties of the plastic molecules [1].

In general use, the term 'plastic' is replicated to the term polymer. Nevertheless, all plastics are polymers but not all polymers are plastic. Polymers can be classified to several groups: a) elastomers (rubbers), b) plastics and c) fibers. This categorization is made depend on their physical properties, elastic modulus and degree of elongation [2].

Plastics are inefficient. So they usually do not combine or make bond with other materials. That's mean that plastics do not decompose normally. It requires thousands of years to decompose or to decay in the environment. Some plastics do not decompose at all. They need to throw in the ocean or in the landfill. Therefore, plastics are producing numerous hurtful problems for humaneness, for environment, for animal in the mainland of even in the oceans. The increase in plastic manufactures and uses are growing daily. So it became an important matter to dominate this issue [3]

Currently, researchers focus more and more on pyrolysis, which has a great benefit over incineration in producing value-added chemicals, even from mixed feedstock.

Pyrolysis process has the ability to breakdown polymer chains into low molecular weight compounds. The output of plastic pyrolysis operation could be used as fuels or chemicals. Two kinds of pyrolysis processes (Thermal pyrolysis, Catalytic pyrolysis) are reported correspondingly.

1-1 Thermal Degradation

Basically, thermal degradation normally produces a mixture of low value hydrocarbons having various products, containing hydrogen to coke. Generally, the increase in temperature of pyrolysis cracking led to produce small quantities of liquid fuels and increase the production of Non-condensing gaseous. The yield and composition of the products gained are controlled by the time of the reaction as well as the temperature [4]. General steps of the thermal degradation are described by the following sequence of procedure [5,6]:

1-Initiation step: happens at random scission of the polymer chain and initiating primary radicals Rp (1)

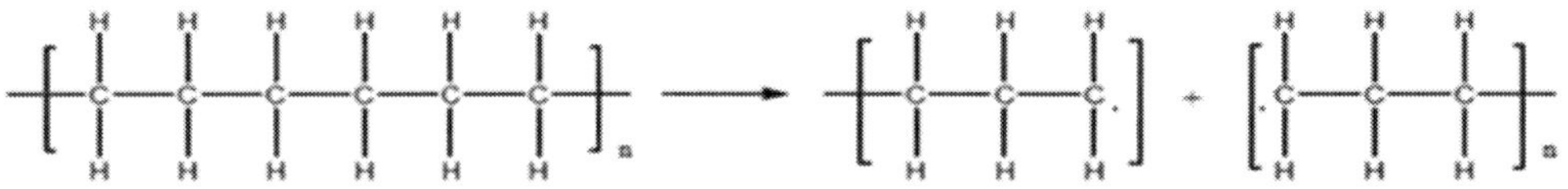

2-Propagation step: olefin monomers are release in this step; it's formed from primary radicals by β-scission which leads to produce predominantly propene (2)

3-Hydrogen chain transfer reaction step: maybe occur as intermolecular or intramolecular (rely upon low or high temperatures) processes.

a. Intramolecular (H) transfers: leads to produce a secondary radical which is more stable. The sequent β-scission of secondary radicals act jointly with the propagation step of the radical chain mechanism since a new primary radical is formed in this process. In addition, the formation of olefins takes place during the process C-C bond hemolysis [1].

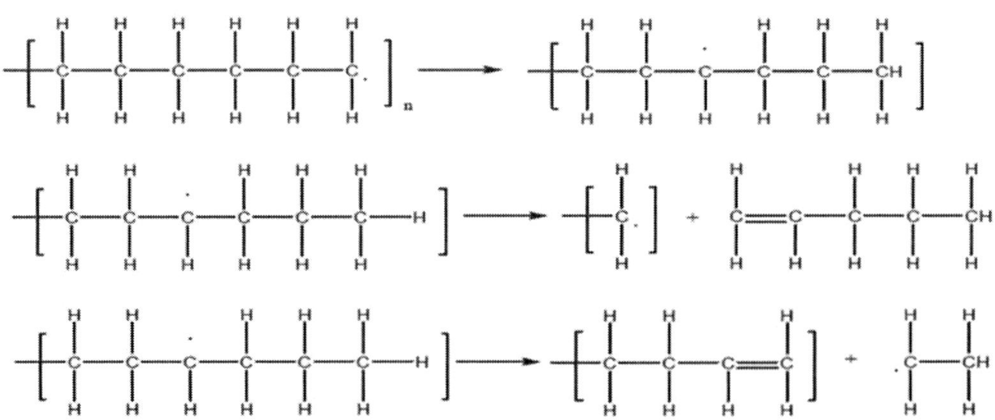

b. Intermolecular (H) transfer: secondary radicals maybe formed through H abstraction by an intermolecular transfer reaction

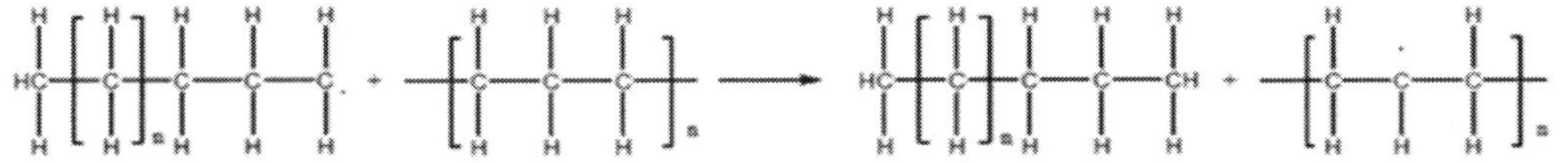

4-Termination step: involves the formation of a neutral molecule by the merging of secondary radicals.

As a result of these reaction steps, the thermal pyrolysis lead to great distribution of hydrocarbon, a C5 C80 range, each fraction being mainly composed of diene, 1-olefin and n-paraffin. Hydrogen formed at high temperature in significant quantity. The products produced from thermal cracking have a finite commercial price, particularly, which is applied as fuel. Where the heavy oils has been proposed, it has been used as a wax [7]. Achieving this great range of products is about one of the main advantages of this technique, which demands temperatures of 500 °C to 900 °C. These factors severely limit its applicability and increase the cost of recycling raw material of plastic waste [8].

The thermal pyrolysis process featured is seem to be a simple process with low costs, but the high energy consumption [9]. A large number of experimental studies have been as listed in Table 1.

Table 1. Recent laboratory experiments on plastics pyrolysis

Author	Experimental steps	Main products and comments
Kiran, Ekinci and Snape (2000) [10]	Thermal pyrolysis of waste and virgin PP plastics & mixtures of waste PP with wastes of polyethylene (PE) and polystyrene (PS)	The liquid yield distribution concerning the aliphatic, mono-aromatic and poly-aromatic compounds varies as the ratio of PP waste increases in the waste plastic mixtures. In addition to this, the alkene/alkane ratio of gas products shows variations depending on the mixing ratio of wastes.
Ballice (2002) [11]	Different LDPE:PP ratios	Maximum product release temperature: 440°C giving straight and branched paraffins and olefins, dienes and aromatics (C1-C27). Increasing ratio of LDPE produced greater amount of C16+ paraffins Produced
Seo & Shin (2002) [12]	Mixed waste plastic pyrolysed at 400-500°C	PE (50-60%), PP (20-30%), PS (10- 20%), PVC (10%) gave pyrolysis oil toluene (49-68%) Aromatics in oil said to be partly due to PS in the feed but likely some other effect in pyrolysis process contributed too.
Faravelli et al. (2003) [13]	PE & PS 370°C and 410°C	HDPE –alkanes, alkenes, dialkenes PS –monomer, dimer and trimer PS degradation independent from presence of PE
Ciliz, Ekinci and Snape (2004) [14]	PE:PP and PS:PP mixes Virgin	polymers gave Ea of 167-181 kJ/mol. PE:PP mix gave high values of gas and residue and low liquid yield. Liquid yield for PE:PP mix reduced with increasing PP, but more C7-C11 aliphatics. Addition of PP increased alkene/alkane ratio for C2 and C3.

1-2. Catalytic Degradation

In this type of pyrolysis a suitable catalyst is used during the pyrolysis to carry out the cracking reaction, the presence of catalyst lead to reduce the temperature of reaction and time. The product distribution of carbon atom number and peak at lighter hydrocarbons in this process is much narrower. To make this process economically feasible, the cost should be further reduced. The use of effective catalysts in lesser amounts and reuse of catalyst can optimize this option. Catalytic pyrolysis can be improved into a cost-effective commercial polymer recycling process for solving the serious environmental problem of plastic waste. It also offers the higher cracking ability of plastics, and the lower concentration of solid residue in the product [15].

Riyadh et al.[16] studied the catalyst pyrolysis of PS, PET and XPET+PS The maximum liquid yields from catalytic pyrolysis of polystyrene, 10%PET +PS, 20%PET +PS and 30%PET +PS was 98.4% at 450 °C, 82.6 % at 450 °C, 72.5% at 450 °C and 55.8 % at 400 °C, respectively. The catalyst pyrolysis of PET lead to acid compounds like benzoic acid, terephthalic acid, ethylene glycol dibenzoate and Dibenzoyl-L tartaric acid anhydride as the main components.

2- Experimental work

2-1. Materials

The plastics utilized in current paper were waste polyethylene terephthalate (PET) and polystyrene (PS) and their mixture. The PET was actual waste plastic from soft drink, and the waste PS was from disposable foam food containers. Additionally, collected waste PET and PS were washed with soap and water and cut into small pieces. The particle size of PET and PS was (5-6) mm as shown in figures (1) and (2). Soft drink bottle and disposable foam food containers were transparent. Nitrogen gas was supplied from Dijlah Factory, with purity 99%.

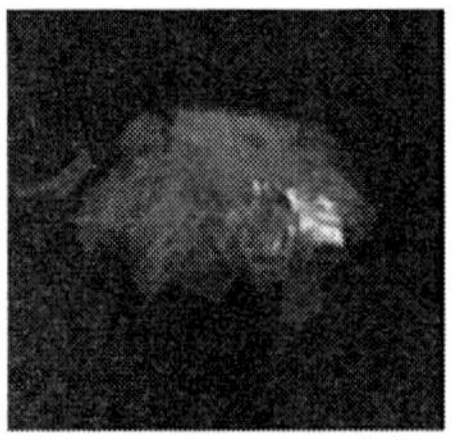

Figure 1. waste PS

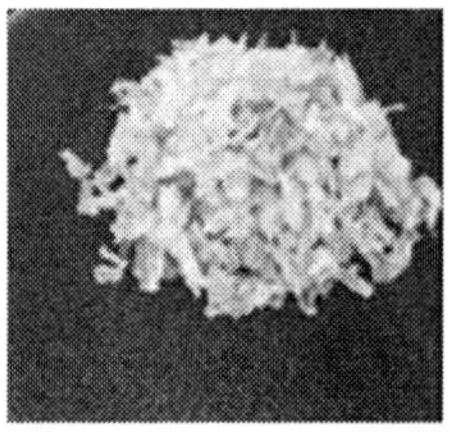

Figure 2. waste PET

2-2.Thermogravimetric analysis of materials

The weight loss verses temperature curve for thermal decomposition of polyethylene terephthalate and polystyrene under a linear heating rate shows that the decomposition was a single step degradation started at 300°C and completed at 463°C for PS while the degradation of PET started at 350°C to 465°C for PET. The TGA analysis was measured by the laboratories of College of Education for pure Science (Ibn Al-Haitham).

2-3.Liquid Product Analysis

The reaction products (liquid) were analyzed in a GC-MS (shimadzy model AOC-20i / AOC-20s) auto injection that is located in the Ministry of Science and Technology. Temperature 100 °C to 260°C.

2-4.Procedure

The equipment used in the thermal pyrolysis of wastes plastic consists of Batch tubular reactor made of stainless steel of about 340 mm length, 40 mm inside diameter and 45 mm outside diameter. Thermo-couple (type K) with digital temperature recorder, connected to the reactor, used to measure the temperature inside the reactor. The heat was supplied to the reactor by 3600 Watt external electrical heaters to get the required reaction temperature. The condensate products are collected from the bottom of the condenser and the gaseous products leave the system at the top of the condenser. The batch reactor was first cleaned with naphtha. 50 g of a certain plastic particles (PS or PET) were weighted by using a four decimal balance ,both of weight plastic particles (PS, PET, 10%PET+PS, 20%PET+PS, 30%PET+PS) were fed to the reactor. Nitrogen was supplied several times to purge the air out of the reactor before the experiment started. The pressure fixed at 1 atmospheric. The electrical heater was turned on to reach the required reaction temperature; the accuracy of temperature readings was about 3°C. zero time was recorded ,when the gases were started to get out of reactor through the condenser. Then, the cooling water was supplied by chiller at temperature 3 to 5°C the condenser was effective enough to drop the temperature of the pyrolysis product to less than 30 °C. The product separated in the condenser to non-condensable gases has left from the top of the condenser to get out of laboratory. Liquid (contains wax) was collected from the bottom of condenser in sealed cylinder. The liquid and the wax weight by four dismal balances. The gas weight was calculated by making difference between initial weight of plastic and summation of products weight (liquid, wax and coke). The time of experiment ends when no further production of liquid and wax (end of plastic inside the reactor). The schematic diagram and photographic of the experimental apparatus are shown in Figures 3 and ٤, respectively.

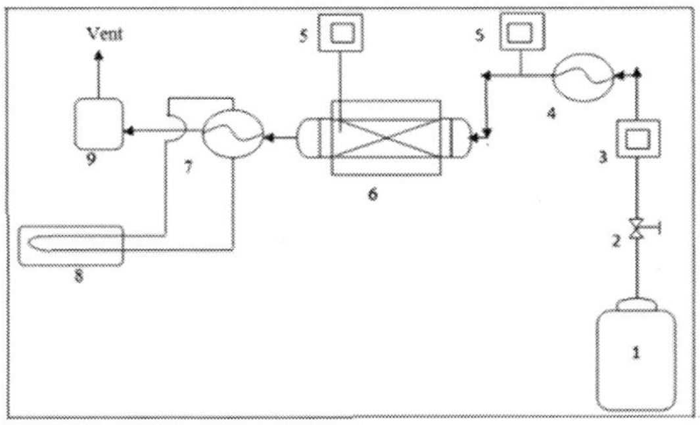

Figure 3. Schematic diagram of the experimental apparatus1) Nitrogen cylinder, 2) Gate valve,3) Gas flow meter,4) Preheater, 5)Temperature reader,6) Reactor, 7)Condenser, 8)Cooling system, 9) Liquid collector.

Figure 4. Photograph of the experimental

2-5. Experimental Design

The following Table 2 shows the thermal pyrolysis of PS and PS+xPET.

Table 2. Experimental procedure of Thermal pyrolysis

Material	Time (min)	Temperature °C
PS	90	350
		400
		450
PS+ 10%PET	90	350
		400
		450
PS + 20% PET	90	350
		400
		450
PS + 30% PET	90	350
		400
		450

3. Results and discussion

3-1 Effectiveness of reaction Time on Thermal Pyrolysis of PS

The degradation of plastic products has been affected by reaction time. In this study, the test of optimum reaction time was studied between 30 min to 120 min for the pyrolysis of PS by using the maximum decomposition temperature (450◦C). The results of the trials are shown in Figure 5. It has been noticed that liquid product increased notably when the reaction time was increased from 30 min to 90 min. The maximum yield was 77% at 90 min. A further increase in reaction time showed no clear change in the yield of liquid products.

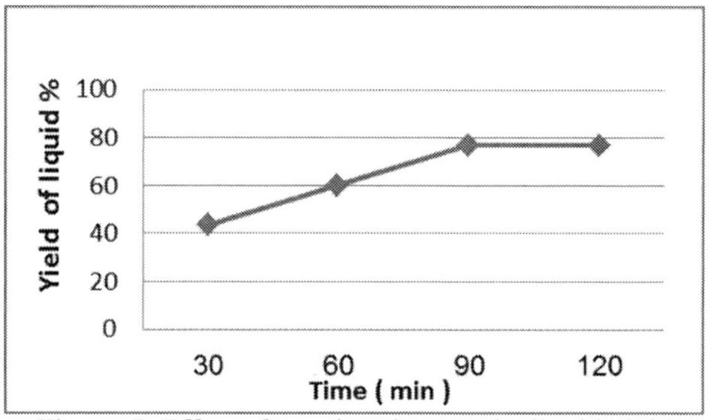

Figure 5. Effect of reaction time on the liquid yield

3-2. Effect of Temperature on Thermal Pyrolysis

Temperature has a strong effect on the characteristics of pyrolysis liquid. The effect of pyrolysis temperature, on the thermal pyrolysis of PS, 10%PET + PS, 20%PET + PS and 30%PET+PS, and PET was studied to know the temperature that gives maximum yield. The liquid produced from waste polystyrene increased from 58.2% to 67.8 %, while the gases and residue decreased from 27.6% to 22 % and 14.2 % to 10.2 %, respectively when the temperature was increased from 350°C to 400°C and the liquid further increased from 67.8% to 77% and further decreased in gases and residue from 22 % to 16 % and 10.2 % to 7% respectively when the temperature was increased from 400°C to 450°C.

For the thermal pyrolysis of 10% PET + PS the yield of the liquid and wax increased from 54% to 58.6 % and from 0.46 % to 0.92 % respectively, while the gases and residue decreased from 22.14 % to 20.5 % and from 23.4 % to 19.98 %, respectively when the temperature was increased from 350 °C to 400 °C. When the temperature increased from 400°C to 450, liquid and wax products further increased from 58.6 % to 67.56 % and from 0.92 % to 1.58 %, respectively and further decreased in gases and residue from 20.05 % to 19.6 % and 19.98% to 11.26%, respectively.

In case of 20% PET + PS, the amount of the liquid and wax increased from 44.6% to 51.4% and from 0.84% to 1.52%, respectively, while the gas and residue decreased from 22.16% to 19% and from 32.4% to 28.08%, respectively when the temperature was increased from 350°C to 400°C, and then the liquid and wax further increased from 51.4% to 54% and from 1.52% to 1.64%, respectively and further decreased in the gases and residue from 19 % to 17% and 28.08% to 27.36% respectively when the temperature increased from 400°C to 450°C. The amount of liquid and wax produced from 30% PET + PS were increased from 27.8% to 37.48% and from 1.96% to 2.02%, respectively, while the gases and residue decreased from 30.04% to 28.9% and from 40.2% to 34% respectively, when the temperature was increased from 350°C to 400°C, but the temperature further increased reduced the yield of the liquid and residue from 37.48% to 30.4% and 34% to 28.9%, respectively and increase of the yield of gases and wax from 26.5 % to 37.46 % and from 2.02 % to 3.24 %, respectively. Fig. 6 show the effect of temperature on the products produced from thermal pyrolysis.

The yield of liquid products reported in the previous study for the catalytic pyrolysis of PS & PS+XPET was calculated in terms of wt.% and the results were compared the thermal pyrolysis, the comparative data are compiled in Table 3.

Table3. Comparison of liquid yield produced from Plastic conversion in absence and presence catalyst

No.	plastic	Temp., °C	Liquid yield, %	
			Thermal pyrolysis	Catalytic pyrolysis [16]
1	PS	450	77	98.4
2	10 % PET + PS	450	67.56	82.6
3	20 % PET + PS	450	54	72.5
4	30 % PET + PS	450	30.4	50.8

To give an illustrating The thermal degradation of PET lead to thermally degradable more oxygen atoms release and that has an effect on polystyrene degradation and produce more non-condensable gases when the temperature increase from 400°C to 450°C in case of 30% PET + PS.

Polystyrene pyrolysis gives higher amount of liquid than 10% PET + PS, 20% PET + PS and 30% + PS. Polystyrene is mainly converted into stable aromatic components as liquid phase and this is clear in the mixture state (polystyrene addition increases the oil yield), Buekens [17] reported the same notice.

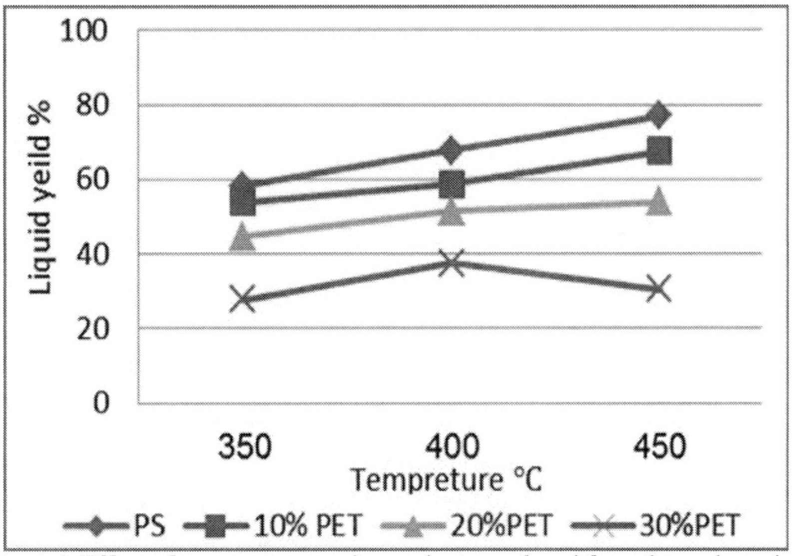

Figure 6. Effect of temperature on the products produced from thermal pyrolysis

3-3. GC-MS analysis

The liquid product obtained from the thermal pyrolysis of polystyrene (PS), and polyethylene terephthalate (XPET), optimized conditions were specified by GC_MS. Figures 7,8,9,10 show the GC-MS spectra of the liquid products derived from PS and PS+XPET. The liquid product gained from the pyrolysis of waste PS, and XPET+PS gave mainly aromatic hydrocarbons. The analysis aimed to study the effect of temperature on the main components distribution. Table (4) gives the main component appear in GC-MS analysis for the liquid produced from thermal pyrolysis of PS and XPET+PS.

The GC-MS analysis for liquid produced from PS and 10%PET + PS shows that the main products were styrene (C_8H_8), toluene(C_7H_8), and ethylbenzene (C_8H_{10}), 1,3 Diphenylpropane ($C_{15}H_{16}$), α-methylstyrene (C_9H_{10}) and 4-phenyl-1 butene ($C_{10}H_{12}$). The monomer styrene is the dominate product. The results of liquid produced form 10%PET + PS approximately which is the same as that found with 100% PS because the PET percentage is low therefore the effect of PET on PS degradation is low.

From GC-MS analysis new chemical compounds appeared in liquid produced from thermal pyrolysis of 20% PET + PS and 30%PET + PS such as Benzaldehyde (C_7H_6O), 1-Phenylnaphthalene ($C_{16}H_{12}$), 1,3-Diphenyl-1-butene ($C_{16}H_{16}$), 1,4-Diphenylbenzene ($C_{18}H_{14}$) as a results of PET percentage increase.

Table4. The major component appear in GC-MS analysis for the liquid produced from thermal pyrolysis of PS and XPET+PS

PS	10%PET+ PS	20%PET+ PS	30%PET+PS
styrene (C_8H_8)	styrene (C_8H_8)	styrene (C_8H_8)	styrene (C_8H_8)
toluene(C_7H_8)	toluene(C_7H_8)	toluene(C_7H_8)	toluene(C_7H_8)
ethylbenzene (C_8H_{10})	ethylbenze-ne (C_8H_{10})	ethylbenzene (C_8H_{10})	ethylbenzene (C_8H_{10})
1,3 Diphenylpro-pane ($C_{15}H_{16}$)	1,3 Diphenylpropane ($C_{15}H_{16}$)	1,3 Diphenylpropane ($C_{15}H_{16}$)	1,3 Diphenylpropane ($C_{15}H_{16}$)
α- methylstyre-ne (C_9H_{10})	α- methylstyrene (C_9H_{10})	α- methylstyrene (C_9H_{10})	α- methylstyrene (C_9H_{10})
4-phenyl-1 butene ($C_{10}H_{12}$)	4-phenyl-1 butene ($C_{10}H_{12}$)	4-phenyl-1 butene ($C_{10}H_{12}$)	4-phenyl-1 butene ($C_{10}H_{12}$)
------	------	Benzaldehyde (C_7H_6O)	Benzaldehyde (C_7H_6O)
------	------	1-Phenylnaphthalene ($C_{16}H_{12}$),	1-Phenylnaphthalene ($C_{16}H_{12}$),
------	------	1,3-Diphenyl-1-butene ($C_{16}H_{16}$),	1,3-Diphenyl-1-butene ($C_{16}H_{16}$),
------	------	1,4-Diphenylbenzene ($C_{18}H_{14}$)	1,4-Diphenylbenzene ($C_{18}H_{14}$)

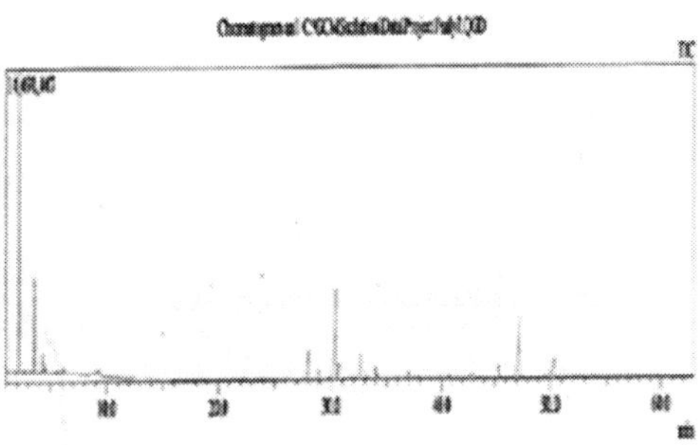

Figure7. GC-MS of liquid produced from Thermal pyrolysis of PS at 450°C.

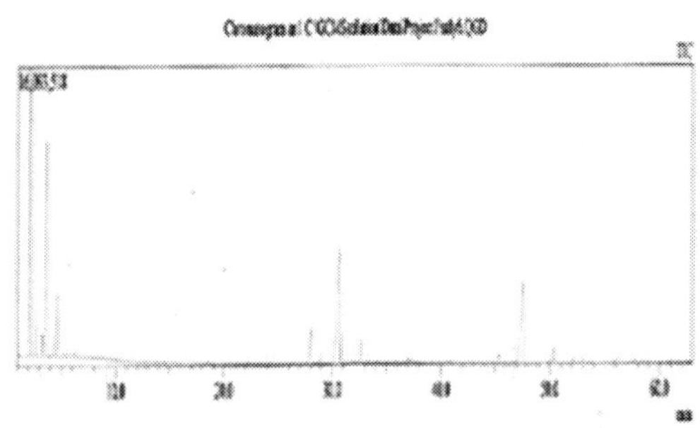

Figure 8. GC-MS of liquid produced from Thermal pyrolysis of 10%PET+PS at 450°C

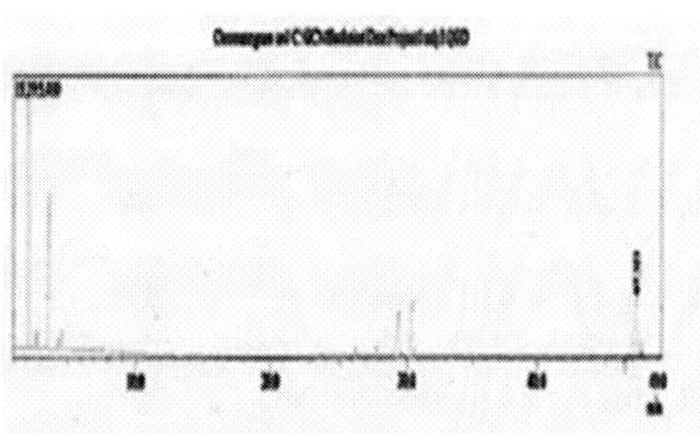

Figure 9. GC-MS of liquid produced from thermal thermal pyrolysis of 20%PET+PS at 450°C.

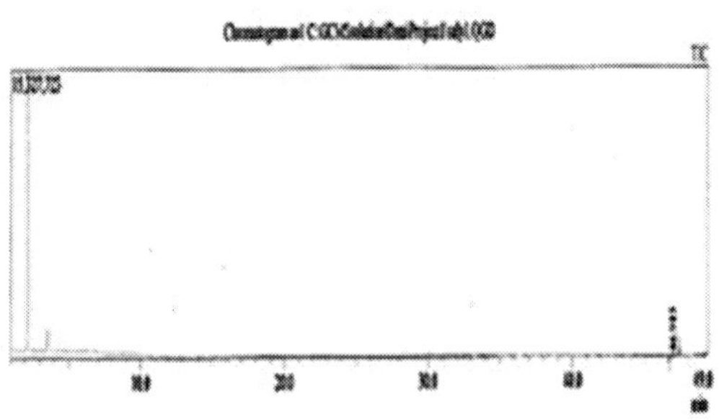

Figure 10. GC-MS of liquid produced from thermal pyrolysis of 30%PET+PS at 450°C.

4. Conclusions

The maximum liquid yields from thermal pyrolysis of polystyrene, 10%PET +PS, 20%PET +PS and 30%PET +PS was 98.4% at 450 °C, 82.6 % at 450 °C, 72.5 % at 450 °C and 55.8 % at 400 °C, respectively.

Mostly, the liquids yield products decreased with the increase in the amount of polyethylene terephthalate. GC–MS analysis of liquid produced from polystyrene and XPET + PS showed the formation of single ring aromatic hydrocarbons (C6–C9) fraction. Styrene monomer was the major product. The increasing concentration of aromatics (C_{12}_C_{24}) and due to increasing of polyethylene terephthalate percentage many new oxygenated hydrocarbons is formed. The physical property of the produced liquid was closely to gas oil specifications.

Acknowledgements

This study was carried out on the chemical engineering department- university of technology, GC-Ms analysis has been done at ministry of science and technology.

References

[1] Ćwik A 2014 *Fuel from Waste Catalytic degradation of plastic waste to liquid fuels* MSc. Thesis University of Science and Technology
[2] Fried J 2014 *Polymer Science and Technology* (Massachusetts: Prentice Hall) ed. 3 p 637
[3] Sohel R 2012 *Feasibility study of Plastic to Diesel factory at Dhaka in Bangladesh* MSc. Thesis Arcada University of Applied Science
[4] Scheirs J and Kaminsky W 2006 *Feedstock Recycling and Pyrolysis of Waste Plastics: Converting Waste Plastics into Diesel and Other Fuels* (West Sussex: John Wiley & Sons) p 816
[5] Baljit S B and Sharma N 2008 *Polymer Degradation and Stability* **93** 561
[6] Khan M I 2014 *Influence Of Some Heterogeneous Catalyst On Pyrolytic Conversion of Model Polyproplyene into Fuel Like Products* Ph.D. Thesis University of Peshawar
[7] Almeida D and Marques M D F 2016 *Polímeros* **26** 44
[8] Lin H T, Huang M S, Luo J W, Lin L H, Lee C M, and Ou K L 2010 *Fuel Processing Technology* **91** 1355
[9] Jahirul M I, Rasul M G, Chowdhury A A and Ashwath N 2012 *Energies* **5** 4952
[10] Kiran N, Ekinci E and Snape C E 2000 *Resources, Conservation and Recycling* **29** 273
[11] Ballice L 2002 *Fuel* **81** 1233
[12] Seo Y H and Shin D H 2002 *Fuel* **81** 2103
[13] Faravelli T, Bozzano G, Colombo M, Ranzi E and Dente M 2003 *Journal of Analytical and Applied Pyrolysis* **70** 761
[14] Ciliz N K, Ekinci E and Snape C E 2004 *Waste Management* **24** 173
[15] Patni N, Shah P, Agarwal Sh, and Singhal P 2013 *ISRN Renewable Energy* **2013** 1
[16] Almukhtar R S and Hammoodi S I 2018 *Engineering and Technology Journal* **36** 27
[17] Buekens A G. and Huang H 1998 *Resources, Conservation and Recycling* **23** 163

IOP Conf. Series: Materials Science and Engineering **579** (2019) 012019 doi:10.1088/1757-899X/579/1/012019

Tracking boil off gas generation into liquefied natural gas supply chain using HYSYS simulator.

S Z Naji[1], A A Abd[2] and A S Hashim[2]
[1]Petroleum Engineering Department, University of Kerbala, Iraq.
[2]Water Recourses engineering college, Al-Qassim Green University, Iraq.

Corresponding author's e-mail address: samah.alrashid@uokerbala.edu.iq

Abstract. Liquefied natural gas (LNG) is becoming one of the prominent clean energy sources with its abundance, high potential energy, and low emission and price. Unavoidable reality is that boil off gas (BOG) generates, because of low temperature of LNG, about -162.5 °C at 101.3 kPa for storage and transportation of liquefied natural gas. As natural gas productions increasing globally, BOG and handling problems become serious critical subject in both economic and environment regulations. In this work, liquefied natural gas storage facilities and loading facilities are simulated to investigate the generated BOG by calculating the heat leaks through cargo tank, pipelines, storage tank, and the generated heat by equipment's. Later, boil off gas rates are estimated for each sector into LNG supply chain with different operational conditions, parameters and voyage time. Factors that affect BOG such as operation temperatures and pressures, methane content, and nitrogen content are presented. Current study would help to properly store and transport LNG to minimize the venting of BOG into supply chain, and thus reducing greenhouse emissions, save energy, and reducing waste of natural gas.

1.Introduction

Liquefied natural gas (LNG) is the fastest growing energy sector due to increasing the demands of clean energy. Worldwide, the natural gas trade increased by 50 % between 2006 and 2014 which is almost double the last 10 years ago. Natural gas exporting countries increased from 6 to 26 countries in the period from 2000 to 2014 [1]. United States Energy Information Administration states that the world natural gas trade, by both pipeline and shipment in the form of LNG, will be poised to increase tremendously in the future [2]. 297 million tons per year is the operating capacity of natural gas recorded in 2014 [3]. New liquefied natural gas plants are under construction will promote LNG trade with up to 125 million tons per year [4].

Economically, natural gas transport in liquefied phase for long distance reduces cost because Liquid phase can reduce volume up to 600 times less than that for gas phase for same mass. However, to liquefy natural gas huge energy is required due to its bubble point below -161 °C. Large difference in temperature of liquefied natural gas and its surrounding can result heat leaks into in spite of carful and efficient insulation. This seep heat leads to evaporate part of LNG into different sectors of the LNG supply chain to generate boil off gas. Boil off gas (BOG) is considered real problem in natural gas trade which reduces the quality of LNG, changes the compositions percent, and increases the pressure of storage tank. Methane and nitrogen are the lighter components of LNG, while ethane, propane, and others represent heavy components. Therefore, methane and nitrogen evaporate first due to their lower boiling point. Generally, boil off gas realses to environment to keep pressure in tank at acceptable range. However, releasing methane to environment is more dangerous than carbon dioxide CO2 because it has radiative efficiency 26 times higher than (CO2) [5]. CO2 emission from unused natural gas from oil production industry was 250 million of tons in 2011 [6], while methane was the second contributor in the greenhouse gas emission between 1990 to 2011 [7]. Therefore, it is important to reduce the flaring and venting of natural gas.

Liquefied natural gas supply chain includes loading and unloading, shipping, and LNG storage. Liquefied natural gas evaporates continuously due to its low temperature and lost as boil off gas through the sectors of supply chain. The rate of BOG depends on the operation and construction conditions which are around 0.12% per day [8]. This rate is considered as losses and due to more intensive global competitions and stricter environmental regulations, venting BOG is unacceptable. Recently, part of boil off gas burns to power the steam turbines in order to reduce the pressure in tank and reducing the flaring, efficiency of this process is estimated around 30% [9].

In addition, BOG applications are fuel in loading stage, re-liquefied in case of large rate, burn or sent to the re-gasification unit, and/or sent back to the ship's tank. Several factors have direct effects on the generation of BOG which are depressurization of LNG, heat ingress through the pipelines and storage tanks, and heat generated by some equipment like pumps [8] [10] [11].

The BOG rate is affected by design and construction of storage tanks, operating conditions, the compositions of LNG itself [2]. Boil off gas rate strongly depends on the mass heat of vaporization and density of LNG [12]. In what follows, we first simulate cargo tank to estimate total leaking heat for a variety of operation The BOG rate is affected by design and constructional conditions, and voyages time. In addition, the leaking heat through the pipelines and storage tank has been estimated with different scenarios. Later, we use these quantitative of heats to determine the total amount of boil off gas generation at each stage. Analysis study has been conducted to test the effect of varying temperatures and pressures of storage tank on boil off gas rate. Due to their low boiling point, the effect of methane and nitrogen content on the boil off gas has been summarized. The study would help proper handling of BOG problems and prediction of the positions that BOG can consist and at what rate. As well as, the study can help designers to plan a suitable method to deal with the generated BOG rate based on the operation conditions.

2. Process description

After liquefication process, natural gas sends to the storage tank at export country until ships arrive to transport the product. Fig. 1 illustrates the process of transport liquefied natural gas from export region to the import region. The ships are usually powered with steam engines, while the speed of ship depends on its capacity typically to travel with 19 knots. The loading plant almost needs 12 to 14 hours to load ship with vessel capacity up to 130000 m^3. Then, ship sails to transport LNG which takes around 12 days in total to travel 5000 miles. Almost, unloading process needs between 10 to 12 hours to unload LNG at receiving terminal. LNG sends to offshore storage tank with storing capacity up to 250000 m^3 using cargo pumps. Liquefied natural gas continuously evaporates because of heat leakage during storing, shipping, and loading/unloading process. These vapor rates called boil off gas are considered a problem. The process of transport LNG simulated using HYSYS program V8.8. Aspen HYSYS (or simply HYSYS) is a chemical process simulator used to mathematically model chemical processes, from unit operations to full chemical plants and refineries. HYSYS is able to perform many of the core calculations of chemical engineering, including those concerned with mass balance, energy balance, vapor-liquid equilibrium, heat transfer, mass transfer, chemical kinetics, fractionation, and pressure drop. The thermodynamic property fluid package used for this simulation was Peng-Robinson. The calculations of heat losses in the pipelines follow Begges and Brill correlations which is available by HYSYS and storage tank calculations use equations of heat modes. Some assumptions have been made to simplify the complexity of simulation. The cargo tank is assumed to be normal storage tank, the unloading and loading pipelines are assumed to have same diameter. Methane percent in this simulation in range 80 % to 99%, 0.1% to 2% nitrogen content and the rest are ethane, propane, i-butane and n-butane [1].

Figure 1. LNG supply chain.

3. Results and Discussion

3.1Heat leaking calculations

Heat leak calculations through storage tank and transport facilities are necessary to study boil off gas generation. Heat transfers from environment to liquefied natural gas by three modes conduction, convection, and radiation.

3.1.1 Heat leaks to cargo tank

Cargo tank is moss type spherical tank with capacity up to 143000 m^3. During the voyage, cargo tank gains some heat from ambient, this heat warms up LNG temperature and generates amount of vapor depending on the operational conditions and voyage time. BOG generation in cargo tank depends on the operational conditions and the sea conditions, which is around 0.1 to 0.15 % [13]. The process of transporting LNG from export to import terminal called laden voyage, whereas the return voyage of the tanker after the unloading process called ballast voyage. During ballast voyage small amount of LNG stays in the tanker to keep the temperature of inside around -162°C which can be considered as losses as well. Possible reasons behind evaporation into cargo tank are the sloshing of cargo in partially filled tanks due to the action of waves, friction on the inner wall of cargo tank creating an additional thermal effect and temperature difference between ambient and cargo tank. The ships usually travel with speed up to 9.7 m/s and by assuming that the distance of journey 4000 km which needs 5 days of traveling. Ship's tank translates up to 3600 kg per one hour, **fig. 2** shows the varying of BOG rate with voyage time. The boil-off gas can be sent as fuel to the propulsion system or re- liquified to change it into liquid phase again.

IOP Conf. Series: Materials Science and Engineering **579** (2019) 012019 doi:10.1088/1757-899X/579/1/012019

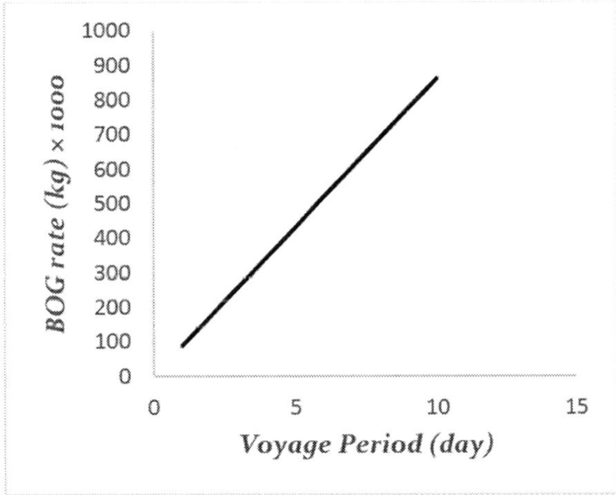

Figure 2. Voyage time against BOG rate

3.1.2 Heat leaks to pipelines

Loading pipes are two parallel lines come with diameters o.6096 and 0.6604 meter or one line with diameter up to 0.762 meter. Tank pipelines are input and output of liquefied natural gas, condensed liquefied natural gas input, perlite input and BOG output pipelines. All pipelines paths are through the top of storage tank, which can be easily checked from platform. Due to direct contact with cryogenic liquid, the pipes distribution and location should be designed in reasonable way. Otherwise, the following situations might be happened (1) shock chilling of line contraction can produce stress concentration or bending deformation; (2) because of heat gasification, LNG can flow into the adiabatic section structure, and then press the vapor back inside the container, it is also known as the penetration phenomenon. Therefore, pipelines usually are designed as S type in the interlayer in storage tank. Pipelines are settled in the same side of the tank at the top of the outer tank, which is easy for installation and maintenance. Whereas, inside tank pipelines should be arranged vertically. Pipelines material of construction is stainless steel 304, thickness and design diameter depend on the capacity rate of liquefied natural gas. LNG pipelines precooling is important as action before loading or unloading to reduce boil off gas rate [13]. Analysis study has been conducted to predict the total amount of heat leaks through pipes with different lengths. **Fig.** 3 shows that total heat leaking through pipes increases, as the length of pipes increases with different ambient temperatures. Whereas, the heat leaks increase at constant rate when temperature of surrounding increases.

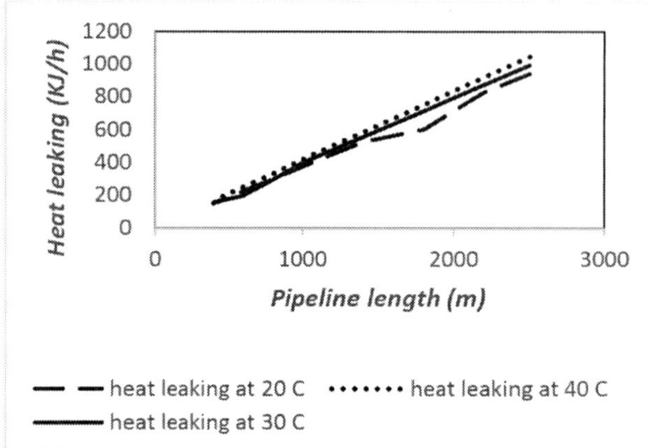

Figure 3. Pipelines length with heat leaking for different ambient temperature.

3.1.3 Heat by pumps energy

Pumps are considered as main element in the unloading/loading process, where LNG transfers from cargo to storage tank by pumps located on the ships. The required energy for each pump depends on the required capacity which is usually in the range of 1200 m^3/h to 1400 m^3/h for the large one. Where the small pump, which is called spray pump, comes with a capacity between 40 to 50 m^3/h [14]. Typically, the spray pump is used to provide LNG to the spry ring to keep the entire of tank in cold state. The simulation shows that the required energy for the pump system is at least 4630 kW for 200000 m^3 storage tank. Where most of this energy converts into heat which raise the temperature of LNG by almost 0.5 °C. The evaporation process can be reduced by increasing the operating pressure of the tank [15].

3.1.4 Heat leaks to storage tank

Storage tanks at both production plant and receiving terminal are stored LNG in atmospheric pressure and temperature around -162.5 °C with rate of boil off gas called tankage boil off gas (TBOG). This gas at LNG plant usually is compressed and exported to the plant fuel system. While, it is either flared or sent to the regasification plant using BOG compressors at receiving terminal. Generally, the designers try to reduce the number of tanks and increase the capacity of storage to minimize the cost. Type of tanks choice of single, double, and/or full containment depends on the cost, land availability, and safety purposes.

3.1.4.1 The limitation of heat leakage for storage tank

Although, large insulation thickness around the offshore storage tank stills, heat transfers to the liquefied natural gas. Usually, heat leaks in two forms: continuous and transient, where continuous leaking time shorter than that for transient type [16]. **Fig.** 4 shows the main parts of liquefied natural gas storage tank. Analysis study has been conducted to estimate amount of leaking heat by three main parts top roof, insulation wall and bottom slab. Generally, total heat leakage limitation is 0.25 volume % per day and can be calculated based on eq. 1;

$$Q_{total} = \Delta G \times q = n \times G_o \times q \qquad (1)$$

where n is boil off gas rate (vol%/day), G_o, is mass of LNG in tank (kg), and q is LNG heat of evaporation about 509 kj/kg. Value of accepted limit of leaking heat for this simulation estimated to be 1250 kw.

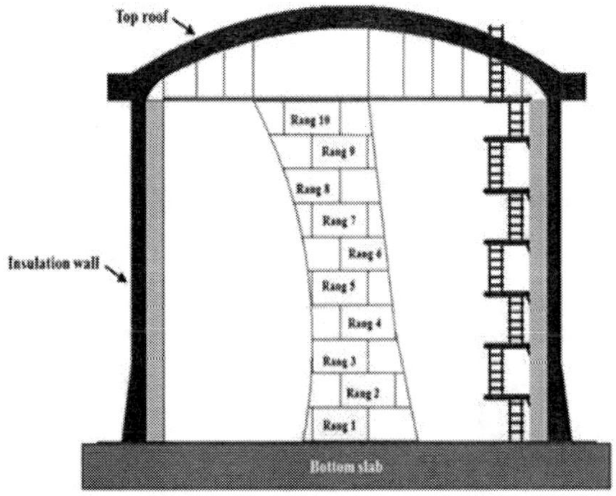

Figure 4. LNG storage tank.

3.1.4.2 Heat leakage- from isolation part

Leaking heat through the insulation part depends on thermal conductivity of construction material and thickness of insulation as shown in **fig. 5**, which can be estimated using eq. 2.

$$Q_a = \lambda_c \times A_m \times \frac{\Delta T}{\delta} \qquad (2)$$

Where λ_c is the effective thermal conductivity which is $0.09\frac{w}{m.k}$ for this simulation, A_m is surface area of inner tank which is 9917 m^2, δ is thickness of insolation part, and ΔT is temperature difference between outer and inner wall. Value of leaking heat is estimated to be 152 kw.

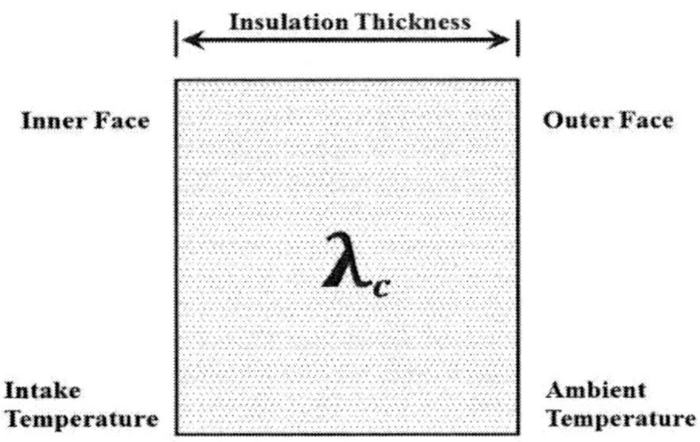

Figure 5. The insulation offshore tank

3.1.4.3 Heat leakage- from bottom part

Bottom slab consists of more than one layer with different thicknesses and materials of construction. The bottom slab layers are foundation layer comes with thickness up to 2000 mm, concrete slab layer with thickness 900 mm and thermal conductivity 0.79 w/m. k, cellular glass with thickness 450 mm and thermal conductivity 0.056 w/m. k, and perlite concrete with thickness 70 mm and thermal conductivity in range 0.09 to 0.22 w/m.k as shown in **fig. 6**. The thickness and material of construction can be manipulated to keep leaking heat through at lower possible value. Total leaking heat through bottom slab can be estimated using eq. 3, where φ is average heat intensity and V is total volume of bottom slab.

$$P = \varphi V \quad (3)$$

While, total heat intensity is 5.97 w/m³ calculated by eq. 4, using 273 k of top concrete slab temperature and the bottom temperature of inner tank is 110 k.

$$t = -\frac{\varphi}{2\lambda}x^2 + \left(\frac{t_{w2} - t_{w3}}{\delta_{tot}} + \frac{\varphi\delta_{tot}}{2\lambda}\right)x + t_{w3} \quad (4)$$

where λ is average thermal conductivity, x is thickness without foundation and concrete slab, and δ_{tot} is total thickness with foundation. Finally, total heat leaking inside the offshore storage tank from bottom slab is 132.7 kw.

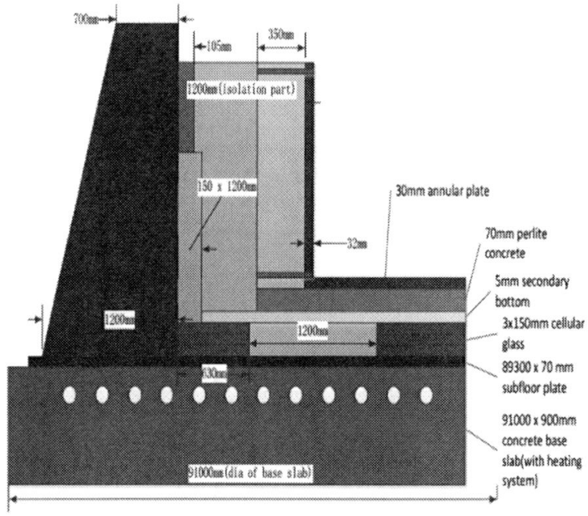

Figure 6. Bottom of tank and corner protection.

3.1.4.4 Heat leakage from roof

Top roof is another possible section to ingress heat, where heat transfers with environment by conduction, convection, and radiation. Conduction heat flux can be calculated using eq. 5:

$$Q_{conduction} = K_{concret} * \frac{\Delta T}{t_c} \quad (5)$$

Where, $K_{concret}$, is thermal conductivity of concrete which is about 2.324 $\frac{W}{m.k}$, t_c is the thickness of the concrete [17]. While, convection heat transfer per unit area can be calculated using eq. 6:

$$Q_{convection} = h.\Delta T \quad (6)$$

here, h is the convection heat coefficient of concrete which is 12.78w/m². k. The heat transfer at the roof comes in two phases conduction for the upper and bottom part of the roof, and radiation in-between as shown in **fig. 7**. Therefore, the heat transfer equations will be as below:

$$Q_{conduction.1} = K_{concret} * \frac{(T_a - T_b)}{t_c} * A_{roof}$$
$$Q_{radiation.2} = F * \varepsilon * \sigma * ({T_b}^4 - {T_c}^4) * A_{roof} \quad (7)$$
$$Q_{conduction.3} = K_{deck\,insulation} * \frac{(T_c - T_d)}{t_d} * A_{Deck}$$

Where k_{deck} is the thermal conductivity of the deck insulation which is $0.038\frac{W}{m.k}$, t_d is deck insulation with value 0.05 m, and t_c is thickness of concrete, A_{roof} and A_{deck} are the areas of the roof and suspended deck with 7467.4 m² and 6647.6 m² respectively, F is the form factor assumed to be 1, ε is emissivity, and σ is Stefan Boltzmann constant with value $\sigma = 5.67 \cdot 10^{-8}$ W/m² ·K⁴. Temperatures of environment is assumed to be 288.13 k, and temperature inside the tank assumed to be 110.15 k. The emissivity should

IOP Conf. Series: Materials Science and Engineering **579** (2019) 012019 doi:10.1088/1757-899X/579/1/012019

be calculated as a resultant for both carbon steel liner emissivity which is 0.66 and deck insulation emissivity which is 0.96 as shown in eq. 8 [17].

$$\varepsilon = \frac{1}{\frac{1}{\varepsilon_{\text{Carbon steel}}} + \frac{1}{\varepsilon_{\text{Deck insulation}}} - 1} = 0.64 \quad (8)$$

At equilibrium conditions, amount of heat transfer modes through the roof can be assumed to be equals:

$$Q_{\text{conduction.1}} = Q_{\text{radiation.2}} = Q_{\text{conduction.3}} \quad (9)$$

$$K_{\text{concret}} * \frac{(T_a - T_b)}{t_c} * A_{\text{roof}} = F * \varepsilon * \sigma * (T_b{}^4 - T_c{}^4) * A_{\text{roof}}$$

$$= K_{\text{deck insulation}} * \frac{(T_c - T_d)}{t_d} * A_{\text{Deck}}$$

By solving these equations, the temperatures distribution can be evaluated to be $T_b = 284.27$ k and $T_c = 280.78$ k. Now the heat transfer through the roof into the storage tank can be calculated to be 368.5 kw. The assumption comes with only 5% of this heat will consider, therefore the total heat that will seep through the top roof is 18.425 kw.

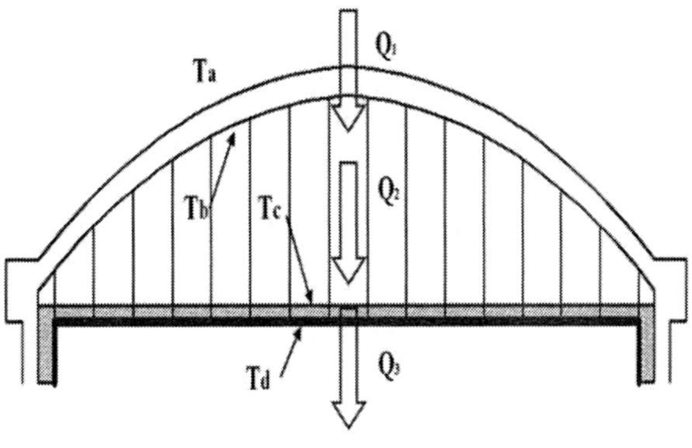

Figure 7. Heat distribution on top roof

3.1.4.5 Total Heat leakage through offshore tank

The total leaking heat through top roof, insulation wall, and bottom slab storage tank can be calculated by using eq. 10:

$$Q = Q_a + Q_b + Q_c = 303.125\text{Kw} < 1250\text{Kw} = Q_{\text{total}} \quad (10)$$

Which means the overall heat seeps inside the offshore tank is inacceptable range. Based on this heat, boil off gas rate per day produces in offshore tank based on the calculated amount of heat leaking can be estimated by eq. 11.

$$\text{Boil off Gas rate}\left(\frac{\text{kg}}{\text{h}}\right) = \frac{\text{Rate of heat leakage}}{\text{Heat of vaporization}} \quad (11)$$

3.2 Boil off gas quantities

Boil off gas generation depends on liquefied natural gas compositions, liquid level in the tank, surface area of the tank, and operation conditions like pressure and temperature. In addition, the construction materials of the supply chain units are another factor that can affect the BOG generation because of their thermal conductivity property. The pressure drop between cargo tank and storage tank increases the BOG generation where there is almost 0.1 °C change in temperature for 1 kPa.

According to this design and based on estimated leaking heats, the amount of generated BOG at the cargo tank depends on voyage time which is almost 3600 kg/hr. boil off gas rate that produces by pumps energy is 20000 kg/hr. Whereas, boil off gas generation in pipelines is varying with length of pipelines and material of construction which is almost 1.6 kg/hr for 2500 m with steel 304. While, BOG generation at both storage tanks is almost 3287.9 kg/hr, where one tank produces 1643.95 kg/hr which is equal to 3.56 m^3/hr (85.44 m^3/day) with 461.2 kg/m^3 LNG density. For tank with capacity 200000 m^3, this is equivalent to 0.043 vol.% which is in range of normal storage tank design specifications. Whereas, the total boil off gas generation for this simulation is 26889.5 kg/hr.

3.3 Storage tank pressure

Usually, the pressure of LNG storage tank is about 101.3 kPa, however pipelines that handle the LNG to storage tank may cause a pressure drop. Therefore, analysis study has been conducted to check the effect of varying storage tank pressure on boil off gas rate. While. The temperature of storage tank is assumed to be constant at -162.5 °C. The compositions of LNG for this study are 90% methane, 1% nitrogen, 5.9% ethane, 2.7 % propane, and small amount of n-i butane. Fig. 8 shows that as the pressure of storage tank increases, the boil off gas rate decreases to reach zero at 114.3 kPa.

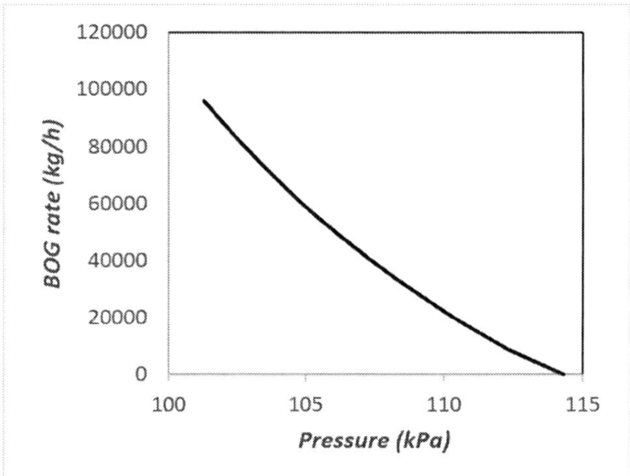

Figure 8. Pressure of storage tank against BOG rate

3.4 Storage tank temperature

During the holding mode which is the period between unloading and loading liquefied natural gas tankers, Heat ingresses from ambient to the tank where the driving force for heat transfer is the temperature difference between environment and inside tank. Well, to keep the temperature of LNG constant and in balance with pressure, LNG cools itself with process known as auto-refrigeration by evaporating part of LNG and it consists of boil off gas. Therefore, temperature can play key role in management the generated rate of BOG. Analysis study has been conducted to test the effect of varying tank temperature on the generated amount of BOG. The compositions of liquefied natural gas in this simulation are 90% methane, 1% nitrogen, 5.9% ethane, 2.7 % propane, and small amount of n-i butane. **Fig. 9** shows that BOG rate starts increases as temperature increases at different rates with constant pressure at 101.3 kPa.

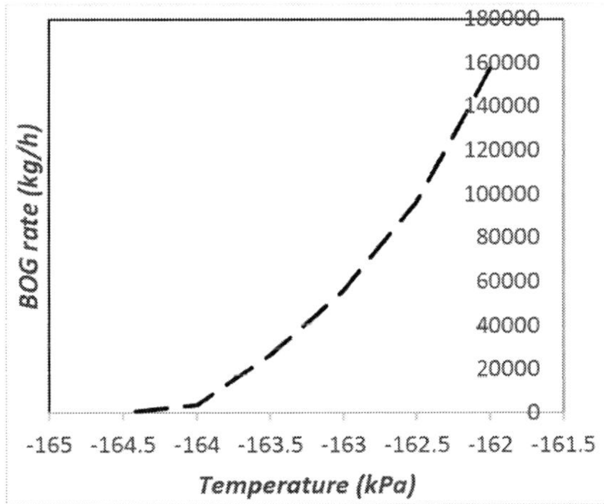

Figure 9. Temperature of storage tank with BOG rate

3.5 Methane and nitrogen content

Nitrogen and Methane considered as volatile components come with evaporation rate higher than less volatile components such as ethane, propane and other higher hydrocarbons. Analysis study has been conducted to examine the effects of varying methane percent in range 88% to 99% on the boil off gas rate at pressure 101.3 kPa and temperature -162.5 °C with 1% nitrogen. **Fig. 10** shows that increasing methane percent in liquefied natural gas causes increasing boil off gas rate. Whereas, **fig. 11** shows the effect of varying nitrogen content in storage tank on the boil off gas rate at pressure 101.3 kPa and temperature –

162.5 °C with 90% methane. The results are supported by fact that the high volatile components produce high evaporation rate.

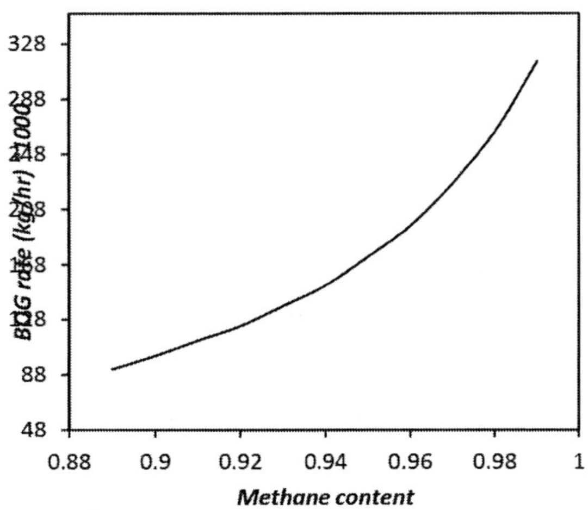

Figure 10. Methane content against BOG rate in storage tank

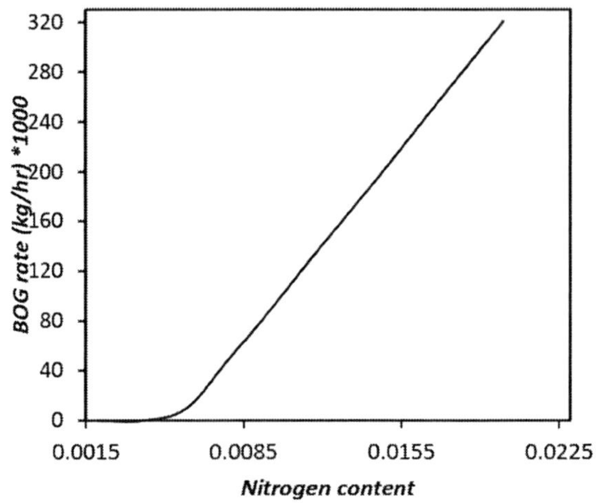

Figure 11. Nitrogen content against BOG rate in storage tank

4. Conclusions

This paper provides comprehensive analysis study to estimate the heat leaks for each part of LNG supply chain and discusses boil-off gas in regarding to this heat. Elements causing boil off gas are presented and the produced amount of BOG at each sector through the supply chain has been estimated. Effects of temperature, pressure, nitrogen content, methane content are tested to conclude best range of conditions that can reduce BOG to low rate. The assumptions for this steady simulation were cargo tank assumed to be normal storage tank, the unloading and loading pipelines assumed that have same diameter. Methane percent in this simulation in range 80 % to 99%, 0.1% to 2% nitrogen content and the rest are ethane, propane, i-butane and n-butane. Fig. 2 displays the generated BOG variation with voyage time, the results show BOG rate increases over time. Fig. 3 explains the effects of pipelines length on the generated BOG for different ambient temperature, the results show that BOG rate increases as pipelines length increases for all range of temperature. Fig. 8 and fig. 9 explain the effects of temperature and pressure on BOG rate, the results show that the best storage tank temperature to stop BOG is -164.5 °C with 101.3 kPa pressure, whereas 112.3 kPa is best pressure to stop BOG. In regards of methane and nitrogen contents, the results show that BOG rate increases as nitrogen and methane increase this is supported by fact that as more volatile components rates increase, the evaporation rate increases. To conclude, boil off gas generation represents a problem facing LNG trade and attention should be paid to reduce the BOG rate. Many factors can be manipulated to reduce the BOG generation into LNG supply chain such as reducing the voyage time, precooling will reduce the heat ingress to pipelines, and the insulation thickness and material of construction. The study would to help properly handling of BOG problems and prediction of the positions that BOG can take and at which rate. As well as, the study can help designers to plan a suitable method to deal with the generated BOG rate based on the operation conditions. It is also a solid foundation for future dynamic modelling to minimize BOG generation.

References

[1] Xiang D and Lawley C 2019 *Ener. Econ.* **80** 206
[2] Wang T, Zhang D and Broadstock D 2019 *Ener Econ.* **80** 707
[3] Oglend A, Kleppe T S and Osmundsen P 2016 *Ener. Econ.* **59** 138
[4] Yan P, William Y, Liu Q, Zhang X and Cao W 2017 *Natur. Gas Ind. B* **4** 25
[5] Lacroix A V 1993 *Chemosph.* **26** 507
[6] Jones D G, Beaubien S E, Lister T R, Graziani S, Finoia M G, Barkwith A K, and Lombardi S 2017 *Ener. Procedia* **114** 3824
[7] Colombini S, Zucali M, Rapetti L, Crovetto G M, Sandrucci A and Bava L 2015 *Agricul. Syst.* **136** 106
[8] Hwang C and Lim Y 2018 *J. Ocea. Eng. and Tech.* **32** 372
[9] Ekanem Attah E and Bucknall R 2015 *Ocea. Eng.* **110** 62
[10] Dimopoulos G G and Frangopoulos C A 2008 *Inter. J. Therm.* **11** 195
[11] Licari F A and Weimer C D 2011 *J. Los. Prev. Proc. Ind.* **24** 736
[12] Hasan M M F, Zheng A M and Karimi I A 2009 *Ind. Eng. Chem. Resear.* **48** 9571
[14] Brenn J, Soltic P, and Bach C 2010 *Ener. Build.* **42** 904
[15] Chen G H and Cheng S B 2007 *Natur. Gas Ind.* **27**
[16] Abd A A, Naji S Z and Lafta F 2018 J. Univ. Babyl. Eng. Sci. **26** 363
[17] Sinnott R K, Coulson J M and Richardson J F 2005 *Coulson & Richardson's chemical engineering. Vol. 6, Chemical engineering design* (Oxford: Elsevier Butterworth-H) ed 2 p 954

IOP Conf. Series: Materials Science and Engineering **579** (2019) 012020 doi:10.1088/1757-899X/579/1/012020

Modeling and simulation of adsorption of methane, ethane, hydrogen sulfide and water from natural gas in (FP)YEu Metal–Organic Framework.

Thaer M. Al-Jadir[1,2,#] and Flor R. Siperstein[2]

[1]Petroleum Technology Department, University of Technology-Iraq, Baghdad, Iraq.
[2]Chemical Engineering and Analytical Science, University of Manchester, U.K.

[#]Corresponding author's e-mail address: thaermunem@gmail.com

Abstract. Metal Organic Frameworks (MOFs) are novel type of hybrid porous materials that have been proposed for natural gas separation and purification due to the possibility of tailoring their capabilities by changing the organic linkers, tuning the length of the organic linker and changing or mixing the metal ion. Molecular simulation helps to understand experimental results by providing a detailed picture on the molecular scale that is not easily accessible from experimental and can be used to study the applicability of classical force fields.

In this work, FPYEu MOF [1] [2] was carefully chosen as an example of a model MOF for studying applicability of force fields, cutoff radius besides the effect of different distributions of metal ions in FPYEu MOF on adsorption of CH_4, C_2H_6, H_2S as well as H_2O.

Grand Canonical Monte Carlo (GCMC) simulations were used to perform gases adsorption on FPYEu MOF cells containing different distributions of Yttrium and Europium. All simulations in this work have been carried out using Towhee code [3].

The applicability of different force fields available in the literature was tested and some of them were applicable like H_2S, where good agreement was found with experimental results, while others were not applicable, like CH_4, C_2H_6 and H_2O, where the calculated isotherms "over or under" estimated the experimental ones. In addition, eight different configurations having Eu atoms arbitrarily located in the system were explored to evaluate the sensitivity of distribution of metal ions.

1. Introduction

Adsorption is one of the most promising and widely used for natural gas separation and purification [4] [5] [6][7]. Different types of adsorbents have been reported for adsorption of gas such as activated carbons, metal oxides, zeolites and metal–organic frameworks (MOFs) [8] [9] [10] [11] [12] [13] [14] [15] [16] [17] [16] [18] [19] [20] [21] [1] [2] [22]. The advantages of MOFs over other porous materials are that MOFs have attractive properties such as, large internal surface areas, synthetic advantages, organic-inorganic hybrid nature, and versatile applications in gas separations, sensors, gas storage and gas purification.

MOFs are a special type of hybrid porous materials constructed by the assembly of organic linking units with metal clusters [23] Figure 1. The numbers of linkers that can be connected to a metallic centre depend on the coordination of the metal and its oxidation state, which result in three-dimensional structures with different topology [23].

The (FP)YEu MOF, with molecular formula of $[(Y_{1.9},Eu_{0.1})\ (phen)_2(fdc)_3\cdot 2(H_2O)]$ as reported by Shi *et al.* in 2014 [1], is an example of a photoluminescent MOF with luminescent properties, 3-dimensional structure and the capability to withstand high temperatures up to 350–400 °C in air, displaying both porosity and selectivity, with the potential to be used as a hydrogen sulfide sensor at low pressure [1] [2].

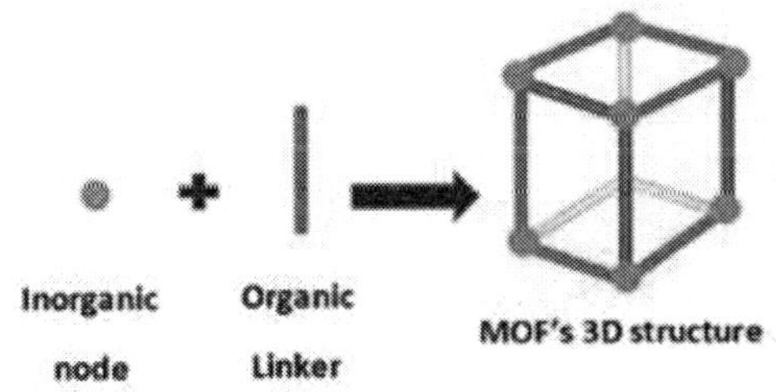

Figure 1. Diagram illustrations the construction of MOFs.

Molecular simulation helps to understand experimental adsorption results by providing a detailed picture on the molecular level that is not possible from experimental methods [24] [25] [26]. Molecular simulation is quicker than experiment and reducing the time and capital investment on experimental work [27] [24] [25] [26]. One of the most popular simulation methods is Monte Carlo (MC) simulation which is widely used to calculate adsorption isotherms of a given MOF by quick study of the impact of molecular scale properties such as surface area, pore size or pore shape on the adsorption performance for specific applications [22] [26] [28] [26] [29].

In this work, MC used to study the effect of the distribution of different metals in the system on the adsorption isotherm. In accordance with the previous work have been studied CH_4, C_2H_6, H_2S and H_2O adsorptions in FPYEu MOF using molecular simulation and the applicability of the force fields focusing on the configuration in which the Europium atoms are close in a model of FPYEu MOF [2]. It was found that force field applicability needs to be treated carefully, as in all cases large deviation with experimental results are obtained [2]. Notably, these force fields were not developed for a specific MOF. In current work, different combinations of force fields were explored to assess their validity [30]. The influence of cutoff radius on adsorbed amounts, nonpolar gases adsorption isotherms were also explored using cutoff radius of 10 Å [30]. In addition, eight different configurations having Europium atoms arbitrarily located in the system were explored to evaluate the sensitivity of distribution of metal ions on adsorption of gases.

The characterisation of FPYEu MOF presented in this work such as the geometric pore volume, crystal density and pore size were computed using Poreblazer package proposed by Sarkisov and Harrison [31]. For the geometric pore volume, the Universal force field (UFF) [32] were used to describe the Lennard-Jones (LJ) interactions of Helium and the framework atoms. The properties of FPYEu MOF are summarized in Table 1. The difference in geometric pore volume between experimental and theoretical is about 10%. This is duo to using different guest molecule to run these experiments and simulations.

Table 1. Theoretical and experimental properties of FPYEu MOF.

ρ_{cyrst} [a] (g/cm^3)	V_{pore} [a] (Å^3)	ρ_{cyrst} [b] (g/cm^3)	V_{pore} [b] (Å^3)	Pore limiting [b] (Å)	Max. Pore [b] (Å)
1.573	8808.3	1.516	9833.4	2.61	5.23

[a] From reference [1];
[b] Computed using Poreblazer v3.0.2, proposed by: Lev Sarkisov, 2012 [31];

2. Methodology

2.1 Models

The crystallographic information of the FPYEu MOF atoms were adopted from Shi *et al.* [1]. This study builds on the previous work where 3-unit cells with lattice: 28.96×28.96×30.50 Å, with 48 metal atoms have been used, having 46 Yttrium atoms and 2 Europium atoms Figure.2.

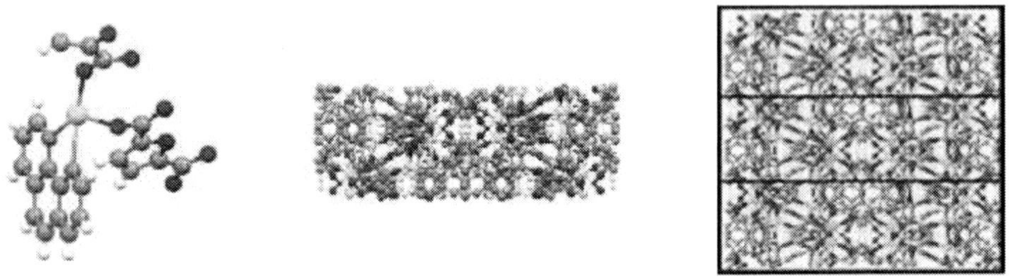

(a) (b) (c)

Figure 2. (a) The asymmetric unit cell of FPYEu MOF (yttrium/ europium: light blue, carbon: grey, oxygen: red, hydrogen: white and nitrogen: blue) [1] [2]. (b) FPYEu MOF unit cell with dimensions 28.9603, b = 28.9603 and c = 10.5023 Å [1] [2]. (c) FPYEu MOF simulation box has 3-unit cells with dimensions: 28.9603×28.9603×30.5023 Å [1] [2].

Moreover, the focus was on close configuration in the previous work Figure 3 [2]. In this study, eight different configurations having Europium atoms arbitrarily located in the simulation box Figure 4 were explored to evaluate the sensitivity of distribution of metal ions on adsorption of gases. Noteworthy, the locations of Europium atoms have been randomly chosen in configurations.

Figure 3. Close configuration, FPYEu with two close europium atoms [2].

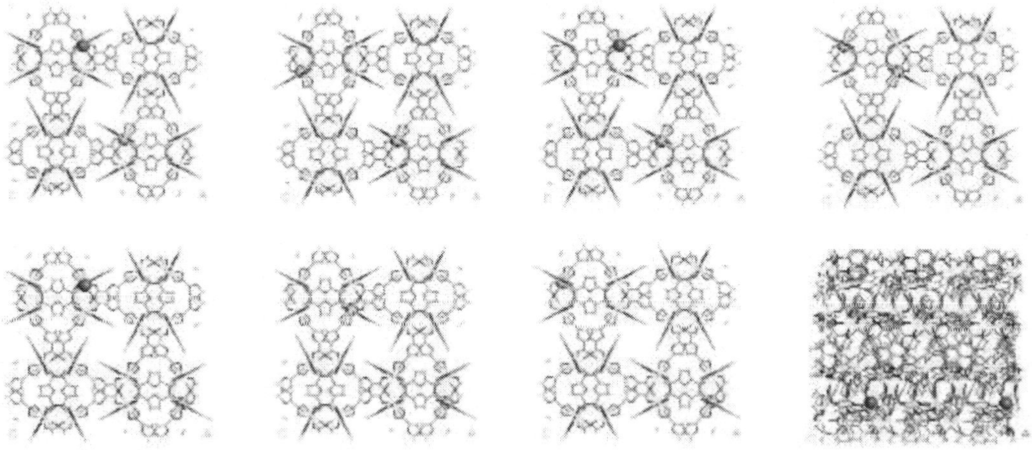

Figure 4. Eight different random configurations, FPYEu with two europium atoms randomly placed, (yttrium: light blue, europium: green, carbon: grey, hydrogen: white, nitrogen: blue and oxygen: red).

Table 2. The partial charges for FPYEu MOF obtained from Qeq method adopted from our previous work [2]. Atom locations can be identified in Figure.5 [2].

atom	Y	Eu	N1	N2	O1	O2	O4	O5	O6	C18	C20
q	0.910	1.088	-0.235	-0.219	-0.444	-0.437	-0.456	-0.481	-0.424	0.067	0.072
atom	O7	C1	C2	C3	H3	C4	H4	C5	C6	H18	C21
q	-0.441	0.068	0.085	0.085	0.068	0.068	0.061	0.053	0.059	0.078	0.082
atom	C7	C8	C9	H9	C10	H10	C11	H11	C12	C19	O3
q	0.066	0.068	0.090	0.079	0.068	0.051	0.068	0.065	0.055	0.076	0.072
atom	H12	C13	C14	H14	C15	H15	C16	C17	H17	H19	O8
q	0.048	0.055	0.060	0.061	0.090	0.075	0.074	0.069	0.077	0.061	-0.109

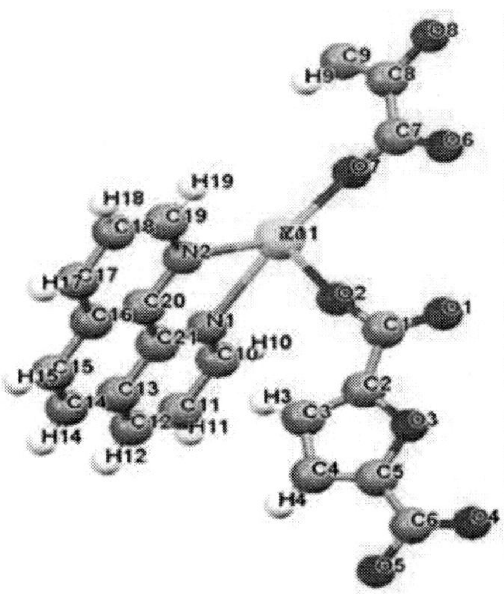

Figure 5. A symmetric unit cell displays atoms labels used to recognise partial charges in table 2 [2].

Several sets of LJ parameters have been derived for methane which are used to describe the methane interactions in all materials, of which DREIDING, TraPPE-UA and Goodbody *et al* are most commonly applied in studies of adsorption [37] [30] [38] [39] [34] [36] [22]. Methane in these force fields is represented by a united-atom model with one site. The different between these force fields is only in epsilon ε, for DREIDING is 149 K and for TraPPE-UA or Goodbody *et al* is 148 K.
In this work, TraPPE-UA force field [40] was used to represent methane and the results were compared with previous work [2] where was used DREIDING [33] model for methane.
TraPPE-UA force field [40] was also used to model ethane and the results were also compared with our previous work [2] where we used DREIDING [33] model for ethane. The different between these force fields for ethane is also only in epsilon ε, for DREIDING is 125.8 K and for TraPPE-UA is 98 K.
Kamath *et al.*[41] was used to represent hydrogen sulphide. Such a model has been successfully used to reproduce the experimental vapor–liquid phase equilibrium data and has been validated in previous studies [34] [22]. Simulations were also performed with the OPLS-UA force field [42] to compare with our previous work [2].
TIP4P-Ew force field [43] was used to model water molecules. Simulations were also carried out with the SPC/E force field [44]. The parameters for framework and the adsorbate or gas molecules are listed in Table 3 and Table 4.

Table 3. LJ potential parameters for the atoms of FPYEu MOF.

Atom type	Force field	σ (Å)	ε (K)
Y	UFF [45]	2.980	36.231
Eu	UFF [45]	3.111	4.025
N	DREIDING [33]	3.262	38.948
C	DREIDING [33]	3.473	47.859
O	DREIDING [33]	3.033	48.161
H	DREIDING [33]	2.846	7.6490

Table 4. Lennard-Jones parameters and partial charges of the gas molecules.

Molecule	Force field	Atom type	Partial charge	σ (Å)	ε (K)
H_2S	Kamath *et al.*[41]	H	0.190	0.0	0.0
		S	−0.380	3.72	232
	OPLS-UA [42]	H	0.235	0.0	0.0
		S	-0.47	3.7	125.80
H_2O	SPC/E [44]	HW	0.4238	0.	0.
		OW	-0.8476	3.1656	78.197
	TIP4P-Ew [43]	HW	0.52422	0.0	0.0
		OW	0.0	3.1643	81.8994
		M	-1.04844	0.0	0.0
CH_4	DREIDING [33]	CH4	0.0	3.7327	149.92
	TraPPE-UA [40]	CH4	0.0	3.7300	148.00

C_2H_6	DREIDING [33]	CH3	0.0	3.699	125.80
	TraPPE-UA [40]	CH3	0.0	3.750	98.00

The adsorption equilibrium between the gas phase and the adsorbed phase was modelled with Grand Canonical Monte Carlo (GCMC) simulations using Towhee code [3] [22]. The descriptions of GCMC simulations are available in several references [3] [46] [47] [28]. In the Grand Canonical Monte Carlo (GCMC) ensemble, the chemical potential μ of each component, the temperature T, and the volume V are fixed as in experiments [22]. The adsorption isotherms were calculated at temperature 298.15 K for hydrogen sulphide, methane and ethane using 5 million Monte Carlo steps for equilibration and 4 million Monte Carlo steps for production [1]. Simulations for water were equilibrated at T= 303.15 K, for 50 million Monte Carlo steps for equilibration and 50 million Monte Carlo steps for production [1]. The translation, rotation, and insertion or deletion of simulation moves was 10%, 20%, and 70% respectively [1]. The distribution of moves is motivated by the desire to have good convergence behaviour in the simulation. In other words, how fast the simulation converges to a stationary distribution to achieve fast equilibration times, fast relaxation times and correlation times. The difference in the temperatures for the calculated isotherms was to enable easy comparison with experimental and simulation data published in the literature [1] [2]. Isotherms were calculated in a range of pressures for comparison with the experimental data [1] [2].
The chemical potentials μ required in the GCMC simulation are taken from our previous work [2]. The influence of cutoff radius on adsorbed amounts [30] [39], nonpolar gases adsorption isotherms were explored for Lennard Jones interactions using cutoff radius of 10 Å and 13 Å. The long-range electrostatic interactions were calculated by the Ewald methodology [22] [28] [48].

3. Results and Discussion

3.1 Adsorption of methane and ethane

The calculated isotherms of methane and ethane in FPYEu MOF compared to the previous work [2] are shown in Figure 6. Although using different combinations of force fields, the calculated isotherms for CH_4 and C_2H_6 are still overestimated the experimental ones. Using TraPPE-UA or DREIDING model gives very similar results.
The results also show that using cutoff radius of 10 Å or 13 Å overestimated the experimental results. Nevertheless, the differences with experimental data for both gases (CH_4 and C_2H_6) indicate different non-accessible volumes [2]. The differences with experimental data could be due to the lack of applicability of Lennard-Jones parameters, pore accessibility or accuracy of the experimental isotherms.

The results also show that the simulations boxes with different Eu distributions do not affect the adsorption of different rare earth metals distributions.

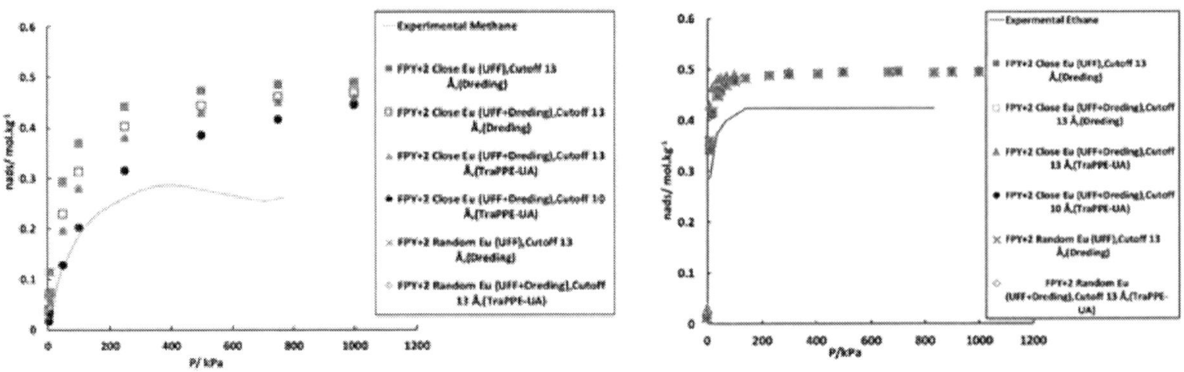

(a) (b)

Figure 6. (a) Methane (b) Ethane adsorption isotherms at temperature = 298.15 K on FPYEu with close and random distributions. Results computed by using additional combinations of force fields and cutoff 10 Å. The blue line displays the experimental as reported by Shi *et al.* [1]. The error bars of the calculated loading are smaller than the symbol size.

3.2 Adsorption of hydrogen sulfide and water

The computed adsorption isotherms for hydrogen sulfide and water with different force fields and sets of charges are shown in Figure 7. It can be observed that the selection of the force field for framework can have a dramatic impact on the simulated data. The agreement between experimental and computed data for hydrogen sulfide is excellent over the whole range of pressure by using DREIDING and UFF for FPYEu and Kamath for hydrogen sulfide, whereas UFF for FPYEu and Kamath for hydrogen sulfide overpredict the experiment data. Nevertheless, using different combinations of force fields for FPYEu has no influence on the simulated data of water. It is possible that modifying the charges of the MOF and/or the force field parameters suggested by Pérez-Peritello [10] and Fairen-Jimenez [49] may satisfactorily address this shortcoming, although it cannot be modified without verifying the pore volume of the MOF.

Interestingly, the calculated adsorption isotherms are not sensitive to the Europium atoms distribution in the system.

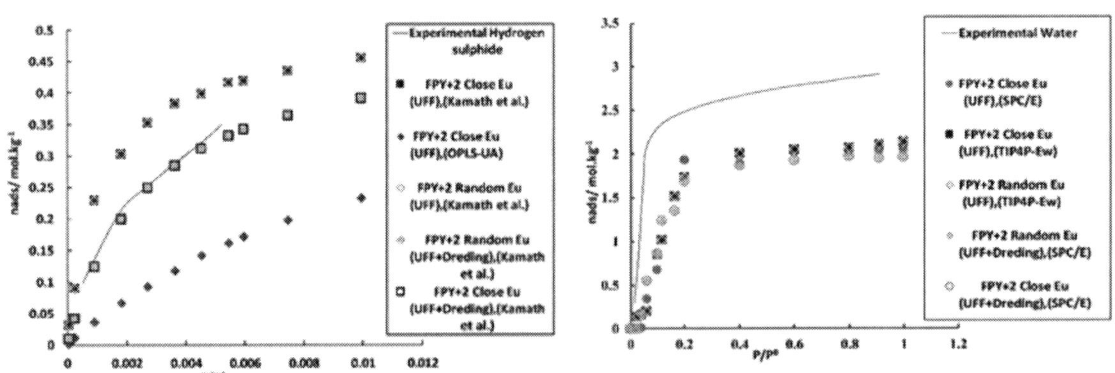

(a) (b)

Figure 7. (a) Adsorption isotherms of hydrogen sulfide at temperature =298.15 K, (b) Adsorption isotherms of water at temperature =303.15 K on FPYEu structures with close and random distributions. Results computed by using additional combinations of force fields. The blue line presents the data of experimental as reported by Shi *et al.* [1]. The error bars of the calculated loading are smaller than the symbol size.

4. Conclusions

The current work has shown that simulation tools can be used to study the applicability of classical force fields and the sensitivity of the calculated adsorption isotherm to the metals atoms distribution in the system. Different force fields available in the literature were used to model the adsorption of gases in FPYEu and some of them were applicable, like H_2S, where good agreement was found with experimental results, while others were not transferable, like CH_4, C_2H_6 and H_2O, where the calculated isotherms "over or under" estimated the experimental ones. While some of the disagreements with experimental data may be due the lack of applicability of Lennard-Jones parameters, pore accessibility, partial charges or accuracy of the experimental isotherms. It is possible that the modifying the charges of the MOF and/or the force field parameters [10] [49] may satisfactorily address this shortcoming, although this was not explored in the present work.

Interestingly, the results also show that the simulations boxes with different Europium atoms distributions do not affect the adsorption of different rare earth metals distributions. The eight different configurations having Europium atoms arbitrarily located in the system give the same results. This also indicates that modelling different ratio of Eu:Y is not necessary.

Acknowledgements

T. AL-Jadir is grateful to the Higher Committee for Education Development in Iraq (HCED) for a postgraduate research scholarship. T. AL-Jadir also acknowledges support from University of Technology, Baghdad, Iraq. The authors acknowledge the assistance given by IT Services and the use of the Computational Shared Facility at The University of Manchester.

References

[1] Shi F, Pinto M, Ananias D and Rocha J 2014 *Microporous Mesoporous Mater* **188** 172
[2] Al-jadir T M., and Siperstein F R 2016 *J. Chem. Eng. Data* **61** 4209
[3] Martin M G, 2013 *Mol. Simul* **39** 1212
[4] Daiminger U, Lind W 2004 *Adsorption Processes for Natural Gas Treatment* (Iselin: Engelhard) p 24
[5] Kohl A and Nielsen R 1997 *Gas purification* (Houston: Gulf Publishing) ed 5 p 1414
[6] Abdel-Aal H K, Aggour M A and Fahim M A 2003 *Petroleum and Gas Field Processing* (New York: Marcel Dekker) ed 4 p 357
[7] Hammer G, Lübcke T Kettner R, Pillarella M R, Recknagel H, Commichau A, Neumann H, Paczynska-Lahme B 2006 *Natural Gas. Ullmann's Encyclopedia of Industrial Chemistry*. (Weinheim: Wiley-VCH Verlag GmbH & Co. KGaA) ed 1 p 54
[8] Eddaoudi M, Kim J, Rosi N, Vodak D, Wachter J, O'Keeffe M and Yaghi Omar 2002 *Science* **295** 469
[9] Farha O K, Bae Y-S, Hauser B G, Spokoyny A M, Snurr R Q, Mirkin Ch A and Hupp J T 2010 *Chem. Commun.***46** 1056
[10] Pérez-Pellitero J, Amrouche H, Siperstein F R, Pirngruber G, Nieto-Draghi C, Chaplais G, Simon-Masseron A, Bazer-Bachi D, Peralta D, Bats N 2010 *Chem. - A Eur. J.* **16** 1560
[11] Dincă M, Dailly A, Liu Y, Brown C M, Neumann D A and Long J R 2006 *J. Am. Chem. Soc.* **128**, 16876
[12] Blomqvist A, Araújo C M, Srepusharawoot P and Ahuja R 2007 *Proc. Natl. Acad. Sci. U. S. A.*, **104** 20173
[13] Bae Y-S, and Snurr R Q, *Microporous Mesoporous Mater* 2010 **132** 300
[14] Thallapally P K, Tian Jian, Kishan M Radha, Fernandez C A, Dalgarno S J, McGrail P B, Warren J E and Atwood J L 2008 *J. Am. Chem. Soc.* **130** 16842
[15] Britt D, Furukawa H, Wang Bo, Glover T G and Yaghi O M 2009 *Proc. Natl. Acad. Sci. U. S. A.*, **106** 20637
[16] Keskin S, van Heest T M and Sholl D S 2010 *Chem Sus. Chem* **3** 879
[17] Amrouche H, Aguado S, Pérez-Pellitero J, Chizallet C, Siperstein F R, Farrusseng D, Bats N

and Nieto-Draghi C 2011 *J. Phys. Chem. C* **115** 16425
[18] Bastin L, Bárcia P S, Hurtado E J, Silva J A C, Rodrigues A E and Chen B 2008 *J. Phys. Chem. C*, **112** 1575
[19] Burtch N C, Jasuja H and Walton K S 2014 *Chem. Rev.* **114** 10575
[20] Øien-ØDegaard S, Bouchevreau B, Hylland K, Wu L, Blom R, Grande C, Olsbye U, Tilset M and Lillerud K P 2016 *Inorg. Chem.* **55** 1986
[21] Hamon L, Serre Ch, Devic Th, Loiseau Th, Millange F, Ferey G, and Weireld G 2009 *JACS* **53**, 8775
[22] Al-Jadir Th M and Siperstein F R 2018 *Microporous Mesoporous Mater.* **271** 160
[23] Rouquerol F, Rouquerol J, Sing K, Llewellyn Ph and Maurin G 2014 *Adsorption by Powders and Porous Solids Principles, Methodology and Applications* (Oxford, UK: Academic Press) ed 2 p 467
[24] Al-Janabi N 2017 *Engineering Novel Porous Materials for Carbon Capture and Storage* Ph.D. Thesis University of Manchester UK
[25] MacGillivray L R 2010 *Metal-Organic Frameworks: Design and Application* (NJ, Hoboken: John Wiley & Sons) ed. 1 p 363
[26] Düren T, Bae Y-S, Snurr R Q 2009 *Chem Soc Rev* **38** 1237
[27] Leach A R 2001 *Molecular modelling: principles and applications* (Edinburgh, UK: Pearson Education) ed 2 p 773
[28] renkel D and Smit B 2002 *Understanding Molecular Simulation: From Algorithms to Applications* (London, UK: Academic Press) ed 2 p 661
[29] Fairen-jimenez D, Seaton N A and Düren T, 2010 *Langmuir* **26** 14694
[30] Lennox MJ, Bound M, Henley A and Besley E 2017 *Molecular Simulation* **43** 828
[31] Sarkisov L and Harrison A 2011 *Molecular Simulation* **37** 1029
[32] Rappe A K, Casewit C J, Colwell K S, Goddard III W A and Skiff W M 1992 *J. Am. Chem. Soc.* **114** 10024
[33] Li Z, Liao F, Jiang F, Liu B, Ban Sh, Chen G, Sun Ch, Xiao P and Sun Y 2016 *Fluid Phase Equilibria* **427** 259
[34] Yang Q, Guillerm V, Ragon F, Wiersum A D, Llewellyn Ph L, Zhong Ch, Devic Th, Serrec Ch and Maurin G 2012 *Chem. Commun.* **48** 9831
[35] Wu D, Maurin G, Yang Q, Serre C, Jobic H and Zhong C 2014 *Journal of Materials Chemistry A* **66** 1657
[36] Rosenbach N, Jobic H, Ghoufi A, Salles F, Maurin G, Bourrelly S, Llewellyn Ph L, Devic Th, Serre Ch, and Ferey G 2008 *Angewandte Chemie* **47** 6611
[37] Vandenbrande S, Verstraelen T, Gutiérrez-Sevillano JJ, Waroquier M and Van Speybroeck V 2017 *Journal of Physical Chemistry C* **121** 25309
[38] Düren T, Sarkisov L, Yaghi O M and Snurr R Q 2004 *Langmuir* **20** 2683
[39] Martin M G and Siepmann J I 1998 *Journal of Physical Chemistry B* **5647** 2569
[40] Mayo S L, Olafson B D and Goddard, III WAG 1990 *Journal of Physical chemistry* **101** 8897
[41] Kamath G, Lubna N and Potoff J 2005 *Journal of Chemical Physics* **123** 0
[42] Jorgensen W L, Maxwell D S and Tirado-Rives J 1996 *J. Am. Chem. Soc.* **118** 11225
[43] Horn H W, Swope W C, Pitera, Madura J D, Dick Th J, Hura G L, and Head-Gordon T 2004 *Journal of Chemical Physics* **120** 9665
[44] Berendsen H J C, Grigera J R and Straatsma T P 1987 *Journal of Physical Chemistry* **91** 6269
[45] Mooij G C A M, Frenkel D and Smit B 1999 *Journal of Physics: Condensed Matter* **4** 255
[46] Deng L-Y 2005 *ACM Transactions on Modeling and Computer Simulation* **15** 1
[47] Darden T, Perera L, Li L and Lee P 1999 *Structure* **7** 55
[48] Fairen-Jimenez D, Galvelis R, Torrisi A, Gellan A D, Wharmby M T, Wright P A, Mellot-Draznieksb C and Düren T 2012 *Dalton Transactions* **41** 10752

IOP Conf. Series: Materials Science and Engineering **579** (2019) 012021 doi:10.1088/1757-899X/579/1/012021

Study the effect of organic promoters on thermodynamics behaviour for refrigerant gas clathrate hydrate.

Shurooq Talib ALHemeri, Riyadh Sadeq ALMukhtar and Mustafa Saadi Mohammed[#]
Department of Chemical Engineering, University of Technology-Iraq, Baghdad, Iraq.

[#]Corresponding author's e-mail address: 080549@student.uotechnology.edu.iq

Abstract. The present work consists of an experimental and theoretical investigation dealing with hydrate formation for the binary system (water+ refrigerant gas) and ternary systems (water+ refrigerant gas+ promoter) at constant initial conditions (pressure and temperature). Isochoric method (constant volume) search was used in current study to measure the formation data (pressure, temperature) for hydrate. Refrigerant R134a gas and different organic promoters such as benzene and cyclohexane were used in this study. For ternary systems at the different concentrations of promoters shifted hydrate formation data when benzene and cyclohexane decree pressure and increasing temperature. Many objective functions were obtained from kinetic models such as the amount of gas consumed (Δn), the growth rate (r (t)), and conversion of the water to hydrate. The gas consumed (Δn) of the binary system increased lead to increase in hydrate formation, also the hydrate growth rate (r(t)) and increase water conversion to hydrate. The gas consumed in the presence of cyclohexane was higher than at using benzene as a promoter. The experimental results show that the presence of different promoters in the ternary systems have an effect on thermodynamic promotion of refrigerant hydrates formation. The promotion effect increased with increasing the concentration of promoter in the select range. Also, the results manifested that benzene and cyclohexane had a clear effect in reducing the pressure and raising the temperature of the hydrate formation. The storage capacity was investigated. It has been noticed that the existence of the promoters in the hydrate formation system led to a slight increase in the storage capacity with maximum value about (78.66) V/V.

1. Introduction

A clathrate is a complex molecule formed by the contact between gas and water at high pressure and low temperatures. The cavities within hydrate formation are formed by the hydrogen bonds which occupied by the gas molecules that are the guest molecules as shown in (Figure.1) [1].

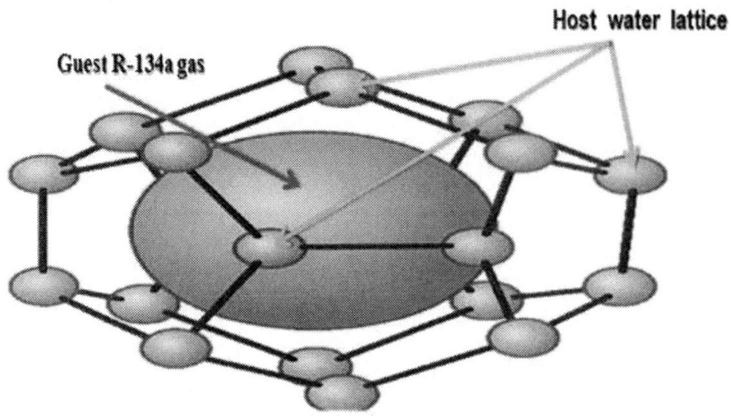

Figure 1. Clathrate hydrate structure

For many years, clathrate hydrates were considered as harmful in oil and gas industry because of their annoying tendency to plug the pipelines [2]. In recent years and as a result of policies to reduce environmental pollution, they entered other fields where the research included: water desalination, separation processes [3]. After the positive aspects of clathrate hydrate technology, enhancing the hydrate formation is one of the main objectives of gas hydrate technology.

The slow formation of hydrates and energy costs involving the formation of hydrates at low temperature and high pressure are considered industrial impediments. Promoters should be used to rapidly form hydrates under the same operating pressure and temperature [4].

Organic component (cyclopentane, cyclohexane, benzene, acetone, etc.) can reduce the required hydrate formation pressure that is considered as a thermodynamic promoter [5]. For the specific case of gas in the cavities, it has been suggested that the enhanced hydrogen bonding and increase their number could be due to the physical tension on the water lattice caused by the relatively gas molecule present in the cavities [6].

The hydrate formation enhancement is not easy to achieve because of the hydrate formation layer volume depends on factors such as promoter choice, guest choice, initial pressure, temperature and concentration of promoter [7].

The main purpose is to study the phenomena of hydrate formation of refrigerant R134a for the ternary system (water +R134a + promoter) for different types of the promoter (Cyclohexane and Benzene) at different concentrations.

2. Experimental

2.1 Materials

Refrigerant gas (R-134a) with a high purity equal to 99% was utilized to form a hydrate with distilled water. In experiments, promoters were used were Benzene and Cyclohexane that were immiscible with water.

2.2 Apparatus

(Figure.2) shows the experimental schematic diagram used in this study. The most important part in the system is the glass cell with a total volume of 1500 cm^3 which consist of the internal diameter of 14 cm and a height of 14.5 cm, the cell was equipped with a magnetic stirrer that used in this experiment and it mainly modified electric stirrer which consists of two shafts, the first shaft connected to electric motor from top and to the other end attached to the magnet. The second shaft contains a magnet (corresponding to the magnet at the first shaft), which is inside a closed chamber to ensure no leakage, the second end of the second shaft is extended to the inside of the cell and at the end of the blender. This

IOP Conf. Series: Materials Science and Engineering **579** (2019) 012021 doi:10.1088/1757-899X/579/1/012021

process is done to ensure that there is no leakage of gas. A thermocouple with a division scale of 0.1 K and a digital pressure gauge with a division scale of 0.00005 bar were used to measure the temperature and pressure.

The cell connected to the system through pipes and controlled by a valve for gas injection, water input, and drain. The cell surrounded by transparency plastic vessel as heat jacket to maintain the temperature inside the cell, moreover cooling water was supplied from chiller type Julabo F10-VC which was which contains a cooling medium made of ethylene glycol/water. The temperature of the medium is adjusted by the controllable circulator.

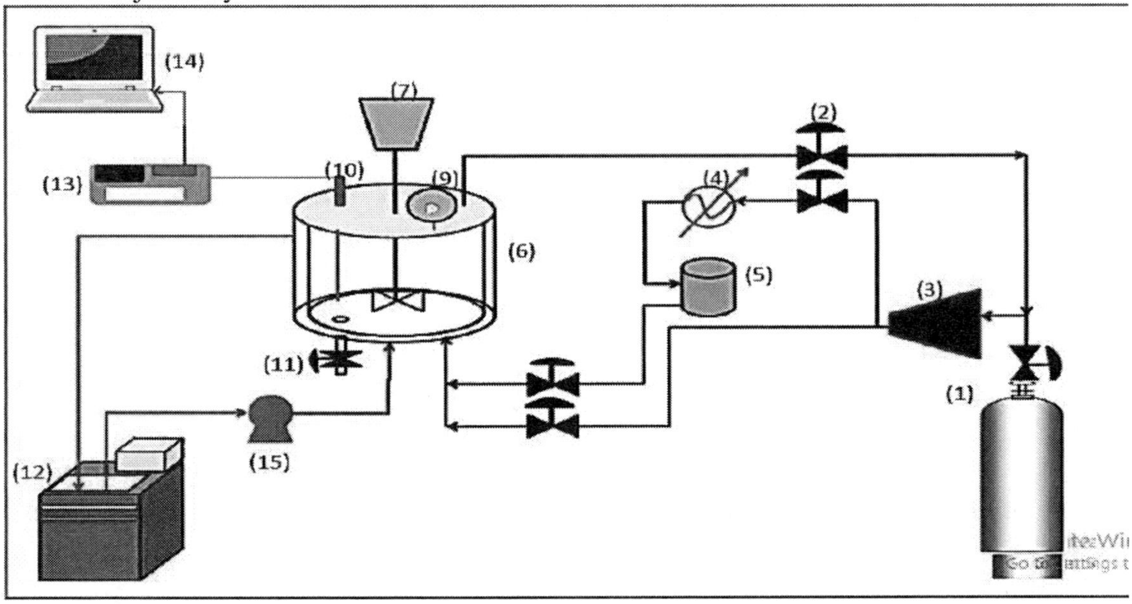

Figure 2. (1) gas cylinder; (2) needle valve; (3) gas compressor; (4)condenser; (5) accumulator of gas; (6) Experimental cell; (7) magnetic stirrer; ; (8) water pump; (9) digital pressure gauges; (10) temperature sensor; (11) drain valve; (12) chiller; (13) interface system; (14) computer.

2.3. Experimental method

The constant volume (isochoric) system search strategy was utilized to investigate hydrate formation in this study. At the start, distilled water was used to wash the cell then it is washed with acetone to flush the cell. Washed water with acetone was drainage. The cell was emptied to vacuum pressure by the compressor for a time of 30 min to guarantee that there were no traces of contamination.

Before starting the experiment, the cell is examined in terms of venting using air at a pressure of 10 bar and using foam solution sprayed on the cell, joints, and valves and on the basis of which it is sure that there is no leakage that can be treated. Repeating this procedure periodically every 10 minutes during the experiment to make sure there is no leakage.

The air is evacuated from the cell by using the compressor then 200 ml (13% vol of the cell) of water solution mixture was f((8) side the cell. A specific measure of R-134a was acquainted of the cell (86% vol of the cell) with achieving the coveted pressure of 0.45 MPa (4.5 bar) at a temperature of 313.15 K (10 °C).

The mixer was rotated at 250 rpm after the gas that is pressurized to the cell is completed. The pressure of cell reduced consistently because of hydrate formation until the point wh(:hieving the steady state condition.

The stirring speed is set at 250 rpm and initial temperature and pressure at 10 °C a respectively with selecting Benzene and cyclohexane at different concentrations (0, 2.5, 5.0, 7.5 and 10.0) Vol%.

IOP Conf. Series: Materials Science and Engineering **579** (2019) 012021 doi:10.1088/1757-899X/579/1/012021

Hydrates are separated by water discharge (non-converted to hydrates) by the pressure inside the cell and thus the hydrate is extracted.

3. Mathematical Model

3.1 Gas Consumed

The moles of the hydrate former consumed throughout the formation of hydrate are obtained by applying the real gas law as follows:

$$\Delta n_H = n_{g,0} - n_{g,t} = \left(\frac{PV}{ZRT}\right)_0 - \left(\frac{PV}{ZRT}\right)_t \quad (1)$$

The "0" Subscripts refer to the initial conditions where "t" refers to conditions at time, t, of the system, P pressure and T temperature at the phase equilibrium state, R was the universal gas constant, and Z was the gas compressibility determined by the Pitzer's correlation [8]:

$$Z=1+B^0\frac{Pr}{Tr} + \omega B^1\frac{Pr}{Tr} \quad (2)$$

Where $$B^0 = 0.083 - \frac{0.442}{T_r^{1.6}} \quad (3)$$

And $$B^1 = 0.139 - \frac{0.442}{T_r^{4.2}} \quad (4)$$

The hydrate number (number of water molecules) is identified by the partial filling the large and small cavities. The simple hydrate number is calculated [1]:

$$M = \frac{136}{(8\theta_s + 16\theta_L)} \quad (5)$$

where θ_S and θ_L are the fractions of linking large and small cavities occupied by guest molecules respectively, as follows [1]:

$$\theta_i = \frac{(C_i \times f_i)}{1+(C_i \times f_i)} \quad (6)$$

C_i is the Langmuir constant for large cavities (C_L) and small cavities (C_S) for R134a hydrate. Where f_i is the fugacity of (R134a) gas which is calculated by *the* following equation:

$$f_G=\varphi * P \quad (7)$$

Where, φ is the coefficient of fugacity, equation (4-8) can be utilized to estimate φ through Peng-Robinson equation [8];

$$\ln(\varphi) = (z-1) - \ln(z-B) - \frac{A}{2\sqrt{2}\,B}\,\mathrm{Ln}\left(\frac{z+(1+\sqrt{2})B}{z+(1-\sqrt{2})B}\right) \quad (8)$$

In which A and B are adjustable parameters for R134a gas.
The Langmuir constant of R-134a Ci is formulated as follows[9]:

$$Ci = \frac{Ai}{T}\exp\left(\frac{Bi}{T}\right) \quad (9)$$

IOP Conf. Series: Materials Science and Engineering **579** (2019) 012021 doi:10.1088/1757-899X/579/1/012021

where Ai and Bi are constants with values 0.00575 and 4908 for C_L and constants equal to zero Cs for (R134a) gas [10]. The gas hydrate volume is not like the aqueous solution volume, To estimate the volume of R134a gas hydrate inside a cell the next equation is used [11] :

$$V_t = V_{cell} - V_{S0} + V_{RWt} - V_{Ht} \quad (10)$$

where V_{cell} is the cell volume with value equal 1500 cm^3; V_{S0} is the initial aqueous solution volume with value equal to 200 cm^3; V_{Ht} is the produced hydrate's volume and V_{RWt} is the reacted water's volume that enters the structure of the hydrate structure which calculated by using the following equation [11]:

$$V_{RWt} = M \times \Delta n_H \times V_w^L \quad (11)$$

The hydrate molar volume, V_{Ht}, can be estimated by the next equation [11]:

$$V_{Ht} = M \times \Delta n_H \times V_w^{MT} \quad (12)$$

The gas hydrate molar volume can be assumed equal to the empty hydrate lattice molar volume.

Klauda and Sandler illustrated that the v_W^{MT} is the volume occupied by empty hydrate lattice (in K temperature and m3/kmol) as a function of temperature and pressure by the following equation [12]:

$$v_W^{MT} = (17.13 + 2.249\times10^{-5}\times T + 2.013\times10^{-6}\times T^2 \times \frac{10^{-30} N_A}{136} - 8.006\times10^{-9}\times P + 5.448\times10^{-12}\times P^2) \quad (13)$$

T and P are the temperature and pressure given in K and MPa units, respectively.

3.2 The Growth Rate of Hydrate Formation

The growth rate of hydrate formation is presented by using the following equation [11]:

$$r(t) = -\frac{1}{n_{wo}}\frac{dn_R}{dt}\Big|_{ti} = -\frac{1}{n_{wo}}\frac{\Delta n_R}{\Delta t}\Big|_{ti} = \frac{(n_{R134a,i-1} - n_{R134a,i+1})}{(t_{i+1} - t_{i-1}) * n_{wo}} \quad (14)$$

$n_{R134a,i-1}$ And $n_{R134a,i+1}$ present moles of R134a gas which is hydrate former at t_{i-1} and t_{i+1}respectively. n_{wo} is the initial number of moles of water.

3.3 Water to Hydrate Conversion

The water is converts to hydrate for each mole of water feed can be estimated by relation the following [13]:

$$\text{The Water conversion} = \frac{M \times \Delta n}{n_{wo}} \times 100\% \quad (15)$$

3.4 The Storage Capacity (SC)

The storage capacity is a volume of gas stored in per volume of hydrate which is calculate by following equation [13]:

$$SC = \frac{V_{STP}}{V_{Ht}} = \frac{\Delta n \times (\frac{R \times T_{STP}}{P_{STP}})}{V_{Ht}} \quad (16)$$

IOP Conf. Series: Materials Science and Engineering **579** (2019) 012021 doi:10.1088/1757-899X/579/1/012021

where subscript STP stands for the standard conditions and V_{Ht} is the gas hydrate volume at the end, which is calculated by Eq. (12). (Figure.3) shows the algorithm of the previous calculation for calculating the gas consumed (Δn), hydrate growth rate (r(t)), conversion of water to hydrate, and storage capacity of R-134a gas.

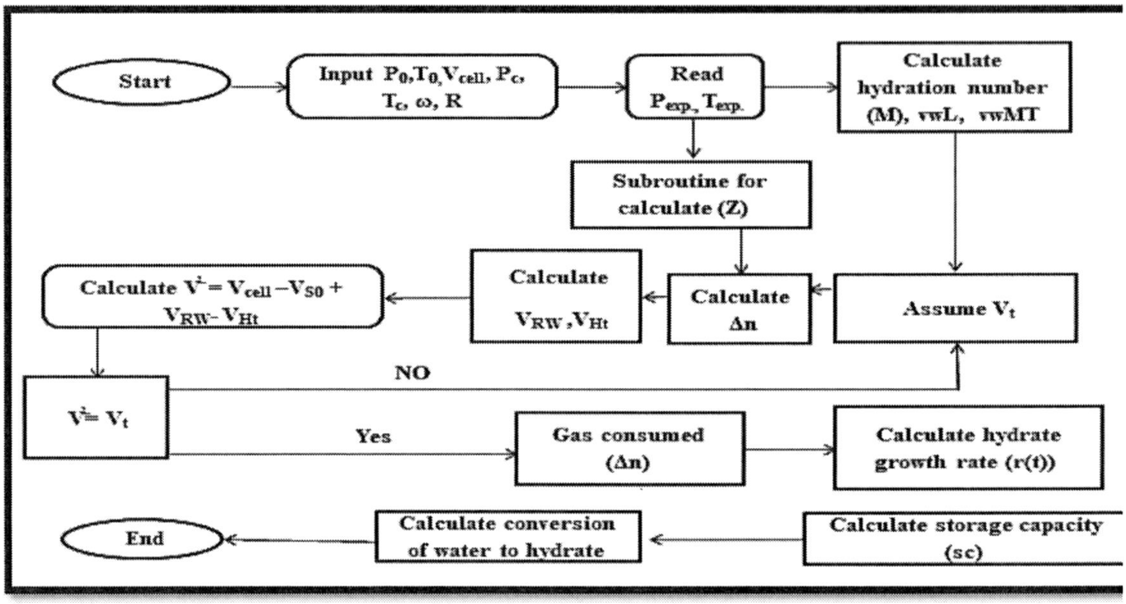

Figure 3. The suggested algorithm to calculations.

4. Result and discussion

4.1Thermodynamic Behavior study

Thermodynamic behavior of clathrate hydrate formation was a function of pressure and temperature with time for refrigerant gas type R-134a with different promoters concentrations. Isochoric search method is used in this study. The measurements of the hydrate phase equilibrium for the pure water and different promoters with refrigerant gas were performed through three phases: hydrate (H), liquid water (L_w) and vapor.

The addition of benzene and cyclohexane as an insoluble hydrocarbon component in water can also lead to a decrease in pressure and an increase in temperature as described by [13,14]. The highest effect of benzene and cyclohexane is achieved at concentration 10 Vol% and this effect is gradually reduced due the concentration that is increase as shown in (Figure.4) and (Figure.5).

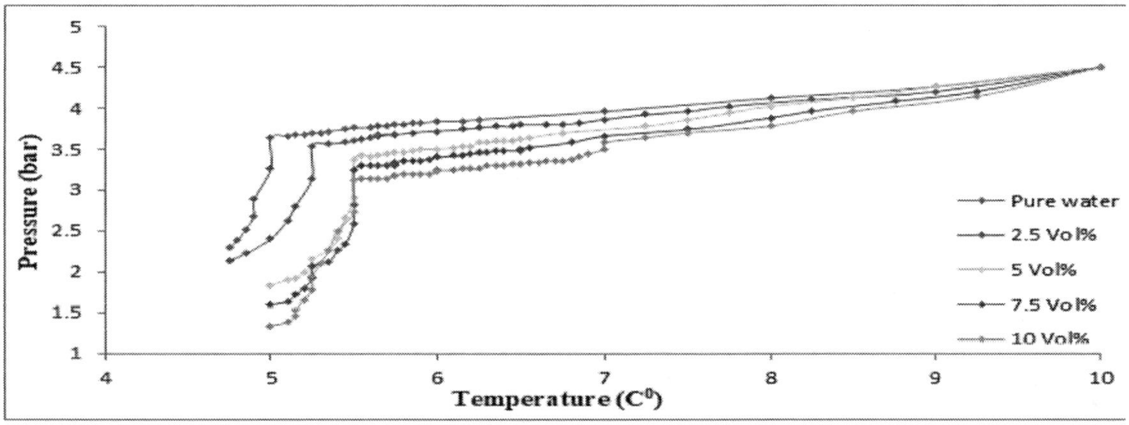

Figure.4 Effect of different Benzene concentrations on the pressure and temperature path during formation of R-134a hydrate.

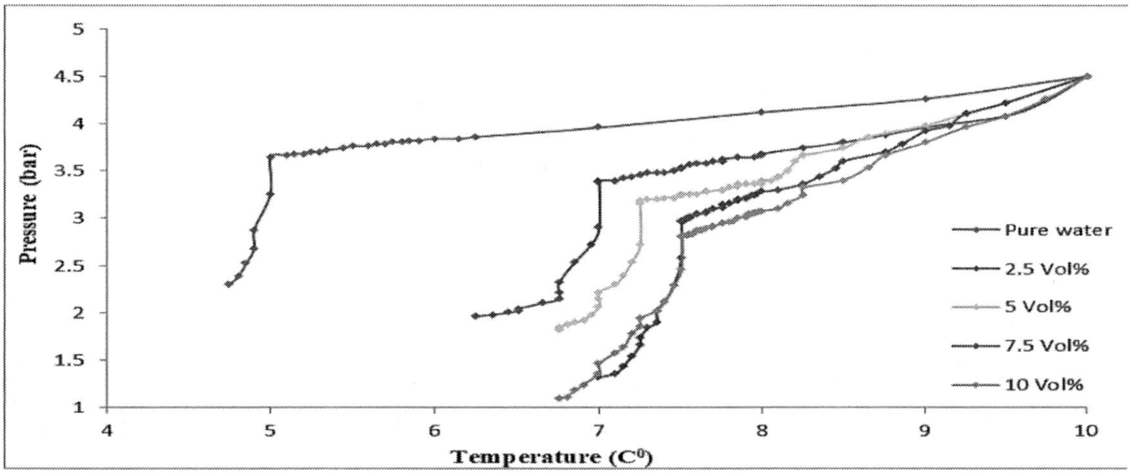

Figure.5 Effect of different Cyclohexane concentrations on pressure and temperature path during formation of R-134a hydrate.

Cyclohexene has a significant effect on the conditions of formation of hydrate gas, which is greater than in benzene in reducing pressure and raising the temperature as described by [6].

In general, the addition of hydrocarbon components (benzene, and cyclohexane) to the solution leads to a decrease in pressure. By comparing benzene, and cyclohexane, cyclohexane was thermodynamically most stable hydrate promoter.

4.2 Gas Hydrate Formation Visual Observations

The experiments were done in the isochoric pressure method. After the addition of water, the gas is injected (until it reaches the required pressure while ensuring the initial temperature stability). In the dissolution stage, The temperature decreases slowly and is accompanied by the dissolving of the gas in the water until it reaches a low dissolving stage and the pressure stabilizes in the nucleation stage. After the end of the nucleation stage and reaching the point of turbidity that recognized in the cell. The turbidity point was a significant sign for hydrate nucleation that is accompanied by formation.

The hydrate appeared in the form of a light layer and are accompanied by a significant reduction in pressure. The hydrate layer thickness is increasing and it grown gradually until it reaches a stage where the pressure drop as described by [6].Hydrate formation was visually observed throug the cell side as the (Figure.6).

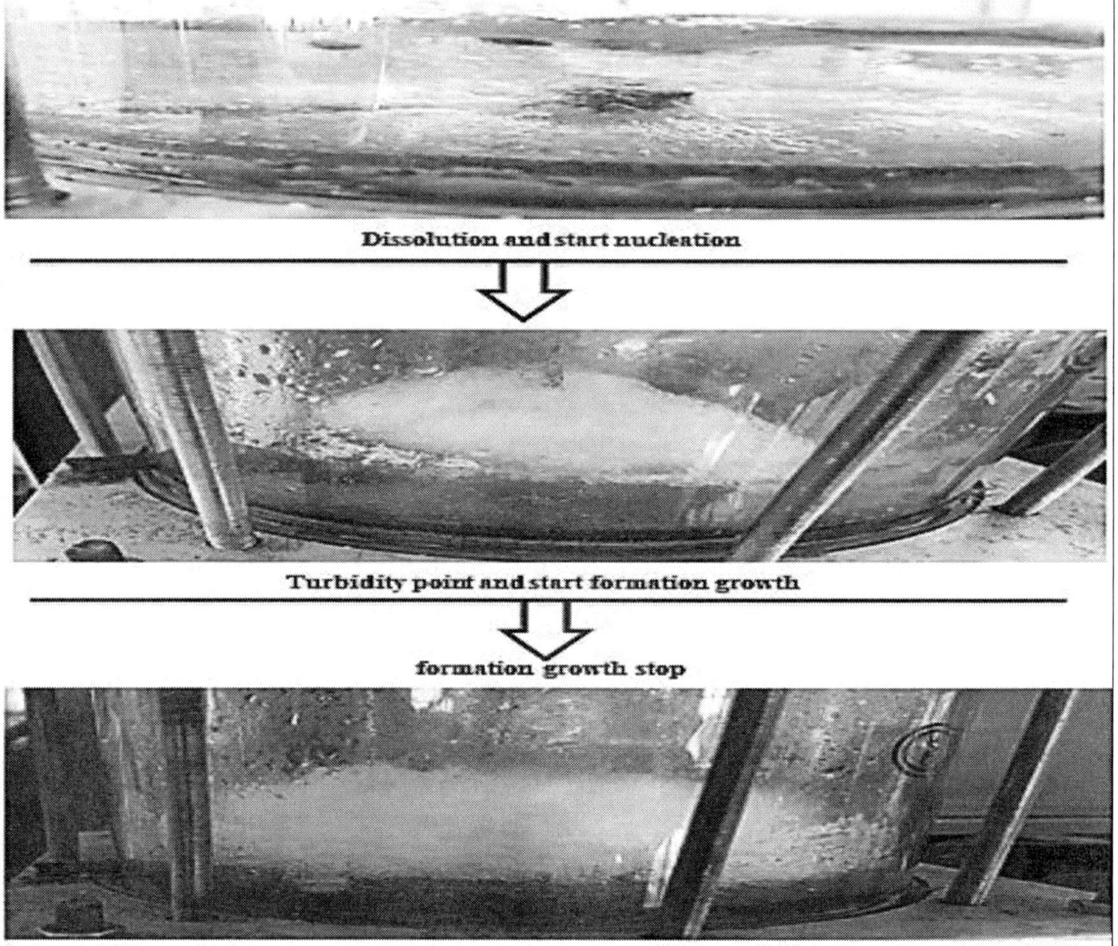

Figures 6. typical sequence for R134a clathrate hydrate formation shows the photographic picture through formation stage.

4.3 Kinetic Model Results

The suggested model mentioned in section 4 was used to show the result of the experimental and obtain the gas consumed, formation rate and other parameters of gas hydrate ternary system.

4.3.1 Gas consumption

The gas feed is 0.2760 mole from eq.(1), the gas consumption was calculated at various concentrations of different promoters through the gas hydrate formation for water + R134a + promoter system. The effect of benzene and cyclohexane on gas consumption in the experiments are shown in (Figure.7) and (Figure.8) The gas consumed was increased slowly at the initial time and the moles of gas consumed increase with a drop in the pressure increase and temperature inside the cell due to the growth of the hydrate layer. The gas consumed was increased slowly at the initial time and the moles of gas consumed increase with a drop pressure increase and the temperature inside the cell due to the growth of the hydrate layer.

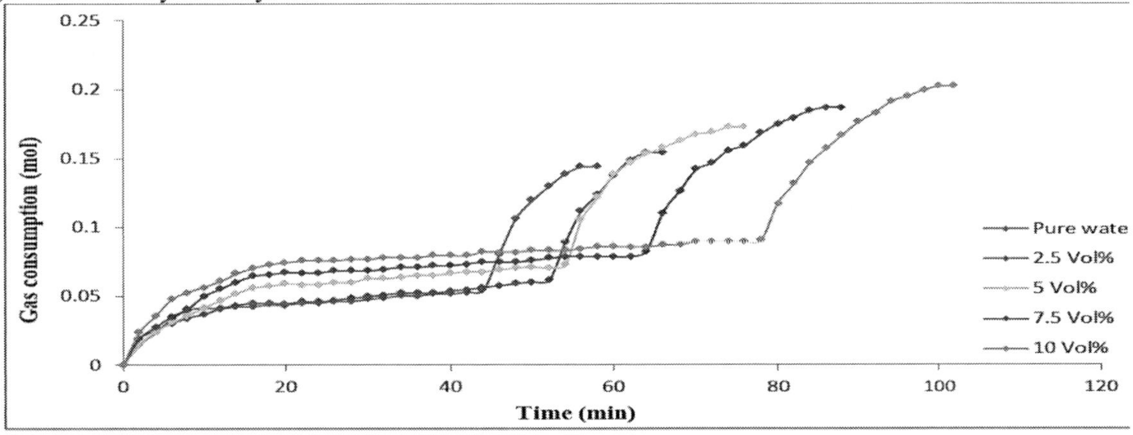

Figure 7. The effect of benzene on R134a gas consumed versus time during the hydrate formation process for the pure water + refrigerant gas R-134a system.

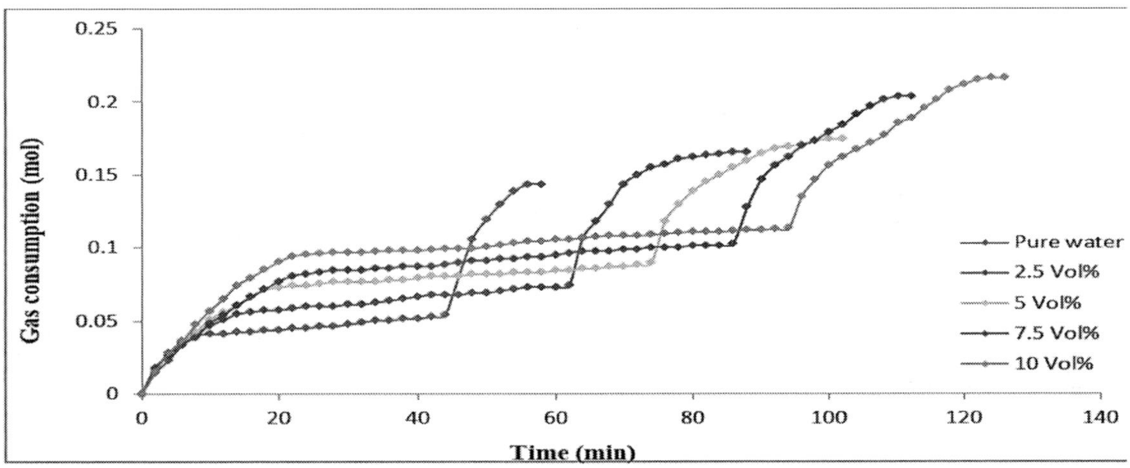

Figure 8. The effect of cyclohexane on R134a gas consumed versus time during the hydrate formation process for the pure water + refrigerant gas R-134a system.

As shown in the plots, with the increase in concentration, the consumed gas raised with time because the promoter concentration has in effect on the number of water cages increase due to the effects of promoter in the solution.

This result indicates that there is a higher driving force represented by the difference in the fugacity of gas phase that depended on pressure of formation of gas hydrate when concentrations is increased. During hydrate formation, the gas consumed enhancing with various concentrations of different promoters and gas consumed higher than in binary system as obtained by [15].

4.3.2 Growth rate of R134a hydrate formation

The rate of hydrate is mostly expressed as the molecules of R134a that are converted to hydrate. The growth rate during the growth period depended on various concentrations of promoters with constant initial pressure. The different promoters have a different influence on the growth of hydrate.

The growth rate in the binary and ternary system first increases gradually during the dissolving gas stage and then becomes semi-stable during the nucleation stage where the particles of hydrates begin to grow gradually until it reaches the formation point and start hydrate growth with an increase in pressure drop.

This is followed by a rapid decline in the value of the growth of hydrate rate and then return to rise again as a result of the dissolving of a new quantity of gas to replace that went to the top. The growth of hydrate equal zero due to stop hydrate formation that represented in (Figure.9) for benzene and (Figure.10) for cyclohexane .

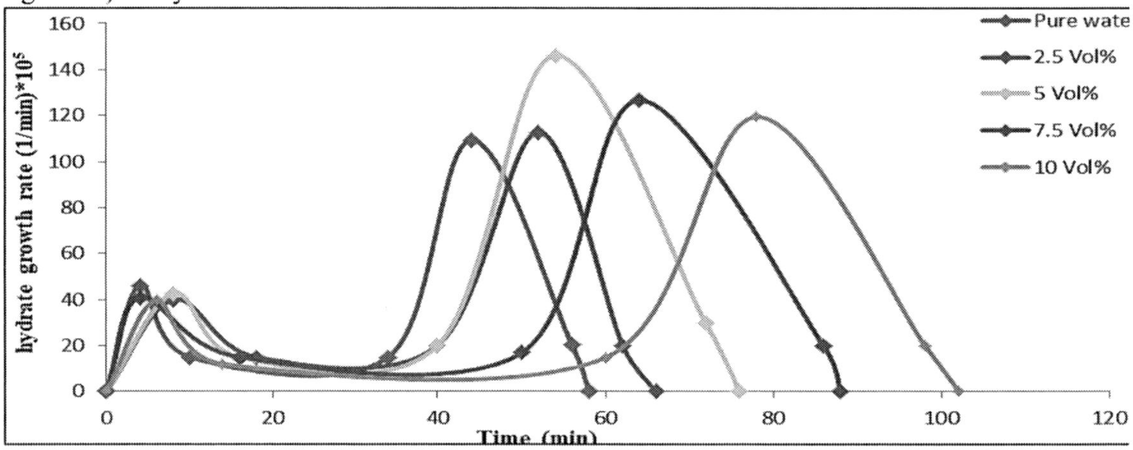

Figure 9. The R134a hydrate growth rate for a ternary system of Benzene with different concentration against time.

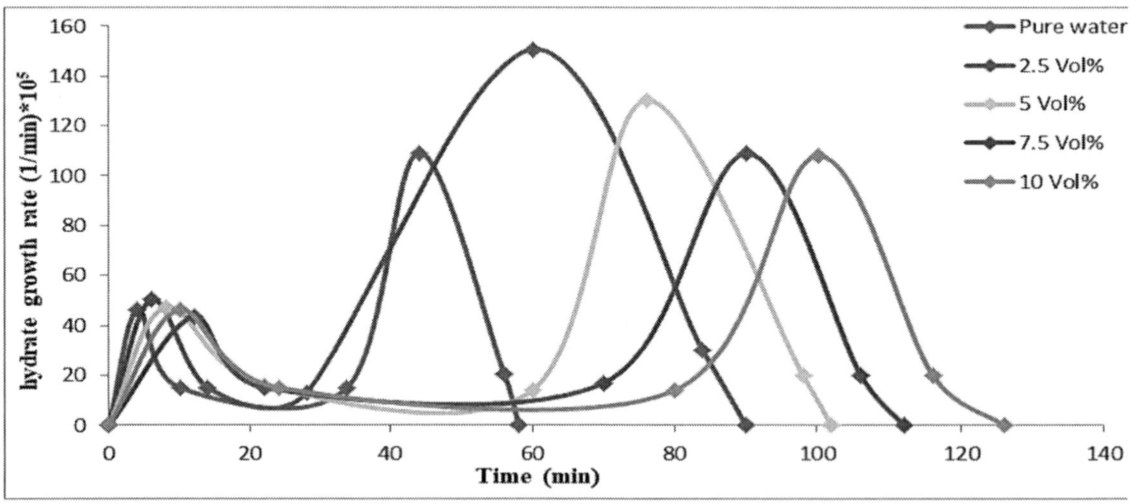

Figure 10. The R134a hydrate growth rate for a ternary system of Cyclohexane with different concentration against time

As mentioned previously, the highest consumption of gas is achieved with the highest concentrations at 10 vol.% However, in calculating the hydrate growth rate, the value of the gas consumed will be divided by the time it takes to grow and the growth rate is inversely proportional to the time taken that shown in **Table.1**.

Table.1. The maximum hydrate growth rate for different promoters used in this study at different concentrations.

Promoters	Concentrations Vol.%	The growth of hydrate rate (1/min)*105
No promoter	-	109.2
Benzene	2.5	112.75
Benzene	5	146.07
Benzene	7.5	126.6
Benzene	10	119.42
Cyclohexane	2.5	150.73
Cyclohexane	5	130.23
Cyclohexane	7.5	108.82
Cyclohexane	10	107.9

The highest value is when the adding of benzene and lower than when adding cyclohexane because the drooping in pressure took a relatively long time that increases the driving force with time. It can be summarized that benzene gave the highest Growth rate at 5 vol.% and the lowest at 10 vol. %, cyclohexane gives the highest Growth rate at 2.5 vol.% and the lowest at 10 vol.%.

4.3.3 The conversion of water to hydrate

The moles of water transform to hydrate during the hydrate growth process were calculated for the binary system and ternary system. For a different type of promoter, the water to hydrate conversion increase with increase the promoter concentration because of adding the promoter to solutions increase hydrate formation.

The ternary system in presence of promoters has the same behavior conversion of water to hydrate which was higher than the conversion of the binary system that is shown with benzene and cyclohexane in the (Figure.11) and (Figure.12).

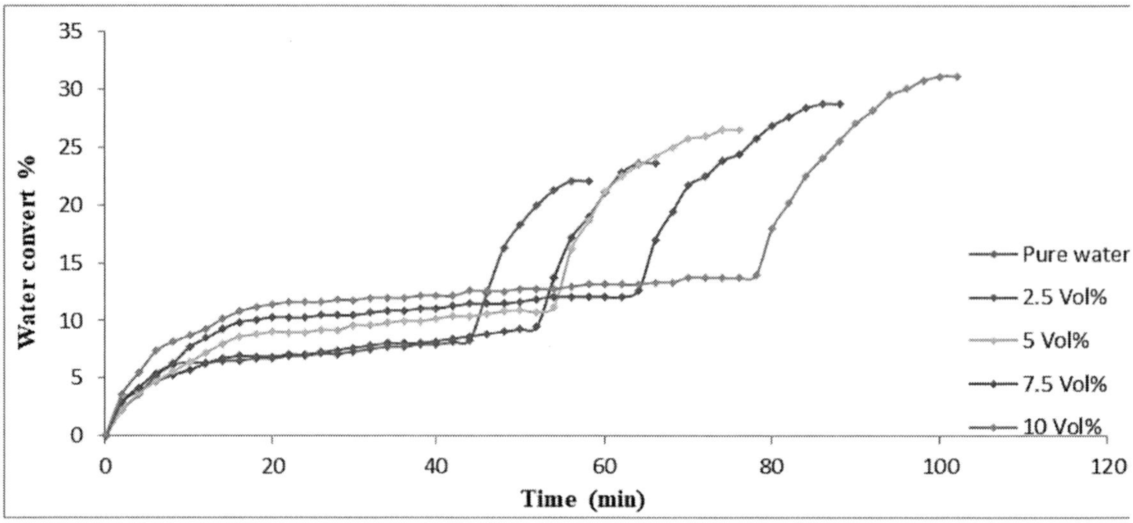

Figure 11. Effect of Benzene on the water to hydrate conversion as a function of time for the ternary system.

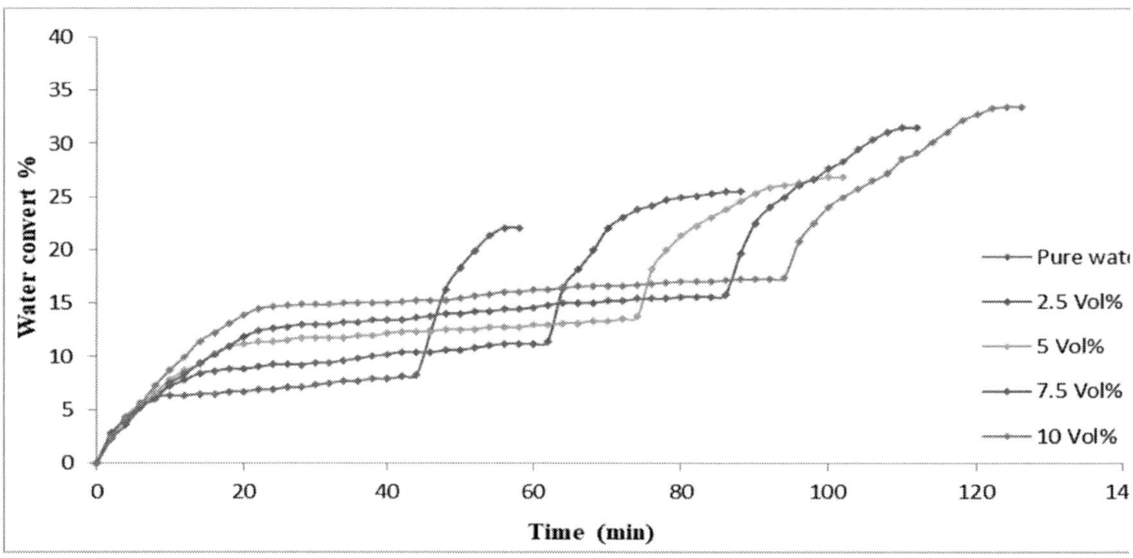

Figure 12. Effect of Cyclohexane on the water to hydrate conversion as a function of time for the ternary system.

The water conversion increase because the higher pressure drop because depending on the pressures and driving forces those obtained by [7]. **Table.2** shows the maximum value of the water to hydrate conversion for a ternary system that was represented as a function of time for promoters at different concentrations for each type of promoter.

Table.2 Maximum water conversion value for promoters at different concentrations.

Promoters	Concentrations	Water conversion%
No promoter	-	22.08
Benzene	2.5	23.63
Benzene	5	26.51
Benzene	7.5	28.75
Benzene	10	31.15
Cyclohexane	2.5	25.48
Cyclohexane	5	26.84
Cyclohexane	7.5	31.46
Cyclohexane	10	33.45

4.3.4 The storage capacity

The storage capacity for the binary system consider as a function of time at the appropriate pressure and temperature. The promoters were added to improve the gas storage capacity in a quiescent system by increasing the pressure drop when adding to the water solution that leads to increases the final storage capacity of the gas hydrate compared to pure water.

Promoters increase the 134a solubility in water hydrates could be formed both at the interface between the gas and the liquid and in solutions. (Figure.13) for benzene and (Figure.14) for cyclohexane reveal that the gas storage capacity could be improved with the presence of promoters.

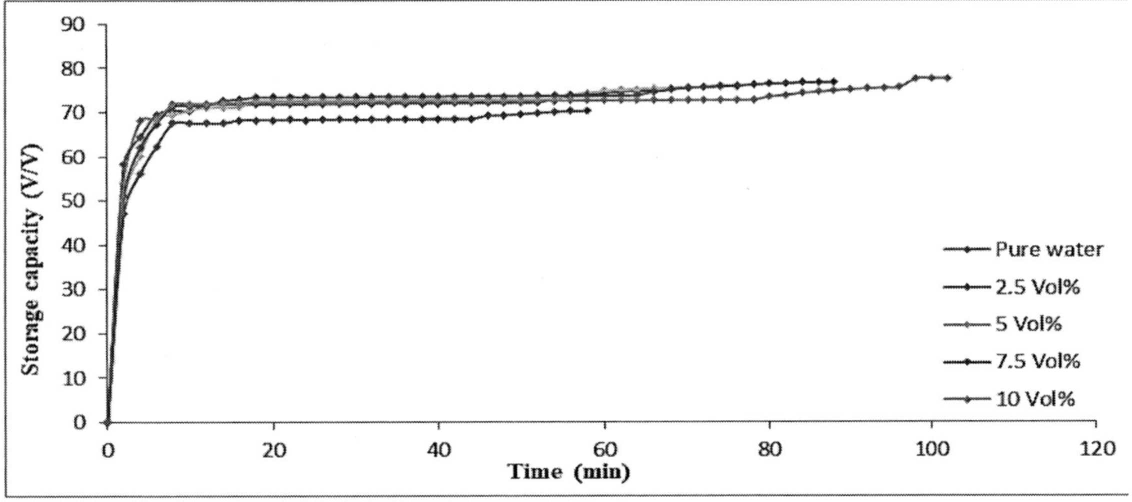

Figure 13. Effect of Benzene Storage Capacity during R134a hydrate growth at constant initial conditions (temperature and pressures).

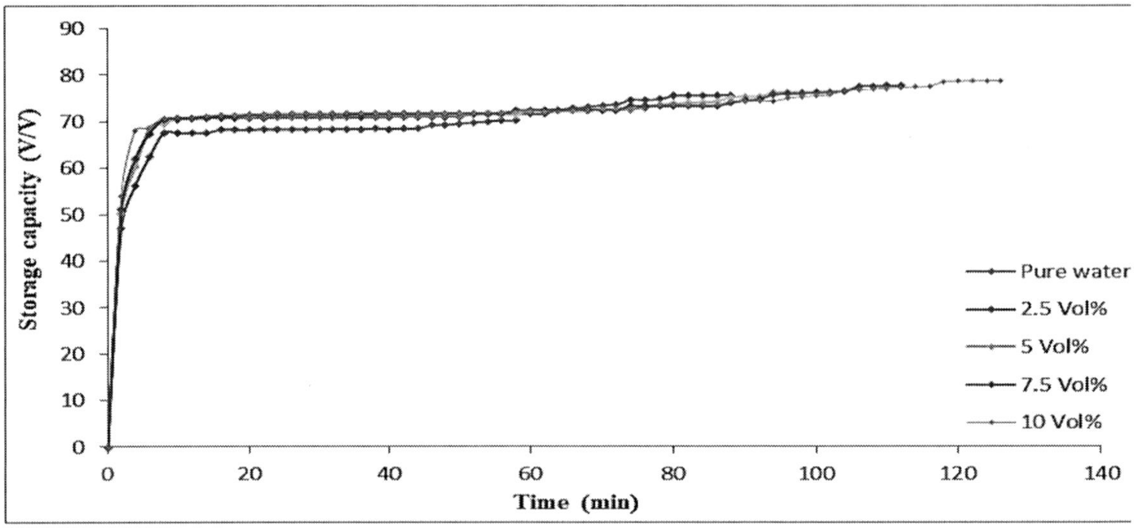

Figure 14. Effect of Cyclohexane Storage Capacity during R134a hydrate growth at constant initial conditions (temperature and pressures).

The final storage capacity of R134a at different concentration of promoters is apparent in **Table.3**. To that end at each tested concentrations, the addition of promoter has effect on the final storage capacity of R134a hydrate. However, it is noticeable that the pressure drop is increased when adding the promoters to the water solution.

Table.3 Final storage capacity value for different promoters used in this study with different concentrations.

Promoters	Concentrations (Vol.%)	Storage capacity (V/V)
No promoter	-	70.16
Benzene	2.5	74.66
Benzene	5	75.69
Benzene	7.5	76.59
Benzene	10	77.47
Cyclohexane	2.5	75.57
Cyclohexane	5	76.33
Cyclohexane	7.5	77.59
Cyclohexane	10	78.66

5. Conclusions

- R134a refrigerant gas was used to form hydrate in a binary system (water+ R134a) and ternary system (water +R134a+ promoter). Two types of the promoter were used in this study (Benzene and Cyclohexane) with different concentrations. The addition of a promoter to the system resulted in shift formation conditions data to high temperatures and low pressures.

- The results of the promoter on the consumed gas, rate of growth and water to hydrate conversion of was studied by kinetic hydrate formation depicted that when the concentration increased, the rate of hydrate formation or the moles of R134 consumed and the water conversion increased with increasing concentration. The promoter in the hydrate formation system slightly increases in the storage capacity with maximum value about (78.66) V/V.
- The best value achieved for R134 consumed is 0.2163 mol., the water conversion is 33.45% and storage capacity 78.66 V/V. All these values achieved with cyclohexane. The best value achieved for the growth rate of hydrate $150\times10^{\wedge-5}$ 1/min, with benzene.

Acknowledgment
The authors wish to thank the Department of Chemical Engineering, University of Technology, Baghdad, Iraq for facilitating this work.

References

[1] Sloan E D 2008 *The clathrate hydrate of natural gases* (New York: Marcel Dekker Inc.) ed 3 p130-150.

[2] Pierce, B S and Collett T S 2004 *Energy resource potential of natural gas hydrates 5th* Conference and Exposition on Petroleum Geophysics Hyderaba India p 899

[3] Shin W, Park S, Ro H, Koh DY, Seol J and Lee H 2012 *Chemistry–An Asian Journal* **7** 2235

[4] Mandal A and Laik S 2008 *Energy & Fuels* **22** 2527

[5] Zhang J, Yedlapalli P and Lee J W 2009 *Chemical engineering science* **64** 4732

[6] Herslund P J, Thomsen K, Abildskov J, Von Solms N, Galfré A, Brântuas P, Kwaterski M and Herri J M 2013 *International Journal of Greenhouse Gas Control* **17** 397

[7] Ishida Y, Takahashi Y and Ohmura R 2012 *Crystal Growth & Design* **12** 3271

[8] Smith A 2015 *Phase Equilibria of Refrigerant Gas Hydrate Systems in the Presence of Sucrose* Ph.D. Thesis University of KwaZulu-Natal Durban

[9] Munck J, Skjold-Jørgensen S and Rasmussen P 1988 *Chemical Engineering Science* **43** 2661

[10] Ngema P T, Naidoo P, Mohammadi A H, Richon D and Ramjugernath D 2016 *Fluid Phase Equilibria* **4** 92

[11] Hashemi H, Babaee S, Naidoo P, Mohammadi A H and Ramjugernath D 2014 *Journal of Chemical & Engineering Data* **59** 3907

[12] Klauda J B and Sandler S I 2000 *Industrial & engineering chemistry research* **39** 3377

[13] Hashemi H, Babaee S, Mohammadi A H, Naidoo P and Ramjugernath D 2015 *The Journal of Chemical Thermodynamics* **82** 47

[14] Zhong D L, Wang Y R, Lu Y Y, Wang WC and Wang J L 2016 *Journal of Natural Gas Science and Engineering* **35** 1566

[15] Verrett J, Renault-Crispo J S and Servio P 2015 *Fluid Phase Equilibria* **38** 160

IOP Conf. Series: Materials Science and Engineering **579** (2019) 012022 doi:10.1088/1757-899X/579/1/012022

Syngas formation by dry and steam reforming of methane using microwave plasma technology

Nabil Majd Alawi[1,2], Gia Hung Pham[1], Ahmed Barifcani[1], Minh Hoang Nguyen[1] and Shaomin Liu[1]

[1]Department of Chemical Engineering, Curtin University, Perth, Western Australia 6845, Australia
[2]Petroleum Technology Department, University of Technology, Baghdad, Iraq
Corresponding author's e-mail address: n.alawi@postgrad.curtin.edu.au

Abstract. The combined dry and steam reforming of methane at atmospheric pressure was experimentally studied by using microwave plasma technology. The effect of the process parameters such as total feed gas flow rate, steam concentration and input microwave power on the synthesis gas H_2/CO ratio was investigated using a commercial microwave reactor system. In order to minimise the carbon formation and plasma instability, the concentration of methane and carbon dioxide in nitrogen plasma were kept at a low level in this study. The long-term test results show that at the flow rate of 0.2 L min^{-1}, 0.4 L min^{-1} and 1.5 L min^{-1} for CH_4, CO_2 and N_2 respectively, the carbon formation was not detectable at the input power of 700 W. This reaction condition offers an opportunity to study the effect of adding water to the feed on the syngas ratio H_2/CO. The test results show that a higher CH_4 conversion (82.74%), H_2 selectivity (98.79%) and yield (81.73%) were achieved compared with those of the dry reforming at the same operating conditions. With the steam addition, the desired H_2/CO ratio for the Fischer-Tropsch synthesis process can be reached.

1. Introduction

Nowadays, the increasing energy demand of the growing population has led to the rapid consumption of fossil fuels, inevitably releasing carbon dioxide (CO_2) which is a prime contributor to global warming and climate change. For this reason, significant efforts have been devoted to developing innovative and cost-effective technologies to produce synthesis gas (syngas) from CO_2 [1]. Syngas, a mixture of hydrogen (H_2) and carbon monoxide (CO), is a versatile intermediate to the synthesis of a variety of valuable chemical feedstock and liquid fuels such as ammonia, methanol, ethanol, acetic acid, methyl format, dimethyl ether, synthetic gasoline and diesel via Fischer-Tropsch process [2-9]. Syngas can be produced from a variety of primary feedstock such as coal, natural gas (NG) and biomass. Among them, NG is the cheapest and cleanest source [10]. Many reforming technologies can be used to produce syngas including steam reforming, dry reforming (CO_2 reforming) and partial oxidation [11-13]:

Steam Reforming of Methane (SRM)

$$CH_4 + H_2O \rightleftarrows CO + 3H_2 \quad (1)$$

$$\Delta H = +206\ k\ /m$$

Dry Reforming of Methane (DRM)

IOP Conf. Series: Materials Science and Engineering **579** (2019) 012022 doi:10.1088/1757-899X/579/1/012022

$$C\ _4 + C\ _2 \rightleftarrows 2C\ \ + 2H_2 \quad (2)$$
$$\Delta H = +247\ k\ /m$$

Partial Oxidation of Methane (POM)

$$C\ _4 + \frac{1}{2}O_2 \rightleftarrows C\ \ + 2H_2 \quad (3)$$
$$\Delta H = -36\ k\ /m$$

SRM is the most important process for syngas production at large scale, because of it stability. The drawbacks of this highly endothermic reaction are the high energy consumption and syngas ratio H_2/CO, which is not suitable for the Fischer Tropsch synthesis process (F-T process). POM is an exothermic reaction (Reaction 3), which produces the synthesis gas at the desired syngas ratio of 2 for the F-T process. Unfortunately, during the reaction carbon is developed and covers the catalyst surface. This leads to the shortening the catalyst life time. In recent years, DRM (Reaction 2) has become an important technique that produces synthesis gas from two main greenhouse gasses: natural gas (CH_4) and carbon dioxide (CO) [7, 14]. The reason for this is that it has environmental, economical, technical and industrial advantages [15]. The drawback of DRM is the low syngas ratio H_2/CO, which can not be used to feed to the F-T process without H_2 addition. From the above reasons, the combined SRM and DRM, abbreviated as CSDRM (Equation 4), also known as bi-reforming of methane, has recently appeared as a promising technique since it is capable of generating a suitable syngas for F-T process by using the greenhouse gasses and water [16, 17].

$$3CH_4 + 2H_2O + CO_2 \rightarrow 8H_2 + 4CO \quad (4)$$
$$\Delta H = +712\ kJ/mol$$

The CSDRM can produce the synthesis gas with flexible H_2/CO ratios of 2.0 [18-20]. The H_2/CO ratio of the synthesis gas produced via the CSDRM can be controlled by changing the composition of the feed gas (H_2O, CO_2, and CH_4) [21, 22].

CSDRM is a complex process, because many side reactions can occur simultaneously as follows:

Water gas shift (WGS):

$$CO + H_2O \leftrightarrow CO_2 + H_2 \quad (5)$$
$$\Delta H = -41\ kJ/mol$$

Reverse water gas shift (RWGS):

$$CO_2 + H_2 \leftrightarrow CO + H_2O \quad (6)$$
$$\Delta H = +41\ kJ/mol$$

Methane decomposition (Methane Cracking Reaction):

$$CH_4 \leftrightarrow C + 2H_2 \quad (7)$$
$$\Delta H = +75\ kJ/mol$$

Disproportionation Reaction (Boudouard Reaction):

$$2CO \leftrightarrow C + CO_2 \quad (8)$$
$$\Delta H = -172\ kJ/mol$$

Carbon Gasification:

$$C + H_2O \leftrightarrow CO + H_2 \quad (9)$$
$$\Delta H = -131\ kJ/mol$$

Nowadays, plasma technology has been used widely in conversion of CO_2 and CH_4 to produce synthesis gas [23]. A plasma process is divided into two main methods; the first is non-thermal (or cold) plasma discharge including dielectric barrier discharge (DBD), corona discharge (CD), atmospheric pressure glow discharge (APGD), gliding arc discharge (GAD), microwave discharge (MWD) and spark discharge [24].

The second is thermal plasma including direct current (DC), alternating current (AC) arc torch and radio frequency (RF) [25, 26].

The microwave plasma technology has a number of advantages that will make it useful in the industry at present. The main advantage of microwave technology is that the energy transfer is rapid and uniform heating by irradiation with microwave [27]. Additionally, the use of microwaves offers technical advantages such as smooth operation, electrodeless reactors and high plasma density [28, 29].

In recent days, some research results on methane reforming by carbon dioxide and water by using microwave plasma are reported in the literature [21, 22]. From these studies, it is found that the adding steam into the dry reforming of methane is effective and leads to increasing the conversion of CH_4, selectivity and yield of H_2. Unfortunately, some knowledge gaps about the combination effect of the steam concentration and microwave power on the process performance and product quality have been remained unknown. Therefore, the main aims of the present work will study the combination influence of the process parameters (total feed gas flow rate, steam concentration and input microwave power) on the H_2/CO ratio under microwave irradiation at atmospheric pressure.

2. Experimental Set up

The schematic diagram of the experimental set-up is shown in Figure 1; microwave plasma system fundamentally consists of a gas cylinders, mass flow controllers, gas mixer, feed gas system, steam vapour unit, plasma reactor, microwave generator, and gas chromatographic (GC-MSD and GC-TCD) analysis system. The three feeding gases (CH_4, CO_2, and N_2) were controlled by a mass flow controller (MFC Alicat Scientific, MCS-Series) and sent into the gas mixer. Deionised water was sent into a vaporiser, which works at the set temperature of 300°C, by using a HPLC pump (Shimadzu LC pump). These gases are joined with the steam in a final feed mixer before entering the plasma reactor. To avoid water recondensation before the reactor, the heated pipe line was used. The quartz reactor has a size of 1.68 mm wall thickness, 25.5 mm outer diameter and length of 126 mm, as shown in Fig. 2. However, the more experimental details were presented elsewhere [30].

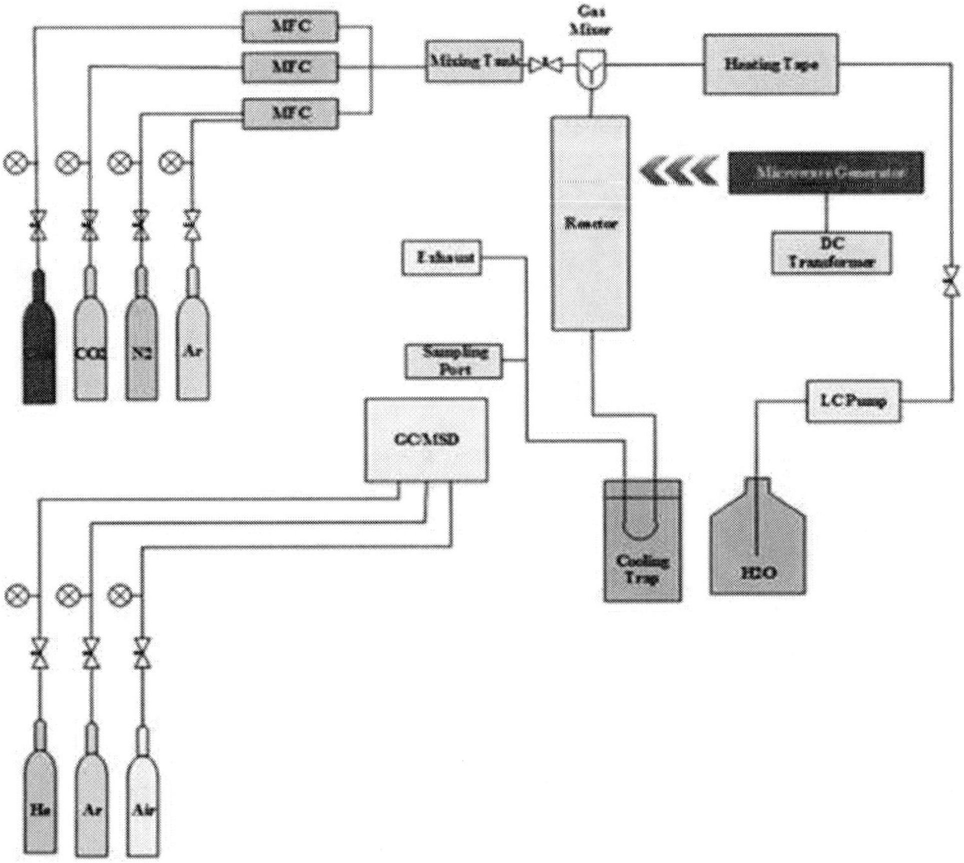

Figure 1. Schematics Diagram of the experimental Process

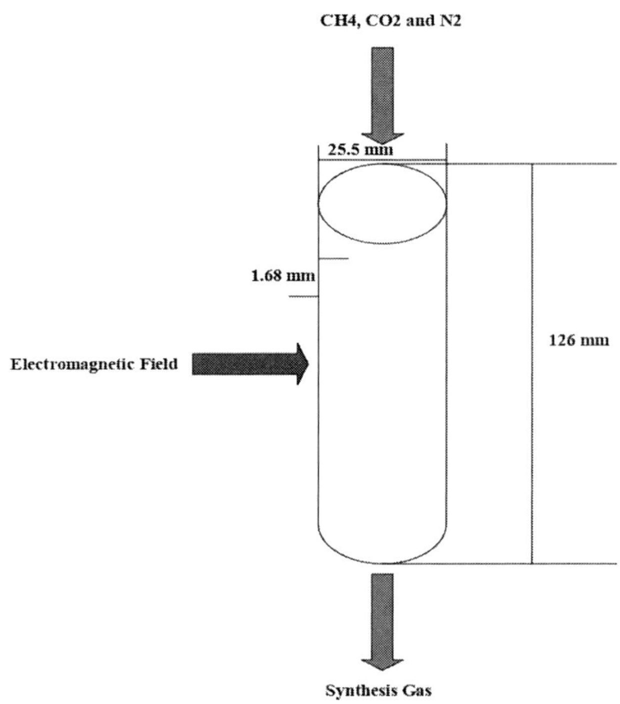

Figure 2. Quartz tube reactor

3. Results and Discussions

3.1 Microwave Plasma Dry Reforming of Methane

3.1.1 Effect of CO_2/CH_4 molar ratio

The CO_2/CH_4 molar ratio has a significant effect on the conversions of CH_4 and CO_2, the selectivity's and yields of H_2 and CO and the H_2 to CO molar ratio, as shown in Figure 3(a-d). In our experiments, keeping the CH_4 and N_2 flow rates constant of 0.2 L/min and 1.5 L/min and microwave power of 700 W. The CO_2/CH_4 molar ratio was changed from 2/1 to 5/1. Figure 3(a)-(d) depict the result of the influence of the CO_2/CH_4 ratio on the conversions of CH_4 and CO_2, the selectivity's and yields of H_2 and CO, and CO and the H_2/CO ratio was investigated by varying the CO_2/CH_4 ratio. As shown in Figure 3(a) and (b), the conversions of CH_4 and the selectivity to CO escalated when increasing CO_2/CH_4 ratio. Conversely, the increase of CO_2/CH_4 ratio leads to decreasing the conversion of CO_2, the selectivity's and the yields of H_2, CO and H_2/CO ratio, as shown in Figure 3(a)-(d), respectively. The decreased in CO_2 conversion could be explained because the main reaction of CH_4 and CO_2 under microwave plasma at atmospheric pressure was the plasma reforming reaction (see Equation (2)). When CO_2 amount exceeds the reaction stoichiometry of the main reaction, the conversion of CO_2 decreased while the conversion of CH_4 increased [31]. The conversion of CH_4 is higher than that of CO_2 under all testing conditions.

In addition, the induction of high amount of CO_2 in the reactor resulted in larger amount of carbon atoms, which are cracked by electron collision. These radicals are reacted with oxygen atoms to form carbon mono oxide (see Equation 6) while the hydrogen is low because of the formation of water as a consequence of the reaction of hydrogen atoms with oxygen atoms (see Equation 6). So, the selectivity's and yields of H_2, CO and the H_2/CO ratio significantly decreased when the CO_2 content in the gas mixture increased, as shown in Figure 3(b)-(d), respectively [32].

The conversion of CH_4 was always higher than those of CO_2. That can be attributed to the nature of the gas molecules, where these gases have a different molecular structure with different chemical bonds.

Methane has covalent bonds, and a large amount of energy may be released when they are broken [33].

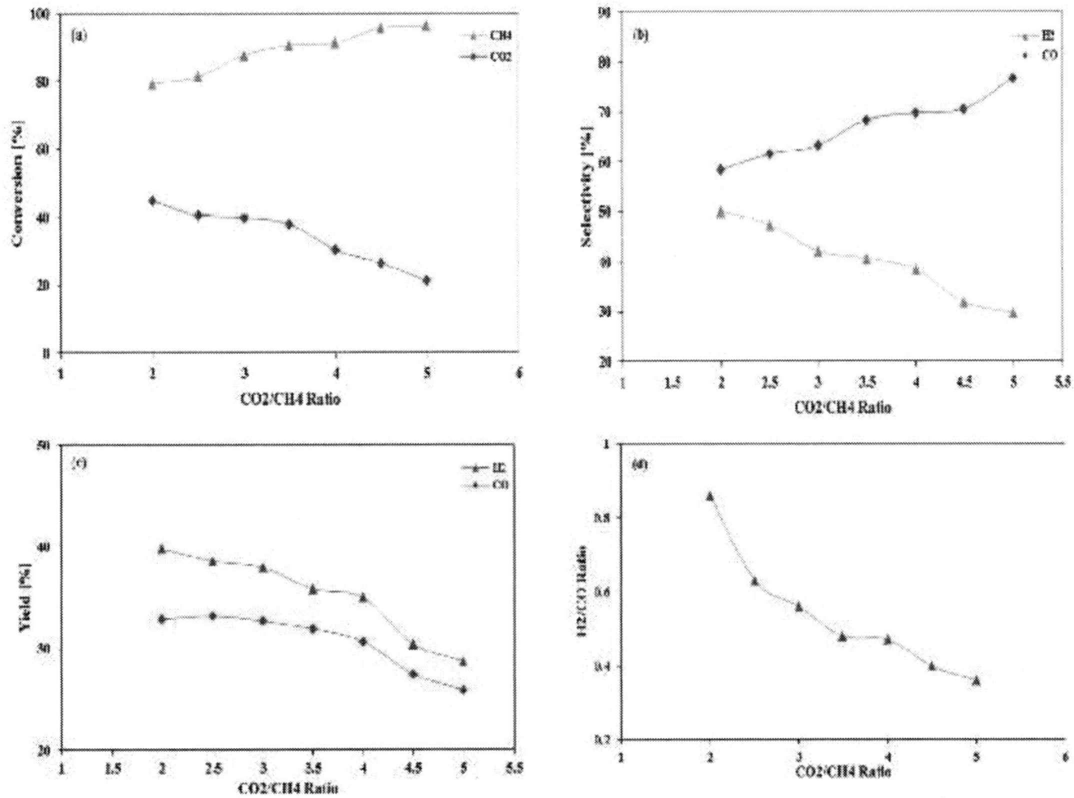

Figure 3. Effect of CO_2/CH_4 Ratio on the Performance Microwave Plasma Dry Reforming Reaction; (a) Conversion of CH_2 and CO_2; (b) Selectivity of H_2 and CO; (C) Yield of H_2 and CO; (d) H_2/CO Ratio (input microwave power at 700 W).

Additionally, the rate of dissociation of methane molecule depends on the initial supplied energy. On the other hand, the rate of thermal dissociation of CO_2 was lower due to its dependence on both temperature and the initial concentrations of CO_2 [34] which adversely affect the conversion of CO_2. N_2 gas is dissociated due to applied microwave energy, so that it is expected the side reaction, which consumes N_2 may occurs. However the results and discussions on N_2 consumption are excluded from this work as they are not the aim of this study. At these parameters, the plasma reaction was stable for more than four hours with the homogeneous plasma flame in the reactor.

3.1.2 Effect of total feed flow rate

The effect of the total feed flow rate on the conversions of CH_4 and CO_2, the selectivity's and yields of H_2 and CO and the H_2/CO ratio are shown in Figure 4(a)-(d), respectively. Experiments were conducted with a wide range of the flow rates of carbon dioxide (0.1-0.4 L min^{-1}) methane (0.05-0.2 L min^{-1}) and nitrogen (0.3-1.5 L min^{-1}). The molar ratio of CO_2/CH_4 of 2/1 and the input power of 700 W were kept constant for all experiments. As seen from the Figure 4(a)-(c), the conversions of CH_4 and CO_2, the selectivity's and yields of H_2, CO and the H_2/CO ratio slightly decreased with increasing total flow rates. This is because of the reduction in conversion of CH_4 and CO_2 could be related to the residence time of the gases in the microwave discharge zone, and it is reduced with the increasing of gas feed flow rate and led to the shorter treatment time [3].

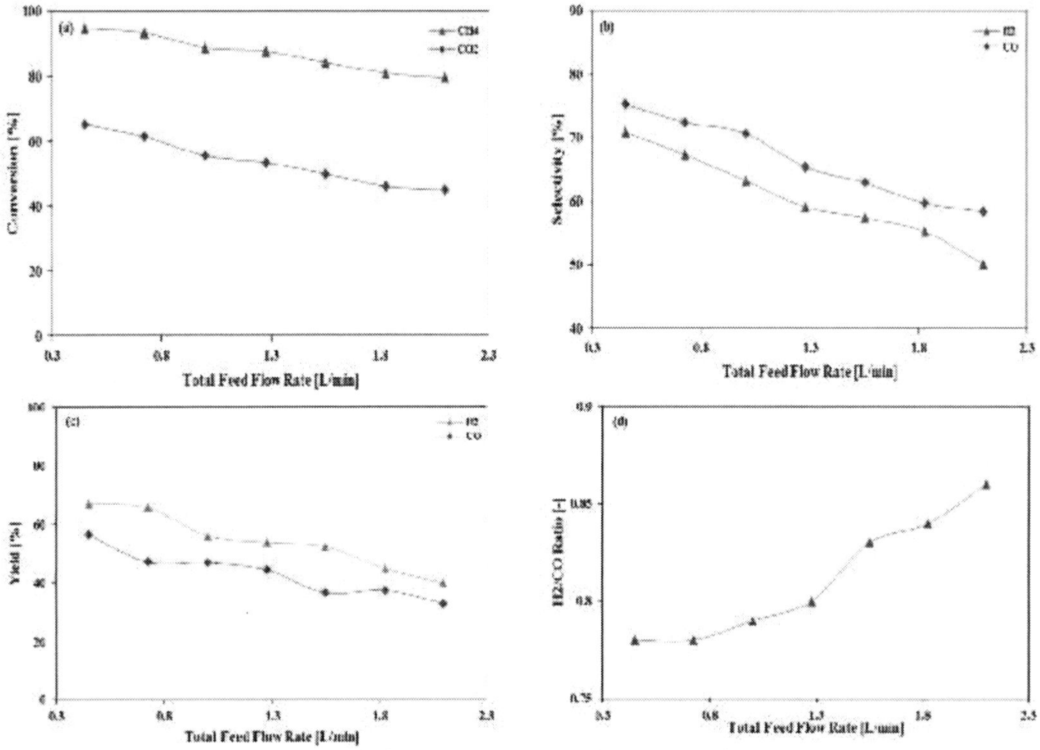

Figure 4. Effect of Total Gas Feed Flow Rate on the Performance Microwave Plasma Dry Reforming Reaction; (a) Conversion of CH_4 and CO_2; (b) Selectivity of H_2 and CO; (C) Yield of H2 and CO; (d) H_2/CO Ratio; (CO_2/CH_4 ratio at 2/1; input microwave power: 700 W).

3.1.3 Effect of input microwave power

It is well known, the input power is an important factor, which affects the dry plasma reforming of methane as it supply the requested energy for plasma formation. In other words, the input power will influence the density of the active species for the reaction. The conversions of CH_4 and CO_2, the selectivity's and yields of H_2 and CO, and CO and the H_2/CO ratio as a function of input microwave power under CO_2/CH_4 molar ratio of 2/1, and the CH_4, CO_2, and N_2 flow rates of 0.2, 0.4 and 1.5 L min^{-1} respectively, as shown in Figure 5(a)-(d), respectively. Experiments were performed in wide range of input power from 700 to 1200 W to get better understanding of the effect of input power on the conversions of CH_4, and CO_2, the selectivity's and yields of H_2, CO and the molar ratio of H_2/CO. Figure 5(a)-(c) exhibit the CH_4, and CO_2 conversions, the CO selectivity and yield increased, while the selectivity and yield of H_2 and the ratio of H_2/CO decreased with increasing the input power (Figure 5(b)-(d)). The increase of input microwave power could provide more energy to dissociate the CH_4 and CO_2 molecules and then leads to cracking of molecules deeply of CH_4 and CO_2 [35]. On the other hand, at high microwave power a side reaction, which consumes H_2 (see Equation 6), occurs and provides enough energy to activate CO_2 to form CO. As a consequence, the product ratio H_2/CO decreases with the increasing microwave power [36].

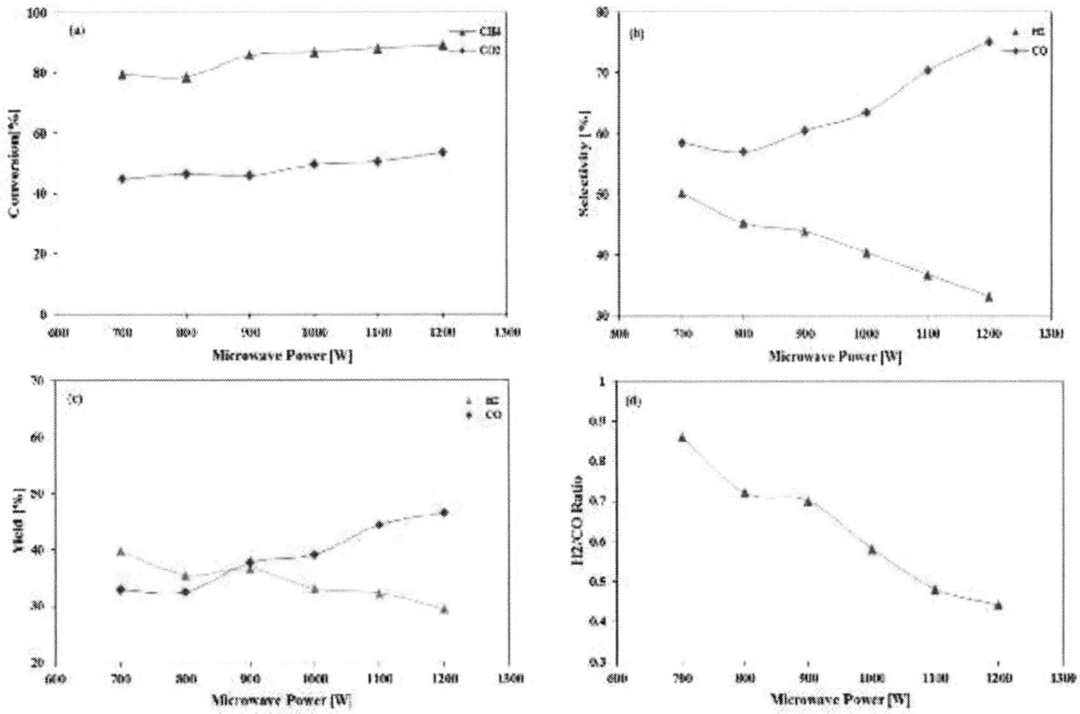

Figure 5. Effect of Input MW Power on the Performance Microwave Plasma Dry Reforming Reaction; (a) Conversion of CH_4 and CO_2; (b) Selectivity of H_2 and CO; (C) Yield of H_2 and CO; (d) H_2/CO Ratio (CO_2/CH_4 ratio at 2/1; the feed gas rate at 2.1 L min^{-1}).

3.2 Microwave Plasma Steam Combined with Dry Reforming of Methane
3.2.1 Steam Concentration

To understand the reaction pathway of CSDRM, the steam concentration was varied at a fixed feed flow rate of CH_4, CO_2 of 0.008, 0.016 mole min^{-1}, CO_2/CH_4 ratio of 2/1 and input microwave power of 700 W. Figure 6(a)-(d) show the effects of steam concentration on the CH_4 and CO_2 conversions, H_2 and CO the selectivity's and yields and the H_2/CO ratio. As shown in Figure 6(a)-(d), CH_4 conversions, selectivity and yield of H_2 and H_2/CO mole ratio increased from 79.35%, 50.12%, 39.77% and 0.86 to 82.74%, 98.79%, 81.73% and 5.23 respectively, whereas CO_2 conversion and CO selectivity and yield slightly decreased from 44.82%, 58.42% and 32.89% to 19.23%, 18.87% and 9.04% respectively with increasing the steam concentration. This is because the steam has more reductive and oxidative radicals (H, OH, and O) in the CSDRM process, and these radicals lead to increasing CH_4 conversion, H_2 and CO selectivity, H_2 and CO yield and H_2/CO ratio [37].

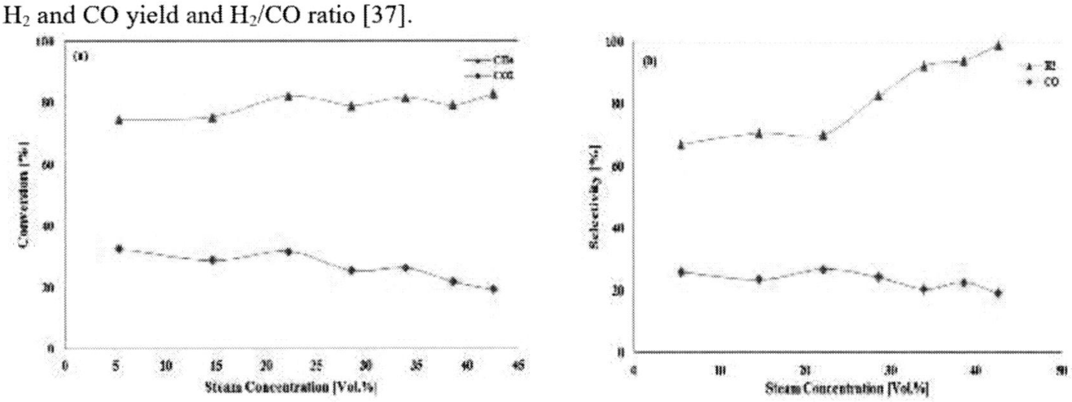

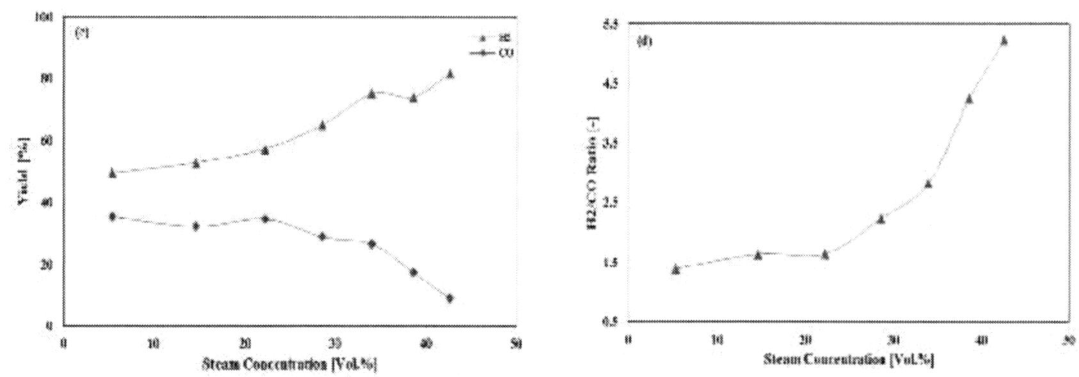

Figure 6. Effect of Steam Concentration on the Performance Microwave Plasma Dry Reforming Reaction; (a) Conversion of CH_4 and CO_2; (b) Selectivity of H_2 and CO; (C) Yield of H2 and CO; (d) H_2/CO Ratio; (feed flow rate of CH_4, CO_2 at 0.008, 0.016, 0.062 mole min^{-1}, CO_2/CH_4 ratio at 2/1 and input microwave power at 700 W).

3.2.2 Effect total feed flow rate

The steam additions have a noteworthy effect on the conversions of CH_4, and CO_2, the selectivity's and yields of H_2 and CO and the H_2/CO ratio are shown in Figure 7(a)-(d), respectively. The total feed gas flow rate of H_2O, CO_2, CH_4 and N_2 ranged from 0.02 to 0.15 mole min^{-1}, as keeping constant the input power of 700 W. As is expected, CH_4 conversion, the product selectivity and the yields of H_2 and the H_2/CO ratio are enhanced with increasing the total feed flow rate, pointing to the fact that combined reforming by H_2O and CO_2 is stronger than dry reforming (see Figure 7(a)-)(d)). This is because the steam molecules could dissociated into OH and H radicals, reacting with CH_4 and resulting in the lower probability of CH_4 recombination due to the production of CH_3 and other by products [38, 39]. The CH_4 reacted with H_2O instead of CO_2 with increasing amount of steam in the feed stream due to the more stable nature of CO_2. On the contrary, the increase of steam content caused a considerable decrease of the conversion of CO_2, CO selectivity, as shown in Figure 7(a)-(c). Although higher steam amounts resulted in the more CH_4 conversion, H_2 selectivity, H_2 yield and the H_2/CO ratio, increasing steam content did not favour CO_2, selectivity and yield of CO due to the fixed CO_2/CH_4 molar ratio in the feed. The conversions of CO_2 and H_2O were dampened as a result of the RWGS reaction (see Equation (6)).

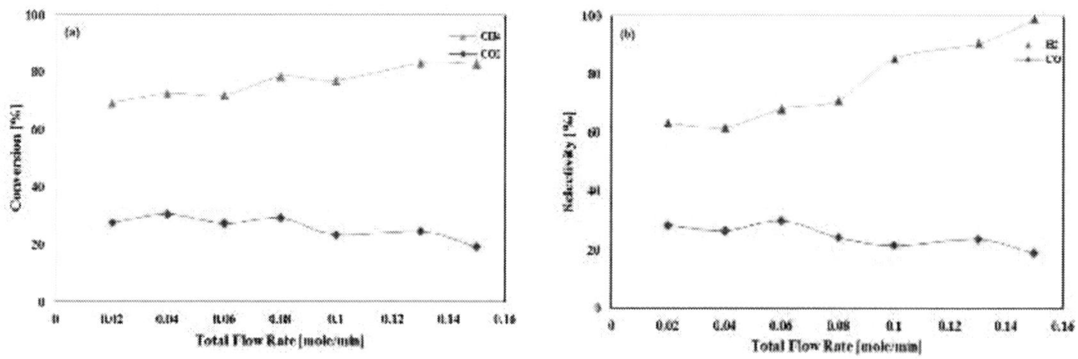

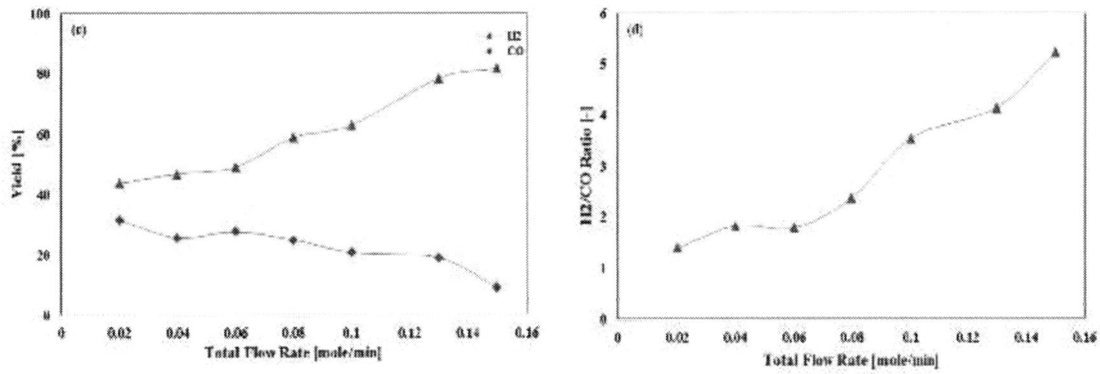

Figure 7. Effect of Total Gas Feed Flow Rate with Water on the Performance Microwave Plasma Dry Reforming Reaction; (a) Conversion of CH_4 and CO_2; (b) Selectivity of H_2 and CO; (C) Yield of H2 and CO; (d) H_2/CO Ratio; (input microwave Power at 700W).

3.2.3 Effect of input microwave power

The effect of microwave power on the conversions of CH_4 and CO_2, the selectivity's and yields of H_2 and CO, and CO and the H_2/CO ratio, as shown in Figure 8(a)-(d), respectively. The input microwave power changed from 700 to 1200 W. The CO_2/CH_4 molar ratio was 2/1, and the CH_4, CO_2 and steam flow rate were 0.008, 0.016, and 0.065 mole min^{-1}, respectively. Figure 8(a)-(c) depict the conversions of CH_4 and CO_2, the selectivity of CO and yields of H_2 and CO increased, while the ratio of H_2/CO decreased with increasing the input power (Figure 8(d)). The reason for this has been explained previously. The selectivity of H_2 did not change with increase the microwave power (Figure 8(b) and (c)). In this study, we observed the amount of water, and the amount of solid carbon powder out the quartz tube low compared with the CO_2 reforming.

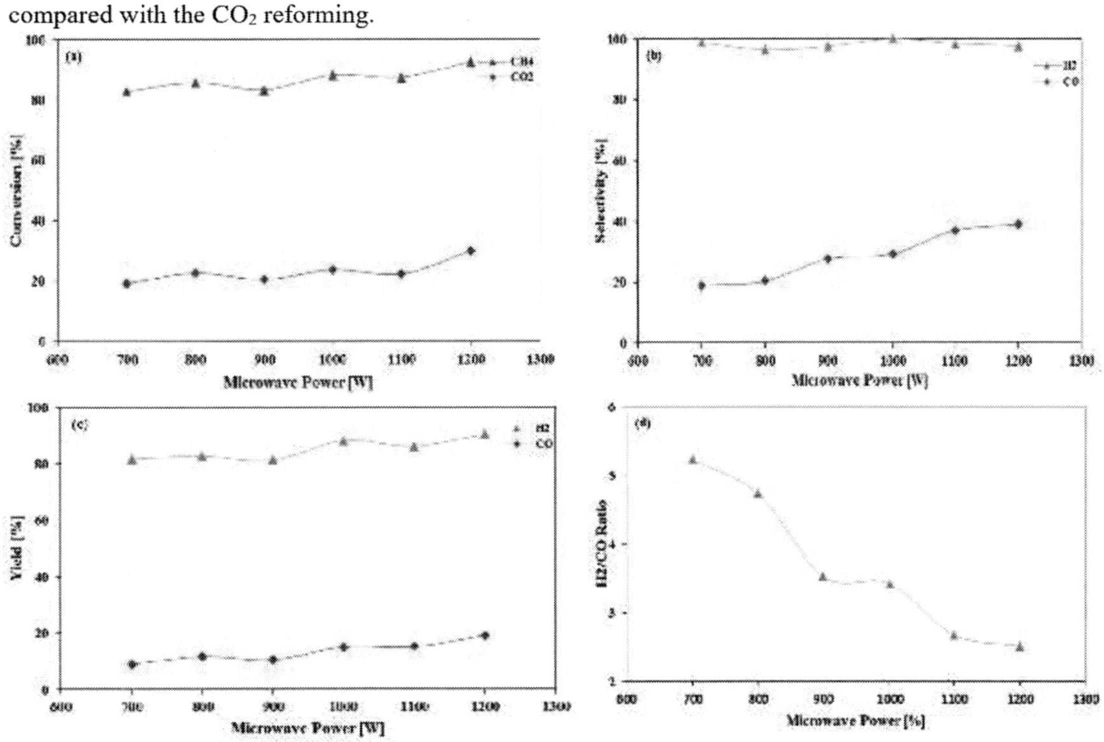

Figure 8. Effect of Input MW Power with Water on the Performance MW Plasma Dry Reforming Reaction; (a) Conversion of CH_4 and CO_2; (b) Selectivities of H_2 and CO; (C) Yield of H2 and CO; (d) H_2/CO Ratio (total feed gas rate at 0.15 mole min^{-1}).

IOP Conf. Series: Materials Science and Engineering **579** (2019) 012022 doi:10.1088/1757-899X/579/1/012022

The combined effects of microwave plasma methane dry reforming and methane steam with dry reforming were compared and displayed in Figure 9(a)-(d). The conversions of CH_4 and CO_2 over dry reforming of methane were about 79.35%, and 44.82% respectively, the selectivity's and yields of H_2 and CO, and the H_2/CO ratio were 50.12%, 58.42%, 39.77%, 32.89% and 0.86, respectively. When the steam adding with dry reforming of methane, the conversion of CH_4, selectivity and yield of H_2 and synthesis ratio were improved apparently, as shown in Figure 9(a)-(c), respectively. The conversions of CH_4 and CO_2 were 82.74% and 19.23% respectively. The selectivity's, yields to H_2 and CO and the H_2/CO ratio were 98.79%, 18.87%, 81.73%, 9.04% and 5.23, respectively. The conversions of CH_4 was 82.74% under combined steam and dry reforming of methane, which was higher than using the dry reforming of methane. Similar tendency the selectivity and yield of H_2 and H_2/CO ratio were also improved under combined of steam and dry reforming of methane. However, it was observed for conversion of CO_2 decreased with adding the steam. The selectivity of H_2 was 98.79% and that of H_2/CO ratio was 5.23, which was even higher than those obtained in dry reforming of methane.

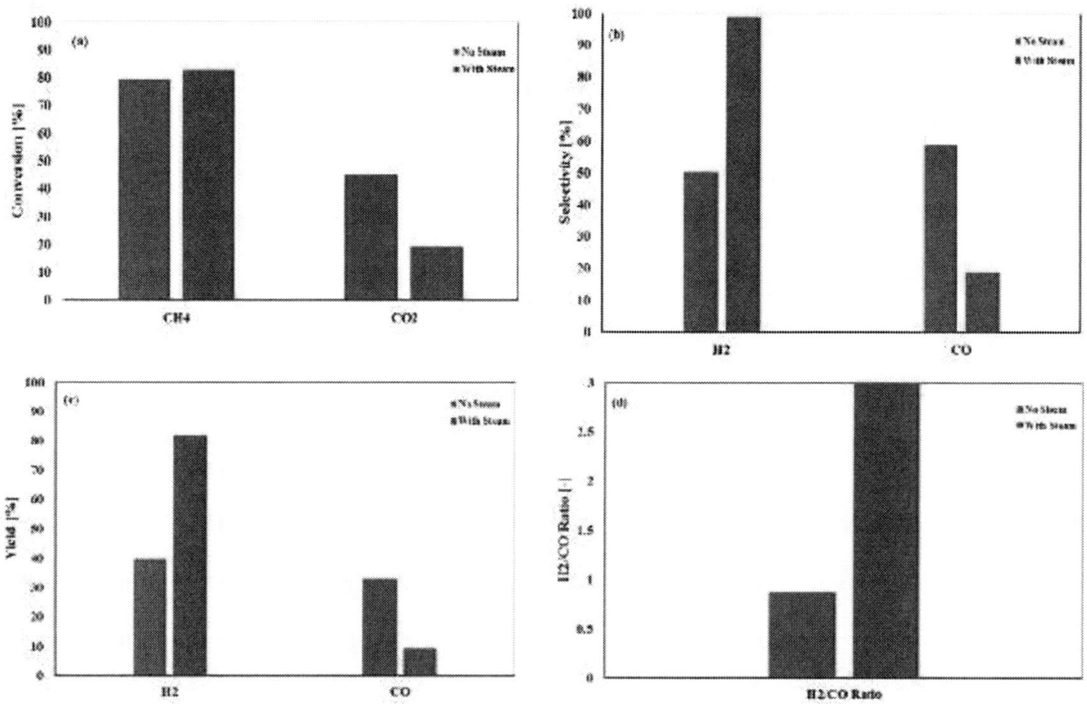

Figure 9. Effect of H_2O on equilibrium (a) CH_4 and CO_2 Conversions; (b) H_2 and CO Selectivities; (c) H_2 and CO Yields and (d) H_2/CO Ratio for combined DR-SR as a function of Microwave Power at 700 W (CO_2/CH_4: 2/1; the total gas feed rate at 2.1 L min^{-1} and the total feed steam rate at 24.5 L min^{-1}).

4. Conclusions

In this paper, the effect of steam adding on the syngas formation has been studied in a microwave plasma reactor at atmospheric pressure for syngas production. Based on the measured results the following conclusions can be made:

1. With H_2O addition, the high conversions of CH_4 and CO_2 were 82.74% and 19.23% of the steam reforming of methane can achieved. At these conversion levels, the selectivity's of H_2 and CO, the yields of H_2 and CO and H_2/CO ratio were 98.79%, 18.87%, 81.73%, 9.04% and 5.23, respectively.
2. The CH_4 conversion, H_2 and CO selectivity, H_2 and CO yield, and H_2/CO ratio were affected by the H_2O concentration in the feed significantly.
3. The combined SRM and DRM process is very sensitive with the operating conditions such as the input

microwave power and total feed flow rate.
4. The desired syngas ratio H_2/CO for F-T process can be reached by optimising the H_2O concentration in the feed and input microwaves power.

Acknowledgements

The first author would like to gratefully acknowledge the Ministry of Higher Education & Scientific Research in Iraq for their sponsorship for his PhD study at Curtin University.

References

[1] X. Tu, J.C. Whitehead, Plasma dry reforming of methane in an atmospheric pressure AC gliding arc discharge: co-generation of syngas and carbon nanomaterials, international journal of hydrogen energy 39(18) (2014) 9658-9669.
[2] J. Ross, A. Van Keulen, M. Hegarty, K. Seshan, The catalytic conversion of natural gas to useful products, Catalysis Today 30(1-3) (1996) 193-199.
[3] D. Pakhare, J. Spivey, A review of dry (CO 2) reforming of methane over noble metal catalysts, Chemical Society Reviews 43(22) (2014) 7813-7837.
[4] M.M. Ghouri, S. Afzal, R. Hussain, J. Blank, D.B. Bukur, N.O. Elbashir, Multi-scale modeling of fixed-bed Fischer Tropsch reactor, Computers & Chemical Engineering 91 (2016) 38-48.
[5] A. Kasht, R. Hussain, M. Ghouri, J. Blank, N.O. Elbashir, Product analysis of supercritical Fischer-Tropsch synthesis: utilizing a unique on-line and off-line gas chromatographs setup in a bench-scale reactor unit, American Journal of Analytical Chemistry 6(08) (2015) 659.
[6] J.R. Rostrup-Nielsen, New aspects of syngas production and use, Catalysis today 63(2) (2000) 159-164.
[7] M. Usman, W.W. Daud, H.F. Abbas, Dry reforming of methane: influence of process parameters—a review, Renewable and Sustainable Energy Reviews 45 (2015) 710-744.
[8] B. Abdullah, N.A.A. Ghani, D.-V.N. Vo, Recent advances in dry reforming of methane over Ni-based catalysts, Journal of Cleaner Production 162 (2017) 170-185.
[9] W. Wang, S. Wang, X. Ma, J. Gong, Recent advances in catalytic hydrogenation of carbon dioxide, Chemical Society Reviews 40(7) (2011) 3703-3727.
[10] P.L. Spath, D.C. Dayton, Preliminary screening-technical and economic assessment of synthesis gas to fuels and chemicals with emphasis on the potential for biomass-derived syngas, National Renewable Energy Lab Golden Co, 2003.
[11] J. Pacheco, G. Soria, M. Pacheco, R. Valdivia, F. Ramos, H. Frías, M. Durán, M. Hidalgo, Greenhouse gas treatment and H 2 production, by warm plasma reforming, international journal of hydrogen energy 40(48) (2015) 17165-17171.
[12] S. Rowshanzamir, M. Eikani, Autothermal reforming of methane to synthesis gas: Modeling and simulation, international journal of hydrogen energy 34(3) (2009) 1292-1300.
[13] G. Centi, E.A. Quadrelli, S. Perathoner, Catalysis for CO 2 conversion: a key technology for rapid introduction of renewable energy in the value chain of chemical industries, Energy & Environmental Science 6(6) (2013) 1711-1731.
[14] J. Hubert, M. Moisan, A. Ricard, A new microwave plasma at atmospheric pressure, Spectrochimica Acta Part B: Atomic Spectroscopy 34(1) (1979) 1-10.
[15] S.T. Oyama, P. Hacarlioglu, Y. Gu, D. Lee, Dry reforming of methane has no future for hydrogen production: comparison with steam reforming at high pressure in standard and membrane reactors, international journal of hydrogen energy 37(13) (2012) 10444-10450.
[16] G.A. Olah, A. Goeppert, M. Czaun, T. Mathew, R.B. May, G.S. Prakash, Single step bi-reforming and oxidative bi-reforming of methane (natural gas) with steam and carbon dioxide to metgas (CO-2H2) for methanol synthesis: self-sufficient effective and exclusive oxygenation of methane to methanol with oxygen, Journal of the American Chemical Society 137(27) (2015) 8720-8729.
[17] M.M. Noureldin, N.O. Elbashir, M.M. El-Halwagi, Optimization and selection of reforming approaches for syngas generation from natural/shale gas, Industrial & Engineering Chemistry Research 53(5) (2013) 1841-1855.

[18] A.N. Pour, M. Mousavi, Combined reforming of methane by carbon dioxide and water: Particle size effect of Ni–Mg nanoparticles, International Journal of Hydrogen Energy 40(38) (2015) 12985-12992.
[19] A.R. Kim, H.Y. Lee, D.H. Lee, B.-W. Kim, C.-H. Chung, D.J. Moon, E.J. Jang, C. Pang, J.W. Bae, Combined Steam and CO2 Reforming of CH4 on LaSrNiO x Mixed Oxides Supported on Al2O3-Modified SiC Support, Energy & Fuels 29(2) (2015) 1055-1065.
[20] V.R. Choudhary, K.C. Mondal, CO2 reforming of methane combined with steam reforming or partial oxidation of methane to syngas over NdCoO3 perovskite-type mixed metal-oxide catalyst, Applied Energy 83(9) (2006) 1024-1032.
[21] D. Czylkowski, B. Hrycak, M. Jasiński, M. Dors, J. Mizeraczyk, Microwave plasma-based method of hydrogen production via combined steam reforming of methane, Energy 113 (2016) 653-661.
[22] L. Li, H. Wang, X. Jiang, Z. Song, X. Zhao, C. Ma, Microwave-enhanced methane combined reforming by CO 2 and H 2 O into syngas production on biomass-derived char, Fuel 185 (2016) 692-700.
[23] R. Snoeckx, Y. Zeng, X. Tu, A. Bogaerts, Plasma-based dry reforming: improving the conversion and energy efficiency in a dielectric barrier discharge, RSC Advances 5(38) (2015) 29799-29808.
[24] W.-C. Chung, M.-B. Chang, Dry reforming of methane by combined spark discharge with a ferroelectric, Energy Conversion and Management 124 (2016) 305-314.
[25] P. Jamróz, W. Kordylewski, M. Wnukowski, Microwave plasma application in decomposition and steam reforming of model tar compounds, Fuel Processing Technology 169 (2018) 1-14.
[26] M.S. Aw, I.G.O. Crnivec, P. Djinovic, A. Pintar, Strategies to enhance dry reforming of methane: Synthesis of ceria-zirconia/nickelecobalt catalysts by freeze-drying and NO calcination, international journal of hydrogen energy 39(12636) (2014) e12647.
[27] M. Nüchter, B. Ondruschka, W. Bonrath, A. Gum, Microwave assisted synthesis–a critical technology overview, Green chemistry 6(3) (2004) 128-141.
[28] J. Menéndez, A. Arenillas, B. Fidalgo, Y. Fernández, L. Zubizarreta, E.G. Calvo, J.M. Bermúdez, Microwave heating processes involving carbon materials, Fuel Processing Technology 91(1) (2010) 1-8.
[29] S. Heijkers, R. Snoeckx, T.s. Kozák, T. Silva, T. Godfroid, N. Britun, R. Snyders, A. Bogaerts, CO2 conversion in a microwave plasma reactor in the presence of N2: elucidating the role of vibrational levels, The Journal of Physical Chemistry C 119(23) (2015) 12815-12828.
[30] N.M. Alawi, A. Barifcani, H.R. Abid, Optimisation of CH4 and CO2 conversion and selectivity of H2 and CO for the dry reforming of methane by a microwave plasma technique using a B ox–B ehnken design, Asia-Pacific Journal of Chemical Engineering (2018) e2254.
[31] A. Serrano-Lotina, L. Daza, Influence of the operating parameters over dry reforming of methane to syngas, International Journal of Hydrogen Energy 39(8) (2014) 4089-4094.
[32] C.-H. Liao, R.-F. Horng, Experimental study of syngas production from methane dry reforming with heat recovery strategy, International Journal of Hydrogen Energy 42(40) (2017) 25213-25224.
[33] J.W. Moore, C.L. Stanitski, Chemistry: The molecular science, Cengage Learning2014.
[34] S. Rayne, Thermal carbon dioxide splitting: A summary of the peer-reviewed scientific literature, (2008).
[35] A. Indarto, J.-W. Choi, H. Lee, H.K. Song, Effect of additive gases on methane conversion using gliding arc discharge, Energy 31(14) (2006) 2986-2995.
[36] A. Aziznia, H.R. Bozorgzadeh, N. Seyed-Matin, M. Baghalha, A. Mohamadalizadeh, Comparison of dry reforming of methane in low temperature hybrid plasma-catalytic corona with thermal catalytic reactor over Ni/γ-Al 2 O 3, Journal of Natural Gas Chemistry 21(4) (2012) 466-475.
[37] T. Jiang, Y. Li, C.-j. Liu, G.-h. Xu, B. Eliasson, B. Xue, Plasma methane conversion using dielectric-barrier discharges with zeolite A, Catalysis Today 72(3) (2002) 229-235.
[38] J. Sutherland, M.C. Su, J. Michael, Rate constants for H+ CH4, CH3+ H2, and CH4 dissociation at high temperature, International Journal of Chemical Kinetics 33(11) (2001) 669-684.
[39] A. Bonard, V. Daële, J.-L. Delfau, C. Vovelle, Kinetics of OH Radical Reactions with Methane in the Temperature Range 295– 660 K and with Dimethyl Ether and Methyl-tert-butyl Ether in the Temperature Range 295– 618 K, The Journal of Physical Chemistry A 106(17) (2002) 4384-4389.

IOP Conf. Series: Materials Science and Engineering **579** (2019) 012023 doi:10.1088/1757-899X/579/1/012023

Modify Chock Performance Equation for Tertiary Reservoir Wells in Khabaz Oil Field.

Haider A. Jumaah

North Oil Company,Kirkuk, Iraq

Corresponding author e-mail address: haiderashour@yahoo.com

Abstract. Gilbert correlation has been modified to determine the performance of multiphase fluid-flow through the wellhead and choke. The modification present two sets of new correlations based on statistical analysis of 33 production tests data from 12 wells produce from Tertiary Reservoir in Khabaz oil field, first correlation is modified of Gilbert equation to predict liquid flow rates as a function of wellhead flowing pressure, gas-liquid ratio and choke size, the second correlation takes the effect of water cut and sediment (BS & W) as an effective parameter to minimize error. A comparison between the results of each correlation with measured date has been made to select the best correlation to predict flow rate in newly drilled wells. The oil flow rates predicted by the new correlations show excellent agreement with the measured rates and the second modified Gilbert correlation are found to be closest to all ranges of flow rate variables with an average error of 10 % and R2=0.9493.

1.Introduction

Pressure is an important energy in the Oil and gas production system, either choke performance to control the flow rate. Thus, different studies have been published tacking present theories and correlations to introduce simultaneous liquid and gas flow through an orifice. The first relationship was present by Gilbert in 1954 based on field data collected from the Ten Section field of California [1], Ros 1960 have also presented relationships that are often used [2]. Gilbert's approach was widely used by petroleum engineers more than any other approach. The goal of this paper is a recalculation of observed available data to modify Gilbert correlation parameters correspond to the tertiary Khabaz oil field, then calculate flow rate by using modified parameters and compare the results with a flow rate estimated from original Gilbert approach Therefore, production data from 12 oil wells in the Khabaz oil field as in Table (1) were used to check correlations.

1.1 Area of Study

The Khabaz oil field is multi-pay zones Carbonate oil fields like most of the carbonate oil fields in the north of Iraqi. It is located in North West of Kirkuk city and far about 12 km from Kirkuk city centre as shown in Figure. 1.

IOP Conf. Series: Materials Science and Engineering **579** (2019) 012023 doi:10.1088/1757-899X/579/1/012023

The field is about 18 km length and its width is about 4 km. It consists of single small subsurface asymmetrical anticline dome sweeping to south direction with depth. Its northeast flange dipper than the southwest flange with general axis extends from northwest to southeast and parallel with the Kirkuk oil field axis in Cotton –Kushkai region. First well (Kz-1) has been drilled in August 1976, since that time, drilling operations have continued.

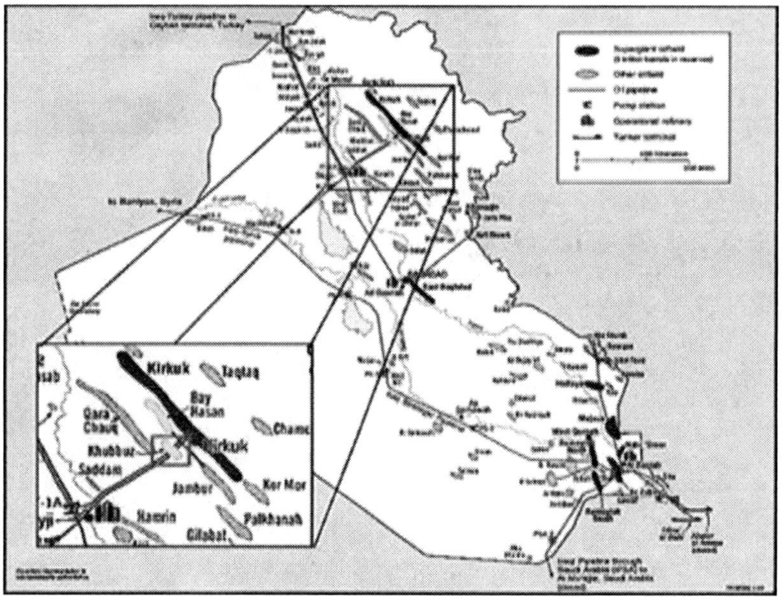

Figure 1. Khabaz oil field location.

The Khabaz oil field consists of three main hydrocarbon reservoirs.

1. Tertiary Reservoir.

2. Cretaceous Upper Qamchuqa (Mauddud) Reservoir.

3. Cretaceous Lower Qamchuqa (Shuaiba) Reservoir.

As shown in Figure. 2.

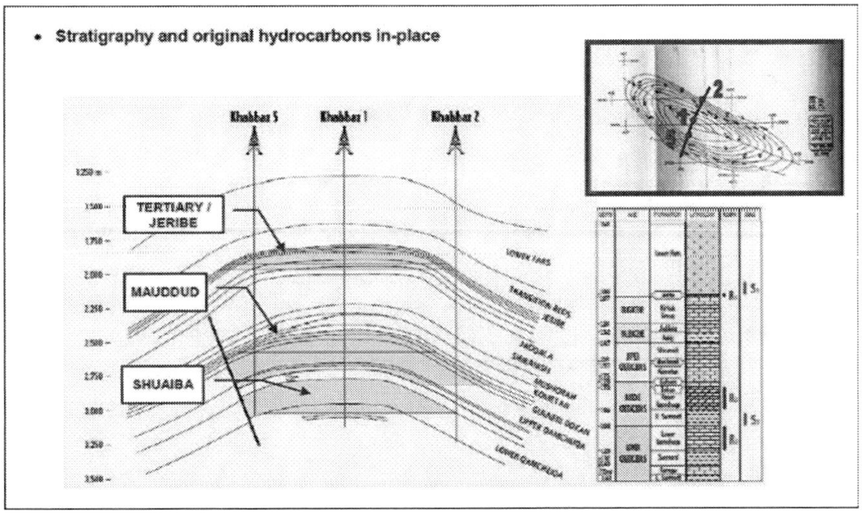

Figure 2. Khabaz oil field cross Section.

All the 12 wells which have been selected in this paper are produce from the Tertiary Reservoir; the important reservoirs in Khabaz field, so the paper confines only to tertiary producer wells.

1.2 Basic Equations

To estimate the relationship between the production rate and wellhead flowing pressure for two-phase critical flow, an empirical correlation has been developed. These correlations are presented in a similar form to eq.1.

$$PWF = c * \frac{GOR^{a} * Q}{D^{b}} \tag{1}$$

Where

Pwf = tubing-head pressure, (psi)

R = gas/liquid ratio, (Scf/STB)

Q = gross liquid rate, (BBL/D)

D = chock size. (Inch)

a, b.and c= constants,

Gilbert's Approach based on the assumption of a knife-edge chokes and simplifying some assumptions that related to the pressure and flow characteristics of the oil and gas through the chock. Theoretically Gilbert's equation assumes that the actual mixture of fluid velocities through the bean should be passed the sound speed; in order to the downstream, flow line pressure, no effect on the rate and pressure of upstream. The fluid velocity reaches the sound speed when the upstream pressure is at least twice greater than the downstream pressure. However, Gilbert, renowned, that his equation works well when the downstream pressure is less than 70% of the upstream pressure. Ross\ (1960) and Baxendell (1958)

IOP Conf. Series: Materials Science and Engineering **579** (2019) 012023 doi:10.1088/1757-899X/579/1/012023

developed another correlations similar to that of Gilbert (1954) correlation but with different correlating parameters. The parameters of each equation are given in the table (1).

Table 1. Correlation Parameters

Correlation	a	b	c
Gilbert (1954)	$3.86*10^{-3}$	0.546	1.89
Baxendell (1958)	$3.12*10^{-3}$	0.546	1.93
Rose (1960)	$4.26*10^{-3}$	0.5	2

1.3 Nonlinear Regression

The Nonlinear regression: is a type of regression analysis of observational data that integrated by a function which is a nonlinear grouping of the model parameters and depends on one or more independent variables. The data is close-fitting by a method of successive approximations [3].

1.4 Correlation Coefficient

A correlation coefficient has measured the strength and what of a linear association between two variables. It ranges from 1 to 0. The stronger relationship is close to the absolute value of 1, and A zero correlation coefficient specifies that there is no linear relationship between the variables. The coefficient can be either negative or positive [4].

1.5 Data acquisition

Recently, the total number of wells in tertiary reservoir reached 26 wells, divided into working, new drilled not connected to Degas station observation, and water injection wells. Although the number of wells selected is 12, but they represent most of the working well and provided data from all parts of the reservoir. It can be adapted to modify the correlation depending on estimate surface flow rates in newly drilled wells as shown in the figure 3.

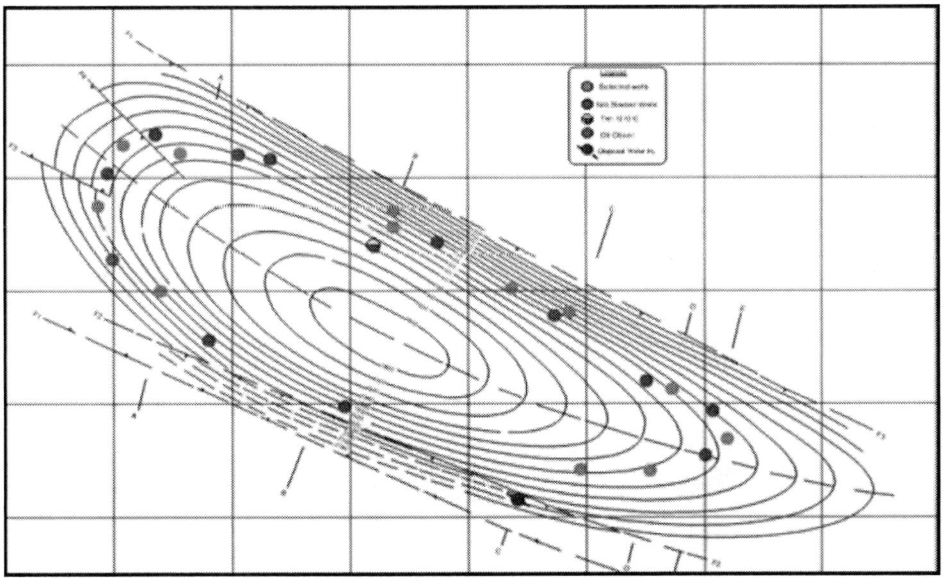

Figure 3. Selected wells locations.

2. Results and Discussion

33 test data of production are gathered from 12 wells, including the flow rates, gas-liquid ratio, choke Size, upstream pressure, downstream pressure, and water cut, lay within the range 400–2900 (STB/day), 847–2595 (SCF/STB), 16–42 (1/64th-inch), 445–1854 (Psi), and (0-4.7) % respectively as shown in the in Table 2.

Table 2. Production Test Data.

No	DATE	Chock size in	Pressure psi	Q bbl/d	GOR scf/bbl	WT %	flow path
1	15/07/2000	0.438	446.5	1000	1112	1	Tubing
2	05/05/2002	0.375	445	850	873	0.5	Tubing
3	22/07/2000	0.375	1102	1300	913	1.3	Tubing
4	01/05/2002	0.375	1233	1150	1290	2.3	Tubing
5	03/09/2000	0.250	1711	400	2595	0	Tubing
6	16/10/2001	0.313	1377.5	450	1976	0	Tubing
7	17/08/2002	0.250	1850	400	2500	0	Tubing
10	15/10/2001	0.375	1116.5	1600	927	4.5	Tubing
11	04/05/2002	0.375	1130	1250	949	4.7	Tubing
12	18/07/2000	0.375	1203.5	1200	1174	0.2	Tubing
13	21/07/2000	0.313	1232.5	850	1221	0.3	Tubing
14	05/09/2001	0.313	1261.5	1100	1078	0.9	Tubing
15	20/10/2001	0.250	1305	750	989	0.8	Tubing
16	10/10/2001	0.563	1102	2600	1027	1.2	Tub.
17	07/07/2000	0.250	1450	500	1780	0	Tubing
18	10/05/2002	0.250	1855	450	1976	0	Tubing
19	16/07/2000	0.500	1160	2250	1121	0	Tubing
20	20/07/2000	0.438	1189	1700	1047	0	Tubing
22	21/05/2002	0.438	1090	1700	1134	0	Tubing
23	04/08/2002	0.500	1190	2200	1079	0	Tubing
24	27/10/2001	0.375	1276	1100	1483	0	Tubing
25	08/05/2002	0.375	1275	1000	1483	0	Tubing
26	29/06/2000	0.656	783	2900	921	0	Tubing
27	01/07/2000	0.563	899	2150	966	0	Tubing
28	03/07/2000	0.500	928	1550	957	0	Tubing
29	23/10/2001	0.250	1363	650	1369	0	Tubing
30	24/08/2002	0.250	1025	700	847	0	Tubing
31	23/07/2000	0.297	1232.5	950	1249	0.8	Tubing
32	17/09/2001	0.297	1322	1250	1068	0	Tubing
33	22/10/2001	0.266	1421	850	1134	0	Tubing

The first correlation for Khabaz field is Gilbert modified equation based on regression analysis of equation (1) as flowing:

$$Q = \mathrm{PWF}\frac{D^a}{c*\mathrm{GOR}^b} \tag{2}$$

Where,

a = 0.000275, b = 0.9058, c = 1.7634

The second correlation has been developed considering a parameter which has not been covered in the previous correlations; water cut (Wct.) measured volume percentage of the production stream, in addition to other parameters which had been added to the correlation, in order to reduce error in the field condition to a minimum, as represented in the following form.

$$Q = \mathrm{PWF}\frac{D^a}{c*\mathrm{GOR}^b} * \left(1 - \frac{Wct.}{100}\right)^d . \tag{3}$$

Where,

a = 0.0000486, b = 1.159, c = 1.733 and d= 1.3936

These new correlations applied to estimate flow rate and compared with Gilbert correlation which is recently used to estimate flow rate in the Khabaz oil field. The results show that flow rate estimated by new correlations is closer to the actual rate in testes than Gilbert correlation as shown in the figure 4.

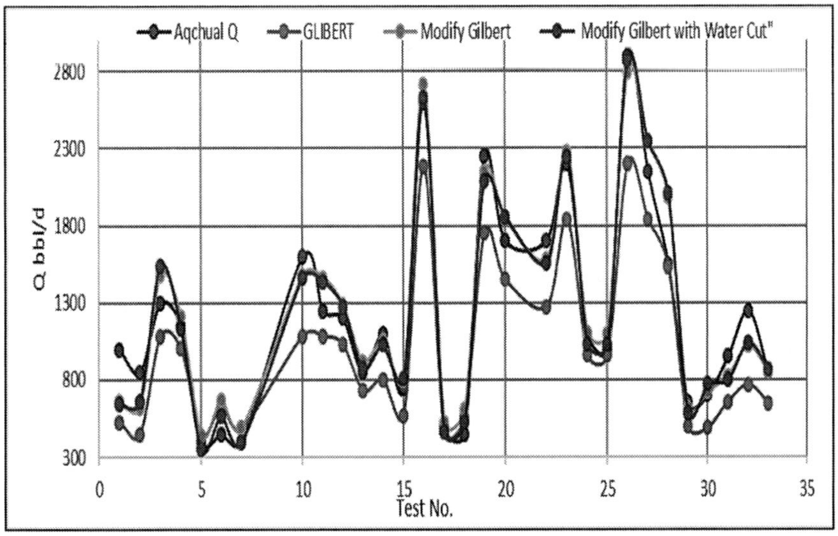

Figure 4. Compares estimated flow rate by correlations with actual rates.

Also, R2 and absolute error techniques are used for the comparison between correlations to predict the stronger correlation that can be applied and guess how far to estimate the flow rates from actual test flow rate data. The absolute errors are calculated using the following equation: -

$$E\ ABS = \left| \frac{Qtest - Qe}{Qtest} \right| \times 100 \qquad (4)$$

According to the results the absolute average errors of the three correlations, Gilbert first modified Gilbert and second modified Gilbert equation, (21%), (12%) and (10%) respectively as tabulated in Table 3. A good match between the measured and the predicted data is estimated by second modify Gilbert. With a high accuracy of predicted production rates of (R2) 0.9493 obtained. As evident from Figure 5.

Table 3. Results in the average absolute errors of the three correlations.

Correlation	Absolute Error%
Gilbert	21
First Modify Gilbert	12
Second Modify Gilbert	10

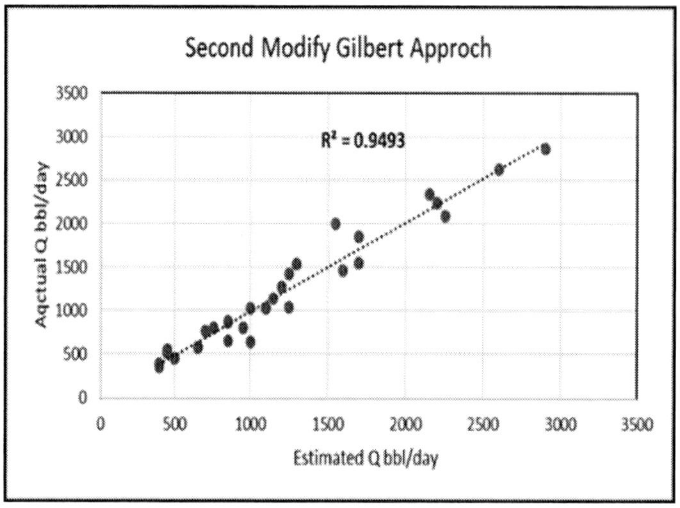

Figur 5. R2 of second modify Gilbert approach.

3. Conclusions

Depending on 33 real production tests data and two sets of newly-established choke performance correlations for the tertiary reservoir of Khabaz field in Iraq. The following conclusions are drawn:

1. The first predicted correlation; the modified Gilbert equation, has given an average error of 12 %, while original Gilbert correlation is giving average errors of 20%.

2. In order to minimize the error of field condition, a new parameter was added to Gilbert correlation ; (free water and sediment (BS & W)), which was not included in the previous Gilbert correlation and other published correlations. This addition of new parameters in this work; the newly predicted correlations (the second modify Gilbert equation) minimizes the average error to 10 %.

References

[1] Gilbert W 1954 Conf on Drill. And Prod. Prac. (New York: API) p 143
[2] Essen I E, Appah D, Udoaka M G 2017 *International Journal of Engineering Research and Reviews* **5** 14
[3] Ritzema H P 1994, *Drainage Principles and Applications* (Wageningen, Netherlands: International Institute for Land Reclamation and Improvement) ed 2 p 1125
[4] Boddy R, Smith G 2009 *Statistical Methods in Practice: For Scientists and Technologists* (Chichester, U.K: Wiley) ed 1 p 246

IOP Conf. Series: Materials Science and Engineering **579** (2019) 012024 doi:10.1088/1757-899X/579/1/012024

Evaluation of Petrophysical Properties Using Imaging Techniques

Ahmad A. Ramadhan#, Hussien I.Z. Al-Sudani, Fadhil Sarhan Kadhim, Ayman Tariq, Muhammed Lutfy

Petroleum Technology Dept./University of Technology/Baghdad/Iraq

#Corresponding author's e-mail address: 150073@uotechnology.edu.iq

Abstract The petrophysics science is concerned with the study and characterization of oil and gas reservoirs for the hydrocarbon potential availability and productivity through the quantitative of rock properties. Porosity is one of the most important property in petrophysics. Estimation of porosity values are classified into two techniques, direct and indirect methods. The direct method has been used to estimate the porosity from core analysis, because the well log records are considered as indirect technique, which depend on the measurments of the physical features related with the porosity values. In this study two methods are used to evaluate the porosity of carbonate reservoir. The first one has been applied to the well logging interpretation, while the second method uses imaging technique interpretation based on core slides. The result of the second method shows a more accurate results of porosity in comparison with first method. The porosity values are varied from 2% to 3% by using Neutron log, while the density and sonic logs recorded about 9%. The optical image results of the porosity are ranging from 6.2% to 19.1% at 40 X magnifications, and from 1.8% to 4.6% at 100X. The SEM image interpretation results are: 12.2%, 10.4, % 4.1%, 5.9% at 400X, 1.99KX, 5.01KX, and 10KX respectively. Finally, the results showed that accuracy of porosity values depends on the examined area of sample that the area of sample is changing with the magnification of microscopes types.
Key words: petrophysics, porosity, image process, well log

1. Introduction

The petrophysical properties and reservoir characterization are important issues in the study of hydrocarbons reservoirs. The specialty of the subject and its relationship to the identification of reservoirs and productivity wells. Therefore, the study of porosity is one of the fundamentals that based on the results of all other factors that are involved in the construction of the geological model of the reservoir which is used to estimate the reserve of reservoir and estimation of productivity. The uncertainty associated with the estimation of porosity will be reflected negatively on the decision makers in making the appropriate decision in the process of economic investment of the reservoir.

Porosity or void fraction is a calculation of the void spaces in a rock, ranging from 0 to 1, or a percentage pore from 0% to 100%. [1] There are many techniques to identify porosity in a rock, such as new technologies of imaging and developed microscopes. Porosity can be estimated from well logs results in addition to results from core data correlated to reservoir situation. [2] Porosity must be estimated precisely because it is required for the estimation of saturation and permeability. Lithology of rocks that affect the porosity estimation is needed to be computed as part of the lithology identification. [3, 4]

While permeability is the ability of rocks to allow fluid to pass, and measured in Darcy. The rocks of a geological formations that pass fluids readily, for example sandstones, can be called permeable. [5] The formations, composed from fine grain sediments, the finer grained be less interconnected voids are called impermeable. Absolute permeability is the estimation of the connected void, when a single fluid is present in the rock. The permeability called effective when it is able to flow or pass a especially fluid in a porous rocks when other not unassailable fluids are exist in the reservoir. [6]

The computation of relative permeability gives the ability to correlate the various capabilities of liquids to pass in exist materials, since exist of more than one fluid generally stop the flow. [7]

Permeability concept is important to estimate the characteristics of hydrocarbons flow in reservoirs. A rock is considered to be exploited hydrocarbon reservoir without motivation when the permeability greater than approximately 100 md. Rocks with permeability significantly lower than 100 md can form efficient seals. Friable or loss sand may have permeability of over 5000 md. [6]

1.1 Porosity Measuring Methods

Conventional petrophysics is related to the characterization of reservoirs to assess the hydrocarbon availability and productivity, by the quantitative study of the properties of rock. The first interest is how naturally interconnected of voids and fractures that control movement and accumulation of hydrocarbon. The main groups methods used for measurements are, (a) core or crushed rock, (b) well logging results and (c) seismic prospecting survey results. Two groups will be reviewed concern, the core and well log that drive to evaluation of porosity from voids and fractures. [8]

In group (a) the methods is subdivide into direct measurements from imaging of a rock core, and group (b) methods are subdivided into indirect measurements from well log results. (Figure 1)

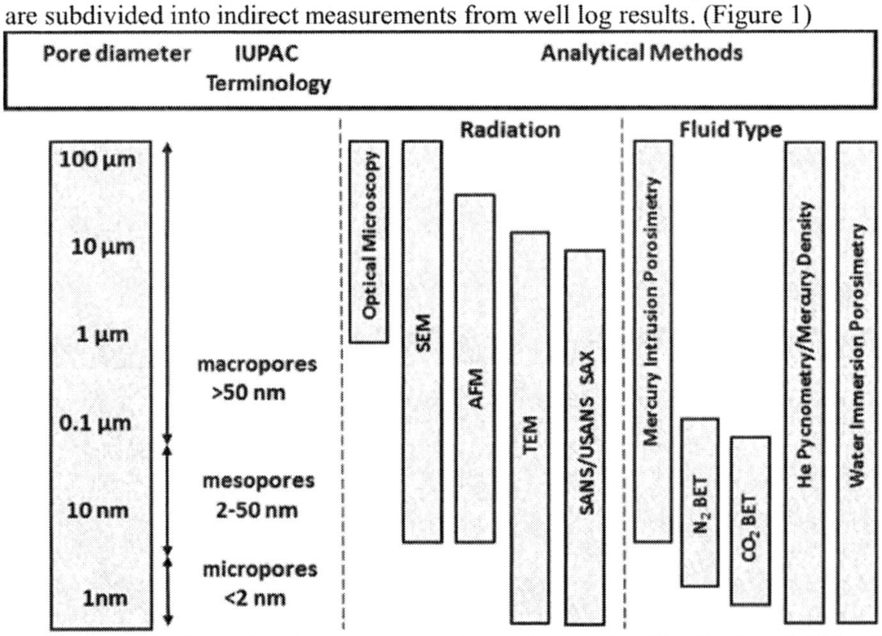

Figure 1. Methods used to determine porosity and pore size distribution (PSD). [8]

1.1.1 Conventional Methods

Several methods and techniques can be applied for direct measurement of rock porosity: (1) saturation or imbibition, (2) buoyancy, (3) gas expansion (He porosimeter), (4) gas adsorption (BET) and (5) mercury intrusion porosimetry [1, 2]. The previous methods can measure the connected voids of the sample, like in a way to the contrast-match technique which is used in neutron scattering to evaluate connected voids. [8]

1.1.2 Imaging Microscopy

There are several techniques of imaging exist to identify the behavior of porosity and its related minerals in rock. For example, optical microscopy, SEM, NMR, FIB SEM.

1.1.2.1 Optical Image Microscope

Optical petrology is the most depending technique, precise and can be repeated to evaluate the void system and lithology of the reservoir rock outcrop, core and cutting samples. For better diagnosis pores and fractures, it is required to prepare the sample for testing by using epoxy solution, for imbibe the rock to identify the pore system. The slide is viewed by plane-polarized and cross-polarized or ultra-violet light to test the slid of cross section through a rock, to identify the lithology, fabric and texture by point counting or image analysis of the slide. In addition to texture, diagenesis, pore system and reservoir quality. However, despite the large amount of information available. [9]

The used of optical techniques in petrology is to study the minerals and rocks by computed their response to optical properties. It is used to evaluate the lithology composition of rocks in order to help identifying their origin and development. [6]

1.1.2.2 Scanning Electron Image Microscope

Through the application of SEM, geologist is now able to go up one step forward the slide analysis to study the void system. [6]. SEM have an advantage over optical petrography that it eases of preparation of sample, greater resolution, and a higher magnification range. When testing an SEM image for the first time, the scale is a major problem. However, with practice and training, the user can soon identify properties previously observed in slide thin section. The SEM replenish slide analysis by providing different type of information which when used with other techniques gives a new information to help characterized void features in sample. [10]

The electrons react with atoms in the rock sample, and send many signals that carry out an information about the sample's surface include topography and composition. The electron beam type is scanned in a raster scan pattern, and the beam's position is combined with the detected signal to produce an image. SEM can achieve resolution better than 1 nanometer. Samples situation can be detected in high and low vacuum or in dry or wet conditions in variable pressure or environmental SEM, and at a wide range of low or high temperatures with specialized instruments [8]. The most common SEM mode is detection of secondary electrons emitted by atoms excited by the electron beam. The number of secondary electrons that can be detected depends, among the other things, on the surface of sample. By scanning the sample and detected the secondary electrons that are released using a special detector, an image has been created representing the topography of the surface.

Historically, William Nicol had created Nicol prism, started to prepare slices of materials, and Henry Thronton Maire Witham (1831) applied his techniques by plant petrifactions study. This technique used in petrology, Sir David Brewster and many other used optical searches for thin sections of crystals. [8]

McMullan, presented the use of SEM. Although Max Knoll takes an image with a 50 mm showing the steering contrast using of an electron beam scanner. Manfred von Ardenne in 1937 created a microscope with high magnification by scanning a very small scaning with a de-magnified and finely focused electron beam. Ardenne used the scanning principle to achieve magnification in addition to purposefully eliminate the chromatic aberration otherwise inherent in the electron microscope. After that he had discussed the various detection modes, possibilities and theory of SEM together with the construction of the first high magnification SEM. Further work was reported by Zworykin's group, followed by the Cambridge groups in the 1950s and early 1960s headed by Charles Oatley, all of which finally led to the marketing of the first commercial instrument by Cambridge Scientific Instrument Company as the "Stereoscan" in 1965. [12]

2. Methods and Materials

Two methods have been applied in this research as follows:

2.1 Log Interpretation Method

The study was made across the Upper Cretaceous Mishrif Formation at Bazergan field. Interactive Petrophysics software IP, V3.5 is used to determine the porosity and permeability from well log data.

2.2 Imaging Interpretation Method

The first step in this method is sample preparation for image in microscope. Three samples of limestone were prepared for imaging by (1) optical microscope and (2) scanning electron microscope, the samples were prepared in three roughness of surface to explain the effect of roughness on porosity calculation. The samples are prepared according to Stutzman P.E., and Clifton J.R. 1999. [13]

3. Results and Discussions

3.1 Log results

The Interactive Petrophysics (IP) software is used to calculate the porosity and permeability from log results after environment corrections. These results will be used to be compared with the results of imaging method which will be used in this study. Fig. 2 and Fig. 3 show that the input data window which is used for porosity and permeability calculations. Fig 4 shows the results for those petrophysics. The results of porosity and permeability as follow:

Porosity at depth 3280m:

1- PHI Density: 9%

2- PHI Neutron: 2-3%

3- PHI Sonic: 9%

4- PHI Effective: 9%

Permeability at depth 3280m: By 3 defaults (Timur, Morris Oil, Schluberger) all nearly the same readings: 500 md.

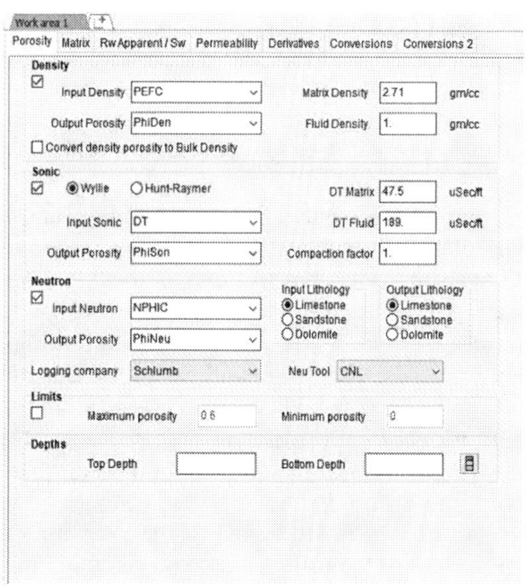

Figure 2. Porosity input for IP.

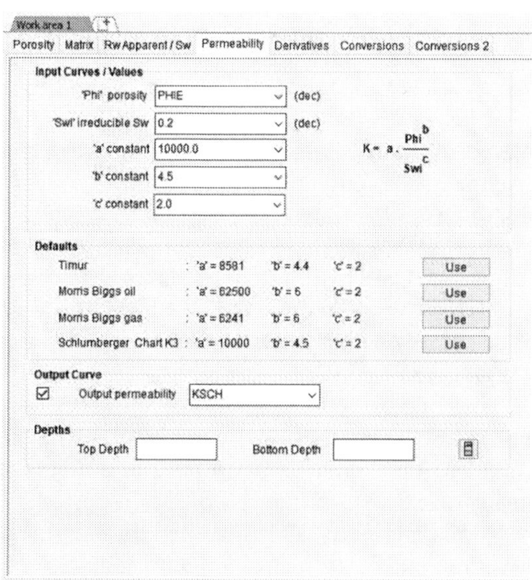

Figure 3. Permeability input for IP

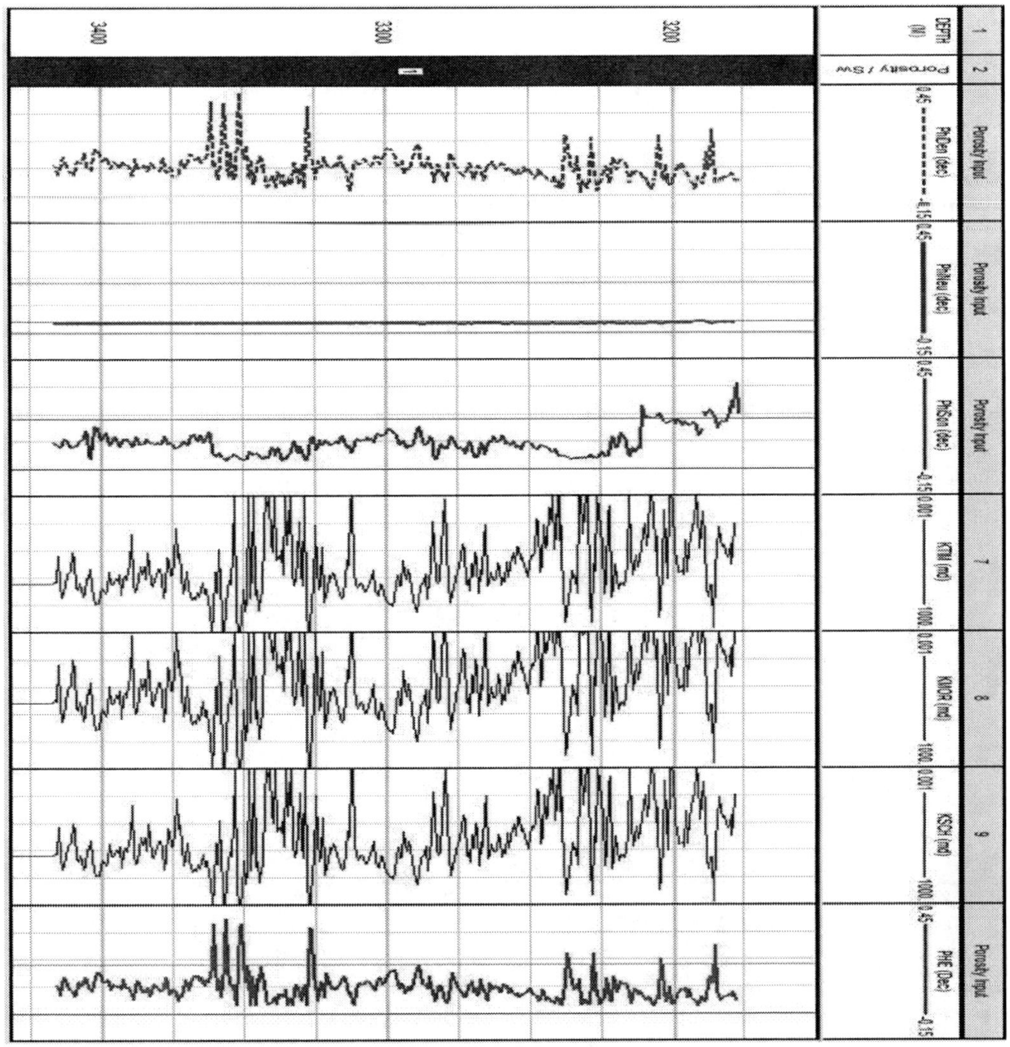

Figure 4. Porosity & Permeability Results from IP software after environment corrections.

3.2 Imaging Methods Results

The 3 sample cores are used in this study from Buzurgan oil field, Mishrif formation. The depth of samples is at 3280m. Those samples are prepared as explained in section 3. And prepared in three roughness's, the sample grinding to a flat surface, using 100, 260, and 600 mesh grit diamond paper.

3.3 Optical Results

Figure 5 shows the original image which taken from Optical Microscope before process in two magnifications lenses at 40X, and 100X. Figure 6 show the threshold image after processed by (Image J) software to calculate the porosity. Figure 7 shows the processed image by image j software to show mapping of porosity. And the results are shown in table 1.

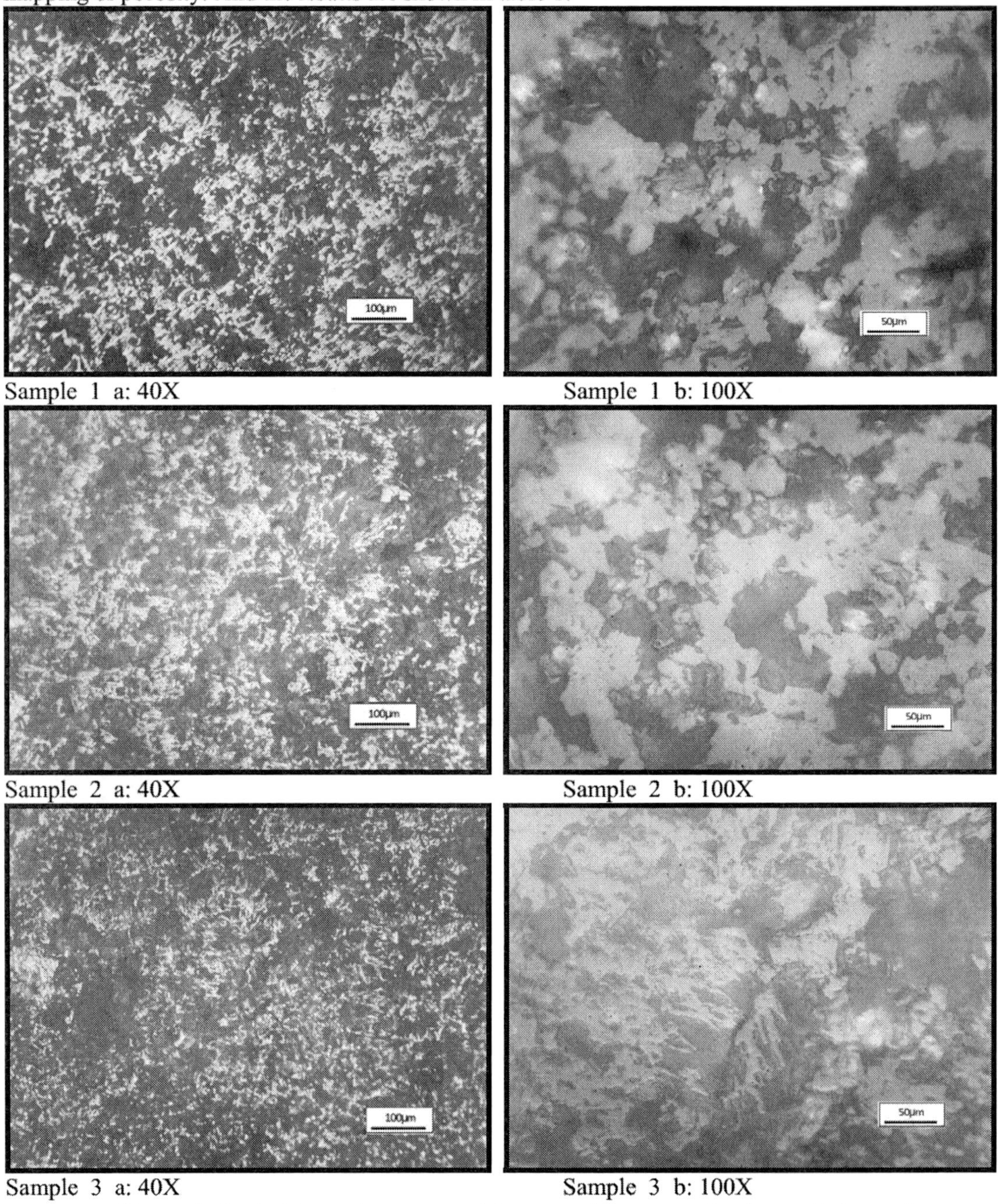

Sample 1 a: 40X Sample 1 b: 100X

Sample 2 a: 40X Sample 2 b: 100X

Sample 3 a: 40X Sample 3 b: 100X

Figure 5. Optical image in two magnification enlargements for the studied samples.

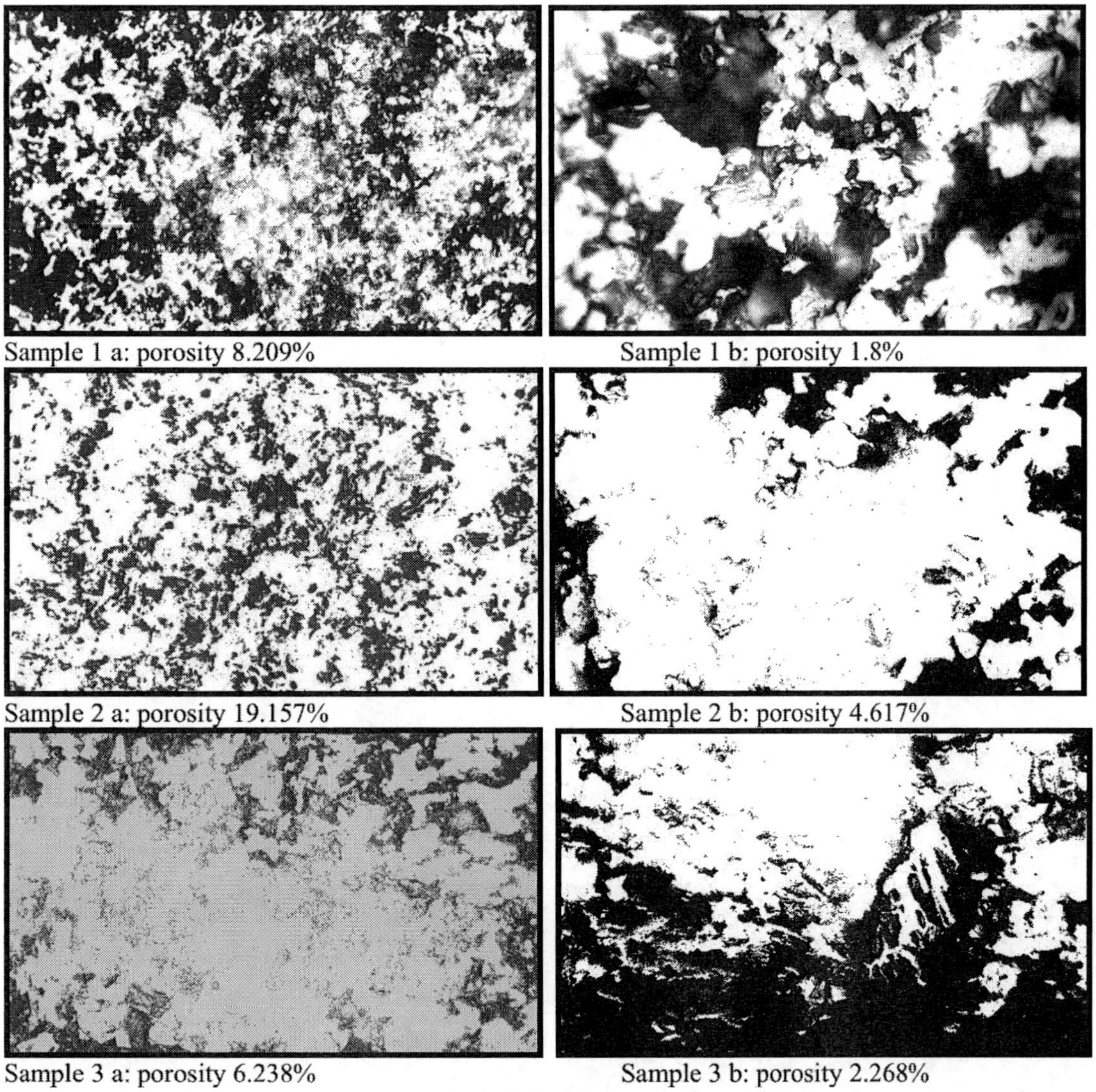

Sample 1 a: porosity 8.209% Sample 1 b: porosity 1.8%

Sample 2 a: porosity 19.157% Sample 2 b: porosity 4.617%

Sample 3 a: porosity 6.238% Sample 3 b: porosity 2.268%

Figure 6. Processed Optical image to determine porosity.

IOP Conf. Series: Materials Science and Engineering **579** (2019) 012024 doi:10.1088/1757-899X/579/1/012024

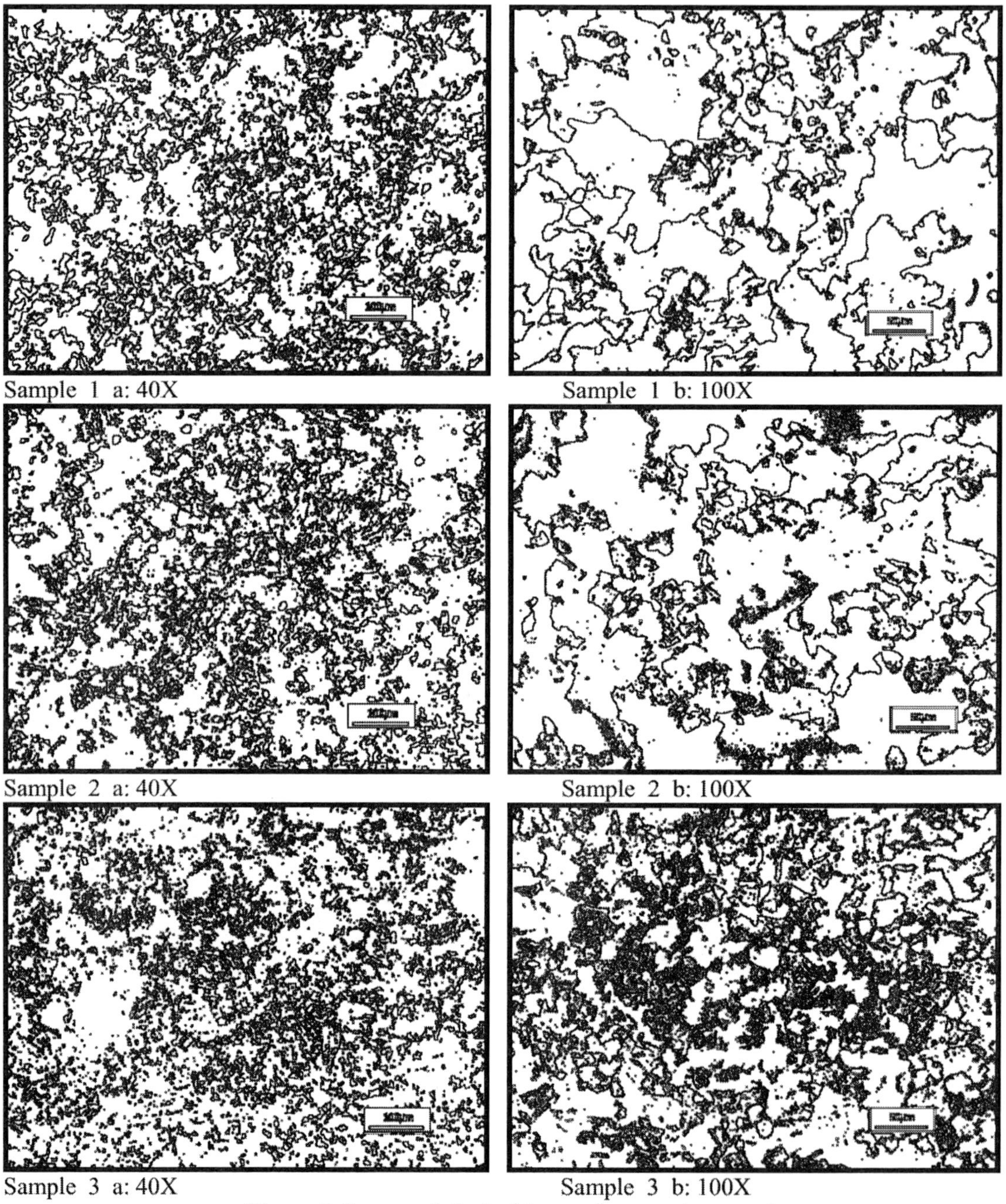

Figure 7. Processed Optical image for porosity mapping.

IOP Conf. Series: Materials Science and Engineering **579** (2019) 012024 doi:10.1088/1757-899X/579/1/012024

Table 1 Results of Optical image processing for porosity.

Sample No.	Count	Total Area (μm)	Average Pore size (μm)	% Area (porosity)
Sample 1 a	1072	61305.296	57.188	14.155
Sample 1 b	504	2201.507	4.368	1.821
Sample 2 a	1631	88226.807	54.094	19.157
Sample 2 b	848	6830.234	8.055	4.617
Sample 3 a	3381	11503.446	3.402	6.238
Sample 3 b	515	2011.027	3.905	2.268

3.4 SEM Results

Figure 8 shows the original image from SEM at magnification enlargement of 228X, 1.0 KX, 4.97 KX, 10 KX, 20 KX. And this magnification is applied to reach the nano scale porosity. Figure 9 shows the porosity of samples after being processed by image j software. Also, the SEM-EDS system provides the chemical composition of sample as $CaCO_3$, as shown in figure 10, and the results are shown in table 2.

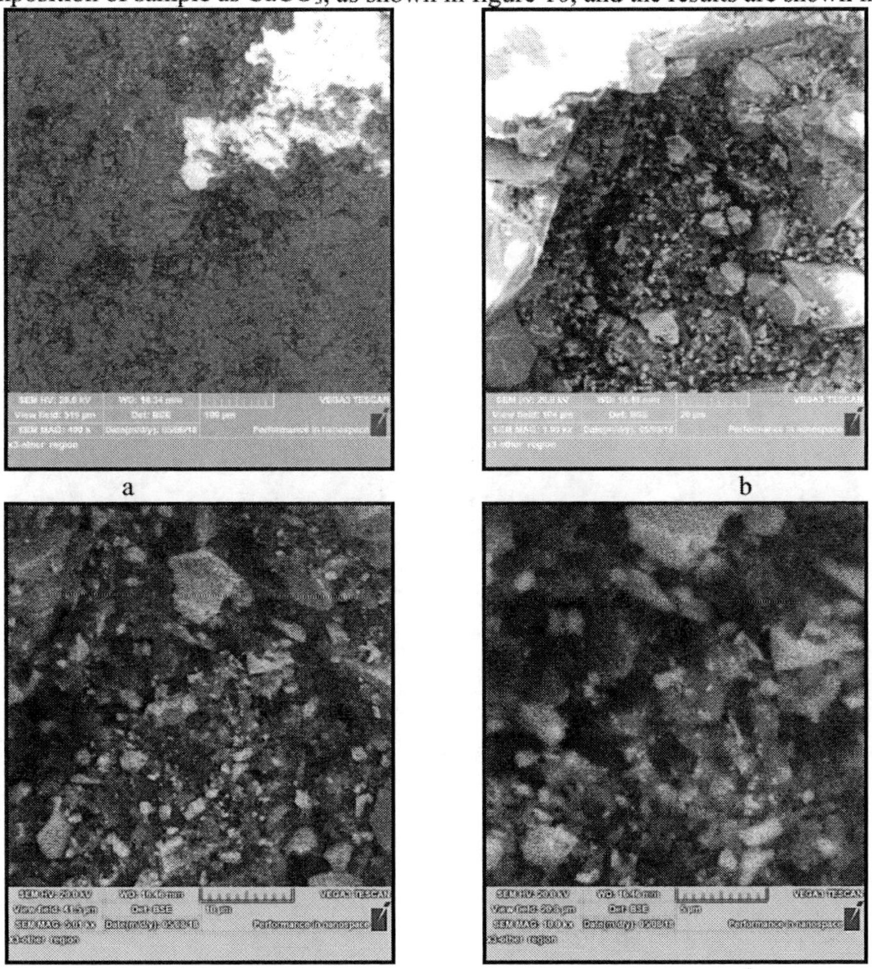

Figure 8. SEM image for Sample 1 in five magnification enlargement 228X, 1.0 KX, 4.97 KX,10 KX, 20 KX.

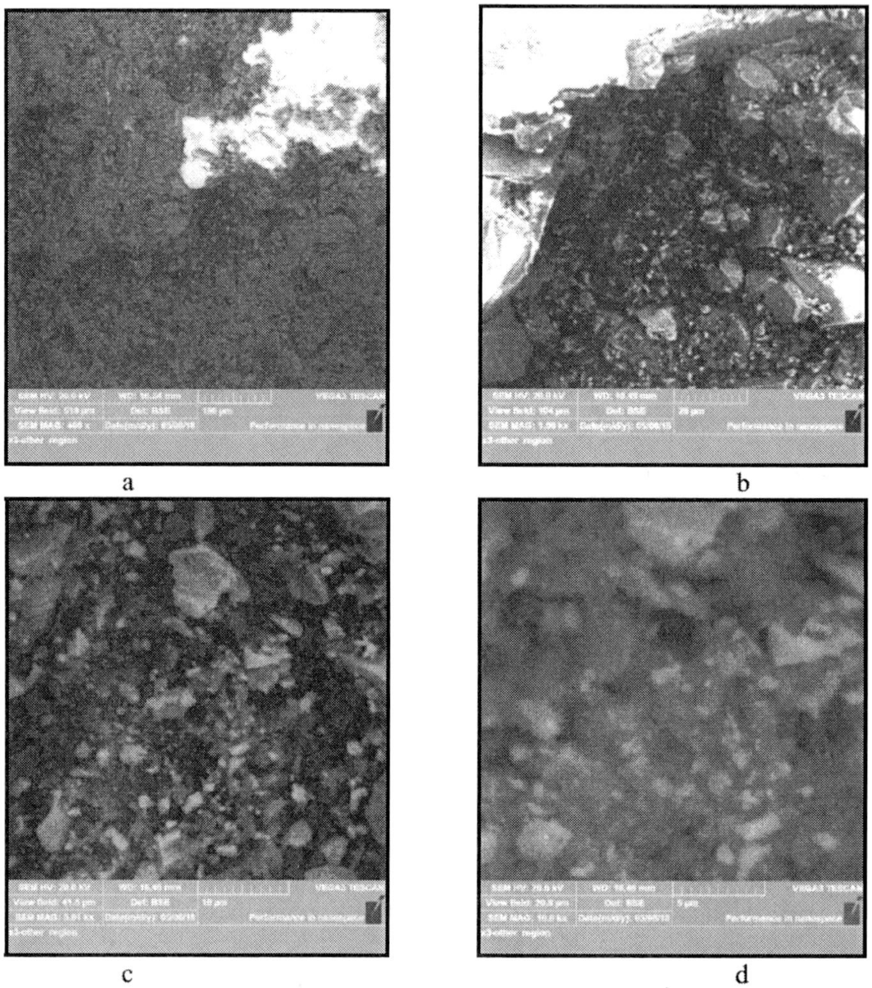

Figure 9. SEM image and processed for Sample 1 in five magnificationenlargement a: 400X, b: 1.99KX, c: 5.01KX, d: 10KX.

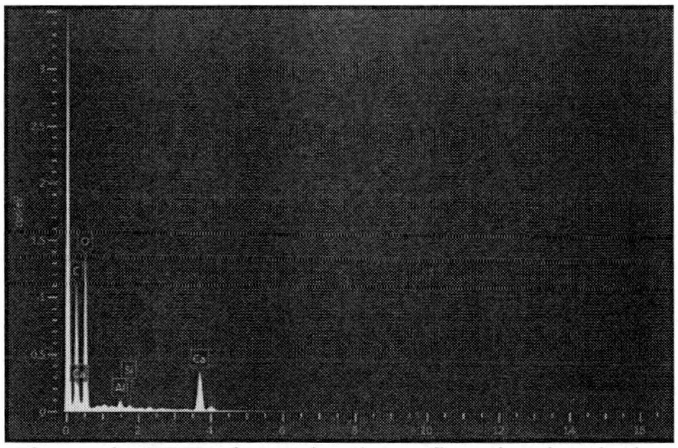

Figure 10. Shows the EDS of samples that reflect the chemical composition.

Table 2. Results of the SEM image processing for porosity.

Sample No.	Count	Total Area (µm)	Average Pore size (µm)	Area% (porosity)
Sample 1 a	7770	85916	11.057	12.186
Sample 1 b	2896	73402	25.346	10.411
Sample 1 c	1984	28916	14.575	4.101
Sample 1 d	2565	41734	16.271	5.920

3.5 Dissections

The result above has shown clearly that the use of image methods to calculate the petrophysical properties such as porosity is useful and accurate in determining the core samples properties and in addition to the pore structures. Table 3 represent the correlation between the IP results with results of imaging methods. These results are logical, but in SEM image the porosity has shown higher value because this technique has calculated the smallest size of pore.

Table 3. Correlation between the IP results with results of imaging methods.

IP	Optical image	SEM image
PHI Density: 9%	6.2-19.1% at 40X	12.186%
PHI Neutron: 2-3%	1.8-4.6 % at 100X	10.411%
PHI Sonic: 9%		4.101%
PHI Effective: 9%		5.920%

4 Conclusion

It can be concluded:

1-The use of different techniques is important to verify the results of petrophysical properties.

2-The porosity are different at using different magnifications of Microscopes because of inhomogeneity of rock materials.

3- Porosity of SEM in sample (1) is greater in larger magnification than Optical image, due to the detailed surface of the SEM.

Acknowledgments
We extend our thanks and appreciation to the Department of Production and Metallurgy Engineering/University of Technology for providing support through the use of SEM and Optical Microscopes to complete measurements of this study.

References
[1] Tiab D and Donaldson EC 2004 *Petrophysics: Theory and Practice of Measuring Reservoir Rock and Fluid Transport Properties* 2nd Edition. *Elsevier*
[2] Sondergeld C H Newsham K E, Comisky JT, Rice M C and Rai CS 2010 Petrophysical considrations in evaluating and producing shale gas *Soc Petrol Eng SPE* 131768
[3] Clarkson C R, Freeman M H L, Agamalian M, Melnichenko Y B, Mastalerz M, Bustin R M, Radlinski A P and Blach T P 2012a Characterization of tight gas reservoir pore structure using USANS/ SANS and gas adsorption analysis, *Fuel* **95** 371–385 DOI: 10.1016/j.Fuel.2011.12.010
[4] Thomas, E C 1992 50th Anniversary of the Archie Equation: Archie Left More Than Just an Equation, *Log Anal* **199** 12-90
[5] Horgan, G W 1998 Mathematical morphology for soil image analysis, *European Journal of Soil Science* **49** 161-173
[6] Noiriel C 2015 Resolving time-dependent evolution of pore-scale structure, permeability and reactivity using X-ray microtomography, *Rev Mineral Geochem* **80** 247–285
[7] Clarkson C R, Wood J M, Burgis S E, Aquini S D, Freeman M 2012b Nanopore-structure analysis and permeability predictions for a tight gas siltstone reservoir by use of low-pressure adsorption and mercuryintrusion techniques. *Soc Petrol Eng, Reservoir Eval Eng SPE* **155537** 648–661
[8] Anovitz L M and Cole D R 2015 Characterization and Analysis of Porosity and Pore Structures, Reviews in Mineralogy & Geochemistry, *Mineralogical Society of America*, **80**:61-164.
[9] Beckingham L E, Peters C A, Um W, Jones K W, Lindquist W B 2013 2D and 3D imaging resolution trade-offs in quantifying pore throats for prediction of permeability, *Adv. Water Resource* **62** 1–12
[10] Zhang S, and Klimentidis R E 2011 Porosity and permeability analysis on nanoscale FIB_SEM 3D imaging of shale rock, *Presentation at the International Symposium of the Society of Core Analysts* held in Austin, Texas, USA 18–21
[11] Welton J E 2003 SEM Petrology Atlas: methods in exploration series No.*4 The American Association of Petroleum Geologists* Tulsa, Oklahoma 74101, U.S.A.
[12] Stokes D J 2008 *Principles and Practice of Variable Pressure Environmental Scanning Electron Microscopy* (VP-ESEM), Chiohester: *John Wiley & Sons*
[13] Stutzman P E and Clifton J R 1999 Specimen preparation for scanning electron microscopy, *Proceedings of the Twenty-First International Conference on Cement Microscopy*, L Jany and A Nisperos, Eds., April 25-29, 1999, Las Vegas, Nevada, USA, 10-22

Application of Geophysical Logs to Estimate the Source Rock Quantity in Ratawi Formation, Southern Iraq: A Comparison Study.

Rami M. Idan
Al-Karkh University of Science – College of Remote Sensing and Geophysics – Department of Geophysics. Baghdad, Iraq.

Corresponding author's e-mail address: ramisc3@gmail.com

Abstract. Quantitative detection of the parameters of source rock that targeted Ratawi Formation in the south of Iraq is achieved via the Geophysical log data, precisely the resistivity, porosity, and gamma ray log of the well X in the northern part of Mesopotamian basin, which is so called Euphrates subzone. Total organic carbon and source rock productivity is detected by mathematically interpreting of the log parameters and that lead to assigning and identification of the content of organic matter and source rock intervals. The responses of log curvatures are relevant to the rise of total organic carbon, which can be identified via logs indicators. Thereby, the organic carbon can be expected with a surge in values of sonic log, gamma-ray values, neutron or total porosity log, and resistivity as well as a decrease in the density log. The adoption of sonic/resistivity overlay technique for the determination of total organic carbon content has indicated the likely lithology of the intervals of interest where the logs initially could indicate the lithology. As such, high organic carbon content is significant to environments that have relatively low energy.

1. Introduction

Typically, source rock is described as very fine particles, which is mainly Shale rocks. The source rock comprises considerable quantity of organic matter (OM) that is frequently called Total Organic Carbon "TOC" which possibly will yield hydrocarbon within maturity or Catagenesis phase [1]. The evaluation of source rocks require to develop the applications of logs to augment the information, especially when geochemical data is limited. This research is a complementary and comparison study with [2], which focused on the calculation of the source rock quantity parameter from geophysical logs to Zubair Formation with the underlying Ratawi Formation. This formation underlain Zubair Formation and in the same depositional cycle. The former represents the deeper environments. The research aims to compare the equations of the (TOC) prediction that has been used to determine the TOC of Zubair Formation on the Ratawi Formation. These equations are derived from [3] with modification to lithology and temperature in respect to the new parameters. In addition, due to the importance of TOC parameter to expect the source rock efficiency in the interested interval, this complementary research focuses on how to calculate these values and provide a prior indication about the TOC in the same succession representing by Zubair and Ratawi formations [4] and [5].

Consequently, accumulation of hydrocarbon depends on many factors, like kerogen type, burial rate, and geothermal gradient in the studied locations. Numerous techniques have been used to evaluate the source rock. The favourable technique to be used are organic matter identification, Rock - Eval

IOP Conf. Series: Materials Science and Engineering **579** (2019) 012025 doi:10.1088/1757-899X/579/1/012025

pyrolysis, and Infrared analysis [6]. Besides logs are used to evaluate source rock, containing sonic, gamma ray (GR), and resistivity logs, in addition to neutron and density. This study represents a feasibility effort to evaluate OM from log data depending on TOC properties. Organic rich intervals have low density values, while high sonic transit time, and high resistivity values [7].

The L. Cretaceous Ratawi Formation is represent the deposits of inner shelf environments [8]. Ratawi Formation consists of two members, the lower composes of detrital limestone from the inner shelf environments during a high stand system tract (HST). Whereas the upper shale member deposited in deep inner shelf environments within the low stand system tract (LST) facies [9], and this division is not fixed in the Iraqi studies concerning Ratawi Formation. The lower and upper contacts in the studied well are conformable with Yammama and Zubair respectively [10]. The formation is approximately 140 m thickness and increase toward the south and the southwest area, as well as in the central Iraq near Baghdad. From the geological point of view, the field of study is located in the Mesopotamian zone of Iraq in Euphrates subzone, Figure 1.

Figure 1. The Tectonic map of Iraq, illustrating the tectonic divisions, and illustrate the study area in Euphrates subezone [8].

2. Materials and methods

Δ Log R is an applied technique, for detecting TOC in organic-rich intervals that established by well logs analysis. The method uses the overlay of sonic log (Delta t) curve with the induction latero-log (ILD) in water-saturated, organic-lean intervals. These two curves are corresponding to each other and can be correlate in the same depths, which each 50 μsec/ft from Delta t equal to one logarithmic

IOP Conf. Series: Materials Science and Engineering **579** (2019) 012025 doi:10.1088/1757-899X/579/1/012025

resistivity cycle.
The separation of the Delta t and ILD curves in organic rich intervals takes place as shown and detailed in Figure 2. Furthermore, the same separation also takes place in reservoir rock but in different pattern. The differentiation between these two cases is depending on the low or high values of GR in reservoir or organic rich intervals respectively [3].

The organic rich intervals separation occurs due to the decrease in velocity and density values, [3]. In immature organic-rich intervals, the separation is mainly due to voids occurrence. While in hydrocarbon bearing intervals, the resistivity increases due to the presence of liquids, in addition to pore space activity. The curve separation in non-reservoirs depends on the TOC and thermal maturity, and this method is worthy in the geochemical exploration especially when data samples are not available. This technique results in providing a TOC assessment close to reality in all lithology types and maturity. Finally, the baseline values characterize theoverlapping interval between the ILD and Delta t logs in immature source rocks based on relatively high values of GR.

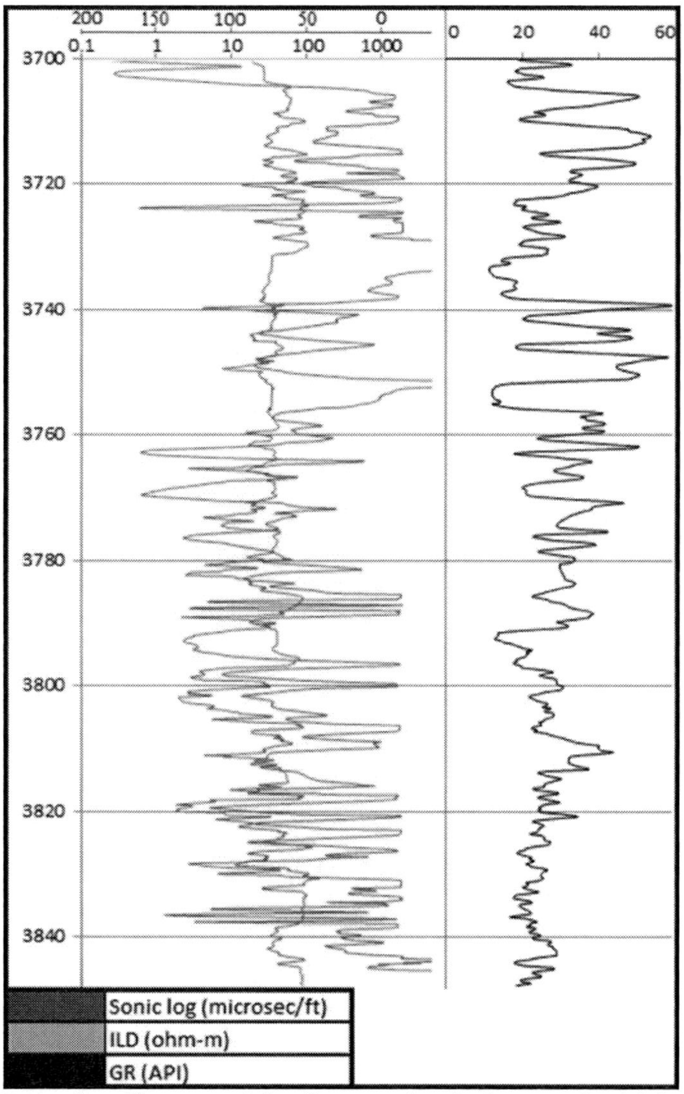

Figure 2: Delta t/ILD overlap, illustrates Δ log R separation in the organic rich intervals. Gamma Ray is contemporaneous to find out the shale base line in the studied case.

IOP Conf. Series: Materials Science and Engineering **579** (2019) 012025 doi:10.1088/1757-899X/579/1/012025

Initially, logs of resistivity, porosity, gamma ray, and calibre are prepared in digital data file to use [12]. The logs of ILD and Delta t are overlay to be in one harmonic track.

GR curve is concurrent to detect the shale base line. As shown in [13] in Figure 2, the ILD and Delta t baselines have been determined and fixed to equal 324 ohm-m and 88 μsec/ft respectively at the interval 3800-3820 m. Then the equation 1 ran, as below:

$$\Delta \log R = (Res / Res\ baseline) + p\ (Delta\ t - Delta\ t\ baseline) \quad \ldots (1)$$

Where:

- Res is the resistivity from log.
- Delta t is the sonic log data.
- Res baseline value is calculated and equalize to 324 ohm-m, at 3800-3820 m.
- Delta t baseline value is calculated and equalize to 88 μsec/ft, at 3800-3820 m.

p is the relationship of 50 μsec/ft to each logarithmic cycle of resistivity log.

The results of Δ log R are equivalent to each 0.25 m or less from the formation. These results from the above equation 1 are submitted to calculate the TOC as shown in equation 2, as follow.

$$TOC = (\Delta \log R) * 10\ exp\ (2.297 - 0.1688 * LOM) \quad \ldots\ldots\ldots (2)$$

Where:

- TOC is the total organic carbon measured in wt % for each 0.25 m or less from the targeted interval.
- Δ log R values are calculated in equation 1.

LOM is the level of organic metamorphism. Vitrinite reflectance Ro values compared with [14], to predict the LOM value that was nine to this work as shown in Figure 3, which the mean of Ro had ranged from 0.6-0.7 as calculated by [15].

- Other numbers are constants.

As well as, equations 3 and 4 to calculate TOC from the density and neutron logs respectively were ran in the same method of calculation.

$$\Delta \log R = (Res / Res\ baseline) + p\ (pb - pb\ baseline) \quad \ldots\ldots\ldots (3)$$

$$\Delta \log R = (Res / Res\ baseline) + p\ (\Phi N - \Phi N\ baseline) \quad \ldots\ldots\ldots (4)$$

Where:

- Res is the resistivity from log.
- pb is the density from log.
- R baseline value in the Δ log R.
- pb baseline value in the Δ log R.
- ΦN is the porosity from log.
- ΦN baseline value in the Δ log R.

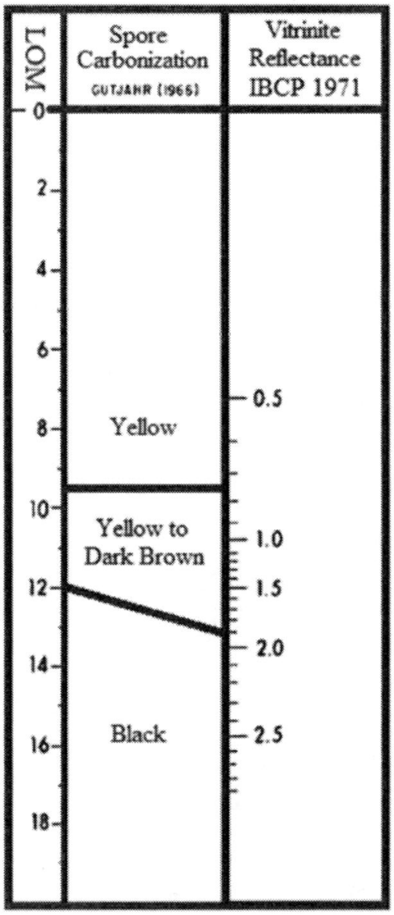

Figure 3: Level of Organic Maturity or Metamorphism (LOM) chart, from [3].

3. Results and Discussion

Typically, this technique operates as a function to the different porosity logs interaction with resistivity logs. The calculated TOC from log are compared with Rock-Eval TOC from thirteen (13) of cutting samples, Fig. 4. Calculation of Δ Log R depends on the basis of porosity - resistivity technique was exercised.

The results of this complimentary work summarize as the TOC can be calculated by the effective integration between porosity and resistivity logs. These results are supported by comparison with TOC from the Rock-Eval results. Logs response, as well as TOC are good indicators to the litho-facies of the studied interval. As a result, the studied section is divided into two main lithology in respect to the TOC and porosity logs response. These two rock packages start at 3700 to 3760 m (60 m), and 3760 to 3840 m (80 m) from top of Ratawi Fn. To the end of the interval. The recognized two intervals of Ratawi Formation represent a succession of layers of limestone intercalated with shale at the bottom, turn into overlapping sand layers with shale at the top of the formation [16]. The relationship of the TOC results from log and Rock-Eval analyses is important to fix and demonstrate the accomplishment of the log techniques in organic geochemistry evaluation. In well X, the correlation between the calculated and the measured TOC reflects a noticeable similarity between them, Figure 4. Thus, the overlay proved real tools for quantitative determination of TOC in this well. On the other hand, resistivity / RHOB and NPHI logs method was used (equations. 3 and 4) and the results are also illustrated in Figure 4. The comparison between these results detects that Sonic / resistivity integration is the more accurate results, as revealed in the Figure 4, in comparing to Rock-Eval TOC from cutting samples.

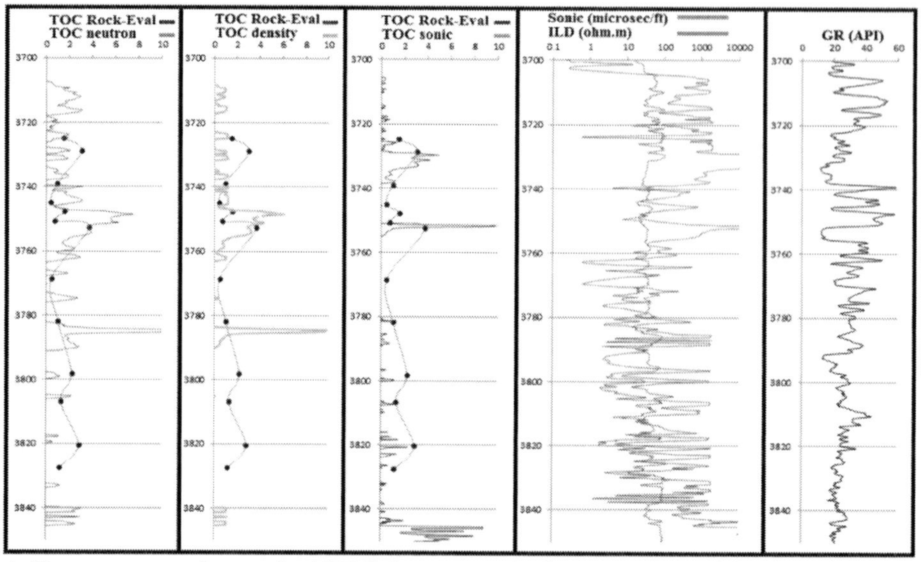

Figure 4: Demonstrates the resulted TOC from porosity logs overlapped by the TOC from Rock - Eval pyrolysis, to compare the results.

Moreover, the petroleum system analysis took place to indicate the maturity of the Ratawi Formation. Ratawi Formation have a transformation ratio ranged between 69 – 83.5 % in the study area, Fig. 6. While the temperature within the formation ranged between 107.36 – 110.28 C°. These results can indicate that Ratawi Formation is within the peak of oil generation [17]. This indication give an idea about the source rock type and its ability to produce hydrocarbons. The additional section of petroleum system analysis is a complimentary section to illustrate the elements of petroleum system, that they are quantity, quality, and thermal maturity. In this research, the quantity and thermal maturity are detected from wireline log data and simple information from final geological report of the well of interest.

4. Conclusions

Geophysical logs evaluation to geochemical properties is considered as the up-to-date tool to identify and quantify source rock. The assessment method predominantly starts by detecting the reactions of the GR curvature, Delta t, ΦN, pb and ILD, to TOC increment. Raising of GR, Delta t, neutron, resistivity and decreasing of density may indicate the increase of TOC but this is not definitely true at all times in each case, see Figure 5.

Porosity/resistivity tool show that log curves can be used to indicate organic-rich intervals within the formations. As in this case, the calculated TOC ranged from 1-10 wt %; and result is close to the Rock-Eval analysis which means that the studied interval is considered as source-seal rock in the studied section, particularly in the part that ranged 3700-3760 m from the top of the formation. Detecting TOC from log analysis may consider as a useful tool of TOC determination. It is indispensable to calibrate with Rock-Eval pyrolysis data. Results from the overlay show a general accepted compatible with core data in estimating TOC in this area Figure 4. Evaluation of results indicates sonic/resistivity combination provides the most correct or acceptable results as explained in the Figure 4, compared with neutron/resistivity and density/resistivity logs methods. As a result, these calculations can be used in intervals that lack of geochemical data to obtain an overview in exploration.

Petroleum system analysis showed that Ratawi Formation is within the peak of oil generation. This means the source intervals within the Ratawi Formation is generating petroleum regardless the type of the generated petroleum.

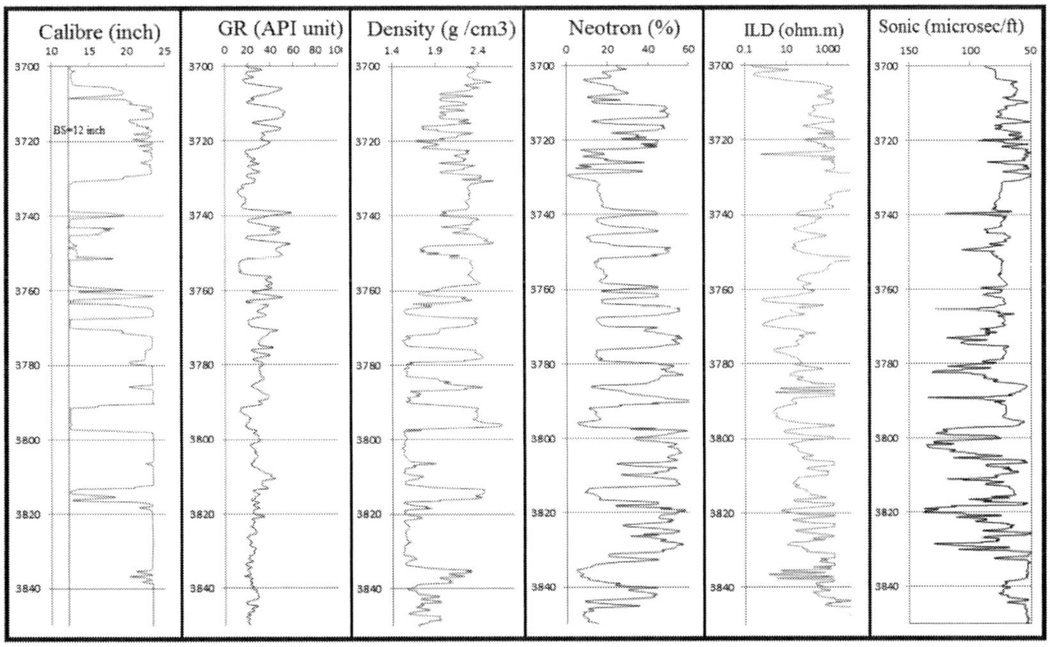

Figure 5: Full set of resistivity, porosity, GR, and caliber of the interested interval, showing the petrophysical properties of the formation. Note that the interval below 3760 m behaves relatively different from upper part.

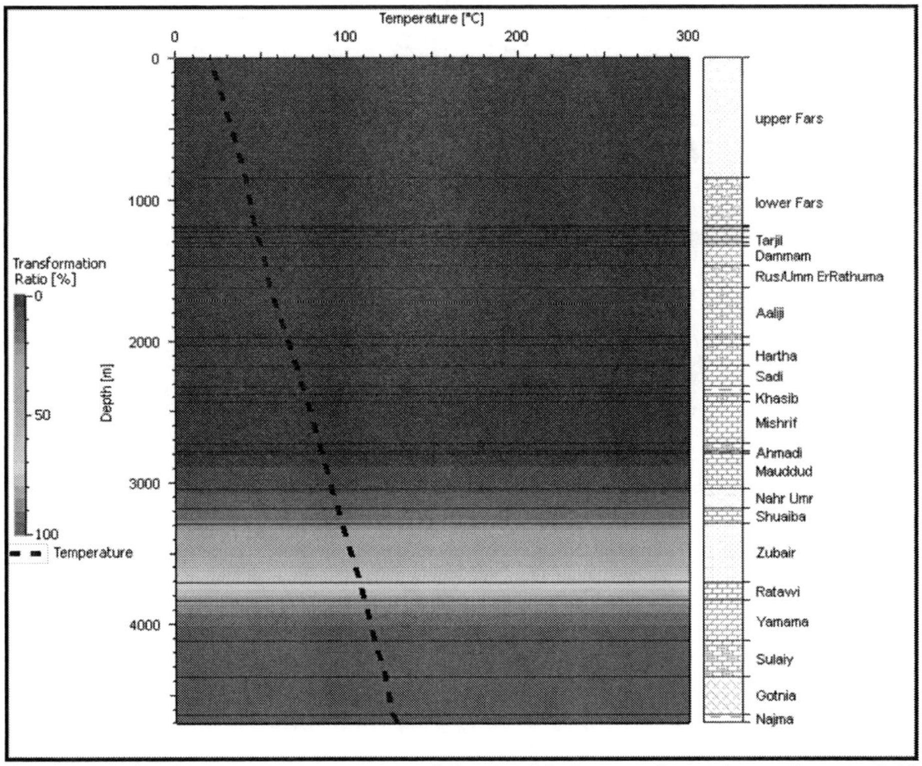

Figure 6: A cross section between temperature and depth with the transformation ration overlaid versus the formations succession.

Acknowledgements

Al-Karkh University of Science is acknowledged due to the all facilities. South Oil Company is also acknowledged for their help, especially the participants of Reservoir Evaluation Department who provide the data. Acknowledgments are also due to the reviewer and proofreading committees in the First International Conference of Petroleum Technology and Petrochemicals, 2019 in University of Technology.

References

[1] Tissot B P and Welte D H 1984 *Petroleum Formation and Occurrence: A new approach to oil and gas exploration* (Berlin: Springer) ed 2 p 702
[2] Idan R M 2017 *Bull. Iraq Nat. Hist. Mus.* **14** 185
[3] Passey Q R, Creaney S, Kulla J B, Moretti F J and Stroud J D 1990 *AAPG Bulletin* **74** 1777
[4] Leckie D A, Kalkreuth W E and Snowdon L R *AAPG Bulletin* **72** 820
[5] Dymann T S, Palacos J G, Tysdal R G, Perry W J and Pawlewicz M J 1996 *AAPG Bulletin* **80** 1177
[6] Idan R M, Al-Rawi D, Nasser M E and AlMashaekhy D 2015 *Arabian Journal of Geosciences* **8** 773
[7] Meyer B L and Nederlof M H 1984 *AAPG Bull.* **68** 121
[8] Jassim S and Goff J 2006 *Geology of Iraq* (London: Geological Society of London) p 352
[9] Sharland PR *et. al.* 2001 *Arabian Plate Sequence Stratigraphy. GeoArabia* (Bahrain: Gulf Petrolink) ed 2 p 371
[10] Idan R M, Faisal R F, Nasser M E, Al-Ameri T K and Al-Rawi D 2015 *Arab J Geosci* **8** 805
[11] Al-Musawi F A, Idan R M and Salih A L M 2019 *1st Springer Conference of the Arabian Journal of Geosciences* on The Structural Geology Contribution to the Africa-Eurasia Geology: Basement and Reservoir Structure, Ore Mineralisation and Tectonic Modelling (Tunisia/ Springer) p 121
[12] Shayesteh M 2011 *2nd South Asian Geoscience Conference and Exhibition* (New Delhi/ Geo India) p 12
[13] Hood A, Gutjahr C C, Heacock R L, 1975 *AAPG Bull.* **59** 986
[14] Jarvie D M 1991 *Chemical Geology* **93** 79
[15] Faisal R F, Idan R M and Salih A L M 2016 *Iraqi Journal of Science* **57** 2258
[16] Al-Ameri Th K, Al-Jubouri N M, Isa M J and Al-Azzawi R E 2013 *Arab J Geosci.* **6** 3725

IOP Conf. Series: Materials Science and Engineering **579** (2019) 012026 doi:10.1088/1757-899X/579/1/012026

Calculating heterogeneity of Majnoon Field/Hartha Reservoir using Lorenz Coefficient method.

J H Al-Joumaa[1#] and M S Al-Jawad[2]

[1]Petroleum Technology Department, University of Technology-Iraq, Baghdad, Iraq

[2]Iraq, Baghdad, Baghdad University, Petroleum Engineering Department

[#]Corresponding author`s e-mail address: 150078@uotechnology.edu.iq

Abstract: One of the principle concepts to understand any hydrocarbon field is the heterogeneity scale; This becomes particularly challenging in supergiant oil fields with medium to low lateral connectivity and carbonate reservoir rocks.

The main objectives of this study is to quantify the value of the heterogeneity for any well in question, and propagate it to the full reservoir. This is a quite useful specifically prior to conducting detailed water flooding or full field development studies and work, in order to be prepared for a proper design and exploitation requirements that fit with the level of heterogeneity of this formation.

The main tool used for these purposes is the application of the famous Lorenz coefficient method for calculating the degree of heterogeneity for any well. The starting point for this kind of complicated studies needs to start from the basics, in order to understand the big picture and be able to plan properly for the scope to be delivered, hence, utilizing analytical tools like the ones mentioned above becomes quite necessary, if not crucial, to the success of full field modelling and choosing an optimum water flood pattern and design.

This paper covers the methodology for quantifying and calculating the level of heterogeneity in a given reservoir (Majnoon Field/ Hartha Reservoir) using three out of four original wells of this reservoir, and how this value impacts and directly feeds into planning for water flooding projects and selection of an optimum flood pattern.

This methodology has been tested successfully in the stated super giant oil field, in which the reservoir in question is a carbonate rock formation. It is important to note that the chosen reservoir is giant areally with moderate thickness. An example of this is covered in the paper. It was concluded that utilizing a heterogeneity calculation method before conducting a detailed reservoir simulation study can save a lot of time and effort by providing guidance to the path which needs to be followed, and sheds light on the critical elements to be looked after. This has also helped to uncover many insights on the reservoir itself , hence allowing the engineer to plan

IOP Conf. Series: Materials Science and Engineering **579** (2019) 012026 doi:10.1088/1757-899X/579/1/012026

for the necessary voidage replacement and water injection rates to sustain the reservoir pressure and pattern development.

The suggested method, in combination with geological and petro physical information available, can be applied to majority of the reservoirs. This combination is paramount to ensure optimum time and planning is followed for each reservoir development study that involves water flooding.

1. Introduction

Heterogeneity is a very important factor in determining the recovery from petroleum reservoirs. Thus, heterogeneity calculations can be classified into static and dynamic techniques. Dykstra Parsons method provide the most excellent means to determine it.

Heterogeneity is the quality and situation of being heterogeneous. It was first defined in 1898 as the difference or diversity in kind from other kinds. Other definition consiss of parts or things that are very different from each other. In petroleum studies it is referred to as the isotropy and anisotropy. Heterogeneity can be named as; complexity, deviation from norm, difference, discontinuity, randomness, and variability.

Many researchers noted that the differences between homogeneous and heterogeneous is relative, and it is based mainly on economic considerations. This shows how heterogeneity has a variable concept which can be changed and re-defined to describe any situation arises during production from a reservoir, and is mainly based on the researchers` experiences and expectations. Other scientists state that heterogeneity is defined as the complexity and variability of the system property of interest in three-dimensional space.

These mentioned definitions clarify that heterogeneity does not refer to the overall system, or individual rock or reservoir unit, but instead it deals separately for each individual unit, properties, parameters and measurement types.

.2. Lorenz Coefficient

The original Lorenz technique was developed as a measure of the degree of inequality in the distribution of wealth across a population. In 1950 it was modified (the Lorenz Curve as used in petroleum engineering) by generating a plot of cumulative flow capacity against cumulative thickness, as functions of core measured porosity and permeability. Fitch et al. investigated the application of the Lorenz technique directly to porosity and permeability data.

Taking a known thickness of the reservoir and the well with most clear core data. Table 1 clarifies the method of calculations if the reservoir was divided into 10 layers only with porosity value of 0.33 and table 2 shows the area calculations, while Table.3 clarifies the calculation for the whole reservoir at each available property with porosity value of 0.18 (Because the calculations are for the permeability variation then the porosity must be of a fixed value).

Where:

κ: permeability (md), Φ: Porosity (%) and the probability was calculated by NORMSINV of excel; normal cumulative distribution.

the approximate area = (1st+ 2nd fracture of cum. κ h)/2*(1st– 2nd fracture cum. Φ*h) … etc.

h* gr = gross thickness- thickness of the rest of the layer (under that point)

IOP Conf. Series: Materials Science and Engineering **579** (2019) 012026 doi:10.1088/1757-899X/579/1/012026

Table 1. Lorenz Coefficient calculations 1

Thickness (h)	κ	Φ	κ*h	cumulative κ*h	Fracture of cum. κ*h	Φ*h	Cum. Φ *h	Fracture of cum. Φ *h
1.00	818.36	0.33	818.36	818.36	0.39	0.33	0.33	0.10
1.00	504.99	0.33	504.99	1323.34	0.62	0.33	0.66	0.20
1.00	311.61	0.33	311.61	1634.96	0.77	0.33	0.99	0.30
1.00	192.29	0.33	192.29	1827.24	0.86	0.33	1.32	0.40
1.00	118.66	0.33	118.66	1945.90	0.92	0.33	1.65	0.50
1.00	73.22	0.33	73.22	2019.12	0.95	0.33	1.98	0.60
1.00	45.18	0.33	45.18	2064.30	0.97	0.33	2.31	0.70
1.00	27.88	0.33	27.88	2092.18	0.99	0.33	2.64	0.80
1.00	17.20	0.33	17.20	2109.38	0.99	0.33	2.97	0.90
1.00	10.62	0.33	10.62	2120.00	1.00	0.33	3.30	1.00
10.00	212.00	0.33	2120.00	2120.00		3.30	3.30	
	avg	Avg	Sum	Max		sum	max	

Table 2. Lorenz coefficient calculations 2

approx. area	Ln_k	h*gr	%h*gr	variation
0.019	6.71	0.000	0.000	
0.051	6.22	1.000	0.100	-1.28155
0.070	5.74	2.000	0.200	-0.84162
0.082	5.26	3.000	0.300	-0.5244
0.089	4.78	4.000	0.400	-0.25335
0.094	4.29	5.000	0.500	0
0.096	3.81	6.000	0.600	0.253347
0.098	3.33	7.000	0.700	0.524401
0.099	2.85	8.000	0.800	0.841621
0.100	2.36	9.000	0.900	1.281552
0.80				
Sum				

The approximated area was calculated from the figure obtained and the probability is calculated by Excel@

From these calculations, the area under layer data and the area under the diagonal (always 0.5) was founded, then the Lorenz coefficient was calculated. Knowing that 0 value means a very homogenous reservoir and 1 value means a very heterogenous reservoir.

Based on the technical approach of calculating the Lorenz coefficient, figure-1 was deduced

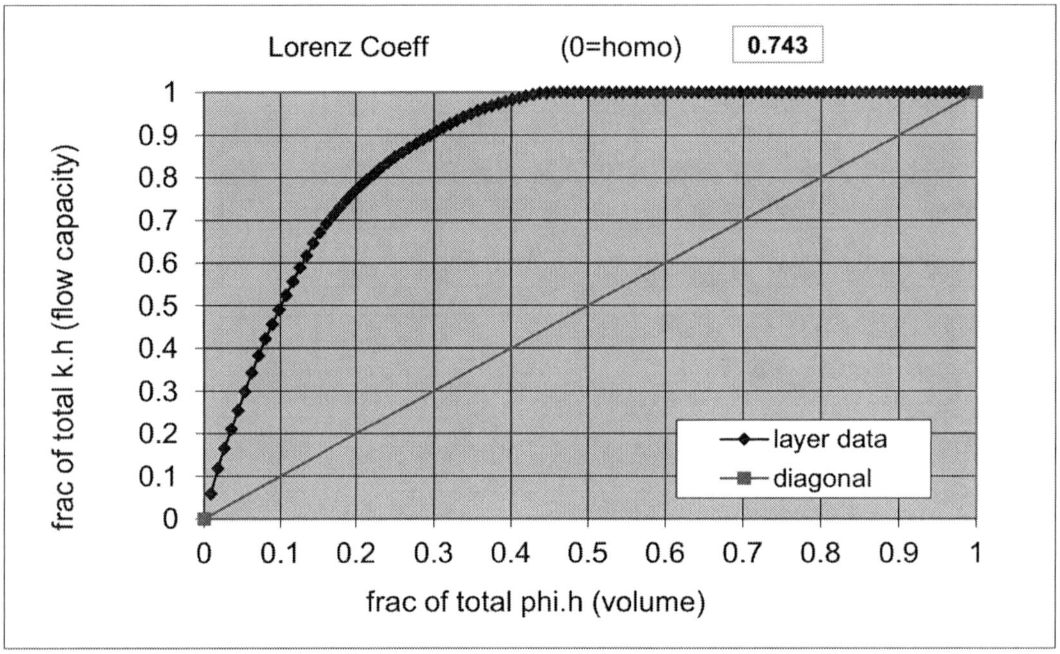

Figure 1. Lorenz Coefficient MJ-1

And the results are as follows:

- Area under layer data = 0.871
- Area under the diagonal = 0.5
- Then the Lorenz coefficient = (0.871-0.5)/0.5 = 0.743 (very heterogenous)

3. Conclusions

1- Determining the heterogeneity of any reservoir is very important before any simulation process
2- Lorenz method is a straight forward and an easy to calculate heterogeneity efficiently and accurately
3- Hartha reservoir – Majnoon Field is a very heterogenous reservoir with a complicated nature.

4- Having a heterogenous reservoir leads to the necessity of making a very detailed simulation model in order to denote all the possible changes in the reservoir.

Acknowledgment
The authors wish to thank the Petroleum Technology Department at the University of Technology - Iraq for facilitating this work.

References

1- Simpson J and Weiner E 1989, *The Oxford English Dictionary* (Oxford: Clarendon Press) ed 2 p 720
2- Okano H, Pickup G E, Christie M A, Subbey S and Sambridge M 2006 SPE Europec/EAGE Annual Conference and Exhibition (Vienna/ *SPE)* P 6
3- Fitch *et. al.* 2013 *Society of Petrophysicists and Well-Log Analysts* **54** 5
4- Atul Arya 1988 *Society of Petroleum Engineers* **3** 4

IOP Conf. Series: Materials Science and Engineering **579** (2019) 012027 doi:10.1088/1757-899X/579/1/012027

Increasing confidence in full field modelling and water flood planning for a giant reservoir under primary depletion through Material Balance modelling.

J H Al-Joumaa[1#] and M S Al-Jawad[2]

[1]Petroleum Technology Department, University of Technology-Iraq, Baghdad, Iraq

[2]Iraq, Baghdad, Baghdad University, Petroleum Engineering Department

[#]Corresponding author`s e-mail address: 150078@uotechnology.edu.iq

Abstract: One of the principle inputs to project economics and all business decisions is a realistic production forecast and a practical and achievable development plan (i.e. waterflood). Particularly this becomes challenging in supergiant oil fields with medium to low lateral connectivity. The main objectives of the Production Forecast and feasibility study for water injection are:

1- Provide an overview of the total expected production profile, expected wells potential/ spare capacity, water breakthrough timing and water cut development over time
2- Highlight the requirements to maintain performance, suggest the optimum development pattern
3- Increasing confidence in business decisions to develop the reservoir in question

The main tool used for these purposes is a sophisticated reservoir simulation software, namely CMG©, since it can predict reservoir behavior, honor physical constraints and capture the heterogeneity within the reservoir to accurately predict performance. However, the starting point for this kind of complicated studies needs to start from the basics, in order to understand the big picture and be able to plan properly for the scope to be delivered, hence, utilizing analytical tools like MBAL becomes quite necessary, if not crucial, to the success of full field modelling and choosing an optimum water flood pattern and design.

This paper covers the methodology for building the reservoir component utilizing a Material Balance model, of which the results will be used as an input to reservoir simulation to evaluate and accurately predict reservoir performance, which directly feeds into planning for water flooding projects and selection of an optimum flood pattern.

A Tank model was built at first to assess and understand the driving forces (energies) of the reservoir in question, utilizing pressure and production data from legacy wells, the prepared model is also supported by geological and petro physical studies to give representative results. Acquired Static Bottom Hole Pressures (SBHPs) in wells were used as anchor points for the tank pressure and to test the validity of the history match. Multiple analytical methods to QC the results and STOIIP volume were conducted, e.g. the Havlena-Odeh method.

This methodology has been tested successfully in the stated super giant oil field, in which the reservoir in question is a carbonate rock formation. An example of this is covered in the paper. It was concluded that utilizing a history matched and coherent MBAL model before conducting a detailed reservoir simulation study can save a lot of time and effort by providing guidance to the path which needs to be followed, and sheds light on the critical elements to be looked after. This has also helped to uncover the driving mechanisms and energies in the reservoir, hence allowing the engineer to plan for the necessary voidage replacement and water injection rates to sustain the reservoir pressure and pattern development. Another technical advantage of the described method is the higher sustainability of the model.

The suggested method, in combination with geological and petro physical information available, can be applied to majority of the reservoirs. This combination is paramount to ensure optimum time and planning that is followed for each reservoir development study that involves water flooding.

1. Introduction

The main reason for constructing a material balance reservoir model is to Provide insight into the production characteristics of the reservoir, help in history matching and to find the reservoir drive mechanisms and off course to determination of initial oil in place

MBal program depends on the material balance equations in calculations. The data and the sorting and using steps are follows by using Linearized form of material balance equation used to estimate the initial oil in place (the intercept of the straight line) - Havlena and Odeh procedure,

It is of great importance to have history matched Material Balance models for any reservoirs targeted for any reservoir study in order to have a better understanding about the drive indices for the reservoir and be able to make decisions and plans with more confidence. The Material Balance model presented here were built using the three wells of interest mentioned earlier. Aquifer modelling for the reservoir was suggested based on the latest geological realizations as well as the field and regional understanding.

2. MBal model

Table1. Summary of Hartha Reservoir rock and fluid properties that were used as input in preparing the Tank (Material Balance) model

STOIIP (MMSTB)	1973
Porosity (frac)	0.19
SWc (frac)	0.1876
Net thickness (meter)	88
Radius (meter)	5710
Temperature (oF)	186
Initial Pressure (psi)	3555.5
Bubble point pressure (psi)	2652.7
Compressibility (1/psi)	4*10-6

Table2. Summary of suggested aquifer parameters

Aquifer model	Hurst-Van Everdingen modified
Outer /Inner radius ratio (frac)	3
Encrochement angle (degree)	270
Aquifer permeability (md)	108

figure below from MBAL software shows the assumptions used as input parameters:

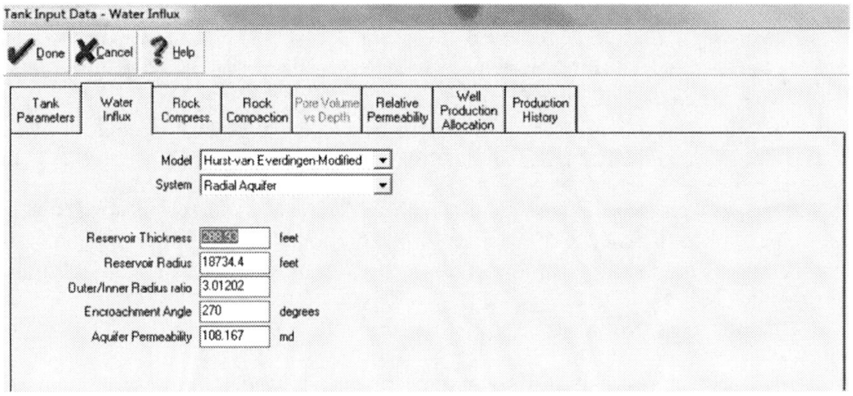

Figure 1. Assumptions of MBal input

As the Hartha formation is water-bearing in all of the surrounding fields; it was evident that the Aquifer size would fall on the high side when history match. Considering outer to inner radius ratio = 3 , and the reservoir radius would be 18734 ft which is a reasonable assumption.

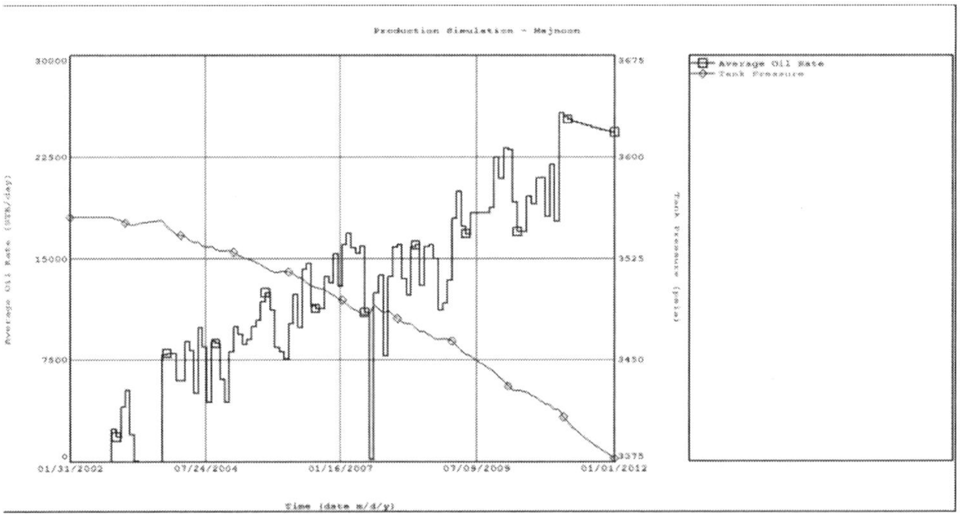

Figure 2. Production rates and Pressure behaviour for Hartha Reservoir.

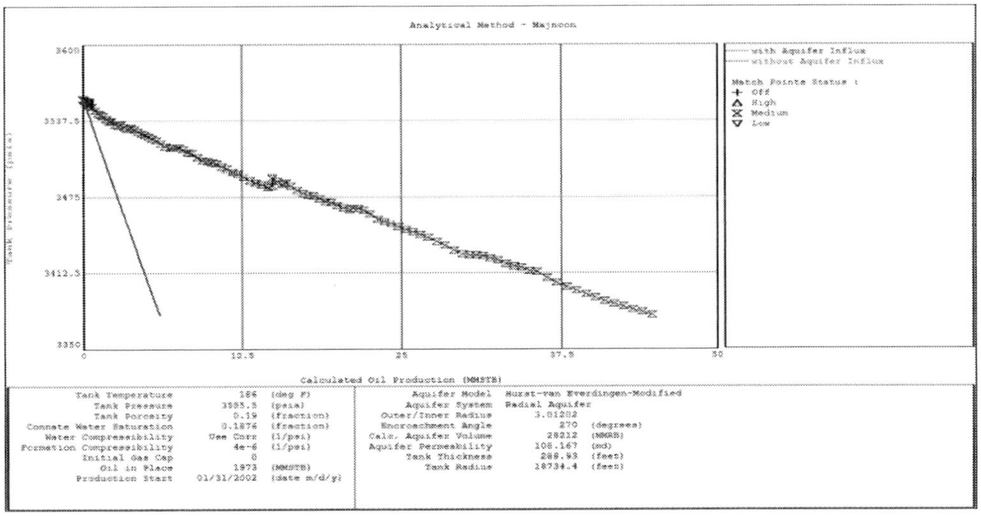

Figure 3. History Matching using The Analytical Technique for Hartha Reservoir.

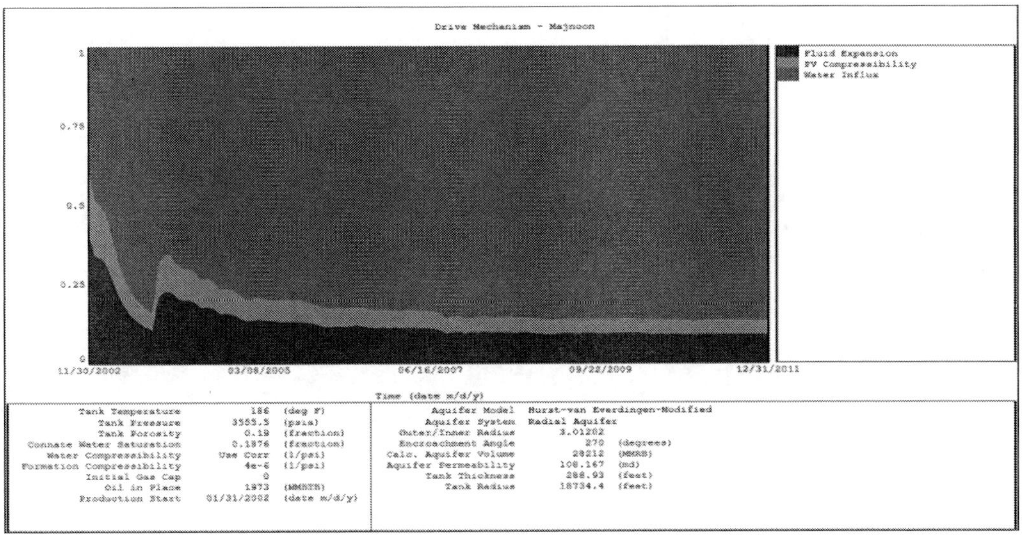

Figure 4. Energy Plot for Hartha Reservoir.

IOP Conf. Series: Materials Science and Engineering 579 (2019) 012027 doi:10.1088/1757-899X/579/1/012027

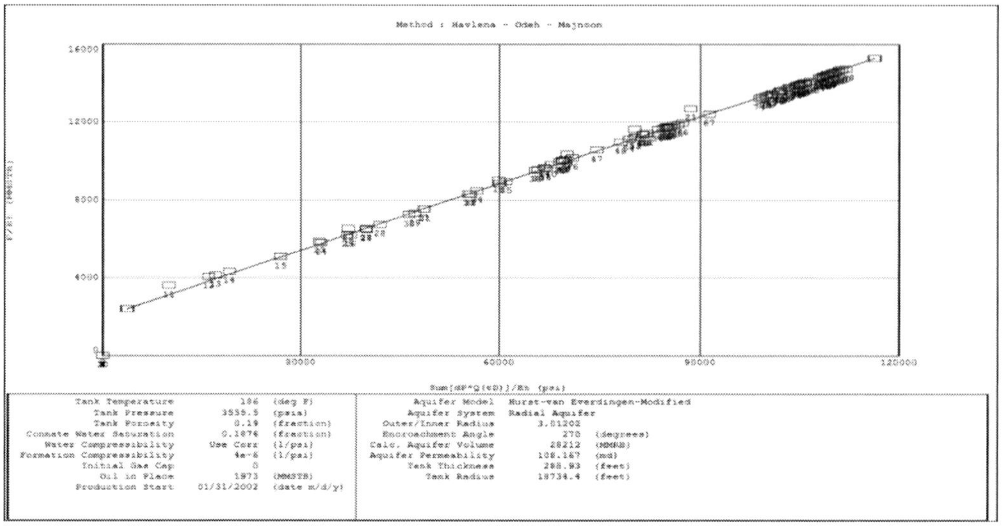

Figure 5. Havlena-Odeh QC plot for Hartha Reservoir.

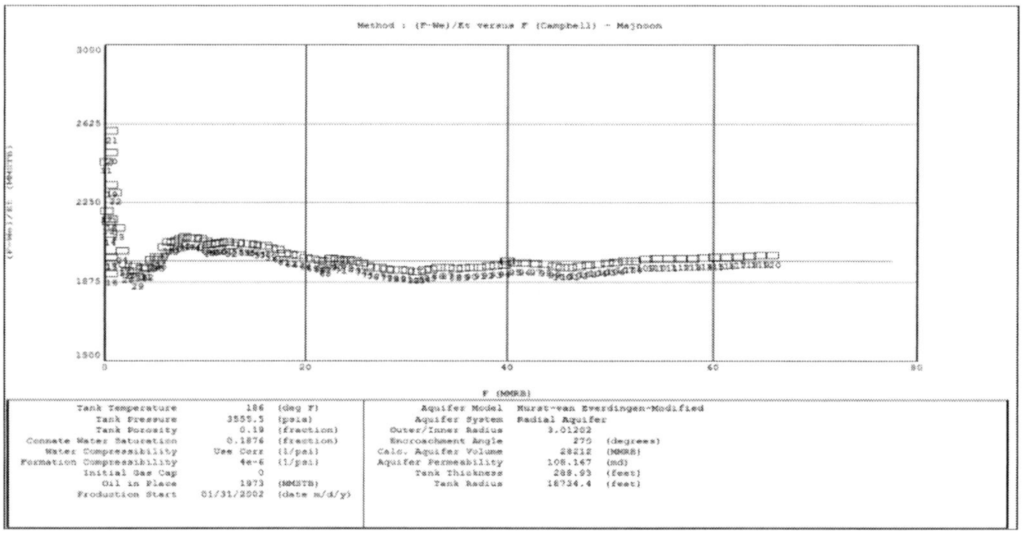

Figure 6. Campbell QC plot for Hartha Reservoir.

IOP Conf. Series: Materials Science and Engineering **579** (2019) 012027 doi:10.1088/1757-899X/579/1/012027

3. Conclusions

the following are the results of the MBal model for the Hartha reservoir;

1- The Analytical method History Matching plot shows an exact match between the historical simulated data and the proposed model
2- The Simulated and History Pressure and Cumulative oil vs. time plot suggests a steady drop of pressure
3- The Energy plot suggests that the driving energy at first is mainly the pore volume expansion; with aquifer support coming into the picture after some depletion
4- Havlena-Odeh and Campbell plots suggests that the STOIIP input and the energies are in line and matching in the later periods
5- The model shows a hydraulic connectivity which means we are dealing with a tank unit that is similar in properties, that means we can use the same pvt tables for any well in building the reservoir models.

Acknowledgment

The authors wish to thank the Petroleum Technology Department at the University of Technology - Iraq for facilitating this work.

References

1- Mogbolu E, Okereke O, Okporiri C, Ukauku I, Esharegharan O, Taiwo I and Sukubo I 2015 *SPE Nigeria Annual International Conference and Exhibition* (Lagos, Nigeria/ SPE) p 2
2- Idogun I, Jeboda O, Charles D and Ufomadu H 2015 *SPE Nigeria Annual International Conference and Exhibition* (Lagos, Nigeria/ SPE) p 6
3- B. Havlena and A.S. Odeh, 1963 *Journal of Petroleum Technology* **15** 898

IOP Conf. Series: Materials Science and Engineering **579** (2019) 012028 doi:10.1088/1757-899X/579/1/012028

Dynamic model of Tertiary reservoir in Khabaz Oil field

Mohamad B. Mousa# and Nabaz Qader
Petroleum Engineering department, North Oil Company, Kirkuk, Iraq

#Corresponding author's e-mail address: mahamad_bandar@yahoo.co.uk

Abstract. This study deals with building a reservoir model for Tertiary reservoir in Khabaz oil field and calculate oil and gas in place by volumetric method using Petrel RE software and by Dynamic method through material balance by using MBAL software ,testing the reservoir dynamic model through history match over 26 years, and predicted (4) case of reservoir and well performance over 29 years start at the end of history match (1/1/2016) until (1/1/2045) by natural depletion for case-1 Do nothing (current producing wells) , Case-2 Base case (work over non-producing wells) case-4 new wells (drill 6 new wells) and add water injection for case-3 (drill 7new water injector wells).

1-Introduction

Khabaz Field is located in the North East of Iraq at approximately 12 km North West of Kirkuk city. Discovered in 1955, the first exploration well (Kz-1) was drilled on April/1976, the structure consists of an elongated asymmetrical anticline, 15 km length and 5 km width, with a NW-SE axis and faulted mainly on its West flank. It lies between Jambur and Bai Hassan structures and South West of the Baba dome (Kirkuk Field). Being more deeply buried than the Jambur and Bai Hassan fields by respectively 600 and 1000 m, Tertiary series are not outcropping and unlike its neighbours, the Khabaz anticline cannot be detected from the surface5.

The field is consist of 2 main carbonate reservoirs: -
1-Tertiary Reservoir.
2-Cretaceous reservoir which consist of two reservoirs:-
A-Upper Qamchuqa Reservoir (Mauddud).
B-Lower Qamchuqa Reservoir (Shuaiba).

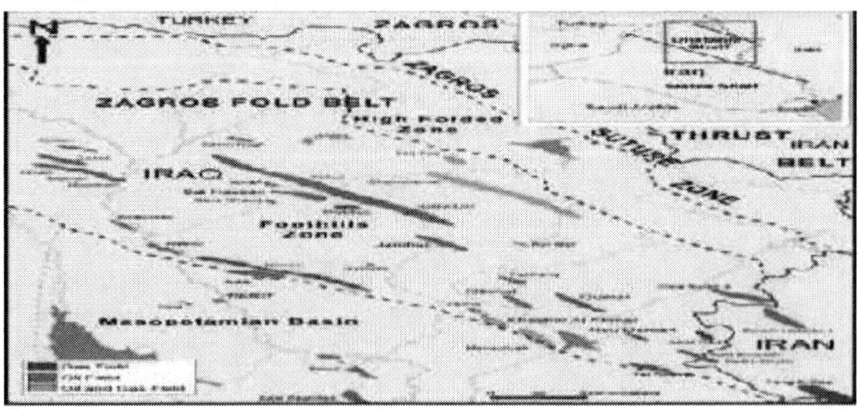

Figure 1. Khabaz Field location

Since the discovery Kz-01 well, 20 wells were drilled on the Khabaz oil tertiary reservoir and the oil production started in 1993-94 with 13 oil producers on Tertiary interval. At the end of 2016, about 89.8 MMSTB were produced representing less than 7.89 % of the overall present reservoir accumulation. The total field oil production reaches approximately 30000 STB/D in 2000, but dramatically dropped down since 2004. Nearly half of the producing wells are now shut down due to well plugging by asphaltene, gas / water break through or even blow-out.

The Tertiary carbonate formations 150 m thick Top reservoir 1794.5 MBSL Saturated oil 36° API with CO_2, no H_2S,the GOC around 1925 MBSL and the WOC around 2050 MBSL at least the top part (JERIBE) of the reservoir is fractured.

The Reservoir production is characterized by good productivities due to the fracturing, early water and gas break-through and problems linked to asphalting precipitation.

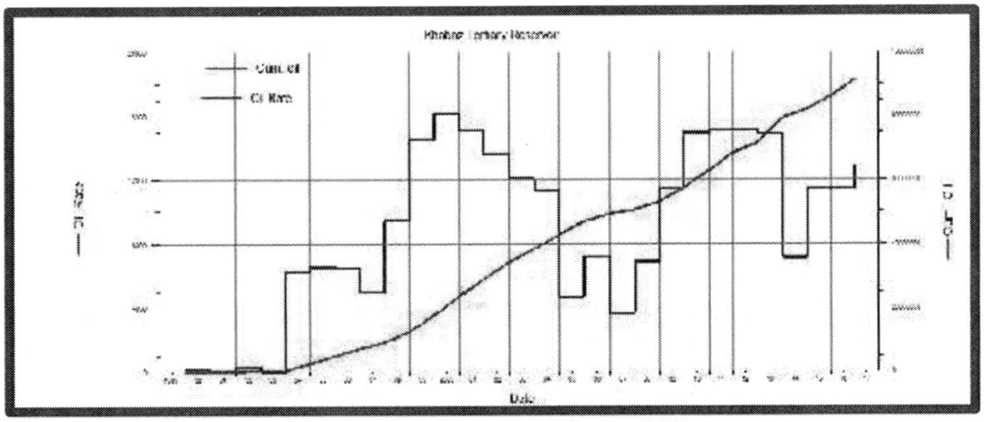

Figure 2. History Production profile for the Reservoir

2-Literature Review

A Reservoir Simulator is a computer generated model that aims to represent the full geological extent and structure of a reservoir. The reservoir simulator gives the reservoir engineer the ability to study and analyze the performance of the reservoir under various operating conditions in order to adopt an efficient and maximum profit-driven development strategy for producing hydrocarbons from the reservoir. Petrel RE is the pre and post processor of Eclipse reservoir simulation interface that is well known in the Oil and Gas Industry. With Petrel RE a user is able to develop reservoir simulation cases that aid in studying and analyzing the performance of reservoirs under various operating conditions to facilitate final decision making for reservoir management.

3- Layering- Reservoir model Grid

The reservoir model layering derives from the geological model. The layering and the horizontal gridding have been defined to allow a good description of the vertical and horizontal flux with a reasonable number of cells to allow an acceptable simulation time. Grid mesh/175 x 75 (100m x 100m), Layering/ 19 layers, Facies/ (dolomite, dolomitic limestone, limestone), the total grid contains / (211584) cells 210238 of which are active in the model, the NTG model, it was calculated in the geological model with (0.06) md as cut off permeability, and (45%) was determined as a cut off for V Clay, where the (NTG) calculated using the formula below: -

NTG=If (VCL<=0.45, If (PHIE>=0.06,1,0),0)) The Geological model divided the reservoir rocks into three types (DOLOMITE), (DOLOMITIC LIMESTONE) and (LIMESTONE).

4-Petrophysic properties

The petrophysical parameters (porosity - permeability - Net to gross) used in the ECLIPSE reservoir model are the Those parameters were exported from the PETREL geological model. The permeability are derived from the plugs permeability/porosity plots, they only reflect the matrix characteristics and thus do not take into account the fracturing of the rock.

5-Buildup Pressures Analysis:

The permeability depended in the geological model represents the permeability of the matrix neither fractures. And because some of the wells in the tertiary reservoir are produced at good rates, have a medium productivity index (P.I), the permeability obtained from their completion tests and mud loss data in the production formations all of that indicate to fractures.

The permeability of the model has been increased in order to allow the wells to produce the rates required by the production history. It is believed that the increase of permeability between the geological model (from Phi K correlations) and permeability's issued from testing reflects the importance of the fracturing in the drainage area of the wells. The adjustment of the permeability has also allowed (combined to adjustments on other parameters) the match of the static pressures and GOR.

In order to stay in a realistic range, this increase of permeability has been based on observed data (permeability derived from well tests).A complete re-interpretation of the available pressure build-up data has been made, 16 well pressure build up were re-interpreted over the 20 wells producing this reservoir using Kappa software (Saphir) for the purpose of obtaining the total permeability of the rocks.

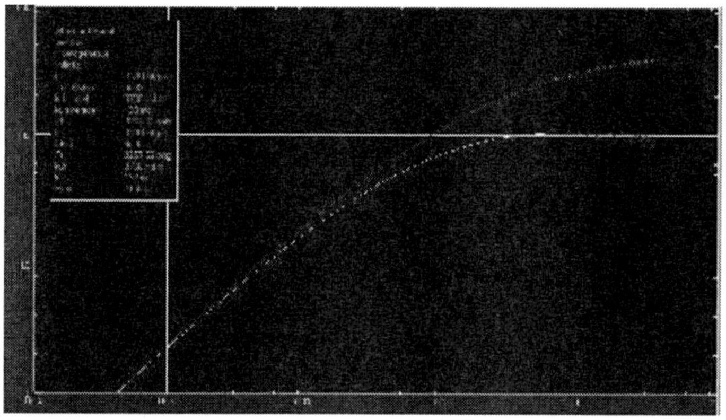

Figure 3. Pressure Transit Analysis Log-log plot:dp and dp'(psi) vs dt(hr) for Kz-10

Most of the build-up were stopped before the end of the well bore storage period leading to poor estimation of the K: very short time for build-up on. The flow rates were only estimated from the pressure drop measured on the flare line (in the build-up interpretation, the K is directly proportional to the average rate). Therefore the resulting K cannot be considered as a 100% reliable data but as an approximation. (Table 1) includes the total permeability values obtained from the analysis process.

IOP Conf. Series: Materials Science and Engineering **579** (2019) 012028 doi:10.1088/1757-899X/579/1/012028

Table 1. total permeability values obtained from the analysis process

WELL	Permeability Well Test (md)	WELL	Permeability Well Test (md)
KZ-03	14.3	KZ-25	314
KZ-08	15	KZ-26	132
KZ-09	104	KZ-27	33.5
KZ-10	133	KZ-32	120
KZ-15	50	KZ-33	55
KZ-17	235	KZ-34	6.4
KZ-19	90	KZ-37	16.6
KZ-22	64		

6-Special core analyses: -

The data of capillary pressure and relative permeability entered into the SCAL software to perform quality control (such as curve smoothing), grouping the data according to lithological parameters or porous unit and transform the laboratory data into rock curves suitable for input to the simulator.

6.1 Capillary pressure data:

The data available from the Centrifuge and Restored State experiments were depended on 85 samples obtained from some of wells of formations (Jeribe, Anah, Anah/Azqand and Azqand). The data were converted from laboratory conditions to reservoir conditions then normalized by using J-Function as shown below:

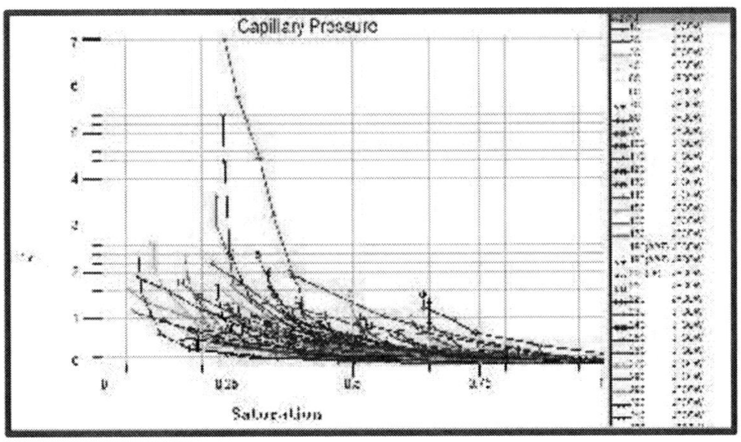

Figure 4. J-Function Curves

The curves were classified into three groups based on the initial water saturation, which was consistent with the number of rock types that were approved in the construction of the geological model.

1- DOLOMITE FACIES which have high permeability, with an initial water saturation 15%.
2- DOLOMITIC LIMESTONE FACIES, which are medium permeability, with an initial water saturation ranging from 15 to 35 %.
3-LIMESTONE FACIES characterized by its low permeability, with an initial water saturation ranging from 35 to 65 %.

Then each group have been averaged as shown in the figure below.

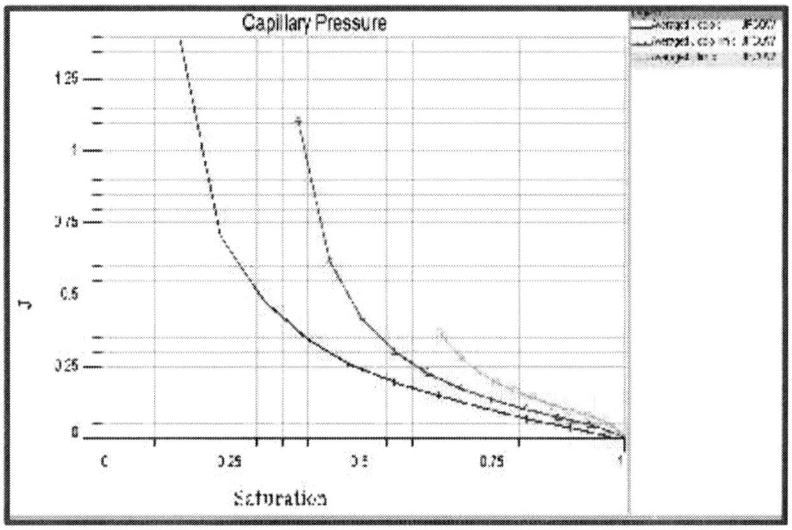

Figure 5. Average J-Function Curves

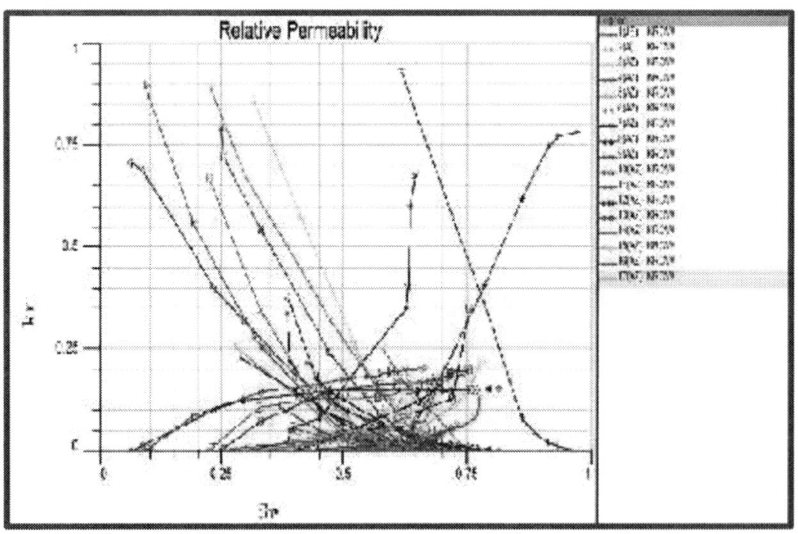

Figure 6. Water-Oil Relative permeability curve for all samples

6.2 Relative permeability data Water-Oil (krw-kro): -

The relative permeability calculations of the water / oil system were based on 17 samples data, as shown in the figure below from water flooding experiments, the samples were classified into groups according to reservoir rock types based on the previous method in capillary pressure calculations.each groups Normalization , averaged and De-normalization to extract the final curves for each group, as shown in the figure below, where the program calculates the modified properties of Sw, Krw, Kro.

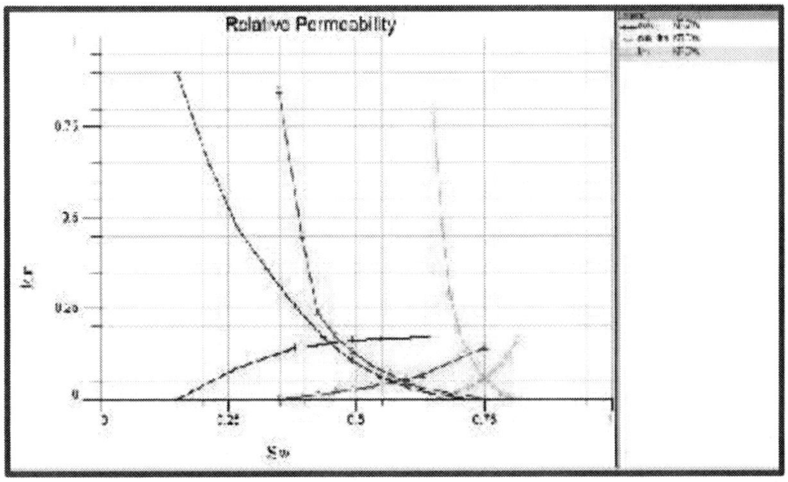

Figure 7. Average water-Oil Relative permeability curves

6.3 Relative permeability Gas-oil (krg-kro): -

The relative permeability calculations of the gas / oil system were based on 11 samples data as shown in the figure below, from Gas flooding experiments,

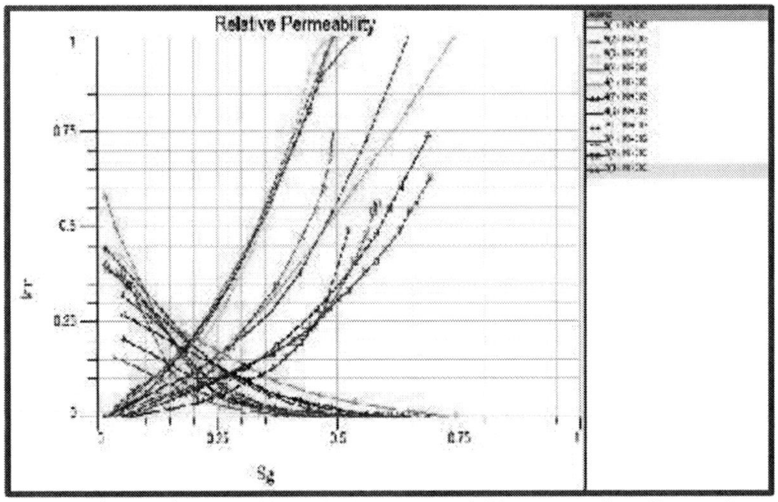

Figure 8. Gas-Oil Relative permeability curve for all samples

The samples classified into three groups according to reservoir rock types based on the previous method in capillary pressure calculations and relative permeability of water-oil,each groups Normalization, averaged and De-normalization to extract the final curves for each formation, as shown in the figure below, where the program calculates the modified properties of Sg, Krog, Kr.

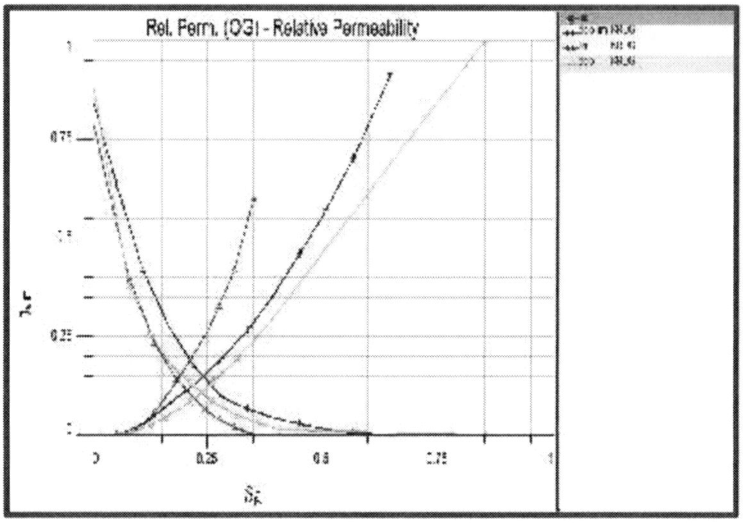

Figure 9. Average water-Oil Relative permeability curves

Tables below represent the final results of these the end point curves for the three rock types which were exported to the Petrel RE software.

Table 2.

FACIES	SWI	SOR	SGC	KRW max	KRO max	KRG max
DOLOMITE	0.15	0.29	0.02	0.17	0.9	1
DOLOMETIC-LIMSTONE	0.35	0.25	0.05	0.14	0.85	0.91
LIMSTONE	0.65	0.19	0.07	0.6	0.8	0.25

7-Fluid Modelling Black oil PVT.

Podbielinak analysis data from Kz-18 table (2) are used to build a pvt model and applied to the whole field by matching them with lab data using regression to obtain the best parameters for the equation of state (Peng Robesion),Then black oil PVT table have been derived and exported to (Eclipse , Prosper , Saphir and MBAL) softwares.

Table 3. Podbielinak analysis data from Kz-18

Well	KZ-18 2nd	
Report date	1989	
	Weight %	Mole %
Methane	9.09	49.61
Ethane	2.84	8.14
Carbon Dioxide	1.99	3.94
Hydrogen Sulphide	-	nil
Propane	2.44	4.81
I-Butane	0.5	0.79
n-Butane	0.9	1.31
I-Pentane	0.24	0.26
n-Pentane	0.23	0.26
Hexanes+	81.8	20.88
Total	100	100
S.G of c6++ @ 60/60	0.8354	
Mwc6++	232	

First separation stage 203 psi 104°F Second separation stage 58 psi 104 °F and Storage tank 14.5 psi 60 °F representing the existing installation process. The main parameters for the reservoir fluid are listed below:

Table 4. Main Oil parameters for the reservoir fluid

GOR SCF/STB	1065
Bo (BBL/STB)	1.63
API	38.2
Oil Viscosity@ Psat. (cp)	0.77

8-Well modelling

The typical well for Tertiary reservoir is KZ-25. IPR established using last test with available GOR on KZ-25 in august 2002.-VLP have been delivered to Eclpise. Oil flow rate in case of blow out when = 8000 STB/D. Absolut opening flow (AOF) = 9800 STB/D.

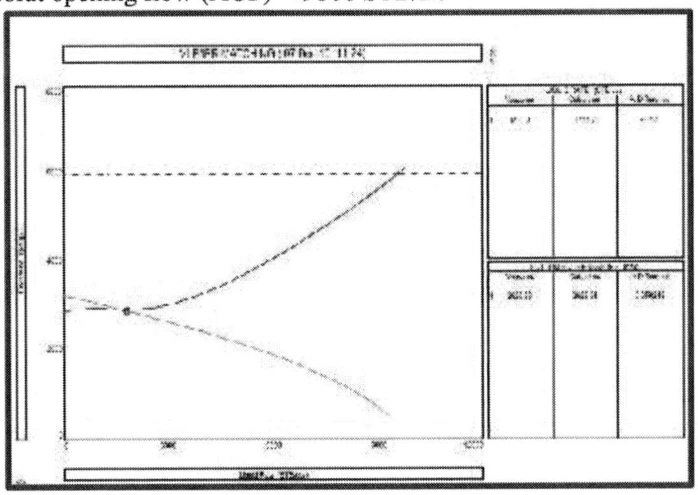

Figure 10. Calibrated Model

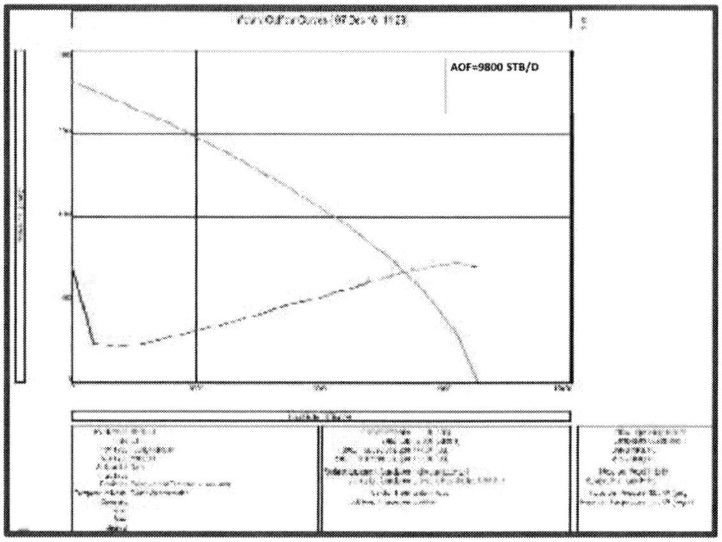

Figure 11. Well flow rate estimation in case of blowout

9- MATERIAL BALANCE

9.1 Material Balance Theory

Schillthuis (1936) first presented a method of reservoir estimation using the material balance. Material balance is a volumetric balance which states that since the volume of a reservoir (as defined by its initial limits) is constant, the cumulative observed production, expressed as an underground withdrawal, must equal the expansion of fluids in the reservoir resulting from finite pressure drop which is the governing principle for the MBAL software i.e.. Underground withdrawal (rb) = Expansion of oil (rb) + originally dissolved gas cap (rb) + reduction in HC pore volume (HCPV) due to connate water expansion (rb) 7

9.2 The MBAL Software Tool

A classic material balance study was conducted to estimate the initial hydrocarbons in-place. The tank model was built using Petroleum Experts Integrated Production Modelling IPM-MBAL software and calibrated by history matching using production data, reservoir pressure reservoir rock data, and reservoir fluid data.

9.3 Data Requirements and Input

The following data are required for Material Balance Analysis using MBAL software:

1- PVT Data
2- Initial Reservoir Pressure
3- Reservoir Average Pressure History
4- Production History
5- All Available Reservoir and Aquifer Parameters

(Table 5) illustrates the data that have been used in the material balance equation

Table 5. data used in the material balance equation

Reservoir	TERTIARY
Res. Temp. F	170
Initial Pressure psig @GOC	3273
Porosity Fraction	0.15
Connate. Water Saturation Fraction	0.3
Initial Gas cap volume Factor	0.238
Rock Compressibility 1/Psi	0.238E-6
Water Influx (BBL/PSIG/DAY)	49.5

Black oil PVT table has been used that derived from equation of state depend on Podbielinak analysis data for Kz-18.Preparation of data and consistency checks were performed on the production history, average reservoir pressures, reservoir and aquifer data. The History Matching Process begins when there is consistency in the data inputted.

9.4 History-Matching

History matching involves a trial and error approach to provide a best fit dimensional level. It comprises the functions of the graphical method, the analytical method and simulation tests. History matching is used to determine and identify sources of reservoir energy and their magnitude, the value of OOIP, OGIP, aquifer type and strength etc. History matching in MBE is the most effective way to determine the aquifer model that best fits the observed data. Comparison between the observed data and the calculated data on a zero- An iterative non-linear regression is used to automatically find the best mathematical fit for a given model, a simulation of production can be run to check the validity of the result of the above two techniques as shown in (figure 12).

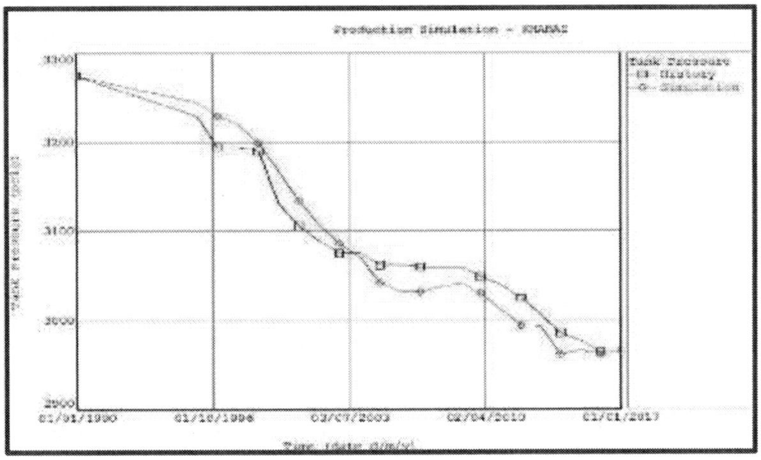

Figure 12. Reservoir history matching

9.4.1 Analytical Method.

The analytical method allows for regression on all reservoir model parameters. Regression is used to adjust the reservoir model to minimize the difference between the observed/measured and the model production. It is used to assess the effects of varying parameters such as formation compressibility that cannot easily be assessed using graphical methods. The quality of the regression match is expressed as the standard deviation between model and measured values. The analytical plot was regressed to compute the oil in place, and the aquifer diffusivity as shown in figure below.

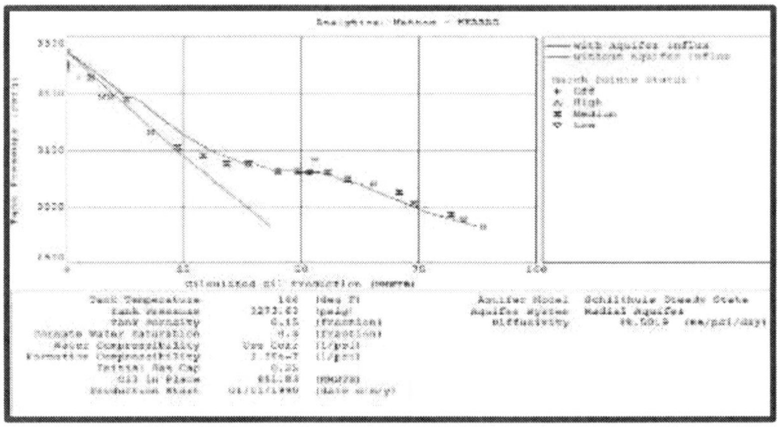

Figure 13.

9.4.2 Graphical Method

The first step taken was to plot (F-We)/Et versus F the withdrawal) known as Campbell's plot with no aquifer defined initially. If there is no other source of reservoir energy other than the total fluid expansion Et, then this Campbell's plot will be a horizontal straight line with a Y axis intercept equal to the original oil in place (OOIP). Any "turn-up" in the plot (i.e deviation from the theoretical horizontal straight line) indicates another source of energy present (due to a source (injector) or an aquifer influx). Sensitivities were conducted using various aquifer models, the best technical case estimate of the initial oil in-place was 850 MSTBO as in (Figure 11).

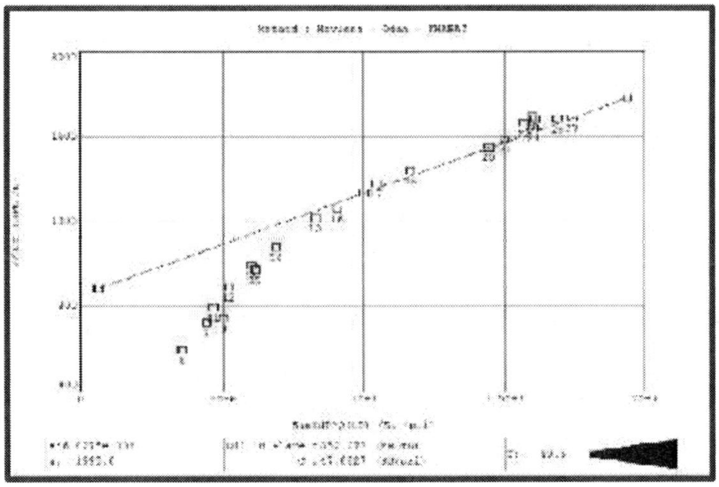

Figure 14. STOOIP by Material balance

9.5 Energy plot

The plot describes the prevalent energy system present in the reservoir; water influx, pore volume compressibility. Fluid expansion, ingestions e.t.c. It describes the fractional contributions of these energy systems present in the reservoir and the most prominent at various dates. The figure below illustrates the drive mechanisms for the reservoir.

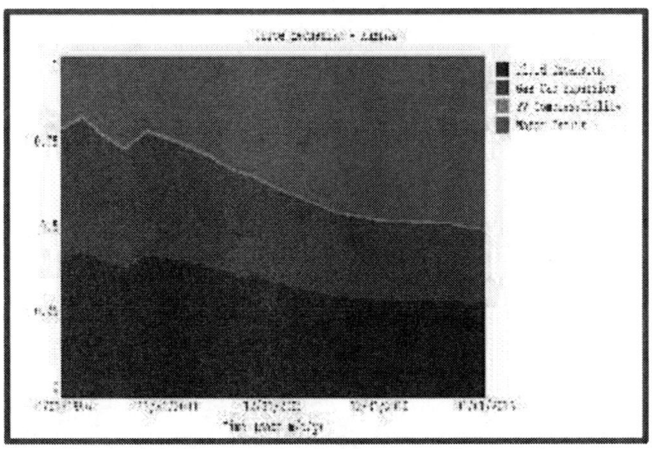

Figure 15. Energy plot for drive mechanism in the reservoir

10- Reservoir dynamic simulation

The tertiary reservoir model has been developed with the petrel RE version 2016.

The model is based on the Dynamic Synthesis and the geological PETREL model, the purpose of the Model is to design a redevelopment plan. The model represents the Tertiary carbonate reservoirs with reservoir fluids are saturated oil, and initial gas cap. The production started in 1993. A total of 20 vertical wells were drilled in the reservoir.

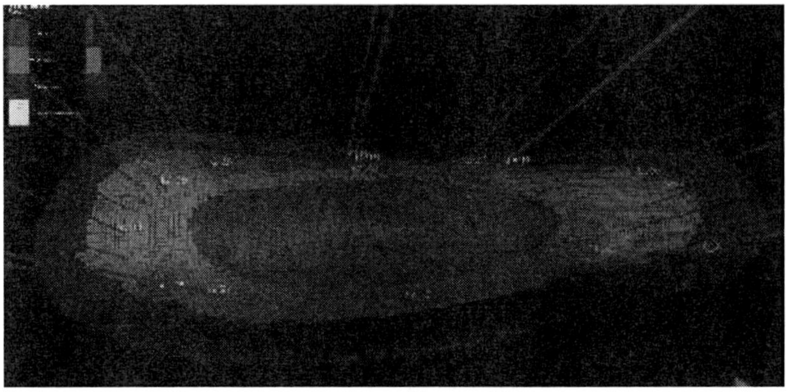

Figure 16.

11- INIALIZATION PARAMETER & OOIP / OGIP

The oil in place was determined by the volumetric method by using data generated from geological and petro physical evaluation (areal extent, formation sand thickness, porosity and the saturation e.t.c) and computing the initial oil in place from the general formula. The governing equation for the volumetric estimation of oil in place is given as:

N=7758 Ah (1-swc)/Boi

N=STOIIP, Ah=Bulk(rock)volume, =porosity(Fraction)

,swc=connate water saturation(Fraction),Boi=Formation volume factor dimensionless.

(Table 6) contain the parameter used in the initialization.

Table 6.

Initial pressure @ datum depth (psi)	3330
Datum depth (MBSL)	1990
Reservoir Temperature (F) @datum depth	170
Reservoir Temperature (F) @GOC	166
saturation pressure @ GOC (PSI)	3273
Saturation Pressure @ reservoirs Temperature (Psi)	3282
GOC (MBSL)	1925
OWC (MBSL)	2050
BO (BBL/STB)	1.636
BG (CF/SCF)	0.0049
Oil Viscosity C.P	0.79
GOR (SCF/STB)	1060

(Table 7) summarizes the estimation of the oil and gas volume by ECLIPSE initialization.

Table 7. Estimation of the oil and gas volume by initialization the model

Zones	STOOIP STB 6810	OGIP SCF 9810
Jr-1	8	8
Jr-2	6	13
Jr-3	43	45
Anah	12	19
Anah/Azkand	45	41
Az-1	94	89
Az-2	248	179
Az-3	372	92
Azkand/Ibrahim	225	3
Total	1053	489

12-History matching

12.1- Model status before matching

On the first run performed before history matching, three main problems were encountered: The permeability issued from the geological model was too small for the wells to reach the observed rates, most wells did not respect the history oil rate required due to a poor productivity. The GOR was not correct. The gas production was generally too low. Gas break through were not represented correctly. The water cut was generally too high, therefore to reduce the water cut ,two region of oil water contact have been used 2065 MBSL for North east region and 2050 MBSL for south west region, the initial oil in place depending on new contact level was 1138 MM STB and initial gas in place was 489 MMM SCF

12.2- Monitoring - Production data

The production history has been matched until 1/1/2017. Modification made on position of the OWC contact, and the transmissibility. monthly oil well production rates for the period 1/1/1990 to 31/12/2016.well shut in dates file and causes for shut-in well test data including oil rate, GOR and WCUT, manifold pressure, well head pressure, type of production (tubing or tubing + annulus) static pressure data issued from pressure build-up ,as no production profile for total gas and water rate is available, the model could not be constrained with a reservoir rate (oil + water + gas rate at down hole condition) as it is normal practice for history matching, it was constrained only with the oil rate .The water-cut and the GOR were matched to the observed well test data.

12.3- Static pressure matching

There was no need for increasing / decreasing the aquifer size in the model nor activating analytical aquifers to match the depletion. There was no attempt to match the pressure observed during the production period, because there was no guaranty on the reliability of those measurements (correct stabilization or extrapolation). Matching was more emphasized on pressure points recorded during long shut in periods.

12.4- Well Head Flowing Pressure matching.
Pressure drop calculations were produced with the PROSPER software on well KZ-25. It was developed for the prediction runs after matching where the production can be constrained by a minimum well head pressure. They represent only the production through tubing, the production through annulus was not considered. The VFP was used to match the well P. I in the late period of the history (100% production though tubing). The matching of the surface flowing pressure was only possible when the gas rates were correctly matched (GOR close to Rs). On high GOR wells the quality of the GOR match does not allow the WHP to be matched correctly. P.I adjustments made in the model for WHP matching:

12.5- GOR matching
Very high increases of GOR have been observed on some wells, while on the other wells the GOR stayed close to the Rs. Those GOR increases were interpreted as fingering of gas from the gas cap, therefore it was tried to adjust the GOC position and locally the permeability field for matching. On KZ-19, due to the low position of the perforations, the breakthrough of gas observed from 2001 cannot be reproduced by the model unless a strong connection to the gas cap. The horizontal anisotropy. The permeability in the vertical direction was increased to improve the communication with the gas cap. The global horizontal permeability was kept: vertical permeability equal to 0.05 of horizontal permeability.

12.6- Water Cut matching
The match was mainly realised by adjustment of the WOC depth on the two previously defined zones. The Kro/Krw favourable to oil was selected in order to minimise the change on WOC depth and OOIP.In the north east zone the WOC was lowered from 2050 MBSL to 2065 MBSL. In the South west zone the WOC was 2050 MBSL. A good match was obtained for most of wells.

13- Development Strategy
Four development cases have been simulated for reservoir and well performance over 29 years start at the end of history match (1/1/2016) until (1/1/2045), the natural depletion for Do nothing case (current producing wells), Base case (work over non-producing wells), add water injection wells for water injection case (drill 7new water injector wells) and Additional drilling of vertical wells is proposed to extend the duration of the production plateau at maximum production capacity (15000 STB/D) for final case (drill 6 new wells) with the existing wells assuming that basic remedial works are made.

The manifold is repaired and group control for oil production of 15000 STB/D, and gas rate 35 MMSCF/D. Gas compression operational on the two stages of separation. The wells are controlled by the manifold outlet pressure (minimum 217 psi), the pressure drop, tables used no Water Cut limit set in the model. The wells stop when they lose their productivity at minimum manifold pressure. The well GOR is controlled by cut-back the liquid rate 20% for GOR in excess of 3000 SCF/STB. No gas flaring nor re-injection.

13.1- Do Nothing Case: -
The DO-NOTHING run is designed as a reference case to evaluate the effect of new investments. The number of producing wells is (13), (Table 8) show the well flow rate.

Table 8. Well flow rate.

wells	Kz-2	Kz-3	Kz-9	Kz-10	Kz-20	Kz-22	Kz-25	Kz-27	Kz-28	Kz-32	Kz-34	Kz-37	Kz-41
Flow rate STB/D	850	1100	400	700	2100	1000	1850	700	350	1100	500	950	700

The plateau at 12750000 STB/D lasts for 1 year, cumulative oil production was 180 MMSTB, RF equal to 10% water cut equal to 19% and reservoir pressure drop equal to 640 psi the, as shown in figure below. Maximum gas rate is reached on year two; the production declines rapidly due to the loss of productivity caused by

- Water breakthrough
- Drop of static pressure.
- Liquid rate cut back due to gas arrivals.

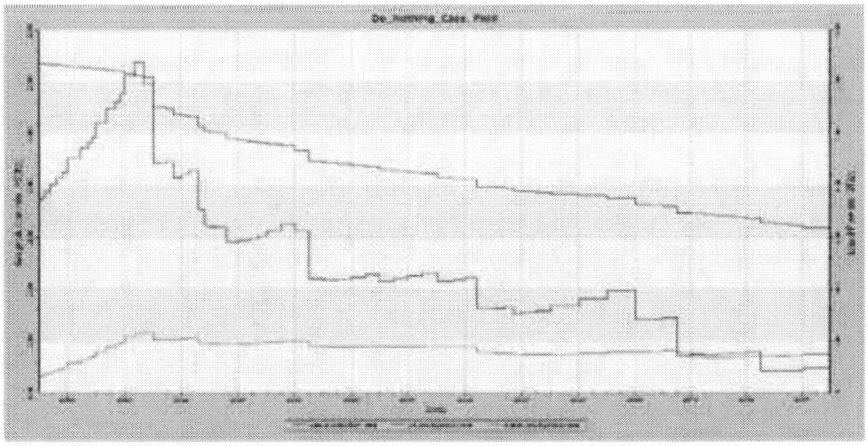

Figure 17. Production prediction profile for Do-Nothing case

13.2- Base case: -

The base case run has been built to evaluate the possible profiles with the same hypothesis of the DONOTHING and with additional production wells were work over in order to maintain the production plateau for (6) years at full capacity (15000 STB/D) as long as possible, the number of producing wells is (20). (Table 9) show the date of putting the additional wells to the work.

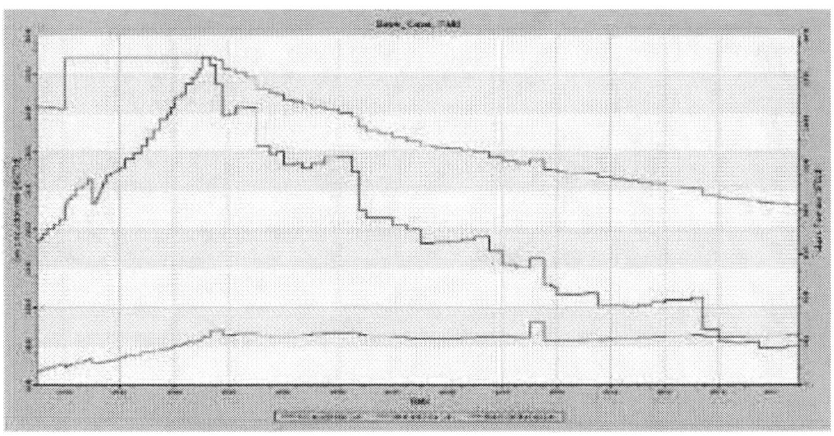

Figure 18. Production prediction profile for Base Case

Table 9. Additional wells time schedule

wells	kz-8	kz-15	kz-18	kz-26	kz-33	kz-39	kz-42
Flow Rate, B/D	500	1200	1500	500	750	500	500
Date on Production	2019/01/01	2018/01/01	2018/01/01	2019/01/01	2018/01/01	2018/01/01	2019/01/01

The plateau at 150000 STB, cumulative oil production 208 MMSTB,RF equal to 18.8% water cut equal to 21% and reservoir pressure drop equal to 730 psi as shown in figure below .

13.3- Water Injection: -

The plateau was for (6) years at 150000 STB, cumulative oil production 205 MMSTB,RF equal to 18% water cut equal to 30% and reservoir pressure drop equal to 430 psi as shown in figure below . The number of producing wells is (20) with (8) water injection wells as shown in figure below.

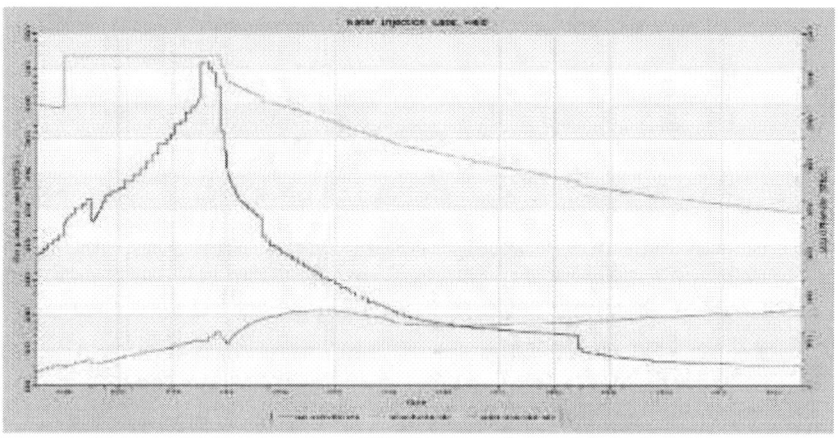

Figure 19. Production prediction profile for Water Injection Case

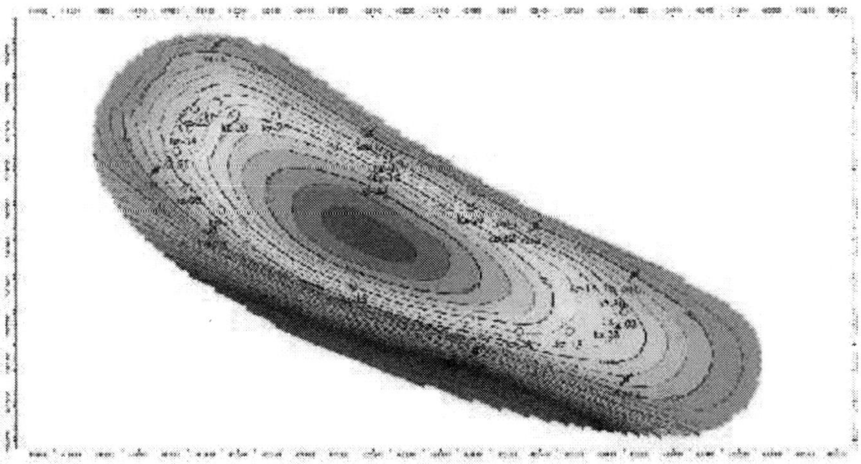

Figure 20. Contour map for top of Jeribe formation with new injection wells location Case 3

13.4- Drill New Wells

The final selected design includes: 26 production wells, 6 of which are new wells drilled the model predicts 14 years plateau at 15000 (STB/D) and cumulative oil production 233 MMSTB in 2045. The expected final recovery is 20%, water cut equal to 18% and the total reservoir pressure drop equal to 800 psi as shown in figure below.

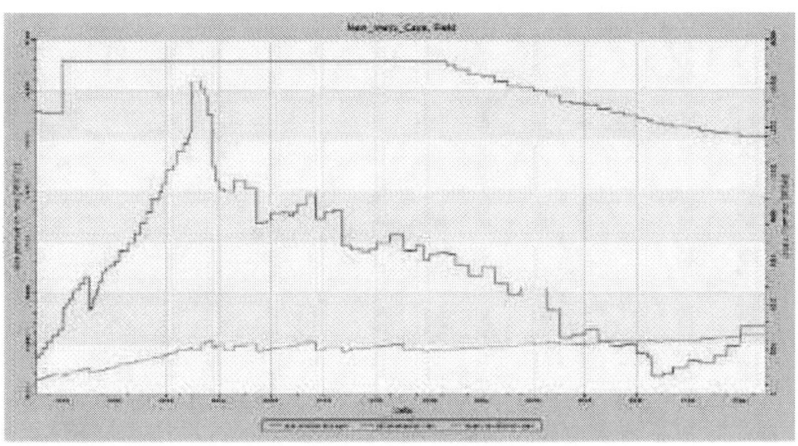

Figure 21. Production prediction profile for Drill New Wells Case

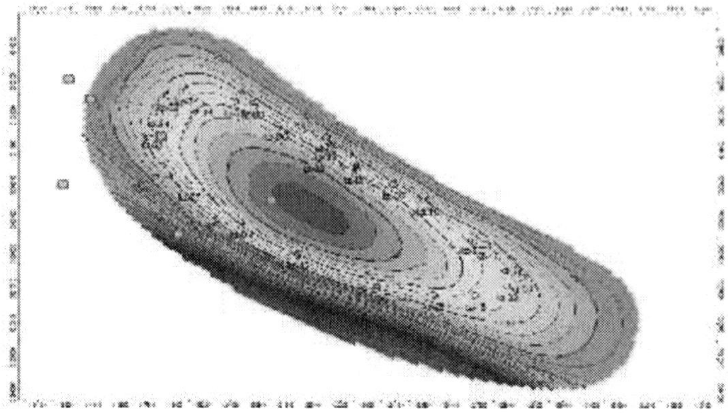

Figure 22. Contour map for top of Jeribe formation with new production wells location Case 4

14- Material Balance vs . Simulation Average Pressure

This subsection compares the average reservoir pressure versus time profile computed from ECLIPSE and MBAL models against the measured pressure at datum depth (1990 mBSL) calculated from pressure measurement surveys for wells(kz-8,kz-9,kz15 and kz-30) as shown in figure below. The average pressure calculations for MBAL model was based on yearly flow rates.

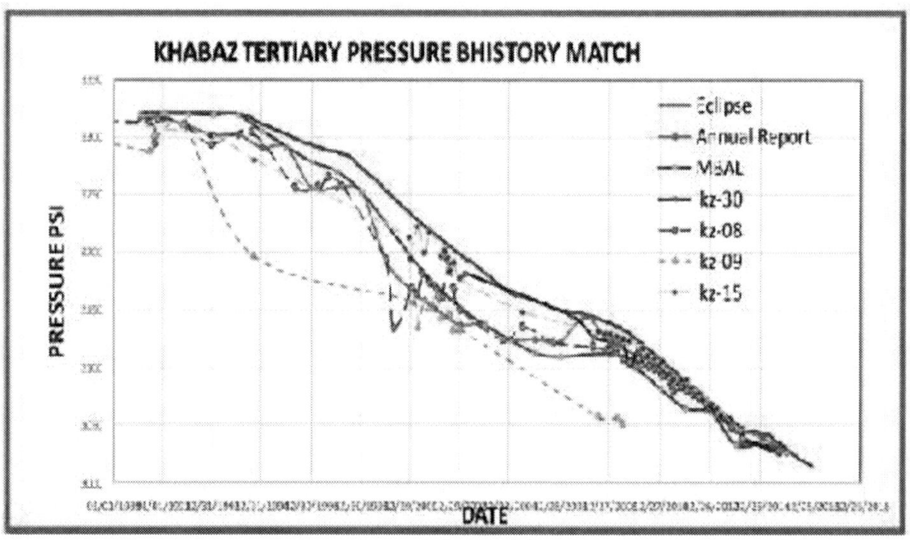

Figure 23.

The MBAL results reflect the theory underlined this model, such as, for example any pressure change is instantaneous and uniform throughout the tank. This implies that the MBAL reservoir pressure depletion associated with fluid production dropped faster compared to ECLIPSE one. The most noticeable features are:

1. The calculated reservoir pressure MBAL and ECLIPSE simulator matched perfectly after 01/2009, the reason is the pressure measurement from kz-30 which is observation wells represent stabilized static reservoir pressure, before that the discrepancy has started between the ECLIPSE and MBAL because the pressure measurement depend on wells kz-8, kz-9and kz15 were are producing wells, the shut in time was not sufficient to reach the static reservoir pressure
2. Comparably, the two calculated pressure curves from ECLIPSE and MBAL, they have the same trend.

15-Conclusions: -

1-Initial Oil in Place (1138 × 106) STB while the Initial Gas in Place reached (489 × 109) SCF, Compared with the calculated from the geological model (937 × 106) STB of oil and (610 × 109) SCF.
2-The Initial Oil in Place calculated by material balance Eq. about (900 ×10^6) STB, which is less than the value calculated by the volumetric method in the reservoir model.
3-The volumetric method estimates the STOIIP based on its static nature and net pay basis by utilizing petro physical and geological properties obtained from static model. With the knowledge of the area and the sand thickness generated from Petrel software, which takes into account the net-to– gross thickness, porosity (effective), the estimate is supposed to be same with that obtained from the MBAL. The difference observed from the estimate (about 286MMSTB) suggests the possibility of a fault which MBAL is not seeing and was not accounted for in the estimate with MBAL tool as it assumes the reservoir to be a tank.
4-Good history matching of pressure achieved in most wells, which have measurements of bottom hole pressure except for some wells and this is due to the incomplete bottom hole build up pressure.
5-Average pressure calculated at the datum level from the model was compared with the measured reservoir pressures at the same level and found they are taking same behaviour.
6-Fourth case forecast (Drill new wells) which is recommended case showed the possibility of sustaining the production of the Tertiary reservoir at a rate of (15000) STB/D for a period of 14 years, assuming that the gas / oil ratio is allowed to rise to 3000 SCF/STB and the lowest pressure on the head of the well (350) psi through (26) producing wells, With RF (20%) with cumulative oil production of (233 × 106) STB at the end of 2044.

Recommendations: -

Having concluded on results gotten, the following recommendations for application towards enhancing the estimates gotten from reservoirs for project planning and development.
1- Wells Production tests had been stopped since 2004, and production rates for wells estimated from empirical correlation from completions tests, last GOR measured for wells were in 2002, Accurate value for oil rate and producing GOR important to perform good history match.
2- Laboratory tests need on core samples focus on (Wettability Restoration) to reflect the case of the correct original wettability of reservoir rocks.
3- Very useful to do a few experiments depended on Steady State method so that a set of integration relative permeability data can be achieved in terms of the endings and curves.

Nnomenclatures

Bg: Gas formation Volume Factor
Bo: Oil formation Volume Factor
BSL: Below Sea Level
OOIP: Original Oil in Place
OGIP: Original Gas in Place

GOC: Gas Oil Contact
GWC: Gas Water Contact
Krg: Relative Permeability gas
Kro: Relative Permeability oil
Krw: Relative Permeability water
SWC: Connate water saturation
SOR: Residual oil saturation
SGC: Critical gas saturation
MBAL: Material Balance
N: Stock Tank Oil in Place
PVT: Pressure, volume, temperature.

References: -

[1] Core analysis report for wells (2,3,4,8,9,29)-North Oil Company
[2] Physical properties report for wells Kz-18 -North Oil Company.
[3] Final well Report for wells (2,3,6,8,9,10,15,17,18,19,20,22,25,26,27,28,30,32,33,34,37,39,41,42)
[4] Evaluation test report of well Kz-5.
[5] Reservoir report for tertiary and cretaceous reservoir of Khabaz field
[6] Field development research directories(1988)
[7] Applied petroleum reservoir engineering (second edition), (B.C. Craft And M. Hawkins)

IOP Conf. Series: Materials Science and Engineering **579** (2019) 012029 doi:10.1088/1757-899X/579/1/012029

Low salinity water flooding in carbonate reservoirs – dissolution effect

Y A Hadi, F Hussain and F Othman

The University of New South Wales,Sydney, Australia

Author's e-mail address: hadi.unsw@gmail.com

Abstract. One of the most significant research subjects in the industry of oil is the low salinity water injection techniques (LSWI). This is because there are possible benefits of raising recovery in oil as opposed to the conventional injection of seawater. Recent research has been done to find out the causes of less salinity with the injection of water in carbonated rocks presented significant findings. The primary mechanisms that are thought to lower salinity water injection are carbonate rocks dissolution, fines migration and alteration of wettability to more water wet. However, many researchers have elected to oppose the mechanism of dissolution and instead attributed high recovery of oil to be the primary wettability modification mechanism. Additionally, previous researchers provided no evidence characterizing about the rock and fluids produced in a detailed manner. This research work provides a detailed description of the underlying mechanisms, the core flood laboratory and the spontaneous imbibition in carbonate rocks. In the experiment, we carried out a series of steps using a limestone core with low permeability. This core was obtained from a nearby oilfield in Iraq and used it to study how dissolution and fines migration mechanisms take place. Moreover, in the process of carrying out the experiment and before its starting, SEM imaging analysis was used. In this case, as we carried out the experiment, we had to inject two-phase flow of sea water (SW), de-ionized water (DW) as well as soltrol-130. This was done in a sequential manner at 50 °C as well as 0.02 ml per minute. As the experiment progressed, the concentration of ions and pH value of the extracted water were measured in a continuous manner. The results obtained indicate that the rate of change in concentration of Ca^{2+} ions is directly proportional to the effluent pH value. By just balancing the mass, we were able to easily see the rate at which limestone dissolved as the experiment progressed. The maximum increase in pH was registered when fresh water was injected (pH=9.2). An increase in pH value can have a positive effect on altering wettability in that it can increase the alkalinity of flood water. To notice fines migration, the Scanning Electron Microscope (SEM) and Energy Dispersive Spectroscopy (EDS) are used to scan the images of the inlet and outlet of the core plugs after completion the experiment. This method can indicate that dissolution can induce fines dislodgement for the surface of the rock. The two mechanisms of fine migration and carbonate rock dissolution can significantly increase recovery of oil and at the same time inject water low in salinity.

IOP Conf. Series: Materials Science and Engineering **579** (2019) 012029 doi:10.1088/1757-899X/579/1/012029

1. Introduction

One of the mechanisms of altering wettability in carbonate reservoirs, which has improved oil recovery techniques, is low salinity water injection (LSWI). Based on increased efficiency level in the displacement of medium and light gravity crude oils, it has become very popular. Other than for this reason, water availability and affordability, low capital and low cost have raised its popularity, which eventually leads to its favorability compared to other methods of IOR or EOR.

However, this technique has a significant challenge when it comes to water disposal and water sourcing.

According to Al-Shalabi et al. (2017) [1], other names that are used for this technique of water injection is LoSalTM by BP, Smart Water Flood used by Aramco Saudi Arabia, Shell's Designer Waterflood and ExxonMobil's Advanced Ion Management (AIMSM). Following the successful injection of low salinity water in carbonate rocks, research has been taken a notch higher, both in the laboratory and in the field scale. This technique has recently been implemented on sandstone rocks. Statistics have it that carbonate oil reservoirs are greater than 50% [2] . This has necessitated the investigation of the validity of LWSI in carbonate rocks. Previously, water-wetting of carbon rocks was done at the deposition age and oil-wetting applied to the aging oil. Moreover, approximately 90% of limestone is either intermediate in wetness or oil-wet [2]. Several factors have to be considered when a rock is being transformed from water-wet to oil-wet. These factors include mineralogy of the rock, base and acid number of the crude oil, the composition of brine in it and the pH value with the reservoir's temperature being no exception.

Carbonate rocks share one unique characteristic, which is their low oil recovery in comparison to the sandstone rocks, which have a system that is oil-wet. This means that they have a lower permeability to oil compared to water-wet rocks. It is not possible to significantly increase oil content in a reservoir by injecting more water into the carbonate rock in a secondary recovery stage, not having undergone the initial treatments of pressure enhancement. This, therefore, shows how important the investigation of the effect of saline brine is. According to (Al-Shalabi et al., 2016) [3], several methods can be followed to lower the salinity of the water, and these are; brine dilution, a decrease in the concentration of cations and the elimination of non-active ions such as Cl- and Na+ . Compared to other methods such as Enhanced Oil Recovery (EOR), reduced water salinity has several advantages and these are; first, it is readily available and affordable. Secondly, it is not difficult to inject water into reservoirs. Thirdly, the cost of operation and the capital required are also low. Fourthly, there is efficiency in the displacement of the medium to light gravity of oil. Fifthly, this technique can ultimately recover oil at a lower cost compared to the EOR method.

Finally, the approach can be easily applied at initial and later stages of the reservoir's life. Morrow (1999) and Lager et al. (2006) [4] argued that the LSWI in limestone lacks validity, because limestone is low in terms of the amount of clay it contains. Additionally, heterogeneity of carbonate rocks makes it difficult to apply LSWI. According to Austad el al. (2005) [5], several studies have been conducted in the field as well as in the laboratory and the majority of these studies have confirmed and given positive feedback on low salinity injection. In fact, these findings have seen their translation in oil recovery in the secondary as well as the tertiary injection modes. In our laboratory, we presented in this work dissolution and fines migration. Moreover, core preparation, experimental procedure, setup and the results of two phase flow, including pH values, effluents concentration, oil recovery and pressure drop, are reported. SEM-images and the EDS-analysis are reported as the last part of the results, as discussed in the sections below:

IOP Conf. Series: Materials Science and Engineering **579** (2019) 012029 doi:10.1088/1757-899X/579/1/012029

2. Core Preparation

The core preparation is one of the most crucial parts of the research because cores are usually heterogeneous and have a low permeability of contents of dead crude oil reservoirs. In the cleaning section, we ensured that pumping solvent was avoided through the cores as it may lead to alteration of the interior structure of the pore of the cores as well as fines migration may occur courtesy of drag force emanating from the injected solvent. Therefore, to ensure that this trend does not recur, the Soxhlet method was utilized as a method of cleaning. To clean the core plugs, both Toluene and Methanol were used. So, the core plugs were cleaned in the regular technique with Toluene and Methanol. Anderson (1986) [6] stated that the effectiveness of Toluene is that it can remove hydrocarbons amongst them asphaltenes and other weak polar compounds. On the contrary, stronger polar methanol (CH_3OH) can eliminate the compounds responsible for the alteration of wettability. In this regard, we used Toluene before the emergence of transparent and clean solvent, which emerged in the apparatus that was used. We also combined the solvent with methanol up to such a point when another explicit solvent reappeared. This process was repeated for both methanol and toluene. The process of cleaning took close to one month as we had to ensure that the cores were cleaned and ready for the experiment even though the core had low permeability.

3. Procedure of the Experiment

From our Iraqi field of Rumali, it was obtained a core of heterogeneous limestone with an average of absolute permeability 0.86 mD as well as porosity of 15.67%. In the absence of clay, from the XRD and XRF results It was found out that the limestone contents were made up of 3.4% dolomite $CaMg(CO_3)_2$ and 96.6% Calcite $CaCO_3$ although these measurements are a deficit of clay. The core had original dimensions of 60.44 mm in length and 2.5 mm in its diameter. Because the sample length was limited regarding utilizing the Scanning Electron Microscope Instrument, the core was cut to half. The two plugs are used in the experiment. From most of the laboratory data available, it is not abnormal to find that the core has selective permeability. Table-1 list the core petrophysical properties are articulated after carrying out experiment. Both soltrol and fluid properties used during the experiment listing in tables 2 and 3.

Table-1, Limestone core plugs used in the experiment.

Core Plug No.	Diameter (mm)	Length (mm)	Porosity (%)	Pore Volume (ml)	Absolute Sea Water Permeability (mD)	Swirr	Sor @ SW	Sor @ DW
1	25.5	29.306	15.165%	2.261	0.86	0.31	0.318	0. 195
2	25.5	31.136	16.172%	2.574	0.86	0.31	0.305	0.194

Table-2, Oil properties used in the experiment

Soltrol Properties	
Viscosity (cp) @ 25°C	1.176
Density (gm/cc) @ 25°C	0.7608

Table-3, Composition of SW and DW.

Ions \ **Brine**	**Sea water (mg/l)**	**de-ionized water (mg/l)**
KCL	860	N/A
$MgCl_2$	5,875	N/A
Na_2SO_4	4,880	N/A
$CaCl_2$	1,385	N/A
NaCl	31,580	N/A
TDS	44,580	N/A
pH	5.85	6.5
Viscosity at 50°C (cp)	0.634	0.548
Density (km/m^3) at 50°C	1032.7	987.683

4. Scanning Electron Microscope (SEM)

Having completed the cleaning step, it was time to take the Scanning Electron Microscope images for the composite core's inlet and outlet. The division of the interface took place realizing four quarters where each quarter had many SEM images selected for it. SEM images from before and after the experiment were compared to study the dissolution impact on the changing geometry of the rock. This can also help in the examination of the possibility of fines migration induced by the dissolution of limestone.

5. Experimental Setup

In the investigation of dissolution mechanism that results from low salinity water, de-ionised water and sea water were used. In a bid to achieve better and higher oil recovery; de-ionised water is used to be the best source of water based on its richness for magnesium, calcium and sulfate ions, which are potential determinants towards carbonated rocks. To obtain higher oil recovery such as wettability alteration, these ions are crucial although not for the dissolution mechanism. Therefore, we prepared sea water synthetically in the lab and utilized it to flood composite core and added de-ionized water (DW). The composition of seawater and de-ionized water are listed in table 3. Then limestone core plugs were saturated with sea water using a Desiccator and later left them in the sea water for 16 hours to allow for full saturation before the commencement of the experiment. Pore volume and porosity of the cores were measured by using the weight difference between wet and dry core. We conducted the experiment at 122 °F reservoir temperature with 0.02 cc/min rate of injection and confining pressure set at 300 psi. 44,580 mg/l total dissolved solids (TDS) of sea water was used in the experiment whereas the DW has zero TDS as shown in table 3. This research was carried out with the aim of examining the dissolution mechanism while the injection of two phase flow of Soltrol-130, high salinity and low salinity water are being tested. In our experiment, we ensured that at the outlet no filter was used in a bid to make sure that the core was the sole origin of the fines and ended up in the effluent container. Moreover, this was to ensure the concentration of ions in the effluent, which typically happens while core flooding takes place. We then used a water pump to flood the core composite with a low flow rate of about (0.02 cc/min) because the core is less permeable. In ensuring that water vaporization was avoided, the experiment was carried out at a temperature of 50 °C and to incorporate the effect of temperature; the reservoir temperature was mimicked. There was a drop in pressure, and we ended up setting final pressure at 300 psig. The figure 1 shows how the experiment was set up. There are two pumps connected into the Hassler core holder and these pumps were controlled by valve in order to use one for the water injection and the other for soltrol injection. Initially sea water was injected into the core plugs which were mounted in a Hassler core holder at least 10 PV and also 20.16 psi pressure drop across the core. Then, the absolute permeability of the composite core was calculated by the pressure drop. Soltrol was then injected at 50 ˚C with low flow rate (0.02 cc/min) by using oil pump in order to establish the irreducible water saturation experimentally. After that, brine was flooded into the core to 2 pore volumes at the same rate and temperature but using the water pump. We also sampled the effluent after an interval of breakthrough and measured the pH instantly. A closed lid was used to test the effluent to ensure that the ion concentration does not change as water vaporizes. It is important to note we stirred brine with evacuation to make sure that there was no air bubble retained in the brine. Accordingly, that was done to ensure minimization of carbon dioxide in the core before flooding the core. Postma and Appelo (2004) [7] argued that CO_2 can increase dissolution of the calcite as it can form carbonate acid through lowering its pH. Next, repeating the soltrol step with the same flow rate and pore volume injected to achieve the same initial water saturation. Then, we switched to de-ionized water flooding so the residual oil saturation was introduced at distilled water. We measured the volumes of produced oil and water at time. The oil recovery factor was measured as % OOIP. The next step also entailed analysis of effluent water samples to quantify pH and the concentration of Ca^{2+}, Mg^{2+}, SO_4^{2-}, Cl^- and Na^+ ions using the ion chromatography and inductively coupled plasma optical emission spectrometry (ICPOES). Soxhlet method was again used as cleaning method to remove the soltrol from the composite core by using 100% isopropanol for four hours only. The core plugs were then dried at 60 ˚C in the oven for 42 hours. The final step took images of SEM and EDS for the outlet and the inlet of the composite core. These were done after seawater injection was completed as well as injection of oil and de-ionized water. Figure-1 shows the experimental setup for the recovery.

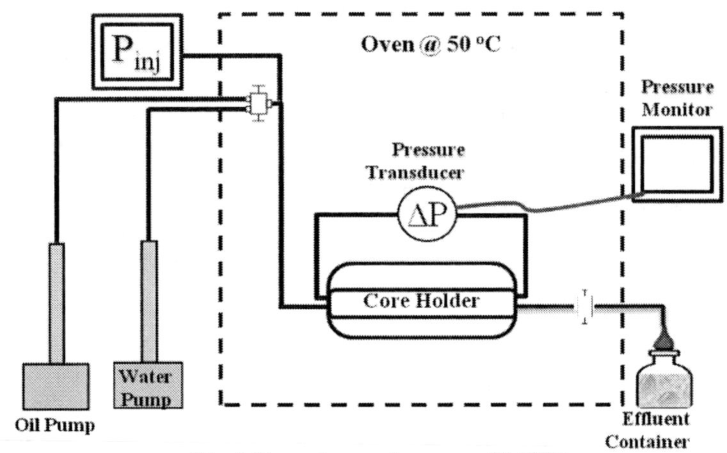

Fig.1 Experimental setup of LSWI.

6. Results and Discussion

pH, effluents concentration, oil recovery and pressure drop results are shown in Figs. 2 – 9 for both the sea water and de-ionised water.

6.1 pH Measurements

The measurements of pH, active and non-active ions concentrations in the effluent and pressure drop curve examined the dissolution of calcite in this experiment. From figure-2 initially we have only one point due to two phase flow. In two phase flow when we did waterflooding, initially we only produce oil and to measure pH. We need to get a sample of water. During the sea water imbibition, the produced water was collected after breakthrough so in two phase flow experiment we cannot have lots of pH measurement. Moreover, the initial pH of the produced sea water recorded around 7.8 which was higher pH value than the injected SW. It can be clearly seen that $CaCO_3$ react with water throughout the core and kinetic dissolution of it was happening. In contrast, the pH went up to 9.2 through low salinity waterflooding but at the end of de-ionized water injection slightly decreased to around 8.8 which was not that much difference. This difference might be due to the dissolution was happening more at the beginning. According to the following equilibria

$$\mathbf{CaCO_3 + H_2O \leftrightarrow Ca^{+2} + HCO_3^{-} + OH^{-}}$$

Calcite ($CaCO_3$) reacts with water, such reaction makes Ca^{2+} with OH^- and OH^- is the reason why pH is going up as well as the reason why we produce more cations, especially Ca^{+2} in the effluent due to calcium hydroxide is liquid. As a result, effluent pH of de-ionized water was higher than the effluent of SW due to active and non-active ions were presented in the injected SW. Sea water might illustrate increasing in the rate of kinetic dissolution.

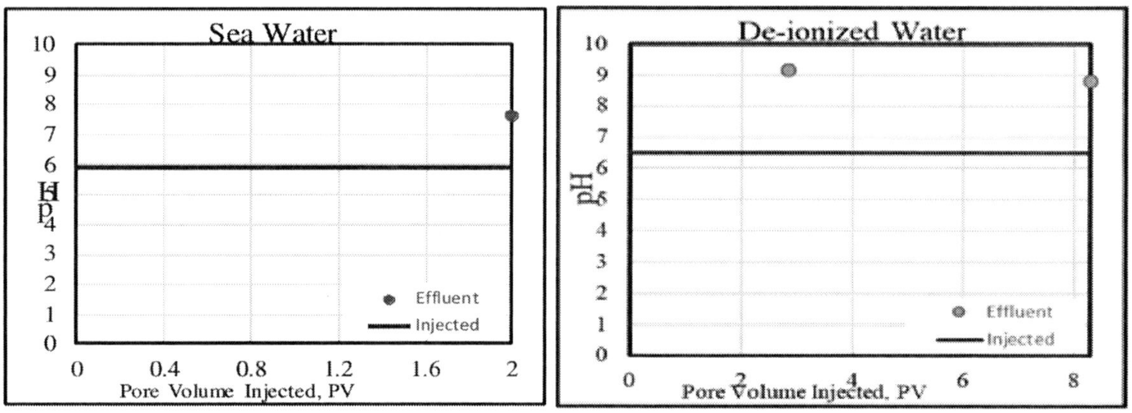

Fig.2 Pore volume injected Vs pH of SW&DW production.

6.2 Effluent Concentration Data

The results obtained from the effluent concentrations for Ca^{2+}, Mg^{2+}, SO_4^{2-}, Cl^- and Na^+ are given in figures 3-7. Furthermore, cations especially Ca^{2+} and pH have increased in the effluent due to the dissolution. The concentration of Ca^{2+} in the effluent is recorded after breakthrough for both flooding SW and DW that is given in Fig-3. In the sea water there are some dissolution inhibitors that work to decrease the rate of dissolution in the limestone reservoirs like SO_4^{2-}, Mg^{2+}, Si, Sr, Ba, PO_4^{3-}, metal ions and Ca^{2+} by itself (Morse and Arvidson, 2002) [8]. Preventing the interaction between the seawater and the surface of calcite molecules could have happened when the metal ions are presented. However, The Ca^{2+} ion concentration in the effluent of sea water is recorded higher than that of de-ionized water due to the higher rate of dissolution in the sea water. Hence, when the initial Ca^{2+} ion and/or high salinity conducted, the kinetic of calcite dissolution takes place faster. It can be seen that the increase of pH in the effluent of low salinity water flooding could pertain not only to the dissolution but to other interaction took place inside the rock and/or some CO_2 consumption related to the de-ionized water which was not vacuumed before flooding into the composite core. The difference between injected and effluent value of Ca^{2+} ion for SW was (150 mg/l) whereas in DW recorded less than the sea water by almost double (Fig-3). It can be concluded that dissolution in carbonate rocks occur higher during high salinity waterflooding than that in low salinity waterflooding. From Fig-4, the concentration of Mg^{2+} ion in the effluent and the injected one was identical in the sea water due to the small percentage of dolomite which represents only 3.4% in our sample. We did not have seen any change in Mg^{2+} ion concentration throughout high salinity water injection. Therefore, dolomite dissolutions definitely did not happen during high salinity injection. In contrast, in the low salinity the injected concentration of Mg^{2+} was zero but we still produced a lot of Mg^{2+} because there is dissolution of Mg^{2+}. An anion might adsorb into the surface of rock at the same time reduce the positive charge of carbonate surface so it will increase the electrostatic attraction. From Figs-5 and 6 during injected sea water more SO_4^{2-} ion was reacted with the negatively charged carboxylic acid group at the carbonate surface. This reaction between an anion and carboxylic group leads to releasing several of the oil from the surface of carbonate as well as changing the wettability of rock to more water wet. For this reason, the concentration of sulfate and chloride ions in the effluent for high salinity was lower 10 times than the injected level. Therefore, the dissolution of calcite might increase during the sea water injection.

We have seen that Ca^{2+} and pH have increased in the effluent of SW due to the dissolution of calcite whereas SO_4^{2-} and Cl^- concentrations decreased. According to Zhang et al. (2007) [9], reducing of SO_4^{2-} and Cl^- concentrations into 10 times of the injected level caused by the existence of Mg^{2+} which is interacted with SO_4^{2-} at the early stage. It can be concluded that improving oil recovery was because of both the adsorption of ions especially SO_4^{2-} on the carbonate surface and dilution of the injected fluid.

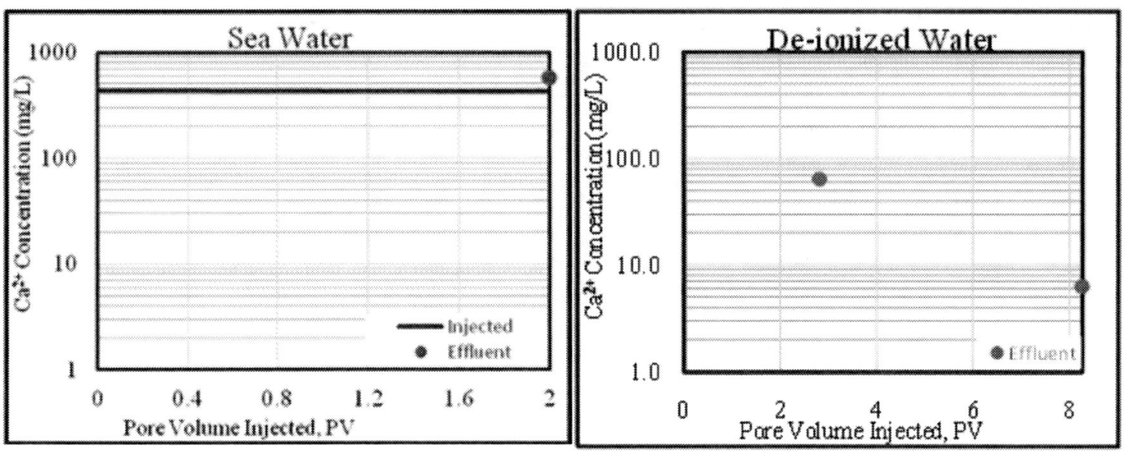

Fig.3, Effluent ion concentration for Ca^{+2} samples of SW & DW production

IOP Conf. Series: Materials Science and Engineering **579** (2019) 012029 doi:10.1088/1757-899X/579/1/012029

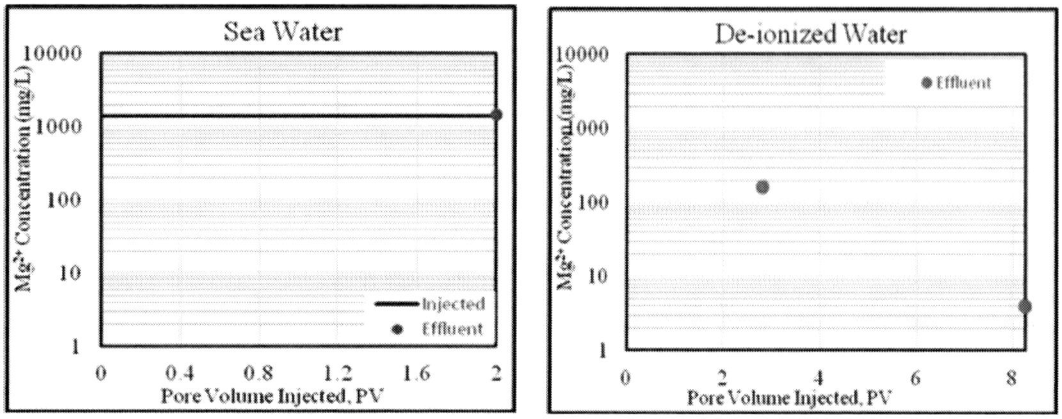

Fig.4, Effluent ion concentration for Mg^{+2} samples of SW & DW production.

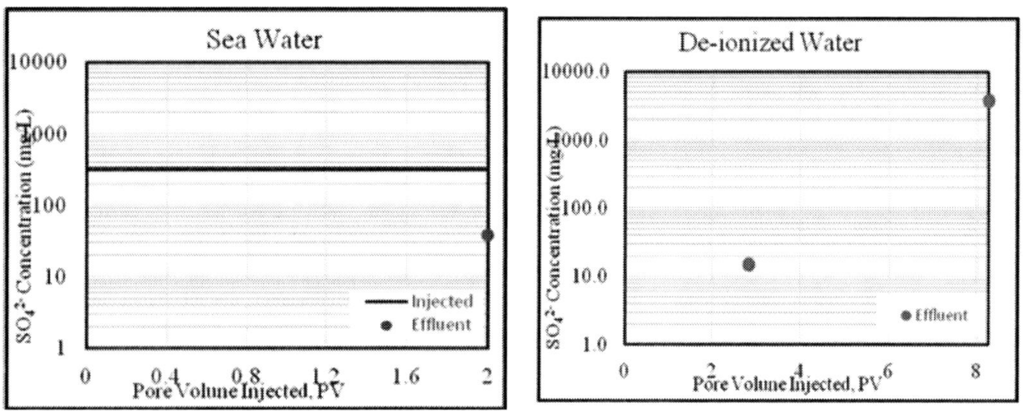

Fig.5, Effluent ion concentration for SO_4^{2-} samples of SW & DW production.

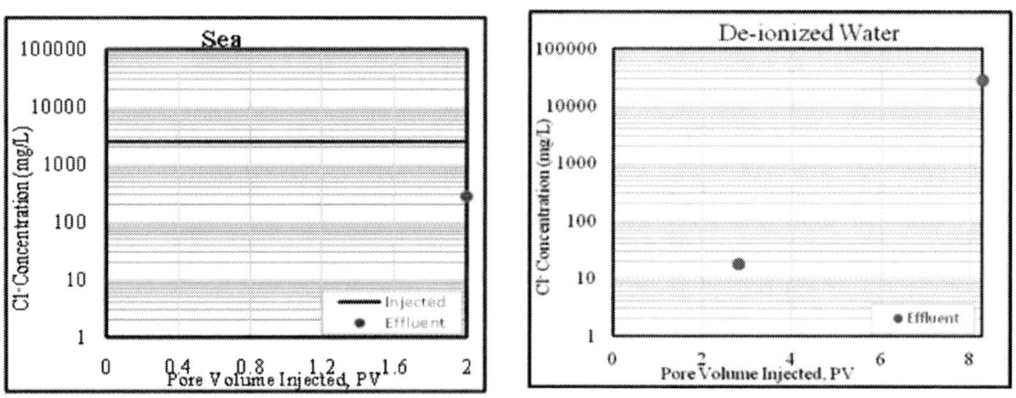

Fig.6, Effluent ion concentration for Cl^- samples of SW & DW production.

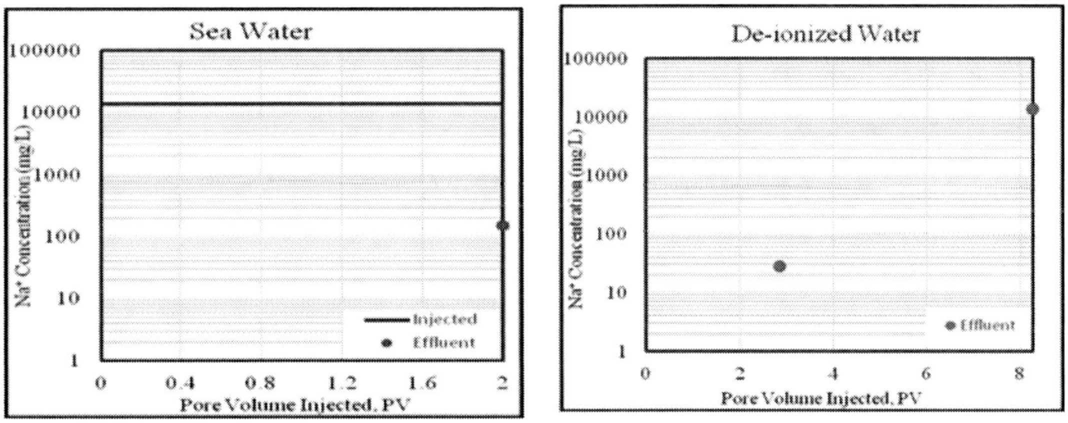

Fig.7, Effluent ion concentration for Na^+ samples of SW & DW production.

6.3 Oil Recovery

The best results of oil recovery data for both high salinity and low salinity waterflooding are given in Fig-8. After flooding 2 PV of sea water, the recovery was recorded as 37% OOIP, while a higher cumulative oil recovery was recorded 50 % OOIP after flooding about 8.26 PV of de-ionized water. There is around 13% difference corresponding to low salinity water injection. There are many reasons why recovery factor different? Initially the increase in pH was significantly higher in the de-ionized water injection than that of in sea water injection because of the dissolution of calcite. Another important reason is that due to the relative permeability is different and the residual oil saturation was less in low salinity by almost 12% than in high salinity. Therefore, we are improving water relative permeability by fines migrations which only happen in water so the oil relative permeability should be decreased in the low salinity waterflooding. As we used heterogenous composite rock in our experiment, the injected water goes to high permeable channels so the low permeable channels remain unsweep. When we reduce the salinity of the injected water, the water go to low permeable channels.

This would improve volumetric sweep efficiency. Hence, the oil recovery also would increase by decreasing the salinity of injected fluid.

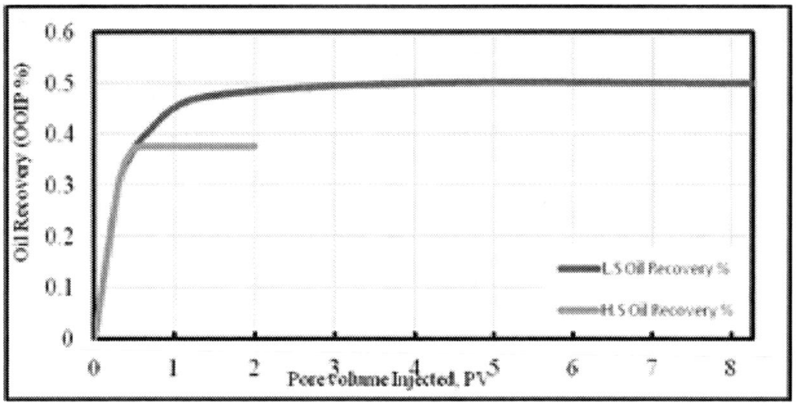

Fig.8, Oil recovery for Sea water and De-ionized water.

6.4 Pressure Drop

The figure 9 down shows the measurements of pressure drop that were done throughout the core plugs with the pore volumes receiving an injection from sea water and de-ionized water. For sea water, the pressure drop decreased and almost stabilized at 11 psi. However, it went up to 22 psi in de-ionized waterflooding. This indicates that fines migration have taken place due to dissolution as shown by the SEM-images. Furthermore, the difference of pressure drop between high salinity and low salinity was almost double. So that's mean relative permeability to water become half which should get better oil recovery. Eventually, despite the increase in pressure, it can be interpreted to be emanating from fines migration and dissolution. Pressure drop decrease in high salinity by dissolution can be compensated by increase in pressure drop by fines migration at low salinity.

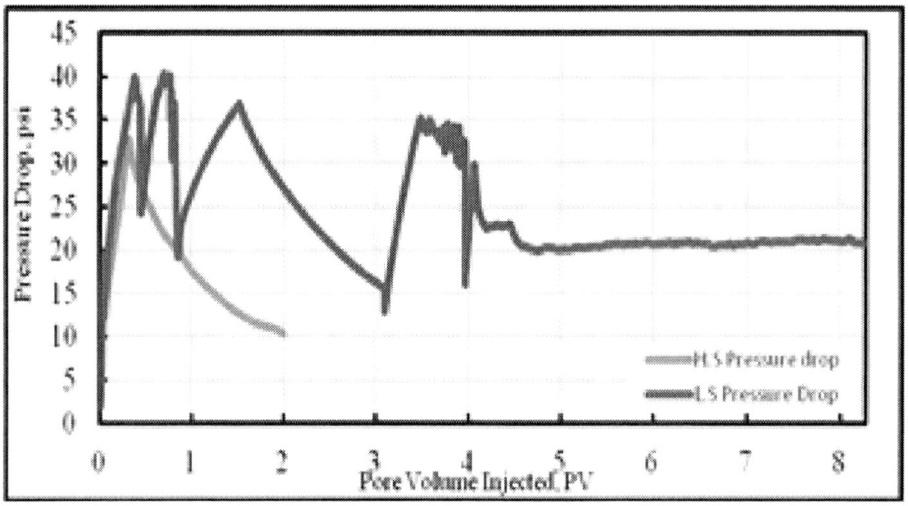

Fig.9, Pressure drop for Sea water and De-ionized water.

6.5 SEM and EDS images

There were about 150 SEM and EDS images of both interfaces of outlet and inlet of the composite core. It is important to note that in most of these pictures, fines migration was detected. However, it was tough to identify and figure out the calcite and dolomite dissolution prior and after the flooding by EDS. As opposed to the discovery of Zahid et al. (2012) [10], dissolution mentioned early was confirmed using the experimental measurement of pH, Mg^{2+} and Ca^{2+} ions concentration in the samples of effluent. From these SEM and EDS images, we have managed to find out some fines migrations, which probably might have happened due to dissolution and therefore occurrence of fines dissolution from the rock. It is important to note that when fines migration occurs, the pressure drop is registered as the fines block some of the pores diverting the flow to a less permeable area. In comparison, pressure drop should be on the decline due to dissolution. Because these phenomena were evident in our experiment, we noticed a significant decrease in pressure. This is associated with enhanced pore throat connectivity. Pressure drop made significant effect and thus cannot be interpreted in an unambiguous manner, because dissolution and fines migration can compensate each other.

When it comes to the recovery of oil, both fines migration and dissolution may end up resulting in a rise in oil recovery in a significant manner. Fines migration blocks the throats of the large pores representing the primary flow of the fluid diverting it to an area with less unswept permeability.

Additionally, dissolution can restore an original water-wet carbonate system before aging of oil dissolves and removes the oil-wet layer on the surface of the carbonate rock [11].

7. Conclusion

With the above findings, the following conclusions can be drawn.

- Both sea water and de-inoized water have the capacity to dissolve carbonate rock as it was evident when the concentration of Ca^{2+} and pH increased. The higher the salinity rate, the higher the rate of dissolution, although to some degree the dissolution of rocks stops after attaining the equilibrium.
- When dissolution induces pH increase, it can have positive effects on how wettability is altered to a more water-wet through conversion of the flood to alkaline, therefore, higher oil recovery was noticed.
- The rate of dissolution of carbonate is a gradual process and equilibrium may not be attained when kinetic dissolution takes place, even at a slow rate of flow (0.02 cc/min).
- Dissolution of carbonate may induce fines dissolution from the surface of the rock and on fines migration as shown in the SEM and EDS images.
- The drop increased in pressure, because of fines migration which was improving water relative permeability.
- Both the dissolution of the rock and the fines migration that emanate from dissolution induction may significantly affect the oil recovery from the carbonate rocks. Fines migration sometimes blocks the throats of the large pore and changes the flow to an area with less permeability. Moreover, the dissolution may end up rendering the surface of the rock as more water-wet. Both mechanisms, however, have the capacity to attain higher recovery of oil.

Appendix

SEM and EDS images:

The Inlet Interface

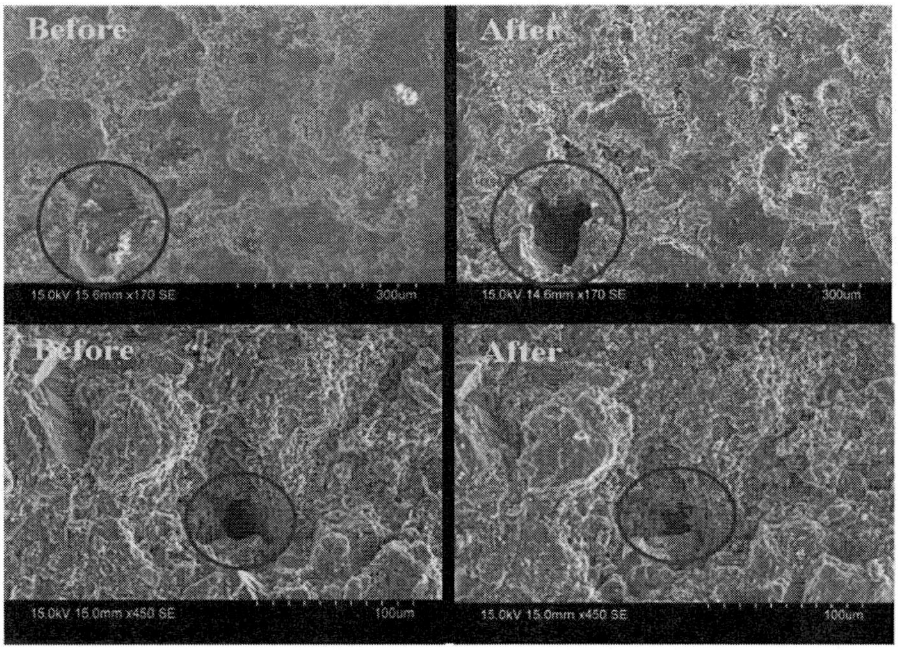

IOP Conf. Series: Materials Science and Engineering **579** (2019) 012029 doi:10.1088/1757-899X/579/1/012029

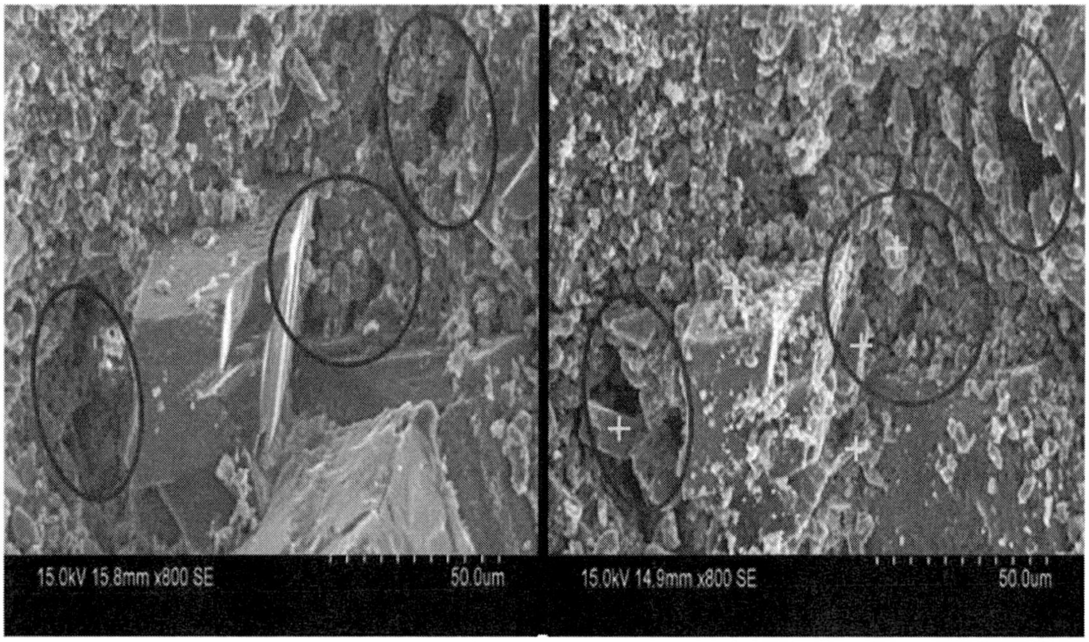

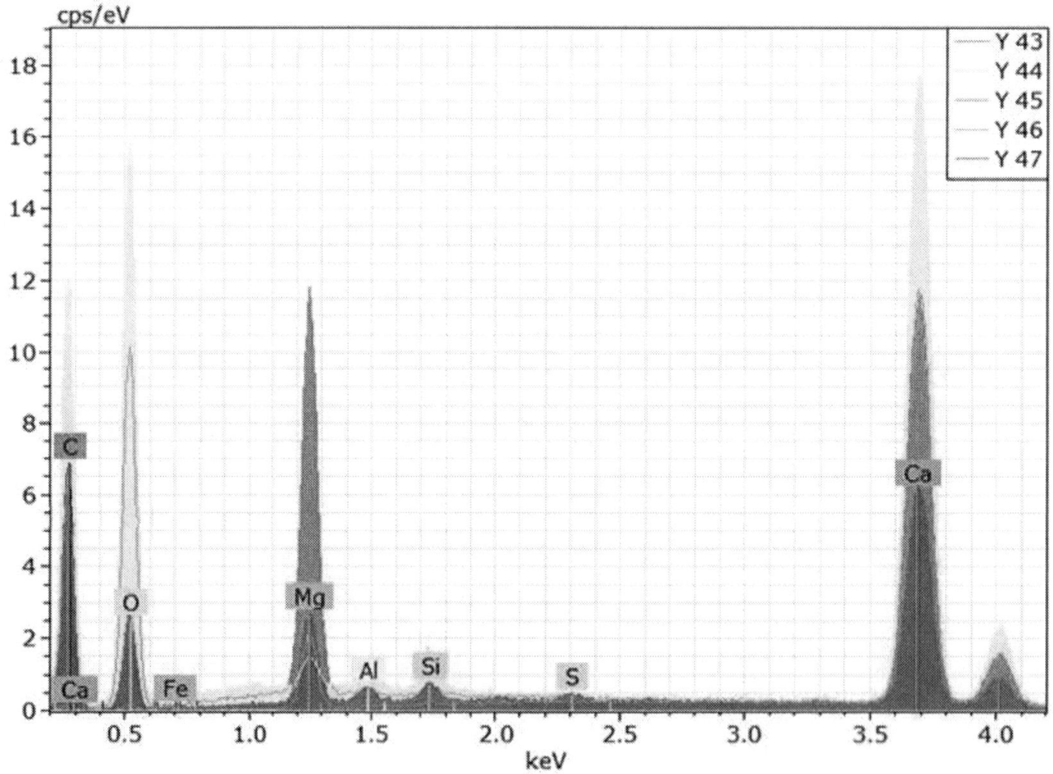

The

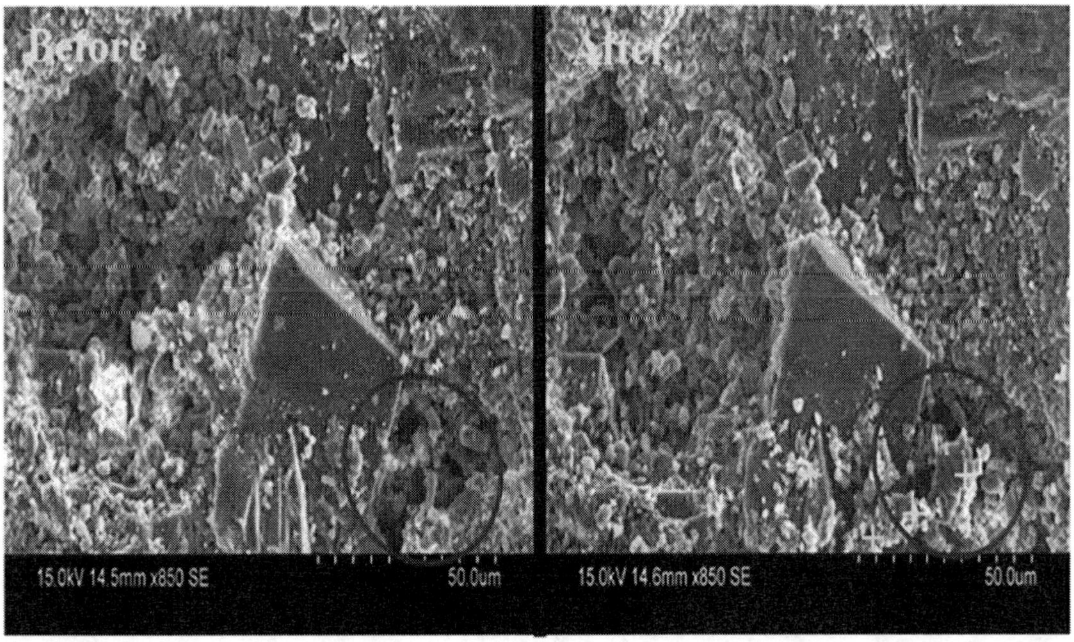
Before
After
15.0kV 14.5mm x850 SE
50.0um
15.0kV 14.6mm x850 SE
50.0um

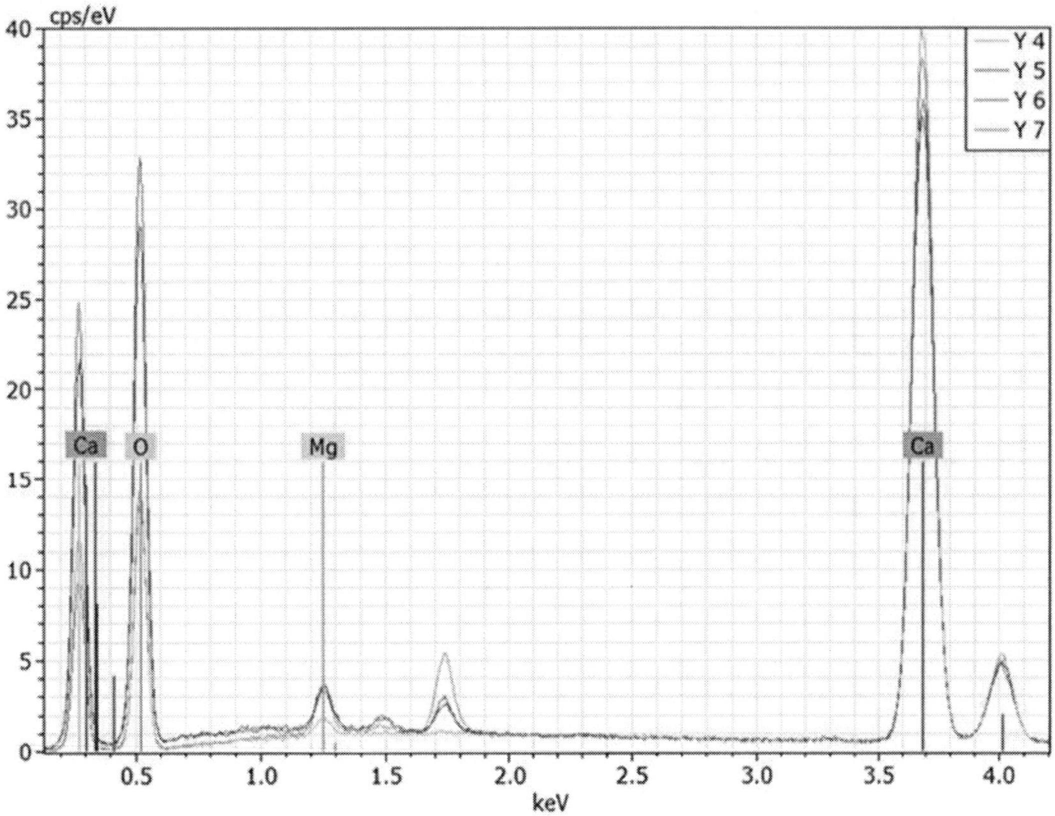
cps/eV
40
35
30
25
20
15
10
5
0
Y 4
Y 5
Y 6
Y 7
Ca
O
Mg
Ca
0.5
1.0
1.5
2.0
2.5
3.0
3.5
4.0
keV

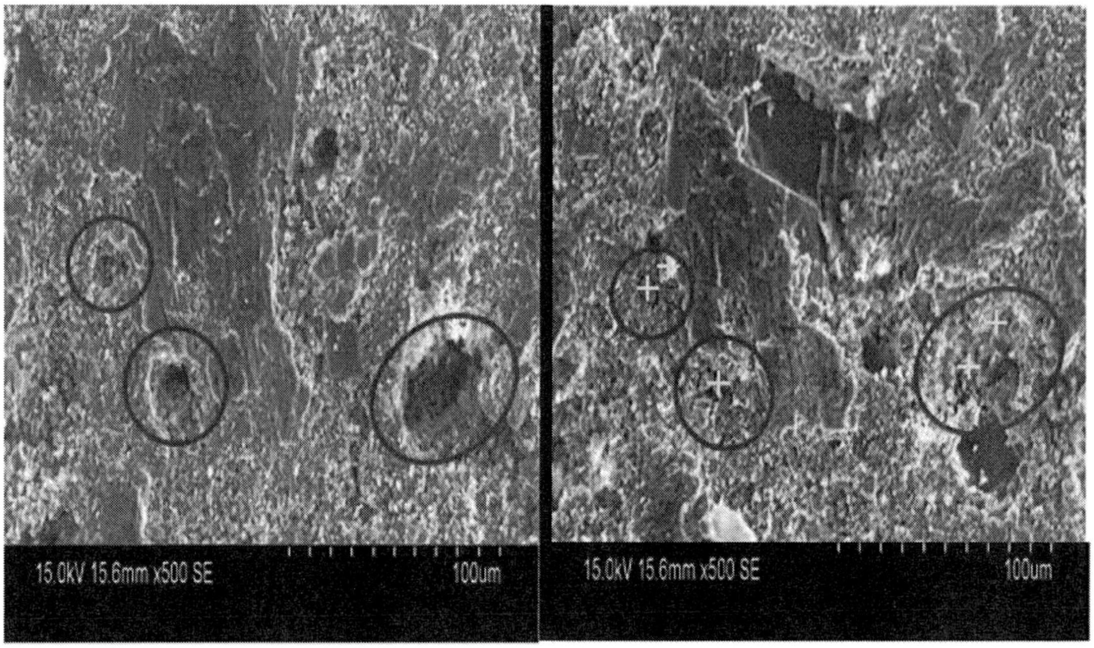
15.0kV 15.6mm x500 SE
100um
15.0kV 15.6mm x500 SE
100um

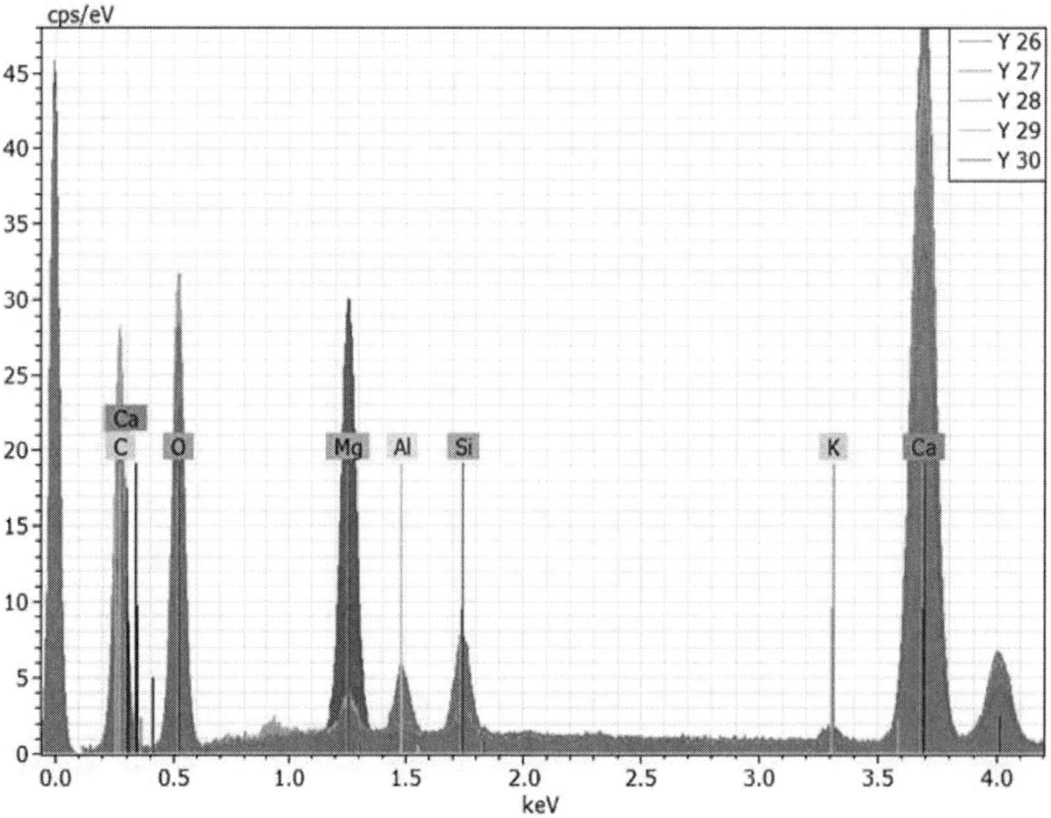
cps/eV
Y 26
Y 27
Y 28
Y 29
Y 30
Ca
C
O
Mg
Al
Si
K
Ca
keV

List of a

Abreviations:

LSWI	Low Salinity Water Injection
EOR	Enhanced Oil Recovery
SW	Sea Water
DW	De-ionized Water
H.S	High Salinity
L.S	Low Salinity
SEM	Scanning Electron Microscope
EDS	Energy Dispersive Spectroscopy
EWI	Engineered Water Injection
OOIP	Original Oil in Place
PV	Pore Volume
TDS	Total Dissolved Solids
So	Oil Saturation
$CaCO_3$	Calcite
$CaMg(CO_3)_2$	Dolomite
Ca^{2+}	Calcium
Na^+	Sodium
Mg^{2+}	Magnesium
SO_4^{2-}	Sulfate
Cl^-	Chloride
Sor	Residual Oil Saturation
Swirr	Irreducible Water Saturation

References:

[1] Al Shalabi E W and Sepehrnoori K *Low salinity and engineered water injection for sandstone and carbonate reservoirs* (Gulf Professional Publishing, 2017)

[2] Akbar M, Vissapragada B, Alghamdi AH, Allen D, Herron M, Carnegie A, Dutta D, Olesen JR, Chourasiya RD, Logan D and Stief D, 2000 A snapshot of carbonate reservoir evaluation Oilfield Review *International petroleum technology conference* 20-21

[3] Al-Shalabi E W and Sepehrnoori K 2016 A comprehensive review of low salinity/engineered water injections and their applications in sandstone and carbonate rocks *Journal of Petroleum Science and Engineering* **139** 137-161

[4] Tang G Q and Morrow N R 1999 Influence of brine composition and fines migration on crude oil/brine/rock interactions and oil recovery *Journal of Petroleum Science and Engineering* **24** 99-111

[5] Austad T, Strand S, Høgnesen E and Zhang P 2005 Seawater as IOR fluid in fractured chalk *SPE international symposium on oilfield chemistry*

[6] Anderson W G 1986 Wettability literature survey-part 1: rock/oil/brine interactions and the effects of core handling on wettability *Journal of petroleum technology* **38** 1,125-1,144

[7] Appelo C A J and Postma D *Geochemistry, groundwater and pollution* (CRC press, 2004)

[8] Morse J W and Arvidson R S 2002 The dissolution kinetics of major sedimentary carbonate minerals *Earth-Science Reviews* **58** 51-84

[9] Zhang P, Tweheyo M T and Austad T 2007 Wettability alteration and improved oil recovery by spontaneous imbibition of seawater into chalk: Impact of the potential determining ions Ca2+, Mg2+, and SO42−," *Colloids and Surfaces A: Physicochemical and Engineering Aspects,* **301** 199-208

[10] Zahid A, Shapiro A A and Skauge A 2012 Experimental studies of low salinity water flooding carbonate: A new promising approach *SPE EOR conference at oil and gas West Asia*

[11] Al-Shalabi E W Sepehrnoori K and Pope G New mobility ratio definition for estimating volumetric sweep efficiency of low salinity water injection *Fuel* **158** 664-671

Enhanced Recovery of Heavy Oil Using A Catalytic Process

M M Hasan[1#] and S P Rigby[2#]

[1]Department of Petroleum Technology, University of Technology, Baghdad, Iraq.
[2]Department of Chemical and Environmental Engineering, University of Nottingham, University Park, Nottingham NG7 2RD, U.K.

[#]Corresponding author's e-mail address: muayad1999@yahoo.com , Sean.Rigby@Nottingham.ac.uk

Abstract. Oil is a major source of energy around the world. With the decline of light conventional oil, more attention is being paid to heavy oil and bitumen, as a good alternative to light oil for energy supplies. Heavy crude oils have a tendency to have a higher concentration of metals and several other elements such as sulfur and nitrogen, and extraction of these heavy oils requires more effort and cost. Toe-to-Heel Air Injection (THAI) is a novel process of enhanced heavy oil and bitumen recovery and upgrading. In this technique, horizontal well concepts are integrated with the reactions of high temperature oxidation to achieve a potentially high recovery ratio. Since the process works through a short distance displacement technique, the produced oil flows easily toward the horizontal producer well. This direct mobilized oil production and short distance are the major features of this method which lead to robust operational stability and high oil recovery. This technique gives the possibility of a higher recovery percentage and lowers environmental effects compared to other technologies like steam based techniques. A novel well design consisting of two horizontal injectors and two horizontal producers was used in different well configurations, to investigate the potential for improved efficiency of the THAI process on the heavy oil recovery. A 3D dimensional model, employing the CMG-STARS simulator, was applied for this process. Two horizontal injectors and producers were utilised in this project, instead of the conventional horizontal injector and producer used in the Greaves model (the base case model), to investigate the effect of the extra injector and producer on the performance of the THAI process. It was found that the locations of the well injections and the well productions significantly affected the oil production. It was found that the amount of the produced oil rose up to about double of the amount of the oil produced by the original model. For the study of the effectiveness of the catalysts in the oil upgrading process, the CAPRI technique has been simulated to investigate the effect of several parameters, such as catalyst packing porosity, the thickness of the catalyst layer and hydrogen to air ratio, on the performance of the CAPRI process. The TC3 model used by Rabiu Ado [1], which was the same model utilised in the experimental study of Greaves et al. [2], was also employed in this study. The silica-alumina catalyst characterised by Hasan [3] was placed around the horizontal producer in this simulation.

1. Introduction

Thermal methods are used frequently for oil production from heavy oil reservoirs. The In-Situ combustion (ISC) technique is one of these methods; it has been used for over 80 years in a lot of fields around the world. ISC is considered the most commonly used of the valuable thermal processes because no external energy source for the process is required. Injected air, used in in-situ combustion, helps to burn part of the crude oil generating heat, which lowers oil viscosity and causes high cracking rates of heavy oil located in front of the combustion zone. The cracking of the crude oil leads to in-situ upgrading of heavy oil. This feature is considered the most important advantage of the in-situ technique compared to the other EOR techniques. However, this method has several disadvantages. One of these drawbacks is that the ignition inside the well can happen spontaneously. Because of this spontaneous ignition, the consumption of the oxygen will occur close to the injector, and the process will be changed by this feature into forwarding combustion [4]. Also, because the oxygen is exhausted near to the injection well, which causes a deficiency in the oxygen supply, it will be difficult to reach a high-temperature combustion mode, which makes it difficult to maintain the combustion [5,6].

In addition, the conventional ISC technique uses a vertical injector and a vertical producer in which hydrocarbons need a long time to travel from the injector well to the producer well. These operational patterns suffer from some considerable problems, such as gravity override of gas, and the production of heavy oil from the reservoir can be affected by these problems. The Toe-to-Heel air injection technique (THAI) is a process in which the heavy oil production rate can be increased dramatically. This method is an advanced ISC technique, which combines the conventional ISC technique with horizontal producer wells, instead of vertical producer wells as in the conventional methods. The reason for this is to avoid the long distance between the injection well and the production well, to keep combustion advanced ahead from the "toe" to the "heel" within the producer well. The THAI process is designed to operate in a way involving stabilization of the drainage by gravity, which is restricted to a narrow mobile oil zone, which will travel along the horizontal production section. This short distance between the injector and the producer causes direct transferring of the mobilized fluids to the production well [7].

The THAI process for enhanced oil recovery follows the same basic mechanism as for the in-situ combustion technique. Part of a heavy residue is burned by air introduced via the injection well, producing heat to increase the temperature of the heavy oil, which causes a high reduction in the oil viscosity; as a result, the mobility of the oil will be increased. The thermal cracking reaction of the heavy oil produces fuel (mainly coke) which will be generated ahead of the combustion advance. The amount of this coke cracked is a very significant parameter to sustain the combustion propagation [8]. Xia et al. [9] have conducted a set of 3D experimental evaluations of THAI for upgrading of Wolf Lake oil with API gravity of 10.5° and Athabasca Tar Sand Bitumen with API gravity of 8°, both originating from Canada, using different well configurations as follows: vertical injector/horizontal producer (VIHP) and horizontal injector/horizontal producer (HIHP) wells in direct line drive, and staggered line drive of vertical injector/two horizontal producer (VI2HP) and two vertical injector/horizontal producer (2VIHP). It was demonstrated that the horizontal injector/horizontal producer (HIHP) was the most effective for obtaining quick start-up, which means that the required time for commencing propagation of the combustion front is very short.

Further previous work using different well configurations includes simulation by Fatemi et al. [10] to evaluate the effects of several parameters regarding injector-producer wells arrangement, namely the distance between the positions of the wells, the depth of the injector and the length of the horizontal producer, on the performance of THAI technique in reservoirs including heavy crude oil. 3D combustion cell, employing the CMG-STARS simulator, has been applied in this study for the heavy oil carbonate reservoir in Iran named Kuh-E-Mond (KEM). It was noticed that significant parameters, such as depth of injection, horizontal producer length, and the distance between injectors, should be taken into account, in order to fulfil the highest performance in the THAT process. The findings revealed that the best configurations forfield scale are the 2VIHP and VIHP schemes.

Experimental work on in-situ combustion (ISC) was performed on the heavy oil of the Orinoco Belt in Venezuela using a combustion tube by Anaya et al. [11], to investigate the technique performance. The laboratory scale results, obtained from the combustion tube experiment, were simulated to scale the laboratory tests up to full field scale. Several different well configurations were considered for the numerical ISC simulation, taking into account distances between the injector and the producer, to determine the most favourable positions of injection and production wells. The tested distances between the injector and the producer were 50 m, 100 m and 200 m, with a constant rate of air injection of ~7063 m^3 min^{-1}. It was found that the suitable distance between the injector and the producer, to operate the experiment in the field, was 100 m. The findings revealed that the model, in which the distance between the injector and the producer was 100 m, gave a high recovery factor and delayed gas breakthrough.

Moreover, a catalytic reactor can be integrated into the THAI technique to be a very effective method to upgrade crude oils prior to the refinery. Catalytic-Processing In-Situ (CAPRI) upgrading of high molecular weight hydrocarbons using ISC provides the possibility of increasing the extent of the improvement process of the heavy oil in situ Abu et al. [12]. The way of using this technique is to attain further upgrading of heavy oil by covering a perforated horizontal producer by a layer of catalyst [8, 13].

In the THAI process, the cracking reaction, created due to the high temperature within the mobile oil zone (MOZ), generates the coke precursor for the CAPRI process [14]. The MOZ involves mobilized oil, water as steam, unconsumed oxygen and gases produced from the combustion process including a small amount of carbon monoxide at high temperatures in the range of 450-650 °C [15]. The gases produced from the combustion reaction operations are given by the following equations [16]:

$$\text{Heavy oil} \rightarrow \text{light oil} + \text{gas} + \text{coke} + \text{absorbed heat} \quad (1)$$

$$\text{Coke} + O_2 \rightarrow H_2O + CO_2/CO + \text{heat} \quad (2)$$

$$\text{Heavy oil} + O_2 \rightarrow H_2O + CO_2/CO + \text{heat} \quad (3)$$

$$\text{Light oil} + O_2 \rightarrow H_2O + CO_2/CO + \text{heat} \quad (4)$$

It is thought that heavy oil is upgraded by a hydroprocessing operation in which carbon-rejection reactions are integrated with the addition of hydrogen at the catalyst surface [14]. According to Hajdo et al. [17], the hydrogen used in the process could be provided via a gasification process and/or the water-gas-shift reaction. The THAI-CAPRI technique is considered environmentally friendly because it makes the heavy oil free of undesirable contaminants such as sulfur, vanadium and nickel. The method is a new technology which offers the significant feature of a combination of enhanced oil recovery process simultaneously with the catalytic upgrading of immobilized to mobilized oil [18].

Experimental THAI-CAPRI work has been conducted by Shah et al. [14], using different catalysts to optimize the catalyst type and to find the favourable conditions for use of this technique. A microreactor was employed to perform this experiment, utilizing different temperatures, pressures and various gas media. Alumina-supported CoMo, NiMo and ZnO/CuO catalysts have been tested. It was found that the improvements of the API gravity, and the reduced levels of the viscosity, are dependent upon the flow rate of the oil and the temperature. It was also noticed that, at high temperature, more upgrading could be achieved, but that led to reduction in the lifetime of the catalyst bed. On the other hand, decreasing the temperature has led to a longer catalyst-bed lifetime, but the upgrading levels of the produced oil went down. Therefore, a balance in the operating temperature, between upgrading performance and the lifetime of the catalyst, is required to achieve acceptable consequences.

Hart et al. [16] have experimentally tested a different type of gas in the THAI-CAPRI method to investigate the effects of the gas environment, using CoMo catalyst supported by γ alumina. Heavy oil has been utilized with API gravity of 14°. It has been found that the reactions of the hydrocracking and hydrogenation occurring in THAI process were promoted when hydrogen gas was used. Employing hydrogen gas gave rise to an API gravity of the produced oil 4° above the original value of the API of the feed, which was the highest number for the different gases level. Further, using hydrogen gas results in a higher reduction in the viscosity and higher conversion of heavier hydrocarbons, with a high boiling point, to lighter fractions, with a lower boiling point. Moreover, it was noticed that hydrogen media leads to convert unsaturated to saturated components as well as enhance the productivity of the saturated hydrocarbon.

A numerical THAI-CAPRI model has been simulated by Rabiu Ado [1] at laboratory scale to investigate the catalyst activity on the heavy oil upgrading, using the same dimensions of the 3D combustion cell used in Greaves model. The effect of the frequency factor of the reaction on several parameters, such as oil rate, cumulative oil production and the API gravity, has been demonstrated. It was found that slight differences in the peak temperature were observed between the various values of the frequency factor, where during the combustion period, the maximum variance in the temperature difference was 40 °C. It is also found that the increase in the frequency factor has led to the increase in the production rate.

Further studies are needed to further understand the THAI-CAPRI technique. In this study, more parameters, such as catalyst thickness, catalyst packing porosity, will be investigated. In addition, the effect of the ratio of hydrogen to air will be taken into account in this project. Additionally, this present work aims to simulate the effect of different well configurations on the performance of the THAI process applied at the laboratory scale. Two horizontal injectors and two horizontal producers (2HI2HP) were employed in different well configurations. Three different locations of the injectors and producers were simulated to look at the effects of the various places of the wells on the THAI process performance. Several parameters such as oil recovery factors, combustion front temperature, oxygen profile and solid phase (coke), have been tested on the THAI technique to choose the best configuration at laboratory scale.

2. Methodology

2.1. 3D combustion cell model

In this work, the Greaves model [2] was adopted, and a numerical CMG STARS model developed to carry out the simulations. A 30 × 19 × 7 (i, j, k) gridblocks experiment scale model was applied using Virgin Athabasca oil sand data, as reported by Greaves et al. [2]. The API of the oil was 8° with specific gravity 1.0143. The measurements of this 3D rectangular combustion cell are 0.6 m long, 0.4 m wide and 0.1m deep, using two horizontal injectors (2HI) and two horizontal producers (2HP) which were positioned in different schemes. Real reservoirs are continuous and all flow factors, which depend on the time, change constantly. Because of this continuous change of fluid properties, a reservoir simulator cannot solve these variables. Therefore, the continuous reservoir has to be divided into finite size elements and this division is indicated as the reservoir discretization. Each element of the group is called a reservoir grid or a cell. The STARS software includes the discretised wellbore (DW), which allows the fluid and the heat flow in the wellbore and between the wellbore and the reservoir to be accurately modelled. All types of equations, wellbore mass, energy conservation and chemical reactions equations are combined together and solved by the software in each wellbore part,using fully implicit finite difference method. However, to solve these equations, several reservoir input parameters are required. The original geological parameters used in this simulation (porosity, vertical permeability and horizontal permeability of the cell) are given in Table 1, and the relative permeability curve can be seen in Figure 1. PVT data used in the THAI process is reported in Table 2. The initial gas and oil saturation were 0.15 and 0.85, respectively, and the initial water content has not been considered. The air was used alone (known as dry combustion) with a flow rate for the air injection of 8000 $cm^3\ min^{-1}$, which corresponded to 12 $m^3\ m^{-2}\ h^{-1}$, at the beginning of the run, when pre-heating was finished. Then, depending on the pre-heating time, the air flow was increased to 10667 $cm^3\ min^{-1}$, which is equivalent to a flow rate of 16 $m^3\ m^{-2}\ h^{-1}$, and continued at this rate until the end of the simulation.

Table 1. Geological parameters of the rocks used in the simulation.

So (%)	Sw (%)	Porosity (%)	Vertical permeability (md)	Horizontal permeability (md)
85	15	34	2300	11500

Table 2. PVT data applied to this project.

Component	Mol fraction (%)	Molecular weight (g mol^{-1})	Critical pressure (kpa)	Critical temperature (°C)	Density (kg m^{-3})
Light component	36.47	170.0	2305.95	425.16	903.8
Heavy component	63.53	878.0	1031.29	780.0	1012.07

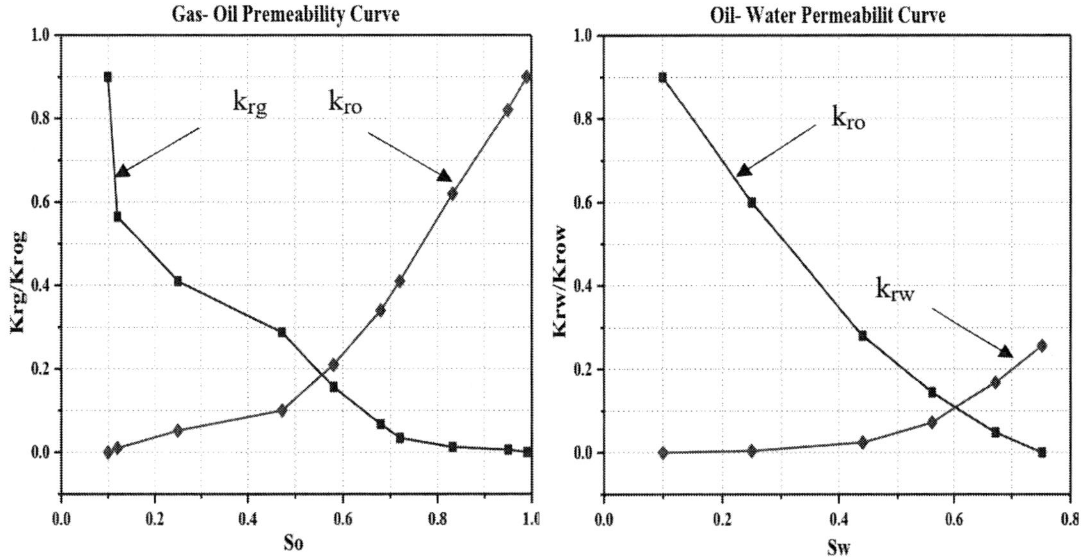

Figure 1. Relative permeability of Athabasca bitumen.

In order to make sure all parameters are consistent with the Greaves model (the original model), the total rate of the injected air for one injector in the Greaves model was equally divided between the two well injections in this study.

2.2 Different well arrangements

A numerical simulation has been carried out in this work using two horizontal injectors and two horizontal producers in different well configurations. In each well configuration, the injectors and the producers have been put in different locations, therefore the different designs of the injectors and the producer's positions are very important in this work. The configurations of the well were arranged as follows:

Greave's model (the original model) is represented, Ao, as shown in Figure 2a. The first well arrangement, A1, was that the first injector, I1, was placed on the top horizontal layer while the second injector, I2, was placed on horizontal layer two rows directly under I1, and the producers, P1 and P2, were positioned on the horizontal layers five and six rows perpendicular to the injectors as shown in Figure 2b. The second configuration, A2, was that the injector, I1, remained at the same position, and the second injector, I2, was moved to the other edge of the reservoir on the top horizontal layer, while the producer, P1 and P2 were centred at the middle of the reservoir as presented in Figure 2c. Finally, in configuration A3, the injectors, I1 and I2, were positioned on the top horizontal layers at the centre

IOP Conf. Series: Materials Science and Engineering **579** (2019) 012030 doi:10.1088/1757-899X/579/1/012030

of the reservoir and the producers, P1 and P2, were fixed on both sides of the reservoir as shown in Figure 2d.

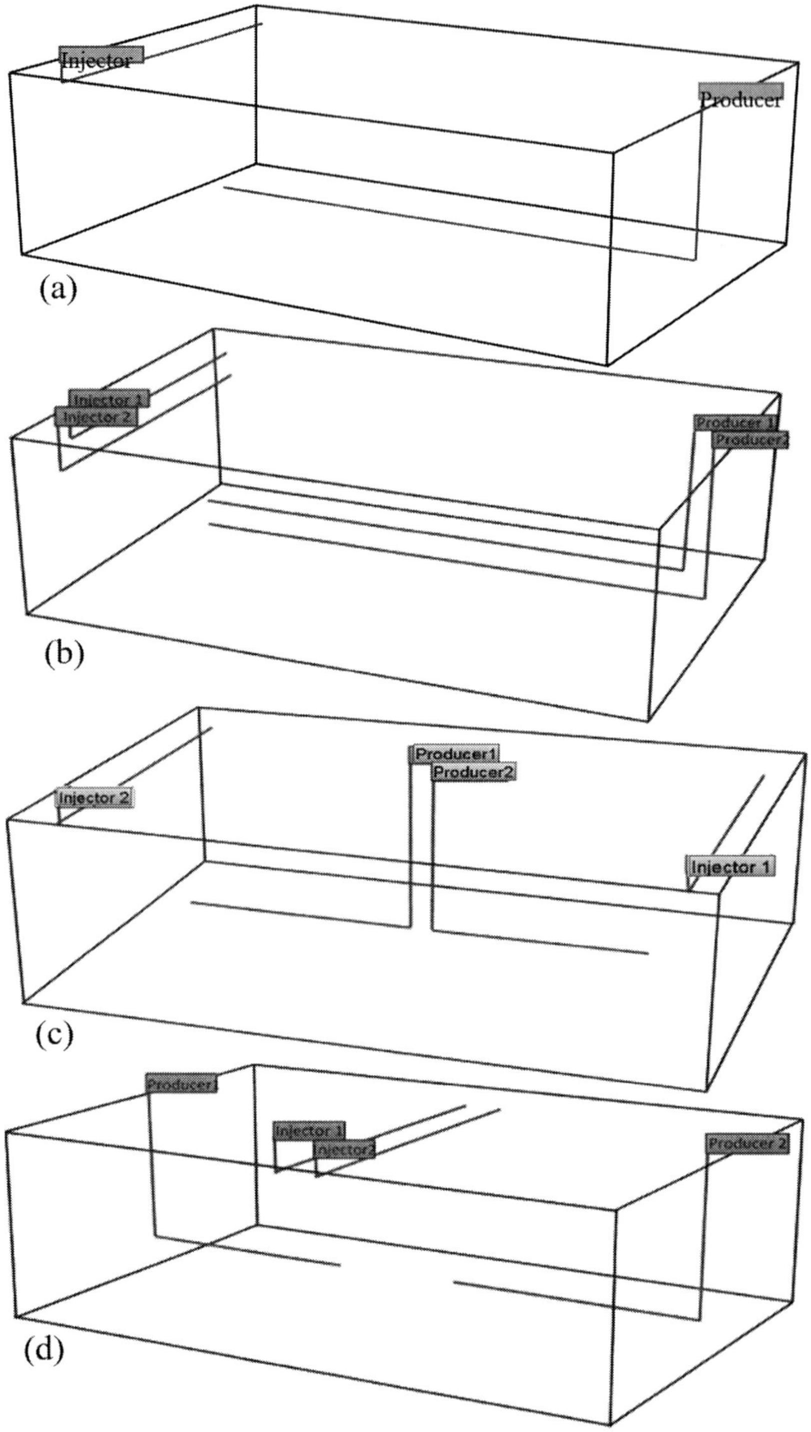

Figure 2. 3D reservoir well configurations in (a) original model, (b) model A1, (c) Model A2, (d) Model A3.

3. Results and Discussion

3.1 Effect of wells configuration on THAI process

Heavy oil of Virgin Athabasca was used in the THAI simulation runs. The oil specific gravity was 1.0143 with 8°API.

3.1.1. Oil rate and production

The oil rate and production are considered the most important indicators to assess the performance of THAI process. Figure 3 shows that no oil was produced during the ~13 minutes of the heating period. This is because no oil exists inside the production well before starting the air to be injected. It can be seen from Figure 3, for all models oil production started earlier in comparison with the Greaves model A0 (the original model). The reason is the presence of two well injectors in the models helps to increase the required amount of heat to pre-heat the inlet zone of the reservoir. As a consequence, it helps to establish a quick communication between the injector wells and the producer wells. It was also observed that, at the beginning, the rate of the oil production dramatically increased to create a high sharp peak. The reason for this is due to the high pressure generated near to the toe of the producer, which is caused by the gaseous and light hydrocarbons produced as a result of the thermal cracking occurring due to the heat provided at the inlet zone of the horizontal injectors. Based on the comparison between all the well configurations, it is obvious that A2 arrangement had the highest production rate and the A3 model then came second; however, the A0 and A1 configurations had the lowest oil production rates. From Figure 4, it can be observed that the cumulative oil recovery, for the A0 and A1 configurations, is almost the same. The highest amount of the produced oil was the A2 model and then the A3 model. It is suggested that the presence of the two well injectors in these models was behind the high amount of the oil production.

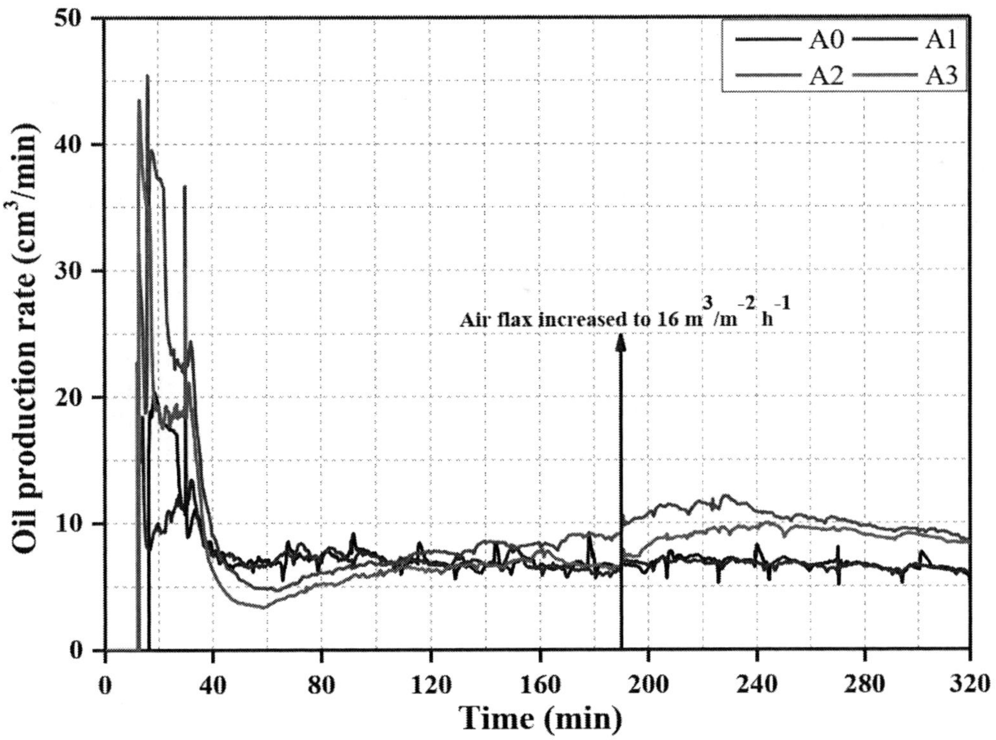

Figure 3. Oil production rates of different models.

IOP Conf. Series: Materials Science and Engineering **579** (2019) 012030 doi:10.1088/1757-899X/579/1/012030

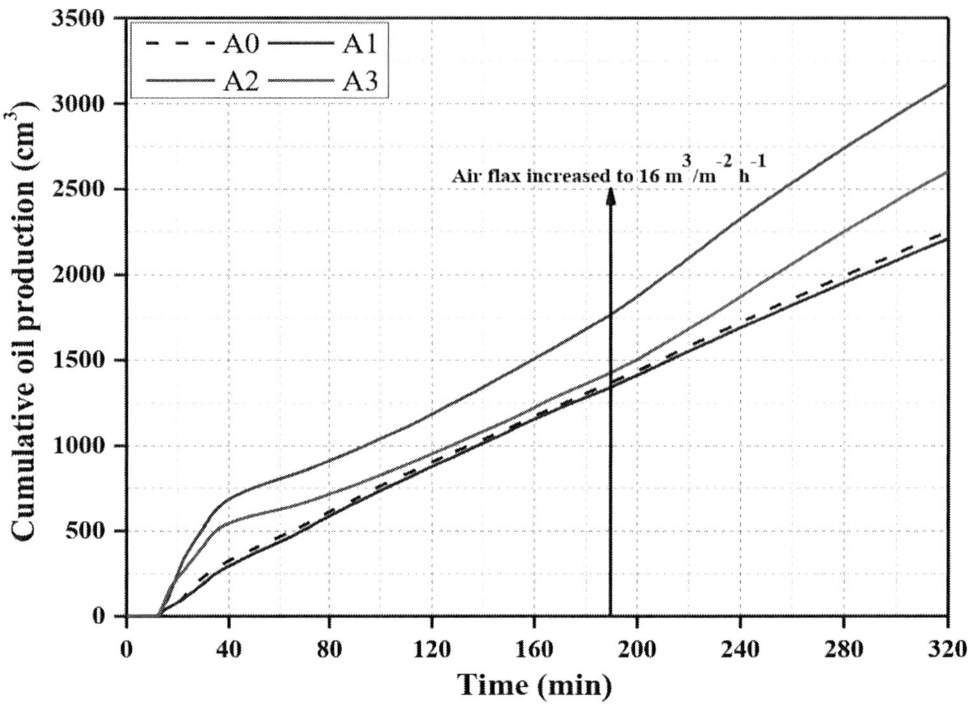

Figure 4. Cumulative oil production.

3.1.2 Combustion front temperature

The most significant feature in the THAI in-situ combustion process is that the advancing combustion front is propagated in a very stable way throughout the distance from the top to the bottom of the heavy oil reservoir. Oxygen is injected from the top of the reservoir to generate this combustion front, which can be sustained by the continued flow of the oxygen-containing gas at the reservoir's top, with low viscosity oil passing toward the production well. High-temperature reactions take place in this process. At the combustion front, most of the oxygen will be consumed by these reactions. From Figure 5, it can be seen that, during the pre-heating time period, the peak temperatures for all the models followed nearly the same trend, except the A3 model which was a little below the others. Then, after ignition is achieved, the A0 and A1 models exhibited almost the same behaviour until the end of the simulation. However, the behaviour of the A2 and A3 models were different, where the A2 model was more stable than the A3 model until the simulation end. As a consequence, the location of the injectors and producers has a significant effect on the stability performance of the THAI process. The small spikes on the more smooth tradition peak temperatures, which have different position depending on the well arrangement, are due to the combustion front reaching a high coke concentration at this point of the peak temperatures inside the producer.

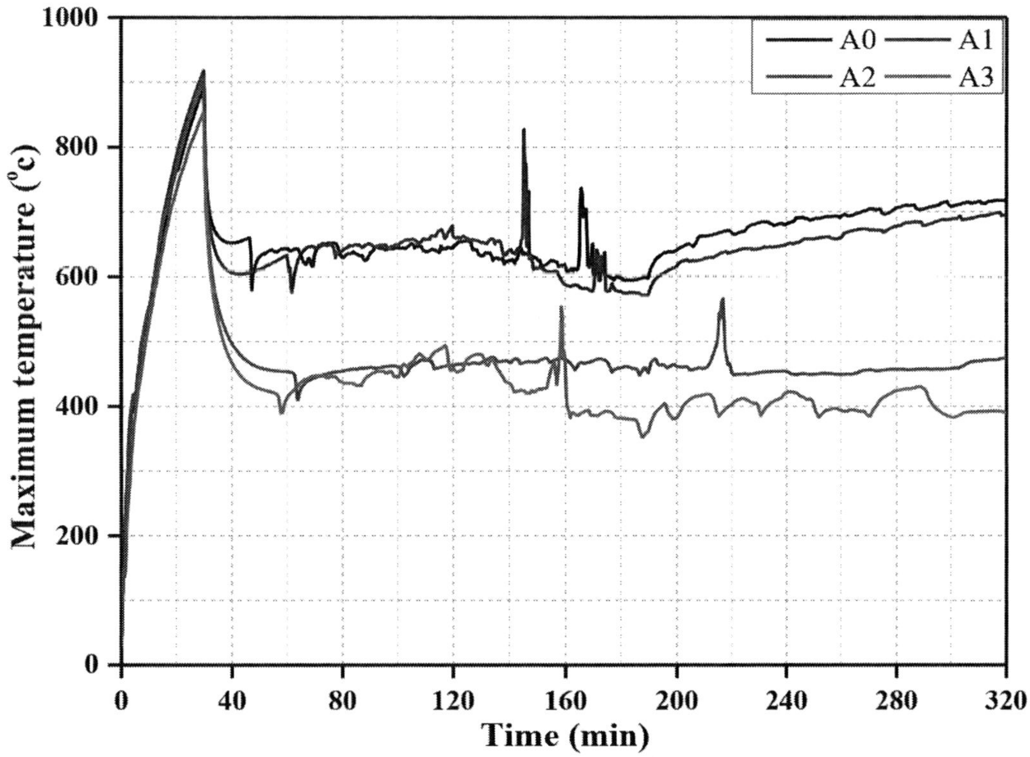

Figure 5. Peak temperatures for different Well Configurations.

3.1.3 Mole fraction of oxygen production

The amount of the oxygen used is one of the most important parameters in the THAI process for enhanced heavy oil recovery. It is considered a good indicator of the performance of the combustion propagation. Therefore, the amounts of consumed and produced oxygen are very significant for the sustainability and the stability of the combustion front. On the other hand, the amount of the consumed oxygen can be calculated from the oxidation process occurring in front of the combustion front because of the high temperature generated through the burning process of the heavy oil. Figure 6 shows the mole fraction of produced oxygen for the different models during the time period of the simulation. It is clear that the oxygen production started early in the A3 model at about 160 min, while it began late in the A1 and A2 configurations. Although the oxygen production commenced late in A2, it behaved more stable than the other models compared to the Greaves model, i.e., that well configuration affects stability. It can be also observed from Figure 7 that oxygen profile in model A2 showed close agreement with that for Greaves model (A0). The concentration of the deposited coke at the toe and the heel of the HP well has a direct effect on the amount of the produced oxygen. In the case of the model A3, where the fuel concentration was low,the oxygen breaks through earlier. In contrast, the concentration of the fuel was more in the A1 and A2 models.

IOP Conf. Series: Materials Science and Engineering 579 (2019) 012030 doi:10.1088/1757-899X/579/1/012030

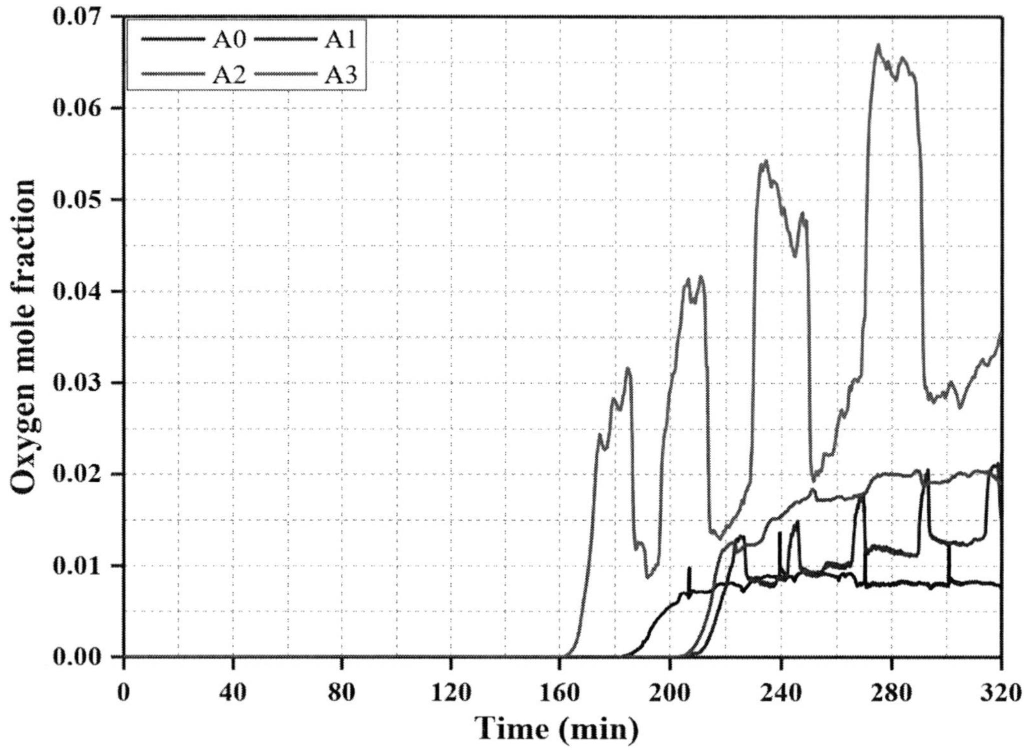

Figure 6. Mole fraction of produced oxygen in the 3D combustion cell.

IOP Conf. Series: Materials Science and Engineering **579** (2019) 012030 doi:10.1088/1757-899X/579/1/012030

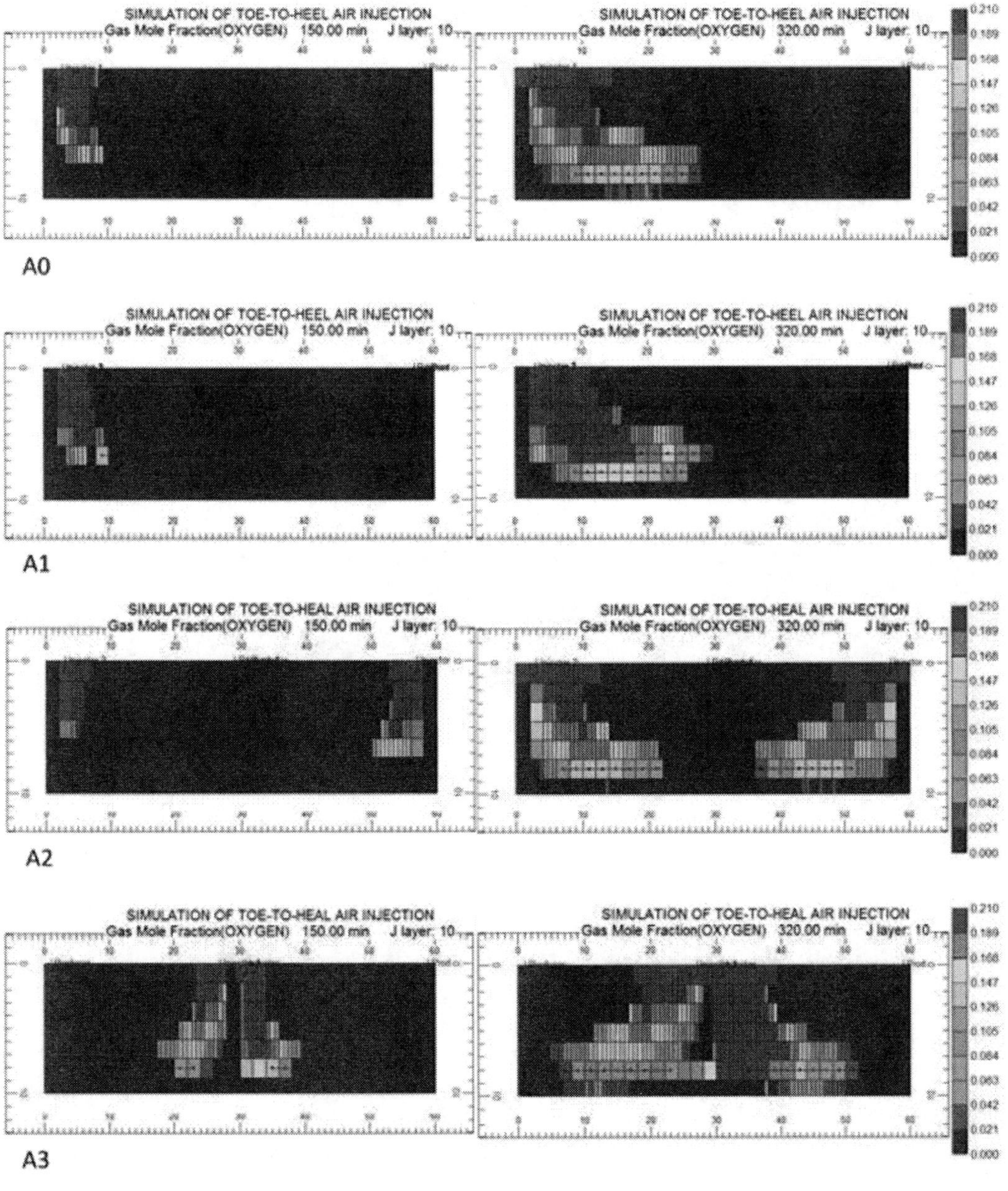

Figure 7. Oxygen profiles along the vertical plane for various configurations at 150 min and 320 min.

3.1.4. Solid Phase (Coke)

The solid phase (coke), which is a result of the heavy oil thermal cracking process, is created in front of the combustion zone in the oil reservoir. This provides the required fuel to sustain the combustion process for the THAI technique. The concentrations of the deposited coke in each of the configurations analysed at the same section and same time, gives a good indication of the efficiency of the volumetric sweep. From Figure 8, it can be seen that the shape of the volumetric sweep in all models is almost the same. It can be noticed that the deposited coke extends over a wide zone gradually from the low concentration at the top reaching the highest concentration at the base of the reservoir down near the horizontal production well. The blue area represents the shape of the propagated combustion front, through the cross-section, which is entirely burned because of the ready supply of oxygen.

IOP Conf. Series: Materials Science and Engineering **579** (2019) 012030 doi:10.1088/1757-899X/579/1/012030

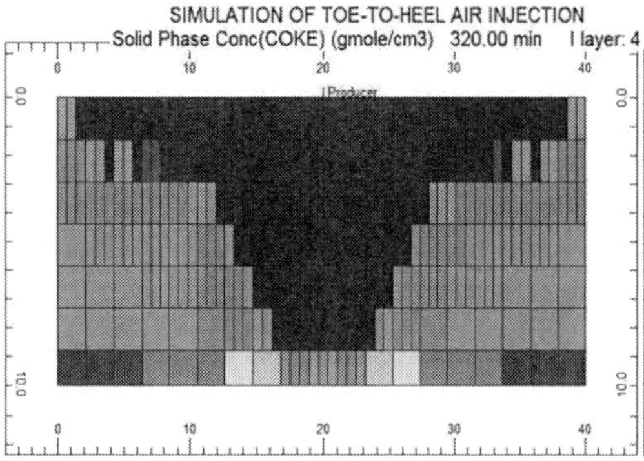

A0 model

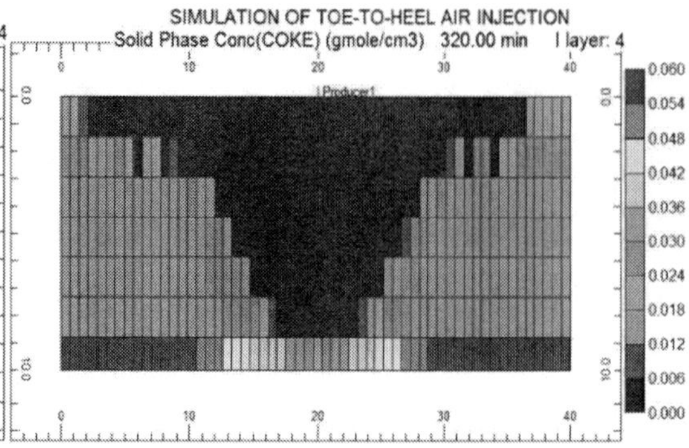

A1 model

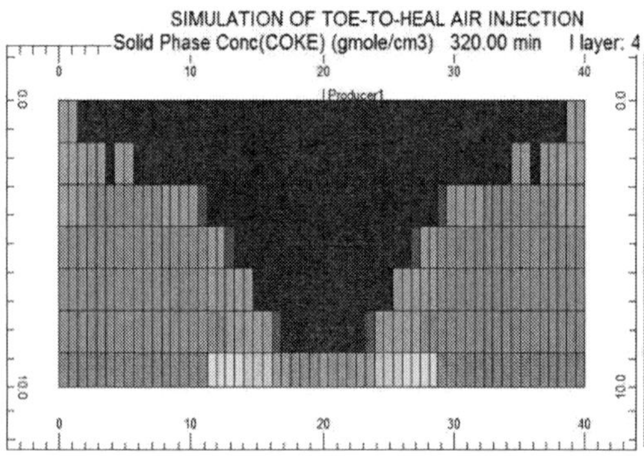

A2 model

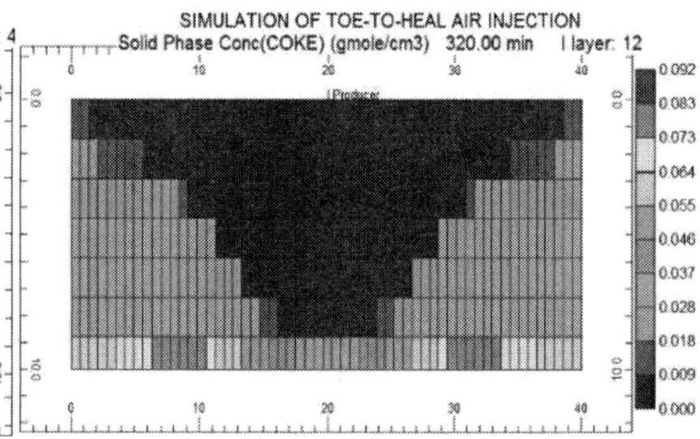

A3 model

Figure 8. Solid Phase (Coke) concentrations for the various THAI well configurations.

To enhance the amount of the produced oil further, the model A2 was used to test the impact of the flow rate of the injected air on the performance of the THAI process. The results obtained from the model A2 were compared to the original model (A00 as shown in Figure 9 and Figure 10. Figure 9 shows the amount of the oil produced in the A0 and the A2 models through the combustion period, at different air flow rates. Quantitatively, it can be observed that oil produced in the model A2 was increased with increasing the flow rate of the injected air. It was clear that the amount of the oil produced in the model A2 had become about twice of that the oil produced by Greaves model (A0) when the air flow rate was increased about 50 % compared to what was injected in the model A0.

Figure 10 clarifies the behaviour of the peak temperatures of the model A2 at different air injection flow rates compared to the peak temperature obtained from the original model. It was noted that the temperature of the combustion cell was significantly affected by the flow rate of the injected air to the reservoir. As a consequence of increasing the flow rate of the air injection in the model A2, the amount of the produced oil doubled, compared to the original model A0. In addition, the temperature of the combustion zone was substantially raised to be very close to that temperature in the original model (A0) as can be noticed in Figure 10.

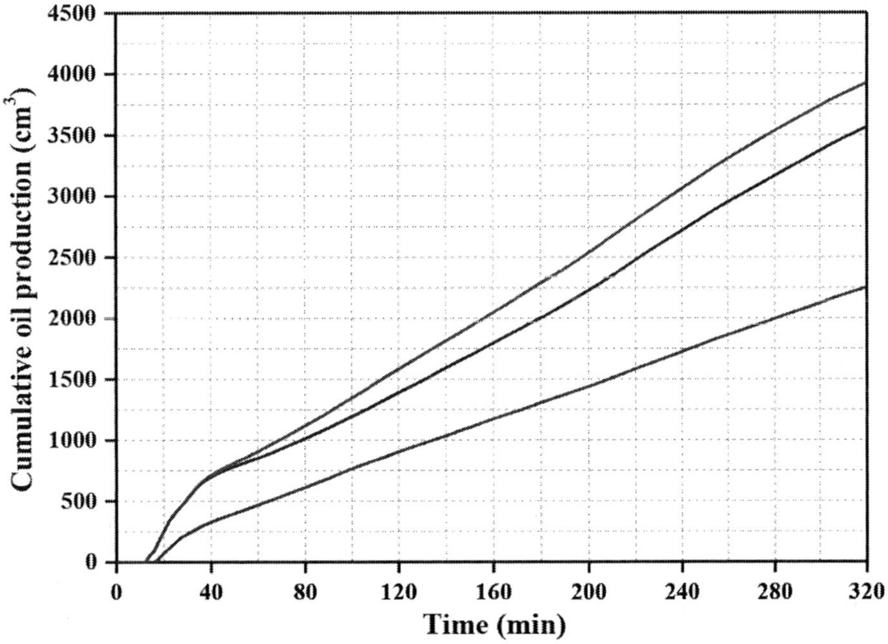

Figure 9. Cumulative oil production for two different well arrangements at different air flow rates, a grey line is the Greaves model at 12 and 16 $m^3\ m^{-2}\ h^{-1}$, a blue line is the A2 model at 15 and 20 $m^3\ m^{-2}\ h^{-1}$ and a red line is the A2 model at 18 and 24 $m^3\ m^{-2}\ h^{-1}$ flow rates.

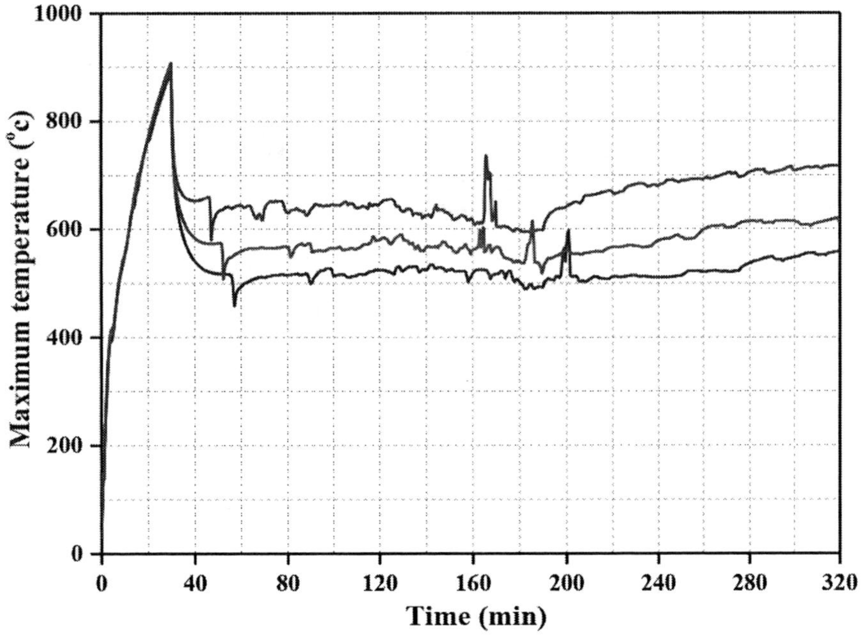

Figure 10. Peak temperatures for two different Well Configurations at different flow rates, a grey line is the Greaves model at 12 and 16 $m^3\ m^{-2}\ h^{-1}$, a blue line is the A2 model at 15 and 20 $m^3\ m^{-2}\ h^{-1}$ and a red line is the A2 model at 18 and 24 $m^3\ m^{-2}\ h^{-1}$ flow rates.

3.2. THAI-CAPRI simulation results

The largest reserves of oil in the world are heavy oil and bitumen which represent about 80% of the World's reserves and are concentrated in Canada and Venezuela. These reserves could compensate for the depletion in conventional oil production [19]. Since heavy oil and bitumen have a high viscosity and are poorly transportable, they need to be subjected to an upgrading process to be transformed to light crude oils. The crude oils then become readily transportable and acceptable to be used as a refinery feedstock [20]. In addition, the upgrading technology adds a significant environmental advantage since it retains heavy metals, such as S, Ni and V, in the reservoir. Rabiu Ado [1] has generated the input parameters to simulate THAI-CAPRI process at laboratory scale. The frequency values of the hydrotreating reactions were adjusted, via trial and error, based on the API values illustrated by Xia and Greaves [8] and Xia et al. [9], until an average value of the API gravity was achieved about 6 points above the original oil. At the beginning of the study, the original frequency factors were applied, creating a base case model to be compared with others having different frequency factors. The models were denoted TC0 (which is a base case) TC1, TC2, TC3 and TC4. In each model, the original frequency value has been multiplied by a variable factor to study its effect on the simulated performance of the CAPRI process. Among these models, it was found that the TC3 model was the best one, in which the frequency factor of the kinetic reactions has been adjusted to give a high cumulative oil production, which has been subsequently adopted in this study. An electrical heater was used to pre-heat the sand pack for 30 minutes, and a gas mixture of air and hydrogen was then injected at a flow rate of 10000 $cm^3\ min^{-1}$, which corresponded to 15 $m^3\ m^{-2}\ h^{-1}$. The initial hydrogen to the air ratio (HAR) used was 1:4 as it was assumed in the study of Shah et al. [14] and Hart et al. [18]. As discussed concerning the standard THAI process, after 190 minutes, gas flow was increased to 13333.3 $cm^3\ min^{-1}$, which is equivalent to a flow rate of about 20 $m^3\ m^{-2}\ h^{-1}$, and it was maintained until the end of the simulation. In this project, the effect of the catalyst thickness has been investigated as well as the effect of hydrogen to the air ratio. In addition, the effect of the catalyst packing porosity has been taken into account.

3.2.1. Model configuration

The TC3 model used by Rabiu Do [1], which is the same model used in the experimental study of Greaves et al. [2], was also used in this study. The model has horizontal injector and producer (HIHP) wells which were configured in a staggered line drive (SLD). In addition, a catalyst layer was placed around the HP section in the CAPRI model as illustrated by the coloured part in Figure 11. Moreover, instead of using 30 × 19 × 7 grid blocks in the original model, the CAPRI model was discretised into 30 × 19 × 9 grid blocks to compensate the area lost due to the added catalyst.

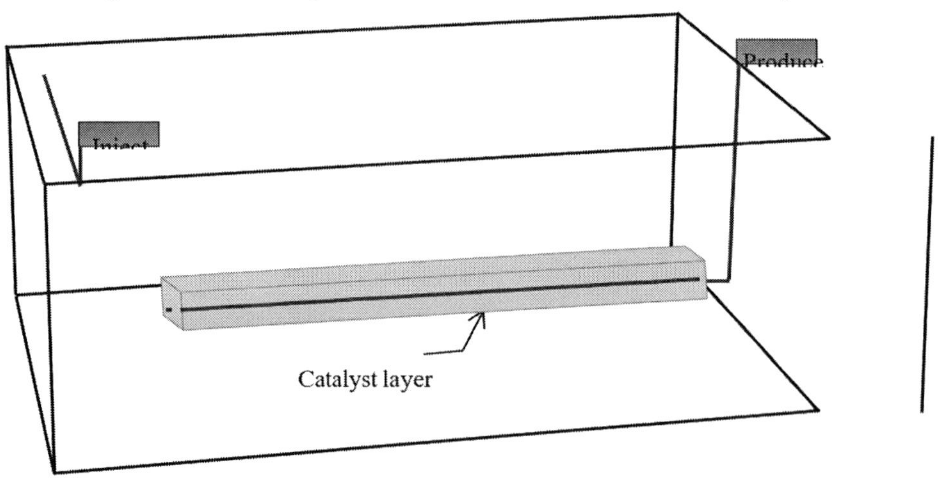

Figure 11. 3D schematic diagram of THAI-CAPRI technique with a catalyst layer placed around the producer.

3.2.2. Physical properties of the upgraded oil
Based on the distillation data obtained by Hart et al. [16] and Rabiu et al. [1] has computed the PVT data of the light and heavy upgraded oil using Aspen HYSYS software as presented in Table 3.

Table 3. Physical properties of the upgraded oil produced [1].

Components	Mole fraction (mol%)	Molecular weight (g/mol)	Density (Kg/m^3)	TC (°C)	PC (kPa)
LUO	20.31	128.01	776.53	353.12	2448.61
HOU	79.69	252.50	850.48	502.19	1523.46

In addition, during the THAI-CAPRI process, catalytic reactions occur beside the different pyrolysis reactions occurring when the THAI process has been tested alone. These catalytic reactions involve mostly hydrodesulfurization (HDS) and hydrodenitrogenation (HDN). Water/gas shift (WGS) and steam gasification (SG) reactions are also observed for the process. Therefore, to simulate the model, further calculations needed to determine the kinetic parameters, the stoichiometry of these reactions, the frequency factor and the activation energy and the fractions of sulphur and nitrogen atoms. All these requirements have been reported in detail by [1], as illustrated in Table 4 and Table 5.

Table 4. Balanced reactions of HDS and HDN with the kinetics parameters.

HDS and HDN reactions	Activation Energy (J/mol)	Frequency Factor (min^{-1})
Heavy oil + 19.932 $H_2 \longrightarrow$ 3.477 HUO + 0.585 H_2S	87×10^3	2.7×10^7
Light oil + 3.286 $H_2 \longrightarrow$ 1.328 LUO + 0.194 H_2S	87×10^3	2.7×10^7
Heavy oil + 1.593 $H_2 \longrightarrow$ 3.477 HUO + 0.189 NH_3	75×10^3	1.0×10^6
Light oil + 0.095 $H_2 \longrightarrow$ 1.328 LUO + 0.011 NH_3	75×10^3	1.0×10^6

Table 5. Elemental compositions of the pseudo-components.

Constituent	Bitumen	Heavy	Light	LUO	HUO
H (wt%)	10.687	10.430	13.005	16.764	12.866
C (wt%)	84.075	84.210	82.859	82.859	84.210
N (wt%)	0.414	0.447	0.112	0.019	0.147
S (wt%)	4.824	4.913	4.025	0.359	2.777

3.2.3. Catalyst preparation

The catalyst used in this study was silica-alumina which has been tested, discussed in the experimental section, by different techniques such as mercury porosimetry, gas adsorption, NMR etc. The catalyst was placed around the producer well so that the produced oil entering the production well has to go through the catalyst first. The layer thickness for the catalyst as used by Rabiu Ado [1] has been adopted in this study with an initial thickness of 0.5 cm to commence with. The catalyst properties obtained from the experiments were used to compute the packing porosity and the concentration of the catalyst used to simulate the model. The elements, and their particular fractions, of the silica-alumina catalyst, can be seen in Table 6. The bulk density of the catalyst, 1361.7 kg m^{-3}, is taken from the empirical results in this study. This density is integrated with the void volume and the packing porosity to calculate the concentration of the catalyst loading, using the following equation taken from STARS manual.

$$\varphi_f = \varphi_v[1 - C_c/\rho_c] \quad (5)$$

where: φ_f is the packing porosity, φ_v is the void porosity, C_c is the catalyst concentration (kg m^{-3} pore volume), ρ_c is the catalyst bulk density (kg m^{-3}). The initial value of the catalyst packing porosity used was taken from Abu et al. [12], which was 45.1%. Two other values of the packing porosity have been investigated to simulate the effect of this feature on the performance of the CAPRI process. These values with the results obtained from the above equation were tabled in Table 7.

Table 6.: Element composition of silica-alumina catalyst.

Element	RMM (g/mol)	Amount/ wt%
Al_2O_3	101.96	40.54
SiO_2	60.08	48.44
Cl	35.45	6.20
TiO_2	79.866	3.03
Fe_2O_3	159.69	1.15
Total	76.89	100.00

Table 7. Calculated catalyst concentrations based on different fluid porosities.

Packing porosity (%)	Void porosity (%)	Bulk density (kg/m^3)	Catalyst loading (kg/m^3 pore volume)	Catalyst loading (mol/cm^3 pore volume)
44.0	0.99	1361.7	756.50	0.009912517
44.55	0.99	1361.7	748.93	0.009813392
45.1	0.99	1361.7	741.37	0.009714267

3.2.4. Effect of packing porosity
The initial porosity is likely the most important parameter, which can be affected by the particle packing. Spherical and cylindrical particles are the most predominantly used in the chemical engineering operations. For cylindrical particles, both the initial porosity and the packing size are dependent on the shape of the particle. Generally, sphericity is considered the most commonly utilized factor to describe the particle shape, which represents the ratio of the sphere's surface area to the surface area of the particle possessing the same volume as the sphere [21]. The packing formation has an effect on many parameters such as fluid behaviour, heat transfer and mass transfer, which affect the macroscopic process. Principally, a packing of cylindrical particles behaves differently from one of spherical particles because cylinders offer directional freedom, and their geometry involves a diversity of surface constituents such as a flat and curved surface in addition to corners [22]. An experimental study has been conducted by Abu et al. [12] to upgrade Athabasca bitumen using packing porosity values of 44 and 45.1 %, which were used in this study as well as 44.55 %.

3.2.4.1. Oil production
Figure 12 shows the amount of oil production at the time of the simulation for several various packing porosities. It can be observed that there is a slight difference in the amount of the produced oil during the CAPRI process. However, an increase in the oil production, with the increase in the value of the packing porosity, can be seen. This is because that the initial porosity of the catalyst increased as more space between the catalyst particles was increased, which allows for more fluid passing through the catalyst into the production well, i.e., the higher porosity has a lower pressure drop.

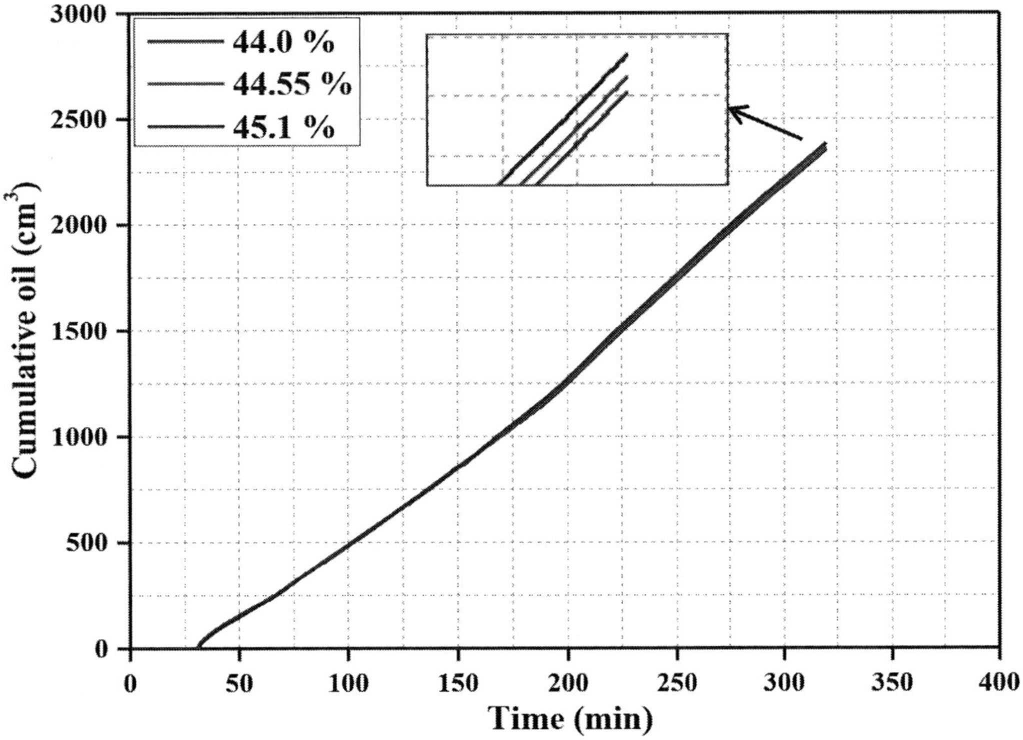

Figure 12. Cumulative oil for different values of packing porosity.

3.2.4.2. Temperature distribution

Figure 13 illustrates the effect of the several values of packing porosity on the temperature distribution with time during the THAI-CAPRI process. Despite the differences in the packing porosity, all models have nearly the same predicted temperature pathway along the combustion front until the simulation end, as can be seen from Figure 13. Therefore, based on the peak temperature characteristics, the change in the packing porosity value has no effect on the temperature distribution. This means that the combustion front propagation is stable, and all models have almost the same combustion front shape as can be noticed in Figure 14. Figure 14 showed that the shape of the combustion front, at 320 minutes, was forward leaning and swept in the upper portion of the reservoir, which is considered as one of the indication parameters of stability.

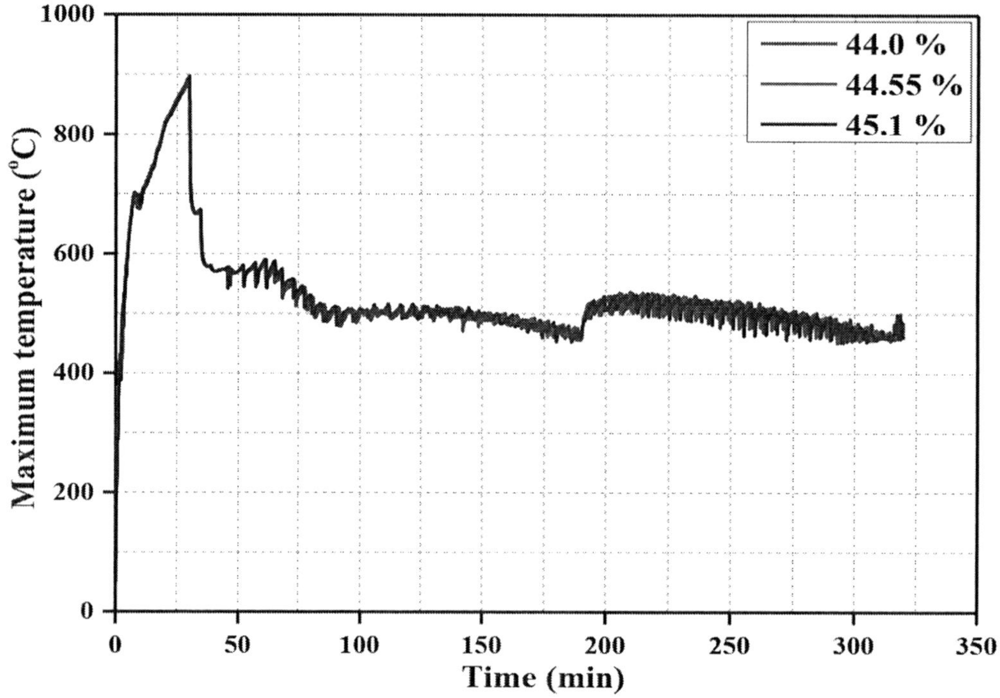

Figure 13. Peak temperature for several different packing porosities.

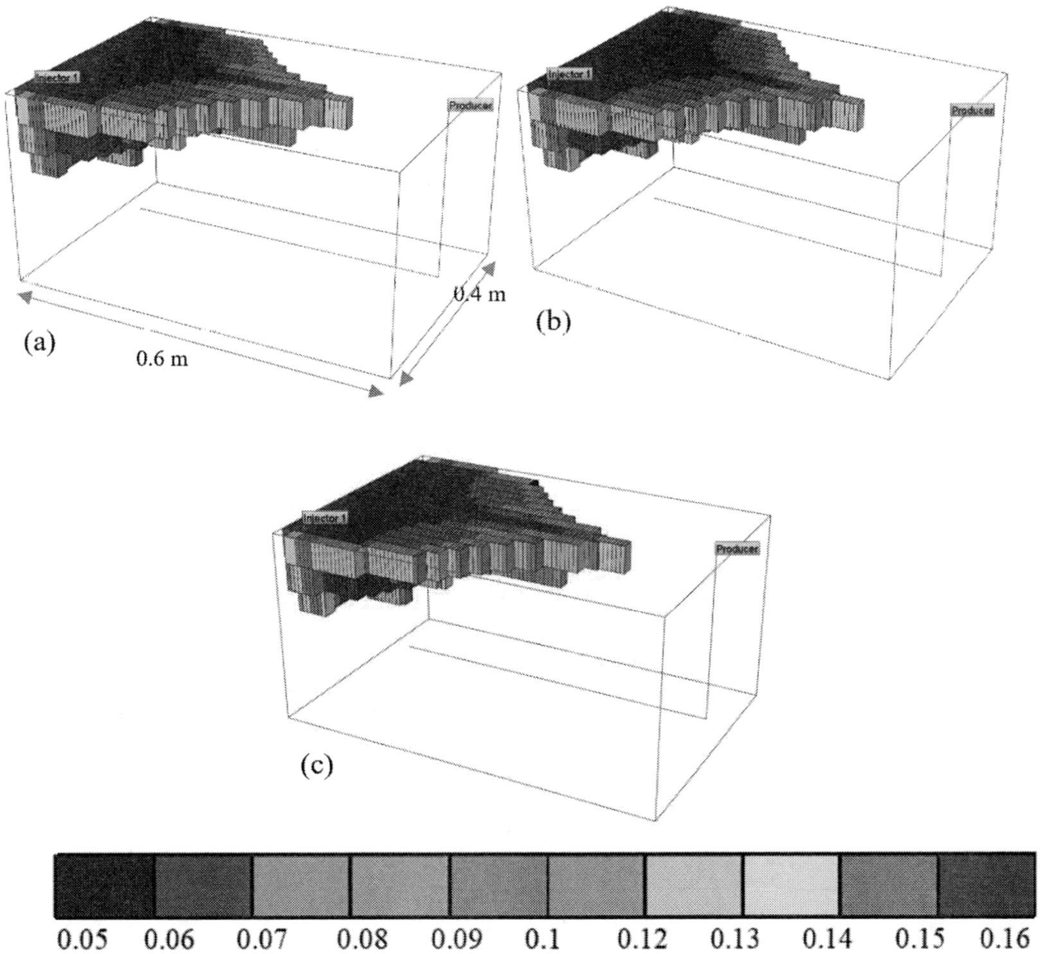

Figure 14. 3D combustion front propagation for CAPRI reactor porosities of (a) 44.0%, (b) 44.55%, (c) 45.1%.

3.2.5. Effect of hydrogen to air ratio

During the THAI-CAPRI process, the existence of hydrogen is a necessary requirement for in-situ catalytic cracking reactions. Hydrogen injected with air in the presence of the catalyst around the horizontal producer assists in the hydrogenation reaction. The hydrogenation process and asphaltene removal result in the upgrading of heavy oil in situ [23]. Another source of hydrogen can be provided by steam gasification (SG) and/or water-gas-shift reactions (WGS) [17], as presented in the following simple equations [1]:

$$\text{WGS: } CO + H_2O \longleftrightarrow CO_2 + H_2 \tag{6}$$

$$\text{SG: } C_iH_a + 2iH_2O \longrightarrow iCO_2 + (2i + a/2)\,H_2 \tag{7}$$

Hydrogen plays an important role in the HDN reaction in which the bond between the carbon atom and the nitrogen is broken and replaced with hydrogen atom, and combination process takes place between nitrogen and hydrogen to form ammonia [12]. Hart et al. [16] have been investigated several types of gas media on the performance of CAPRI upgrading using heavy oil and bitumen. It was found that the addition of hydrogen enhances the reactions of the hydrocracking and the hydrogenation

IOP Conf. Series: Materials Science and Engineering **579** (2019) 012030 doi:10.1088/1757-899X/579/1/012030

processes, whereas, in the case of the other gas media, heavy coke formation was associated with the upgrading process. It is also noticed that hydrogen results in characteristically high API gravity, low viscosity and high distillables in the produced oil. In this study, three different ratios of hydrogen to air have been employed to investigate its effect on the performance of the CAPRI process. These ratios are 1/4, 2/4 and 2.5/4.

3.2.5.1. Oil production

The effect of the various ratios of hydrogen to air on the cumulative oil in the THAI-CAPRI technique has been illustrated in Figure 15. As can be seen from Figure 15, the amount of the produced oil was affected by the value of the injected hydrogen. Where the oil production decreased with increasing of the hydrogen to air ratio. 2366 cm^3 of oil was recovered when the ratio of hydrogen to air was low, and when the ratio was increased, the amount of the produced oil dropped to 2161 cm^3, and a further increase of the hydrogen to air ratio, the recovered oil declined to 2059 cm^3. The high ratio of hydrogen to air means the lower injected air, and this resulted in an increase in the amount of the coke deposited, which in turn led to drop in the oil production.

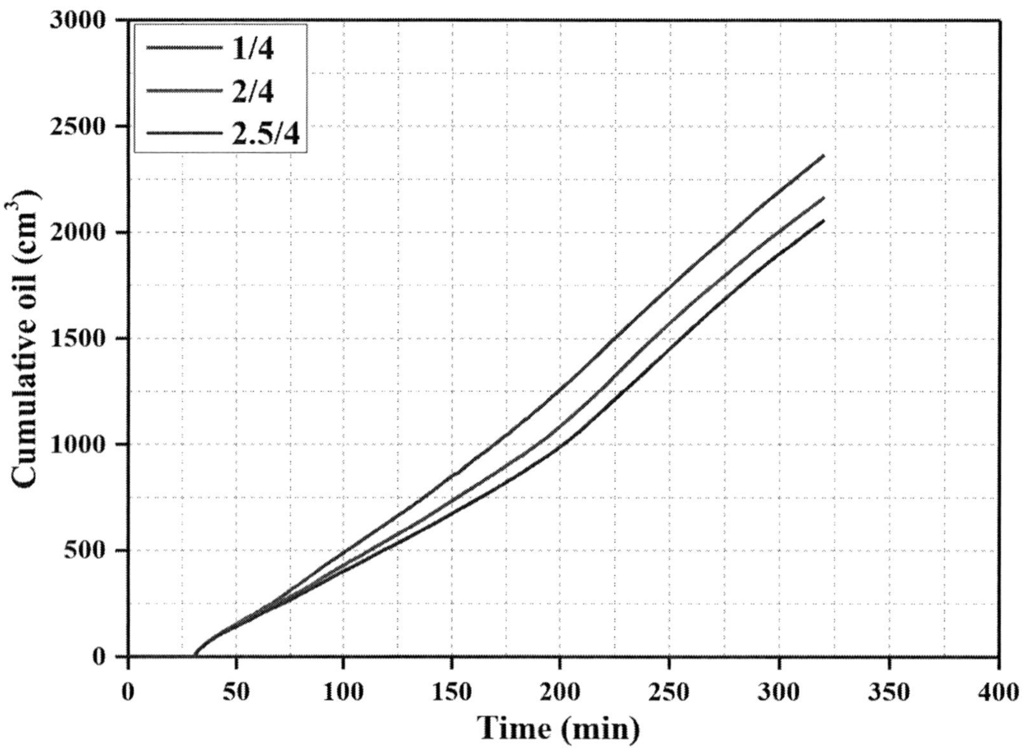

Figure 15. Cumulative volume of produced oil against time using different hydrogen to air ratios.

3.2.5.2. Peak temperature

Figure 16 shows the variation in maximum temperature of the produced oil along the reservoir as a function of time. It is noticed that the peak temperature having the lowest hydrogen to air ratio diverges first and continues to lie above all the other curves up to the end of the process, and then the second one with the middle value and following with the highest ratio of hydrogen to air. This is because of the produced oxygen which started very early in case of the highest ratio of hydrogen to air.

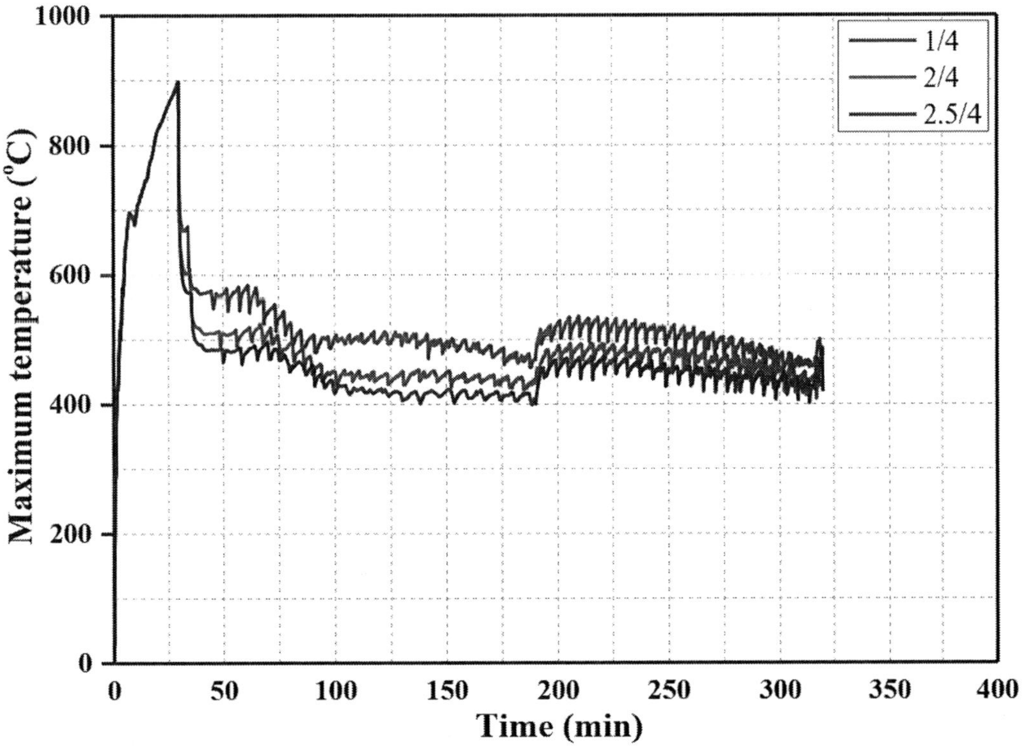

Figure 16. Peak temperature variation with time with different hydrogen to air ratios .

3.2.6. Effect of catalyst thickness

Catalysts are materials that can speed up chemical reactions and enhance the selectivity of the reaction without consumption themselves (Huang et al., 2012). The reaction can be described by the following equation:

$$\text{Heavy oil} + 19.932\ H_2 + \text{CAT} \longrightarrow 3.477\ \text{HUO} + 0.585\ H_2S + \text{CAT} \quad (8)$$

In this part, the effect of the catalyst thickness on the performance of the THAI-CAPRI process is illustrated. Three different thicknesses of catalyst have been investigated to study its effect on the cumulative oil production and the peak temperature. The thicknesses were 0.5, 0.75 and 1.0 cm.

3.2.6.1. Oil production

Figure 17 shows the effect of the catalyst thickness on the cumulative oil during the simulation period. As can be noticed from Figure 17, there is a slight difference in the amount of the produced oil. It is observed that the model having a catalyst with a thickness of 0.5 cm has the highest amount of the produced oil, and this amount decreased with an increase of the catalyst thickness. This is because that most of the thermal cracking has taken place on the surface of the catalyst in the presence of a high temperature, which resulted in a high concentration of the coke. However, as further of the catalyst thickness increased, more coke formed on the catalyst surface, and thereby, the amount of the produced oil reduced.

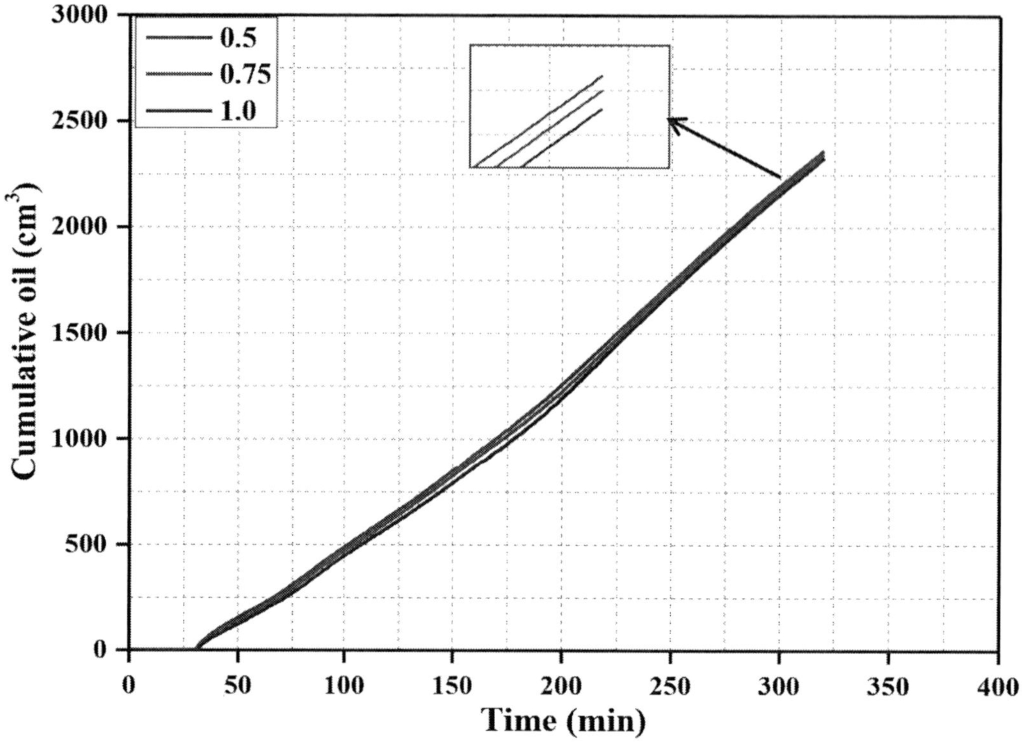

Figure 17. Cumulative oil vs time using different catalyst thickness.

3.2.6.2. Peak temperature

Figure 18 displays the change in the peak temperature along the reservoir with the time for diverse values of catalyst thicknesses. From Figure 18, the peak temperature, in the case of the catalyst thickness of 0.5 cm, rose up to about 900 °C with ~80 to 100 °C higher than that when the thicknesses of the catalyst are 0.75 and 1.0 cm respectively. This is due to the difference in the produced amount of the coke at the end of the preheating time. It is observed that during the lowest air flux from 30 to 190 minutes, the trend of all curves is almost the same.

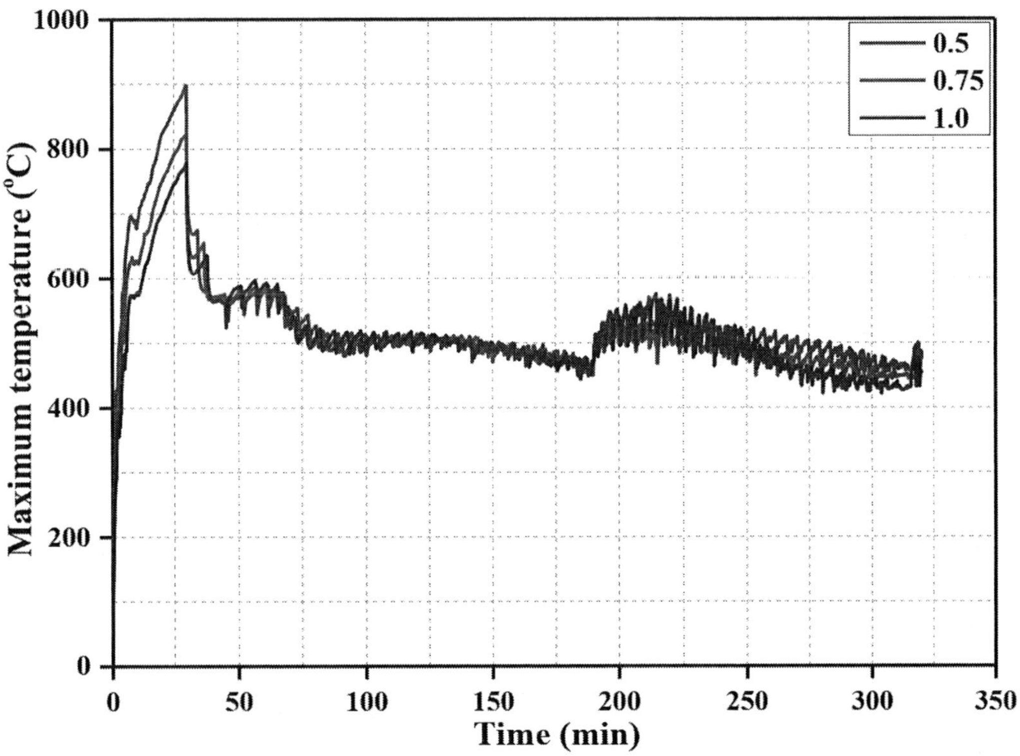

Figure 18. Peak temperature for different catalyst thickness.

However, after 190 minutes when the air flow increased, the curve for the model having a thickness of 1.0 cm was above the other curves up to about 230 minutes, and then they reordered depending on the thickness from the lowest one to the highest until the end of the combustion period. During the period between 190 to 230 minutes, when the flow rate of the injected air was increased, more oil flows through the catalyst layer, and due to the large catalyst thickness, the produced oil remains a longer time within the catalyst which leads to increase the temperature in this zone.

4. Discussions

It has been found that the best model was the A2 model in which the injectors were designed to be on both sides of the reservoir, and the producers were located in the centre of the reservoir. The findings revealed oxygen breakthrough, which is considered a most important parameter, acted in the same trend as the original model of Greaves (A0). This feature gives an indication that, in appropriate reservoir, THAI technique is stable and effective process. However, in the other models, the oxygen production followed an oscillating path to the end of the simulation period. It is also observed that model A2 gave high cumulative oil compared to the other models, at the same flow rate of the injected air used in the A0 model. The reason for this is that the symmetry of the combustion front achieved in A2 model has affected a larger volume of the reservoir and hence resulted in higher volumetric sweep of the reservoir.

Further, when the air injection flow rate was increased in the model A2 by about 50 % of that used first, the amount of the produced oil rose up to about double of the amount of the oil produced by the original model. As a result, about half reduced the time required in the A2 model to recover the same amount of the oil produced by the original model A0.

Moreover, catalytic upgrading using THAI-CAPRI process was simulated for the recovery of Athabasca Heavy Oil to investigate the effect of several parameters on the performance of the CAPRI technique. The change in the packing porosity values had a slight effect on the oil production, and no effect mention on the peak temperature. This small impact could be due to the difference in the particles arrangement from one to another. Similar, the effect of the catalyst thickness on the performance of the CAPRI process, where the amount of the cumulative oil increased with decreasing the thickness of the catalyst placed on the producer. This because of the coke deposited on the surface of the catalyst on which the thermal cracking occurred. Since the cracking reaction takes place on the solid surface, the high catalyst thickness obstructs the flowrate of the oil which leads to a high concentration of the coke on the catalyst surface.

However, the oil production was obviously affected by the ratio of hydrogen to air injected by the injectors to the reservoir. The low ratio of hydrogen to air produced the higher oil cumulative, compared to the high ratios. The high amount of hydrogen implies a low amount of injected oxygen and that resulted in an increase in fuel availability which substantially affects the recovery of the oil.

5. Conclusions

3D models using CMG-STARS have been employed for simulations of the THAI and THAI-CAPRI processes to investigate an effect of several parameters on the performance of the technique. The influence of variables such as different well configurations using two horizontal injectors and producers have been studied for the THAI technique. Moreover, the packing porosity, catalyst thickness and hydrogen to air ratio have been tested to illustrate its effect on the THAI-CAPRI process. Three different models, with varying locations for the injectors and producers, have been simulated for the THAI process, using Virgin Athabasca oil sand. Compared to the Greaves model (the original model), the second configuration, in which the injectors are located on both sides of the reservoir while the producers are centred in the middle of the reservoir, was the best scheme for the THAI technique, since it gave the highest cumulative oil, and, it also gave rise to the same behaviour as the Greaves model in terms of the temperature distribution and the oxygen breakthrough. In addition, the combustion front shape for the presented simulation was forward leaning, like the combustion shape in Greaves model, which indicates that the process was perfectly stable. However, in the other models in which the injectors are designed to be near to each other, and this could be the reason for that the amount of the produced oil was less than that in the case of the second model.

In terms of the THAI-CAPRI process, the model used by Rabiu Ado (2017) has been adopted in the present study, but using silica-alumina catalyst whose properties have been obtained from the practical part of this study. Several parameters such as packing porosity, hydrogen to air ratio, and catalyst thickness were investigated as an extension to the Rabiu Ado (2017) work. The CAPRI simulation presented here provides a new prediction of the required amount of the hydrogen used in the hydrotreating process during THAI-CAPRI technique for the upgrading of heavy oil and bitumen. The findings found that the amount of the produced oil was affected by the value of the hydrogen to air ratio. The simulation also displays that the combustion front propagation was entirely restricted to the upper portion of the sandpack, indicating that the process was completely stable. It is also that the results showed that the oil production was slightly influenced by the packing porosity and the thickness of the catalyst placed around the horizontal producer.

References

[1] Rabiu Ado, M 2017 *Numerical simulation of heavy oil and bitumen recovery and upgrading techniques* (University of Nottingham)

[2] Greaves, M, Dong, L & Rigby, S 2012 Simulation study of the toe-to-heel air injection three-dimensional combustion cell experiment and effects in the mobile oil zone *Energy & Fuels,* **26** 1656-1669

[3] Hasan M 2018 *Enhanced Recovery of Heavy Oil Using A Catalytic Process* (University of Nottingham)

[4] Karimi, G & Samimi, A K 2010 In-situ combustion process, one of IOR methods livening the reservoirs *Petroleum & Coal* **52** 139-147

[5] Thomas, S 2008 Enhanced oil recovery-an overview *Oil & Gas Science and Technology-Revue de l'IFP* **63** 9-19

[6] Shah, A, Fishwick, R, Wood, J, Leeke, G, Rigby, S & Greaves, M 2010 A review of novel techniques for heavy oil and bitumen extraction and upgrading *Energy & Environmental Science* **3** 700-714

[7] Greaves, M, Xia, T X & Ayasse, C. 2005 Underground upgrading of heavy oil using THAI-'toe-to-heel air injection' *SPE International Thermal Operations and Heavy Oil Symposium* (Society of Petroleum Engineers)

[8] Xia, T & Greaves, M. 2001 3-D physical model studies of downhole catalytic upgrading of Wolf Lake heavy oil using THAI. *Canadian international petroleum conference* (Petroleum Society of Canada)

[9] Xia, T, Greaves, M, Werfilli, W & Rathbone, R. 2002 Downhole conversion of Lloydminster heavy oil using THAI-CAPRI process *SPE International Thermal Operations and Heavy Oil Symposium and International Horizontal Well Technology Conference* (Society of Petroleum Engineers)

[10] Fatemi, S, Ghotbi, C & Kharrat, R 2009 Effect of wells arrangement on the performance of toe-to-heel air injection *Brazilian Journal of Petroleum and Gas* **3**

[11] Anaya, I, La Cruz, R E, Alvarez, A J, Gutierrez, D, Skoreyko, F A & Card, C. 2010 Simulation Study for Designing an In-Situ Combustion Pilot in the Orinoco Belt of Venezuela: From Laboratory Studies to the Field Scale. *Canadian Unconventional Resources and International Petroleum Conference* (Society of Petroleum Engineers)

[12] Abu, I I, Moore, R G, Mehta, S A, Ursenbach, M G, Mallory, D G, Pereira Almao, P & Carbognani Ortega, L 2015 Upgrading of Athabasca Bitumen Using Supported Catalyst in Conjunction With In-Situ Combustion *Journal of Canadian Petroleum Technology* **54** 220-232

[13] Greaves, M 2000 Upgrading athabasca tar sand using toe-to-heel air injection *SPE/CIM International Conference on Horizontal Well Technology* (Society of Petroleum Engineers)

[14] Shah, A A, Fishwick, R P, Leeke, G A, Wood, J, Rigby, S P & Greaves, M 2011 Experimental optimization of catalytic process in situ for heavy-oil and bitumen upgrading *Journal of Canadian Petroleum Technology* **50** 33-47

[15] Xia, T, Greaves, M, Turta, A & Ayasse, C 2003 THAI—a 'short-distance displacement'in situ combustion process for the recovery and upgrading of heavy oil *Chemical Engineering Research and Design* **81** 295-304

[16] Hart, A, Leeke, G, Greaves, M & Wood, J 2014 Down-hole heavy crude oil upgrading by CAPRI: Effect of hydrogen and methane gases upon upgrading and coke formation *Fuel* **119** 226-235

[17] Hajdo, L, Hallam, R & Vorndran, L. 1985 Hydrogen generation during in-situ combustion *SPE California Regional Meeting* (Society of Petroleum Engineers)

[18] Hart, A, Shah, A, Leeke, G, Greaves, M & Wood, J 2013 Optimization of the CAPRI process for heavy oil upgrading: effect of hydrogen and guard bed *Industrial & Engineering Chemistry Research* **52** 15394-15406

[19] Desena, M F M, Rosa, L P & Szklo, A 2013 Will Venezuelan extra-heavy oil be a significant source of petroleum in the next decades *Energy policy* **61** 51-59

[20] Carrillo, J A & Corredor, . M 2013 Upgrading of heavy crude oils: Castilla *Fuel Processing Technology* **109** 156-162

[21] Zou, R P & Yu, A B 1996 Evaluation of the packing characteristics of mono-sized non-spherical particles *Powder Technology* **88** 71-79

[22] Zhang, W, Thompson, K E, Reed, A H & Beenken, L 2006 Relationship between packing structure and porosity in fixed beds of equilateral cylindrical particles *Chemical Engineering Science* **61** 8060-8074

[23] Weissman, J, Kessler, R, Sawick, R, Belgrave, J, Laureshen, C, Mehta, S, Moor, R & Ursenbach, M 1996 Down-hole catalytic upgrading of heavy crude oil *Energy & Fuels* **10** 883-889

[24] Huang, C, Li, C & Shi, G 2012 Graphene based catalysts *Energy & Environmental Science* **5** 8848-8868

IOP Conf. Series: Materials Science and Engineering **579** (2019) 012031 doi:10.1088/1757-899X/579/1/012031

Study of different geostatistical methods to model formation porosity (Cast study of Zubair formation in Luhais oil field)

Jassim M. Al Musawi[1#], Mohammed S. Al Jawad[2]

[1]Petroleum Technology Department, University of Technology, Baghdad, Iraq.

[2]Petroleum Engineering Department, College of Engineering, University of Baghdad, Baghdad, Iraq.
[#]Corresponding author's e-mail address: Jassimm.jaleel@yahoo.com

Abstract: This study is concerned with making comparison in using different geostatistical methods for porosity distribution of upper shale member - Zubair formation in Luhais oil field which was chosen to study.
Kriging, Gaussian random function simulation and sequential Gaussian simulation geostatistical methods were adopted in this study. After preparing all needed data which are contour map, well heads of 12 wells, well tops and porosity from CPI log. Petrel software 2009 was used for porosity distribution of mentioned formation in methods that are showed above. Comparisons were made among these three methods in order to choose the best one, the comparing criteria was according to different statistical information and variograms analysis of entered porosity which was scaled up and modeled. The best method gave porosity distribution in model closest to enter porosity that represent real porosity of Zubair formation.
The comparison proved that sequential Gaussian simulation is the best model of porosity distribution followed by Gaussian random function simulation and kriging methods respectively. Based on the results obtained, it was concluded that the distribution accuracy of the porosity or other petrophysics properties is one of the main factors that affect building correct geological model that represents the base of reservoir model and also pore volume (Oil reserve) value was affected by geostatistical methods variation.

1. Introduction

Geostatistical modeling is accepted in petroleum industry as a technique of heterogeneous reservoir characterization, it preserve variance data instead of just the mean value, as in interpolation [1].
Geostatistical modeling deals with large number of different methods of properties estimation such as different Gaussian and kriging Techniques [2].
Kriging techniques are the estimation methods of petrophysical properties (i.e. porosity) that use variogram for showing the spatial variability [3], so simple kriging (SK) was used as one of the methods that utilized in this case study.
Another method of petrophysical properties distribution is sequential Gaussian simulation (SGS) that beginning with defining values univariate distribution e.g., Evaluation of grade values, performing an original score transformation of values to a standard normal distribution, and assuming multi normality of normal scores [4].
Gaussian random function simulation (GRFS) is subsequently differ from sequential Gaussian simulation in accuracy of distribution, it is faster with additional efficiencies due to its parallel architecture [5].

For many years, variogram had been used for the determination of the spatial phenomena variability, sometimes variogram calculation, modeling and interpretation were performed hastily and the wrong must be skipped with taking in account more attention to establish a durable model of spatial variability (variogram) before the beginning to make numerical reservoir models [6].
By using these methods, porosity value at a location in space was determined, firstly by getting the probability distribution at that location, and then plotting data randomly (i.e. simulated porosity values) from this distribution. The simulated values of porosity reproduced statistical and spatial patterns of the input data [7].
Porosity values at some locations of Zubair formation in Luhais field was entered, scaled up and distributed in all of the above geostatistical methods. The comparison among these models is accomplished according to statistical and variograms analysis.

1.1 Case study area

Luhais oil field that located a way 100 km from north-west Basrah is producing from two reservoirs (Nahr omer and Zubair), Zubair formation of said field was chosen to this study because of the availability of data.
Zubair formation represent the main reservoir that contains high amounts of hydrocarbons in southern Iraq which is deposited in lower cretaceous, generally it is a sequence of sandstone, shale and siltstone layers. Mainly it is divided into three members, upper shale member, sandstone and lower shale members. The current study is for upper shale member, where the oil cumulated, while the rest of the members are water bearing [8] [9].

1.2 Geostatistical review

Geostatisics: is a continuous development branch of math and statistics applications, it offers numbers of tools for spatial variability modeling and its understanding. Geostatisics was found previously in mining industry, new estimation method was developed by D.G krige, a South African engineer, and H.S sichel, a statistician at early of 1950s where classical statistics became inappropriate for the estimation of disseminated ore reserves, Krige's concept was expanded and formalized within a single framework, and coined the word "kriging" in recognition of Krige's work, by The French engineer Georges Matheron.
Kriging technique was developed to solve problems of ore-reserve estimation, in 1970 where the high-speed computer was turn up, Kriging technique was spread to many other areas of geology.
Although gesotatisics techniques have been developed and grown steadily and significant, it did not use in petroleum industry before 1980s.
Engineers and geoscientists are facing complex relationship and huge geographical extension of geological phenomena with scientific requirements for preparing geological model involving the measurement of interest minute volume or area [10]. So, the main purpose of geostatisics is spatial geological systems characterization that are incompletely known [11].

1.3 Variogram theory

Variogram is an important input for studying of geostatistecs, it represents a good tool for investigation and modeling spatial variability of petrophysical properties and lithology. Variogram is used in 90% of geostatistical reservoir characterization studies as a basic geostatical modeling methods [6].

1.3.1 Experimental variogram

It is a special function that calculated from variability measurement between pairs of points at various distances, mathematical representation is [2] [12]:

$$2\gamma(h) = E\{[z(u) - z(u+h)]\}^2 \qquad (1)$$

Semivariogram: is a variogram one half and its mathematical representation is:

$$\gamma\ (h) = \frac{1}{2N(h)} \sum\nolimits_{N(h)}^{n} [\,z(u) - z(u+h)]^2 \qquad (2)$$

1.3.2 Variogram parameters

Sill there is a total variance where the empirical variogram appears to level off.

Range is the distance or lag at which the sill is reached. (i.e the distance after which data are no longer correlated.)

Nugget is an apparent intersection with Y-axis [13].

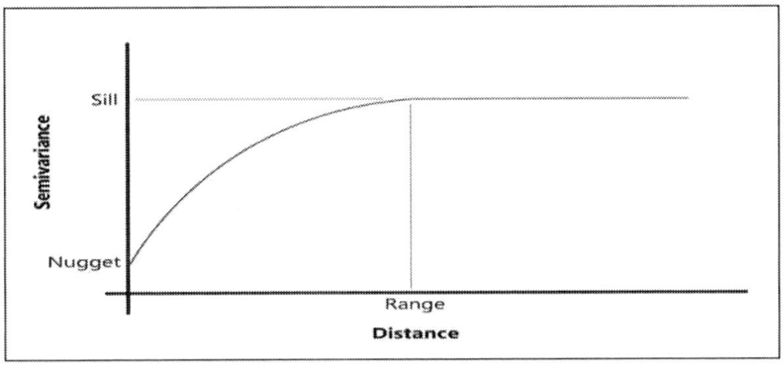

Figure 1. Variogram parameters

1.3.3 Variogram models

Experimental variogram is not adopted directly in geostatistical modeling methods, so its points must be fitted to parametric variogram model [2]. In petrel software, there are three variogram models points:

1) Spherical model

Represented by quadratic modified equation where spatial dependence speared out as the sill and range.

$$\gamma\ (h) = c[\frac{3h}{2a} - \frac{h^3}{2a^3})] \qquad (3)$$

$a \geq h \geq 0$, $\gamma\ (h) = c, h \succ a$

2) Exponential model

It is similar to spherical model in variability with distance reaches to sill gradually. Two samples points relationship is decaying gradually, but at a distance of infinite spatial dependence dissipates.

$$\gamma\ (h) = c[1 - \exp\left(-\frac{3h}{a}\right)] \qquad (4)$$

3) Gaussian model

It is using normal probability distribution curve.

Where phenomena are identical at short distance this model type is used because its progressive rise up the y-axis.

$\gamma(h) = c[1 - \exp\left(-\frac{3h^3}{a^3}\right)]$ (5)

So, the above three equations variables are (c is sill- nugget, h is a spatial vector, distance from central point and a is an effective range (the distance where h equal 95% of sill).
The three variograms models are existing in Petrel showed below in Fig.2 [14].

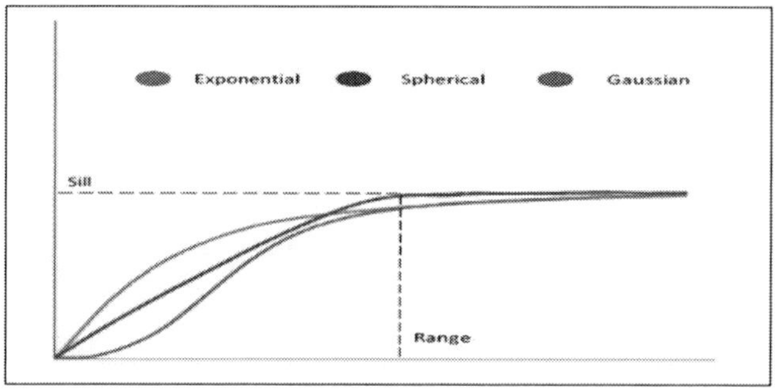

Figure 2. Exponential, spherical and Gaussian theoretical variogram models.

1.4 Kriging

Kriging is a number of popularizing linear regression methods for decreasing a guesses variables from covariance of past model.
All kriging techniques are detailing on the logarithmic basic of linear regression generalization and corresponding estimator.
Simple kriging technique (SK) assumes any set of weight $\lambda\alpha$ (u) are linearizing to zero in calculation and this leads to minimize the estimation variance from solving the equations for these weights [12]. Its expression as:

$$[Z^*_{SK}(u) - m(u)] = \sum_{\alpha=1}^{n} \lambda_\alpha(u)[Z(u_\alpha) - m(u_\alpha)] \quad (6)$$

The classical stationary system of simple kriging:

$$\sum_{\beta=1}^{n} \lambda_\beta(u)C(u_\beta - u_\alpha) = C(u - u_\alpha), \alpha = 1, \ldots, n \quad (7)$$

1.5 Sequential Gaussian simulation

In the reservoir modeling applications, Gaussian sequential simulation (SGS) represents a common approach which was adopted in recent periods. Simply, flexibility and reasonability are main feature of this method.
The following steps show the SGS procedure:

1) Original z data must be transformed to normal distribution.
2) Place transforming data into the model.
3) Seek all neighboring and previously simulated data of random function u.
4) Get kriged guesses and the corresponding kriging by achieving kriging on these above and previously simulated data.

Variance:

$\Upsilon^*(u) = \sum_{\beta=1}^{n} \lambda\beta \,.\, \Upsilon(u\beta)$ (8)

$\sigma^2 sk(u) = c(0) - \sum_{\alpha=1}^{n} \Upsilon\alpha \, c(u, u\alpha)$ (9)

5) Plot a random residual R(u) that follows a normal distribution with mean of zero and variance of $\sigma^2 sk$

(u).
6) Get simulated value by adding the kriged estimate to residual:
$Ys(u) = Y^*(u) + R(u)$ (10)
Ys(u) must be acquired from the drawing normal distribution with mean Y*(u) and variance $\sigma^2 sk$ (u)>
7) Correction of future predictions is tested by adding Ys(u) to data set to ensure that covariance with this value.
8) Revise the location of random order.
9) When model is populated, all data values and simulated values are backed transform.

10) By repeating the above steps with different random numbers seeds, you can create any number of realization [2].

1.6 Gaussian random function simulation
Petrel software introduced a new Gaussian simulation algorithm, geostatistical simulation library and user's guide (GSLIB) showed substantially difference's from the sequential Gaussian simulation:
1) Typically, It is faster than SGS.
2) It is not a sequential algorithm.
3) It has been parallelized.
4) It has a fast co-located co-simulation option.
As showed above, other than SGS, it is not a sequential algorithm and this has permitted to be parallelized algorithm.
It is based on well-known decomposition which stated:
Conditional simulation = Kriging + Unconditional Simulation (11)
So, new parallel kriging algorithm will use as a base.
When correct algorithm is used, unconditional simulations can be very fast. Fourier Transform is a basic of a faster algorithm that is used. By is parallelizability it is showing a good variogram reproduction for a wide class of variograms [3].

2 Methodology
This case study work was started with preparing data of contour map, well heads, well tops, and CPI porosity, and finished with making Zubair surfaced that had distributed porosity in three showed methods.

2.1 Structural modeling
Structural modeling of Zubair formation was built by applying different statistic algorithms in Petrel software such as convergent interpolation on the contour map, inserted well tops and the plotted polygon as a boundary.
Fig.3. is a structural contour map of Zubair formation top, it represents a clear image of formation that has four anticline
folds with axis trending NW- SE, the length of structure is 16.079 km and width is 15.218 km.

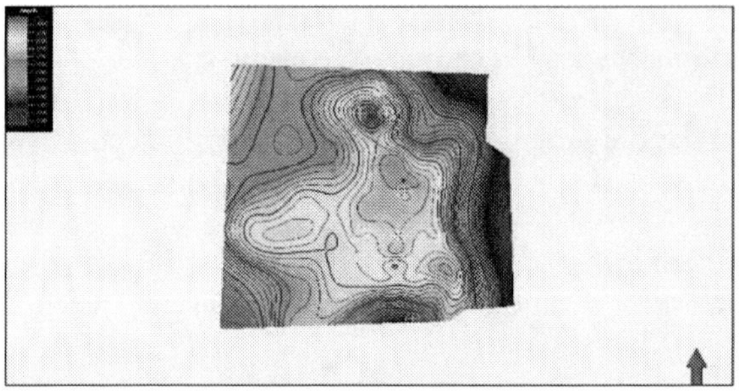

Figure 3. Structural contour map of Zubair formation top by convergent interpolation.

2.2 3D grid construction

Zubair formation was represented by three dimensional grid system of 53 cells along the x- axis and 51 cells along the y- axis and 75 layers along the Z- axis. The size of grids was 300 m in width, 300 m in length and the total number of cells is 202725 cells as shown in fig.4.

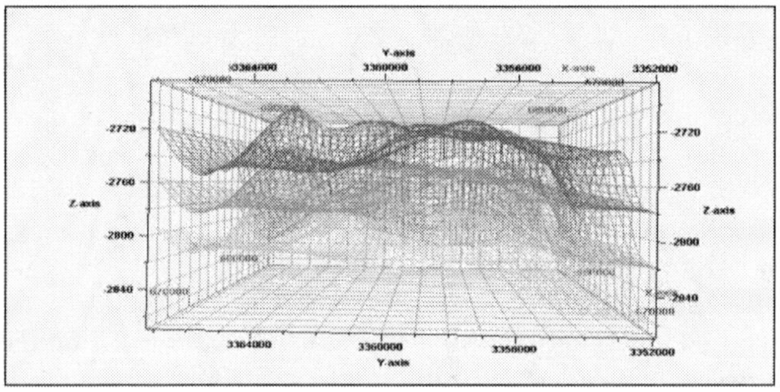

Figure 4. The skeletons of Zubair formation in Luhais field.

2.3 Layering

The average thickness of the mentioned formation in 12 Luhais wells is approximately 60 m. Previous reservoirs studies were divided this formation to seven zones according to the formation evaluation process, but in this case study, it was taken as single zone and divided to 75 layers for distribution porosity at each layer and that for easy interpretation to results of statistics and variograms analysis. Top layer was selected as a sample in the illustrative figures.

2.4 Scale up of well log porosity

Scale up is the process of averaging porosity by any selected algorithm to produce one porosity value that will fall in each grid cell of skeleton. Each grid cell in skeleton will be used as start point for porosity modeling.

Before entering into the grid, porosity (or any petrophysical property) of well log must be scaled up, this process is also called blocking of well logs as shown in Fig.5.

Many statistical methods are used for scale up such as (arithmetic mean, harmonic, geometric mean method and etc….), where arithmetic average is always utilizing for numbers of properties such as porosity, water saturation, and net/gross, so arithmetic average method was used for porosity scale up

[15].

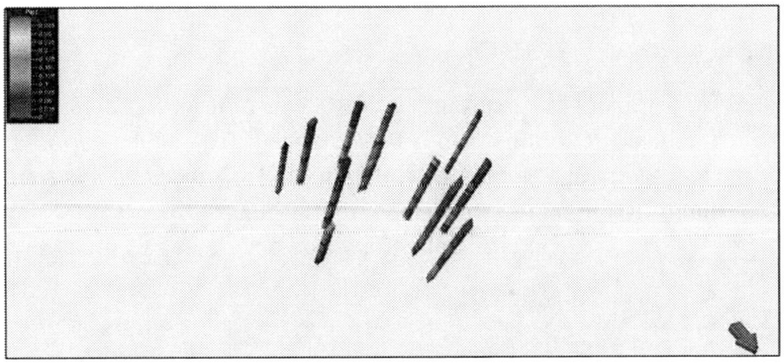

Figure 5. Porosity scale up of Zubair formation

2.5 Histograms

Histogram is a corresponding graph of data sets and its frequency, a frequency records values is showing how often observed values fall within certain intervals or classes [16].
Histograms of entered porosity, scale up and the distribution porosity in all layers by three methods recalled above were shown in Fig.6, 7 and 8, pinky column is well log porosity while the scale up porosity is represented by green column and blue column is the modeling porosity.

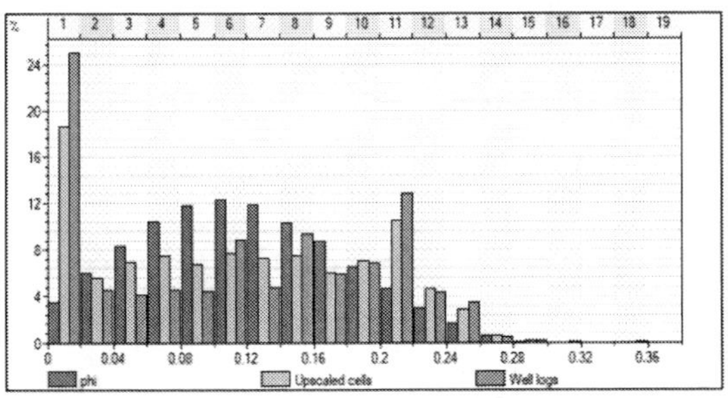

Figure 6. Histogram of entered, scale up and porosity modeled by sequential Gaussian.

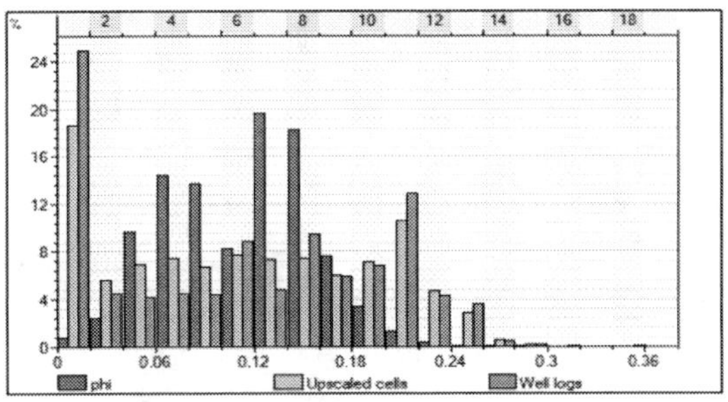

Figure 7. Histogram of entered, scale up and porosity modeled by kriging.

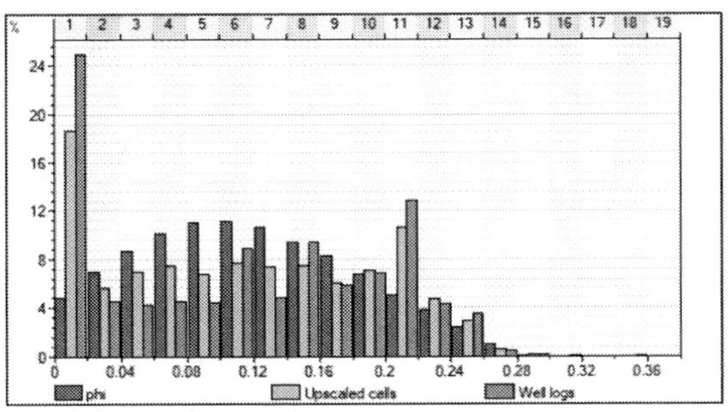

Figure 8. Histogram of entered, scale up and porosity modeled by Gaussian random function.

2.6 Variogram map and directions

Variogram is calculated for a number of directions and distances in reservoir and its values posted on a map called variogram map, the center of this map is lag distance (zero), so the variogram map takes the subject of calculating the variogram in a number of directions and distances, so determination major and minor directions in continuity is the main target of variogram map. Fig.9. shows the 2-D variogram maps that calculated from porosity. Geometric anisotropy that occurs in 2-D areal direction is clear by elliptic shape [17].

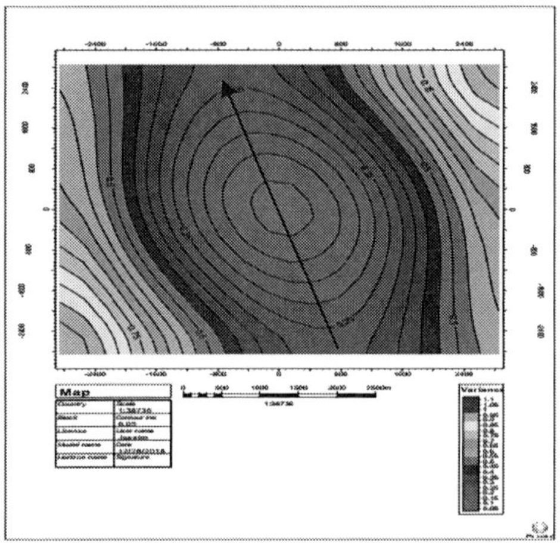

Figure 9. 2D variogram map from porosity data.

Variogram two horizontal direction below in Fig.10 are according to 2D variogram map in Fig.9, the number of lags was determined as 17 lags with 909.1 m distances.
Porosity variability with different scale distances and directions is explained by variogram calculation and interpretation as be shown in results.

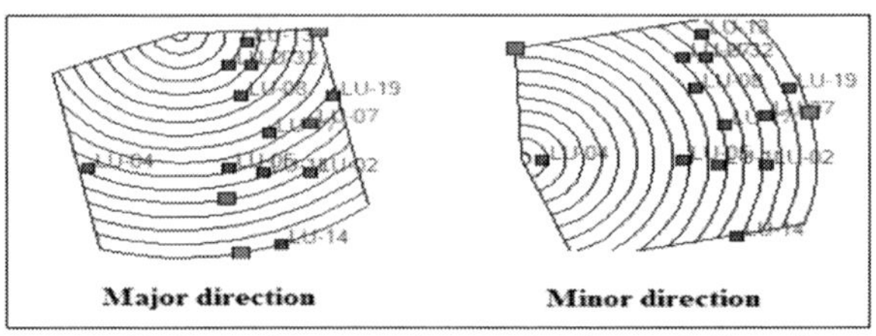

Figure 10. Location map in two horizontal direction

2.7 Reservoir modeling

As clarified above, porosity was distributed in three methods, these distributions were in 75 layers that subdivision of formation. Fig 11, 12 and 13 below show the porosity distribution of Zubair formation top in three mentioned methods and the color scale is also shown in these figures. According to this scale, red and yellow represent higher porosity values, while blue and purple are lower ones.

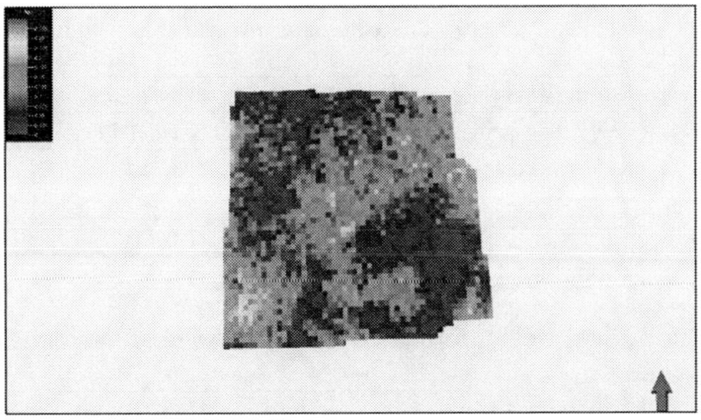

Figure 11. Porosity distribution of Zubair formation top by sequential Gaussian method

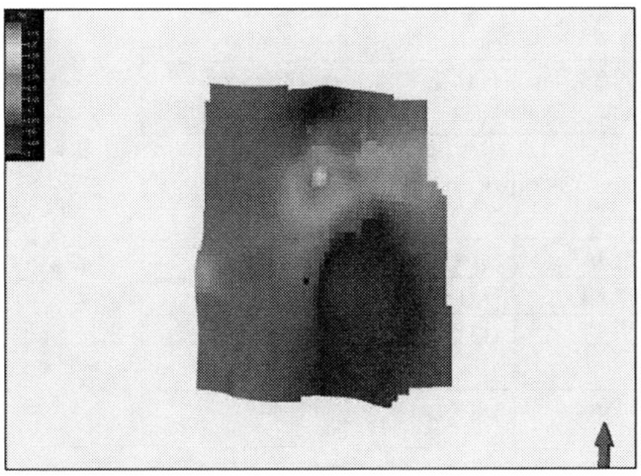

Figure 12. Porosity distribution of Zubair formation top by kriging method.

Figure 13. Porosity distribution of Zubair formation top by Gaussian random function method.

3. Results and Discussion

The results of this case study were obtained from Petrel software after entered all mentioned data and the made model.

The results were statistical information, variograms and pores volumes calculation, Comparison has been made between the results and finished with choosing the best distribution of porosity model that is closest to real reservoir and pores volume calculation in each method.

3.1 Statistical analysis

Statistical analyses were performed for entered porosity log of 12 well, scale up and the distribution porosity by sequential Gaussian, kriging and Gaussian random function models on Zubair formation layers. Tables 1, 2 and 3 involved the estimations of basic univariate statistics including the mean, standard deviation and variance, and these estimation values gave an indication for closest results of sequential Gaussian and Gaussian random function techniques to scale up and entered porosity, While kriging results were further away from them, but these results do not make a decision for what is the best method, so the variogram analysis is a key for determination of the best porosity distribution model.

Table 1. Statistical information results of sequential Gaussian modeling.

Porosity type	Mean	Std.	Variance
Well log porosity	0.0814	0.0871	0.0076
Up scaled porosity	0.0953	0.0897	0.0081
Porosity modeling	0.1095	0.0558	0.0031

Table 2. Statistical information results of Kriging modeling.

Porosity type	Mean	Std.	Variance
Well log porosity	0.0814	0.0871	0.0076
Up scaled porosity	0.0953	0.0897	0.0081
Porosity modeling	0.0963	0.0225	0.0005

Table 3. Statistical information results of Gaussian random function modeling.

Porosity type	Mean	Std.	Variance
Well log porosity	0.0814	0.0871	0.0076
Up scaled porosity	0.0953	0.0897	0.0081
Porosity modeling	0.1194	0.0660	0.0044

As shown in Fig. 6, 7 and 8 the scale up histograms and distributed porosity by random and sequential Gaussian models histograms were showed convergence to entered porosity rather than kriging model.

3.2 Variograms analysis

For the variogram analysis of 12 wells porosity producing from Zubair formation that distributed in three listed methods, was utilized to calculate corresponding experimental variograms.

Firstly, average of experimental variogram was calculated for scale up of entered porosity that represent the real date and reference for comparing the modeling methods, Fig.14 shows the experimental variogram result that fitted to exponential variogram model, there are small error in porosity values that measured by well logging because of the zero-nugget effect except the value nugget (0.251) in major direction (The area of most wells) that somewhat was acceptable. The horizontal experimental variogram changed from positive correlations to negative correlations at a length scale that indicate geologic cycles, these cycles according to varies in rocks properties along formation. In major direction there is some deviation because of the experimental variogram reach to sill before exponential model that happen because of large distance between data (i.e. distance between LU 13 and LU4). The presence of a number of layers extend along large distance of reservoir and the porosity difference of two separated wells by large distance (i.e. LU19 and LU4) made the minor horizontal and vertical Variograms did not reached or reached near to sill and that called areal trends and layering. [6] The exponential variogram model results is shown in table 4.

IOP Conf. Series: Materials Science and Engineering **579** (2019) 012031 doi:10.1088/1757-899X/579/1/012031

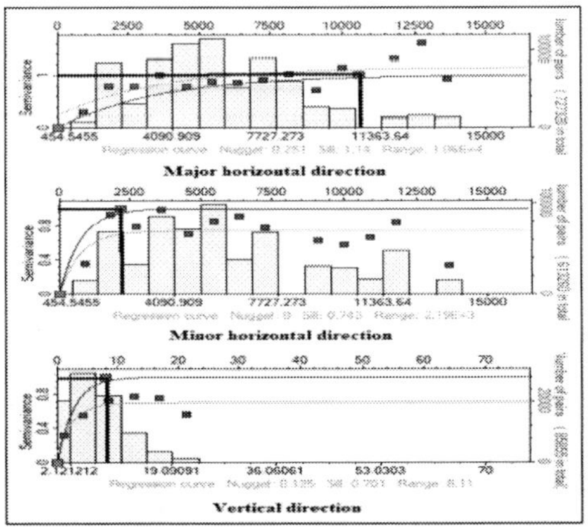

Figure 14. Average experimental variogram of scale up porosity in three direction.

Table 4. Results from exponential variogram model analysis of scale up porosity.

Sill	1
Nugget	0
Major range (m)	10606.1
Minor range (m)	2185.8
Vertical range (m)	8.1

Porosity was transformed to normal score because the using of Gaussian technique required data transformation, Fig.15 shows the experimental variogram of porosity distributed by sequential Gaussian model. It is fitted to exponential variogram model. The comparison of these variograms with scale up entered porosity variograms gave the same behavior and very close results as shown in table 5.

Experimental variograms of porosity distributed in kriging and Gaussian random function were calculated and fitted to exponential variogram model and that clearly was shown in Fig.16 and Fig.17. Comparison between variograms of kriging porosity model with entered scale up porosity showed a spacing in behavior and results, because the variograms of this method was indicated to zonal anisotropy and large deviation as shown in Fig.16 and results in table 5.

The last comparison between porosity Gaussian random function modeling and entered porosity scale up variograms, Fig.17 elucidated the high nugget values that indicating a discontinuity in porosity values and that did not appear in Fig.12, also it showed zonal anisotropy. [12]

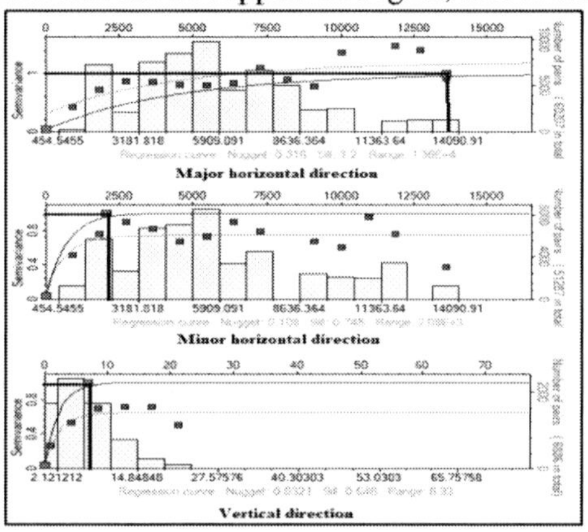

Figure 15. The average of experimental variogram of porosity modeling by sequential Gaussian method of Zubair layers in three directions.

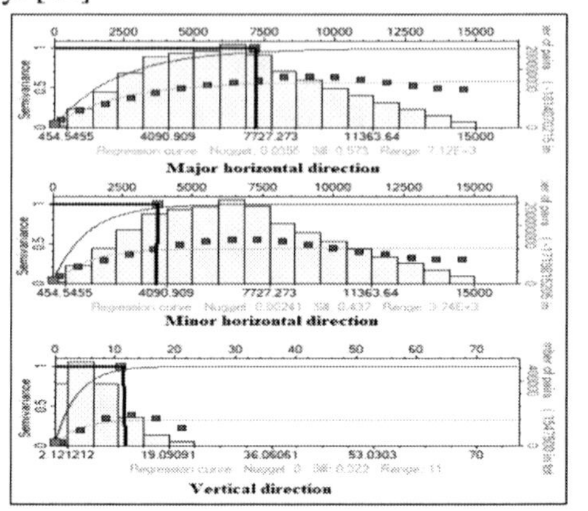

Figure 16. Average experimental variogram of porosity modeling by kriging method of Zubair layers in three directions.

Table 5. Results from exponential variograms models analysis of porosity modeling by three methods.

m	Sequential Gaussian modeling	Kriging modeling	Gaussian random function modeling
	1	1	1
	0	0	0
)	13636.4	7123.1	9697
)	2080.5	3744.7	4695.5
Vertical range (m)	6.9	11	9.4

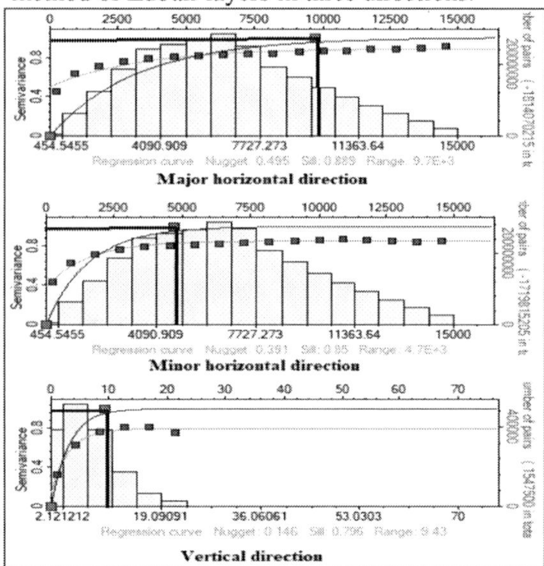

Figure 17. Average experimental variogram of porosity modeling by Gaussian random function method of Zubair layers in three directions.

From statistical analysis, above comparisons and according to the results in tables 4 and 5, the best porosity modeling technique in upper shale member of Zubair formation is sequential Gaussian followed by Gaussian random function and kriging techniques respectively.

IOP Conf. Series: Materials Science and Engineering **579** (2019) 012031 doi:10.1088/1757-899X/579/1/012031

3.3 Pores volumes calculation

Using different geostatistical models for porosity distribution have clear effect on pore volume, where each model was given a different pore volume value as shown in table 6. That means oil reserves values are similarly changed accordingly.

Table 6. Pore volume calculation results at each modeling method.

Modeling method	Pore volume (m^3) 10^6
Sequential Gaussian	294
Kriging	267
Gaussian random function	283

4 Conclusions

For building porosity distributions model, contour map, well heads and tops, and CPI porosity log data must be available. Statistical parameters analysis (one-dimension parameters) is not enough for making decisions about property modeling so any reservoir modeling study must be modeled by variogram (two-dimension parameters) interpretations where the variogram is used throughout geostatistical reservoir modeling as a measure of spatial variability, so it will represent the true reservoir heterogeneity if it modeled. Sequential Gaussian simulation was the best method of porosity distribution of upper shale member in Zubair formation. Accuracy of porosity on any petrophysical modeling means building of a good geological model that represent the structure of reservoir model. Difference using of Geostatical methods in porosity modeling effect on pore volume values which means changing in oil in place.
According to the three models comparisons, Pore volume value (294 e^6 m^3), that calculated by using porosity distributed in Sequential Gaussian model, absolutely it represents the real pore volume value of upper shale member of Zubair formation.

5 Nomenclatures

$2\gamma(h)$ = Variogram.
$z(u)$ = property value at position u.
h = lag distance.
$z(u + h)$ = property value at position u + h.
$N(h)$ = number of pairs for lag.
$Z(u)$ = random function model at u location.
$u\alpha$ = are n data locations.
mu = $E\{z(u)\}$ = RV $z(u)$ expected values that depended on location.
$Z^*sk(u)$ = linear regression estimators also called simple kriging estimators.
$\lambda\alpha(u)$ = S.K. weights.
$C(u\alpha, u\beta)$ = covariance matrix.
CPI= computer program interpretation

Acknowledgements
This study was carried out on Luhais oil field – Upper shale member of Zubair formation under supervision my destiny Prof. Dr. Mohammed Al-Jawad with the support of the Ministry of Oil to provide the necessary data and helping Dr. Abdilalli Al Dabaj and Eng. Ahmed Alher.

References

[1] Society of Petroleum Engineers 2016 Geostatistical conditional simulation (Petrowiki website) https://petrowiki.org/Geostatistical_conditional_simulation
[2] Pyracz M and Deutsch C 2014 *Geostatistical reservoir modeling* (New York: Oxford university press) ed 2 p 116
[3] Schlumberger 2009 *Petrel technical manual.*
[4] Nowaki M and Verly G 2005 *Quantitative Geology and Geostatistics volume 14* (Dordrecht, Netherlands: Springer)
[5] Daly C, Quental S and Novak D 2010 *Geoconvention Conf. On Working with the Earth (Alberta)* (Calgary: Canada) p 1
[6] Gringarten E and Deutsch C. V 1999 *SPE Annual Technical Conference and Exhibition* (Houston) (University of Alberta) p 5
[7] Sahin A and Al-Salem A. A 2001 *SPE Middle East Oil Show Conf.* (Bahrain) p 1
[8] Ministry of Oil 1986 detailed reservoir study Reservoir and Development Department *Ministry of Oil Report*
[9] Mohammed A 2017 *Static and Dynamic Predictive Modeling of Zubair Reservoir in Luhais Oil Field in Iraq* MSc thesis University of Technology- Petroleum Technology Department
[10] SPE 2016 Origin of geostatistics https://petrowiki.org/Geostatistics#Origin_of_geostatistics
[11] Olea R 2009 A Practical Primer on Geostatistics *U.S. Geological Survey Report*
[12] Deutsch C and Journel A 1998 *GSLIB Geostatistical Software Library and User's Guide* (New York: Oxford University Press) ed 2 p 44
[13] GisGeography 2018 https://gisgeography.com/semi-variogram-nugget-range-sill/
[14] Gunnarsson N 2011 *3D modeling in Petrel of geological CO_2 storage site* (Uppsala: Uppsala University, Department of Earth Sciences, Geotryckeriet) p 12
[15] Salman S J 2015 *Three dimensions geostatistical model of the Mishrif carbonate reservoir of Amara oil field in Iraq* MSc thesis University of Baghdad, Collage of Engineering, Petroleum Engineering
[16] Isaaks E and Srivastava R 1989 Applied Geostatistics (New York: Oxford University Press) p 11
[17] Abdullah A A 2014 *Geostatitical of Mishrif reservoir in Noor oil filed in Iraq* MSc thesis University of Baghdad, Collage of Engineering, Petroleum Engineering

IOP Conf. Series: Materials Science and Engineering **579** (2019) 012032 doi:10.1088/1757-899X/579/1/012032

Wells location effect on the underground gas storage in UM ERadhuma Formation-Ratawie oil field.

Hayder S. Fukaa Al asida[1#] and Sameera M. Hamd-Allah[2]

[1]Ministry of oil/ petroleum research and development center (PRDC), Baghdad, Iraq

[2]Petroleum Engineering Department, University of Baghdad, Baghdad, Iraq

[#]Corresponding author's e-mail address: Haideralkhafaje123@gmail.com

Abstract. The aim of this study is to investigate the feasibility of underground gas storage in UM ERadhuma formation /Ratawie oil field. This formation is high permeability bed according to the available coring data, dolomite, and water bearing and located below RUS formation which consists of impermeable anhydrite beds. Interactive petrophysics (IP), Petrel RE and Eclipse 100 softwares are used for well log interpretation, conducting a reservoir simulation model and predicting the reservoir behaviour during storage respectively. A black oil, three dimensional and two phase fluid model has been used. The results showed that upper reservoir of UM ERadhuma formation is good for underground gas storage because of the tightness, of its cap rock and its properties for storing gas. It has been found that the available volume for storage is 14.3 billion cubic feet with a structural closure of 45 m. moreover, two groups of suggested injection wells location are used during this study to investigate the effect of wells location on the injection behaviour and it has been found that high space wells location is more suitable for the gas injection.

1. Introduction

Natural gas is playing an important role in meeting the world's energy demand because of the clean burning and its huge energy. According to the International Energy Agency, in the next twenty years, the average growth of natural gas demand is expected to be over 1.8% per year. Also, there is no doubt that natural gas has a changeable demands according to the months of the year, whether they are cold or hot. [1]

Natural gas storage is used to face the expected future peak demand, maintain the balance between supply and demand and to save the non-invested gas from burning and losing[1].

Large volumes of natural gas don't invested and lost by burning in Iraq because the main type of natural gas is the associated gas and needs high capitals of money to be invested [2]. For example, more than1.5MMMSCF/D were burned in 2016 which means, 7500 ton of liquefied gas, 1.2 MMMSCF of dry gas and 1650 ton of benzene have been lost[2]. Therefore; a new technique for gas storage must be used. In fact, underground gas storage is not very modern but it is highly used by many countries in the world such as Europe, US and Russia [3].

2. Methodology

In current study, two phase and 3D black oil reservoir simulation model have been conducted to present the aquifer behavior during the injection processes. Therefore; Petrel RE and Eclipse softwares have been used to build the reservoir model. The four main stages that have been conducted during this study were:

2.1 Static geological model

Static geological model is very important in reservoir description and represents the beginning for dynamic fluid model preparation. Therefore, during this study a geological model for RUS and UM ERadhuma formation have been built by using Petrel RE software and depending on well logs, core data, well heads and tops of 7 wells of Ratawie field which were; RT-3, 6, 7, 8, 9, 10 and 11.

2.2 Dynamic fluid model

Dynamic fluid model is important to predict the reservoir behavior with time under any process such as; production, injection, stimulation, etc..... Therefore, a black oil 3D dynamic fluid model has been built for RUS and UM ERadhuma formations to investigate their behavior during the injection process.
Dynamic model have been built depending on the static geological model, initial reservoir properties (**Table 1**), relative permeability data [4], capillary pressure data [4], formation water properties [5] (**Table 2**) and injected gas properties [6] that has been used for injection was from zubair oil field because its data was available. Figure 1, 2 and 3 represent the relative permeability data, capillary pressure data and the injected gas properties respectively.

Table 1:- initial reservoir properties.

Reservoir pressure	Datum depth (ft)	Water contact(ft)	T (F^{o})
1450 psi	-3000	-2000	110

Table 2: - formation water properties [5].

Water density(lb/ft^3)	Comp. 1/psi	BW (RB/STB)
67.549	$2.4*10^{-5}$	1.0003

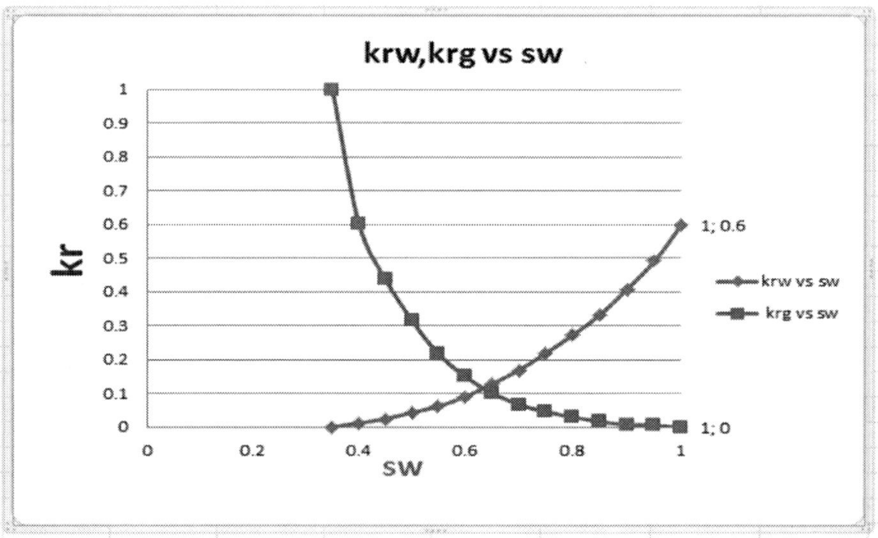

Figure.1:- relative permeability of gas and water for UM ERadhuma formation. [4]

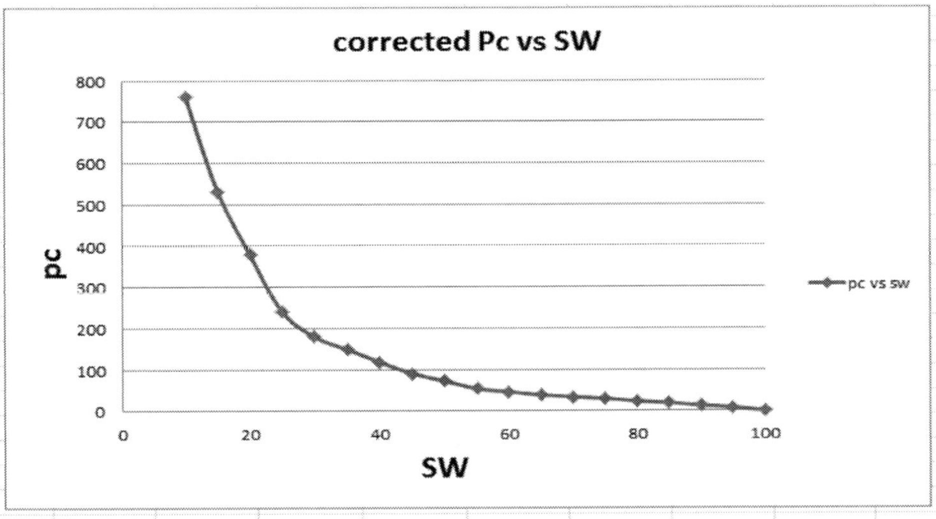

Figure. 2:- capillary pressure curve for UM ERadhuma formation. [4]

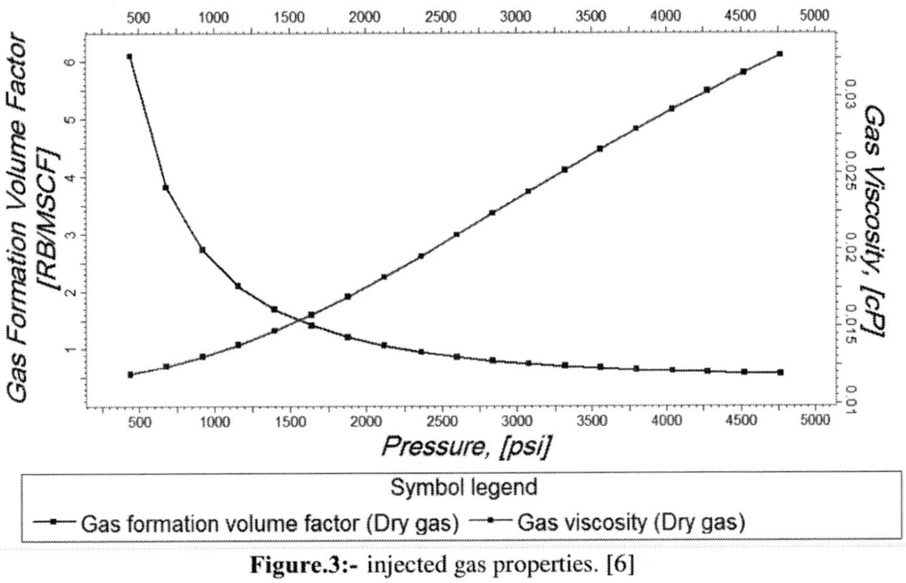

Figure.3:- injected gas properties. [6]

2.3 History matching

History matching is essential to insure that the constructed model represents the reservoir. Because of UM ERadhuam formation is an aquifer and located at shallow depth (800-850 M), no production data are available. However, history matching in this study was depending on 6 days DST tests which are conducted on the well RT-10 in 1984.Figure 4 and 5 represent the obtained history matching of the well RT-10 for water production rate and bottom hole pressure respectively.

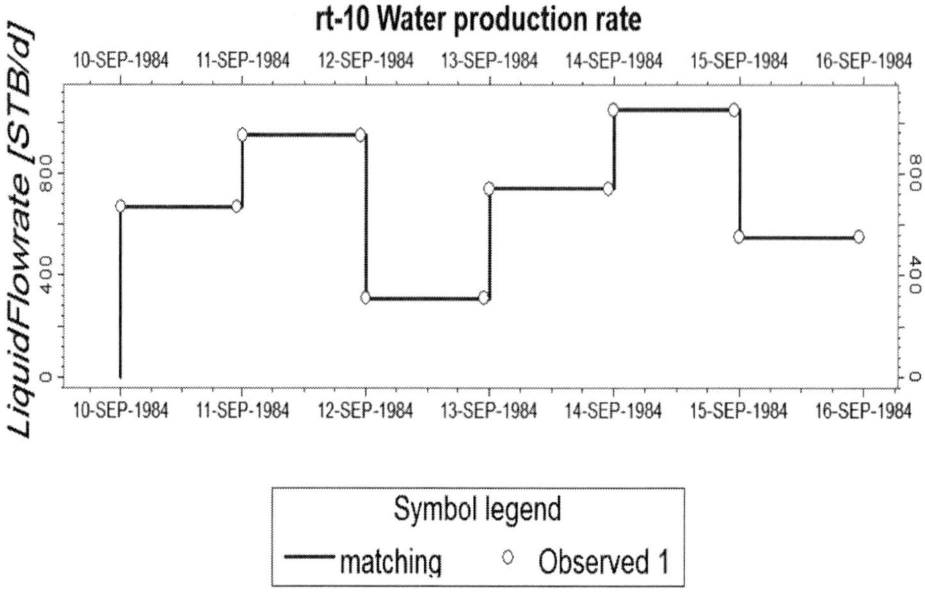

Figure.4: Water production rate history matching of well RT-10.

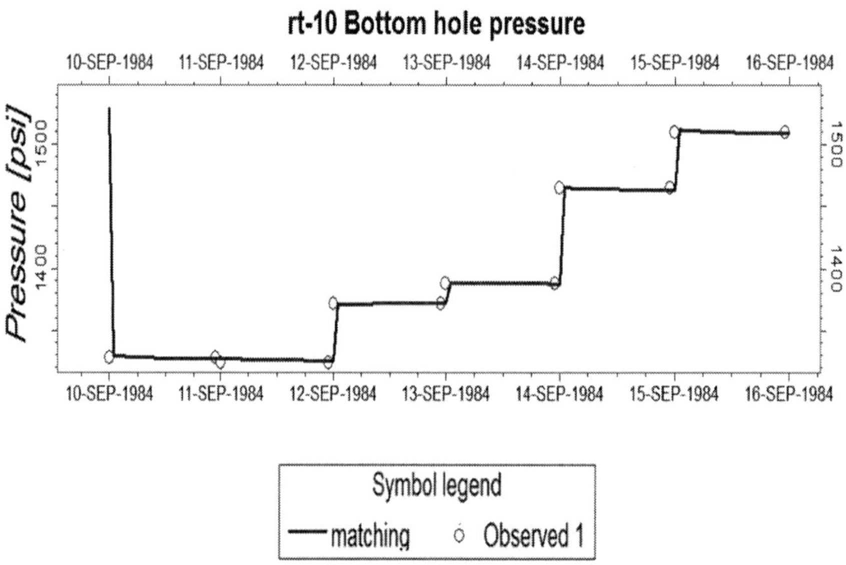

Figure.5:- Bottom hole pressure history matching of well RT-10.

2.4 Prediction of reservoir behaviour during the injection process

In order to predict the reservoir behavior during the injection process, 9 wells have been assumed to be drilled in two groups of wells location (A and B) in addition to the wells RT-8 and RT-10 to use them for the gas injection. These two groups are different in wells location (high spacing and small spacing). Furthermore, the choice of the location is depended on the permeability distribution. Simulation injection period was of 10 years of injection with a perforation interval of 19 m and constant bottom hole pressure (BHP) of 1600 psi which is less than fracture pressure of 1800 psi. Figure 6 and 7 represent the wells location of group A and B respectively.

3. Results

3.1 Available storage volume

Volumetric calculation have been conducted on the geological model to find the total pore volume according to the structural closure which is 45 m. Figure.8 shows the 3D porosity distribution of the upper reservoir unit and its cap rock. However, the result showed that the total pore volume of the structural closure of upper reservoir of UM ERadhuma formation is 22 billion cubic feet. Additionally the available storage volume is equal to 14.3 billion cubic feet which is calculated by using the following equation.

Available volume = total pore volume of the closure area * (1-swc)

Where:-

Swc: - connate water saturation which is equal to 0.35 acoording to the relative permeability data.

3.2 Effect of injection wells location

Two cases have been conducted on the upper reservoir during this step with two different 11 wells locations (group A and B) as shown in Fig.6 and Fig.7 at constant BHP of 1600 psi, perforation interval of 19 m and injection period of 10 years. The injection processes were without injection rate target and without water withdrawal to put the reservoir under high pressure in order to check the cap rock capability to reserve the gas and prevent it from upward movement

The values of BHP have been selected less than the fracture pressure of the cap rock of the upper reservoir which is about 1800 psi. Figure.9 and 10 show the resulted curves of cumulative gas injection, average formation pressure and gas injection rate of the mentioned two cases.

The results showed that there is a high effect of wells location on cumulative injected gas. However, cumulative injected gas was equal to 144.1 MMMSCF at group A wells location and equal to 182.8 MMMSCF at group B wells location because the small space between wells has an effect on the reservoir pressure and pressure interference may occur.

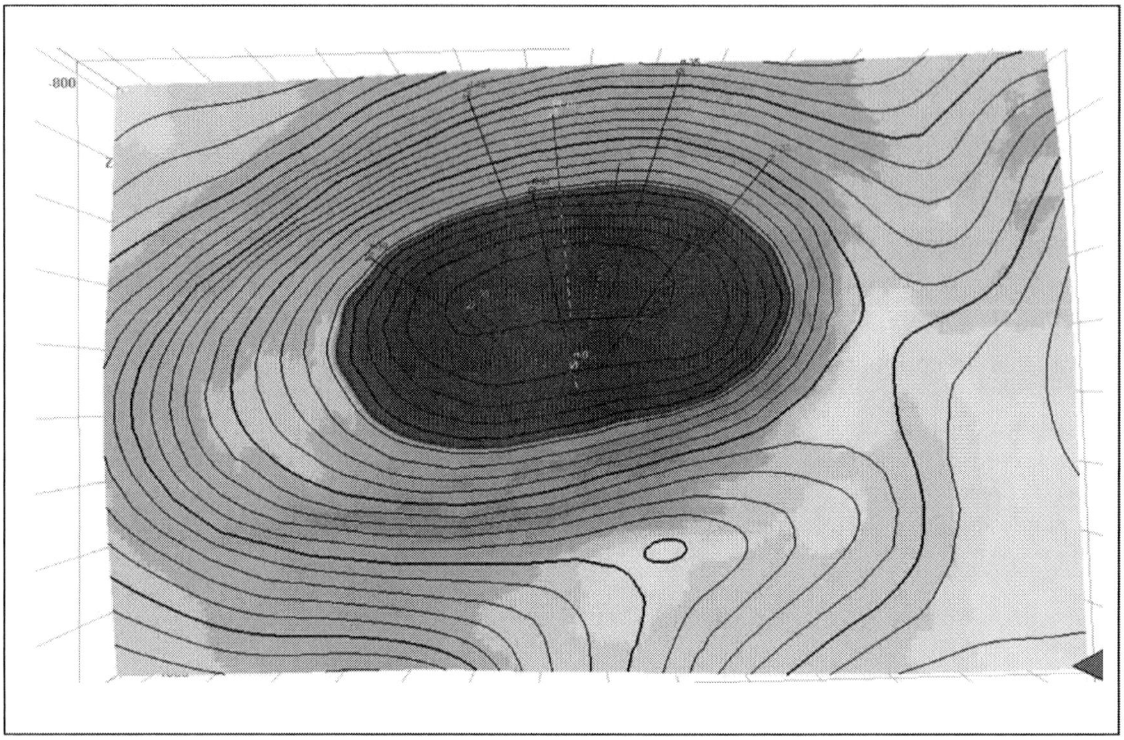

Figure.6 : - Group A wells location.

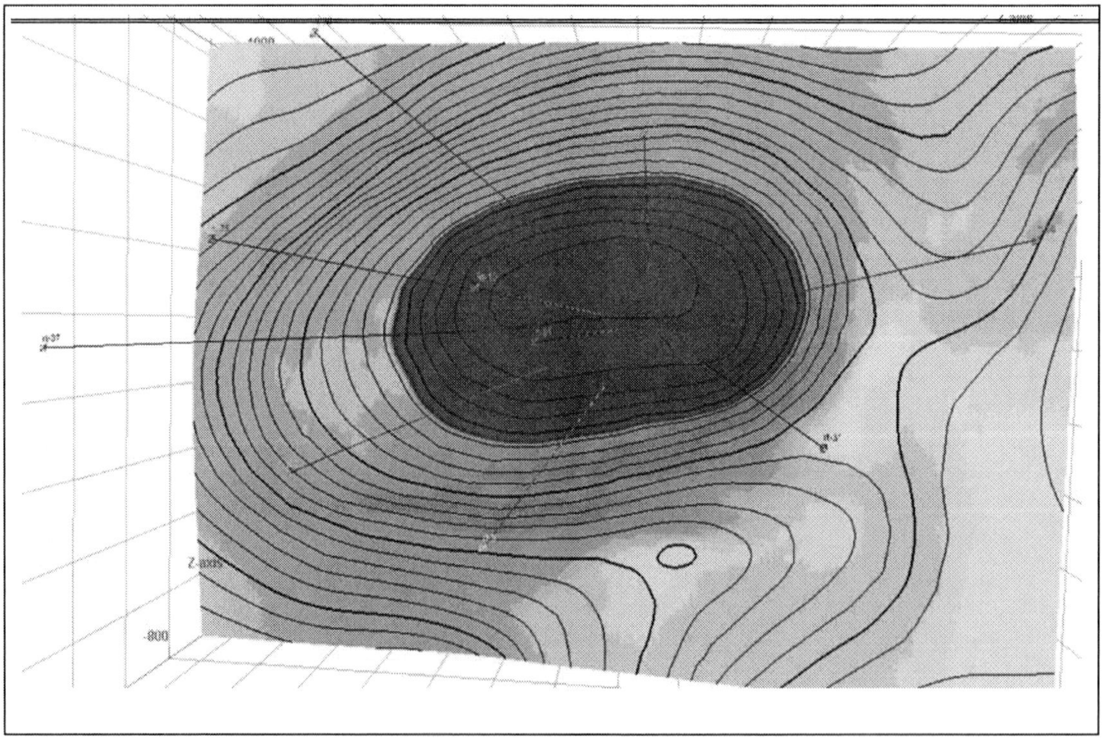

Figure.7 : - Group B wells location.

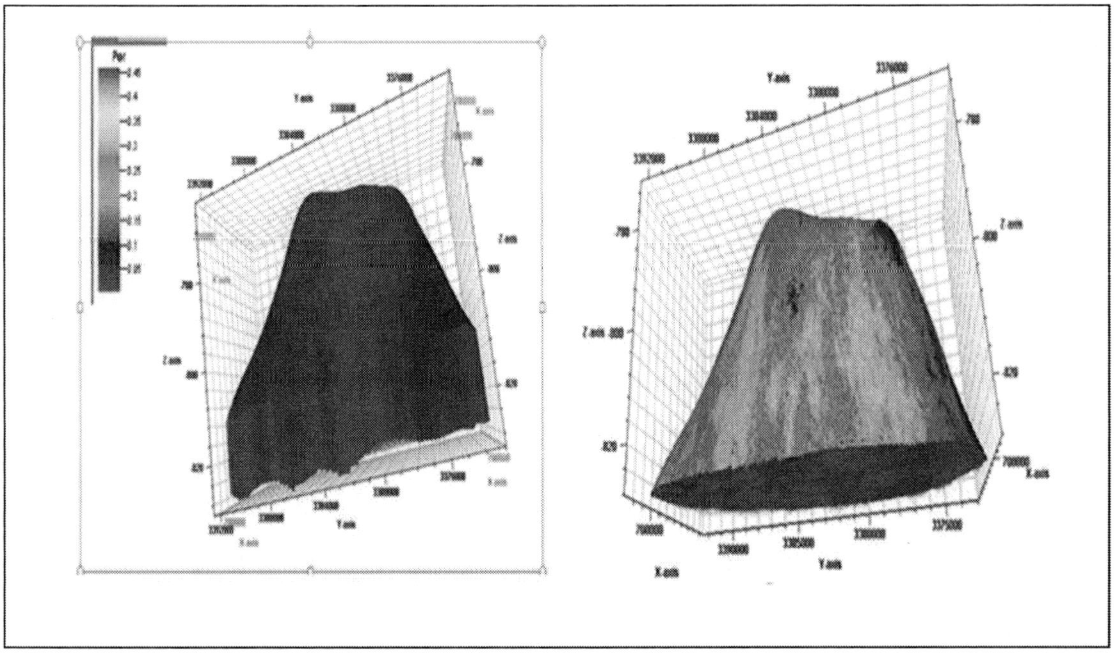

Figure.8:- porosity distribution of the closure area of UM ERadhuma upper reservoir and its cap rock.

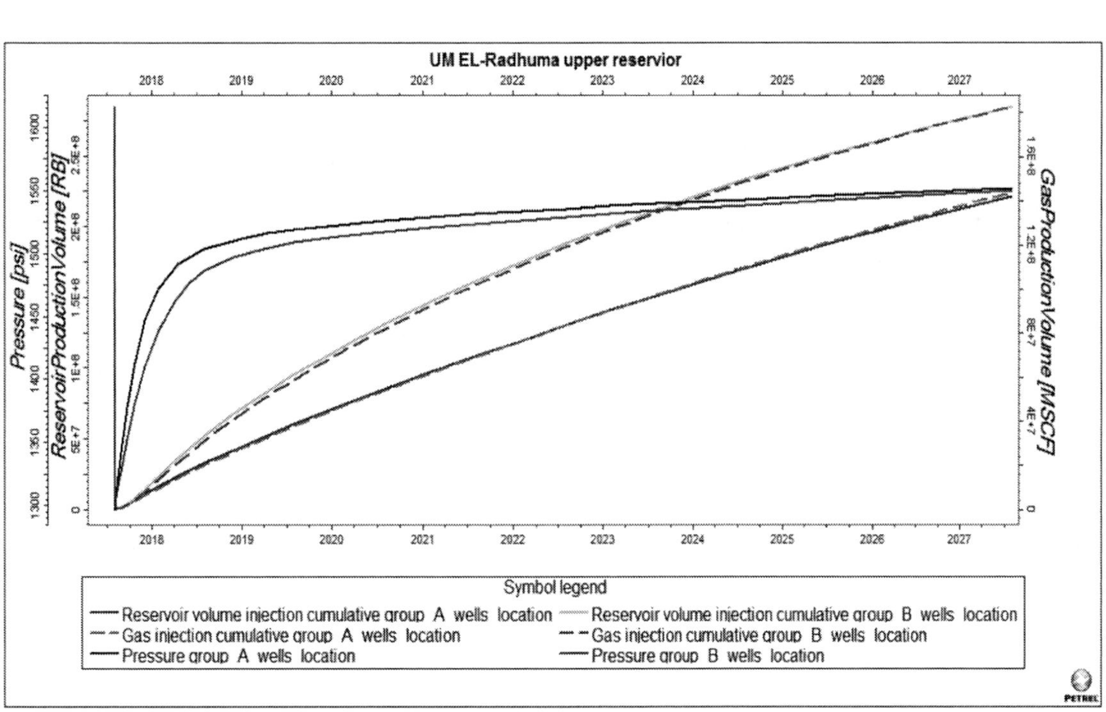

Figure.9 : - Effect of wells location on average reservior pressure and injeceted gas cummulative during 10 years of injection.

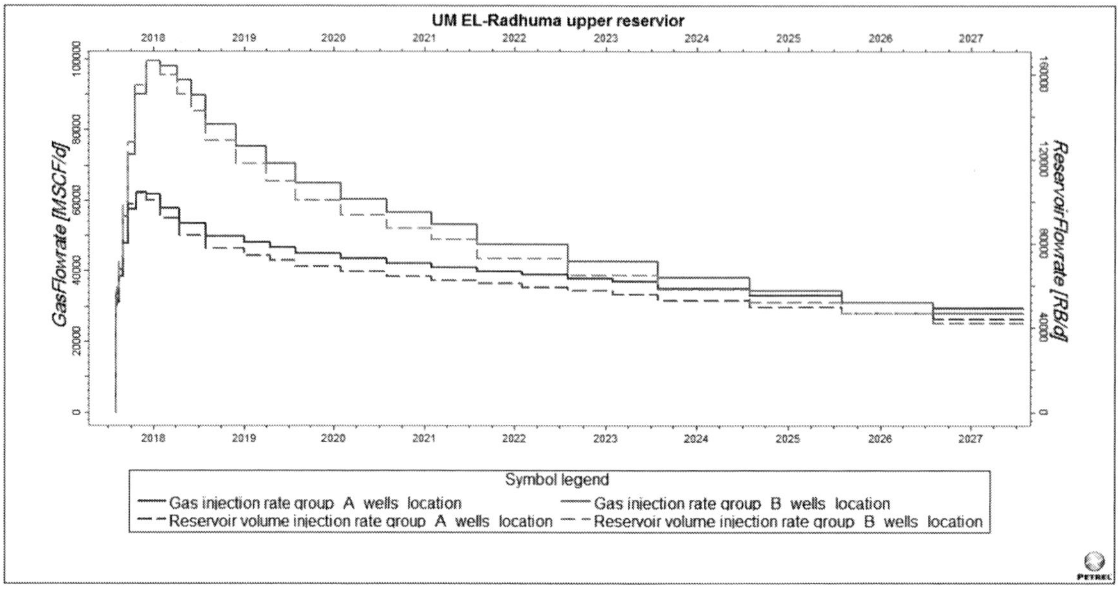

Figure.10 : - Effect of wells location on gas injection rate during 10 years

4. Discussion

During this study, upper reservoir of um ERadhuma formation has been tested for underground gas storage using two wells location groups and find the effect of wells location on cumulative gas injection and reservoir formation pressure during 10 years injection period.

Regarding the injection wells location effect , it can be noticed that injection wells location have a good effect on the cumulative gas injection and formation pressure because of well location effect on formation pressure distribution and an interference between wells will be occurred which has an effect on the differential pressure between bottom hole pressure and formation pressure, therefore; changing in well location from small well spacing (group A) to a high well spacing (group B) between injection wells leads to a high increment in cumulative gas injection because it effects on the reservoir pressure.

As noticed in the injection rate curve (**Figure.10**), injection rate is changing during 10 years period. This change in gas injection rate is due to the increase in formation pressure which leads to a decrease in differential pressure value. Therefore; in order to decrease the increment in formation pressure during the injection, pressure relief is recommended which is done either by water withdrawal from the bottom of the reservoir or by making a shut in period between injection periods.

From gas saturation distribution of UM ERadhuma formation and its cap rock (**Figure.8**), it has been noticed that gas saturation in the cap rock and all the remaining upper part of RUS formation is equal to zero which means that there is no gas escaped above the cap rock which implies that RUS formation is good cap rock and UM ERadhuma is suitable for underground storage purposes.

5. Conclusions

1- Upper reservoir of UM ERadhuma formation is suitable for underground storage because of the tightness of RUS formation that overlying it.

2- High volume for storage is available in UM ERadhuma formation which is equal to 14.3 billion of reservoir cubic feet.

3- The location of gas injection wells is very effective on gas injection rate and cumulative injected gas because it has an effect on pressure distribution. High spacing between injection wells (group B) gave a good result and high cumulative gas injection which is 182.8 MMM (billions of surface cubic feet).

References

[1] Economides M J and Wang X 2013 *SPE Annual Technical Conference and Exhibition* (New Orleans) (Louisiana/SPE)
[2] Musa A 2016 Iraq as producer and exporter of gas *Iraqi Economists Network Report*
[3] Alejano R, Perucho Á, Olalla C, Jiménez R 2014 *Rock Engineering and Rock Mechanics: Structures in and on Rock Masses* (Leiden, Netherlands: CRC Press) p 372
[4] Reservoir & Field Development Directorate Core analysis report of wells (RT-8, RT-9, RT-10, RT-11) *Iraqi Ministry of oil*
[5] Reservoir & Field Development Directorate Water analysis report of UM ERadhuma formation water 1986 *Iraqi Ministry of oil*
[6] Gas chromatographic report of Zubair field /well ZB-71, 2016, *Basrah Gas Company*

IOP Conf. Series: Materials Science and Engineering **579** (2019) 012033 doi:10.1088/1757-899X/579/1/012033

Identifying hydrocarbon and shaly zones of Mishrif Formation in Nasiriya Oil Field by using well logs

Ameer T. Hameed[1], **AbdulAali A. Aldabaj**[2], **Ahmad A. Ramadhan**[2]

[1] Ministry of Oil, Petroleum Research and Development Center, Baghdad, Iraq

[2] University of Technology, Petroleum Technology Department, Baghdad, Iraq

Abstract Well logging is the process of recording physical, chemical, electrical and other properties of rock/fluid mixture penetrated by drilling a borehole into the earth crust. Many of these logs are electrical in measurements. Hydrocarbon may exist in a porous and clean formation. That is, gamma ray and spontaneous potential can identify shaly/clean zones while neutron, density, sonic logs or even NMR logs may be used for porosity estimation. Resistivity logging is used to differentiate between formation filled with salty water (low resistivity) and with those filled with hydrocarbons (high resistivity). Mishrif Formation in Nasiriya Oil Field was chosen as a case study. The Nasiriya oil field structure is an anticline with NW- SE trend. Mishrif Formation is the Majer middle cretaceous carbonate in the stratigraphic column of southern Iraq. The result of well logs interpretation for the first five wells showed that a shaly unit separate upper Mishrif (MA) from the lower one (MB). Also, both units MB1 and MB21 are considered the movable hydrocarbon zones. The CPI data reflect that in which good values of porosity of about 0.18 in unit MB1 and 0.2 in unit MB21, besides low values of water saturation 0.28, 0.45 in units MB1& Mb21, respectively.

1. Introduction

Well logs (electric logs or wireline logs) can be defined as a plot of a well between depth and any rock property like resistivity or formation bulk density. Well log readings can be processed either down hole and transmitted to the surface or processed at the surface. Porosity and fluid saturations are examples of some important rock properties that could be estimated from well logs. In spite of that, well logs data need equation or physical model to convert their recorded data to find these properties [1]. Well logs using different tools plays an important role in geological decision. After interpreting these logs, it is applicable to identify porosity, fluid saturation or even the lithology. That means, that hydrocarbon (oil or gas) and water can be distinguished [2].

2. Area of Study

Nasiriya Oil Field is located in Thi-Qar governorate about (38 km) north-west of Nasiriya city. Include an anticline structure with NW-SE trend. The exploration operation and the exploratory drilling had discovered the major reservoir in Mishrif, Yamama and Nahr Umr formations [3]. The field latitudes is 34 80' -34 60' N and longitudes 57 50' -60 10' E, as shown in Fig. (1). The Mishrif formation is considered heterogeneous formation described as organic detrital limestone with beds of algal, rudist, and coral-reef limestone, capped by limonitic fresh water limestone [4]. Mishrif formation had divided into two main units (upper Mishrif, MA & lower Mishrif, MB). They are separated by thin shale unit within 10 m thickness and lower Mishrif is subdivided into MB1 and MB2 with barrier rocks in some areas of the field [5].

3. Methodology

3.1. Gamma ray log (GR)

It measures the natural radioactivity in a formation. Shale zones has a high radioactive material in which it reflect the high values of GR reading. The opposite is in the shale-free sandstone and carbonate formations so that it can be used to determine the volume of shale through equations (1),(2) and (3)[2].

$$I_{GR} = \frac{(GR_{reading} - GR_{min.})}{(GR_{max.} - Gr_{min.})} \quad ....(1)$$

Then, volume of the shale can be estimated depending on the age of the formation, for cretaceous rocks;

$$V_{sh} = 0.33\,(22 * I_{GR} - 1) \qquad (2)$$

While for tertiary rocks;

$$V_{sh} = 0.83\,(23.7 * I_{GR} - 1) \qquad (3)$$

3.2. SpontaneusPotential log (SP)

It is a log which is used to identify permeable zones and distinguish these zones from shale. SP log based on the self-diffusion of the dissolved ions of fluid in both borehole and formation. It measures in millivolts. A deflection of SP log is a sign for a porous and permeable zones which has different ionic concentration than the mud. It can be used to identify formation water resistivity when formation temperature and mud filtrate are known [6].

3.3. Resistivity log

They are electric logs which are considered important in identification of water & hydrocarbon zones. Hydrocarbon, like rock matrix, are non-conductive in which current can't be transmitted. For that, it can be notice the high resistivity values in hydrocarbon zones in opposite to water zones. Two types of resistivity logs present, induction and electrode logs. The first one is better to be used in non-salt-saturated drilling mud ($R_{mf} > 3R_W$) while the electrode log is used when ($R_{mf} \approx R_W$).An indication of a permeable zone is the invasion in which resistivity in flushed zone (R_{xo}) doesn't equal the uninvaded zone (R_t). These zones are illustrated in Figure (1).Many types of resistivity which are deep in investigation like LL_3, LL_7 or ILD logs. While for flushed zone, we have MLL, MSFL, ML…etc[2]. It should be mentioned that a decrease in porosity may lead to an increase on R_t values for a zone saturated only with water. In this case, a comparison of a separation between R_{xo} (shallow) and R_t (deep) is made between this zone with the detected water zone and it can be said that it may contain hydrocarbon when it is less in separation [7].

3.4. Porosity logs

3.4.1. Density logs

Used to find the porosity in case we have about lithology of a formation. The measured bulk density tries to find the density of a formation with the uniform distribution of fluid pores in a rock matrix [8] as it declared in equation (4).

$$\rho_b = \Phi * \rho_f + (1 - \Phi)\rho_{ma} \qquad (4)\ [6]$$

3.4.2. *Sonic Log:*

It measures the transmitted pulse velocity from the transmission to the receiver. It can be used to transmit porosity when both transit time of matrix and fluid are available as presented in equation (5) [8].

$$\Phi = (DT - DT_{ma}) * (DT_f - DT_{ma})^{-1} ....(5)\ [6]$$

3.4.3. Neutron log:

Its measurements is based on hydrogen which is considered effective in slowing-down the fast neutrons due to the large scattering cross section and small mass of it. Since the existed hydrocarbon or water contains hydrogen, then neutron log can easily be a reflection of porosity values [6].

IOP Conf. Series: Materials Science and Engineering **579** (2019) 012033 doi:10.1088/1757-899X/579/1/012033

4. Determination of formation Water Resistivity

A precise value of formation water resistivity (Rw) is fundamental in order to complete interpretation process. It can be determined when the chemical analysis of the formation water is available .Another approach is from SP curve [9].The concentration of a solution of pure Nacl has the same resistivity of (Rw) at a given temperature because major component of dissolved material in most of formation waters is sodium chloride. Then, Rw is expressed in terms of equivalent Nacl [9]. From water analysis report, the concentration of pure Nacl in well NS2 is (190788 ppm) . Equation (6) is applied to calculate Rw @ 75°F (equal to 0.0454 ohm.m) using Nacl concentration in ppm. Then, it is converted to the bottom hole temperature by using equation (7). Rw value for NS1 at formation temperature (185°F) is equal to 0.02 ohm.m.

$$Rw_{75} = \frac{1}{2.74*10^{-4}*Cp^{0.955}} + 0.0123 \quad (6)\ [10]$$

$$Rw_{fm} = Rw75 * \frac{81.77}{ft+6.77} \quad (7)\ [10]$$

Where, Rw_{75} :Formation water resistivity @ 75°F Rw_{fm} : Formation water resistivity @ formation temperature, °F .CSP: Nacl concentration in ppm

Ppm: part per million It can be mentioned that mud filtrate resistivity is read from well log header reports at a given temperature.

5. Shale Volume Calculation

Gamma ray index in equation (1) was used with equation (2) to calculate shale volume for Mishrif Formation which is considered old rocks (Cretaceous age). Figure (1) represents the shale volume of well NS1 and temperature curve. It can be noted that there is a shale unit reaching (10) m which separates the two main units (MA & MB) of Mishrif Formation in Nasiriya Oil Field.

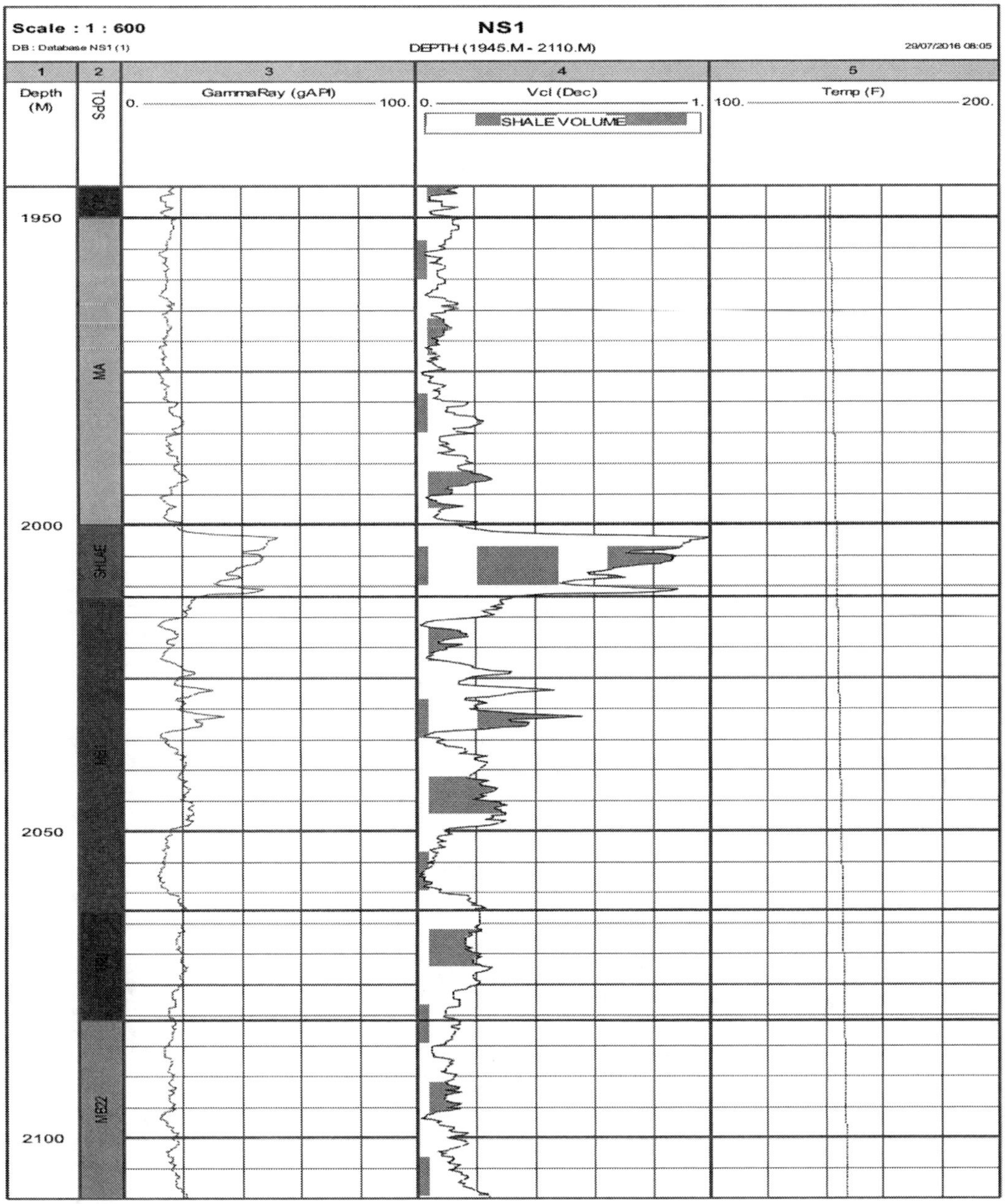

Figure (1): Shale Volume for Well NS1

6. Porosity Estimation

Equation (8) is used to determine the total porosity. The derived porosity from sonic, density, neutron and neutron-density logs are compared with the lab core data. The neutron-density porosity has given best match with the core porosity among other derived porosities and as shown in figures (2) for NS1. The obtained (N-D) porosity is then used to determine the effective porosity after extracting shale volume from equation (9). Secondary porosity index (SPI) can be determined by using equation (10).

$$\varphi_t = \frac{(\varphi N + \varphi D)}{2} \quad \ldots(8)$$

$$\varphi_e = \varphi_t(1 - Vsh) \ldots(9)$$

$$SPI = \varphi_t - \varphi_s \qquad (10)$$

Figure (3) shows the total & effective porosities for well NS1. From the figures mentioned above, it is shown that unit MB has the highest porosity values in the formation which reflects its important role in storage capacity.

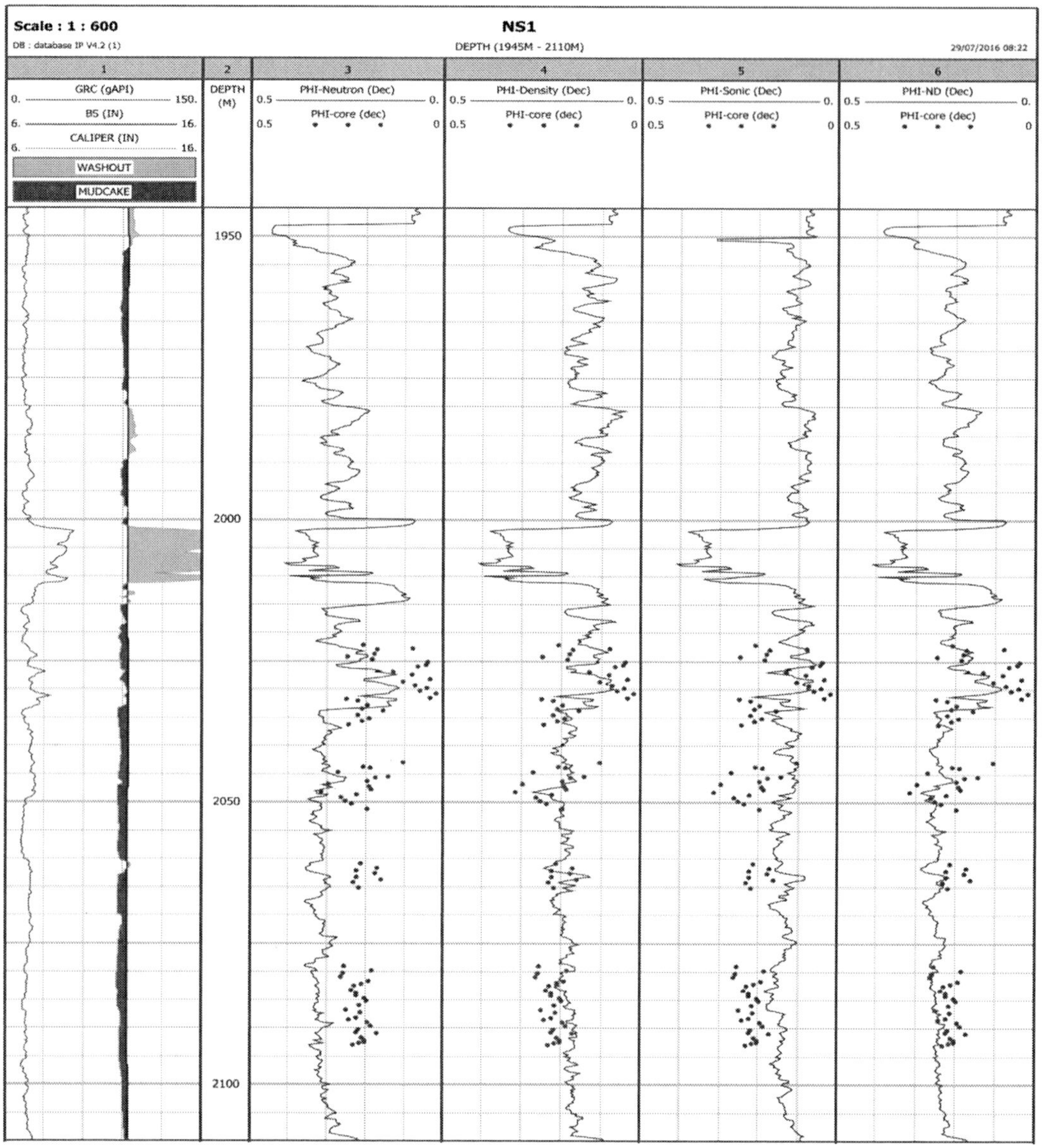

Figure (2): Comparison between the Derived Porosities for well NS1

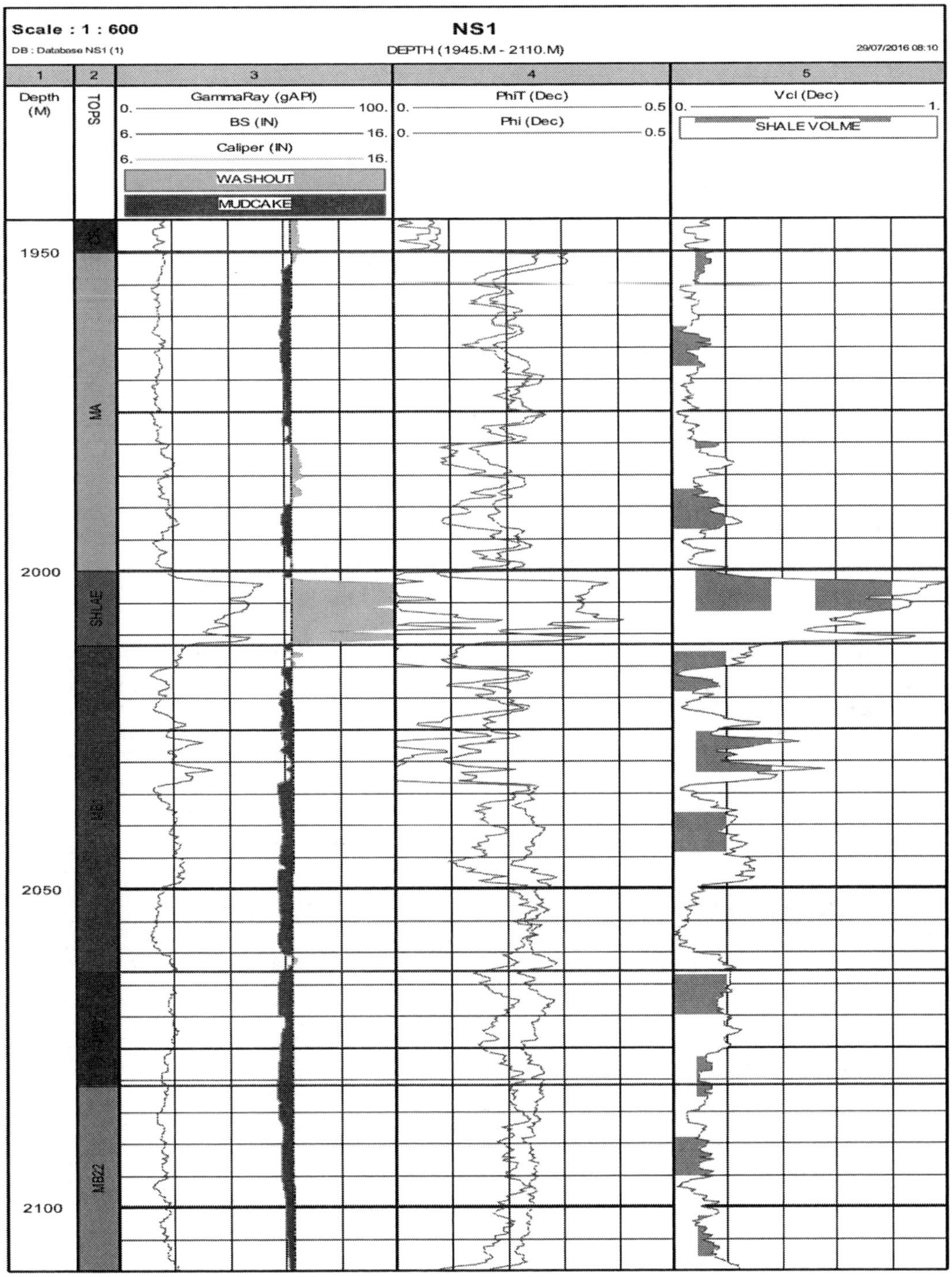

Figure (3): Total and Effective Porosity for Well NS1

7. Estimation of Archie's Parameters for Sw Calculation

Archie's equation (11) is applied to determine water saturation. The heterogeneity of carbonate reservoirs makes it somewhat difficult to apply the equation when its parameters depends highly on carbonate characteristics. Uncertainty analysis for Archie's parameter declared that (a, m & n) parameters have the higher influence on water saturation calculations than the resistivities (R_t & R_w) [11].

$$Sw^n = \frac{a}{\varphi^m} \cdot \frac{R_w}{R_t} \qquad (11)$$

Pickett's method was presented to calculate Archie's parameters by a basic rearrangement of his equation [12].

$$Rt = \frac{F.Rw}{Sw^n}\ldots \quad (12) \quad \text{and,}$$

$$F = a/\varphi^m\ldots \quad (13)$$

And by using logarithms, yields;

$$\log Rt = -m\log(\varphi) + \log aRw - n\log Sw.. (14)$$

In the water bearing zone, equation (14) becomes;

$$\log Rt = -m\log(\varphi) + +\log(aRw) \quad (15)$$

Equation (15) is a straight line on log –log plot with a slope=m and intercept equal to (a.Rw) at (φ=1) where Rw is known previously. It's good to keep in mind that this method should be applied in a clean water formation zones in order to give a good estimation of the parameters. Archie's Pickett's plot is also useful to know (n) parameters; for 100 % water bearing formation (SW=1), equation (14) becomes;

$$aRw = \varphi mRt \quad (16)$$

Equation (11) can be rewritten for irreducible water levels;

$$Swi^n = \frac{a}{\varphi^m}.\frac{R_w}{R_{tirr}} \quad (17)$$

Morris and Biggs in 1968, observed that the multiplication of Sw by φ for irreducible water levels is a constant value:

$(Swi * \varphi = C)$.Coates & Dumanoir in 1974 concluded from the studies of core analyses that the assumption of (n=m) for irreducible levels is fair, so that equation (17) can be reduced with the application of equation (16) [13];

$$(Swi\varphi)^n = \frac{\varphi^m Rt}{R_{tirr}} \quad (18)$$

With a rearrangement and the logarithm to both sides yields;

$$\log Rt = \log(Swi^n.R_{tirr} + (n-m)\log(\varphi) \quad (19)$$

Equation (19) is a log-log plot with a slope of (n-m) and an intercept of $(Swi^n.R_{tirr})$. Exponent (n) can now be easily estimated as (m) in known. Figure (4) represents Pickett's plot for the well NS1. Archie's parameters for NS1 have been estimated in which (a) equal to 1 and the value 1.8 is for (m) and (n).

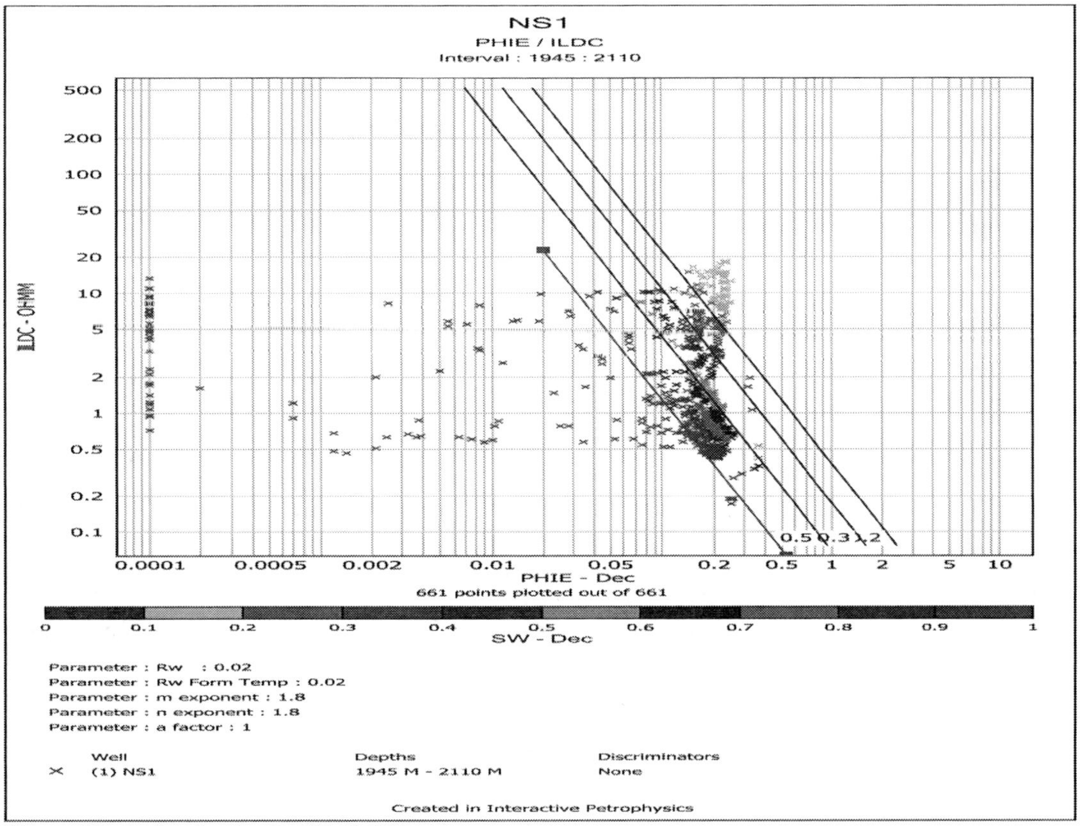

Figure (4): Pickett's plot for Well NS1

8. Reservoir Fluid & formation Analysis

8.1. Fluid Analysis

Archie's equation has been used in Mishrif formation, Nasiriya Oil Field to calculate the water saturation in uninvaded zone. Also, in flushed zone, water saturation can been calculated using equation (20).

$$Sxo^{n} = \frac{a}{\varphi^{m}} \cdot \frac{R_{mf}}{R_{xo}} \qquad (20)$$

Determination of flushed zone water saturation has its advantage where the ratio Sw/Sxo is an indicator of oil movability .When Sw/Sxo=1, then no moved hydrocarbon, whether or not the formation contains hydrocarbon or not. If Sw/Sxo is 0.7 or less, that means that there is an indication of movable hydrocarbon existence [14].

8.2. Bulk Volume Analysis

The bulk volume of water can be identified as a unit volume of porous media occupied by water. The bulk volume of hydrocarbon is the amount of hydrocarbon pore volume.

$$BVW = Sw * \varphi_e \qquad (21)\ [14]$$

$$BVW_{Sxo} = Sxo * \varphi_e \qquad (22)\ [14]$$

The bulk volume of the movable hydrocarbon can be identified by [13];

$$BVmo = (Sxo - Sw) * \varphi_e \qquad (23)\ [15]$$

The computer processed interpretation (CPI) is a set of logs for the results of porosity, saturation, fluid & matrix analysis tracks and other parameters of interpretation.

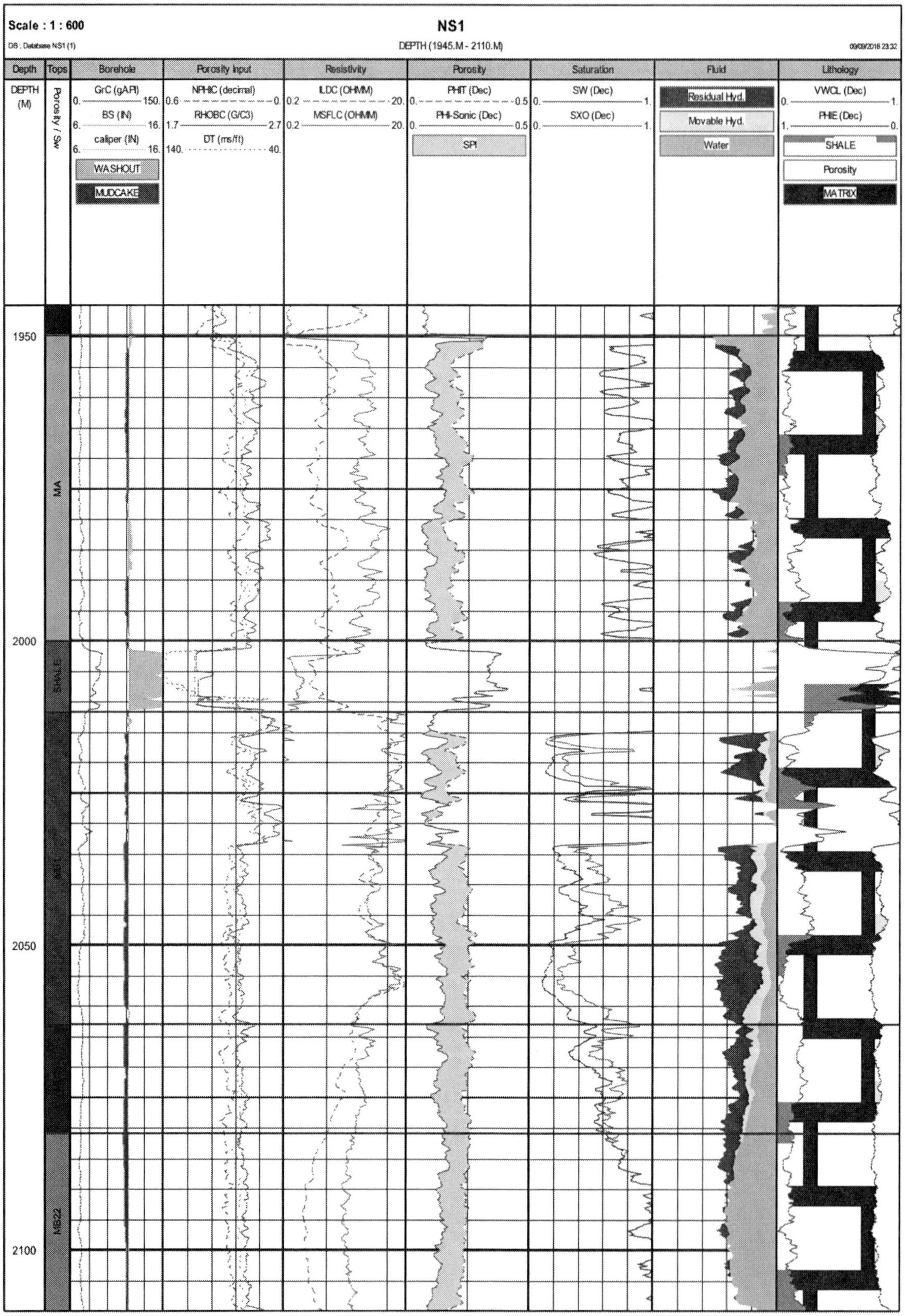

Figure (5): Computer Processed Interpretation for Well NS1

Table (1): Averages of properties for the first five wells

well	unit	Avg. Vcl	Avg. Phie	Avg. Sw	bulk Movable	Moveable
	MB1	0.187384507	0.142007	0.435077	0.029728169	0.209343
1	MB21	0.169784	0.184249	0.524608	0.025228947	0.136928
	MB22	0.115184071	0.193592	0.918077	0.000465487	0.002404
	MB1	0.075618033	0.194516	0.286207	0.073180328	0.376217
2	MB21	0.095752174	0.210604	0.603257	0.0627	0.297715
	MB22	0.0924	0.1988	0.908319	0.01380625	0.069448
	MB1	0.152816889	0.15806	0.444495	0.061167111	0.386987
3	MB21	0.140922667	0.188737	0.607369	0.057174026	0.302929
	MB22	0.138419481	0.183308	0.929418	0.012218182	0.066654
	MB1	0.124071548	0.19845	0.252542	0.066594979	0.335575
4	MB21	0.106473469	0.231465	0.570488	0.045542857	0.196759
	MB22	0.079521374	0.205156	0.874779	0.006782443	0.03306
	MB1	0.134812658	0.218559	0.249219	0.06607173	0.302307
5	MB21	0.1229475	0.225483	0.61860	0.0345225	0.153105
	MB22	0.094671429	0.221271	0.972528	0.001633835	0.007384

From the table (1) we can notice the following important notes:

1. Effective porosity, part of neutron- density porosity, has good values in all units of lower Mishrif which reflect the high hydrogen content, lower density and low values of clay volumes.
2. Unit MB1 has the lowest values of water saturation in lower Mishrif which with porosity values ensure the accumulation of oil. Beside that, unit MB21 has higher Sw values than MB1, but stills important. Mb22 is considered a water reservoir unit.
3. Good moveable hydrocarbon values in both MB1& MB21 units. Also, very low values in MB22 for the high values of water saturation.

9. Conclusion:

From using different well logs, the computer processed interpretation can give the following realities for the well NS. Both upper and lower Mishrif are porous with high index of secondary porosity. MA unit is almost water zone and Unit MB22 is a water zone where oil water contact (OWC) has been reached. Movable hydrocarbon is concentrated in the unit MB1 and in the upper part of MB2 (MB21).A thick shale unit (about 10 m) separate upper Mishrif from the lower one. Borehole diameter from caliper log (CAL) is much higher than BS in shale unit which is an indication of a washout while a mudcake can be seen from CPI in other units.

References

[1] Etnyre L M 1989 *Finding Oil and gas from well logs*, Springer science and business Media New York

[2] Asquith G and Gibson C 1982 *Basic well log analysis for geologists*, methods in Exploration Series, AAPG

[3] Handhel A M 2006 *Reservoir Properties Study for Mishrif Formation in Nasiriya oil field and their relation with oil production*, thesis, Department of Geology, university of Basra in Arabic

[4] Atiaa A A and Handhel A M 2009 A fuzzy Logic approach in infer reservoir permeability from depth and porosity measurements for Mishrif limestone Formation at Nasiriya Oil Field, south of Iraq *Journal of al-anbar university for pure science* **3** No.1

[5] Jreou G 2013 A preliminary Study to Evaluate Mishrif Carbonate Reservoir of Nasiriya Oil Field *International Journal of Engineering and Technology IJET-IJENS* **13** 5 69-82

[6] Ellis D V and Singer J M 2008 *Well Logging for Earth Scientists* 2nd Edition springer science and Business Media B V 692

[7] Hameed A T 2017 *Geostatistical Approach of Petrophysical Properties Distribution for Mishrif Formation in Nasiriya Oil Field* M.Sc. thesis Petroleum technology Dept. University of Technology

IOP Conf. Series: Materials Science and Engineering **579** (2019) 012033 doi:10.1088/1757-899X/579/1/012033

[8] Tittman J 1986 *Geophysical well logging* Technical Consulting Services Danury Connecticut Acadimic press 186

[9] Moore E J Szasz S E and Whitney B F 1966 Determining formation water resistivity from chemical analysis *SPE* **1337** 373-376

[10] Rabanit P M V 1952 *Rock units of Basrah area* BPC unpublished report

[11] Hamada G M Almajed A A Okasha T M and Algathe A A 2013 Uncertainty analysis of Archie's parameters determination techniques in carbonate reservoirs *Journal of Petrol Explor Prod Technol* **3** 1–10 DOI 10.1007/s13202-012-0042-x

[12] Kadhim F S 2016.,Correlation Between Cementation Factor And Carbonate Reservoir Rock Properties, *Researcher* **8** 1 29-65

[13] Schlumberger 1989 *Log Interpretation Principles/Applications*, Houston

[14] Schlumberger 2008 IP-Interactive Petrophysics V-3.5, Manual

[15] Ashour H 2016 *Reservoir Characterization and 3D petrophysical Modeling for Tertiary Reservoir in Khabaz oil field*, M.Sc. thesis, Petroleum engineering, University of Baghdad

Porosity and Permeability Calculation Using NMR Logging in an Iraqi Oil Field

Marwah Dhahir Abdulkadhim
Petroleum Technology Department, University of Technology, Iraq-Baghdad

Corresponding author's e-mail address: 150060@uotechnology.edu.iq

Abstract. This study aims of analysing the NMR data and extracting the petrophysical properties of a well in an Iraqi oil field, T2 analysis forms the basis of more complex analysis for the interpretation of NMR, including wettability analysis, facies discrimination and the identification of reservoir units from NMR data. T2 analysis can also be used to calculate porosity, calculate the geometric mean of the distribution, track the position of the modes (peaks of the distribution), and estimate fluid viscosity. NMR logging data has been used for an Iraqi oil field and interpreted to calculate PHIT, PHIE, BVI, CBW, FFI and K. In this current research the petrophysical properties have been calculated and the moveable fluid zone has been detected, where the porosity and permeability values are significant with a good value of FFI which represents the reflection of high value of T2.

1. Introduction

Basically NMR is an acronym for Nuclear Magnetic Resonance. Nuclear Magnetic Resonance relies on a simple principle that the molecular behaviour of materials can be affected by a magnetic field [1]. The best model for describing this behaviour is to consider nuclei as bar magnets. In order to be able to align certain nuclei with magnetic properties in a magnetic field [1].

These nuclear bar magnets have two important properties:

- A magnetic moment, which determines the strength of the interaction.
- Spin: in a magnetic field, the nuclei spin which causes the nuclear bar magnet to precess around the applied magnetic field. The nuclear spins process at a frequency that is dependent on the strength of the magnetic field. Thus, the nuclear spin possesses a resonance frequency that is related to the strength of the magnetic field, hence the term Nuclear Magnetic Resonance [1]

NMR is seen to be interest to the formation evaluation specialist because the behaviour of the nuclear spins is related to:

- The properties of the fluid—most importantly the chemical composition of the fluid, and the viscosity of the fluid.

IOP Conf. Series: Materials Science and Engineering **579** (2019) 012034 doi:10.1088/1757-899X/579/1/012034

- Where a fluid is contained in a porous medium, the interaction of the nuclear spins with the pore surface perturbs the behaviour of the nuclear spins.

In rocks, the overall magnetization of a fluid is related to pore size and fluid composition. By knowing the pore size distribution, it is possible to derive the bound fluid volume which can be used, in turn, to derive permeability.

The NMR measurement used in the petrophysical analysis of rocks utilizes three different applied magnetic fields:

1. A static magnetic field B0.

2. A magnetic field generated using oscillating radio frequency pulses (B1). This field is applied normally (at right angles) to the static B0.

3. Local magnetic field fluctuations which are generated by unpaired electrons and neighbouring nuclei.

Detection of magnetization decay in the transverse plane is referred to as the "transverse relaxation" or the "T2 relaxation". In the case of formation evaluation, the NMR experiment basically examines the time taken for the protons' spins to relax from B1 to B0 by detecting magnetization in the transverse plane.

2. T1 versus T2 Relaxation

T2 relaxation is related to the detection of the proton spins in the transverse plane which is aligned with the B1 field. T1 relaxation relates the time taken for the proton spins to translate from a random alignment to become aligned in the direction of the B0 field.

The difference between T1 and T2 can be considered as the measurement of relaxation to B0 in two different measurement directions. T1 looks in the direction of B0 and normal to B1. T2 looks in a direction normal to B0 and parallel to B1 (i.e., T1 refers to the y-z plane and T2 refers to the x-y plane). [2]

3. Understanding T2 Distributions

The distribution of exponentials, or T2 distribution, can be interpreted in terms of pore size and the composition of the fluid residing in the pore. [2] Figure 1

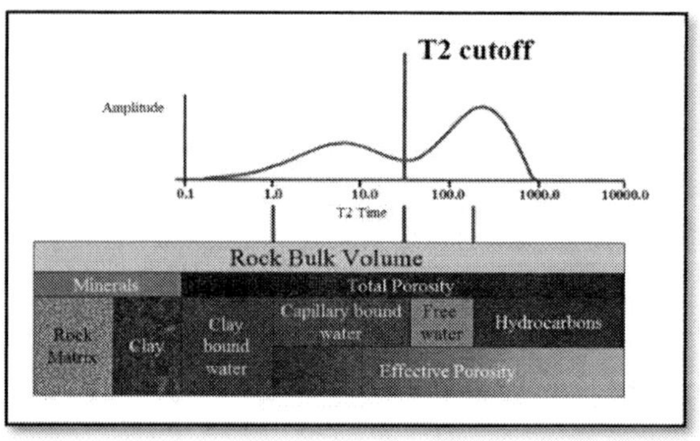

Figure 1. T2 distribution in relation to pore size and fluids [2]

4. Subdivision of the T2 Distribution:

It has been noticed that the T2 distribution is subdivided on the basis of pore size and fluid type. The shortest relaxation times contain clay and capillary bound water, collectively referred to as the bound fluid volume (BFV). Moreover, the BFV can be determined by applying a T2 cutoff. The use of a T2 cutoff relies on the assumption that bound fluid resides in the smallest pores. For sandstones, this is usually quoted as 33ms and for carbonates, 100ms is often used. [3]

Clay bound water (CBW) is often estimated by using a 3ms cutoff. In terms of NMR interpretation, the bound volume irreducible (BVI) is defined as capillary bound water which is

IOP Conf. Series: Materials Science and Engineering **579** (2019) 012034 doi:10.1088/1757-899X/579/1/012034

calculated from BFV - CBW.

The T2 cutoff can be calibrated using NMR measurements on core samples. Core samples are analyzed at 100% brine saturated conditions. De-saturating the core by using a centrifuge or porous late method can be used to determine Swirr. The Swirr value can be used to determine the appropriate T2 cutoff for the brine-saturated rocks.

The desiderated core may be measured by using NMR to confirm that the bound does actually reside in the smallest pores. [2]

Furthermore, the unique information, such as *BVI* and *CBW*, provided by NMR logging can significantly enhance the estimation of resistivity-based water saturation and can greatly assist in the recognition of pay zones that will produce water-free. [1]

In the NMR literature, the concept of bound water (*BVI*) is used in two different ways. The first refers to the water contained in the pore space that will not flow out of the rock and into the wellbore during production. This volume is accurately determined only by a relative permeability measurement, but can be reasonably estimated from a capillary-pressure saturation curve. This water volume is primarily a property of the rock and the wetting conditions. The second use of *BVI* refers to the water that is not displaced by hydrocarbons during the filling of the reservoir. This volume is a function of both the capillary-pressure curve for the rock and the height above free water. In a transition zone, this water volume can include water that can be produced. However, for a sufficient height above free water, capillary forces can drive water saturation to levels well below the point at which the relative permeability to water is effectively zero. [1]

The FFI is the free fluid volume and FFI = φ - BVI. In the other expression, the Mean T2 (or SDR) model, the size parameter enters through the geometrical mean of the relaxation spectra, T2gm. [1]

5. Methodology:

The analysis of T2 distributions aimed at proving estimates of the bound-volume irreducible water and permeability. T2 analysis can also be used to calculate porosity, calculate the geometric mean of the distribution, and track the position of the modes (peaks of the distribution).

1) **Porosity calculation:** The Porosity module calculates porosity for the T2 distribution obtained from the MRIL tool. Figure 2
   a) Set the required input values
   b) Calculate the NMR porosity

2) Estimation of the T2 geometric mean (T2LM):

The T2LM is the geometric mean of the T2 distribution. This parameter is used for calculating the mean pore size and is used in permeability equations. The T2LM is computed by averaging the logarithms of the relaxation times in distributing each weight by its signal amplitude. The T2LM can be calculated for the whole T2 distribution, or part of the T2 distribution, by specifying minimum T2 and maximum T2 values.

3) Calculation of the bound fluid:

The T2 distribution is subdivided on the basis of pore size and fluid type. The shortest relaxation times contain clay and capillary bound water, collectively referred to as the "Bound Fluid Volume" (BFV). The BFV contains both clay and capillary bound water (BFV = BVI + CBW)

Where:
BFV = bound fluid volume
BVI = the capillary bound water
CBW = clay bound water

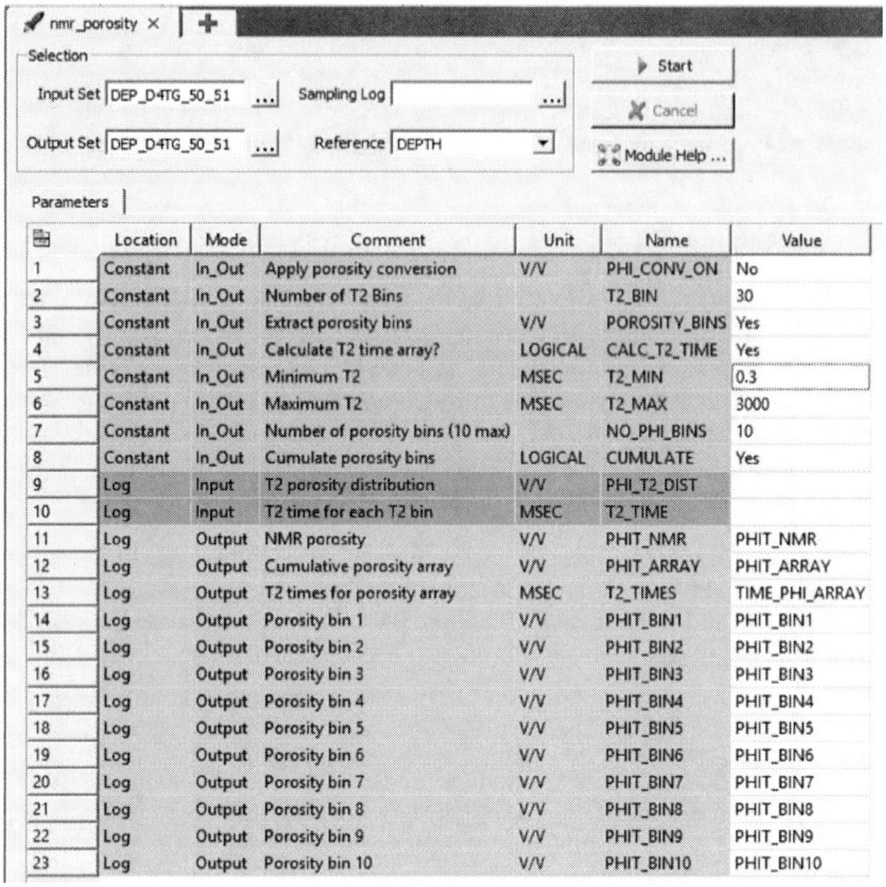

	Location	Mode	Comment	Unit	Name	Value
1	Constant	In_Out	Apply porosity conversion	V/V	PHI_CONV_ON	No
2	Constant	In_Out	Number of T2 Bins		T2_BIN	30
3	Constant	In_Out	Extract porosity bins	V/V	POROSITY_BINS	Yes
4	Constant	In_Out	Calculate T2 time array?	LOGICAL	CALC_T2_TIME	Yes
5	Constant	In_Out	Minimum T2	MSEC	T2_MIN	0.3
6	Constant	In_Out	Maximum T2	MSEC	T2_MAX	3000
7	Constant	In_Out	Number of porosity bins (10 max)		NO_PHI_BINS	10
8	Constant	In_Out	Cumulate porosity bins	LOGICAL	CUMULATE	Yes
9	Log	Input	T2 porosity distribution	V/V	PHI_T2_DIST	
10	Log	Input	T2 time for each T2 bin	MSEC	T2_TIME	
11	Log	Output	NMR porosity	V/V	PHIT_NMR	PHIT_NMR
12	Log	Output	Cumulative porosity array	V/V	PHIT_ARRAY	PHIT_ARRAY
13	Log	Output	T2 times for porosity array	MSEC	T2_TIMES	TIME_PHI_ARRAY
14	Log	Output	Porosity bin 1	V/V	PHIT_BIN1	PHIT_BIN1
15	Log	Output	Porosity bin 2	V/V	PHIT_BIN2	PHIT_BIN2
16	Log	Output	Porosity bin 3	V/V	PHIT_BIN3	PHIT_BIN3
17	Log	Output	Porosity bin 4	V/V	PHIT_BIN4	PHIT_BIN4
18	Log	Output	Porosity bin 5	V/V	PHIT_BIN5	PHIT_BIN5
19	Log	Output	Porosity bin 6	V/V	PHIT_BIN6	PHIT_BIN6
20	Log	Output	Porosity bin 7	V/V	PHIT_BIN7	PHIT_BIN7
21	Log	Output	Porosity bin 8	V/V	PHIT_BIN8	PHIT_BIN8
22	Log	Output	Porosity bin 9	V/V	PHIT_BIN9	PHIT_BIN9
23	Log	Output	Porosity bin 10	V/V	PHIT_BIN10	PHIT_BIN10

Figure 2. Set Porosity Parameters

4) Estimation of T2 bumps:

The amplitude of the modes of the T2 distribution has been calculated. Where amplitude is scaled to porosity, then the porosity of the modes of the T2 distribution has been calculated.

5) Permeability:

There are three permeability models available for use with NMR data, SDR method has been selected for this research.

SDR model: [2]

$$K = C_1 \cdot T_{2_{LM}}{}^{a_1} \cdot \phi_{NMR}{}^{b_1}$$

Where:
T_{2LM} = logarithmic mean T2
φ_{NMR}= NMR porosity (V/V)
C_1, a_1, b_1 = coefficients

6) Tracking the T2 of the modes of the distribution (Peak Tracking):

It is agreed upon the fact that the Peak Tracking module tracks the position of the peak of a T2 Mode based on search limits. The peak is spline fitted for accuracy. [4]

6. Results & Discussion:

Figure 3 below represents data from a Limestone formation in Iraqi oil field. The NMR porosity parameters have been set to calculate PHIT & PHIE, the T2LM has been calculated and used in Permeability equation, the T2 multi cutoff, saddle point and spectral analysis also calculated to find BVI, FFI and CBW.

Track 1 contains gamma ray and calliper logs. Track 3 shows the NMR permeability using SDR equation (green curve), the SDR model will give good estimates of permeability when the porosity is unaffected by HC.

Track 5 contains NMR porosity PHIT & PHIE (Black curve) further BVI, FFI and CBW.
The NMR interpretation results show the logs through a good-porosity (PHIT= 0.15-0.23 and PHIE= 0.10 -0.20), and the permeability ranged (200-850 md).

NMR porosity is Lithology- independent, the change from Limestone in displayed log to Shale has no effect on the accuracy of porosity values. In addition, a comparison with core porosity has been made and it has been found that there is a slight difference between the NMR porosity and core Porosity, for example at depth 3864 m the core porosity was 0.17 while the NMR porosity was 0.20, and at depth 3916 the core porosity was 0.05 and NMR porosity was 0.08 and that because of shale existence (see the Gama log),

Over the depicted interval, the log shows the almost clean Limestone formation, and an intervening shale between the Limestone layers and at the bottom. The zone of oil has been detected between depth (3835-3910) meter furthermore clay bound water, irreducible water, and free fluid, the oil – water contact can be easily identified by using resistivity logs at depth 3915 m.

7. Conclusions:

1. NMR is the most advanced and effected way for formation evaluation which provides an accurate calculation of Petrophysical properties.
2. The NMR represents an advanced tool for petrophysicist to determine as much as possible data during interpretation.
3. NMR is a significant way to calculate the porosity and permeability, further the water volume in reservoir and transition zone.
4. The NMR cost is an effective tool because a reduction in the number of traditional Wireline has been run to get same information.
5. NMR permeability, clay bound water and detection of free fluid zone is the core of this tool, therefore; applying the NMR will be a useful tool for petrophysicist and reservoir engineer to get required information and to achieve their plan.

8. References:

[1] George R C, Lizhi Xiao and Manfred G P 1999 *NMR Logging Principles and Applications, Halliburton Energy Services* (Halliburton Energy Services Publication, USA) p 251.
[2] NMR interpretation Manual (Paradigm TM Geophysical) 2013 chapter 1, 6 and 7
[3] Robert Freedman 2006 *Advances in NMR Logging* JPT (Schlumberger Oilfield Services)
[4] Robert Freedman1 and Heaton1 N 2004 *Fluid Characterization using Nuclear Magnetic Resonance Logging* **45** 241–250

IOP Conf. Series: Materials Science and Engineering **579** (2019) 012034 doi:10.1088/1757-899X/579/1/012034

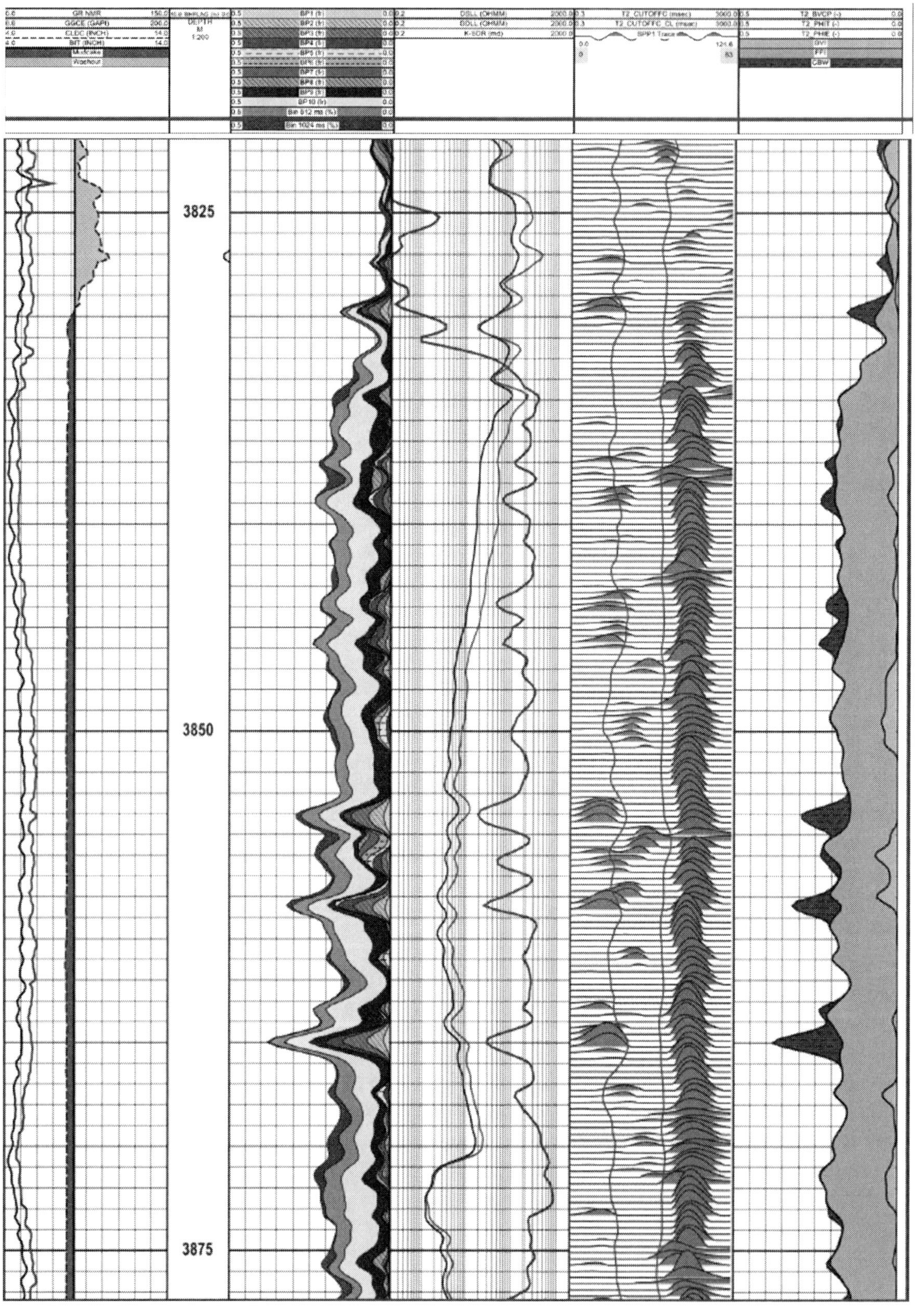

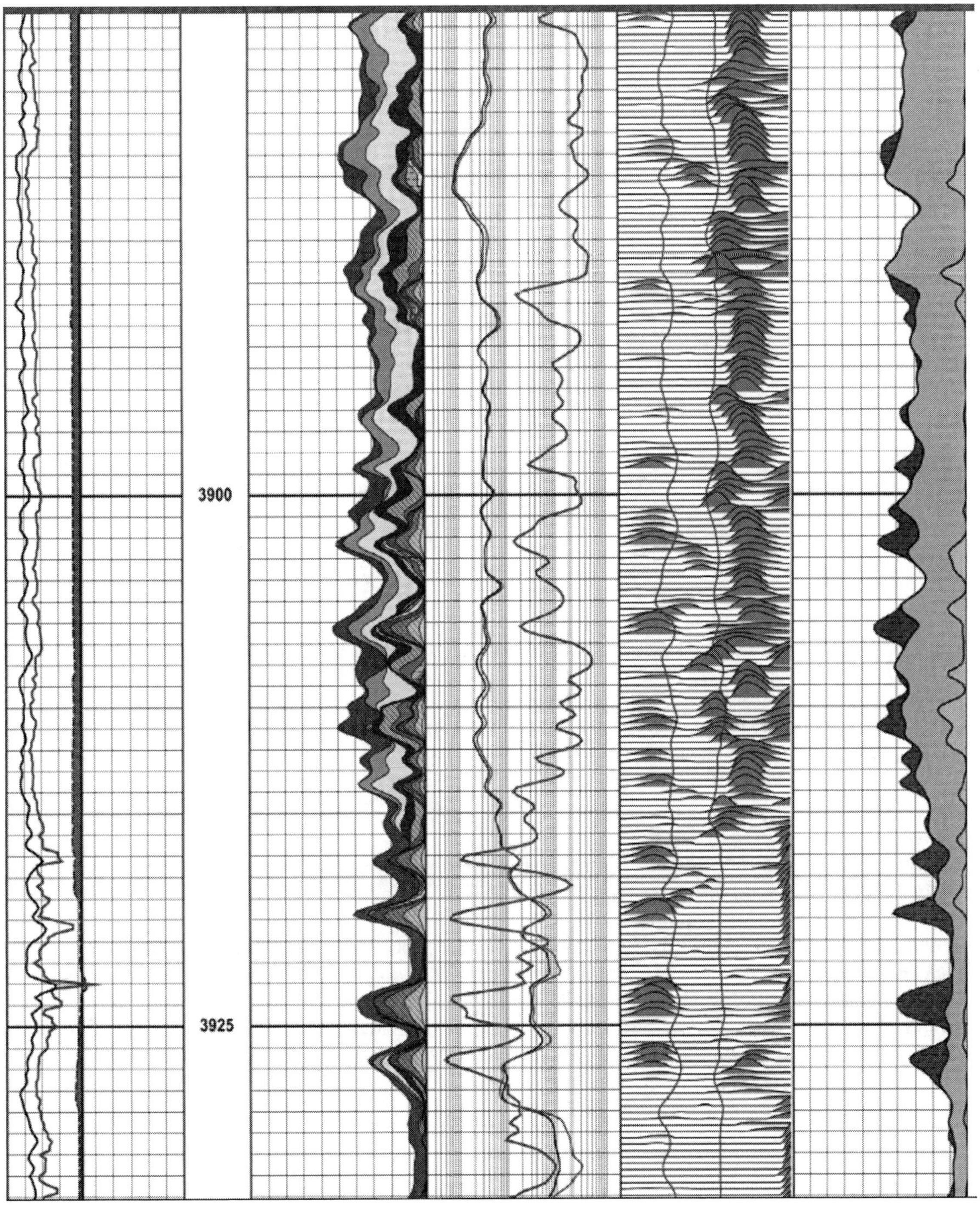

Figure 3. Well log data including NMR

IOP Conf. Series: Materials Science and Engineering **579** (2019) 012035 doi:10.1088/1757-899X/579/1/012035

Application of streamlines for improved waterflood management of iraqi's reservoir.

S al-musawi, M keshavarz moraveji[#1], M Adnan[2] and M Jawad[1]
[#1]Amirkabir University of Technology, Iran
[2]Karbala University, Iraq

Corresponding author's e-mail address: saifalialmosawy@aut.ac.ir

Abstract. In this paper, Using the unique advantages of streamline based (SL) flow simulation, for the first time, we have attempted to simulate waterflood strategies in Buzrgan oil field in southern Iraq and to optimize it. Therefore, in this paper the streamlines are really significant that:

- The numerical calculation of the produced and injection fluid at any given time is possible for each pair of wells.
- Fux pattern maps represent the significant information of the streamlines.
- The runtime in the simulation of the (SL) is much less than finite-difference (FD).

Using this (WAF's), it is possible to calculate the efficiency of injected wells as the proportion of in jected water to the oil delivered at balance wells. With injectedefficiencies (IE) known over the field for every injector, water can be reallocated from low-efficiency to high-efficiency wells, in this way computing the assemblyfor each barrel of water injected, resulting in more sweeping and significant optimization. In addition to this, the IE chart has been given obvious visual of the reservoir management such as (eg, the changes of the wells with low IE to the injection wells.

1. Introduction

In view of the fact that the reservoirs in the middle-east have often come to maturity, or when they are at the stage of reaching their maturity, the use of the secondary recovery operation is suggested/considered. And water flooding is the most common operation for this situation, however due to the limited access to water, it is considered one of the problems of this type of operation.

In this project we have been optimizing the water flooding operation by keeping the water levels constant, in this operation we use the streamline and its unique capabilities to determine the amount of the injected and produced quantities, through the well allocation factors (WAF). The well allocation factors help to calculate the efficiency of the injected wells to improve the productivity of the injection. And the fact that water can be used from the well with low injection efficiency to high efficiency well and, consequently, optimization of the production per injection barrel [2].

2. Streamline simulation

In more than a decade the Streamline based simulation has been transpired as a robust complementary tool to more traditional FD simulation. The streamline simulations statistics can be read in most of the references on the methodology [1, 4, 5, 6, 8, 9, 10, 13, 16, 17]. The information about the streamlines have showed up in the petroleum literature since Muskat's book [12], the streamlines simulations have taken most of its plan to the streamtube manner of the 1960s and 1970s [14, 15]. One of the strong element of the streamline reproduction is the capability to evaluate and imagine supply stream get from well positions and rates, geographical portrayal and congruity, fluid properties, and relative permeability.

The information about the streamlines have showed up in the petroleum literature since Muskat's book [12] and [13, 16]. The streamline simulations have taken most of its plan to the streamtube manner of the 1960s and 1970s [14, 15]. One of the strong element of the streamline reproduction is the capability to evaluate and imagine supply stream get from well positions and rates, geographical portrayal and congruity, fluid properties, and relative permeability.

At split second, streamlines give a depiction of the manner in which the store is associated and how significantly the liquid is dole out between injector/producer sets. Owing to the fact that streamlines are rely upon a numerical reproduction approach, a streamline model can be changed in accordance with historical data and can be represented nonuniform liquid disseminations and multiphase-stream impacts correspondingly as a FD model can.

The perceiving aspect of a full streamline reenactment as opposed to just following streamlines in a FD test system, is that the general rates and the discrete stage rates between well matches can be determined in light of the fact that a streamline test system takes care of a 1D transport issue in each streamline. what's more, the work to the 1D issue that ushers to the definition IEs either for an injector or an injector/producer pair. This information isn't accessible from the tracing alone.

When the IEs are found, it is feasible to promptly obtain a fieldwide metric of the execution of the waterflood, disregarding of the number of well involved. A metric such as that is indispensable for any successful reservoir management.

3. Reservoir simulation

At first it is essential to require the simulation for the reservoir, to manufacture a sample for the reservoir that consist of both the static and dynamics' properties, and the crucial data that we need to obtain such simulation are:

- Contour Map
- Well top
- Well log
- Some elements of the porosity, permeability and the water saturation.
- A few statistics of geology and petro physics that has been achieved in the reservoir simulation.

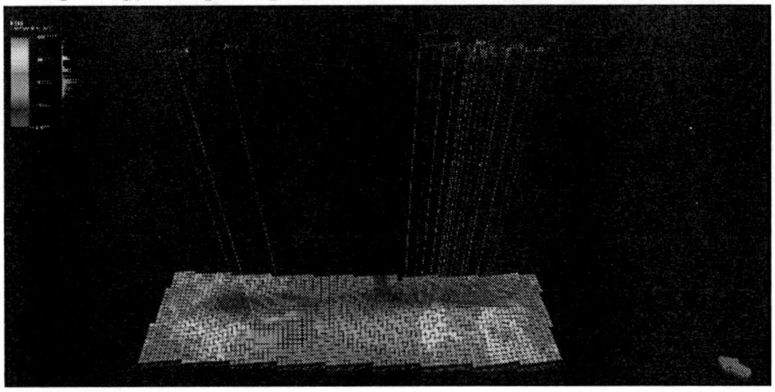

Figure 1: Overview of the designed reservoir's permeability and along with the well's location which is in the southern dome and northern dome of the buzrgan oil field.

Table 1: overview of the designed reservoir's permeability and along with the well's location which is in the southern dome and northern dome of the buzrgan oil field.

Cell number	92×34×14
poroducers	18
Injectorers	3
Rock type	Shaly sand

4. History matching

Every simulation that has been carried out require a history matching to assure that the simulations are in line with the actual reservoir, otherwise the simulation will not be acceptable. As luck would have it the data that is required for implementing the history matching is already available and for that reason we can assemble a case study with above 85% of the history matching, which for this research model, this matching is acceptable.

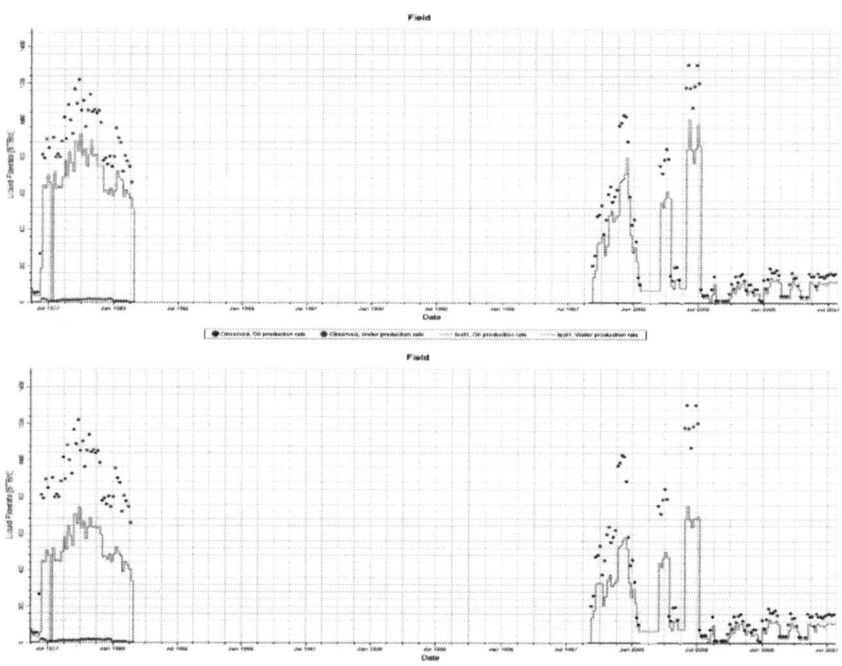

Figure 2: Apparent difference between the field history matching of oil (green line) and water (blue line) production, first illustration after alternation and modification second illustration before the changes, we obtained almost 85% of the history matching.

5. Defined development strategy

After executing the history matching, we could go forth by defining the strategy, we initially designed a production strategy for the following reservoir, we have limited in production data due to the fact of the reservoir's security problems, and for that reason we have obtained the average rate for all the reservoir observed production data that is available and it is about 500 STB/D for most of the wells, after that we have acquired that the information is safe to the majority of the well's production rate, so we used 500 STB/D for every production well. Scheduled time for the injection is when the oil production rate has reached below 600 STB/D and that would be in the near 2047. After this set, we have made an assumption the injection process would be 2700 STB/D for all of injection wells, the injection rate has been deduced from the rule of thumb The injection rate better to be 1.5 times the production rate, which is supposed to be better for waterflooding operation and that is for the following reasons:

- The produced fluid is compressible and the injected fluid is incompressible.
- Waste part of the injection fluids if there are aquifers.
- Leaks portion of the fluid in the adjacent grafts.

We utilize this reservoir to show the fundamental strides of the work process exhibited in this paper.

Note the following points:

- Production toward the finish of 10 years is taken to be the present condition of the field and the minute from which improved injection/production management is to be started.
- The objective is to make the best utilization of as of now (2700 STD/D) and increment oil production by reallocating water among injectors. We expect that there are no limitations in reallocating water in light of surface-facility or well profitability/ injectivity issues.
- We contrast our outcomes with a base case thought to be a "do-nothing" scenario and normal waterflooding.

6. Injection efficiency.

The key idea we use all through our optimization approach is that of an injection efficiency, for our optimization approach We used the concept of "Injection Efficiency", which is defined as follows:

$$I_{\text{eff}} = \frac{\text{off} - \text{set oil production [STB/day]}}{\text{water injection [STB/day]}} \ldots (Eq.1)$$

The accompanying perceptions apply to Eq. 1:

- Being a proportion of rates, Eq. 1 represents to an immediate amount. In any case, the condition likewise can be composed as a proportion of combined volumes, in which case the outcome would be an average efficiency over time.
- It is conceivable to characterize an IE for every active injector in the field. The water-injection rate is known (denominator), while the offset oil production (numerator) must be determined utilizing the data from the WAFs controlled by the streamlines.
- It is additionally conceivable to characterize an injection efficiency (e_{wp}) for every producer/injector pair as we have utilized it (injection efficiency) for the improvement activity which is a noteworthy case in this project. For this situation, both water injection (denominator) and offset oil production (numerator) are processed from the WAFs.
- Although in this paper we center only around water injection and oil production, the meaning of an IE can be stretched out to any sort of injected and produced volumes.

7. Toward improved injection management.

Expanding the oil recuperation for each unit volume of water injected is the purpose of every effective waterflood management scheme, subject to production/injection requirements given by surface facilities, (for example, all out accessible water or most extreme injection/production rates) and local pressure maintenance. In spite of the fact that there are numerous ways to deal with dealing with a waterflood, we recognize two principle choices:

- Maintain an oil-production focus by diminishing injected volumes (water) as much as possible.

- Increase the production of oil as much as possible due to better utilization of available volumes of injection (water).

Both of these purposes can be obtained with the aid of IEs and the workflow presented in this paper. Next, the methodology is applied to a real reservoir with 21 wells, although production rates have been altered for purposes of demonstration and confidentiality. Using the accessible water to increment oil production as contrast to a base case is the strategy that will be used in this case.

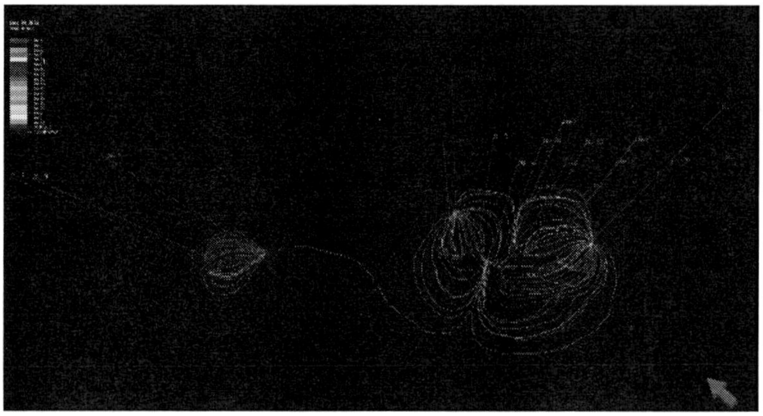

Figure 3: Streamlines after the 5 years of production and for every pair of the production/injection well.

7.1 IE of a well pair

The IE of a well pair (e_{wp}) is calculated in much the same way as the IE of an injector, except that now it is also necessary to extract the injected- water rate related with each heap of streamlines. These information (WAF) are again promptly accessible as a feature of the streamline computations. The subsequent information to compute e_{wp} values are appeared Table 2. The data related with the streamlines appeared in Fig.3.

Table 2: (WAF) data used to calculate the Injection Efficiency (e_{wp}) for each production/injection pairs.

Well name	BU-2	BU8	BU-21
Water in	306.97	569.62	417.71
Oil out	171.09	294.39	190.11
e_{wp}	0.557351	0.516818	0.455124

7.2 Calculate injection efficiencies at the running time.

The initial phase in the work process is to decide the IEs for every injected well, the average IE of the field is determined utilizing the total of the oil produced partitioned by the water injected. For this situation, the average field efficiency is 0.509%. Note that the average field efficiency is the one number that can generally be determined from the deliberate creation information and does not require a recreation display. A vital part of ascertaining IEs is the means by which to represent injection water lost to the aquifer or, then again, oil produced in light of help from the aquifer. Water lost to the aquifer (i.e., an aquifer that goes about as a sink and that both injection water and dislodged oil may spill out to) is incorporated into the IE estimation. Hence, a well that is losing water to the aquifer will see a lower IE than it may somehow or another accomplish if the water were bringing about counterbalanced oil production. Alternately, oil produced in light of a supporting aquifer is excluded in the productivity computations. This is on the grounds that there is no injection over the aquifer, and, in this way, it ought not be incorporated into a metric that is utilized to infer target rates.

7.3 Reallocation of injected water.

Similarly, as the IEs are determined for every injection well, e_{wp} is additionally determined for each well pair. It is the e_{wp} values that are utilized in the work process to compute new purpose rates as an example, and the e_{wp} values for all associations in the model are found in Table2 these estimations are registered naturally by parsing a WAF file made by the streamline simulator. Once e_{wp} is known for every injector/producer association, the focal thought is to reallocate water from wasteful to proficient well associations. How is this done? Consider the injection rate related with every injector/producer pair. The thought is to multiply the old injection rate by a factor 1+w, where w is an element of the e_{wp} of that association. In other words, we are looking for

$$q_i^{new} = (1 + wi) * q_i^{old} \ldots (Eq. 2)$$

Where i alludes to the ith injector/producer pair. In this work, we utilize the average field efficiency as the reference point to choose if the weight w ought to be bigger or littler than zero. For instance, on the grounds that the average field efficiency is 50.9%, we would need the weight to be bigger than zero for the association of BU-2 since that e_{wp} is 55.73% (see Table 2), while we need the weight to be littler than zero for the association of BU-21 since that e_{wp} is just 45.51%. Notice that the decision of utilizing the average field efficiency as a kind of perspective point is subjective. An elective methodology could be to just position all estimations of e_{wp} and allot loads in that way. In this work, we picked the main methodology and utilized Eq. A-1 in the Appendix to ascertain the loads as a component of average field efficiency, e_{wp}, two key components become an integral factor here:

1. The meaning of an "average" efficiency, with the related principle that estimations of e_{wp} with an efficiency higher than the average efficiency should discover a rate increment, while estimations of e_{wp} with a lower efficiency than the average efficiency should discover a rate decline.
2. The immensity of the capacity used to relate e_{wp} with a weight. This function is completely arbitrary.

Table 3:Statistics of the unconstrained & constrained for the old and current injection rates.

well name	e_{wp}	Wi	q_i^{old} STB/D	(1+wi)* q_i^{old} Unconstrained, STB/D	r*q_i^{new} constrained, STB/D
BU-2	55.73%	0.3	900	1170	1167.435
BU-8	51.68%	0.00659	900	905.93	903.9463
BU-21	45.51%	-0.3	900	630	628.6188
			2700	2705.932	2700

*α= 2; e_{min}= 0.4551: e_{max}= 0.5573: w_{min}= -0.3: w_{max}= 0.3

7.4 New Injection-Rate Targets.

Utilizing the estimations of e_{wp} in Table 2. The figuring in Table 3 are accomplished for every injector in the field, bringing about refreshed target rates for each rate in the sixth section for Table 3. This is only a post-handling step once the WAF information have been created by a streamline test system (3DSL 2006). There is no system in the computation displayed so far to guarantee that the new injection rate targets mean the measure of accessible water. In this precedent, we are keen on regarding an all-out injection imperative of 2700 STB/D. To do this, we essentially rescale the unconstrained target rates by a worldwide proportion given by

$$r = \frac{field\ target\ water - injection\ rate}{\sum q_i^{new}\ |unconstrained} \ldots (Eq.3)$$

The new target rates for every injector (unconstrained and obliged) are appeared in the fifth section in Table 3.

7.5 Determining the time intervals for the reinjection

We have done the optimization operation once and we used the time data in 1/1/2053 for the optimization as the (WAF) data base.

Note: As a reminder redoing the operation in the time intervals for multiple times will grant us better results, we need to be extra careful that every time there is a change in the average field efficiency, we need to calculate it separately.

8. Conclusions

1. The waterflooding for the useable actual reservoir are capable for operation, it shows an increase 429,000 STB of oil production over a period of 10 years is given in Fig. 5.
2. IEs on a well-pair premise enable a heuristic yet ground-breaking way to deal with improve reservoir management of (water) floods, by moving injected volumes from wasteful to effective injector/proucer sets, it is conceivable to expand oil production around 46243 STB without expanding volumes of injection as given in Fig.4.
3. This is a new exploratory approach, and to achieve accurate results, there must be an integration between the simulated and the actual reservoir.

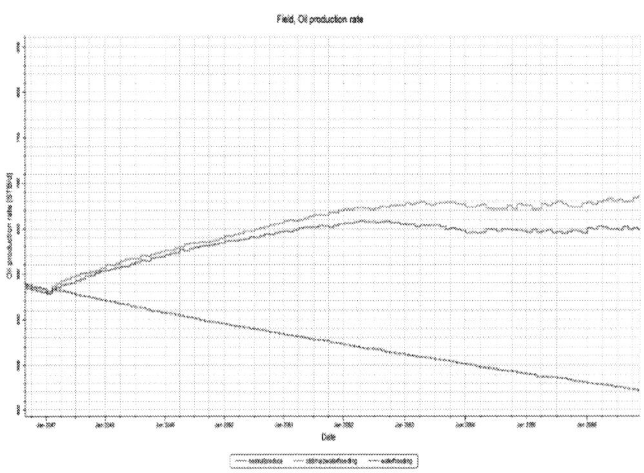

Figure 4: charts for the 3 different case study in production rates, red is for simple production, blue is for the waterflooding operation, and green is for the optimized waterflooding operation.

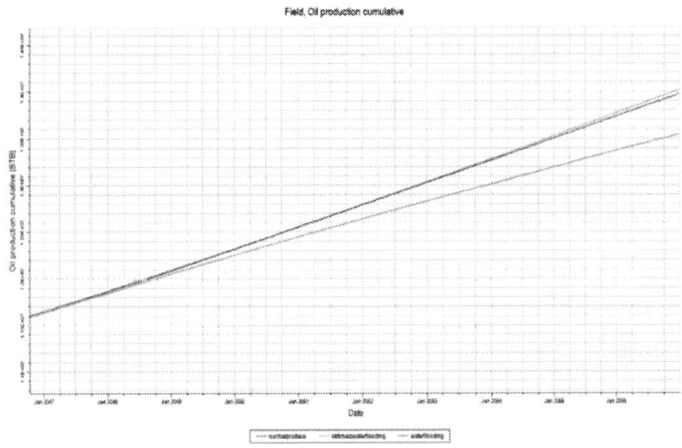

Figure 5: Chart for three case of the cumulative oil, green for the simple production, black for the waterflooding, and blue for the optimization waterflooding.

Appendix

$$e_i > \bar{e}: w_i = min\left[w_{max}; w_{max}\left(\frac{e_i - \bar{e}}{e_{max} - \bar{e}}\right)^{\alpha}\right]$$

...Eq. A-1

$$e_i < \bar{e}: w_i = max\left[w_{min}; w_{min}\left(\frac{\bar{e} - e_i}{\bar{e} - e_{min}}\right)^{\alpha}\right]$$

Where:

e_i *Injection efficiency for well i*
$\bar{e}$ *Average field injection efficiency*
w_i *Average field injection efficiency*
w_{max} *Maximum weight at* e_{max}
w_{min} *Minimum weight at* e_{min}
e_{max} *Upper injection efficiency limit*
e_{min} *Lower injection efficiency limit*
a *Exponent*

References

[1] 3DSL v2.30 *User Manual* 2006 StreamsimTechnologies Inc San Francisco
[2] Marco R Thiele and Rod P Batycky 2006 *Using Streamline-Derived Injection Efficiencies for Improved Waterflood Management* April SPE-84080-PA
[3] Batycky R P, Blunt M J and Thiele M R 1997 *A 3D Field-Scale Streamline-Based Reservoir Simulator* SPERE **12** 4 246–254 SPE-36726-PA
[4] Batycky R P, Thiele M R, Baker R O and Chugh S H 2005 Revisiting Reservoir FloodSurveillance Methods Using Streamlines *Paper SPE 95402 presented at the SPE Annual Technical Conference and Exhibition Dallas* 9–12 October
[5] Bratvedt F, Gimse T and Tegnander C 1996 Streamline computations for porous media flow including gravity *Transport in Porous Media* 25 1 63–78
[6] Baker R O, Kuppe F, Chugh S, Bora R, Stojanovic S and Batycky R 2002 Full-Field Modeling Using Streamline-Based Simulation: Four Case Studies *SPEREE5* 2 126–134 SPE-77172-PA
[7] Thiele M R 2003 *Streamline Simulation.* Keynote address at the 7th Intl Forum on Reservoir Simulation Baden-Baden Germany 23–27 June
[8] Thiele M R, Batycky R P and Blunt M J 1997 *A Streamline-Based 3D Field-Scale Compositional Reservoir Simulator Paper SPE* 38889 San Antonio Texas 5–8 October

[9] Thiele M R, Batycky R P and Kent L T 2002 Miscible WAG Simulations Using Streamlines. Paper presented at the 8th European Conference on the Mathematics of Oil Recovery (ECMOR) Freiberg Germany 3–6 September

[10] Thiele M R, Batycky R P, Blunt M J and Orr FM Jr 1996 Simulating Flow in Heterogeneous Systems Using Streamtubes and Streamlines *SPERE11* 15–12 SPE-27834 PA

[11] Wang P, Litvak M and Aziz K 2002 *Optimization of Production Operations in Petroleum Fields Paper SPE* 77658 presented at the SPE Annual Technical Conference and Exhibition San Antonio Texas 29 September–2 October

[12] Muskat M 1937 *The Flow of Homogeneous Fluids Through Porous Media* (McGraw-Hill Book Co. Inc. New York City) p 770

[13] Samier P, Quettier L and Thiele M 2002 Applications of Streamline Simulations to Reservoir Studies *SPEREE5* 4 324–332 SPE-78883-PA

[14] Higgins R V and Leighton A J 1962 A Computer Method to Calculate Two-Phase Flow in Any Irregularly Bounded Porous Medium *JPT* 14 6 679–683 SPE-243-PA

[15] LeBlanc J L and Caudle B H 1971 A Streamline Model for Secondary Recovery *SPEJ* **11** 3 7–12 SPE-2865-PA

[16] Flanders WA and Bates GR 1987 Optimizing Reservoir Surveillance by Using Streamlines and the Microcomputer *Paper SPE 16482 presented at the SPE Petroleum Industry Application of Microcomputers* Lake Conroe Texas, 23–26 June

[17] Grinestaff G H 1999 Waterflood Pattern Allocations: Quantifying the Injector to Producer Relationship With Streamline Simulation *Paper SPE 54616 presented at the SPE Western Regional Meeting* Anchorage 26–28 May

IOP Conf. Series: Materials Science and Engineering **579** (2019) 012036 doi:10.1088/1757-899X/579/1/012036

3D geological model of Nahr Umr /Ratawi oil field in Basra.

Mohammed Rashad Jemeel[1#], Samahr A.Lazim and Sameera[2] M. Hamdullah[3]
[1,2]Petroleum Technology Department/ University of Technology, Baghdad, Iraq.
[3] Petroleum Engineering Department/ University of Baghdad, Baghdad, Iraq.

[#]Corresponding author's e-mail address: moh0780107801@gmail.com, sameerahamdulla@yahoo.com and Samaher2013@yahoo.com .

Abstract. The study includes building a 3-D geological model, which involves get the Petrophysical properties as (porosity, permeability and water saturation). Effective Porosity, water saturation results from log interpretation process and permeability from special correlation using core data and log data. Clay volume can be calculated by six ways using IP software v3.5 the best way was by using gamma Ray. Also, Water Resistivity, flushed zone saturation and bulk volume analysis determined through geological study. Lithology determined in several ways using M-N matrix Identification, Density-Neutron and Sonic-Neutron cross plots. The cut off values are determined by Using EHC (Equivalent Hydraulic Unit) method. The full grid of the model was 454308 cells. Layering of units depend on the importance of these units and scale up had done using several methods depend on property (Arithmetic for porosity and water saturation and geometric for permeability). Petrophysical properties distributed using Gaussian statistical approach. Original oil in place calculated for Nahr Umr which equal to 460 m3.

1. Introduction

Ratawi oil field was discovered by the Iraqi oil company and the first exploration well drilled in 1948, at depth 3868 m. In 1972 another exploration well, Rt-2 was drilled to depth 3020 m. Ratawi oil field located at 70 km from Basra city and 12 Km in parallel to north Rumaila field [1] (Figure 1). Nahr Umr formation first discovered by Glynn Jones 1948 in Nahr Umr field where the thickness of formation reaches to 360m and in the south of Baghdad reach to 160 m while in Ratawi filed between (200-240m) which in deposited in Cretaceous era with shale sand rocks and thin sand layer in upper part with some clastic rocks. The log interpretation process includes Environmental correction, Estimate the petrophysical properties, fluid resistivity determination, lithology, Archie Parameter and net pay, net to gross determination. In the middle part of the layer consist of layers of sand rocks and shale rocks respectively and followed by the thick layer of sand stone rocks In the bottom of Nahr Umr totally consist of thick sand rocks and one third of upper part of this section consist of layers of sand and shale rocks respectively[2].There are no fault in the Ratawi which may effect on the compartments [3],thirty one wells were drilled in Ratawi Oil Field, just 14 wells were penetrate Nahr Umr formation as well as 8 wells (Rt-4,Rt-16,Rt-17,Rt-18,Rt-20,Rt-21,Rt-22 and Rt-24)were have las files. These las files have been imported to IP software v3.5 (Interactive Petrophysics) in order to get interpretation as shown (Figure 2).

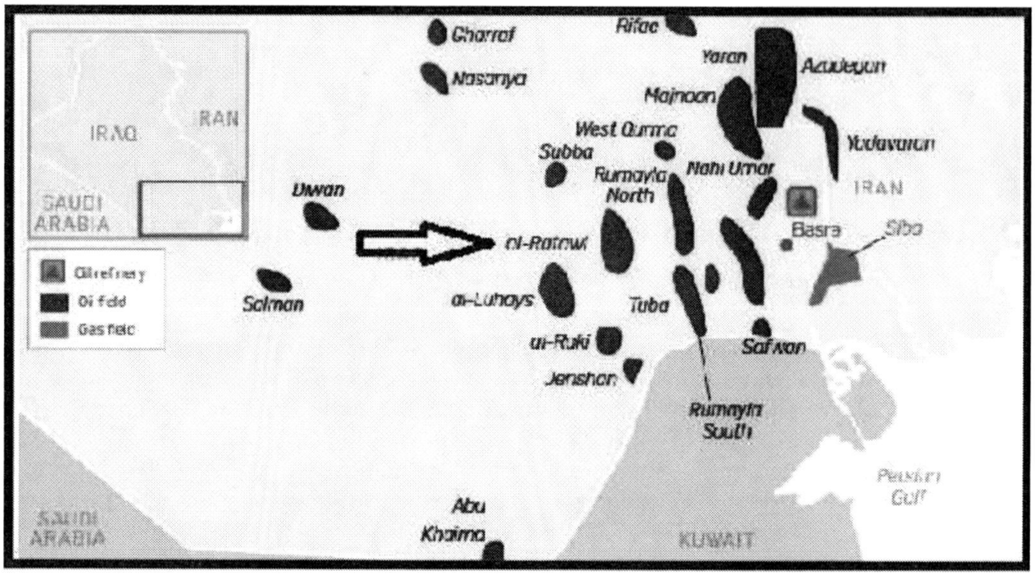

Figure 1. Location map of Ratawi oil field.

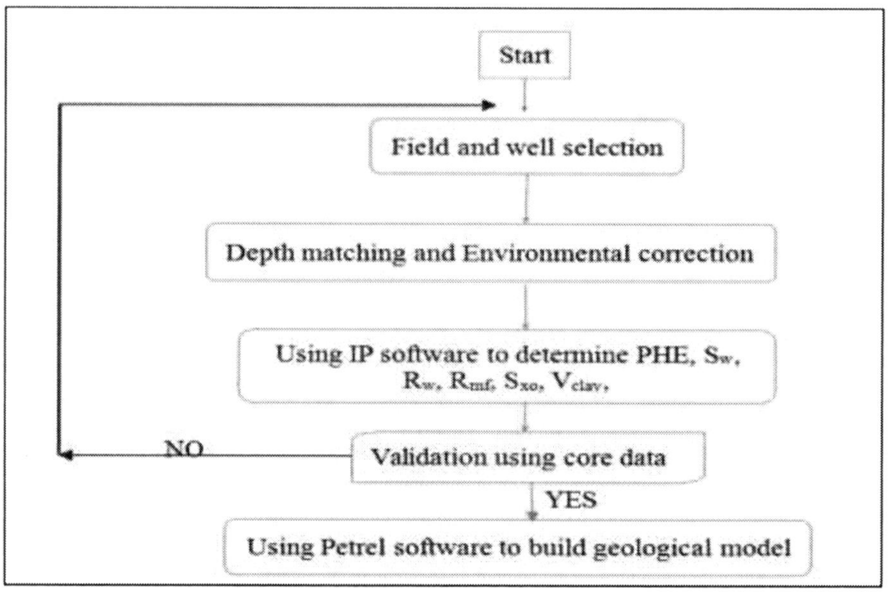

Figure.2. Procedure of interpretation process

2. Environmental correction and Clay Volume

Many charts available to correct the logs like Baker Hughes, Halliburton, Schulmberger…. etc. Schulmberger charts were used to this purpose. Environmental correction was used in order to bring back log into the standard condition. The input data beside the log needed for Gamma Ray, Mud weight and Top, Bottom of each well, density mud weigh. Resistivity logs need temperature curve, hole diameter, mud filtrate resistivity, mud resistivity and temperature of each them. (Figures 3&4) these figures show the correction for density, resistivity curves, gamma ray and Neutron respectively [4]. Clay volume can be calculated by six ways using IP software which include gamma Ray, neutron, resistivity, SP, density neutron and sonic neutron as shown (Figure 5).

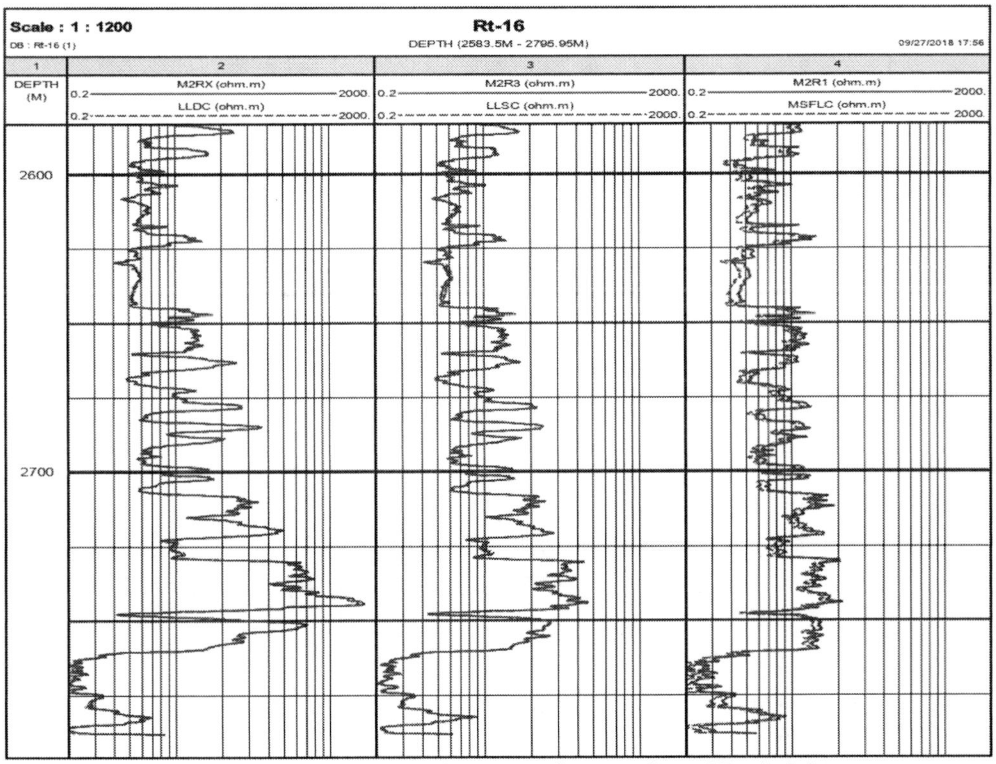

Figure.3. Environmental correction of Resistivity.

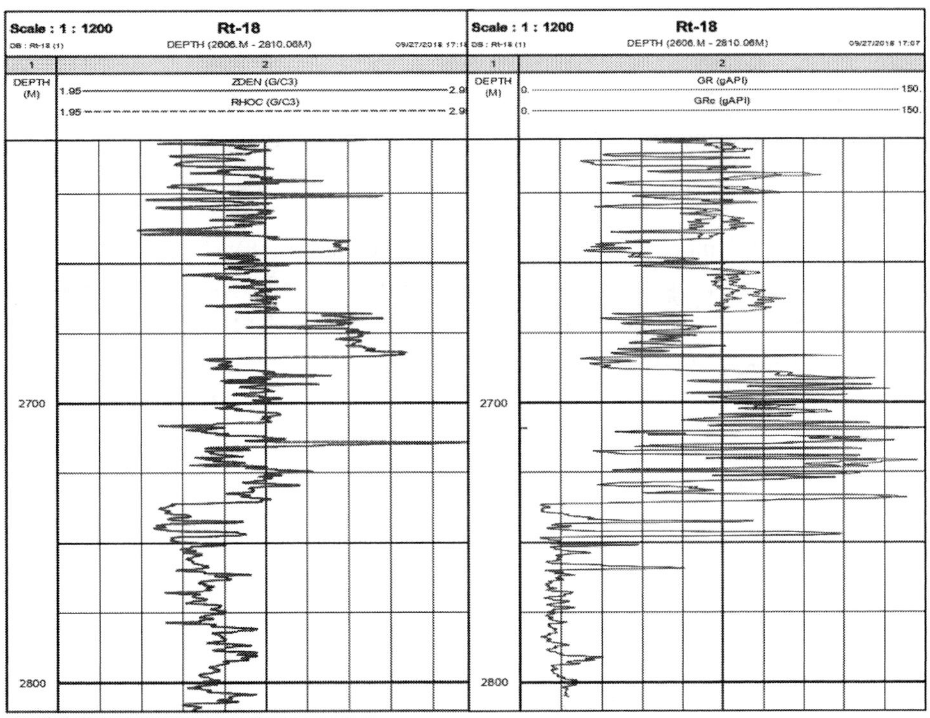

Figure 4. Correction of gamma ray and density

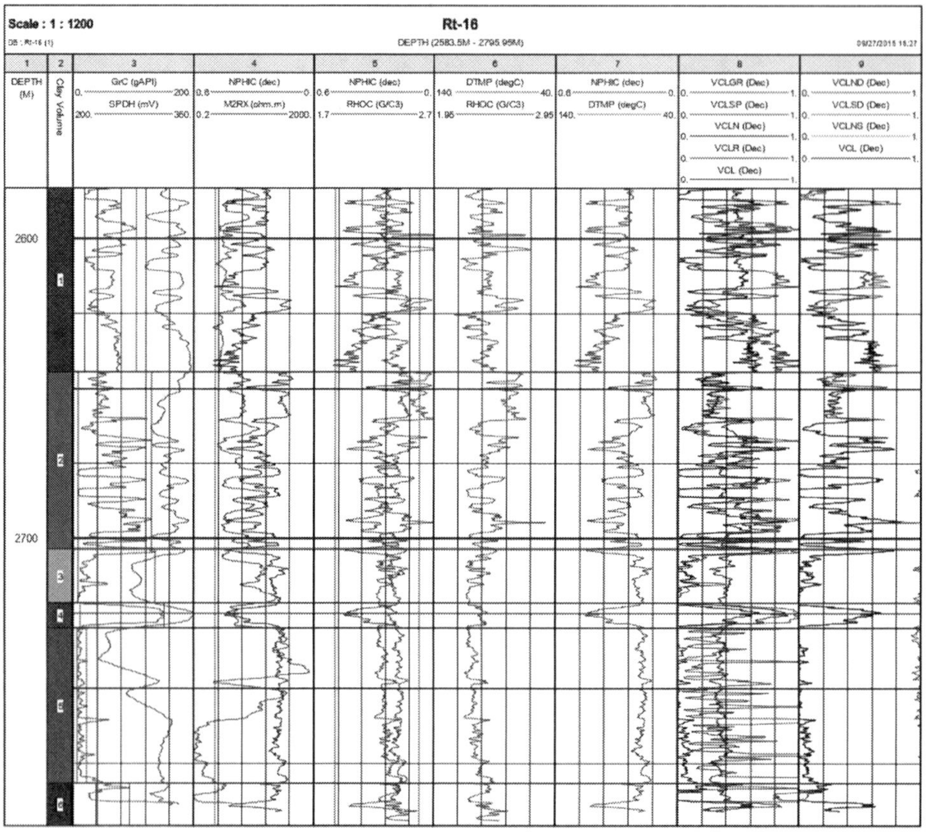

Figure 5. Estimated clay volume.

3. Effective porosity and Lithology

Effective porosity [5], depend on the clay volume as expressed below:

$$\varphi_e = \varphi_t(1 - V_s) \tag{1}$$

Lithology can be determined from different method using IP software such as density neutron cross plot, Neutron sonic cross plot, M-N Mineralogy Identification plot equations (2 &3) can be used to estimate M and N and Matrix Identification (MID) involve estimate apparent matrix density (ρma) and apparent transit time of rock matrix (Δtma) as in equations (4 &5). Lithology of Nahr Umr as mentioned in introduction section composed mainly of sand stone and shale with minor amount of limestone which located mainly in upper part of formation. As shown in (Figures 6,7,8&9).

$$M = \frac{\Delta t - \Delta t_i}{\rho - \rho} * 0.01 \tag{2}$$

$$N = \frac{\varphi - \varphi}{\rho - \rho} \tag{3}$$

$$\Delta tma = \frac{(\Delta tb - \Delta tf . \varphi ta)}{1 - \varphi ta} \tag{4}$$

$$\rho_m = \frac{(\rho_b - \varphi_t . \rho_f)}{1 - \varphi_t} \tag{5}$$

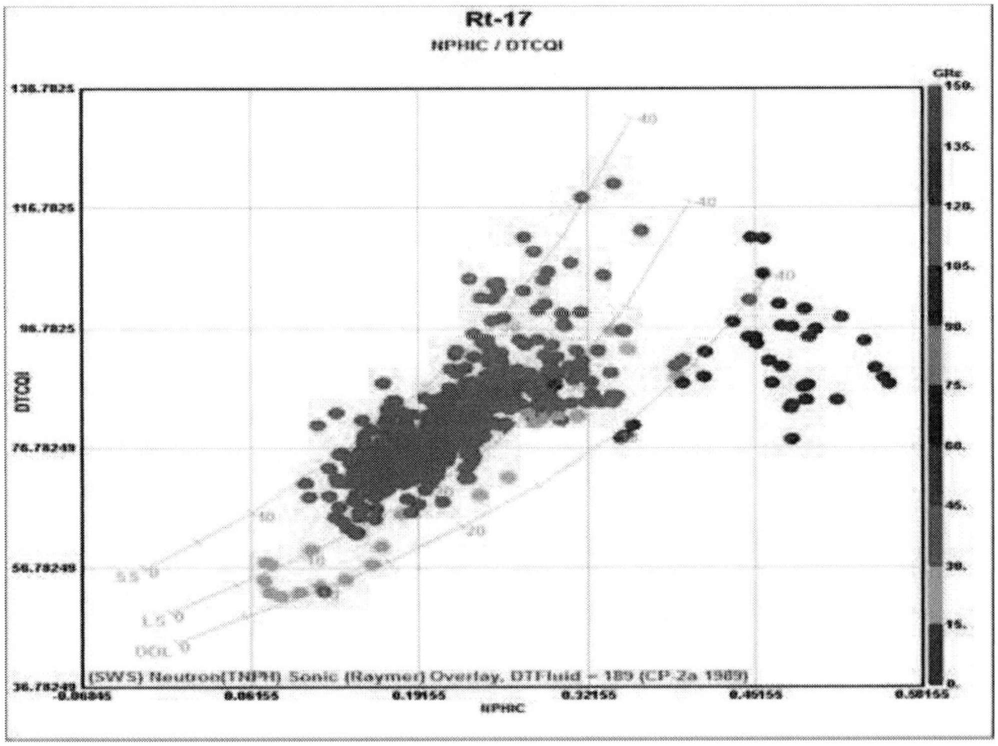

Figure 6. Sonic Neutron cross plot.

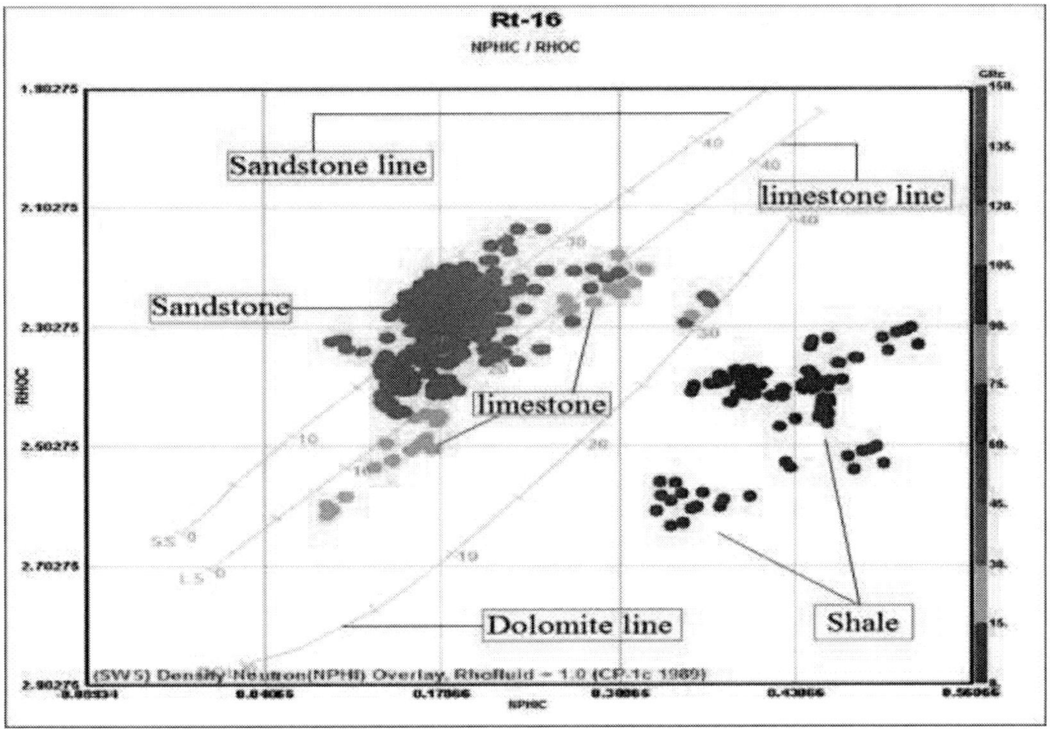

Figure7. Neutron-density cross plot.

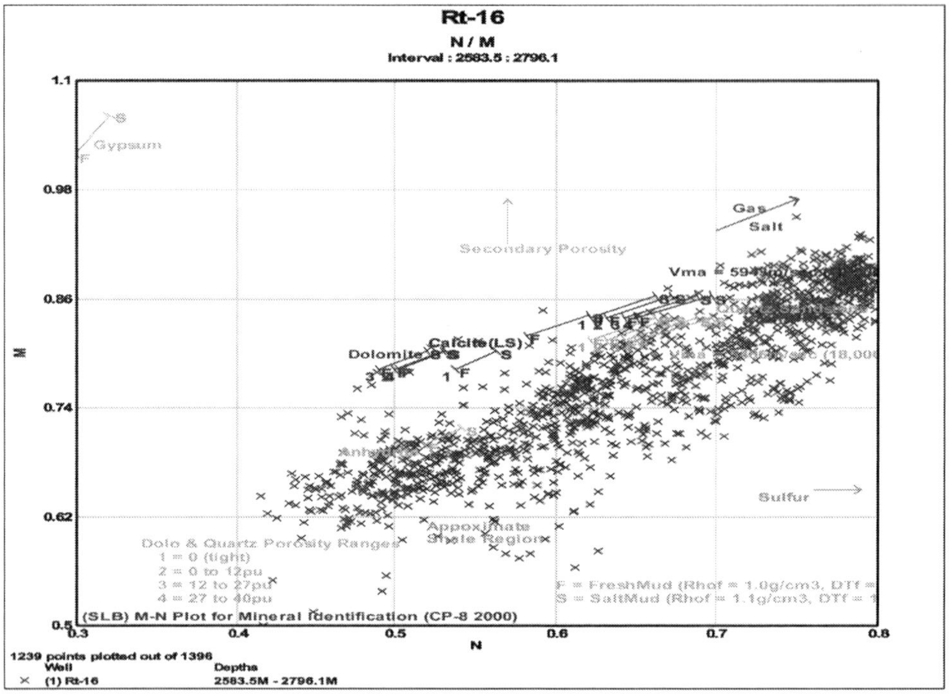

Figure .8. M/N cross plot

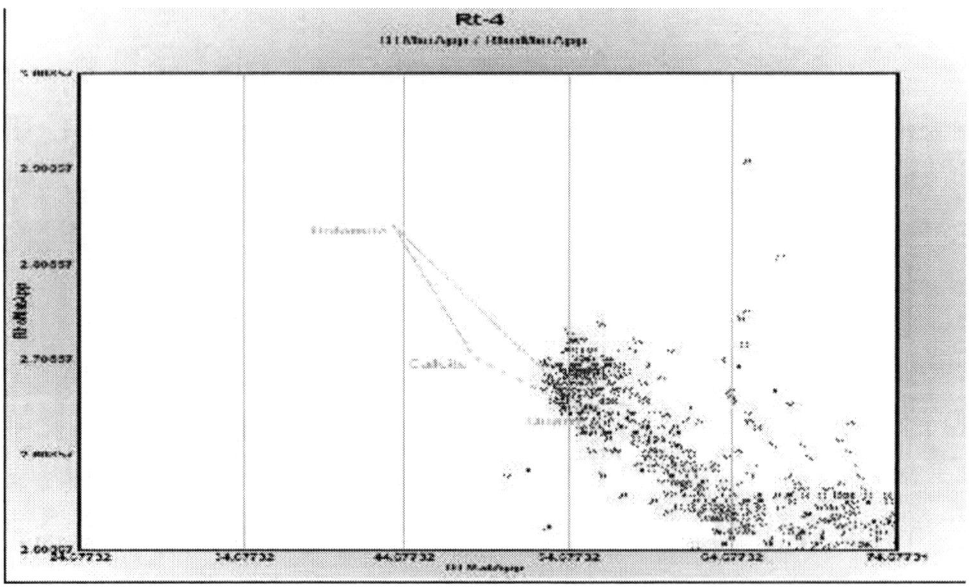

Figure. 9. Matrix density vs. transit time plot.

4. Water resistivity (R_w) determination

R_w is important factor to estimate water saturation beside the Archie parameter. Rabanit 1952 proposed a method to calculate water Resistivity. First must calculate Rw at 75 °F depending on the Salinity concentration then determine the water resistivity in reservoir temperature [5], (Table 1).

$$R_{w7} = \frac{1}{2.7 * 1^{-4} * C^{0.9}} + 0.0123 \quad (6)$$

$$R_w = R_{w7} * \frac{8\ .7}{T_f + 6.7} \quad (7)$$

Table 1. R_w and R_{mf} for each well

Wells	BHT (F°)	Rmf @MT	Rw @ BHT
RT-4	176.5	0.52	0.019589
RT-16	177.8612	0.32	0.01944
RT-17	178.682	0.177	0.019356
RT-18	176.936	0.165	0.0195
RT-20	163.4	0.22	0.02109
RT-21	168.8	0.22	0.02044
Rt-22	165	0.18	0.0209
RT-24	174.2	0.1	0.01983

5. Archie Parameter

Archie Parameter (a, m and n) is necessary to determine water saturation, so that picket plot used for this purpose. By assuming that a=1 and n=2 and By Knowing the Rw values the m (cementation factor) will be determined, (Figure 10) and (Table 2).

Table 2: Archie parameters for different wells.

wells	a	m	n
Rt-4	1	1.98	2
Rt-16	1	1.94	2
Rt-17	1	1.95	2
Rt-18	1	1.92	2
Rt-20	1	1.95	2
Rt-21	1	1.93	2
Rt-22	1	1.98	2
Rt-24	1	1.98	2

6. Fluid saturation determination

There many equations to determine the water saturation one of the most equation was Archie, besides this equation there are many equations like Simandoux, Indonesian and Indonesian. Archie can calculate water saturation for clean zones depend on the effective porosity(φe), water resistivity (Rw), true resistivity (Rw) and Archie parameter (m, n, a) [6].

$$S_w = \left\{\frac{a.R_w}{R_t.\varphi_e{}^m}\right\}^{\frac{1}{n}} \quad (8)$$

$$S_x = \left\{\frac{a.R_m}{R_x\ .\varphi_e{}^m}\right\}^{\frac{1}{n}} \quad (9)$$

Residual oil saturation (S_{or}) and movable oil saturation (S_{hr}) [7] can be determined as:

IOP Conf. Series: Materials Science and Engineering **579** (2019) 012036 doi:10.1088/1757-899X/579/1/012036

$$S \quad = [\varphi(1 - S \quad)] \tag{10}$$

$$S_{hr} = [\varphi(S_x \quad - S_w)] \tag{11}$$

Indonesian equation was used to estimate water saturation for this study because the high shale percent in the Nahr Umr formation [8] as shown below:

$$S_w = \left[\frac{1}{\left[\frac{V_{sh}^{\frac{1-V_{sh}}{2}}}{R_{sh}} + \frac{\varphi_e^{\frac{m}{2}}}{(A * R_w)^{\frac{1}{2}}} \right] * R_t^{\frac{1}{2}}} \right]^{\frac{2}{n}} \tag{12}$$

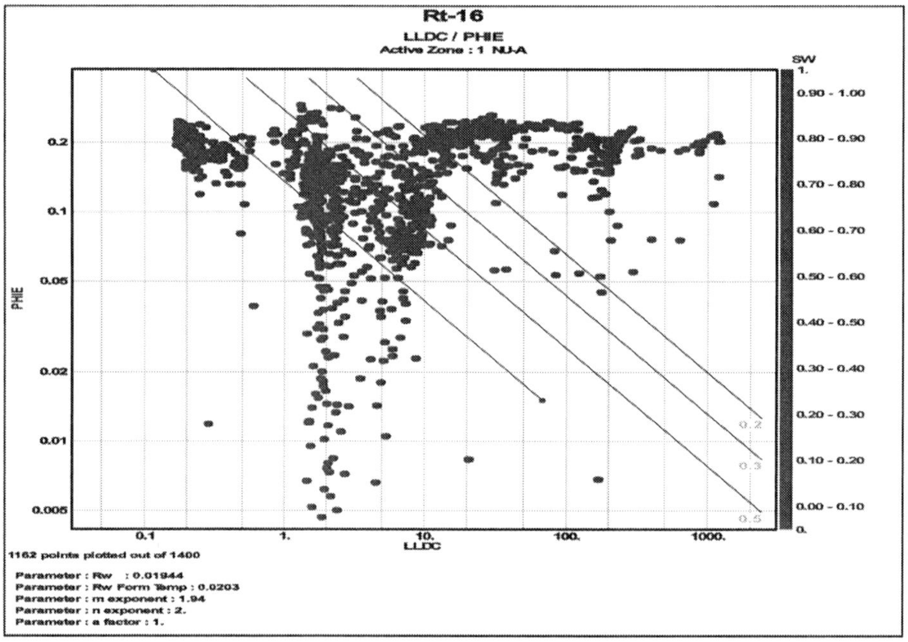

Figure 10. Picket plot.

7. Bulk volume Analysis

In porosity track BVWSXO (Black line) which represent the bulk volume of porosity in flushed zone filled with water as shown in eq.13, BVW (red line) which is bulk volume of porosity in true zone filled with water as shown in eq.13, where area between them represent the movable hydrocarbon and the area between effective porosity (φ_e) and BVW represent the total hydrocarbon. Through this study the result that the percentage of shale in NU-A and NU-B equal to 45% and 44% respectively while V_{sh} in NU-C and NU-D equal to 28%, (Figure 11).

$$BI \quad = \varphi_e * S_x \tag{13}$$

$$BI \quad = \varphi_e * S_w \tag{14}$$

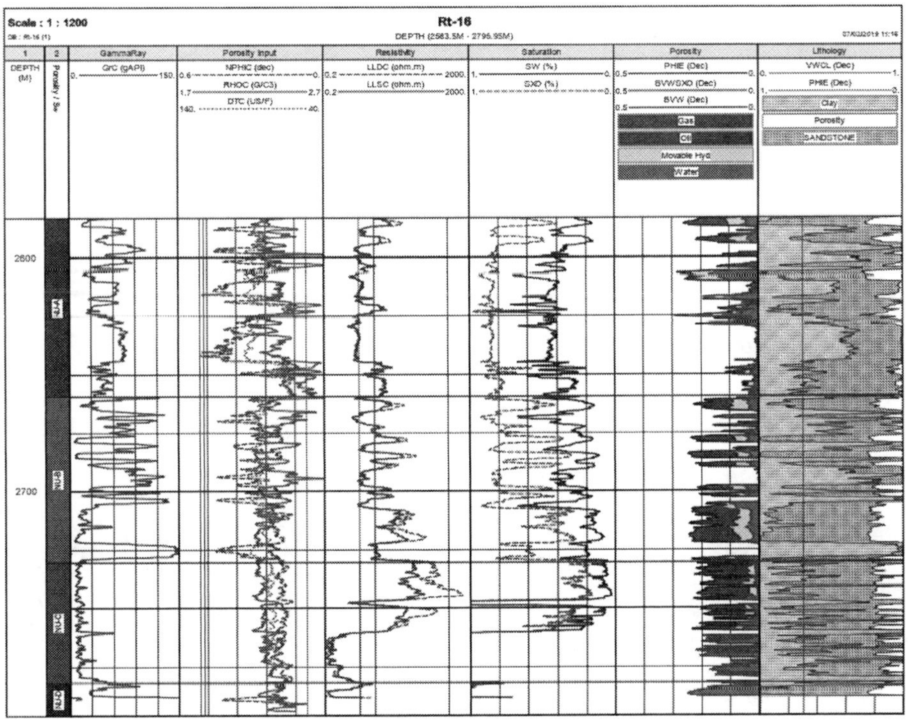

Figure 11. Bulk volume Analysis of Rt-16.

8. Cutoff parameters

There many ways to determine the cutoff values one of these methods was EHC (equivalent hydraulic column) method, eq.14, where this method depend on multiple values of water saturation like (Sw>40%, Sw>50, Sw>60%, Sw>70% and Sw>80%). Porosity and water saturation cutoff were determined (Figure12 and Figure 13). So, the criteria of Nahr Umr will be porosity >8%, water saturation <70%, permeability>0.1 md and shale cutoff <0.3.

$$E \quad = (1 - S_w) * \varphi_a \quad * \sum h_n \tag{15}$$

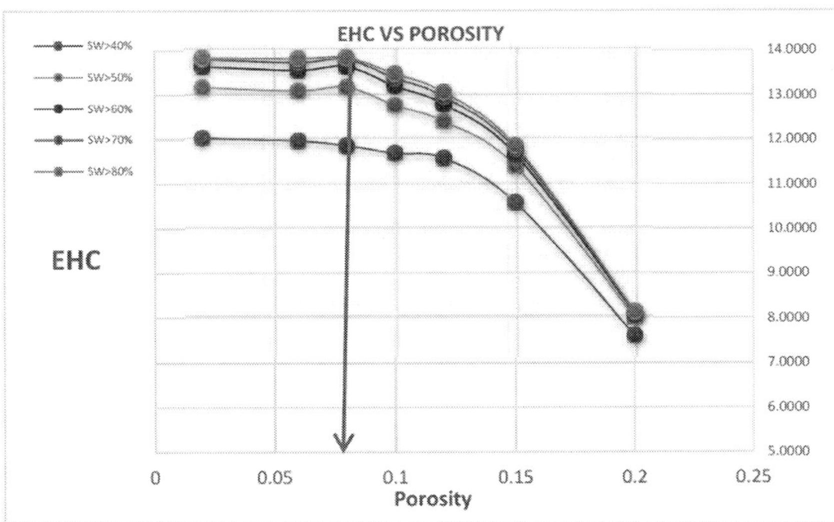

Figure 12. Porosity cutoff determination.

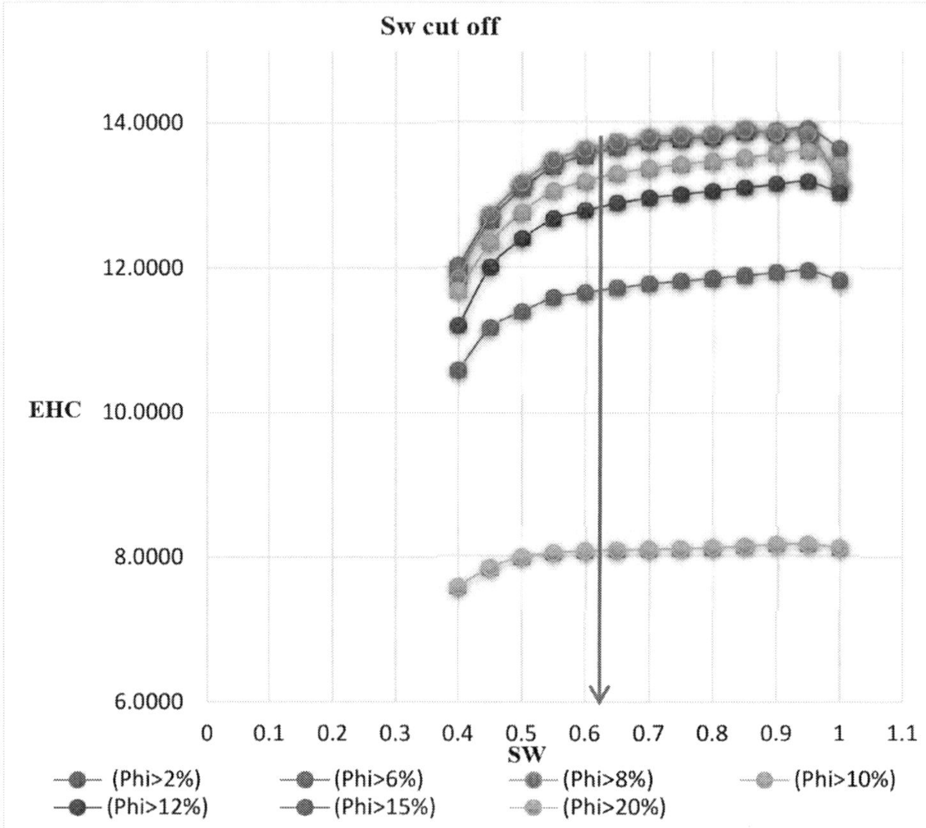

Figure 13. Water saturation cut off Nahr Umr formation.

9. Permeability calculation

Permeability can be predicted using core data in order to create correlation use to convert the log porosity to core porosity for the rest wells as in eq.16, (Figure 14) show the relation between the porosity log and porosity core for Rt-4. The core porosity that result from this correlation can be used in eq.17 which result by draw between core permeability and core porosity for Rt-4 in order to get permeability as (Figure 15).

$$\varphi_c = 0.8731\,\varphi_{lc} + 0.01 \quad (16)$$

$$K_{core} = 0.0005\,e^{62.82\,Phi_{core}} \quad (17)$$

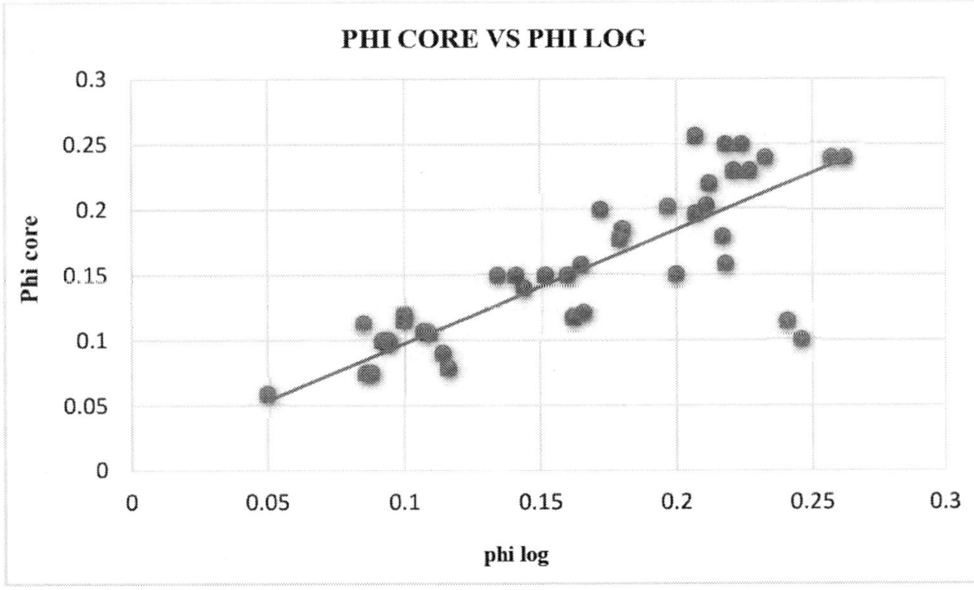

Figure 14. Phi log-core relationship of Rt-4.

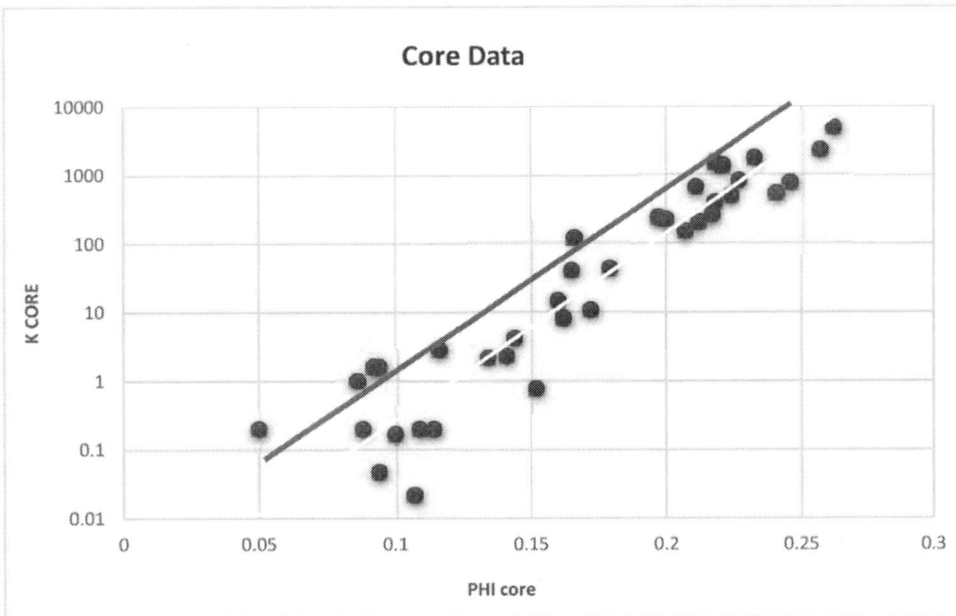

Figure 15. Relation between porosity core and Permeability core of Rt-4.

10. Geological modeling and petrophysical properties distribution

Geological model has been built using Petrel software 2015.1 and by input multiple data that need as well head (where well head must be imported to know the position of the well path as well as the measured depth), well tops (represent a peck point along the well path), well logs (represent the CPI , Net to Gross, shale volume…..etc.), and contour map. Use simple grid to define the structural of the model by define the dimension of grids (ΔX=250, ΔY=250 and ΔZ=4.218). The structural contour map of NU-B depened for this model (Figure 16), so based on this map other maps were constructed, (Figure

17). The skeleton of this model represents the framework which represent the first step of 3D the model, where the Top, Bottom and middle of this skeleton represented in (Figure 18). Layering process used to split the important zones into several layers depending on the importance of these zones. Where NU-A split into 15 NU-B into 20, NU-C into 10 layers and NU-D into 1 layer. In order to get average value of properties to each cell need to make scale up. There are many ways to make scale up such as arithmetic, harmonic, geometric .... etc. [9]. The total number of cells reach to 454308 cells (68 cells in I direction, 131 in the y-direction with 51 number of layers). The Porosity, permeability and water saturation was distributed throughout the model using Gaussian statistical approach [10]. Porosity result from IP v 3.5 as a result of interpretation process. Porosity maximum and minimum of the model was (0.074 to 0.195, 0.071 to 0.259, 0.132 to 0.259,0.099 to 0.256) for NU-A, NU-B, NU-C, NU-D respectively. Water saturation model was used distribute the water saturation that have been gotten from well logs interpretation for four layers mentioned previously as (0.2539 to 1, 0.0872 to 1, 0.01 to 1 and 0.855 to 1). Permeability was supported from porosity-permeability correlation, where permeability model was construct based on the maximum and minimum of NU-A was from 0.00033666 to 6561.55, NU-B from 0.00024531 to 162.18 md, NU-C from 0.00085822 to 7256.9 md and NU-D was 0.00024531 to 1218.53md, shown (Figure 19,20,21).

11. Original oil in place

Water oil contact was determined for Nahr Umr formation from the comparison between the water saturation and depth of well log data where the water oil contact was -2729 m, (Table 3), (Figure 22). Original oil in place has been calculated for Nahr Umr which was 2893.31 MMBBL using Petrel software 2015.1. where NU-B has the biggest reserve (1257 MMBBL) and for NU-A equal to (1062.97 MMBBL), while oil in place in NU-C equal to 572.37 MMBBL.

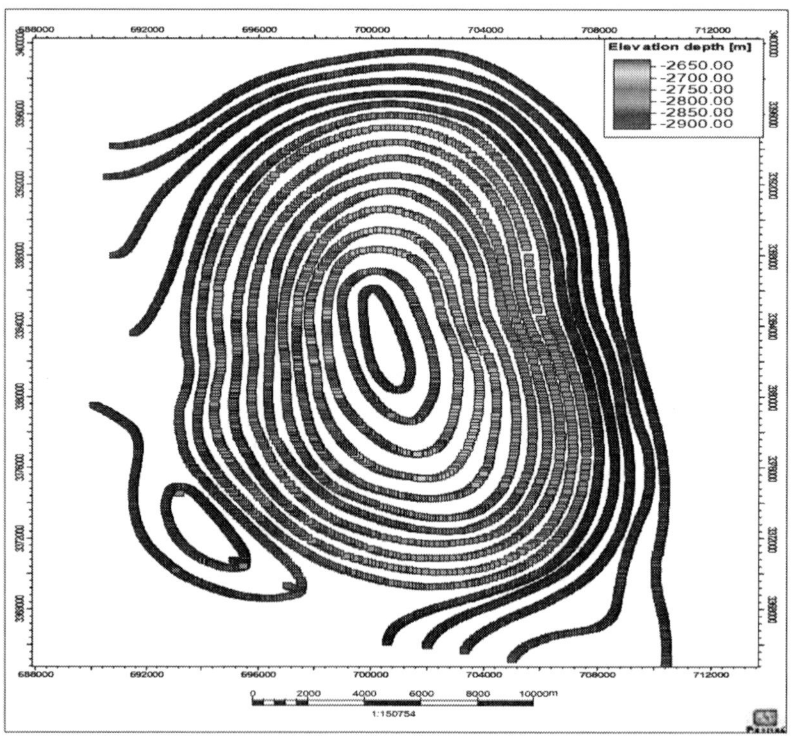

Figure 16. structural Map of NU-B formation.

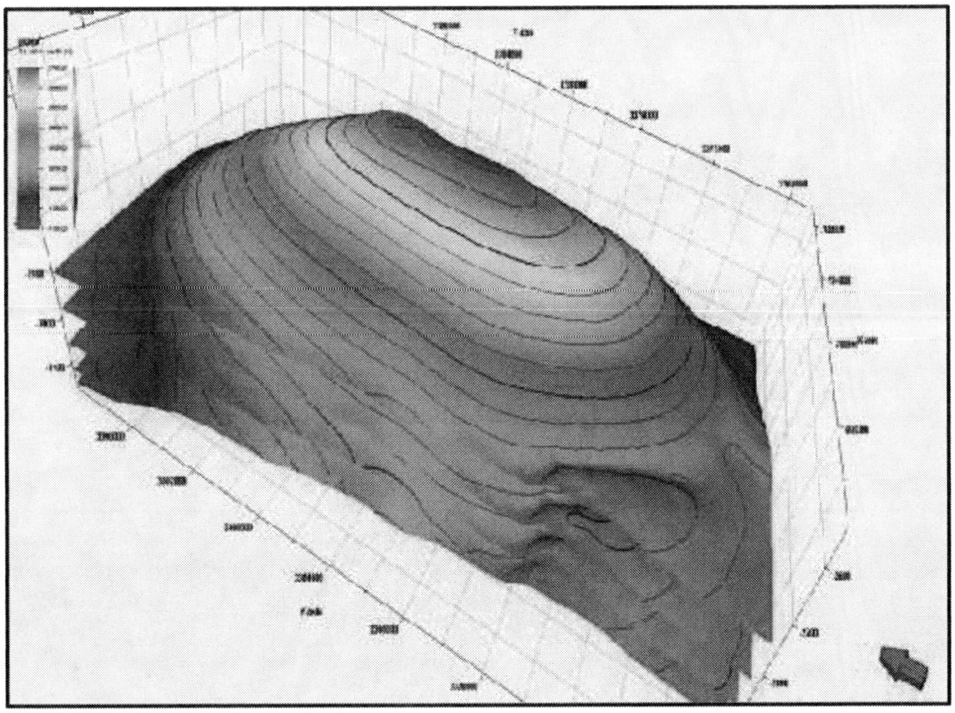

Figure 17. Map surfaces of Nahr Umr formations.

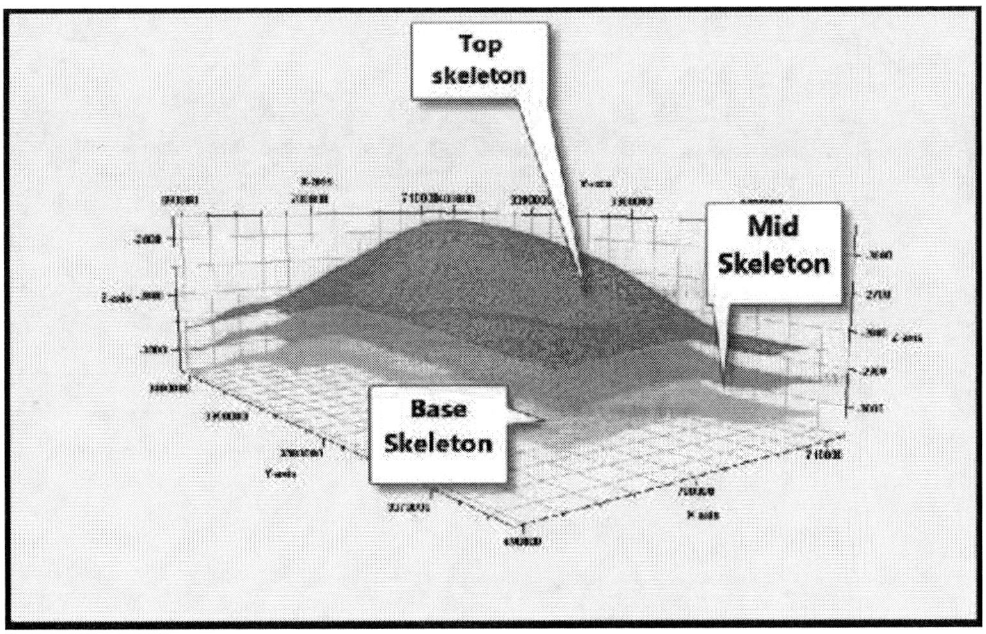

Figure.18. Structural skeleton of Nahr Umr formation.

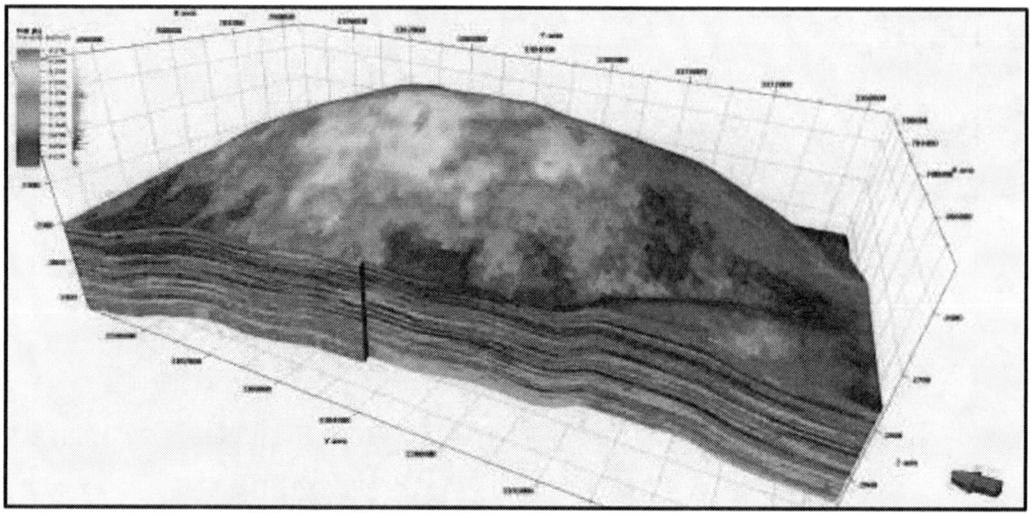

Figure 19. Porosity Model of Nahr Umr formation.

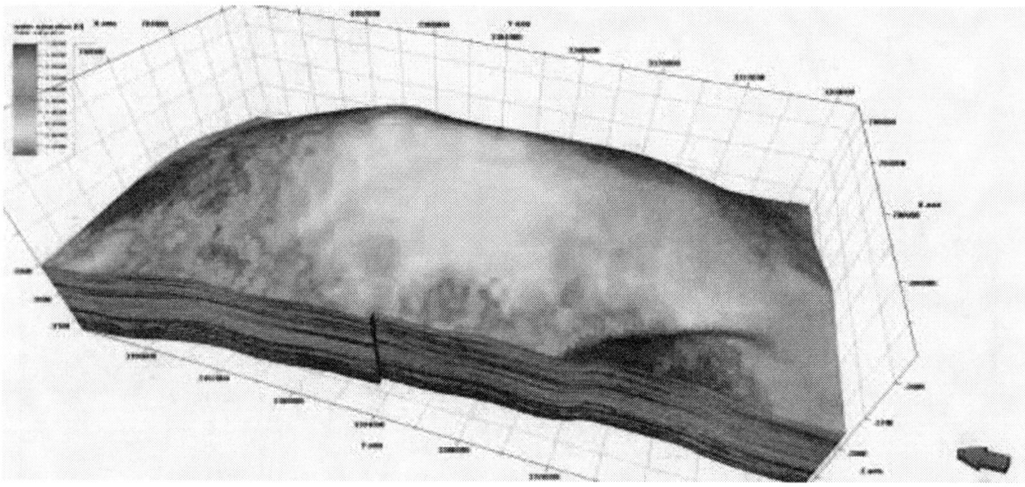

Figure 20. Water saturation Model of Nahr Umr. formation

Table 3: Oil water contact values of different wells of different wells

Well name	OWC (meter)
Rt-16	-2725.6
Rt-17	-2718.9
Rt-18	-2720.39
Rt-20	-2730.3
Rt-21	-2743.03
Rt-22	-2729
Rt-24	-2730.42
Rt-4	-2735.3

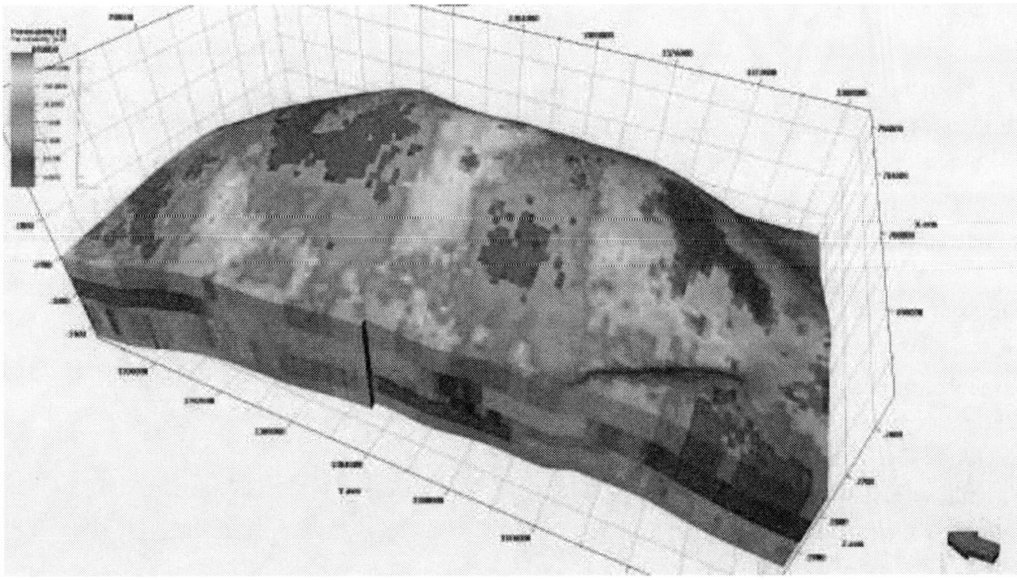

Figure 21 Permeability model of Nahr Umr.

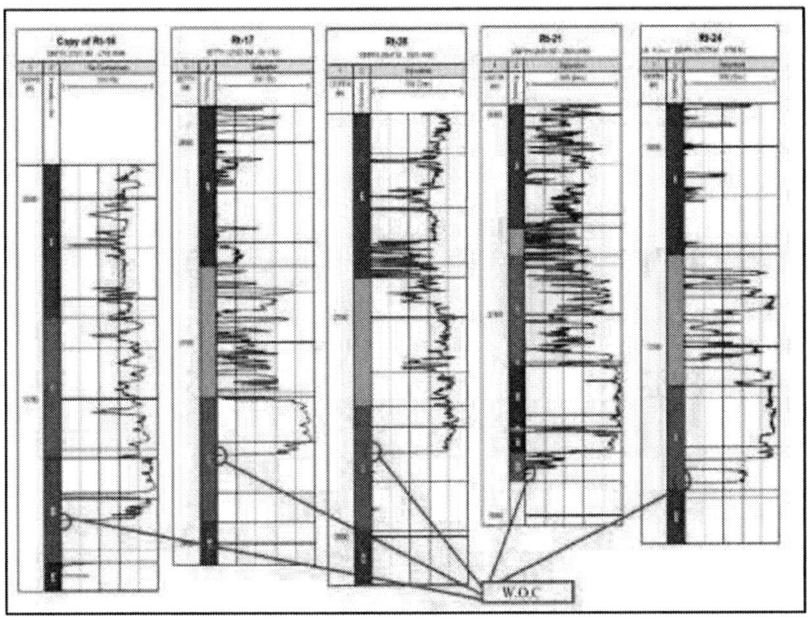

Figure 22. O.W.C determined with depth for different wells

12. Conclusions

Nahr Umr can be divided into four units depending on the composition of each unit.NU-A have large percentage shale about 45% so little perforations recommended in this Unit.NU-B has a biggest oil reserve where NU-B and NU-C represent main reservoir of Nahr Umr. Depending on all sections that mentioned in lithology the main composition of hole formation is sandstone and secondary from shale. EHC method was used to determine the cutoff values. The relation between porosity log and porosity

IOP Conf. Series: Materials Science and Engineering **579** (2019) 012036 doi:10.1088/1757-899X/579/1/012036

core give R^2=0.6224 while the relation between the porosity core and permeability core give 0.8742 R^2. Indonesian equation was used for water saturation calculation. Porosity, water saturation and permeability models have been built using statistical Sequential Gaussian simulation method.

Nomenclatures

a: tortuosity factor;

CSP: Concentration of salinity ppm;

n. Saturation exponent.

m. Cementation factor.

φ_e. Effective porosity.

φ_{ta}: apparent total porosity fraction which can be determined by appropriate neutron density;

$\varphi_{NF}=1$;

K. Permeability, md.;

S_w: Water saturation;

S_{xo}: Invaded zone saturation;

BVW: Bulk volume of porosity filled with water;

R_t: True Resistivity ohm-m.;

R_w: Water Resistivity ohm-m.;

R_{mf}: Mud filtrate Resistivity ohm-m.;

R_{sh}: shale Resistivity ohm-m.;

Rw_{75}: Formation Water Resistivity @ 75°F;

Rw_{fm}: Water Resistivity at formation temperature;

BVWSXO: Bulk volume of porosity in flushed zone filled with water;

S_w: Water saturation;

S_{hr}: movable oil saturation.

S_{or}: Residual oil saturation;

T_f: Formation temperature, °F;

ρ_{ma}: apparent matrix density.

Δt_{ma}: apparent transit time of rock matrix.

Δt_f: 189 usec/ft;

V_{sh}: Shale volume.

Reference

[1]. J. Gergies & A. Mohesin, 1992 *Geological evaluation study for Mishirf, Muddud and Nahr Umr formation* (Baghdad) **3** 70

[2]. Jawad Mohsen Mozan, Abdulzahra Eskander, 2014 *A study on reexplanation of earthquake information of Ratawee Field (stress study)* 1012 (In Arabic)

[3]. JOGEMEC, 2010 *Geological study for Ratawi oil field* (Ministry of oil Iraq) p 5

[4]. Synergy LTD 2008 *Interactive Petrophysics software v3.5 manual.***816** 1662

[5]. Rabanit, P.M.V., 1952 *Rock units of Basrah area. Unpublished, Basra Petroleum Company (BPC), Iraq National Oil Company Archives* (Baghdad, No. BGR) P 8.

[6]. Desbrandes, R., 1985 *Encyclopedia of well logging* (Editions OPHRYS)**12** 460

[7]. Asquith, G. and Gibson, C., 1982 *Basic well log analysis for geologists: methods in Exploration series. AAPG* (Tulsa, Oklahoma) p 216

[8]. D. Marsan &R. Nada Pratma 2015 *Well log interpretation and Petrophysical analysis* (Halliburton) p 4-6

[9]. Bartucz, D., 2009 *Exploration of Geothermal Systems with Petrel Modeling Software* (Doctoral dissertation) p 20-30

[10]. Schulmberger Company, 2009 *Schulmberger Petrel manual 2009 version.*

[11]. Brown D.G., 2003 *Formation evaluation and Petrophysics* (Core Laboratories, Jakarta, Indonesia) p194

IOP Conf. Series: Materials Science and Engineering **579** (2019) 012037 doi:10.1088/1757-899X/579/1/012037

Modeling of Petrophysical Properties and Reserve Estimation of Mishrif Formation-Garraf Oil Field

Ahmed T. Abeed [1], Samaher A. Lazim [2], Ramzy. S. Hamied

1, 2, 3 Department of Petroleum Technology, University of Technology, Baghdad-Iraq
Corresponding author, ahmedtalal284@gmail.com

Abstract. In this study a build-up a 3D geologic model both depend on the information of field and subsurface is a usual task in geologic studies including resource evaluation and hazard assessment. During this study a 3D geologic model for Mishrif Formation in Garraf oil field has been set-up by using petrel 2015 software system. Mishrif Formation represents the foremost important reservoir in Garraf oil field. Four vertical oil wells and one directional well were chosen in Garraf Oil Field so as to set-up structural and petrophysical models delineated by a 3D static geologic model in 3 dimensions. The structural closure is shifted from GA-3 well to GA-A1P well. Mishrif Formation was divided into 10 zones.

Sequential Gaussian Simulation algorithm was used to build-up Petrophysical model (NTG, permeability, water saturation and porosity) for Mishrif Formation keeping with information analyses and also the results from modeling the units (M1.2, L1 and L1.2) are thought-about as prime quality reservoir units because of the high PHIE and low water saturation. Units (L2, L2.2, L2.3 and L2.4) are thought-about as poor reservoirs because of low PHIE and high water saturation, and non-reservoir units include (Top Mishrif, M1 and M2) thus, they represent cap units. Cross sections of petrophysical model were conducted to Illustrate the vertical and horizontal distribution of porosity, permeability, Net to gross and water saturation between wells within the field. This study calculated the reserves of the Mishrif formation of Garraf oilfield by petrel, that alter the issue that the reserves parameters are difficult to properly acquire when estimate complex reservoir's reserves in terms of ancient volume technique.

1. Introduction

Models of reservoir can be formed comparatively quickly with geo-statistical tools, like petrel 2015 software package, one of preferred modeling code in petroleum industry, but often, a limited variety should be chosen for input to flow simulation due to procedure time needs [1]. 3D model is the method of developing a mathematical illustration of any 3– D surface of object via specialized software system. The product is termed a 3D model [2]. The ranking engine has been written as petrel plug-in. This enables petrel users to rank reservoir models in petrel and choose limited representative ones for flow simulation [3]. The objective of this study to build up a 3D geologic model for Mishrif Formation in Garraf oil field using petrel 2015 software system. This model includes structure, Stratigraphic and reservoir properties (porosity and water saturation) in 3 dimensions (X, Y and Z). Geological modeling is an engineering science of making processed representations of parts of the Earth's crust, particularly hydrocarbon reservoirs. Within the petroleum industry, Geological models are needed as input to reservoir simulator software, that predict the performance of the rocks beneath various hydrocarbon recovery situations using reservoir simulation permits reservoir engineers to detect that recovery choices provide the safest and most economic, efficient, and effective development plan for a specific reservoir[4].

Reserve volume is that the comprehensive analysis of oil field exploration and development, as a consequence, the rationality of reserves calculation positively have an effect on the choice of oil field prospecting and exploitation. The reserves calculation method based on the high exactness of tridimensional reservoir geologic model, will be used to influence the issues that reserve parameters are troublesome to correctly acquire when estimating complicated reservoir's reserves in terms of traditional volume method[5].

2. Area of Case Study

The Garraf oil field is located in Thi Qar province regarding eighty-five km to the north of Nassriya city south of Iraq. The Garraf oil field could be a north west south east trending anticline with an area of twenty-four km length and five km width [3]. Three wells were drilled in Garraf oil field throughout. Six vertical wells (GA-1, GA-2, GA-4, GA- A1P, GA-3 and GA-5) and one directional well (GA-B8P) were studied in Garraf Oil Field (Figures 1 and 2) [6].

The Mishrif and Yamama Formation were containing the most oil accumulation within the field area. The second accumulation zones are found within the Ratawi and Zubair Formation [7].

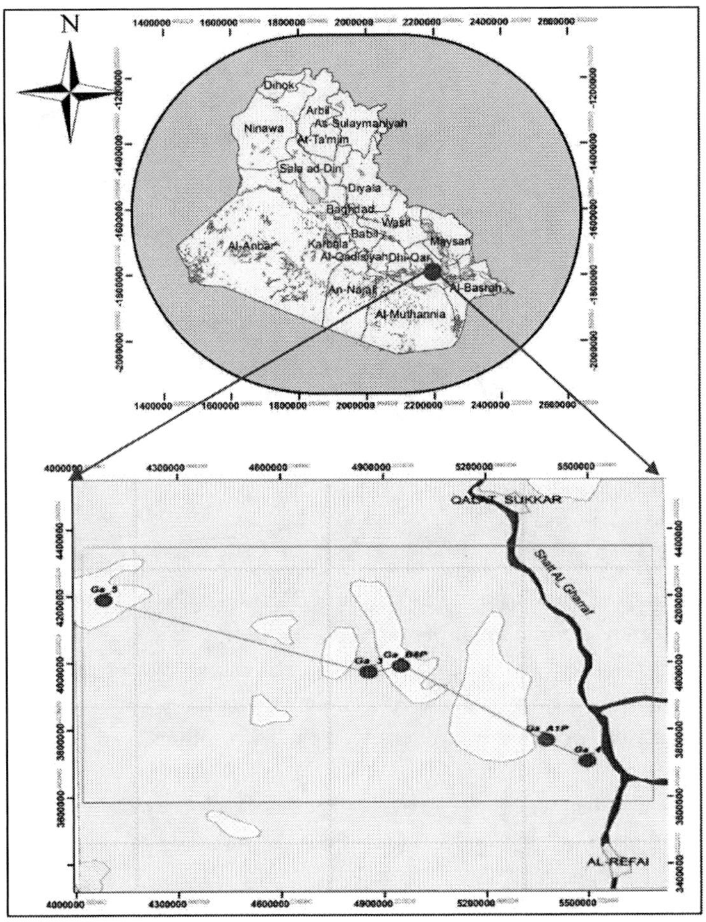

Figure 1 Map of appearance location of the area study (ARCMAP, 2015).

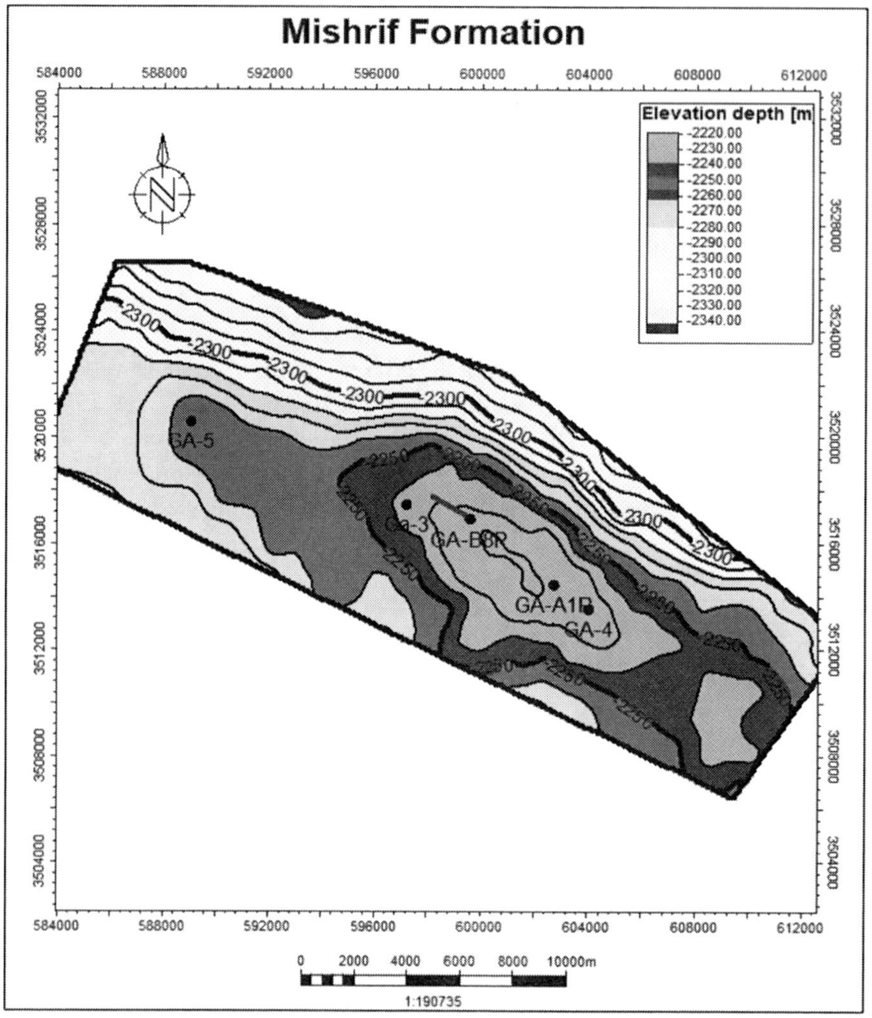

Figure 2 Contour map of top Mishrif in Garraf oil field.

3.Workflow

Petrel software system 2015 has been used to set-up a 3D model. Petrel is a PC- based mostly progress application for underground interpretation and modeling [8].

On this basic of program system required and analysis area characteristics. The input data are imported from files- on file for every data object. These data include:

1. Well head: includes the coordinates of every well in 3-dimentions, and also the measured depth along the path.

2. Well tops: markers represent well picks on the well path, commonly a change in stratigraphy.

3. Well logs: the information of logs include effective porosity, permeability and water saturation values on the well path.

4. Depth maps: structural contour maps obtained from seismic acquisition.

IOP Conf. Series: Materials Science and Engineering **579** (2019) 012037 doi:10.1088/1757-899X/579/1/012037

5. WOC: Water oil contact is a used in several reservoir studies to explain the elevation higher than that fluids apart from water can be found within the pores of a rock.

4. Design of Model

The main steps of set-up a geological model of a crude oil reservoir using petrel 2015 software system are:

- Data import.
- Quality check (Q.C).
- Well correlation.
- Structural modeling, that includes:
  a.Pillar gridding,
  b.build horizons,
  c.Layering.
- Property modeling, that includes:
  a.Scale up,
  b.data analysis
  c.Petrophysical modeling.
- Volumetric calculation
- Uncertainty analysis

5. Quality Check

In any modeling study, the input file need to be pre-processed to create it appropriate for fitting the model. Similarly, the results could need post- process before mapping and coverage. The process includes: editing, removing or adding data, furthermore as logical, mathematical and object connected operations. Therefore, it's necessary to see the statistics whenever any process operations are performed, and it's necessary to continuously compare those statistics before and when the operation to visualize if the results appear logical or not.

6. Well Correlation

Well correlation principles might shed light on the distribution of petrophysical properties, extents and thickness of various lithological units in reservoirs (Figure 3). Well correlations perform by using petrel 2015 software. Well correlation in petrel 2015 permits the chance to observe multiple wells in a well section, produce marker picks and bring up new wells to check with already correlate wells [9]. Figure (4) illustrates correlation section of Mishrif Formation in Garraf oil field. Mishrif Formation in Garraf oil field has been divided into 3 main components, upper, middle & lower. The upper & middle components are divided by marl units. The upper part from top of Mishrif to M1 unit. However, there's oil shows inside this part it's not thought of inside reservoir zone, as a result of it's not producible. Middle & lower components are reservoir units. They extend from M1 to top of Rumaila Formation. The Mishrif Formation (middle & lower) components contain many reservoir units (M1.2, L1, L1.2, L2, L2.2, L2.3 and L2.4) that are sealed by 2 cap layers (M1 and M2).

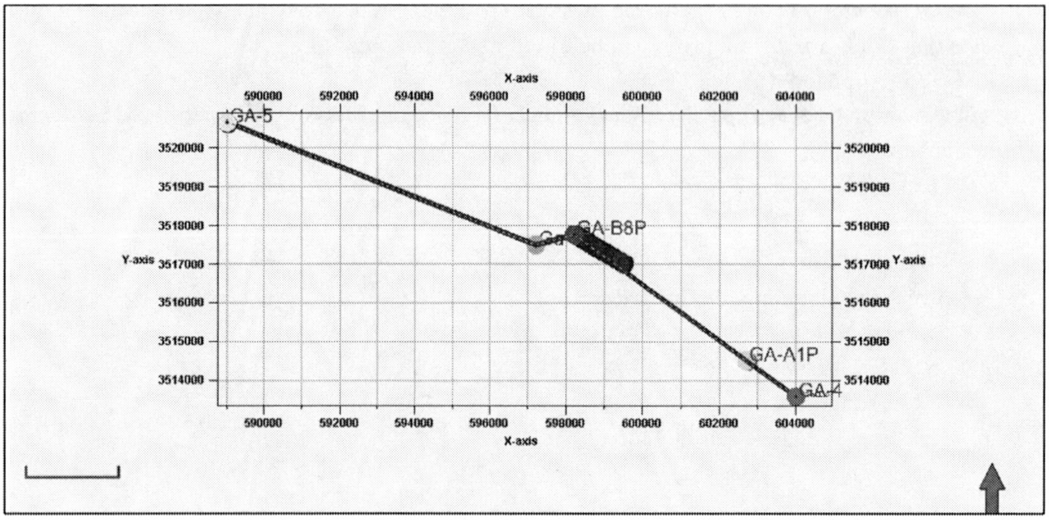

Figure 3 X section for Well Locations of the study area

7. Structural Modeling

Structural modeling represents setting up the structural map for every unit in Mishrif Formation (Figure 5). It's divided into 3 operations as follows: fault modeling, pillar gridding, and layering. All the 3 operations are performed one when the opposite to create one single information model [10]. Simple grid method is used here rather than pillar gridding. The studied formation in Garraf Oil Filed has no fault thus fault modeling method is neglected. Contour maps can be created by computer from the surface data and related to boreholes [11].

8. 3D-Grid Construction

In easy terms, a 3D grid divides a model into boxes, every box is named a grid cell and will have a single rock kind, one value of shale volume, one value of permeability, etc. [10].
A three-dimension grid building is the first step to set up the three-dimension model in easy terms. These are remarked because the cell's properties. This can be a simplification of actuality case, however permits us to come up with an illustration of reality that may be used in calculations, etc. [12].

9.Pillar Gridding

Pillar gridding is the method of creating the grid, that denotes the bottom of all modeling. The skeleton is a grid involving of top, middle and lower skeleton grids (Figure 6).

10. Make Horizons

Make horizons method step was utilized in describing the vertical layering of the three-dimension grid in petrel 2015 software. This presented a real three-dimension methodology within the creating of Two-dimension surfaces that were gridded within the same method, taking the relationships between the surfaces under consideration [10]. The next step in structural modeling is to insert the structural horizons into the pillar grid. It's a step for building the vertical layering of the three-dimension grid in petrel 2015.

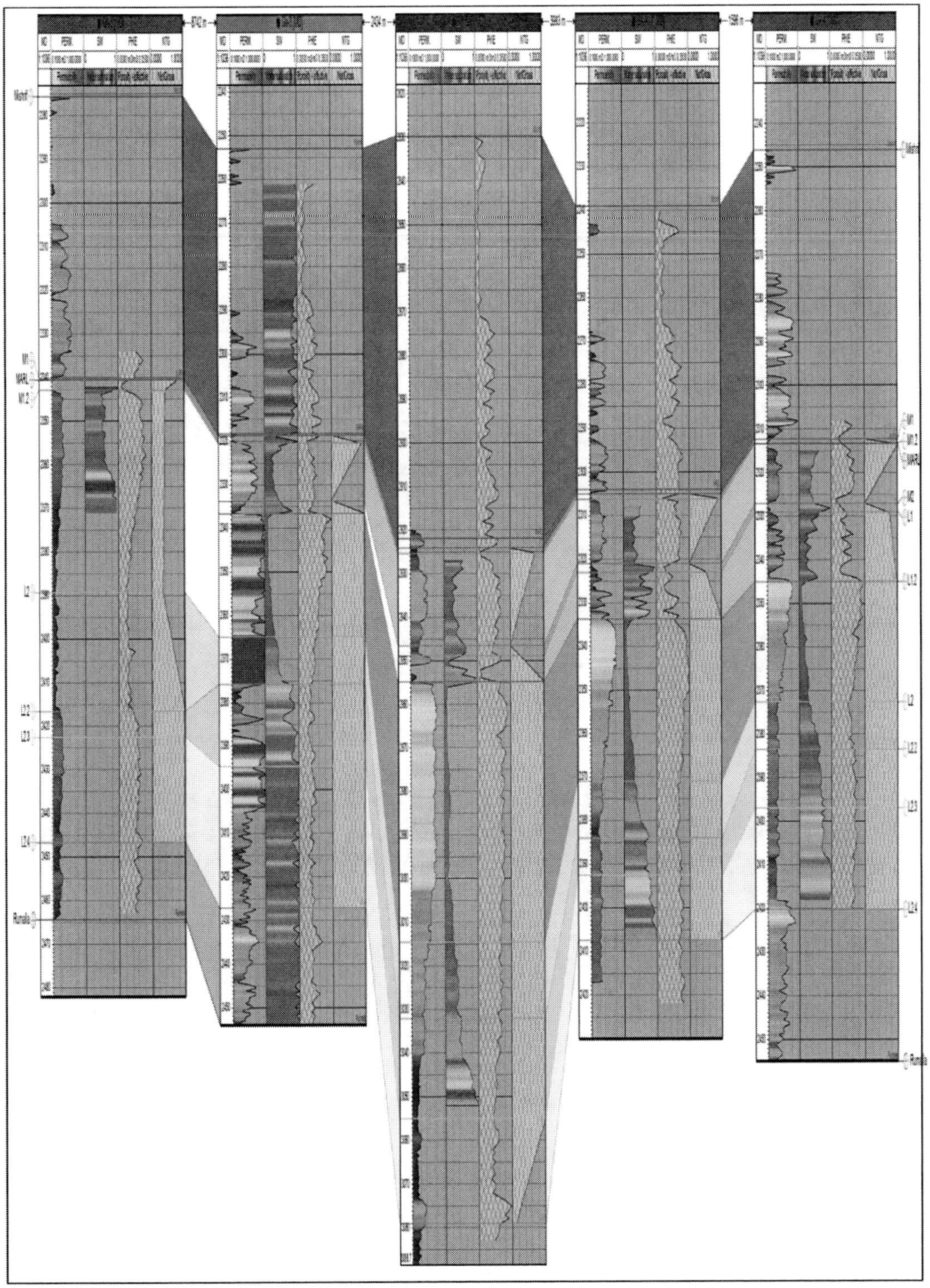

Figure 4 Correlation section of Mishrif Formation in Garraf oil field.

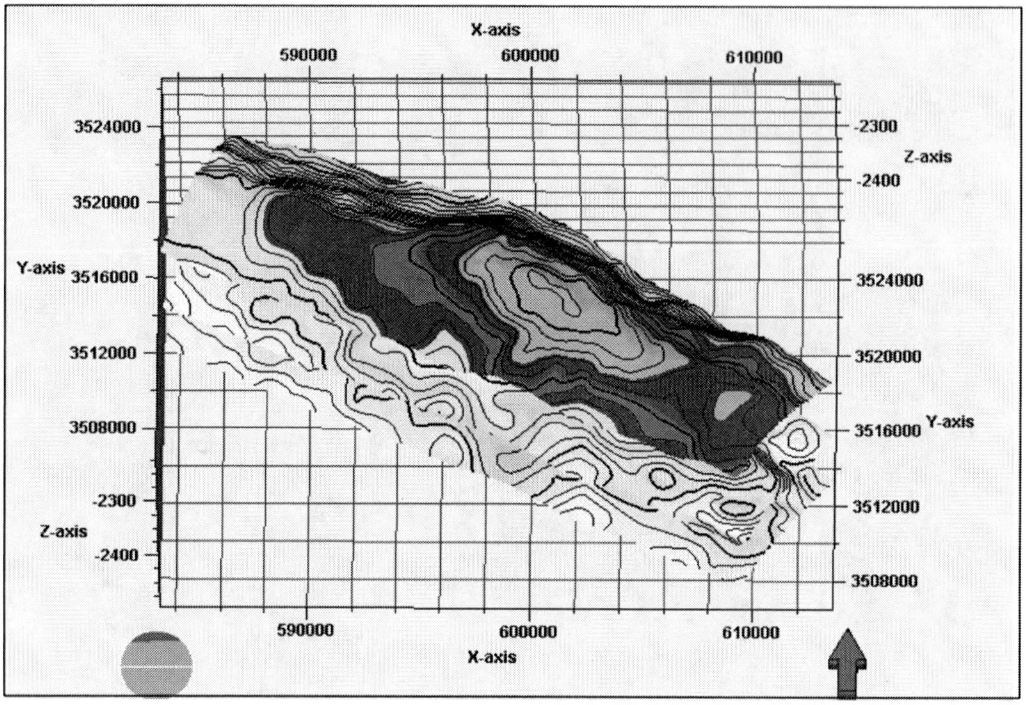

Figure 5 Structured Model of Mishrif Formation in Garraf oil field.

The used grid increment in Mishrif Formation here is (50m*50m) with Conformable horizon rule that horizons are truncated by erosional, base and discontinuous.

The contraction of horizons generates independent geologic horizons from X, Y, and Z input file and it's used to generate extra horizons using relative distance to existing horizons (Schlumberger, 2009). Figure (7) shows horizons of Mishrif Formation in Garraf oil field in a 3D window.

11. Layering

The 3D grid might have as several main layers as number of horizons inserted into the set of pillars. This is shown as horizons within the models window of the petrel adventurer. The create zones and create subzones processes were the 2 last steps used in process the vertical resolution of the 3D grid [10].

The create subzones allowed the definition of the final vertical resolution of the grid by setting the cell thickness or the quantity of desired cell layers. The future step was to start out the interpretation of the well logs to stipulate contacts between totally different facies identification of unconformities. The top and bottom of this specific unit were known inside the cell sections and well tops inserted to be displayed inside the well section in 3D [10]. Layering is the final vertical subdivision of the framework (Fig. 8). The layering is a part of the zone, and cannot have an instantaneous filter just like the zones do; layering but, is outlined because the internal layering reflective the geologic deposition of a selected zone [11].

Zones is added to the model by introducing thickness information within the form of isochore maps, constant thickness, and percentages[13]. Table (1) represents the layering of Mishrif units.

IOP Conf. Series: Materials Science and Engineering **579** (2019) 012037 doi:10.1088/1757-899X/579/1/012037

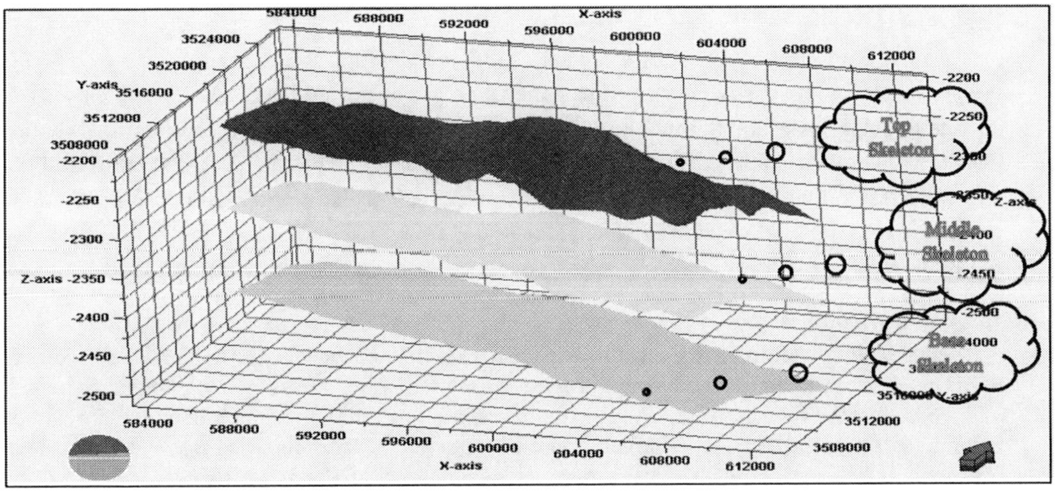

Figure 6 The skeletons of Mishrif Formation in Garraf oil field.

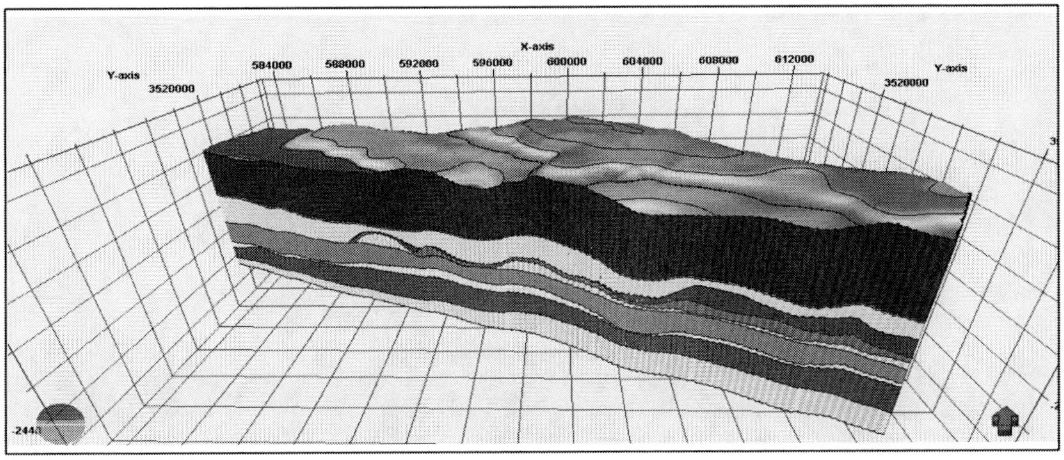

Figure 7 The 3D representation shows main horizons of Mishrif Formation in Garraf oil field.

Table 1 Layering of Mishrif units

Zones	Avg. Thickness(m)	No. of Layer
Mishrif	60.5	20
M1	15.42	5
M1.2	35.49	12
M2	22.31	5
L1	18.177	15
L1.2	49.26	20
L2	62.36	15
L2.2	8.9	10
L2.3	33.9	10
L2.4	30.9	10

12. Scale up Well Logs

The Scale up process averages the values to the cells within the three-dimension grid that are penetrated by the wells. Every cell has one value per up scaled log. These cells are used as a place to begin for property modeling [10]. There are several statistical ways used to scale up like (arithmetic, harmonic, and geometric methods). The shale volume, porosity, permeability and water saturation values within the current model are scaled up using the arithmetic and geometric technique whereas in the facies are scaled up by using most of the ways. Figure (9) represents scale up of PHIE and SW logs for well GA-3 in correlation window.

13. Quality Control

After upscaling well logs, it's sensible to visualize if the necessary units are capture d well or not. This step is related to the layering process during which thickness of layers ought to be set with regard to the intense flow units. If the layers are too thick, too much data could also be lost and as a result, thickness of layers must be adjusted [14]. A comparison between well log information before and when scale up process may be visualized by using histograms. It incontestable that the upscale properties in MB1&MB21 are close to the original log information. Figure (10) represent the quality control histograms of unit MB1 for PHIE, Sw, K and rock kind, respectively.

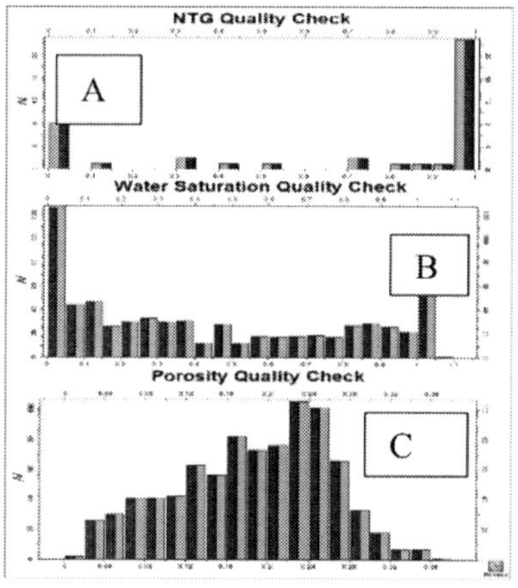

Figure 8 A-Quality histogram for NTG B- Quality histogram for SW C- Quality histogram for PHIE

14. Data analysis

Analysis of petrel data permits interactive analysis of distributions and trends and their relationships across all information varieties. Histogram, function, and stereonet windows-as well because the petrel information Analysis process are provided for analyzing upscaled well information and grid properties. The analysis of variogram includes selections for initial search-cone factor suggestions and fitting the variogram to the regression line, with the power to additionally build nested variograms. Careful analysis is kept for every property for direct use within the modeling processes. Sensitivity plots enable users to match the impact of every unsure variable in a risk-analysis context. Variogram for each unit and property presented in peterophysical modeling.

Figure 9 The Layering of Mishrif Formation in Garraf oil field

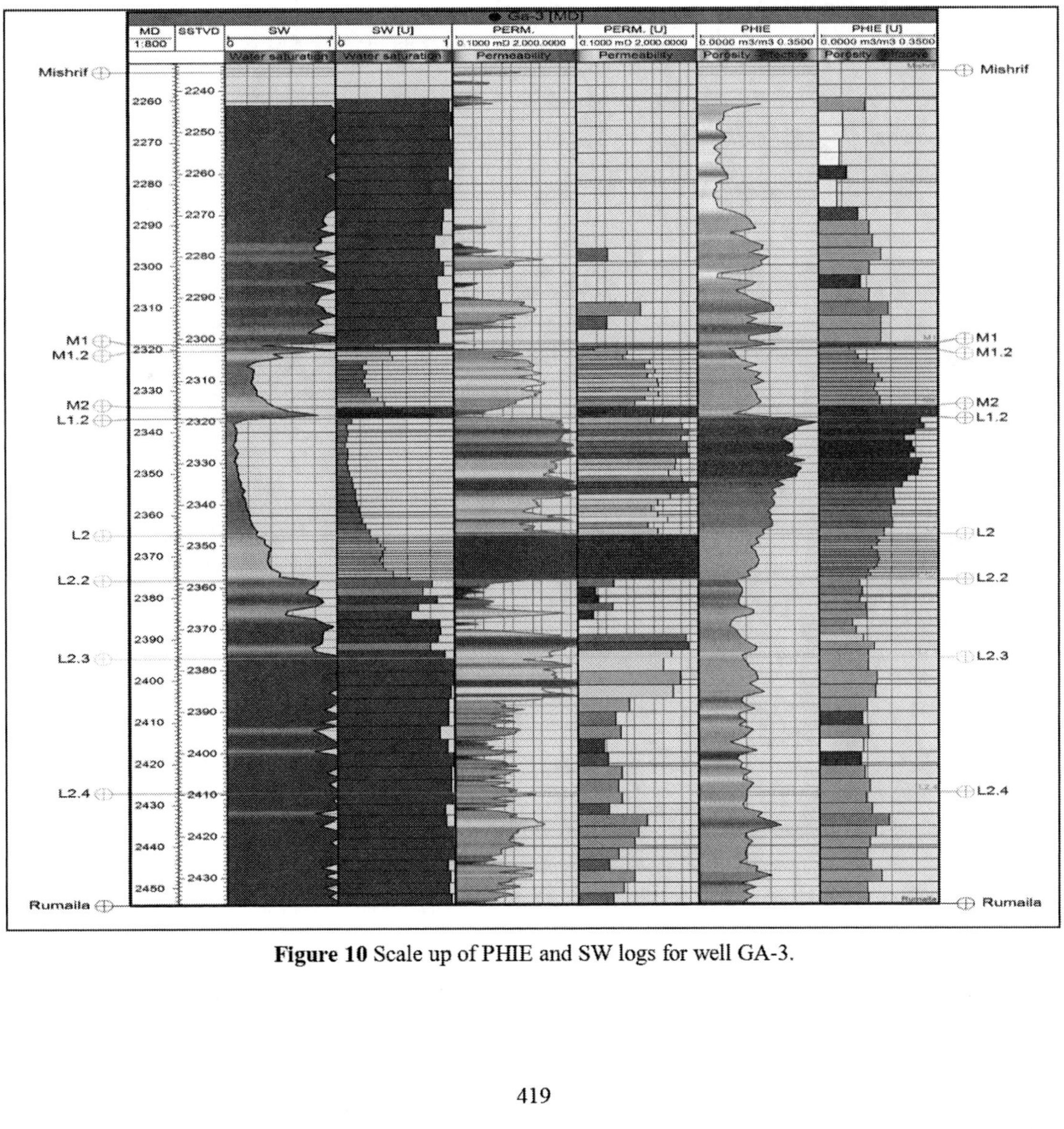

Figure 10 Scale up of PHIE and SW logs for well GA-3.

It is used to outline the behavior of variation during a property wherever it's considered as a key parameter in a lot of geostatistical algorithms. Simply, a variogram may be a plot of variability between semivariance (y-axis) versus separation distance (lag) during a specific direction [15].

Two valuable steps area unit performed in analyzing variogram:

1. Determining the directions of the variogram, major, minor and vertical direction wherever every of them are perpendicular to the opposite.
2. Calculative the experimental variogram and so making the variogram model based on the experimental one for each direction [15].

The 2D variogram map in figure (11), defines the direction of sample points, has 300 azimuthal angle in major direction. The minor search direction is 210 in azimuth (perpendicular on the major one). Figures (12 and 13) show the major and minor directions, respectively.

The necessary parameters that characterize the search core of major and minor direction are the azimuth, lag, range of lag (usually capable the no. of wells) and lag distance (approaches to the distance between wells).

The following parameters also are taken into consideration [15]:

1. Search Radius: that is that the most separation distance utilized in the search for sample pairs (no. of lags * lag distance).
2. Bandwidth: it's 0.5 the width of the search cone used as a cut-off to prevent the search area from turning into too wide at large separation distance.
3. Tolerance Angle: it's the width angle measured from the search cone main axis.

The following table (2) summarizes the most parameters for major and minor direction. The parameters in vertical direction are varied in every unit because of the distinction in thicknesses.

Table 2 Search Cone Parameters

Direction	Azimuth	Dip	Lags
Major	300	0	8
Minor	210	0	8
Direction	**Lag Dist.**	**Radius**	**Band Width**
Major	2612.4	20899	3424
Minor	2511.2	19689	5600

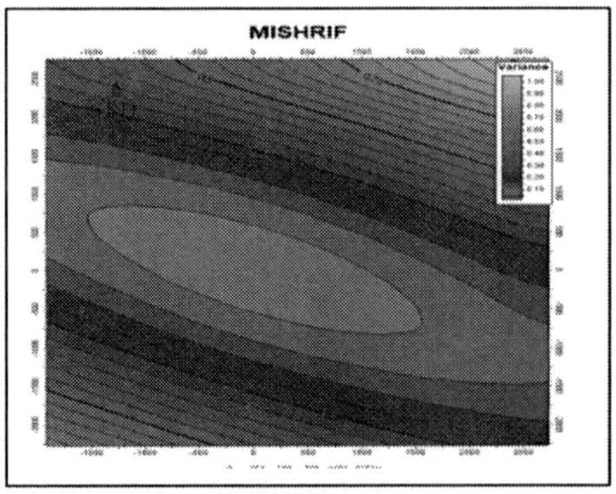

Figure 11 2D Variogram Map for the direction of Mishrif Formation

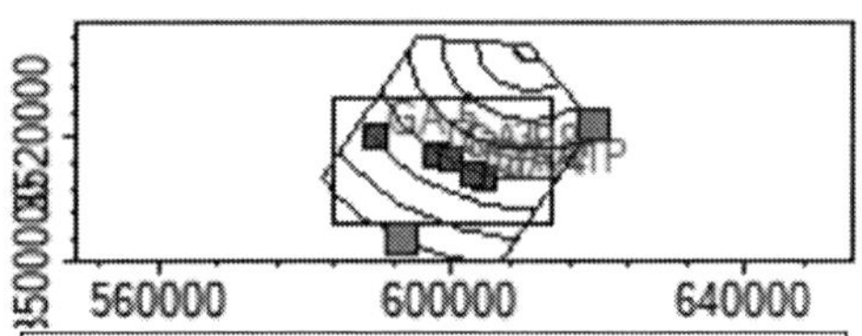

Figure 12 Major Direction Based on 2D Variogram Map

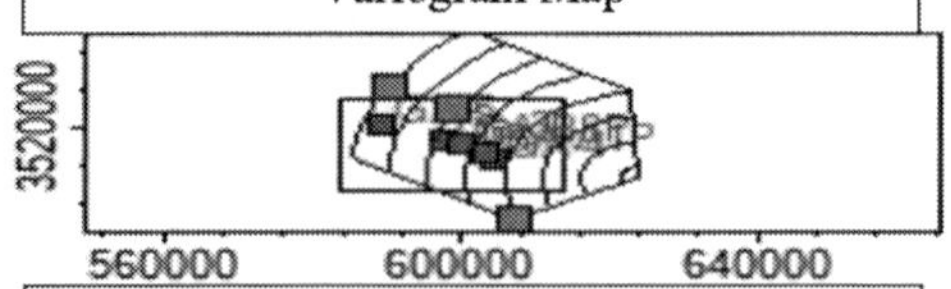

Figure 13 Minor Direction Based on 2D Variogram Map

Variogram analysis is a very important part of geostatistical modeling. Indeed, variogram shows the mean sq. distinction between the 2 values as a function of their increment [16]. The value of variogram is calculated by

$$y(h) = \frac{\sum_{i=1}^{n}(x_i - x_{i+h})}{2n_h} \dots\dots\dots\dots\dots\dots\dots\dots\dots\dots\dots (1)$$

in the above equation, $y(h)$ is named semivariogram and 2 $y(h)$ is named variogram, and n_h is the variety of pairs of points with the distance h from each other participating within the variogram. xi is a grade in point i and x_{i+h} is grade in a point with distance h from the purpose i. By hard variogram for various h values, the $y(h)$ diagram in terms of h are often drawn, and the h value is called lag.

For many modeling, the variogram model begins from a nonzero value and will increase up to vary called effective vary (a), eventually reach to the constant value called sill. the effective vary is that the range in that the data spatial structure has relevancy, and outside the vary, the info impact is independent from one another [17] (Table 3).

Table 3 Parameters of variograms are delineated for porosity and saturation data.

Zones	Nugget	sill	Major	Minor	Vario. Method	Property
Mishrif	0	0.9704	5628	5001	Exponential	Porosity
M1	0	1.273	7342	6274	Exponential	Porosity
M1.2	0	1.085	9401	8899	Exponential	Porosity
M2	0	0.879	7887	6331	Exponential	Porosity
L1	0	0.99	3855	2907	Exponential	Porosity
L1.2	0	0.934	4619	3756	Exponential	Porosity
L2	0	0.894	4976	6293	Exponential	Porosity
L2.2	0	0.96	8750	7008	Exponential	Porosity
L2.3	0	0.997	7560	4472	Exponential	Porosity
L2.4	0	0.995	9056	7640	Exponential	Porosity
Mishrif	0	0.923	4256	3214	Exponential	SW
M1	0	0.958	9754	9558	Exponential	SW
M1.2	0	0996	6648	6038	Exponential	SW
M2	0	0.975	9800	8324	Exponential	SW
L1	0	0.992	4463	2801	Exponential	SW
L1.2	0	0.907	8090	7866	Exponential	SW
L2	0	0.914	6622	6413	Exponential	SW
L2.2	0	0.949	8910	8258	Exponential	SW
L2.3	0	0.997	3836	3158	Exponential	SW
L2.4	0	0.991	3385	3244	Exponential	SW

15. Property Modeling

Petrophysical property modeling is the method of distribution petrophysical property values (porosity, water saturation, NTG, K, etc.) to every cell of the 3D grid.The objective of property modeling is to distribute properties between the accessible wells therefore it realistically preserves the reservoir heterogeneity and matches the well data; so, property

IOP Conf. Series: Materials Science and Engineering **579** (2019) 012037 doi:10.1088/1757-899X/579/1/012037

modeling is the method of filling the cells of the grid with distinct (Rock type) or continuous (Petrophysics) properties [10].The 3D property modeling relies on well logs information. This includes a calculation for resolution advanced mathematical equations involving one or many 3D property models; i.e. SW transforms based on porosity 3D model [13].

The aim of a geologic reservoir model is to produce a whole set of continuous reservoir parameters (i.e. porosity, and water saturation) for every cell of the grid. Many various techniques are used to generate these parameters [15].

Petrel offers many algorithms for modeling the distribution of petrophysical properties in a reservoir model. Petrophysics model was designed using geo-statistical methods. The Petrophysics models include:

1. Porosity Modeling

Porosity model was built based on the results of porosity logs (density, neutron and acoustic) that are corrected and interpreted within the interactive petrophysical 3.5 software system. Sequent Gaussian Simulation formula was used as a statistical procedure, which inserts with the quantity of the accessible information [18]. Figures (13) to (22) show the porosity models of Mishrif reservoir unit of every unit is characterized by porosity distribution.

2. Water Saturation Modeling

Water saturation model was built after the scale up of water saturation that exported from IP software system for every reservoir unit of the Mishrif Formation. A similar geostatistical methodology was used in the porosity models (Statistical Gaussian Simulation Algorithm) in line with the accessible information. Figures (23) to (32) show the water saturation models of Mishrif reservoir unit of every unit is characterized by water saturation distribution.

3. NTG Model

A net pay is a main parameter in analysis of reservoir, because it describing the penetrated geologic sections that have enough reservoir quality and interstitial hydrocarbon volume to perform as important producing intervals. The net pay participates to the calculation of the original oil in place. The net pay demonstrates facilitates reservoir simulation, as a result of non –reservoir rock doesn't get to be characterized for inclusion [19].

Geoscientists have usually been annoyed by the absolute assignment of petrophysical log cutoffs to outline reservoir intervals capable of hosting producible hydrocarbon. Process of the net pay is essential to estimating reserves, this method deserves an exacting level of scrutiny [20].

The net Pay is quantified through the use of petrophysical cut-offs that are applied to well logs. Cut-offs is range values of formation parameters that eliminate non-producing intervals [20]. There's no usually accepted technique of characteristic cut-offs. historically, a porousness of zero.1md is mostly thought of minimum for production, the cutoff values of porosity is used to describe net reservoir, and the cutoff values of water saturation is used to describe the net Pay [20].The figures (33) to (42) show the final net pay model for Mishrif formation within the Garraf field

16. Water Oil Contact

Water oil contact is used in several reservoir studies to explain the height higher than that fluids apart from water can be found within the pores of a rock. In most things within the hydrocarbon industry the term is qualified as being an WOC.

In a traditional hand-excavated water well, the extent at that the water stabilizes represents the groundwater level, or the elevation within the rock wherever air starts to occupy the rock pores. Water oil contact analysis is extremely necessary stage in characterization of reservoir and calculation of hydrocarbon reserve, as a result of it's required for boundary delineation between oil and water zone [21]. For the Mishrif formation oil water contacts at the reservoir units.

17.OOIP Calculation

Estimation of Volume process can accurately estimate the varied volumes (bulk, pore and fluid) during a 3D grid figure (12). Throughout the process of modeling of the reservoir, many three-dimension grids could produce with completely different treatments of the faults and horizons. inside those grids, there is also many versions of a property model and contacts and there are also many representations of the fluids at intervals the model.

Table (4) shows the results of the volume of Initial oil in place (OIIP) in Mishrif formation of Garraf field. As noted from Table (4), the OIIP that obtained by using petrel 2015 software system is 1061 *106 M3 that is simply too near to that calculated mistreatment average properties that equal 1085 *106 M3.

18. Results and Discussion

The correlation of the wells in Garraf field shows that Mishrif zones extent everywhere the field except m2, L1 and L1.2 zones. m2 zone that represents cap rock of L1 and L1.2 extends over the field from GA-4 well within the southeast to the well GA-3 further as L1.2 reservoir zone. Whereas L1 reservoir zone extends from GA-4 well within the southeast to well GA-B8P within the middle of the field. Rom porosity and water saturation models for every unit of Mishrif Formation, the subsequent points can be shown:

• **Top Mishrif Unit:** The thickness of this unit is homogeneous. There are oil shows among this unit it's not thought of among reservoir zone, because it's not producible. It shows low PHIE Figure-16, low NTG Figure-36 and water saturation Figure-26 average values that reach 11% ,0.8 and 93% respectively.

• **M1.2 Reservoir Unit:** The unit thickens towards GA-5 and GA-4 wells. Generally, this unit shows good PHIE Figure-18, good NTG figure 38 and water saturation Figure-28 average values that may reach 80%, 1 and 43%, severally. However, the reservoir quality decreases within the area between GA-5 and GA-3 as indicated by the higher water saturation and lower PHIE values.

• **L1 Reservoir Unit:** The L1 reservoir unit pinches out towards GA-3 well. The direction of thinning is related to decreasing PHIE values and increasing water saturation. the average PHIE Figure-20 is 12%, NTG figure 40 is 0.9and water saturation Figure-30 is 39%.

• **L1.2 Reservoir Unit:** The L1.2 units is characterized by high reservoir properties. This unit thins towards GA-3 and GA-4 wells. In most wells, very little changes in PHIE Figure-21, NTG figure 41 and water saturation Figure-31 are determined. The average of PHIE is 26%, NTG is 1 and water saturation reaches 16%. Therefore, it represents the most effective reservoir unit in Garraf oil field.

• **L2 Reservoir Unit:** The L2 unit has low reservoir quality. The L2 unit thickens towards GA-5 that has lowest reservoir properties. The average of PHIE Figure-22 is 14%, NTG figure 42 is 1 and water saturation Figure-32 reaches 58%.

• **The L2.2 units:** The L2.2 unit has low reservoir quality with 15 % PHIE Figure-23 ,1 NTG figure 43 and 73% water saturation Figure-33. Lowest reservoir properties occur in GA-3 and GA-B8P.

• **L2.3 Reservoir Unit:** The L2.3 reservoir unit is the lower reservoir quality as indicated by the values of PHIE and water saturation in GA-3 and GA-B8P wells. Other wells have higher reservoir quality. In general, the L2.3 unit has poor reservoir quality with 15% PHIE Figure-24, 1 NTG figure 44 and 88% water saturation Figure-34.

• **L2.4 Reservoir Unit:** This unit is comparable to L2.3 reservoir unit. The GA-4 and GA-A1P wells have the lowest PHIE values. Other wells have higher values. However, L2.4 units is taken into account as a poor reservoir because of low average PHIE Figure-25 (13%) 0.95 NTH figure 45 and high water saturation Figure-35 (100%).

Figures (46 to 48) show the porosity, NTG and water saturation models for Mishrif Formation in Garraf oil field that are designed from NTG, porosity and water saturation values using sequential gaussian Simulation rule as a statistical method after scale up of NTG, porosity and water saturation. Finally, the cross sections in NW-SE and NW-SE directions for NTG, porosity and water saturation models were set so as to Illustrate the vertical and horizontal distribution of NTG, porosity and water saturation in every well beneath study. Figures 49, 50 and 51 show that the most effective location characterized by good reservoir properties is between well GA-4 and GA-3 particularly within the units m1.2 and L1.2 and they decrease step by step toward well GA-5 that become bad reservoir properties.

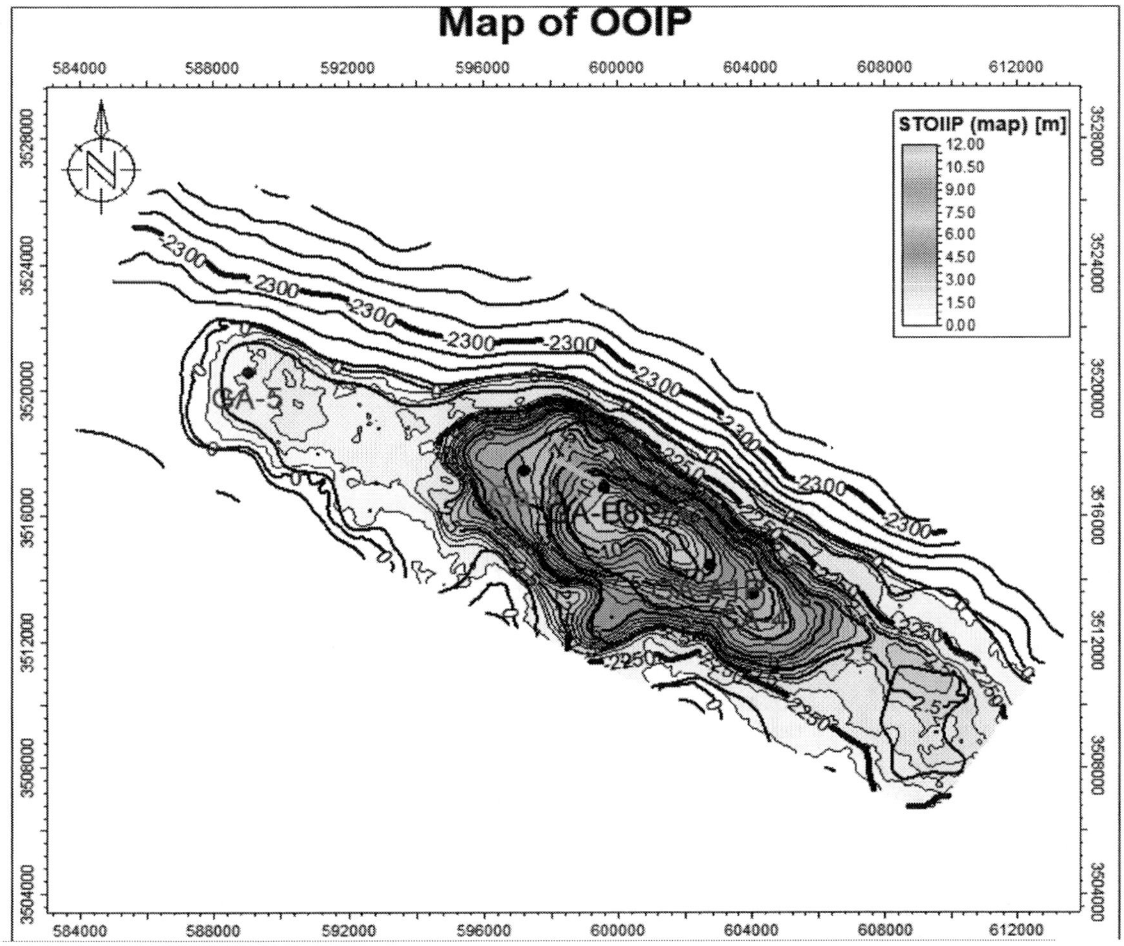

Figure 14 Mapof original oil in place for Mishrif formation in garraf field

Table 4 Volumetric calculation for Mishrif formation in Garraf field

Zones	Bulk volume[*10^6 m3]	Net volume[*10^6 m3]	Pore volume[*10^6 rm3]	HCPV oil[*10^6 rm3]	STOIIP (in oil)[*10^6 sm3]	Type of fluid
Mishrif	16703	6816	868	519	371	Oil
M1	549	260	33	6	4	Oil
M1.2	4347	1944	411	222	158	Oil
M2	363	64	9	3	2	Oil
L1	1136	889	135	82	59	Oil
L1.2	2151	2151	581	494	353	Oil
L2	1233	1233	229	137	98	Oil
L2.2	333	329	63	23	17	Oil
L2.3	0	0	0	0	0	Water
L2.4	0	0	0	0	0	Water
TOTAL	26816	13687	2329	1485	1061	

IOP Conf. Series: Materials Science and Engineering **579** (2019) 012037 doi:10.1088/1757-899X/579/1/012037

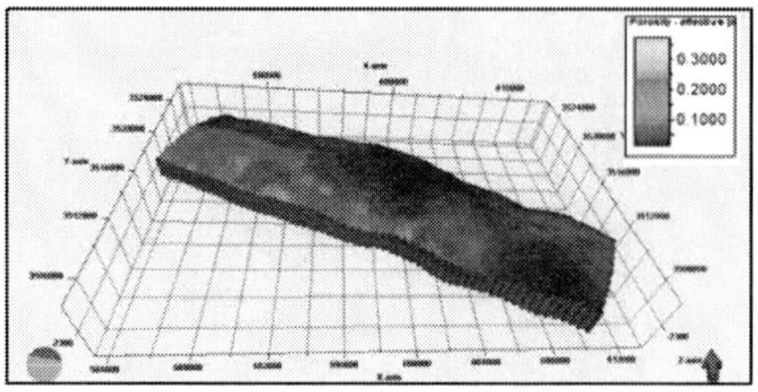

Figure 15 Porosity model of top Mishrif Formation unit in Garraf oil field.

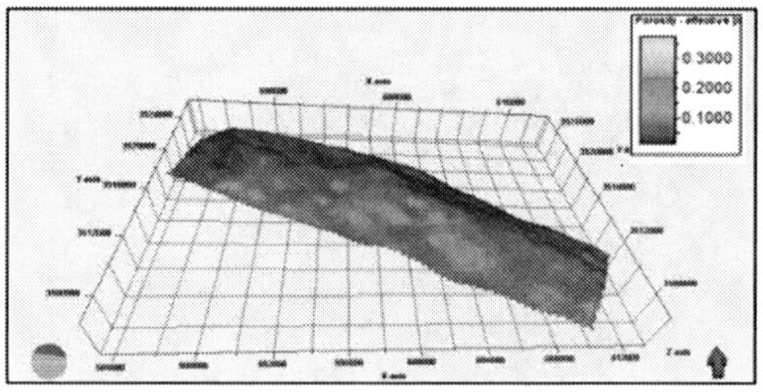

Figure 16 Porosity model of top M1 Formation unit in Garraf oil field

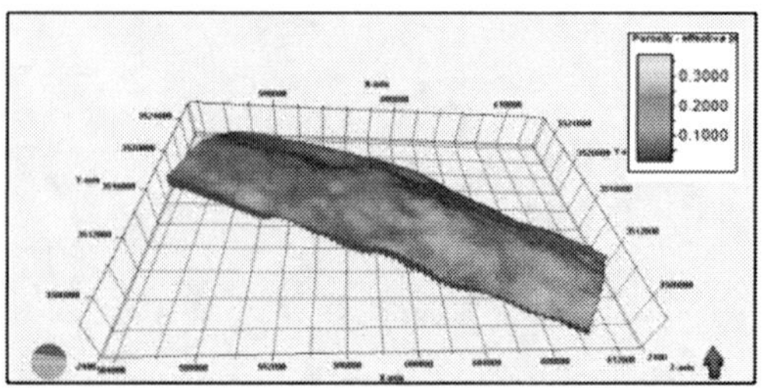

Figure 17 Porosity model of top M1.2 Formation unit in Garraf oil field

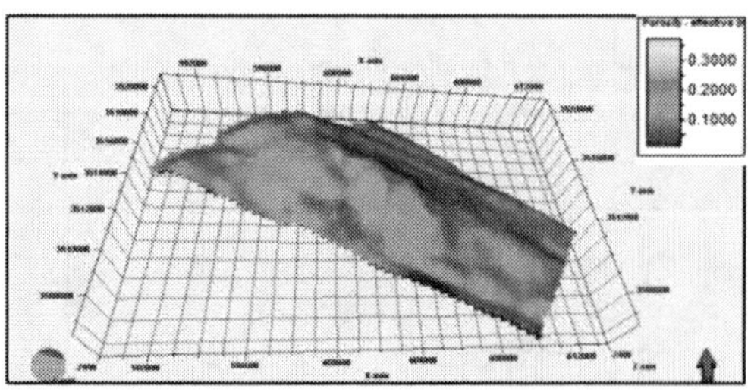

Figure 18 Porosity model of top M2 Formation unit in Garraf oil field

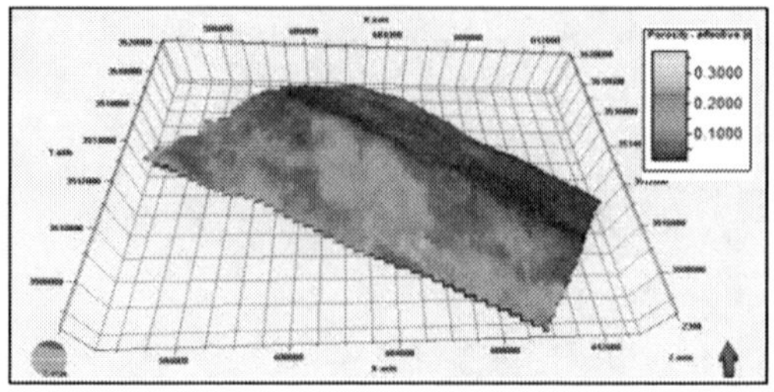

Figure 19 Porosity model of top L1 Formation unit in Garraf oil field

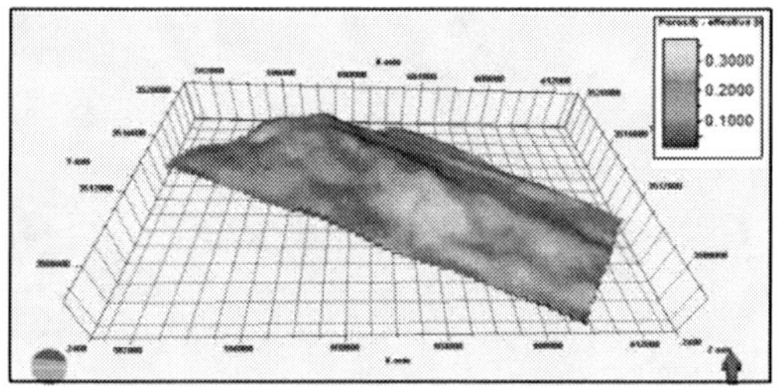

Figure 20 Porosity model of top L1.2 Formation unit in Garraf oil field

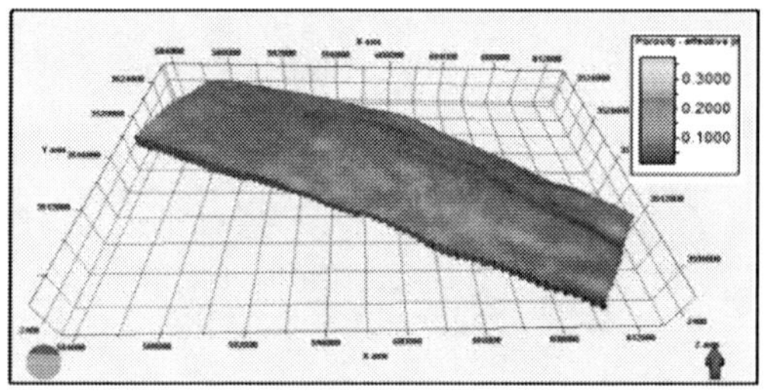

Figure 21 Porosity model of top L2 Formation unit in Garraf oil field

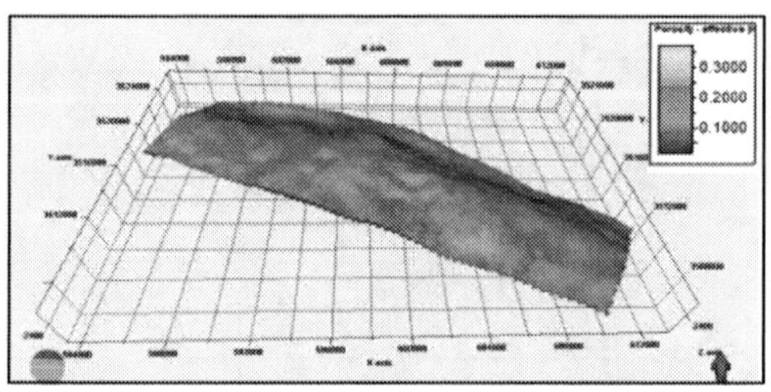

Figure 22 Porosity model of top L2.2 Formation unit in Garraf oil field

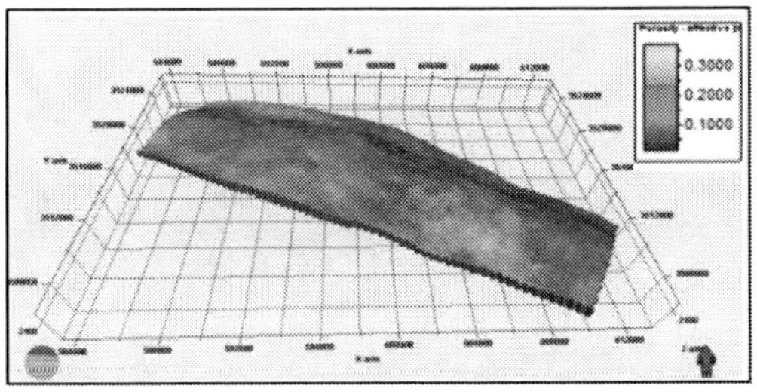

Figure 23 Porosity model of top L2.3 Formation unit in Garraf oil field

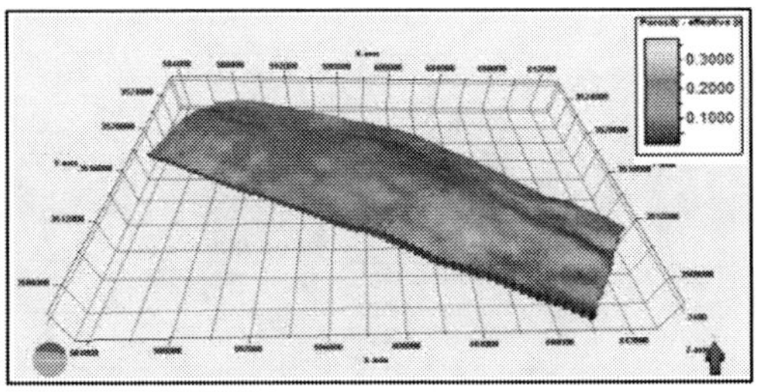

Figure 24 Porosity model of top L2.4 Formation unit in Garraf oil field

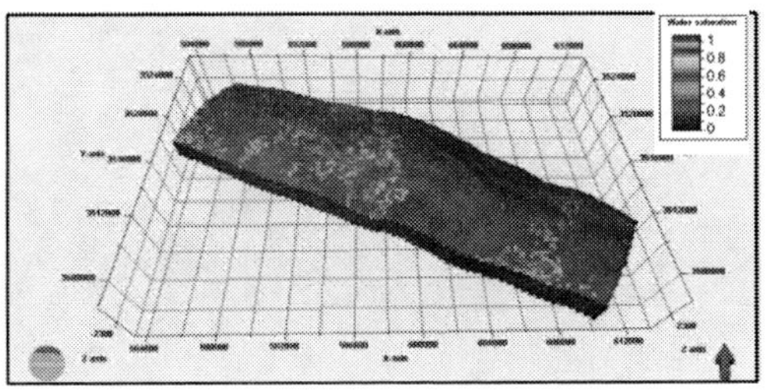

Figure 25 Water saturation model of top Mishrif Formation unit in Garraf oil field

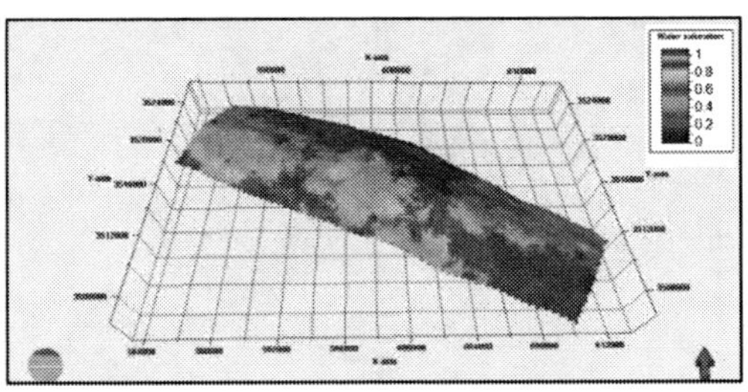

Figure 26 Water saturation model of top M1 Formation unit in Garraf oil field

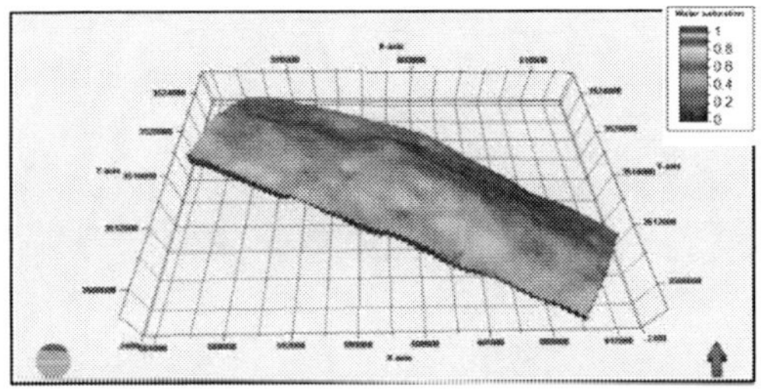

Figure 27 Water saturation model of top M1.2 Formation unit in Garraf oil field

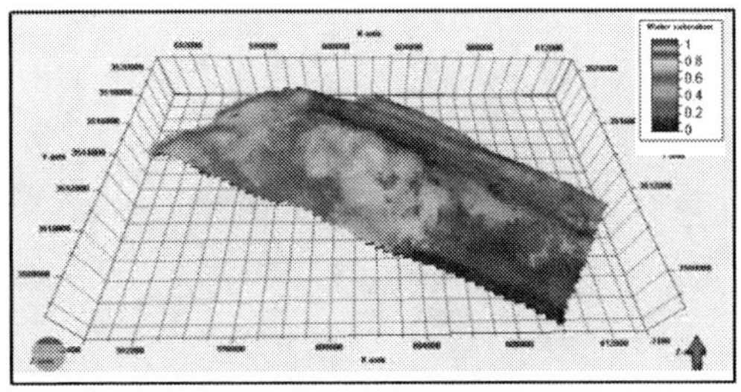

Figure 28 Water saturation model of top M2 Formation unit in Garraf oil field

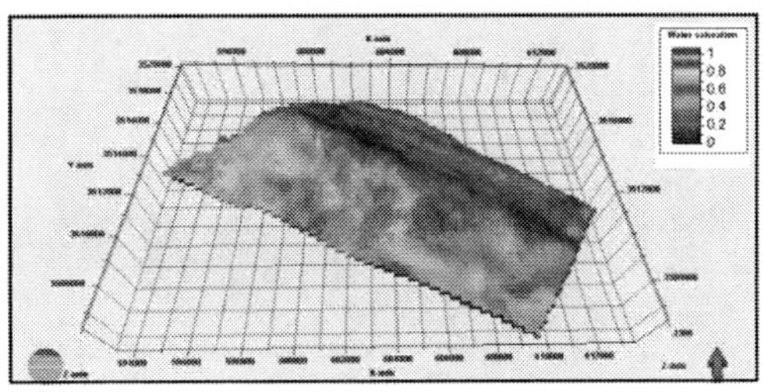

Figure 29 Water saturation model of top L1 Formation unit in Garraf oil field

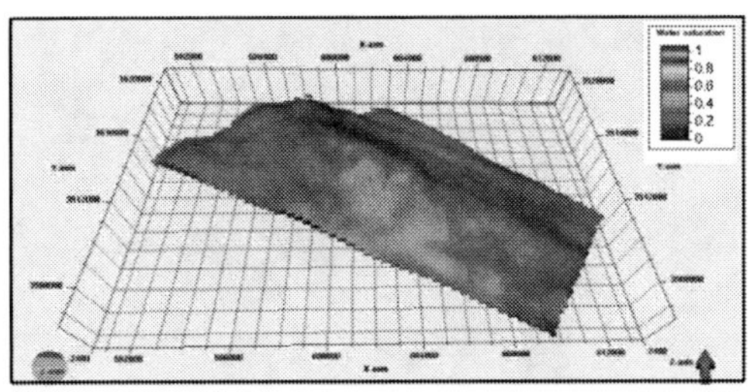

Figure 30 Water saturation model of top L1.2 Formation unit in Garraf oil field

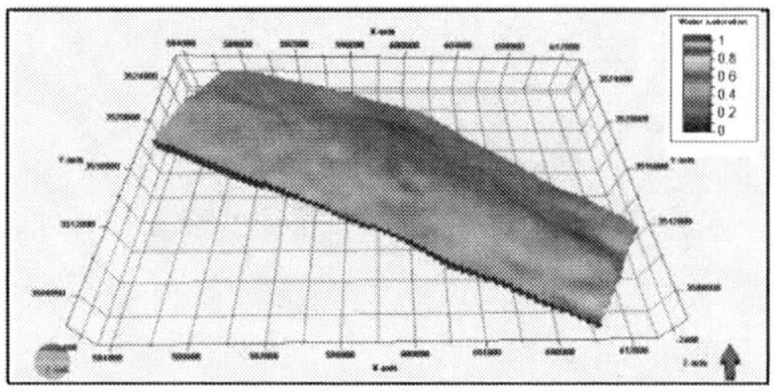

Figure 31 Water saturation model of top L2 Formation unit in Garraf oil field

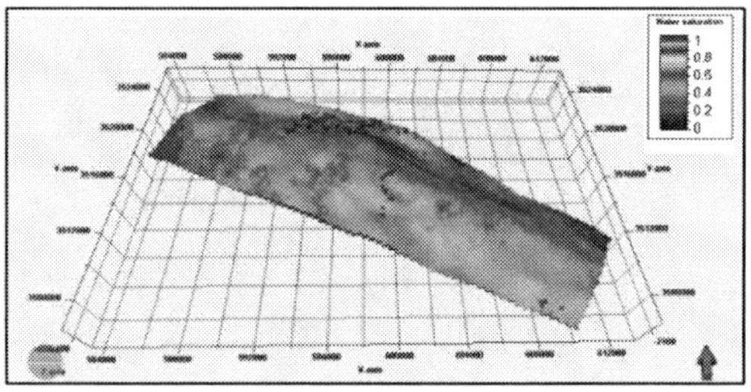

Figure 32 Water saturation model of top L2.2 Formation unit in Garraf oil field

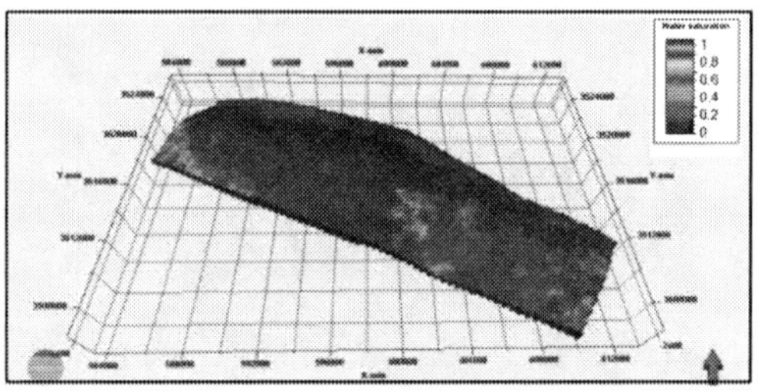

Figure 33 Water saturation model of top L2.3 Formation unit in Garraf oil field

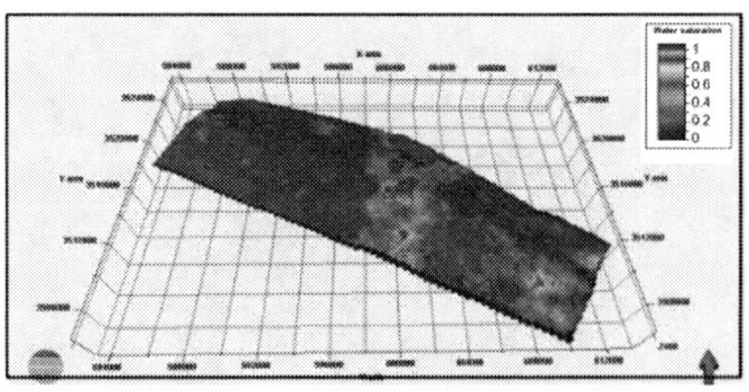

Figure 34 Water saturation model of top L2.4 Formation unit in Garraf oil field

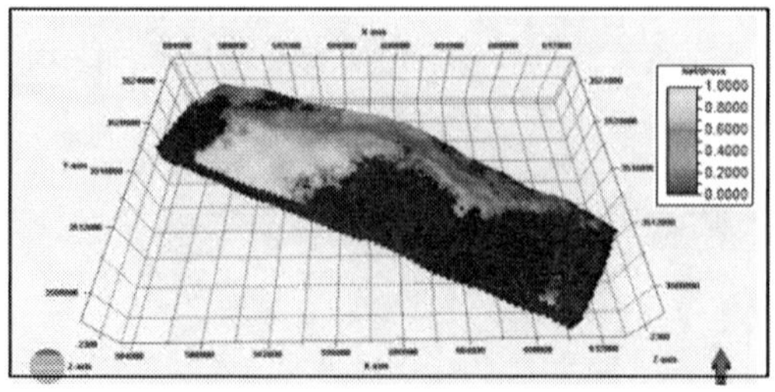

Figure 35 NTG model of top Mishrif Formation unit in Garraf oil field

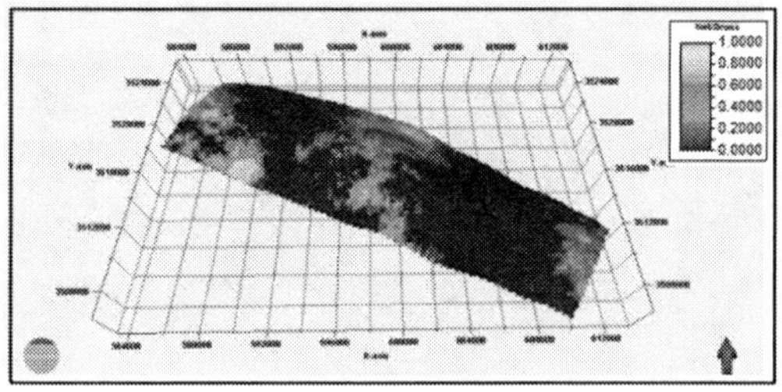

Figure 36 NTG model of top M1 Formation unit in Garraf oil field

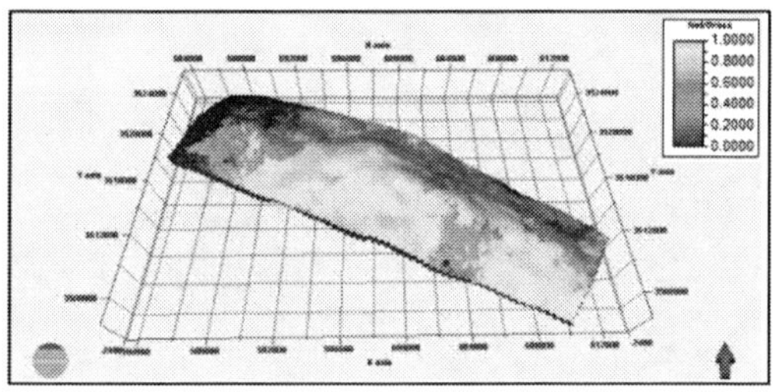

Figure 37 NTG model of top M1,2 Formation unit in Garraf oil field

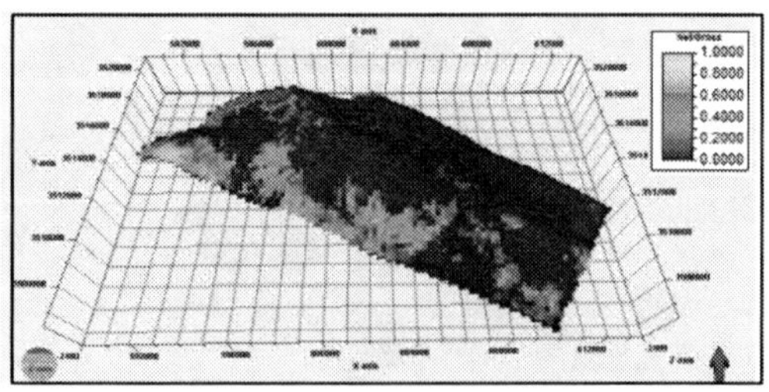

Figure 38 NTG model of top M2 Formation unit in Garraf oil field

IOP Conf. Series: Materials Science and Engineering **579** (2019) 012037 doi:10.1088/1757-899X/579/1/012037

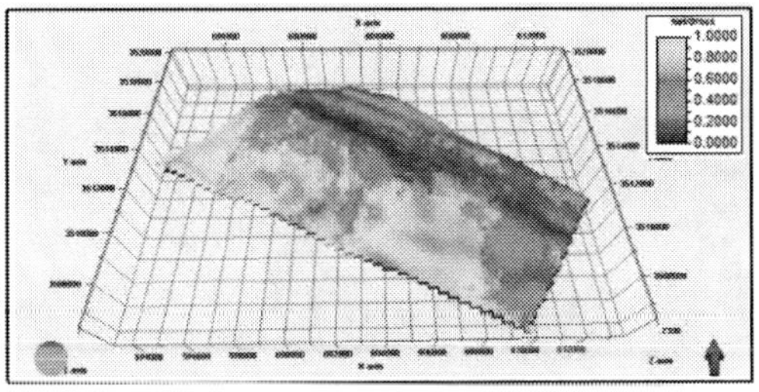

Figure 39 NTG model of top L1 Formation unit in Garraf oil field

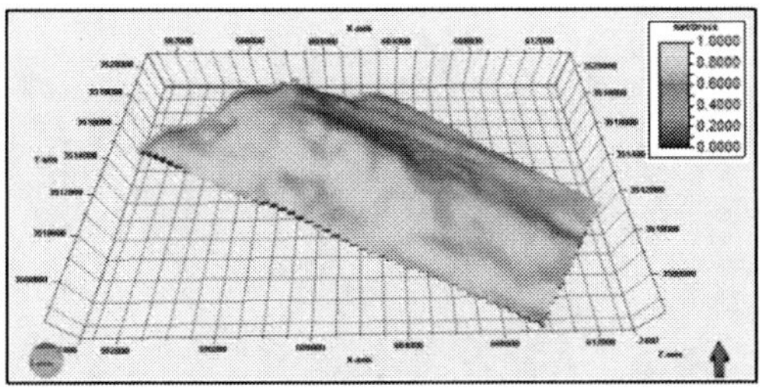

Figure 40 NTG model of top L1.2 Formation unit in Garraf oil field

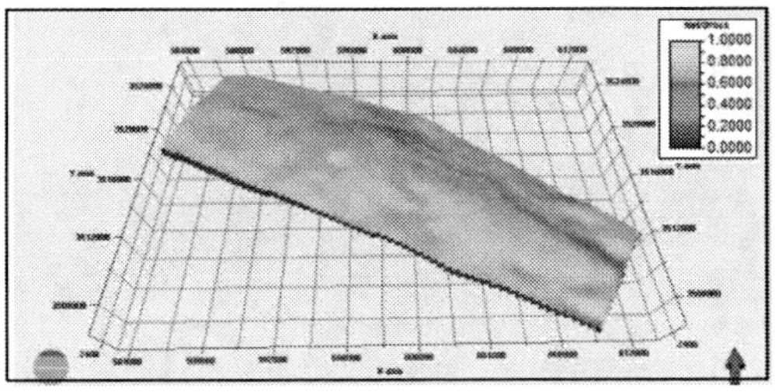

Figure 41 NTG model of top L2 Formation unit in Garraf oil field

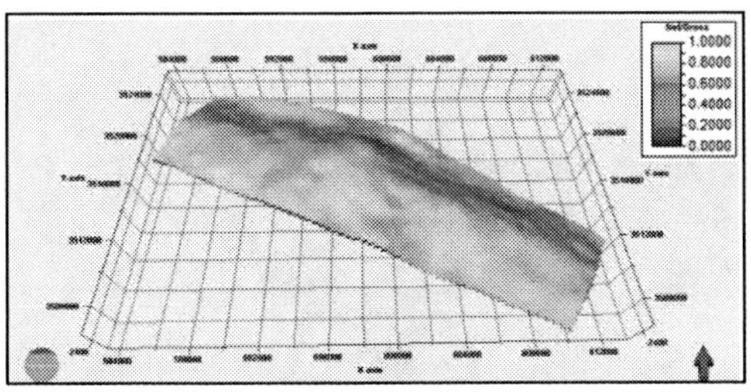

Figure 42 NTG model of top L2.2 Formation unit in Garraf oil field

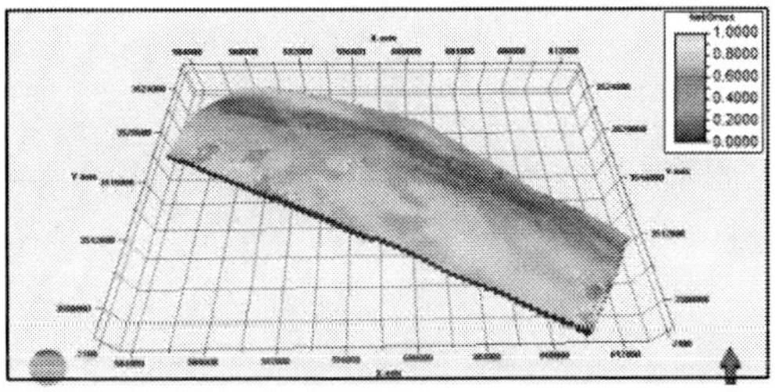

Figure 43 NTG model of top L2.3 Formation unit in Garraf oil field

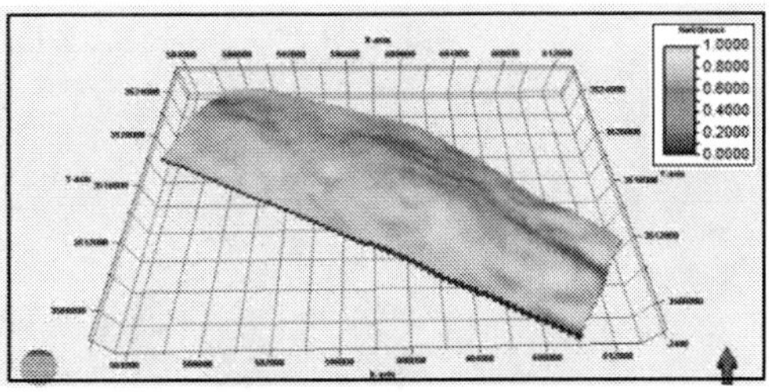

Figure 44 NTG model of top L2.4 Formation unit in Garraf oil field

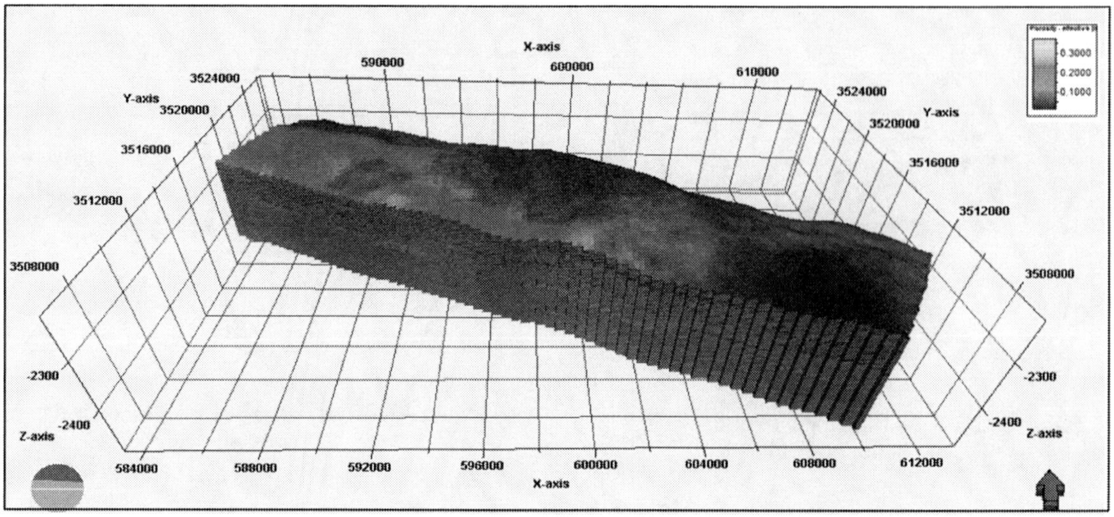

Figure 45 Final porosity model for Mishrif Formation in Garraf oil field.

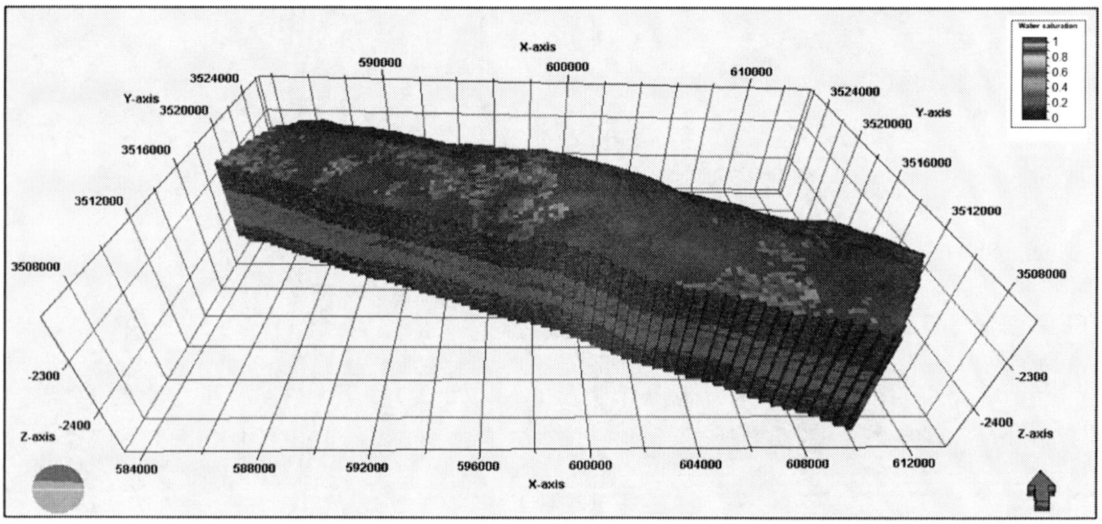

Figure 46 Final water saturation model for Mishrif Formation in Garraf oil field.

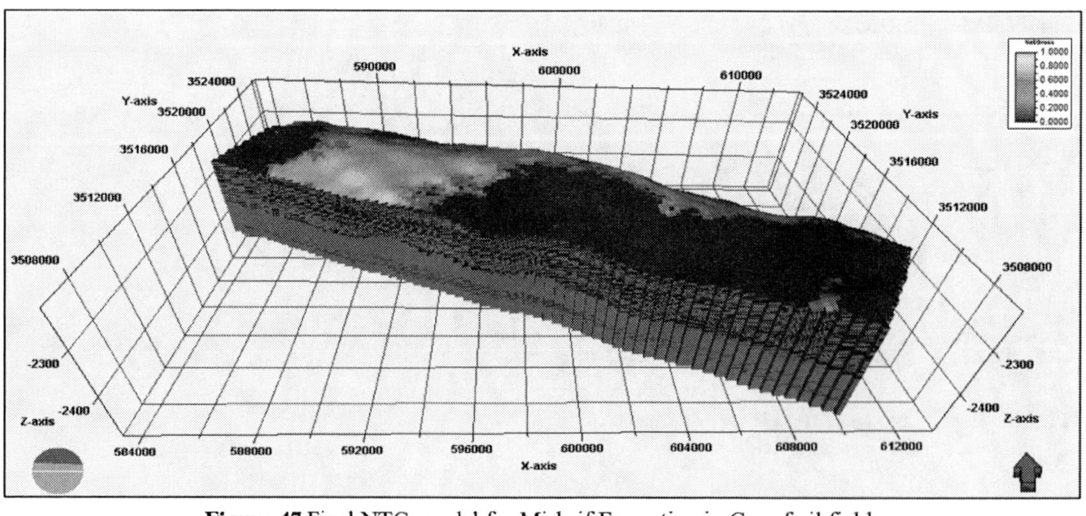

Figure 47 Final NTG model for Mishrif Formation in Garraf oil field.

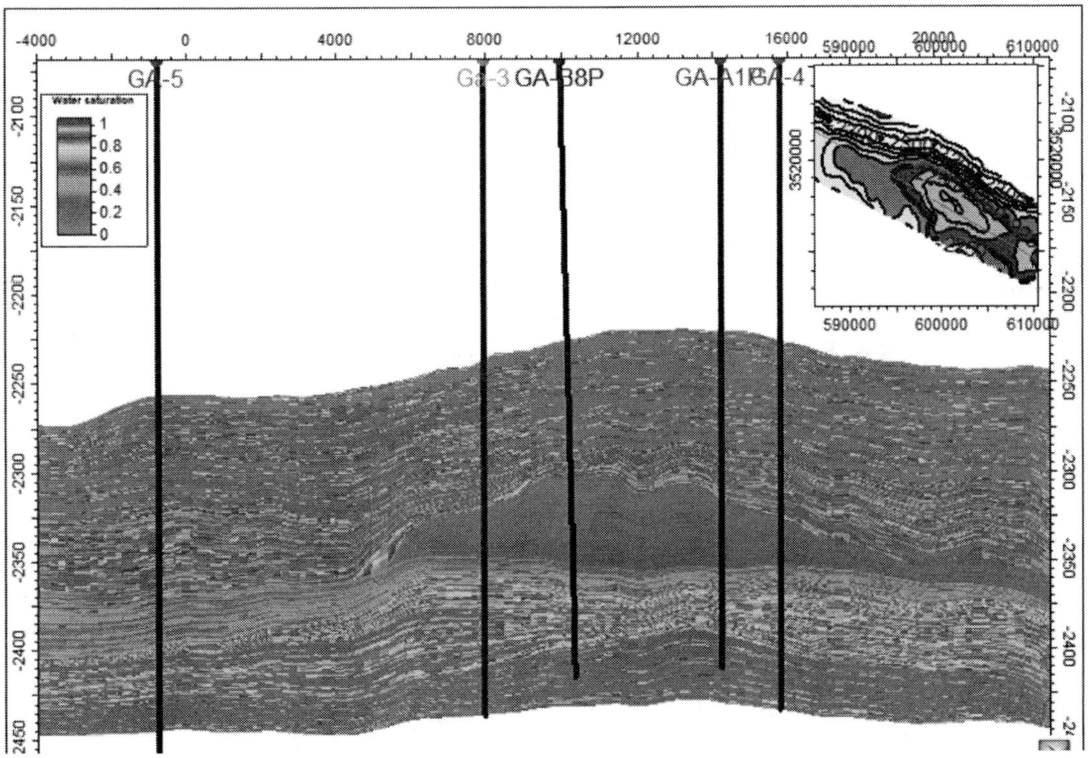

Figure 48 Cross section in direction NW-SE distribution of SW for Mishrif Formation in Garraf oil field.

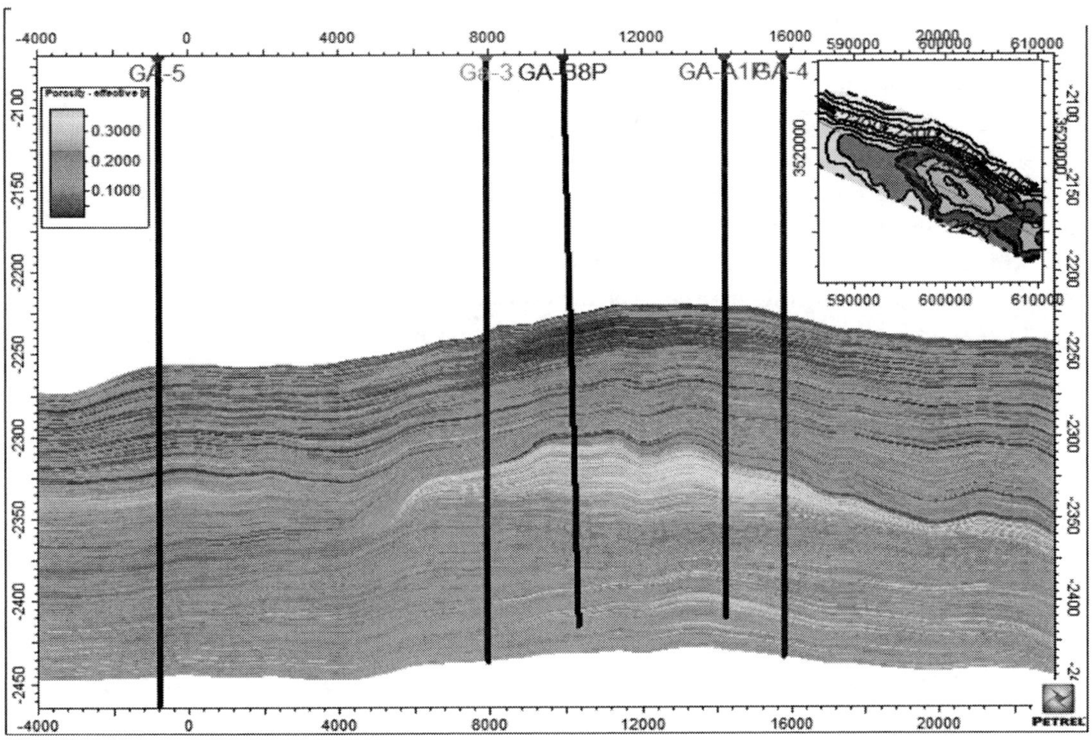

Figure 49 Cross section in direction NW-SE distribution of PHI for Mishrif Formation in Garraf oil field.

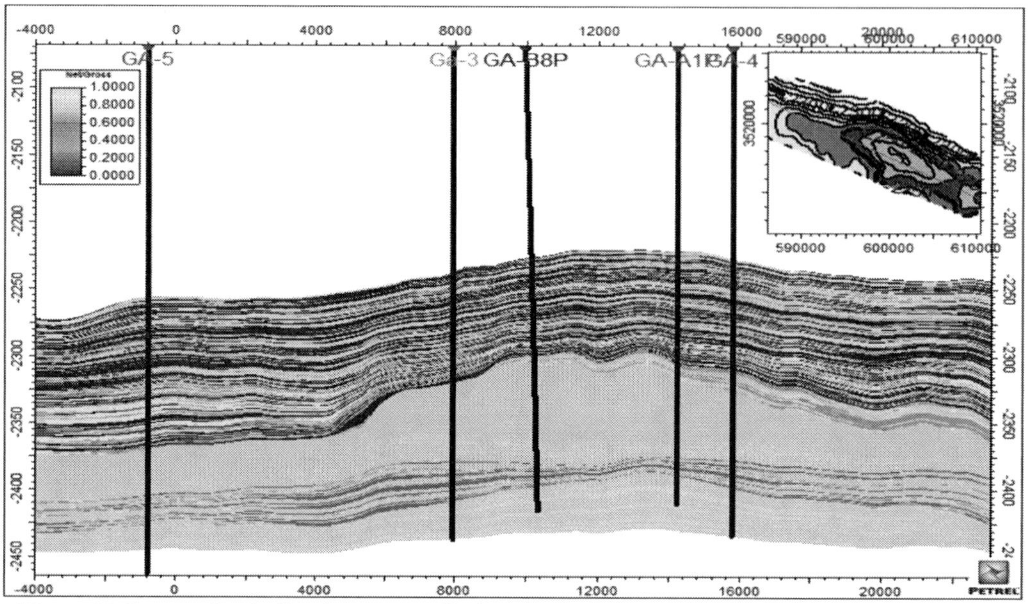

Figure 50 Cross section in direction NW-SE distribution of NTG for Mishrif Formation in Garraf oil field.

19. Conclusions

1. The structural model of Garraf Field created by using petrel software system showed that Garraf oil field represents a domal structure as indicated by the structural maps at prime of reservoir units. The structural closure is shifted from GA-3 well to GA-A1P well.

2. Horizons for Mishrif Formation are divided into three essential elements, further as 10 zones. Layers were designed for every zone based on petrophysical properties.

3. Mishrif zones extent complete the field except m2, L1 and L1.2 zones. m2 zone that represent cap rock of L1 and L1.2 extends over the field from GA-4 well within the southeast to well GA-3 further as L1.2 reservoir zone. Whereas the L1 reservoir zone extends from GA-4 well within the southeast to well GA-B8P within the middle of the field.

4. Model of petrophysical properties (volume shale, porosity permeability and water saturation) for Mishrif Formation in Garraf oil field was designed by using sequential Gaussian Simulation algorithmic rule as a statistical method after scale up of well logs. The model includes prime Mishrif, M1, M1.2, M2, L1, L1.2, L2, L2.2, L2.3 and L2.4. every unit is characterized by totally different reservoir properties. The unit's money supply.2, L1 and L1.2 are thought-about as high quality reservoir units because of the high PHIE and low water saturation values. Units L2, L2.2, L2.3 and L2.4 are considered as poor reservoirs due to low PHIE and high water saturation, and non-reservoir units include prime Mishrif, m1 and m2 so, they represent cap units.

5. Depend on cross sections for porosity and water saturation models that built in NW-SE and WNW-
ESE directions show that the most effective location characterized by good reservoir properties is between well GA-4 and GA-3 and these properties decrease step by step toward well GA-5.

6. The original oil in place (OIIP) within the Mishrif formation of Garraf field is (1061) million cubic meters.

IOP Conf. Series: Materials Science and Engineering **579** (2019) 012037 doi:10.1088/1757-899X/579/1/012037

References

1. Tarek Ahmed, 2005. "Working Guide to reservoir rock properties and fluid flow". Gulf professional published, USA.
2. Konica Minolta, 2012 "3D Scanning Advancements in Medical Science".
3. Li Shaohua, 2009 "Remaining Resource Reservoir Group", Yangtze University, Jingzhou, China.
4. Turner A. K. and Gable C. W., 2008. "A review of Geological Modeling. Colorado School of Mines ", USA, Los Alamos National Laboratory, Los.
5. Schlumberger, 2013.Petrel Geology and Modeling, Petrel Introduction Course, 559pp.
6. 4. Mustafa Baghdad, 2015. ‟Reservoir Characterization and 3D Petrophysical Modeling of Mishrif Formation in Garraf Oil Field‟,pp.2.
7. Modified after Al-Naqib Sargelu,1967. ‟Reservoir Characterization and Stratigraphic Relationships of Mishrif Formation in Gharraf Oil Field , ‟PP.32.
8. Ashour, H., 2016" Reservoir Characterization and 3D petrophysical Modeling for Tertiary Reservoir in Khabaz oil field "M.Sc. thesis, Petroleum engineering, University of Baghdad.
9. Schlumberger Petrel online help,2008, "Petrel 2008 version ".
10. Schlumberger, 2009. "Petrel online help, Petrel Introduction Course Schlumberger ", pp. 560.
11. Schlumberger, 2010(a). "Petrel introduction course. Schlumberger ", 13- 493p.
12. Schlumberger, 2007. "Petrel Structural modeling course Schlumberger ", 105-123p.
13. Ataei, M., 2013. "Log Facies Evaluation and Property Modeling of a Turbidite Reservoir ", the Gulf of Mexico. A thesis submitted to Norwegian University of Science and Technology, Department of Geology and Mineral Resources-Norway, pp. 103.
14. Hector H. Perez, Akhil Datta –Gupta, "The Role of Electrofacies, Lithofacies, and Hydraulic Flow Units in Permeability Predictions from Well Logs: A Comparative Analysis Using Classification Trees", Texas A&M U. and S. Mishra, SPE, 2003
15. Schlumberger "Petrel technical Manual", 2013.
16. N. Cressie and D. M. Hawkins, "Robust estimation of the variogram," Journal of the International Association for Mathematical Geology, vol. 12, no. 2, pp. 115–125, 1980.
17. R. Corstanje, S. Grunwald, and R. M. Lark, "Inferences from fluctuations in the local variogram about the assumption of stationarity in the variance," Geoderma, vol. 143, no. 1-2, pp.123–132, 2008.
18. Schlumberger, 2010 (b). "Reservoir Engineering course ". Schlumberger, 137- 177 p
19. Peter Phillips, Janice Liwanag, 2006 Improving Net-to-Gross Reservoir Estimation", Dew Journal, India.
20. Paul F. Worthington, 2009 "Net pay: what is it? What does it do? How do we quantify it? How do we use it?", Gaffney, Cline & associates, SPE.
21. Jarot Setyowiyoto, Ariffin Samsuri, 2006 "Oil Water Contact and Hydrocarbon Saturation estimation based on Well logging data", University GadjahMada Indonesia and University Technology Malaysia, Regional Postgraduate conference on Engineering and Science, Johore.

IOP Conf. Series: Materials Science and Engineering **579** (2019) 012038 doi:10.1088/1757-899X/579/1/012038

A New Algorithm for Foot-By-Foot Permeability and Irreducible water Saturation Estimation in Iraqi Limestone Reservoirs

Haider Qays M, Ruya Ahmed G

Ministry of Oil – Midland Oil Company MdOC, Baghdad, Iraq

Corresponding author's e-mail address: hydr_kais@yahoo.com

Abstract. This paper discusses the methodology developed for petrophysical modeling and its application to the main limestone formations in the Mid-region of Iraq. The systematic approach presented in this study evaluates not only combination of porosity, permeability and irreducible water saturation at each depth, but also free-water level and capillary pressure curve, such that water-cut of flow tests is closely reproduced in numerical simulation. It is one of the major achievements in this study. The algorithm matched to several sets of core samples: 2,903 porosity, 2,767 permeability, and 95 Irreducible water saturation samples that give the algorithm the robustness for future properties prediction. To develop the prediction Algorithm, it was employed multiple regression analysis (MRA) which mathematically requires independence between explanatory variables. Since possible explanatory variables (porosity, permeability and facies) indicate interdependency, it was investigated when necessary to transform the variables into independent axes. It was found that logarithm of permeability has correlations positive to porosity and FZI and negative to irreducible water saturation. Therefore, foot-by-foot evaluation shall meet the following conditions, (S_w) along wells shall be explained by single free-water level (FWL) for each pressure compartment and Water-cut observed on production test shall also be closely simulated with the FWL, For modeling purposes, Sw is split into irreducible water saturation (S_{wir}) and transitional (mobile) water saturation (S_{wm}) as water-cut is a function of only the latter.

1. Introduction

The determination of the Petrophysical properties in the specific geographical location is a challenging matter since it depends on the geological and rock properties, reservoir characteristics and as well as on the economic parameters. Hence, reservoir characteristics correlation is recognized as one of the most reliable tools for defining the optimum strategy for petrophysical properties prediction in uncored intervals the method proposed could be one of the useful tool to assist the progress in the decision making for the field development and strike towards the optimal and effective hydrocarbon Evaluation.

2. Conventional Core Analysis

This section presents the petrophysical analyses on conventional core measurements. From 28 Khasib and 20 Tanuma wells, 2,903 porosity and 2,767 permeability data are available. But water saturation information is somewhat limited (see Table.1). We conducted the following works using those data in this study.

- Core data classification by Hydraulic Unit concept

IOP Conf. Series: Materials Science and Engineering **579** (2019) 012038 doi:10.1088/1757-899X/579/1/012038

- Construction of property prediction mode

Table 1. Available Data sets for Property modelling.

Formation	Porosity	Permeability	S_{wir}
Tanuma	990 samples	990 samples	38 samples
Khasib	1879 samples	1777 samples	57 samples

3. Hydraulic Unit concept

This concept of rock type classification focuses on pore-structure geometry. Resulted rock type is denoted "hydraulic unit" and is characterized by a value called "flow zone indicator (FZI)", Based on Kozeny-Carman equation:

$$k = \frac{\emptyset^3}{(1-\emptyset)^2} \frac{1}{k_z S_{gv}^2} \qquad (1)$$

Where,
K: Permeability (cm^2),
$\emptyset$: Effective porosity,
K_z: Kozeny's constant
S_{gv}: Surface area per unit grain volume
Eq. (1) can be transformed into the following expression using field unit:

$$0.0314\sqrt{\frac{k}{\emptyset}} = \frac{\emptyset}{(1-\emptyset)} \frac{1}{k_z S_{gv}} \qquad (2)$$

Rock matrix's quality comprises two properties, i.e. storage capacity (porosity) and flow capacity (permeability). An index called "reservoir quality index (RQI)" relates pore scale attributes to macroscopic reservoir parameters as follows:

$$RQI = 0.0314\sqrt{\frac{k}{\emptyset}} \qquad (3)$$

RQI distinguishes between reservoir intervals with different hydraulic characters. Parameters called "flow zone indicator (FZI)" and "normalized porosity ratio ($\emptyset_z$)" are defined as follows:

$$FZI = \frac{1}{\sqrt{k_z S_{gv}}} \quad , \emptyset_z = \frac{\emptyset}{1-\emptyset} \qquad (4)$$

Using Eqs. (3) And (4), Eq. (2) can be rewritten as

$$RQI = \emptyset_z FZI \qquad (5)$$

Or,

$$Log(RQI) = \log \emptyset_z + \log(FZI) \qquad (6)$$

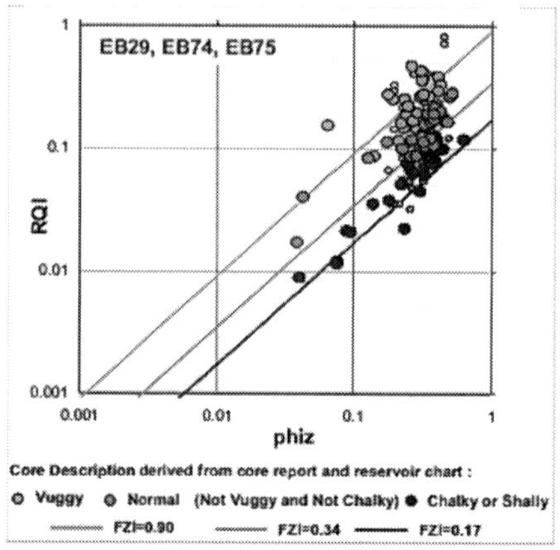

Figure 1. Cross-plots of core Calculated RQI and porosity

As Kz and Sgv stay constant so long as pore structure is similar, relation between RQI and "normalized porosity ratio ($\emptyset_z$)" should form a unit slope in log-log plot. Its y- intercept at x=1 is log (FZI). Samples with similar pore-space attributes are plotted Log (FZI) around same line and thus, RQI and FZI are powerful tools to correlate between different rock properties. Fig. 1 shows a cross plot of RQI against $\emptyset_z$ on some cores. Facies classification is copied from core description. Fig. 2 shows histograms of Log (FZI) by facies. Strong relationship can be observed between facies and FZI in Khasib formation. Strong relationship can be observed between facies and FZI in Khasib formation. Fig.3 and Fig.4 show histograms of porosity and permeability in each hydraulic unit. FZI technique was proved to be robust and was found to be valuable to supplement core data in the prediction of log-permeability in the reservoir modelling in the Middle East[1][2][3].

If FZI value in uncored sections is estimated, corresponding permeability can be computed using the following relation

$$K = 1014\, FZI^2 \frac{\emptyset^3}{(1-\emptyset)^2} \qquad (7)$$

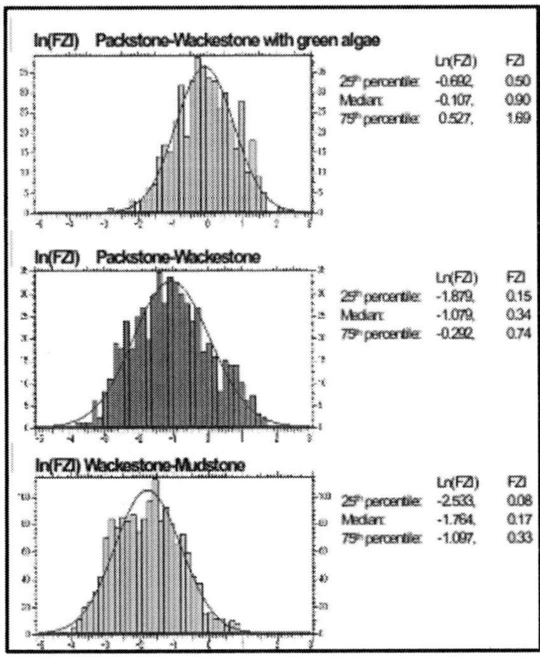

Figure 2. Histogram of Log (FZI) by facies

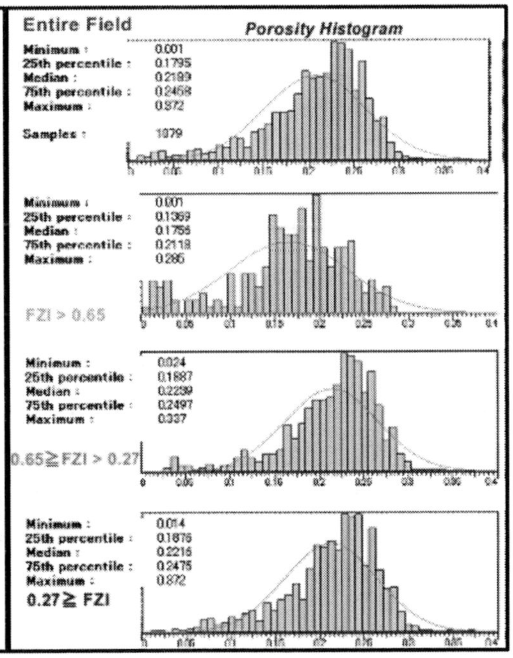

Figure 3. Histogram of Porosity in each hydraulic unit

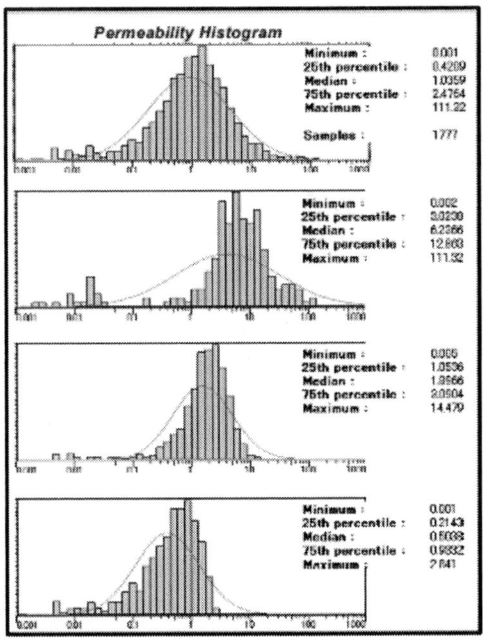

Figure 4. Histogram of Permeability in each hydraulic unit

4. Construction of property prediction model

In order to predict permeability and irreducible water saturation from log data for uncored sections, we constructed a property prediction model. Scatter plots between porosity, permeability and irreducible water saturation on cores are shown in Figs (5, 6) which indicate notable interdependency among those properties. The prediction model was constructed in a way that the correlations are kept.

Permeability, Porosity, and S_{wir} data are derived from Core Flooding test & Capillary Pressure measurements. The correlation between these properties does not honour the facies or rock type therefore the predicted property estimated from these correlation will be tested for facies and production test result to get the more reasonable value.

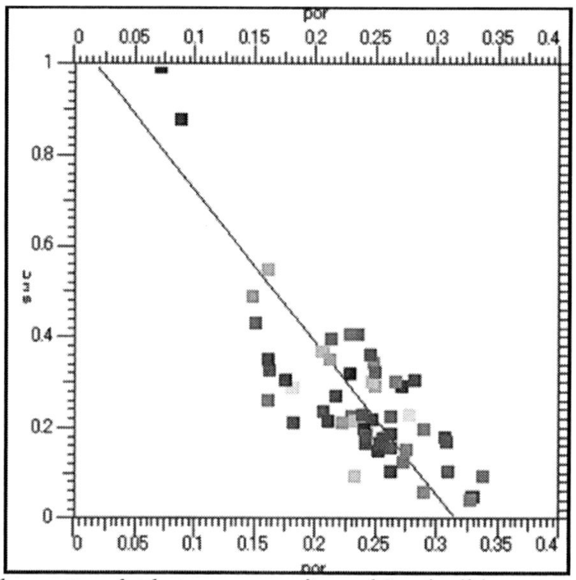

Figure 5. The scatter plot between porosity and Irreducible water saturation

Comprehensive analysis confirmed results that are different from well log for oil and gas, in case of oil and gas are produced from reservoir interval that has a high water saturation but does not produce water during test the interpretation for this phenomena is the existence of irreducible water saturation therefore in many cases the simulator fails to predict the actual water cut due to unrealistic value of S_{wir} ,even though that there are many methods for s_{wir} estimation[5] of Swir , we have used Crude oil displacement method by displacing rock samples of water by crude until the rock samples water content has not changed.

IOP Conf. Series: Materials Science and Engineering **579** (2019) 012038 doi:10.1088/1757-899X/579/1/012038

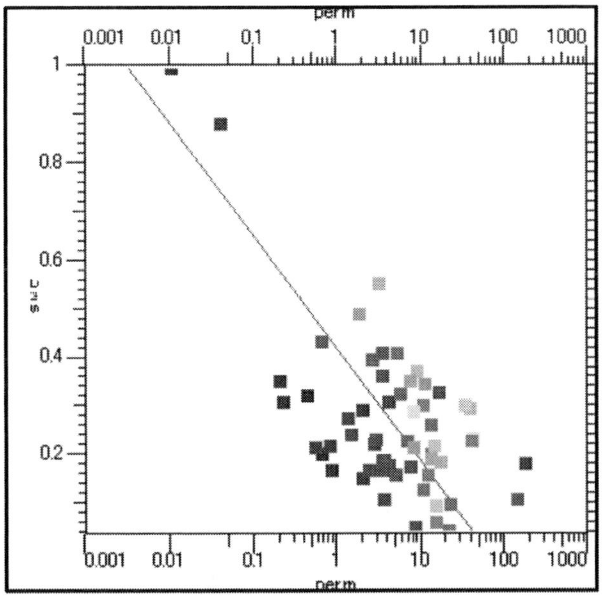

Figure 6. The scatter plot between permeability and Irreducible water saturation

5. Extraction of correlation from core data

To develop the prediction model, we employed multiple regression analysis (MRA) which mathematically requires independence between explanatory variables. Since the possible explanatory variables (porosity, permeability and facies) indicate interdependency (Fig.7), it was investigated when necessary the transform of the variables into independent axes. (Fig. 8 and Fig. 9) Compare Two prediction models using MRA with:

1. "Raw" variables (case MRA) and
2. Principal components of those variables, (case PCA+MRA).

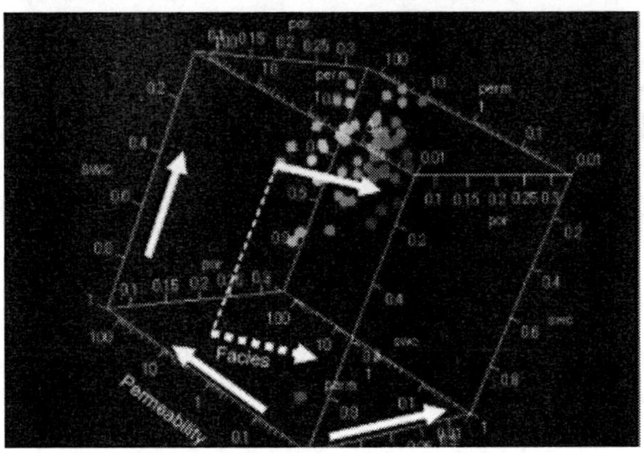

Figure 7. 3-D plots of properties

Correlation coefficients on Swir between the prediction and core measurement were 0.88 and 0.87, respectively for the two models. Both achieve good property Reproduction. However, Swir values predicted by PCA+MRA were too high in low permeability range (compare in Fig. 8 and Fig 9 solid lines). We therefore chose MRA with raw variables for Khasib property prediction with the following expression for Swir hereafter

$$s_{wir} = \exp[-0.1263\ \ln(k) - 4.62762\emptyset + 0.1224] \quad (8)$$

Where,
K: permeability. (md),
φ: effective porosity,
Swir: irreducible Water saturation.

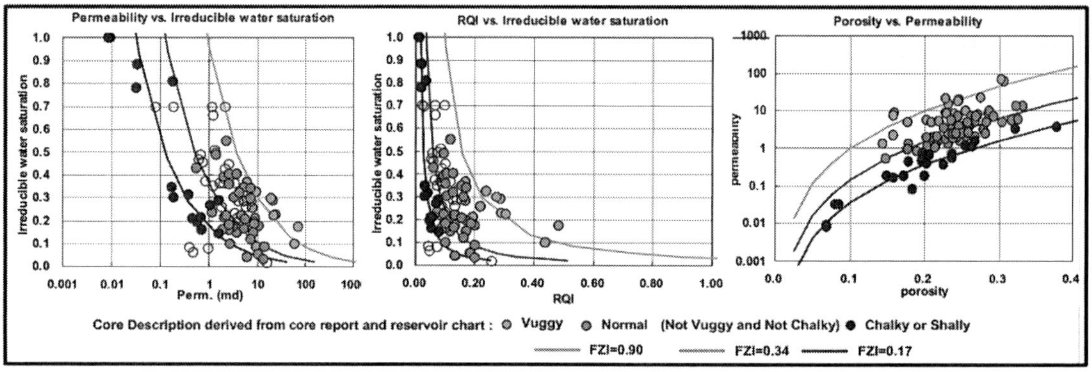

Figure 8 . Prediction model using MRA with “raw” properties (MRA)

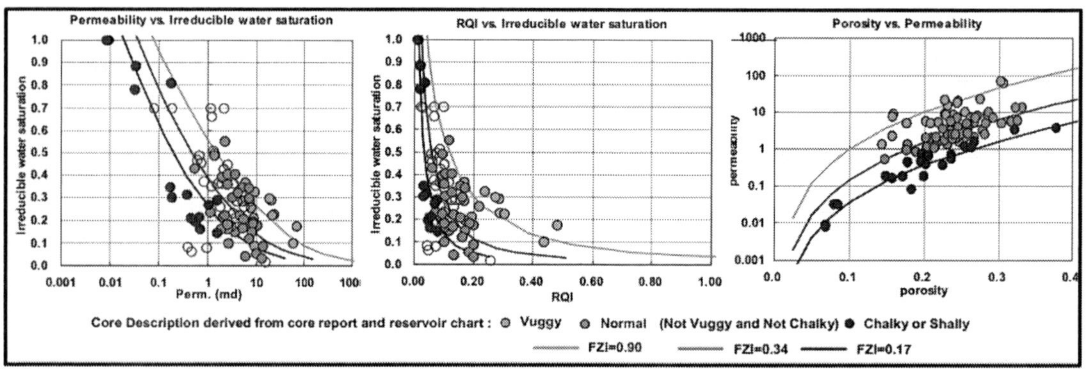

Figure 9. Prediction model using MRA with principal components of properties (MRA+PCA)

6. Capillary Pressure Curves

Capillary pressure data from oil-water system of drain cycle from special core analysis were used. The core data were modeled and averaged for each rock type with the capillary pressure function then, these curves were converted to values of "J" and these were related with water saturation from capillary pressure curves, We took the following J-function approach to extract representative capillary pressure Curves from a number of core measurements.

$$P_c = J(s) \times ST \times \left(\frac{\varphi}{k}\right)^2 \times U_{const} \quad (9)$$

Where:

J(S): J-function, ST: Surface tension, φ: Porosity, K: Permeability,
U_{const}: Constant depending on the unit system =4.61678 (FIELD units, pressure in PSIA, perm in MD)
Derived J-functions for Tanuma and Khasib are shown in Fig.10 and Fig.11 for each facies. Table 2 present the proposed constant input data to J-function equation

$$J(s) = A\,(sw+0.49)^{B} \times U_{const} \qquad \textbf{(10)}$$

Uconst=0.216601,see Table 2. for J-Function equation constants.

The J-function plot shows Moderate initial stage and relative long middle stage indicates the good sorting and seepage conditions. The calculated Sw from the J-function equation for each rock type is calculatred as mobile water saturation (Swm) then compared with swi estimated from wireline log .In case the values are identical the value is considered for fractional flow calculation to match the production test result at the specified depth, however if the estimated value mismatched the log sw then according to the proposed algorthim the FZI is revised to start inner itration of calculating new Swi and K properties. reasonable agreement between predicted and measured water saturation is relying on appropriate estimations permeability and flow unit indicator FZI of the specified depth.

Table 2. for J-Function equation

	Constant A	Constant B
K>100 md	3.1758	-4.4300
Vuggy Limestone	2.4925	-3.8251
Normal Limestone	1.9661	-3.2326
Chalky Limestone	1.5269	-2.6011

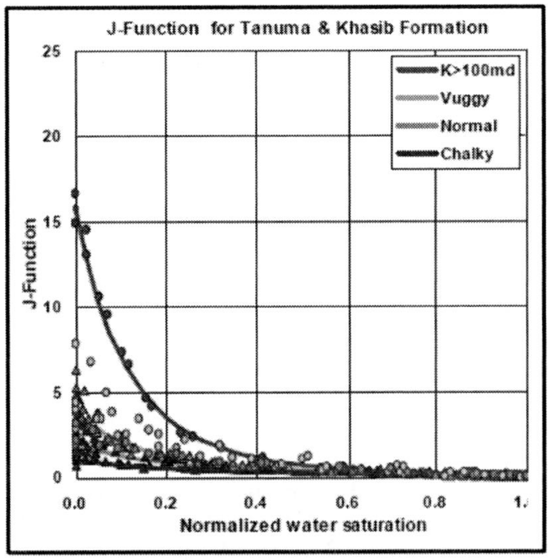

Figure 10. Derived J-functions for Tanuma and Khasib

The proposed formula for water saturation covers the most avaiable limestone formations in the middle region of Iraq and can be utilized for any carbonate formation with rock type similar to our mode in the same area.

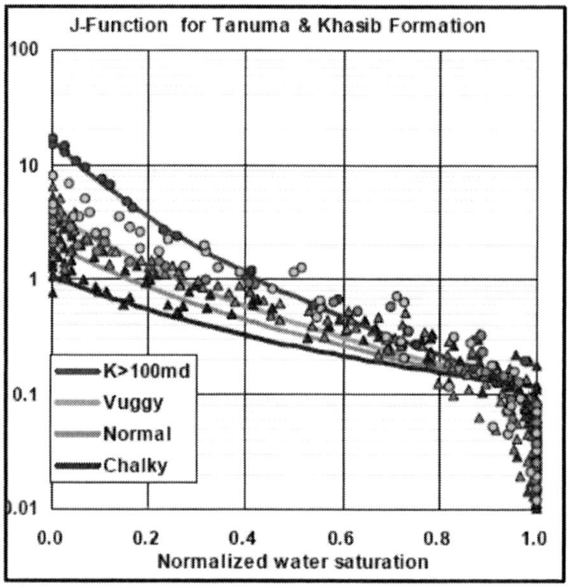

Figure 11. Derived J-functions for T&K Formation in semi-log

7. Foot-by-foot estimation of permeability and irreducible water saturation

Next, it was estimated the permeability and irreducible water saturation foot-by-foot along wells for uncored sections. Three parameters (Porosity , Sw and facies classification) are available from log analyses. Representative J-function curves are also available for each facies (see section 6 for details). It was found that logarithm of permeability has correlations positive to porosity and FZI and negative to irreducible water saturation. Now foot-by-foot evaluation shall meet the following two conditions.

Sw along wells shall be explained by single free-water level (FWL) for each pressure compartment.

Water-cut observed on production test shall also be closely simulated with the FWL.

For modeling purposes, Sw is split into irreducible water saturation (Swir) and transitional (mobile) water saturation (Swm). As water-cut is a function of only the latter,

$$Sw = Swir\,(FZI) + Swm\,(FZI, FWL, J\text{-function})$$

$$WC = function\,(Swm, Kr, Vis)$$

Where, FZI: flow zone indicator, FWL: free water level, Kr: relative permeability, Swir: irreducible water saturation and Vis: fluid viscosity. FZI and FWL are unknowns while the others are known.

We developed a new algorithm that honors available all information and conditions. Fig.12 and Fig. 13 are its flow chart and computation process for foot-by-foot permeability and irreducible water saturation. Via using this procedure, we built permeability and irreducible water saturation "logs" calibrated to production test results. Moreover, free water level, estimated in try-and-error basis, is now endorsed by dynamic data. discussions on pressure compartments are based on the free water levels established in this way.

IOP Conf. Series: Materials Science and Engineering **579** (2019) 012038 doi:10.1088/1757-899X/579/1/012038

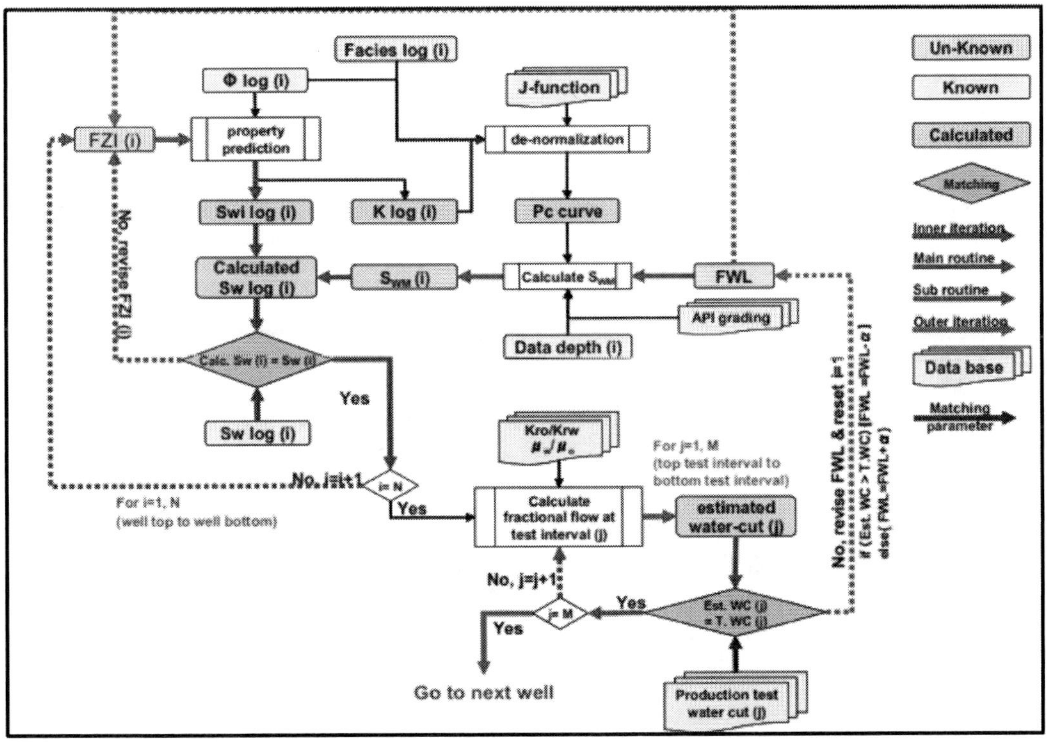

Figure 12. The new Algorithm Calculation flow chart of foot-by-foot permeability and irreducible water saturation

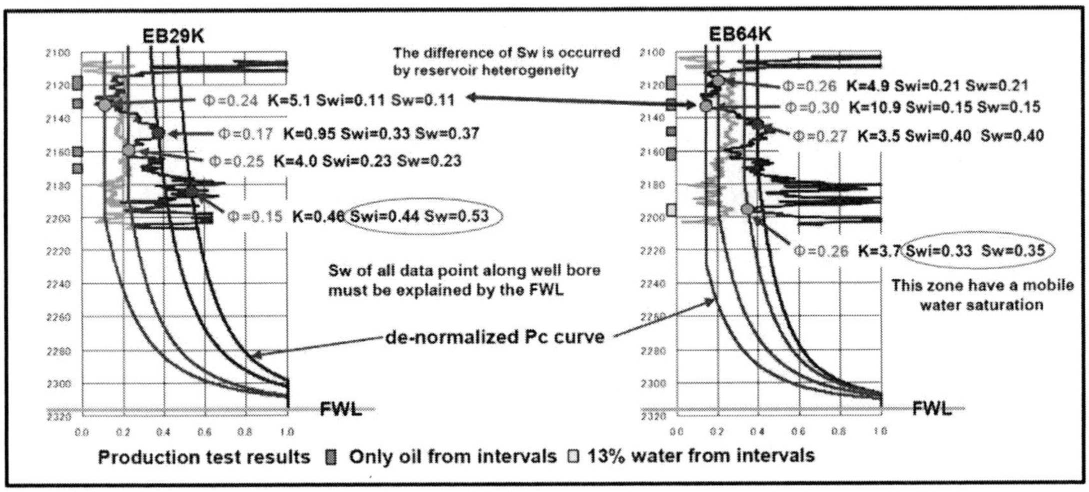

Figure 13. Example calculations: Sw profile along a well is explained by single FWL for each pressure

The computed permeability and irreducible water saturation "logs" are used later in geostatistical reservoir modeling, to characterize reservoir heterogeneity. The implementation of the new algorithm shows the good match between the actual and the predicted petrophysical indices as seen in Fig.13.

Results of this paper contribute to further understanding of the dynamic reservoir characters of Late Cretaceous Formation in Middle of Iraq, and benefit for favorable reservoir predicting and oil producing in this oilfield.

This study algorithm consistently produced correlation superior to the existing models that were compared against and extremely sensitive to possible changed in flow zone indicators within the area of interest

8. Conclusions

The new algorithm honors all available information and conditions such as free-water level (FWL) for each pressure compartment and the production test at each production interval through the fractional flow calculation at each depth.

This study is not exceptional, comparable work has been conducted and is accessible in the literature. What is noteworthy about this study is that this method delivered a significantly improved irreducible water saturation and permeability estimation in limestone reservoirs compared to the standard correlation models, moreover the uniqueness of this study is that it overcomes the data limitation constraint when conducting reservoir studies because it includes large amount of genuine data collected from two main carbonate formations in the middle region of Iraq (total core samples =2903).

The calculated results of porosity and permeability are in accord with the core analysis data, the algorithm model is reliable in petrophysics properties prediction for simulation purposes.

Acknowledgements

I would like to express my special thanks of gratitude to Midland Oil Company MdOC, Secondly I would also like to thank my colleagues who helped me a lot in inalizing this project within the limited time frame.

References

[1] Kharrat R, Mahdavi R, Bagherpour M, and Hejri S, 2009 Manama Rock Type And Permeability Prediction Of A Heterogeneous Carbonate Reservoir Using Artificial Neural Networks Based On Flow Zone Index Approach *Society of Petroleum Engineers* 10.2118/120166-MS,

[2] Shenawi. S, White J, Elrafie E and AL-Kilany K 2007 Permeability and Water Saturation Distribution by Lithologic Facies and Hydraulic Units: A Reservoir Simulation Case Study Manama *Society of Petroleum Engineers*

[3] Pourafshary A Mirzaei-Paiaman Hadi Saboorian-Jooybari, and Peyman 2015 Improved Method to Identify Hydraulic Flow Units for Reservoir Characterization *Energy Technology* **3** 726-733

[4] Cui Y and Kuan Jie Y 2015 The measure of irreducible water saturation of low permeability reservoir and its influencing factors is analysed **42** 17 20-21

[5] Wang Fengqi and Daqing Heilongjiang 2016 China Discussion on the Irreducible Water Saturation and Ultra-Low water saturation *IOSR Journal of Engineering (IOSRJEN)* **06** 2278-8719

Evaluation of Formation Capacity and Skin Phenomena of Mishrif Reservoir -Garraf Oil Field

H A Baker, A Al-Rikaby* and I S Salih
University of Baghdad, Baghdad- * Dhiqar Oil Company, Dhiqar, Iraq

Corresponding author's e-mail address: ha.baker@yahoo.com

Abstract. Mishrif Formation is the main reservoir in Garraf Oil Field. It is divided into three productive units (M1, L1.1, and L1.2), the unit L2 contains a very small percentage of oil with high majority of water so it is considered to be of a non-economic feasibility. Developing of such reservoir requires a study of its capacity and how the effective communication properties changes laterally. Build up pressure testing was conducted for five wells (A16P, Ga-1, A4P, B5P and C43P) in Garraf oil field. The data have been analysed by Ecrin Kappa Saphir program and well behaviour was detected by pressure and pressure derivative responses (or signature) with time. The results show a difference in capacity with good flowing capabilities. All the wells exhibit a negative skin factor due to stimulation.

1. Introduction

In well testing, pressure data interpretation recorded has been used to evaluate reservoir properties. Shut in reservoir pressure used to predict reserve in place during calculations of material balance. Reservoir flowing behavior has been described during transient pressure analysis. There is a proposal for interpretation method based on derivative pressure analysis with respect to Horner/superposition time or natural logarithm time functions [1]. Many methods are presented to construct a type curve derivative pressure. The most commonly used type curve of pressure derivative is based on ideas of Bourdet et. al. [2] who made type curve based on dimensionless pressure derivative with respect to logarithm of dimensionless time that have two advantages above type curve based on a log-log of dimensionless pressure vs. dimensionless time for uses wellbore storage and skin problems. Construction of type curve has an alternative method based on pressure derivative presented by Onur, Veh, and Reynolds [3] that involves plotting both dimensionless times vs. dimensionless pressure and dimensionless pressure divided by logarithm derivative dimensionless pressure for uniform flux and infinite conductivity vertically fractured wells. Pressure buildup behavior must be plotted on semi log coordinate as a straight line for approaches of asymptotic represented by Soliman, Correa and Ramey [4] who represents approximation a long time to Horner solutions that uses only a data portion in period of radial flow if the data are available [5]

2. Brief Idea about the Field

1- Garraf oil field is located in southern of Iraq, approximately 5 km northwest of Rife city in Thi-Qar governorate.
2- Garraf consists of an anticline forward northwest-southeast, 10 km width and 31 km long. [6]

IOP Conf. Series: Materials Science and Engineering **579** (2019) 012039 doi:10.1088/1757-899X/579/1/012039

3- Five wells were used in this study (Ga-1, C43P, A4P, B5P and A16P) for the detection of wells behavior and its characteristics by data analysis through pressure derivative vs. time. Figure 1 illustrates the location of these wells.

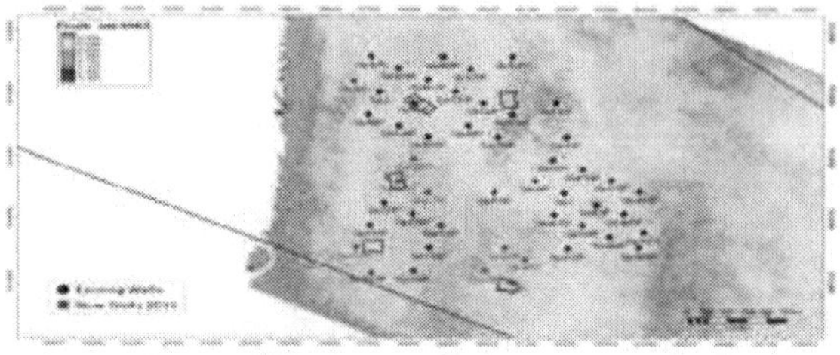

Figure 1. Wells location

3. Bourdet type curve:

This curve represent relationship between dimensionless group which are:, dimensionless pressure(P_D) ,dimensionless time(t_D) ,dimensionless pressure derivative(P'_D) and dimensionless wellbore storage C_D. Mathematically:

$$P_D = \frac{k\ (P_i - P)}{1\ .2q\mu B} \quad (1)$$

$$t_D = \frac{0.0\ \ \ k}{\phi\ C_t\ r_w} \quad (2)$$

$$C_D = \frac{0.8\ \ \ C}{\phi C_t\ h\ r_w^2} \quad (3)$$

$$P'_D = t_D\left(\frac{\partial P_D}{\partial t_D}\right) \quad (4)$$

The type curve is illustrated in figure 2.

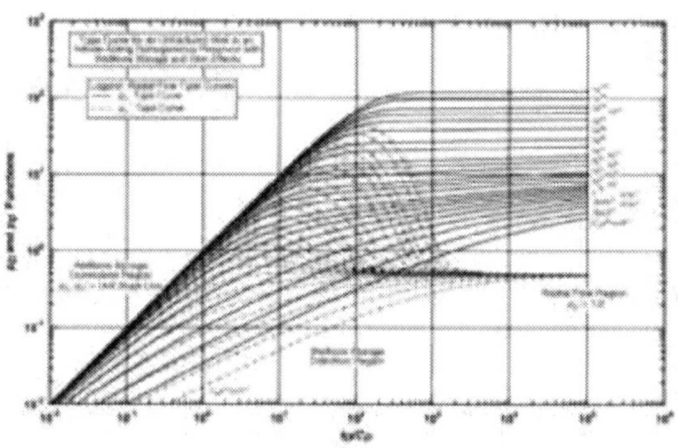

Figure 2. Bourdet type curve.

4- Transient Test Analysis and results

Pressure transient analysis was done in five wells producing from heterogeneous limestone formation. The produced fluid is black oil with API of 24 and GOR of 450 Scf/STB. The layer is considered highly anisotropic where the pressure flow regime changes in signature (shape of pressure derivative) from well to another. The first example is a slightly deviated well (well A16P) of 28 degrees in relatively lower permeability region, the well production rate is 3500 b/d. The pressure build up response shows a homogenous layer of effective permeability of 69.5 m.D and effective thickness of 223 ft. The values of layer capacity, skin factor, storage coefficient is calculated by matching the plot with Burdate type curve which done with the aid of software. Figure 3 shows the pressure response of the well. The curve of green colure is the pressure difference between shut-in pressure and bottom-hole flowing pressure prior to shut-in (Pws-Pwf) versus time, the curve of red colure represents the pressure derivative versus time. The plot shows the absence of unit slop line at the early time of test which gave a good evidence of low damage or stimulation.

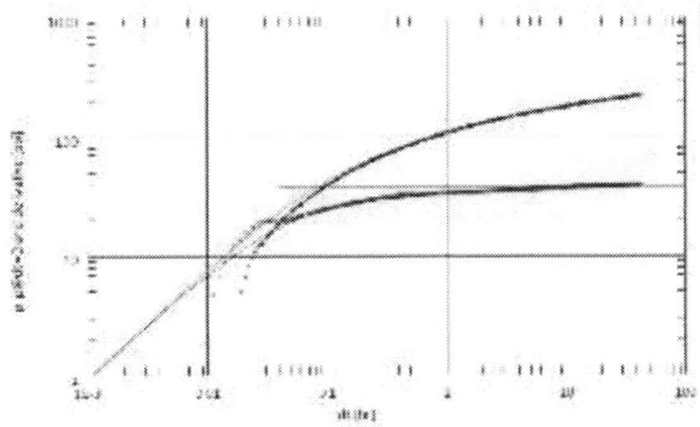

Figure. 3. Pressure build up of well A16P

The skin parameter reported is - 6 which represent a very strong acid stimulation, however, the geometrical skin is positive, this is because the well is partially penetrated and not centered in mid-layer. Figure 3 represent the well architect

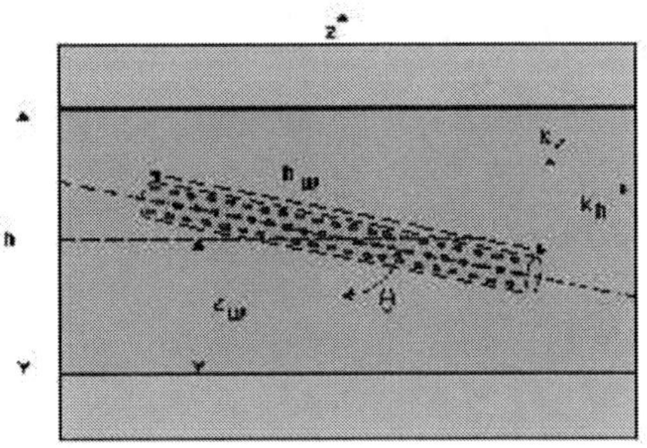

Figure 4. Slanted well architect

The effect of time was studied at constant applied voltage of 10v, mixing speed of 400rpm and initial concentration of 100NTU. Figure 2 illustrates the turbidity as a function of operating time. It is clearly seen that the operating time has an important effect on the turbidity removal.

IOP Conf. Series: Materials Science and Engineering **579** (2019) 012039 doi:10.1088/1757-899X/579/1/012039

There is a high possibility that when dealing with heterogeneous rocks; partial depletion, vertical changes in flow permeability could occur, and because the productive interval (hw) is less than layer thickness (h), there is a chance that only small portion of the perforated interval is well stimulated and gives production more than 90% and the rest of the layers are not strongly stimulated either due to lower permeability or higher pore pressure. Production logging here is requested to confirm whether there is or no uniform acid distribution.

The second well (Ga-1) has similar response. The well penetrating the same layer but here the estimated reservoir capacity (kh) of 3740 m.D-ft. is much smaller than previous well. The stimulation job has a good efficiency in a single layer model. Skin is -4.5.Figure 5 represents the transient behavior of the well during test

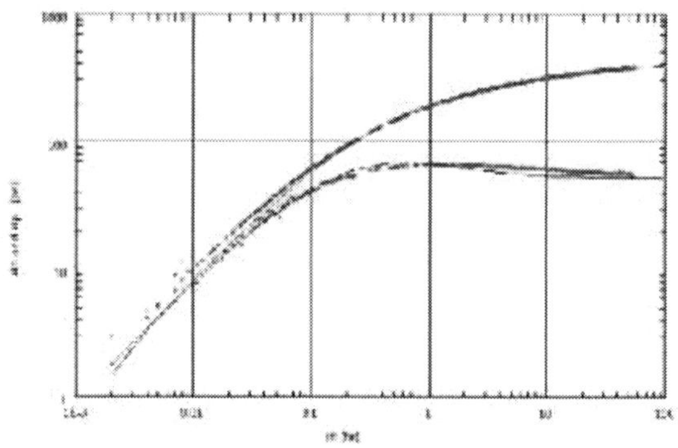

Figure 5: Transient response of well Ga-1

The third well (A4P), is horizontal well penetrating the same layer The flow regimes of a horizontal well is clearly visible starting with early hemiradial flow reached around 0.2 hour and swiftly changed to early linear flow as shown in figure 6 The early radial flow reached less than 1 hour of buildup, which shows a very strong geometrical permeability in the region surrounding the well. The early linear flow regime starts at 1 hour and ends with the build- up end. The build- up time was not sufficient to reach the pseudo radial flow which is expected around 150 hours. The well producing effectively with 267 mD, and geometrical skin of the well is -6.64

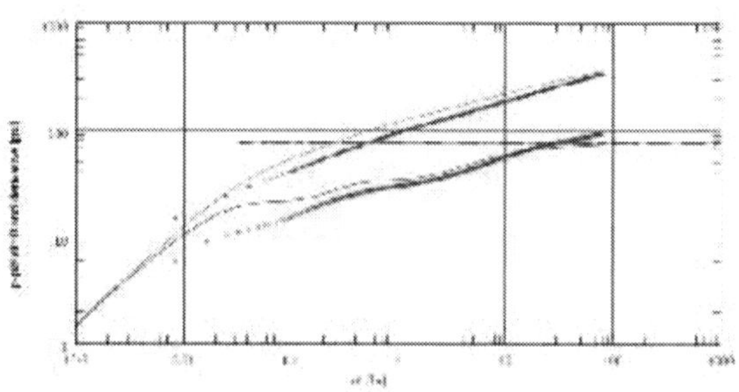

Figure 6. Response of horizontal well.A4P

The fourth case is a vertical well with a clear dual porosity response (well B5P) as shown in figure 7

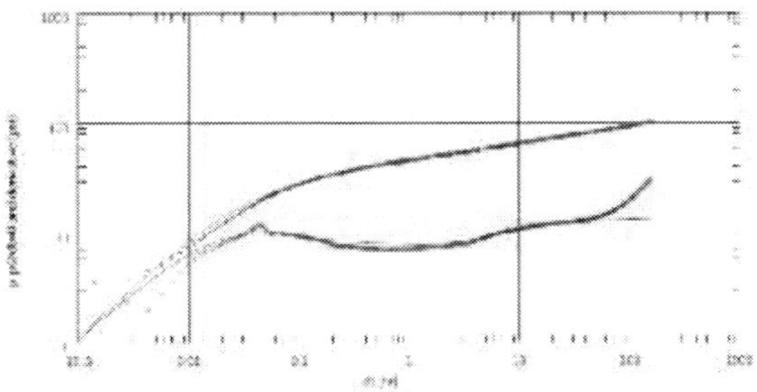

Figure 7. Dual porosity response of well B5P

The after-flow period is approximately ended at Δt=0.03 hr., the pressure response from Δt=0.03 till Δt=3 hr. during buildup represents the transient response of a high permeability layer contributing to high level of production and has as well a very low skin factor. From Δt =3 to Δt =7 hours the derivative shows a ¼ slope due to transition period between the high permeability layer to the overall system flow capacity. The late response on the derivative after 70 hours of buildup correlated to a hydrodynamic well interference boundary, which is clear as the pressure difference increases rapidly.

The fifth well (C34P) shares the same concept yet no second layer stabilization reached as indicated by figure eight.

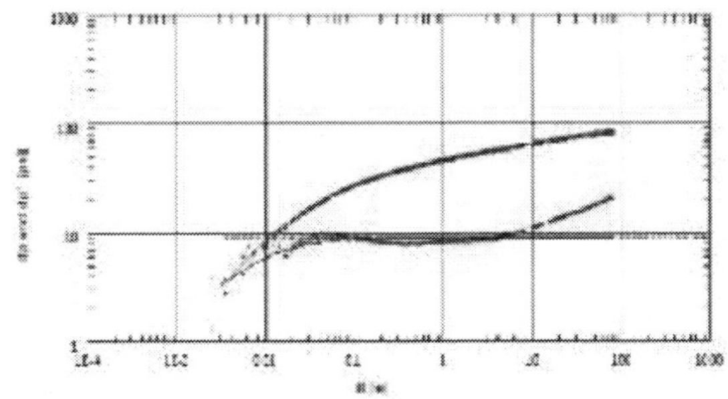

Figure 8 .Transient profile of well C34 P

Only one layer reaches radial flow and the buildup still in transition period to reach total system radial flow the kh here is considered low compared with offset cases 3500 mD.ft. Table 1 summarize the test results.

Table 1: Summary of results of test analysis

Well	kh mD.ft	Total skin
A16P	15500	- 4.84
Ga-1	12500	-4.68
A4P	24200	-6.64
B5P	51496	-5.5
C43P	11400	-3.37

IOP Conf. Series: Materials Science and Engineering **579** (2019) 012039 doi:10.1088/1757-899X/579/1/012039

Nomenclature and Abbreviations

C:Wellbore storage(bbl/psi)
Ct: Total Compressibility,psi-1
dP: Pressure difference, psi
dp': Pressure derivative
dt: time ,hrs
h:layer thickness
hw: length of the productive zone of well
K: Permeability, mD
Kh: Horizontal Permeability, md
Kz: Vertical Permeability
rw: well radius, ft.
S: Skin factor, dimensionless
Φ:Porosity,fraction

References

[1] Bourdet D, Ayoub J A and Plrard Y M 1989 *SPE Form. Eval.* 293.
[2] Soliman M Y 1986 *SPE Form. Eval.* 363.
[3] Onur M Veh N and Reynolds A C 1989 *SPE Form. Eval.* 429
[4] Correa A C and Ramey H Jr 1986 SPE 15579 *SPE Annual Tech. Conf. and Exhi.* (New Orleans) .
[5] Ozkan E Vo D T and Raghavan R 1987 SPE 16811.
[6] PETRONAS Co 2016 Final development plan for Garraf oil field.
[7] Internal report.

IOP Conf. Series: Materials Science and Engineering **579** (2019) 012040 doi:10.1088/1757-899X/579/1/012040

Skin and Reservoir Pressure Profiling Utilizing Production Logging

H A Baker# and I S Salih
University of Baghdad, Baghdad, Iraq

#Corresponding author's e-mail address: ha.baker@yahoo.com

Abstract. Production Log is a key log that directly define the facies and layers behavior. The PLT measures the inflow zone thickness and the flow rate of each phase which is directly function of facies properties in terms of deliverability. On the basis of pseudo steady state flow regime (which is the dominant phase of production during PLT log operations) a mathematical relationship developed to determine skin and reservoir pressure per each inflow zone with reasonable error.

1. Introduction

Evaluation of multilayered system in pseudo steady state is required to plan and monitor the well efficiency taking into consideration the commingled system interaction when various flow capacity coexist [1]. Usually with heterogonous carbonates; stratification of difference facies usually occurs, having in between barriers such as non-permeable layers or same facies with non-quantifiable vertical permeability. With these cases, partial depletion is expected and should be closely monitored as it affects the ultimate recovery of the well [2].

Production logging tool can run periodically to determine the pressure and productivity index for all the commingled system inflow zones[3], however, with such technique, multichoke required, and shut-in period required as well[4].

The limitation of this technique is the operation parameter of the well, as may the reservoir pressure low and unable to startup naturally after shut-in and/or multichoke test may not be easy to perform when the flowing bottom hole pressure is close to saturation pressure hence flow after flow will not be representative[3].

Having these limits in mind, a new technique established based on single choke flowing production log

The Model can take two different analysis trends:

a- Customer provides Modular formation dynamics tester pressure profile to calculate skin for each zone

b- Unknown reservoir pressure and the calculation include part of the skin determination.

Two derivation done for both cases for single phase laminar flow pseudo steady state regime for homogenous reservoir (matrix contribution mainly).

2. Skin Profiling

Based on pseudo steady state flow regime during Production testing with Production Log, Darcy equation that defines the system deliverability expressed as [1]:

$$Pe - Pwf = \frac{141.2QB\mu}{Kmh}\left\{ln\left(\frac{re}{rw}\right) - 0.75 + S\right\} \qquad (1)$$

Where k_m = mean permeability in the pay interval (estimated from well test analysis or from Enhanced

Processed open hole log); h = pay thickness; Q = Total flow rate during production logging; B = formation volume factor; μ = reservoir fluid viscosity; P_e = reservoir pressure; P_{wf} = flowing bottom hole pressure; r_e = reservoir external radius (drainage); r_w = borehole radius; S = skin factor.

Considering now a subinterval with thickness h_i and permeability k_i with production q_i and skin of s_i then equation (1) becomes:

$$Pe - Pwf = \frac{141.2\ qiB\mu}{kihi}\left\{ln\left(\frac{re}{rw}\right) - 0.75 + Si\right\} \tag{2}$$

Combining equation (1) and (2) with respect to skin yields to:

$$S - si = \frac{\Delta P}{141.2\ B\mu}\left\{\left(\frac{Kh}{Q}\right) - \left(\frac{kihi}{qi}\right)\right\} \tag{3}$$

Where:
S: mathematical expression considers the theoretical total skin factor of the well. This number accounts for other deviation in parameters and will not be the same as the actual total skin of the well
Si: layer (i) skin factor corresponds to damage/stimulation of ith layer.
ΔP: Pressure drop between the average reservoir(s) pressure and wellbore flowing pressure (psi)
KH: Total well flow capacity mD.ft
Kihi: layer (i) flow capacity mD.ft
Q: well production rate STB/D
qi : Layer (i) production rate STB/D
B: formation volume factor
μ: fluid viscosity cP

Having the corrected matrix permeability log as an input with the pressure drop per inflow zone (In case of vertical compartment or baffles) then simply the term (S-si) can be determined per inflow zone. The second step of interpretation is to compare the measured and calculated productivity index with iterating on S to determine si per zone where:

From PLT $PI = \frac{qi}{\Delta p*}$

* Δp can be either given from customer or determined simply from shut in pass in condition of multi choke test

The productivity index to be calculated with "S" iteration to match the PLT PI using following equation:

$$\left(\frac{1}{PI}\right) = \frac{141.2B\mu}{kihi}\left\{ln\left(\frac{re}{rw}\right) - 0.75 + Si\right\} \tag{4}$$

Example 1 shows the validity of the derived term.

3-Pressure Drop determination

In case of:

1- no previous knowledge of reservoir pressure
2- expected compartments and/or pressure baffles
3- no shut-in passes performed
4- Injection operation ongoing with offset wells that may disturb the pressure map.

Then, a method derived for pss flow to determine the drawdown value based on constant – rate drawdown test given by Ramey and Cobb [5]

$$P_D\ (t_{DA}) = 2\pi t_{DA} + 0.5\ \left[ln\left(\frac{4A}{e^{\gamma}C_A r_w^2}\right)\right] \quad (5)$$

Where dimensionless pressure drop is:

$$P_D(t_{DA}) = \frac{kh\Delta p}{141.2qB\mu} \quad (6)$$

Pressure drop:

$$\Delta p = P_i - P_{wf}(t)$$

For drawdown and dimensionless time based on drainage are:

$$t_{DA} = \frac{0.0002637kt}{\emptyset(\mu C_t)A} \quad (7)$$

Integrating equation 5 with respect to t_{DA} having

$$P_{DI}\ (t_{DA}) = \left\{2\pi t_{DA} + 0.5\ \left[ln\left(\frac{4A}{e^{\gamma}C_A r_w^2}\right)\right]\right\} + 0.5\ \left[ln\left(\frac{4A}{e^{\gamma}C_A r_w^2}\right)\right] \quad (8)$$

Combining equation 5 and 8 yields to

$$P_{DId} = 0.5\ \left[ln\left(\frac{4A}{e^{\gamma}C_A r_w^2}\right)\right] \quad (9)$$

Equation 9 suggests that when dimensionless pressure integral function, P_{DId} is plotted against dimensionless time, a horizontal line will be obtained during pss flow regime regardless of the reservoir geometry.

To link the derivation with the main objective of this note, means that, during PLT operation, a constant rate drawdown is to achieve the job objective, hence, without any extrapolation of the pressure versus time plot we can consider that the Pwf reached is basically the start of a zeroline slope which will be the start of the calculations for determining the Drawdown per inflow zone.

Expressing equation 9 in real oilfield variables [5], we have:

$$(\Delta P)_{Id} = q_o\ b_{pss} \quad (10)$$

Where:

$$b_{pss} = \frac{141.2B\mu}{kihi}\left[0.5ln\left(\frac{4A}{e^{\gamma}C_A r_w^2}\right)\right] \quad (11)$$

Equation 10 is the basis of the analysis and strictly applicable during pseudo steady state flow regime for a constant rate case.

Two examples will show how to determine skin and reservoir pressure based on above.

Case#1

IOP Conf. Series: Materials Science and Engineering **579** (2019) 012040 doi:10.1088/1757-899X/579/1/012040

This is a simulator case of a well having 6 different reservoirs of different properties, vertical scale represent the depth and for convenience the scale subtracted by factor of 2000:

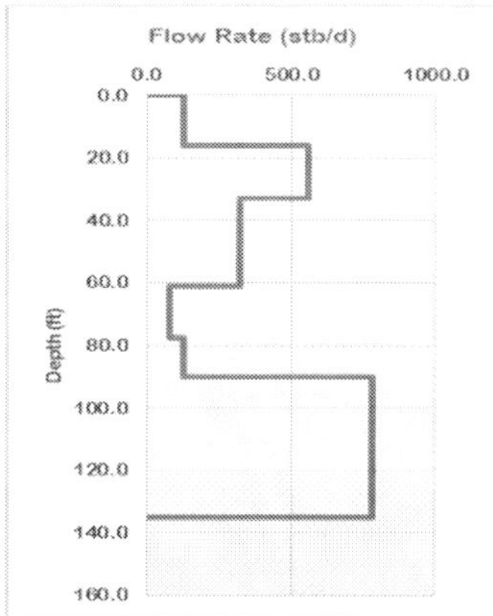

Figure 1. PLT Profile of well with 6 flow zones

Table 1 summarizes the results of simulated PLT response

Table 1. PLT simulated analysis of each flow zone

from simulated PLT		model input	PVT		Constant Pe	model input
h (ft)	qi (stb/d)	ki mD	mu (cp)	Bo	Dp (psi)	skin
16	128.5	10	1.90	1.38	2922	-2.03
17	558.4	38	1.90	1.38	2922	-2.98
28	324.3	16	1.90	1.38	2922	-0.99
16.5	80.4	11	1.90	1.38	2922	6.45
12.5	130.4	21	1.90	1.38	2922	4.59
45	778	33	1.90	1.38	2922	3.89

Table 1 shows:

Track 1: layer inflow thickness (ft) from simulated model input
Track 2: Layer inflow production rate (stb/d) simulated production model output
Track 3: Layer permeability ki simulated model input, mD
Track 4&5: PVT data per zone (formation volume factor and viscosity)
Track 6: The pressure drop per layer (layer reservoir pressure minus borehole flowing pressure) psi
Track 7: input model skin factor per layer

Applying equation 3 for each inflow zone to determine the expression (S-si), the results are shown in table 2

Table 2. Comparison of measured productivity index and skin factor with calculated values

Si	PI from PLT	Re (ft)	darcy PI	E% in PI	E% in Si
-2.0	0.044	56737.5	0.047	6.0%	1.4%
-2.8	0.191	76488.1	0.200	4.5%	6.5%
-0.9	0.111	61458.3	0.116	4.6%	9.9%
6.2	0.028	56973.4	0.028	1.8%	4.1%
4.3	0.045	60976.3	0.045	1.9%	6.8%
3.4	0.266	72880.6	0.269	0.9%	14.5%

Table 2 shows:

Track 1: The skin factor per layer calculated using equation 4

Track 2: Productivity index determined from simulated model output

Track 3: Drainage radius in ft per layer part of linear regression output for all layers

Track 4: Productivity index from Darcy equation utilizing solution of equation 2

Track 5: is the error percent in productivity index between the measured in PLT (track 2) and the calculated (track 4)

Track 6: is the error percent between the calculated skin factor per later (track1) and the actual skin factor (track 7 in table 1)

Case#2

A simulated PLT in a well produced from multilayer system with pressure baffles where cross flow encountered during PLT survey, the result of PLT is illustrated in figure 2

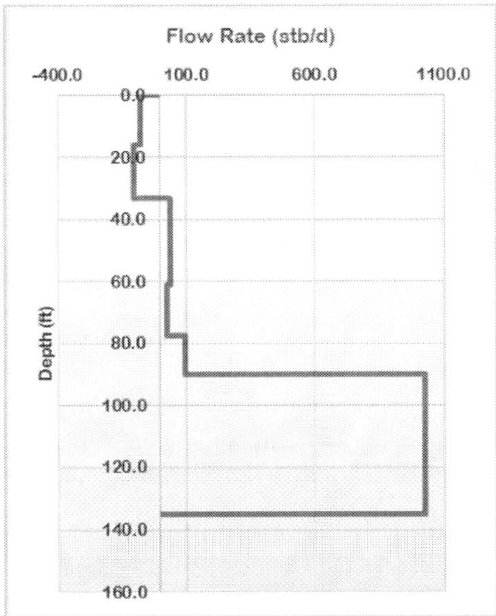

Figure 2. PLT of well producing from 6 layers

IOP Conf. Series: Materials Science and Engineering **579** (2019) 012040 doi:10.1088/1757-899X/579/1/012040

The raw data for further calculation are shown in table 3

Table 3. PLT Analysis of Case #2

From PLT results		From IPTT	PVT	
h (ft)	qi(stb/d)	ki (mD)	mu (cp)	Bo
16.0	-76.5	100.0	1.90	1.38
17.0	-105.1	380.0	1.90	1.38
28.0	36.1	160.0	1.90	1.38
16.5	24.4	110.0	1.90	1.38
12.5	94.2	210.0	1.90	1.38
45.0	1026.9	330.0	1.90	1.38

With this case, both drawdown per layer and skin is unknown, using equation 3 and 10 to determine both as given in table 4

Table 4. Summary of calculation and comparison

dp (psi)	si	q/dp	Re	darcy PI	E% in PI	E% in Si
-297	-2	0.258	56737455	0.268	3.8%	0.0%
-97	-3	1.084	76488135	1.129	4.2%	0.0%
53	-1	0.682	61458301	0.702	3.0%	0.0%
123	6	0.198	56973370	0.203	2.3%	0.0%
303	4	0.311	60976268	0.319	2.6%	0.0%
563	3	1.824	72880580	1.874	2.7%	0.0%

Track 1: the estimated pressure drop (from equation 10 with optimized linear regression to equation 4)

Track 2: model skin factor calculated using equation 4

Track 3: productivity index using the calculated pressure drop with the determined flow rate per zone from the simulated PLT.

Track 4: Drainage radius in ft per layer part of linear regression output for all layers

Track 5: Productivity index from Darcy equation utilizing solution of equation 2

Track 6: is the error percent in productivity index between the measured in PLT (track 2) and the calculated (track 4)

Track 7: is the error percent between the calculated skin factor per later (track1) and the actual skin factor (track 7 in table 1)

4-Conclusion

This research establishes a new method of determining reservoir pressure and skin profile which will have an impact on well managements in terms of:

1- Work over cost
2- Reservoir depletion
3- Productivity monitoring
4- Unnecessary technology utilization
5- Work over optimization
6- Well troubleshooting
7- Work over cost
8- Reservoir depletion
9- Productivity monitoring
10- Unnecessary technology utilization
11- Work over optimization
12- Well troubleshooting

With this value, the study require development for:

1- multiphase flow,
2- gas wells,
3- Deviated and horizontal wells.
4- Fractured reservoirs

Once the research proves its advantage, to be converted into program linked with two main things:

1- PLT interpretation
2- Work over optimization in terms of acid job, plugs necessity, coning forecast, re-perforation, and tabular management

References

[1] Ehlig-Economides C and Joseph J 1987 *SPE Form. Eval.* SPE-14167-PA **2** 261–283.

[2] Aly A Chen H and Lee W 1995 *SPE eastern regional meeting, Charleston* (West Virginia) p 1–12 SPE 29176

[3] Sullivan M 2007 *JPT* 80-87

[4] Yan Pan, Michael Sullivan, David Belanger 2010 SPE 132596

[5] Agarwal R 2010 *SPE Annual Technical Conference and Exhibition* (Florence, Italy) SPE 135804

IOP Conf. Series: Materials Science and Engineering **579** (2019) 012041 doi:10.1088/1757-899X/579/1/012041

Behavior of Swelling Soil Treated by Grid Geocell Columns under a Tank Footing

Raid R. Al-Omari[1] , Mohammed Y. Fattah[2#] , Haifaa A. Ali[3]
[1]Professor, Civil Engineering Department, College of Engineering, Al-Nahrain University, Baghdad, Iraq.
[2]Professor, Building and Construction Engineering Department, University of Technology, Baghdad, Iraq.
[3]Lecturer, Civil Engineering Department, College of Engineering, University of Baghdad, Baghdad, Iraq.

[#]Corresponding author's e-mail address: myf_1968@yahoo.com.

Abstract. Fuel tanks usually induce large settlements due to heavy loading which requires treatment for the underlying soils. In this paper, a method of treating the swelling of expansive soil is presented. The method is simply based on the embedment of a geogrid (or a geomesh) in the soil. The geogrid is extended continuously inside the volume of soil where the swell is required to be controlled and orientated in the direction of swell. Soils with different swelling potentials were employed; bentonite-based Na and kaolinite mixed with bentonite. To investigate the swell, swell- partial shrinkage and other phenomena for untreated soils in comparison with their counterparts of treated soils, laboratory model tests were carried out in 800 x 200 mm box container and 600 mm in height. Swelling and cyclic wetting along with drying tests were carried out on each sample with pore water pressure measurement. Swelling test in the model reveals promising results of the proposed treatment technique. The improvement factors, range between 30% to 60 % depending on the geogrid stiffness, soil plasticity, fill martial (sand or clay) and dry density of sand fill. The geocell has a significant effect on the experimental results. The reduction in swell increases with increasing the geogrid stiffness and when using sand as a fill material, the reduction is apparently due to a strong interference bond which restricts the relative movement between the clay and the grid. The use of geocell filled with the same expansive soil causes a decrease by about 19% and 42 % in the final swelling of soils, whereas filling the cell with sand causes about 35 % and 64 % reduction for the soils, respectively. The treatment method shows the activity of geocell fill material in reducing the matric suction (pore water pressure). When sand is used, a new channel is created for the dissipation of pore water pressure.

1. Introduction

Expansive soils are characterized as problematic as they exhibit swelling with absorption of water and shrinking with adsorption. Such volume changes caused by swelling and shrinkage movements often distress the infrastructure that is not designed to withstand these movements. In addition to swelling, shrinkage-related volume change is critical in influencing the overall volume change properties of the soil. Field conditions that promote shrinkage or shrinkage-induced crack formation include environmental changes, construction process, and surcharge loading. Environmental changes include freezing (the growth of ice lenses), differential swelling (coupled with the weakening of interparticle forces during rapid wetting), and drying (shrinkage of clay mass) [1].

Many factors govern the expansion behavior of soil. The primary factors are the availability of moisture and the amount and type of the clay size particles in the soil. Factors affecting the expansion behavior include the type of soil (natural or fill), the condition of the soil in terms of dry density and moisture content, the magnitude of the surcharge pressure, the amount of non-expansive material (gravel-or cobble-size particles), and the amount of aging [2] [3]. In general, expansive potential increases as the dry density increases and the moisture content decreases. In addition, the expansion increases as the surcharge pressure decreases.

Using the enlarged oedometer, the improvement factor, IF, is defined as the difference between the final swell of unreinforced sample, W_u, and the final swell of the corresponding reinforced sample, W_r, divided by the final unreinforced swell, thus [4]:

$$\mathbf{IF} = \frac{\mathbf{W_u - W_r}}{\mathbf{W_u}} \tag{1}$$

The value of W_r is corrected for the unreinforced length as the soil sample height is slightly larger than the geocell height.

The swell potential and swell pressure are dependent on several factors namely the type and amount of clay minerals present, the initial dry density (void ratio) and water content of the soil specimen, the nature of pore fluid, the type of exchangeable cations, the overburden pressure, and the wetting and drying effects [2] [3] [5]. The swell potential and swell pressure are known to increase with increasing of clay content and dry density and decrease with increasing of initial water content, overburden pressure, pore salt concentration, and exchangeable cation valence [2] [6]. However, clay soils in arid and semi-arid regions are subjected to cycles of wetting and drying in the field due to climatic variations.

Rao et. al., [7] proposed an expansive soil rating system, termed as the "Expansive Soil Index" (ESI). The model was developed as a function of using the soil properties most correlated with shrink–swell potential, swell index, liquid limit, and cation exchange capacity (CEC). The model gave expansive soil potential ratings (ESI) for each soil series.

Zumrawi [8] studied the swelling potential of compacted expansive soil and its relation with the soil state (water content, dry density, void ratio and the surcharge pressure) and the soil type (plasticity index and clay content). The study revealed a linear relationship of the swelling potential and the initial factor represented by soil state and the soil type. The coefficients of the linear relationship (constant and slope) were found to depend on the clay content, plasticity index and surcharge pressure.

In this paper, a method of treatment of swelling of expansive soil is presented. The method is simply based on the embedment of a geogrid in the soil. The geogrid is extended continuously inside the volume of soil where the swell is required to be controlled and orientated in the direction of swell.

Different research parameters are studied like geogrid stiffness, type of geocell fill and swelling potential of clay. Geocell grid columns of diameter 50 mm are inserted through the swelling soils and encased with geogrids of different stiffnesses.

2. Experimental Work, Materials and Procedures

The experimental work that has been conducted to validate the proposed method for treatment of swelling soil can be performed using a large scale model represented by model tests carried out in a test tank manufactured of a steel frame; the container is sufficiently rigid and exhibits no lateral deformation during the preparation of the bed of soil and during the tests. This model is provided with three threaded holes along depth of tank to measure pore water pressure using pore water pressure transducers. Three dial gages were used to measure the swell occurring in three points; one point is lying on the footing and the other two gages measure the movement of two points lying 200 mm away from footing.

Soils with different swelling potentials were employed in this research. Bentonite-based Na and Bentonite-based Ca samples were brought from Karkuk city in Iraq, while the non-expansive soil was Kaolinite. Another type of soil sample was prepared in the laboratory by mixing 30% of Kaolinite with 70% Bentonite-based Na. Standard tests were performed to determine the physical and chemical properties of each soil.

There are two main types of bentonite the Na-bentonite and the Ca-bentonite, the first one has the greatest ability to swell [9].

Details of the properties and the designation are given in Tables 1 and 2. The grain size distribution of the soil used is shown in Figure 1. According to the Unified Soil Classification System; all the soils are classified as (CH) fat clay.

Table 1. Physical properties of the soil used in the experimental work.

Samples Designation	Test Method	Sample 1 (S1)	Sample 2 (S3)	Sample 3 (S2)	Sample 4 (S4)
Description		Bentonite-based Ca	Bentonite-based Na	30% Kaolinite 70% Bentonite-based Na	Kaolinite
L.L%	ASTM D4318-00	125	155	130	45
P.L %	ASTM D4318-00	46	55	48	26
P.I %		79	100	82	19
Max. Dry density (gm/cm^3) by modified compaction test	ASTM D1557-99T	1.46	1.44	1.58	1.83
Max. Dry density (gm/cm^3) by standard compaction test	ASTM D698-99T	1.37	1.35	1.5	1.77
Max. Dry density (gm/cm^3) by static compaction test		1.268	1.275	1.35	1.72
W_{opt}.%, optimum moisture content @ Max. dry density by modified compaction test		23	25	20	12.5
W_{opt}.% @ Max. dry density by standard compaction test		25	28	22	15
W_{opt}.% @ Max. dry density by static compaction test		20	20	20	20
Specific gravity, Gs	BS 1377:1990 [16]	2.75	2.78	2.70	2.63
% Finer than 0.005 mm		95	97	88	65
*Activity %		0.83	1.031	0.91	0.76

$$\text{Activity \%} = \frac{PI}{\text{\% Finer than } 0.005 \text{ mm}}$$

Table 2. Chemical properties of the soils used in the experimental work.

Type of Sample	SO_3 (mg/liter)	Gypsum content (mg/liter)	Total dissolved salts TDS (mg/liter)	pH value	Chloride Cl^{-1} mg/liter
Bentonite-based Ca (S_1)	2.05	4.4075	1650	8.2	350
Bentonite-based Na (S3)	3.1	6.665	1775	8.4	625
30% Kaolinite 70% Bentonite base-Na (S2)	1.675	5.2	1731	8.3	514
Kaolinite (S4)	0.8	1.72	1628	8.2	250
Test method	BS 1377		Earth Manual, E8	BS1377 Test11(A)	BS812:Part4 1976

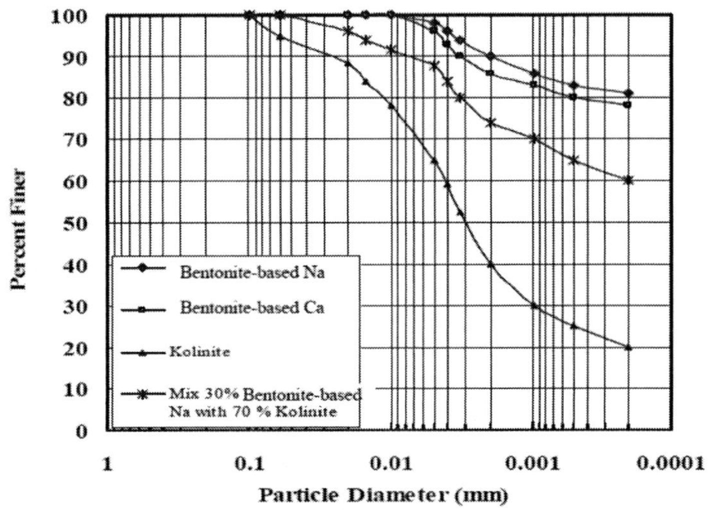

Figure 1. Grain size distribution for the clay soil used.

Chemical analysis was conducted to determine the chemical composition. The qualitative X-Ray diffraction was conducted on samples to get an idea about the mineralogical composition of the swelling soil samples. The mineralogical compositions of the soil, using the X-Ray diffraction method, are presented in Table 3.

Table 3. X-Ray diffraction results of testing soils.

Sample type	Clay Minerals
Sample 1 (S1) Bentonite-based Ca	Calcium Oxide (CaO), Silicon Oxide (SiO_2)
Sample 3 (S2) 30% Kaolinite 70% Bentonite-based Na	Sodium Aluminum Silicate Hydrate, Silicon Oxide
Sample 2 (S3) Bentonite-based Na	Sodium Aluminum Silicate Hydroxide Hydrate $NaO_3AL_2(Si_1AL)_{40}10(OH)_2.2H_2O$
Sample 4 (S4) Kaolinite	Silicon Oxide, SiO_2

For the soils used (S1, S2, S3), the swelling potential corresponding to plasticity indices 79%, 82%

and 100% is found to be very high according to [3] as indicated in Table 4.

Table 4. Typical soil properties versus expansion potential (Day, 1994).

Expansion Potential	Very low	Low	Medium	High	Very high
Clay content (< 2 μm)	0-10%	10-15%	15-25%	25-35%	35-100%
Plasticity index	0-10	10-15	15-25	25-35	> 35

Sand fill

Sand is used in the geocell grid columns. The sand is poorly graded of uniform size. The physical properties of the sand were determined according to ASTM (D422-01) [10]standard as presented in Table 5. The placing density of sand in the model box as a filter or in the hole of the geocell was achieved by raining technique. To obtain the desired density, different values of placing density were given by different heights of sand drop. The height of drop of 15 cm was chosen to maintain a placing unit weight of 15.2 kN/m^3 which is corresponding to a value of the void ratio and relative density of 0.70 and 26 %, respectively.

Table 5. Physical and chemical properties of the used sand fill.

Index property	Test Method	Index value
Max. dry unit weight (kN/m^3)		17.4
Min. dry unit weight (kN/m^3)		14.4
Dry unit weight used (kN/m^3) at D_r =26 %		15.2
e_{min}		0.59
e_{max}		0.82
e_{used} at dry unit weight used		0.70
Specific gravity (Gs)	ASTM D 854-2005	2.63
Coefficient of uniformity (Cu)		3.15
Coefficient of curvature (Cc)		1.07
T.S.S (%)	Earth Manual, E8	15.7
SO_3 (%)	BS 1377	3.121
Organic matter (%)	BS 1377 -Test 8	1.46
Relative density (D_r %)	ASTM D2049-64T	26
Angle of internal friction (□°) at D_r = 26 %	Direct-shear test	28°

Geogrid reinforcement

The geogrids used are polymer meshes commercially known as Netlon CE 121(G1) and Tensar SS2 geogrid (G3). Figure 2 shows the geogrid reinforcement used.

Figure 2. Geogrid tubes.

Testing Program

To investigate the swell, swell and partial shrinkage, loading behavior with pore water pressure measurements and other phenomena for untreated soils in comparison with their counterparts of treated soils, model tests were conducted for this purpose. Each soil bed is compacted at the dry side of maximum dry density and water content using the static compaction.

The optimum moisture content and the dry unit weight of soil are very important for construction specification of soil improvement by compaction. Standard Procter compaction, modified compaction and static compaction are employed using the CBR mould for each swelling soil sample in the experimental work to investigate the effect of compaction method on the values of dry density, water content and the expected swelling results.

In the present work, the compaction characteristics of the soils are obtained using "standard" compaction test according to (ASTM D698-66T), and "modified" compaction test according to (ASTM D698) as well as static compaction.

The results are demonstrated in Table 1 using the Procter compaction mould and CBR mould which indicate that the maximum dry density of the soil reached uniform and repeatable when compacted by static method.

The bed of soil was prepared using static compaction at a water content of 20% and a stress of 300 kg to develop the corresponding maximum swell and maximum dry density.

Secondly, the variation of shear strength is measured using a Load Ring Penetrometer device at different liquidity indices.

According to Figure 3, the bed of soil was prepared at water content of 20 % corresponding to liquidity index in the range of 20 % to maintain an undrained shear strength of 17.5 kPa for sample S_2 and 21 kPa for sample S_3.

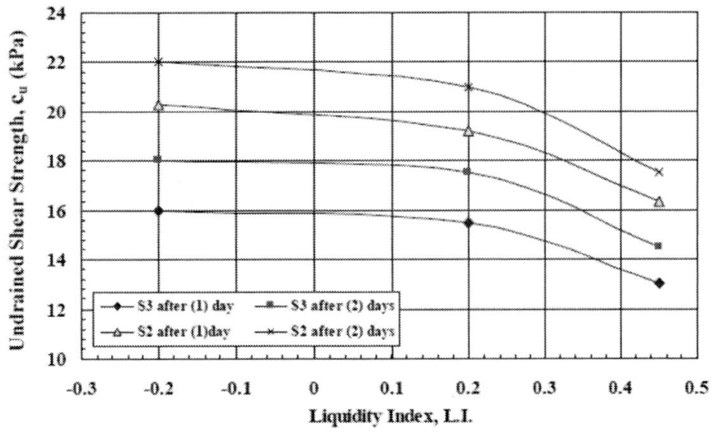

Figure 3. Variation of undrained shear strength with liquidity index at different times.

Placement of the bed of soil

Each 25 kg of soil sample was mixed with enough quantity of water to get the desired consistency. The wet soil was kept inside tightened polythen bags for a period of two days. This period was sufficient to get uniform moisture content. After that, the soil was placed in three layers inside a steel container of (800×200) mm and 600 mm in height. Before that, the sides of the container were coated with silicone grease to minimize the friction effect. After the placement of each layer, it was pressed gently with a wooden tamper of size 75×75 mm in order to remove entrapped air. After completing each layer, the top surface was scraped, levelled and compressed by a steel sheet 790×190 mm loaded from axial loading system to reach the bed of soil and left for a period of two days to regain part of its strength.

IOP Conf. Series: Materials Science and Engineering **579** (2019) 012041 doi:10.1088/1757-899X/579/1/012041

Construction and installation of geogrid encased sand columns
All the sand columns have a diameter of 50 mm, length to diameter ratio (L/D = 6) with spacing of 2 times the footing width (B).

Then, a geogrid cell (Figure 2) was inserted into the column hole using a hallow steel pipe with internal diameter of 50 mm. The sand was poured into the hole in layers and compacted gently by a tamping rod. After pouring all the specific amount of sand, the full depth of the hole was filled with sand at a dry unit weight of 15 kN/m^3.

Testing Procedure
The general view of the testing ring employed in the present research is illustrated in Figure 4. The ring has the capability to sustain the applied load to simulate the stresses occurring under plate or pile load test. An experimental formulation based on an approximate tenth 10th scale for general square footing was adopted in this work. A square footing model with dimensions of 100×100 mm was used. The length of grid geocell columns was 300 mm embedded inside the swelling soil.

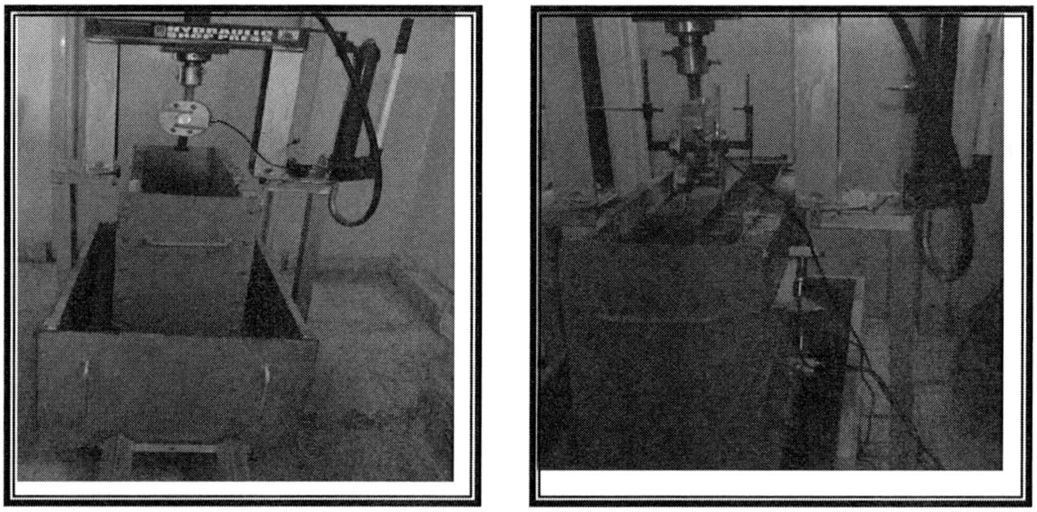

Figure 4. General view of the large scale model.

The pressure applied to the model was measured using a load cell connected to the readout pressure system; the information was recorded at selected intervals of time in a data file in the computer along with the reading signals of pore water pressure transducers. The entire testing process was run with the aid of computer software.

The displacement of the soil model was measured using three dial gages; one is placed on the footing and the others are placed at a distance of 200 mm away from the footing on the geocells located right and left the footing.

The other side of container model has three threaded holes; each is 10 mm diameter in order to accommodate the pore water pressure transducers and the de-airing equipment of the pore water pressure system as shown in Figure 5.

To simplify the observations of the soil samples, geogrid types and grid geocell columns that will be dealt with in the experimental work, symbols are used to characterize each model. Table 6 shows each model and its symbol.

Figure 5. Pore water pressure measurement system.

Table 6. Model types and their symbols.

Sample type	Sample symbol
Bentonite-based Ca sample	S1
Mixture 70% Bentonite-based Na sample with 30% Kaolinite	S2
Bentonite-based Na sample	S3
Geogrid No. 1 (stiffness = 120 kN/m)	G1
Geogrid No. 2 (stiffness = 40 kN/m)	G2
Geogrid No. 3 (stiffness = 240 kN/m)	G3
Geocell filled with the same soil	F1
Geocell filled with sand	F2

3. Results and Discussion

Swell-time relationships

A relatively large scale model was used to estimate the swell-time relationship for different types of soils of different plasticity indices before and after treatment.

Table 7 graphically describes the locations of the measurement points in the model. The measured swell and pore water pressure are plotted against time for samples with and without treatment (using geocell filled with the same soil or sand). The results are demonstrated in Figures 6 to 11.

It can be seen that the expansive soils takes long period of time to complete their expansion, therefore the testing time of 27 days was chosen for all test groups.

Figures 6 to 8 present the results for S3 soil sample model with and without treatment with pore water pressure records while Figures 9 to 11 for S2 sample model with and without treatment. From the results, the effective method for treatment can be summarized by presenting the amount of decrease in the measured swell and the activity of geocell grid columns in restricting the soil swelling due to increase of the interlocking between the geocell with fill material.

It can be noticed that the changes in pore water pressure during swelling process are small; the pore water pressures are negative especially at the top and bottom of the model indicating that matric suction is related to swelling.

The largest value for variations in matric suction occurs near the ground surface of the soil samples. The matric suction measurements indicate low values that can be attributed to the conditions of the soil sample tested; the soil sample is compacted at optimum moisture content during static compaction. In addition, there is no way to dissipate the air pressure along the soil sample depth during soaking of the soil sample from top and bottom sides in the model. This is clear especially in data measured in channel 2.

Table 7. Geocell locations and descriptions of measuring points in the model.

Point symbol	Description of measuring points	Measuring points location details
A	Lies below middle of footing.	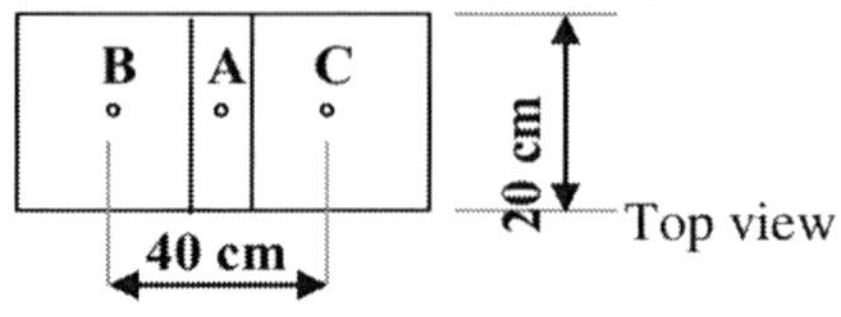
B and C	Lies at 20 cm distance on both sides of point A.	
Channel 1	High point in the box.	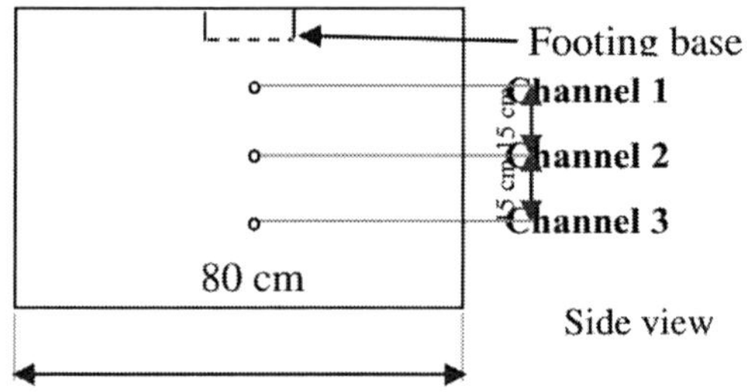
Channel 2	Medium point in the box.	
Channel 3	Low point in the box.	

The effect of treatment is clear from the results which reflect the effectiveness of geocells in restricting the soil movement. This finding is in agreement with the results of [11] who indicated that the displacement field in foundation is changed significantly as a result of reinforcement.

The mineralogical model for the variations of the pore water pressure with time dependent swell behavior may be more appropriate because the ingress of water between the plates in clay crystal is not controlled by effective stress, but by the physicochemical reactions taking place between the plates. These physicochemical reactions are controlled by the osmotic suction caused by the presence of exchangeable cations between the plates, and are therefore only indirectly influenced by the matric suction prevailing in the pores between the clay crystals, [12]. Therefore, the pore water pressure (matric suction) at the top and bottom has considerable difference at the beginning and end of all tests. The method of treatment increased the dissipation of pore water pressure by increasing the void channels and permeability of soil sample especially with the use of geocell cylinders filled with sand.

The same results were obtained for S2 soil sample model. Figures 9 to 11 display the effectiveness of the method of treatment as indicated in the testing results. The figures also demonstrate the effect of soil plasticity on the magnitude of improvement. When geocells are filled with sand which has low plasticity, the treatment method is found to be more effective than the case of geocells filled with clay.

From these results, the method of geocell embedment technique is affirmed to be successful as expressed by the improvement factors listed in Table 8. The improvement factors vary from 19 to 64 based on the plasticity index and type of fill material.

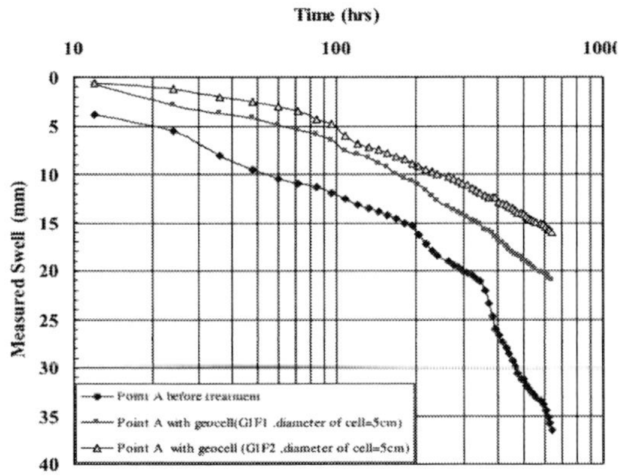

Figure 6. Swell measured at point A before and after treatment of S3 model beneath square footing.

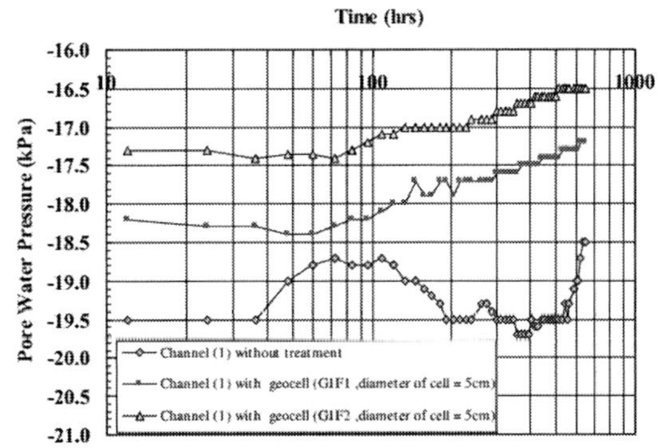

Figure 7. Pore water pressure at channel (1) before and after treatment of S3 model beneath square footing.

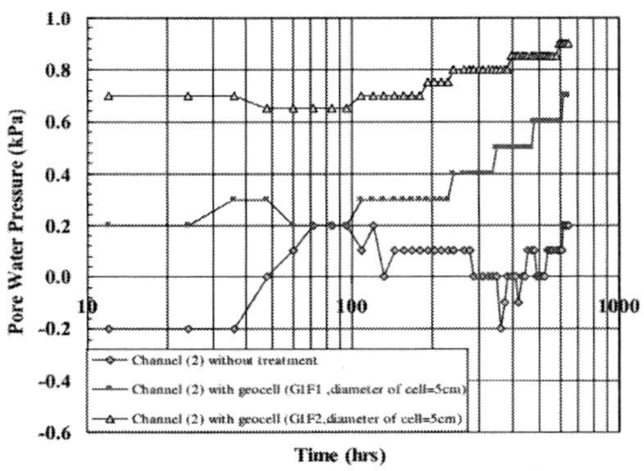

Figure 8. Pore water pressure at channel (2) before and after treatment of S3 model beneath square footing.

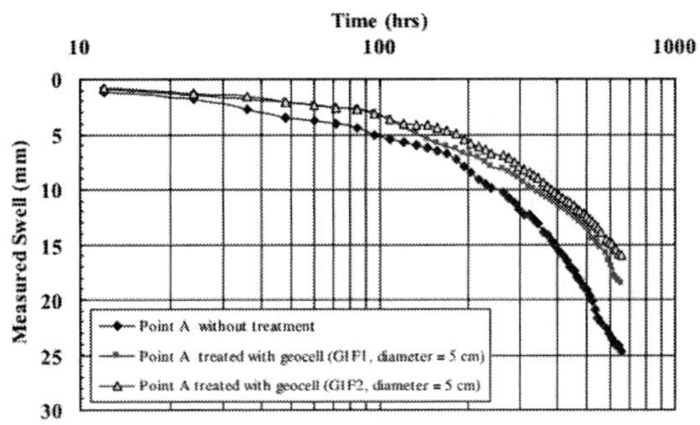

Figure 9. Swell measured at point A before and after treatment of S2 model beneath square footing.

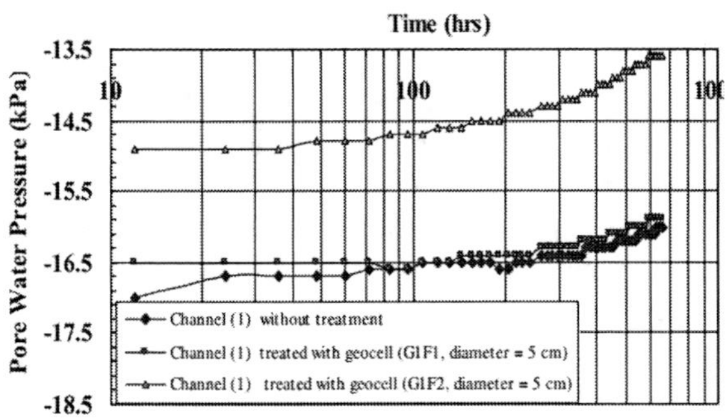

Figure 10. Pore water pressure at channel (1) before and after treatment of S2 model beneath square footing.

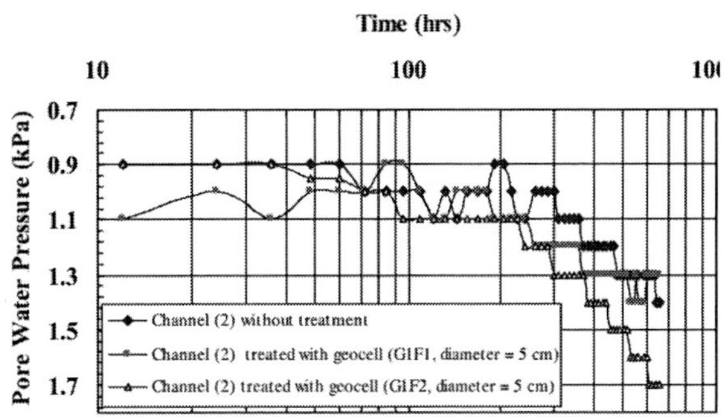

Figure 11. Pore water pressure at channel (2) before and after treatment of S2 model beneath square footing.

Table 8. Improvement factors for different samples treated by using geocells of different fill materials, stiffness of geocell is 120 kN/m.

Sample symbol model	Dry density by static compaction test (gm/cm^3)	Measuring point	Type of fill material	IF (%)
S2	1.35	B	the same soil	19
S2	1.35	B	sand	35.3
S3	1.275	B	The same soil	42
S3	1.275	B	sand	64

Effect of fill material

Figures 12 and 13 are drawn to study the effect of fill material on the effectiveness of the proposed method of treatment. Geocell material creates effective channels which support the basic work of the treatment method by increasing the permeability and reduction in soil matric suction.

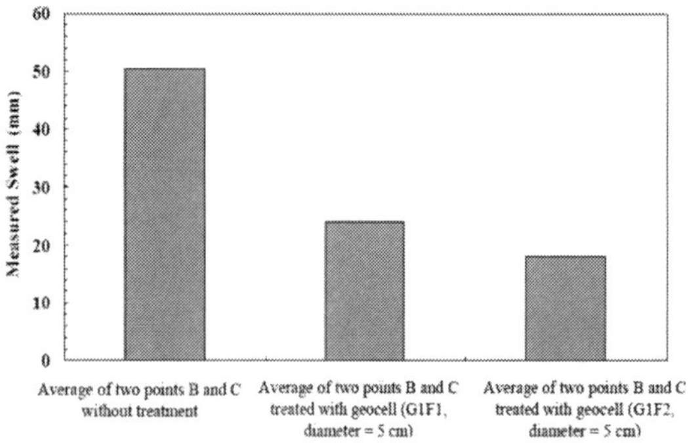

Figure 12. Effect of fill material on geogrid treatment for S3 soil sample model beneath square footing.

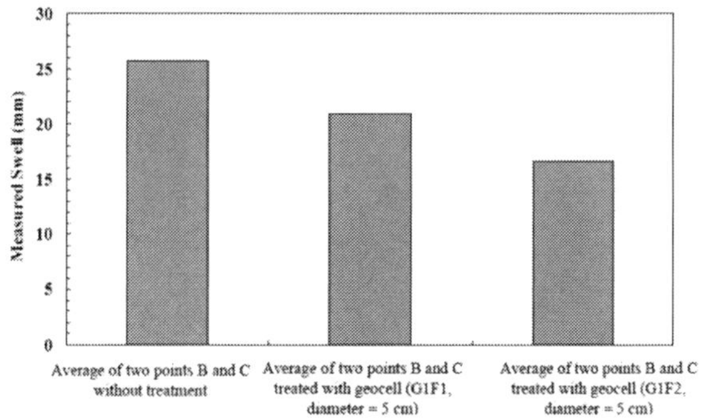

Figure 13. Effect of fill material on geocell treatment for S2 soil sample model beneath square footing.

Table 8 presents the results of the effect of fill material on the recorded results. From the results, the use of geocell filled with the same expansive soil causes a decrease of about 19 % and 42 % in the final swell for S2 and S3 soils, respectively whereas filling the cell with sand causes a decrease of about 35.3 % and 64 % in the final swell reduction for S2 and S3 soils, respectively.

Sand fill material accelerates the flow of water through the soil sample model; therefore the treatment with geocell filled with sand is more effective. In addition, the area of clay in contact with footing is reduced. The sand in the pockets forms a better composite material and the geocell behaves as a stiffer bed that redistributes stress over a wider area giving an increase in the bearing capacity and a reduction in the settlement of the footing as outlined by [13].

For the treated soils, the geogrid reinforced column with fill materials works like a pile structure and sustain the applied pressure for the whole soil. Therefore, the mechanical behavior of the treated soil is different from any chemical additive behavior. The aim of using the mechanical treatment by geogrid reinforced column is to restrict the soil swell in the direction of movement during wetted or dry season, the same principle is controlling when the swelling soil is under tension stress due to swelling pressure or under compression stress by consolidation pressure [14].

Effect of geocell stiffness

To demonstrate the effect of geocell stiffness on the treatment method, G1 and G3 types of geogrid are chosen to show the effect of the variation in stiffness of the geocell on the measured swell.

The relationships of the final swell and pore water pressure with time are plotted in Figures 14 and 15. The results suggest that increasing the stiffness lead to an increase in the restraint and decrease in matric suction of swelling soil.

It is concluded that, the geocell stiffness has a little effect on the pore water pressure, which can be due to no change in the conditions of the model situations as a result of this fact; there is no change in the matric suction. Also the degree of saturation in the model is not homogenous in the vertical direction during the test period of 27 days; the effect of stiffness may be more pronounced if the soil model is allowed to swell until full saturation of the whole soil.

Figures 14 to 17 show a comparison of swelling measured at two points with geocell stiffness variation. improvement factors obtained in the present work are given in Table 9. The amount of improvement factor increases with increasing the grid stiffness. The geocells used in experiments included in this table are 5 cm in diameter filled with sand.

The effect of the elastic modulus of the geogrid on the swelling, pore water pressure, and degree of saturation decreases slightly when the elastic modulus increases. Subsequently, the effect diminishes gradually with the depth. The stiffness of the geogrid leads to the restriction of soil movement and the concentration of stresses within the geocell, hence decreasing the swelling and pore water pressure [15].

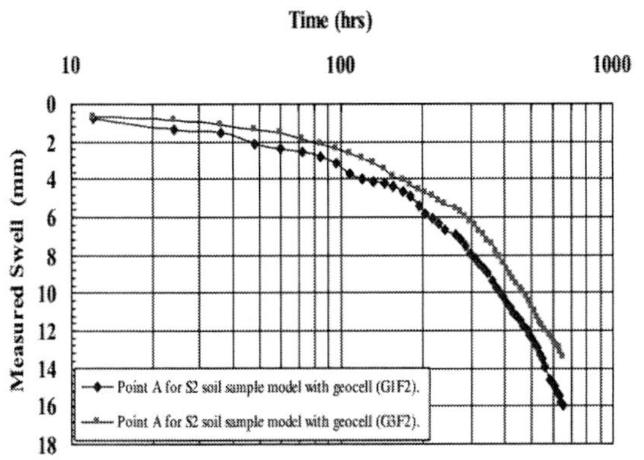

Figure 14. Effect of geocell stiffness on the swelling for S2 soil sample model measured at point A beneath square footing.

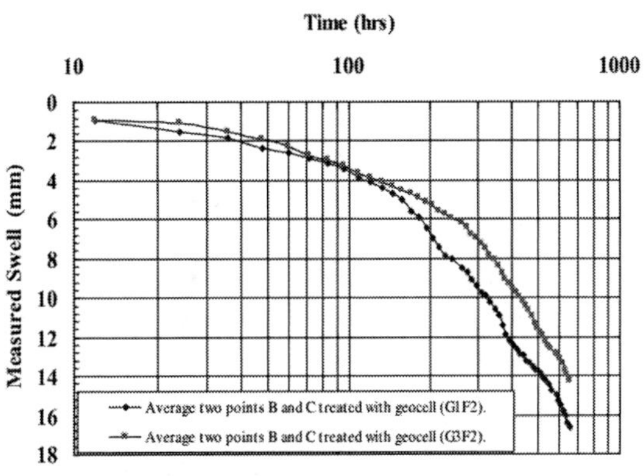

Figure 15. Effect of geocell stiffness on the swelling for S2 soil sample model measured as an average of two points B and C beneath square footing.

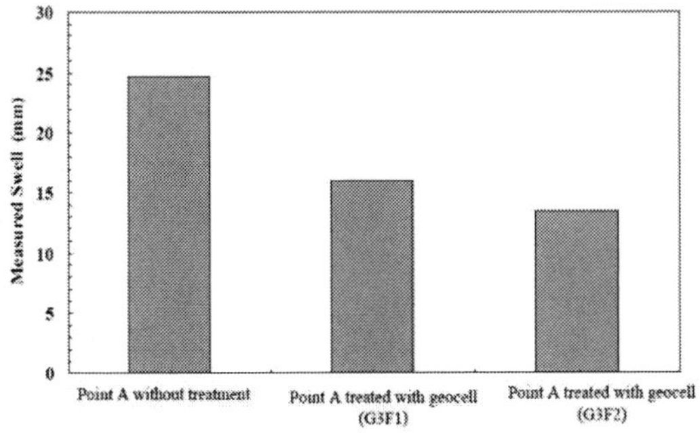

Figure 16. Effect of geocell of stiffness on the swelling for S2 soil sample model at point A beneath square footing.

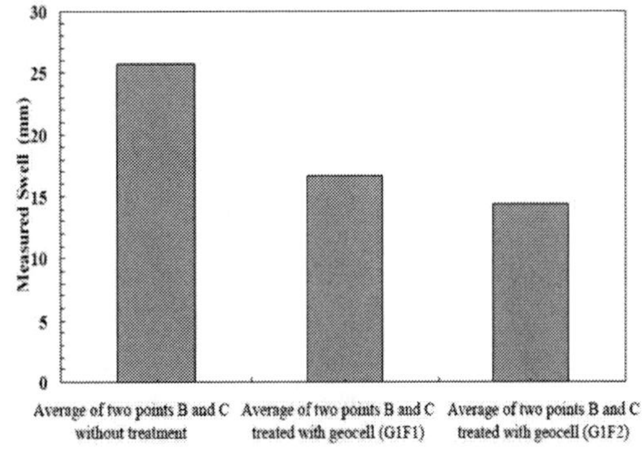

Figure 17. Effect of geocell stiffness on the swelling for S2 soil sample model at average two points B and C beneath square footing.

Table 9. Improvement factors indicating the effect of geocell stiffness on the obtained reduction in swell.

Stiffness of geocell (kN/m)	Measuring points	IF (%)
120	A	35.3
240	A	45.6
120	B	35.3
240	B	44.3

4. Conclusions

The main points concluded from the experimental work can be summarized as follows:-

1. The treatment of swelling soil using grid geocells filled with sand or the same soil is found to be effective. If the geocell is filled with the same expansive soil, it causes a decrease of about 19 % and 42 % in the final swell for S2 and S3 soils, respectively, whereas filling the cell with sand causes a decrease of about 35.3 % and 64 % in the final swell reduction for S2 and S3 soils, respectively.
2. Swelling test in the model reveals promising results of the proposed treatment technique. The improvement factors, IF, range between (30 to 60) % depending on the geogrid stiffness, soil plasticity, fill martial (sand or clay), dry density of sand fill and clay percentage of the fill material.
3. The geocell has a significant effect on the experimental results. The reduction in swell increases with increasing the geogrid stiffness and when using sand as a fill material, the reduction is apparently due to a strong interference bond which restricts the relative movement between the clay and the grid. The sand fill material accelerates the flow of water through the soil sample model; therefore, the treatment with geocell filled with sand is more effective.
4. The treatment method shows the activity of geocell fill material in reducing the matric suction (pore water pressure). When sand is used, a new channel is created for dissipation of pore water pressure. The pore water pressure is slightly changed because the swelling soil needs a long period to complete the swelling and reach a saturation state. This change is sensed near the top surface and near the bottom and both location points are near the flooding water. The geocell fill material increases this change because the sand materials accelerate the water migration.

References

[1] Katha, B. R., 2002 *Shrinkage strain characterization of expansive soils using digital imaging technology, M.Sc. Thesis* (The University of Texas at Arlington, Texas)
[2] Chen, F.H., 1988 Foundations on expansive soils *Developments in Geotechnical Engineering*, Vol. 12 (Elsevier Publications, the Netherlands)
[3] Day, R.W., 1994 Swell-shrink behavior of compacted clay *Journal of Geotechnical Engineering ASCE* **120** 618–623
[4] Al-Omari, R. R., and Hamodi, F. J., 1991 Swelling resistant geogrid-a new approach for the treatment of expansive soils *Geotextiles and Geomembranes* **10** 295-317
[5] Al-Homoud, A. S., Husein M. B., and Al-Bashabshah, M. A., 1995 Cyclic swelling behavior of clays *Journal of Geotechnical Engineering ASCE* **121** 562–565
[6] Mitchell, J.K., 1993 *Fundamentals of soil behavio* (John Wiley and Sons, New York)
[7] Rao, S.M., Thyagaraj, T., and Thomas, H.R., 2006 Swelling of compacted clay under osmotic gradients *Geotechnique* **56** 707-713
[8] Zumrawi, M. E. 2013 Swelling potential of compacted expansive soils *International Journal of Engineering Research and Technology* (ISERT) **2** 1-6
[9] Athanassopoulos, C., 2011 Natural sodium versus activated calcium bentonite in geosynthetic clay liners (GCLs) *Solid Waste and Recycling Conference*, available at URL:http://www.nyfederation.org/PDF2011/04 , Athanassopoulos.

[10] ASTM D422-2001 *Standard test method for particle size-analysis of soils* (American Society for Testing and Materials, West Conshohocken, Pennsylvania, USA)
[11] Wang Wei's, 2000 Study on model tests and mechanism of soft soil foundation reinforced by geotextile *Chinese Journal of Geotechnical Engineering* **22** 750-753
[12] Saxena, K.R., 1994 *Geotechnical engineering-emerging trends in design and practice* (A.A Balkema, Rotterdam) pp. 33-36
[13] Moghaddas, S.N., and Dawson, A.R., 2010 Comparison of bearing capacity of a strip footing on sand with geocell and with planar forms of geotextile reinforcemen *Geotextiles and Geomembranes* **28** 72–84
[14] Fattah, M. Y., Al-Omari, R. R., Ali, H. A., 2015 Numerical Simulation of the Treatment of Soil Swelling Using Grid Geocell Columns *Slovak Journal of Civil Engineering* **23** 9–18, DOI: 10.1515/sjce-2015-0007.
[15] Al-Omari, R. R., Fattah, M. Y., Ali, H. A., 2016 Treatment of Soil Swelling Using Geogrid Reinforced Columns *Italian Journal of Geosciences* **135** 83-94, doi: 10.3301/IJG.2014.54.
[16] B.S. 1377 1990 *Methods of test for soils for civil engineering purposes* (British Standards Institution)

IOP Conf. Series: Materials Science and Engineering **579** (2019) 012042 doi:10.1088/1757-899X/579/1/012042

Improving the Properties of Acrylic by Creating Crack Using Laser Beam

R N Hwayyin[1] and A S Hammood[2]

[1] University of Technology/ Petroleum Technology Department, Iraq-Baghdad.
[2] University of Technology/Factory and Training Center, Iraq-Baghdad.

[1]Corresponding author's e-mail address: 10596@uotechnology.edu.iq

Abstract. The study investigates the improvement of the mechanical properties of the acrylic layer by creating a crack according to polynomial's function by using the CO2- laser and without additives. The cracks were created by the laser at different powers and different speeds. The shape of the cracks was created according to six degree polynomial's function by using CNC-laser machine. The study focused on the effects of creating an edge on both sides of the crack generated from the laser heat helping to change the microstructure on it to resist the failure crack growth. The study adopts a laser beam path according to equation g(x) to show the effect of the shape of the laser beam on improving the properties of the acrylic specimens. The experimental results have shown the effect of laser beam speed and laser beam power on the stress-strain. The experimental results shown increasing in the stress with average rate (12.25%) due to increasing laser beam power at (40, 50 and 60). The experimental results of specimens type (O and A4) shown improving in the stress at average rate (12.87%) and copmaring the experimental results of specimens type (O, A5 and A6) has revealed an incresing in the tensile stress with average rate (14.088%).

1. Introduction

Today's laser applications are widely used in industrial and technological fields such as welding and laser cutting in polymeric materials and other materials. Extensive is use of acrylic material and transparent materials in windows for structures, ships and submarines used in marine exploration and marine life studies as well as geological nature and seabed erosion. Acrylic has also been applied in many tourist buildings with the advancement of technology in the manufacture and improvement of properties of acrylic material.

Konstantin Hofmann et. al. [1] They studied the effect of alloying strategies on the thermal energy input and the metal structure resulting from modified tool surfaces. The cooling behavior of the mixing lines was also of the length of the input unit of different interest; the microstructure was analyzed with respect to the fine hardness as well as the carbide formation. E. Abouda et. al. [2] He studied the mechanical component that operates under high pressure conditions and high temperature conditions and needs enhanced mechanical and cognitive properties using the laser beam. Frank L. Palmieri et.al.[3] The investigation into the laser ablation process was developed to carefully remove the target depth of the resin, approximately (0.1 to 20) μm, of the composite surface of the carbon fiber reinforced epoxy while simultaneously changing the surface chemistry and achieving micro-roughness. Yusuf Z. Akpinar et.al.[4] The study of the effect of the different angles of the laser beam on the strength of the cement cut Y-TZP-resin (SBS) was evaluated. The forty-plate ceramic Y-TZP is randomly selected to be four groups (n = 10).The application ensures the laser pulse of the femtosecond amplifier on the Y-TZP surface at angles (90 °, 75 °, 60 °, 45 °). Md. Saidin Wahab et. al. [5] It made progress in research on the advantage of laser cutting in carbon fiber and reinforced plastics (CFRP) and composite materials used in the space structures. CFRP is a high performance material and is one of the most important materials in the aviation industry. Rajib Kumar Mandal et.al [6] Adopted laser beam welding study (LBW) which has become increasingly used in manufacturing and manufacturing which is due to its distinctive characteristics. LBW offers high energy density around its concentration, making it well suited for welding a certain class of materials that are difficult to weld. M. M. Noor et. al. [7] their study developed a model to predict the roughness of the surface by cutting the laser beam (LBC) into acrylic layers. The using of the Box-Behnken design was depend on the surface response method to predict the effect of laser cutting marks, including the ability, cutting speed and tip distance to surface roughness during operation.

The aim of this study is to improve the mechanical properties of acrylic by creating laser beam crack without additions. The study adopts new investigation field depending on improving the mechanical properties by creating a crack laser beam that effects on micro-stricture of acrylic specimens when determining the concentration of stresses in the test specimens and by using a crack created by laser beam to redistribute the concentration of stress and increase the stress required for failure.

2. Theoretical analysis

Tensile strength is described as a function of the increase in gage length. Where, the relationship between tensile strength versus tensile elongation will be of little value if not compared as ratio of the dimensions of the specimen. Engineering Stress, or nominal stress, (σ), is defined as [4]:

$$\sigma = \frac{F}{A} \qquad (1)$$

Where,

F: is the tensile force

A: is the initial cross-sectional area of the gage section.

Engineering strain, or nominal strain, (e), is defined as

$$e = \frac{DL}{L} \qquad (2)$$

Where,

L: is referring to the initial gage length,

DL: is referring to the change in gage length (Lf- Lo).

Where (Lf) is represented the final length and (Lo) is the initial length.

When force elongation data is translated into stress and geometrical stress, the stress and emotion curve are similar in shape to the longitudinal force curve. The ability to deal with stress versus stress rather than pregnancy versus elongation is that the stress and stress curve is almost independent of the dimensions of the test specimen. Failure is due to fracture results in the separation of the body into two or more parts. This process is accompanied by the formation of new surfaces. For example, with the formation of fracture surfaces, the associated dynamics can not be described in a simple manner. Conclusions are possible and based on experimental results. They show two basic patterns to form fracture surfaces. For a fracture controlled by natural stress, the fracture surface coincides with the cross section to the maximum extent of the main stress, which should be the concentration of tension as shown in Figure 1.a. When the fracture surface is formed by cross sections where the specific shear stress (e.g. , τmax, τoct (octahedral shear stress, etc.) reaches a critical value, this is called broken fracture predominates as shown in Figure 2.b. Depending on the state of stress and physical behavior, both types also occur in different mixed forms.[5].

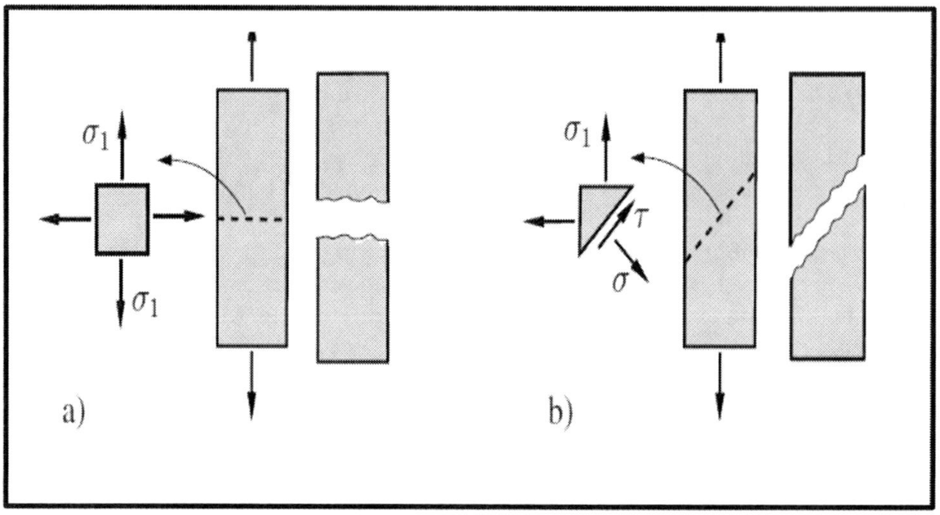

Figure 1. a, b. show the fracture surface

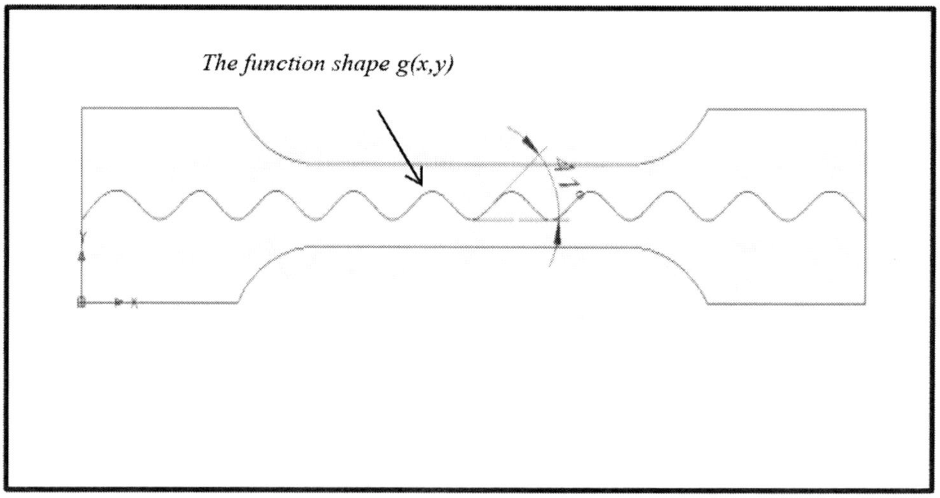

Figure 2. show the fracture surface

Figure 2 shows the crack laser shape which created according the function g(x) that is predicated by curve fitting the draw in AOUTOCAD software. The angle of the crack is created to be less (41o) than the angle of shear stress which is (45o) for redirection of the crack failure expected at that angle. The redirection of crack failure growth by using crack laser is to make the crack growth path to be longer which is requiring greater stress to reach the fracture. Equation (3) represents the laser crack shape:

$$g(x)=17.27265-0.3257*x-0.0113*x^2+0.00092*x^3-2.0E-005*x^4+1.83E-007*x^5-6.073E-010*x^6 \quad (3)$$

The following equation (4) predicates the experimental result's, a relationship of the effect of conditions of laser beam on the stress-strain at laser beam speed (15 mm/sec) and different laser beam powers (40, 50 and 60) watt.

$$F_1(x,y) = y_{a1} + y_{b1} * x \quad (4)$$

Where (y) represents the equations for the curve fitting function of the relationship between constant values in the symmetric limits in equation:

$$ya1= 638.0260304 + 53.85141122 * x - 2.232615722 * x^2 + 0.02113974764 * x^3 \quad (4\text{-}1)$$

$$y_{b1}= -33.62634717 + 1.458127618 * x - 0.01379198272 * x^2 \quad (4\text{-}2)$$

The following equation (5) predicates the experimental results a relationship of the effect of conditions of laser beam on the stress-strain at laser beam speed (30 mm/sec) and different laser beam powers (40,50 and 60) watt.

$$F_2(x,y) = y_{a2} + y_{b2} * x \quad (5)$$

Where (y) represents the equations for the curve fitting function of the relationship between constant values in the symmetric limits in equation:

$$y_{a2} =638.0260304 - 100.2692261 * x + 4.407515589 * x^2 - .04855984697 * x^3 \quad (5\text{-}1)$$

$$y_{b2}= x . \quad (5\text{-}2)$$

The following equation (6) predicates the experimental results a relationship of the effect of conditions of laser beam on the stress-strain at laser beam speed (45 mm/sec) and different laser beam powers (40,50 and 60) watt.

$$F_3(x, y) = y_{a3} + y_{b3} * x \quad (6)$$

Where (y) represents the equations for the curve fitting function of the relationship between constant values in the symmetric limits in equation:

$$y_{a3} = 638.0260304 + 0.405506005 * x - 0.14810265 * x^2 + 0.00222613865 * x^3 \quad (6\text{-}1)$$

$$y_{b3}= x \quad (6\text{-}2)$$

3. Experimental analysis

Table (1). shows the type of specimens according to laser power and its speed.

Spec. symbol	The spec. conditions
O	The original tensile test specimen absence of the effect of laser beam.
g(x,y)	The function relationship of the effect of speed laser beam and different laser beam power on stress-strain
A1	Is refer to speed 15(mm/sec) and laser power 40(watt).
A2	Is refer to speed 15(mm/sec) and laser power 50(watt)
A3	Is refer to speed 15(mm/sec) and laser power 60(watt)
A4	Is refer to speed 30(mm/sec) and laser power 40 (watt)
A5	Is refer to speed 30(mm/sec) and laser power 50(watt)
A6	Is refer to speed 30(mm/sec) and laser power 60(watt)
A7	Is refer to speed 45(mm/sec) and laser power 40(watt)
A8	Is refer to speed 45(mm/sec) and laser power 50(watt)
A9	Is refer to speed 45(mm/sec) and laser power 60(watt)

3.1. The Process of Set-up

The tensile specimens were prepared according to (ASTM D412) standard as shown in Figure 3 by using the drawing program Auto CAD and then using the laser machine to cut the acrylic tensile specimens. The Physical properties of acrylic sheet are shown in table 2. The study uses laser machine as shown in figure 4 to create the crack in the acrylic tensile specimens. The crack in the tensile specimens is created according to the sixth degree. The crack is generated at different laser beam intensities with continuous wave (CW) CO2 at 20 kHz and powers equal to (40, 50 and 60) watt at different speeds (15, 30 and 45) mm/sec. The laser crack shape is drawn by using AUTOCAD software and it is transformed to CNC laser machine which works with dimensions (140 cm *190cm) and maximum power (80 watt) and distance from laser lens to specimen (6.5 mm).

Table (2). shows the physical properties [6].

Quantity	**Value**
Thermal. Expansion	50-90 μ-6/K
Thermal. Conductivity	0.167- 0.25W/m.K
Specific heat	1466 J/kg.K
Glass transition temp	105°C
Density	1170 - 1200 kg/m3
Shrinkage	0.3 - 0.8%
Friction co-efficient	0.54
Refractive Index	1.492

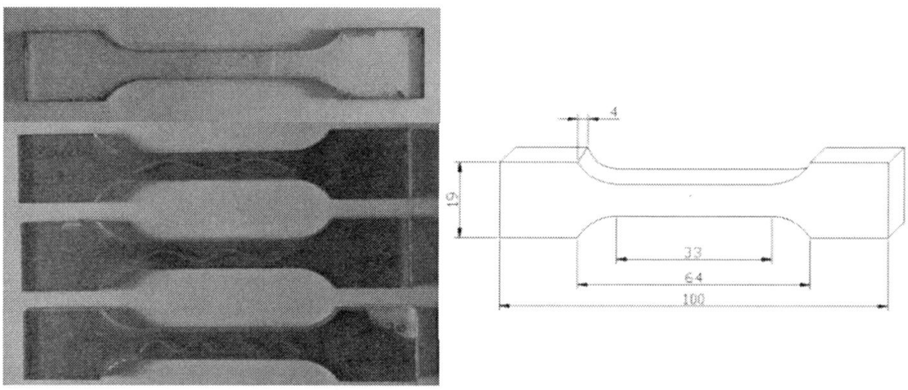

Figure 3. the tensile test specimen type (ASTM D412).

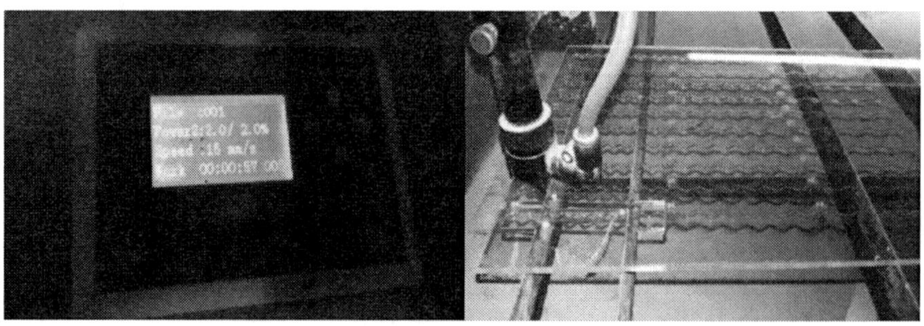

A-Control board of laser machine B-Creating cracks by laser beam

Figure 4. shows the CNC laser machine to create different types of cracks

The microscopic images are obtained by using microscopic type (Bel-X1600) as shown in figure 5. The microscopic images of cracks created by laser show the microscopic structure of the laser cracks at different locations of it as shown in figure 6.

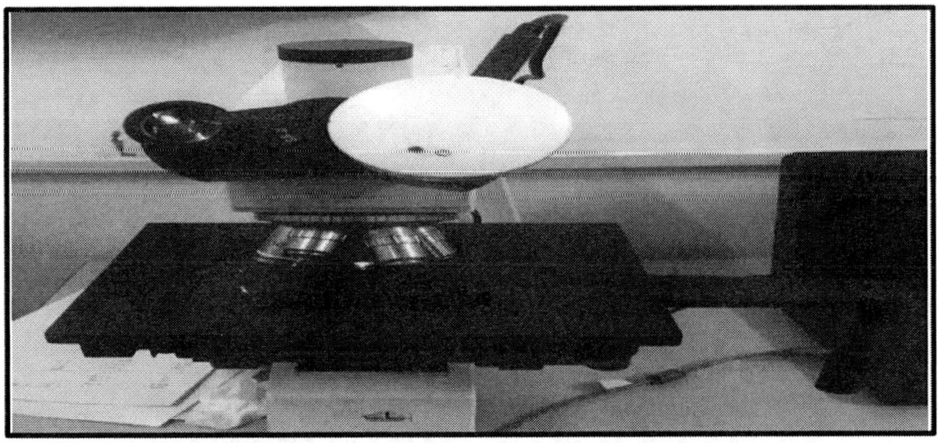

Figure 5. shows microscopic type (Bel-X1600)

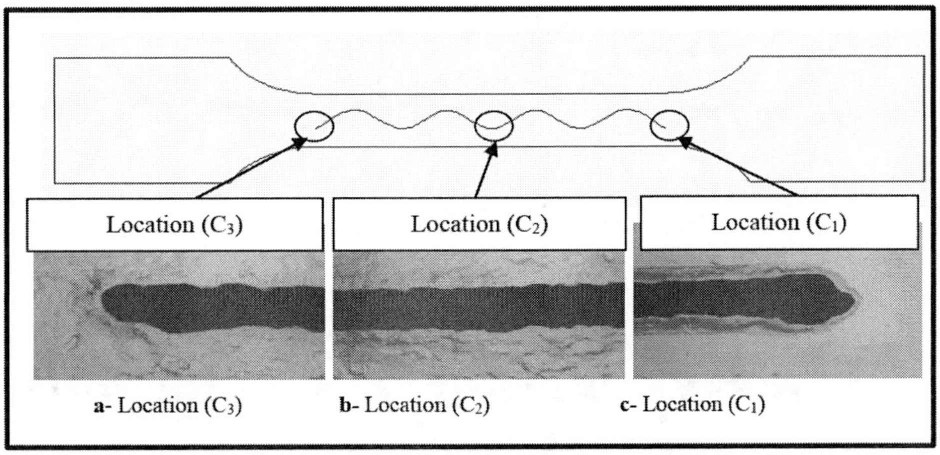

Figure 6. shows microscopic images of the laser crack created at speed (30mm/sec) and laser power (60 watt)

4. The Results and Discussions

4.1. The effect of laser beam on the microstructure

The scanning electron microscope (SEM) shown in figures 7, 8, and 9 of the laser crack created at speeds (40 mm/sec) and (15 mm/sec) show that the lateral surface of the slit at low-energy cutting energies is a uniform and regular surface compared to the surfaces at medium cutoff powers (50 watt) and (60 watt) at speed (15 mm / sec). This is due to the fact that the high temperature generated by the high intensity of the laser beam causes the melting of the cutting area and distortion of the side surfaces of the slit. Therefore, these surfaces will decrease their resistance to crack growth and failure. When surfaces have low energies they are more resistant. The microstructure of the acrylic specimens shows a significant change in the use of a high laser capability that deforms the microscopic structure due to the high temperatures generated by the laser beam.

The shape change of the microscopy is due to the laser beam at the high-energy which creates the cavities on the surface of the laser crack, which are caused by the changing in the stress concentration at applying tensile stress to the specimens due to increase of stress resistance to failure.

IOP Conf. Series: Materials Science and Engineering **579** (2019) 012042 doi:10.1088/1757-899X/579/1/012042

Figure 7. shows scanning electron microscope (SEM) of the laser crack created at speed (15 mm/sec) and laser power (40 watt).

Figure 8. shows scanning electron microscope (SEM) of the laser crack created at speed (30 mm/sec) and laser power (50 watt).

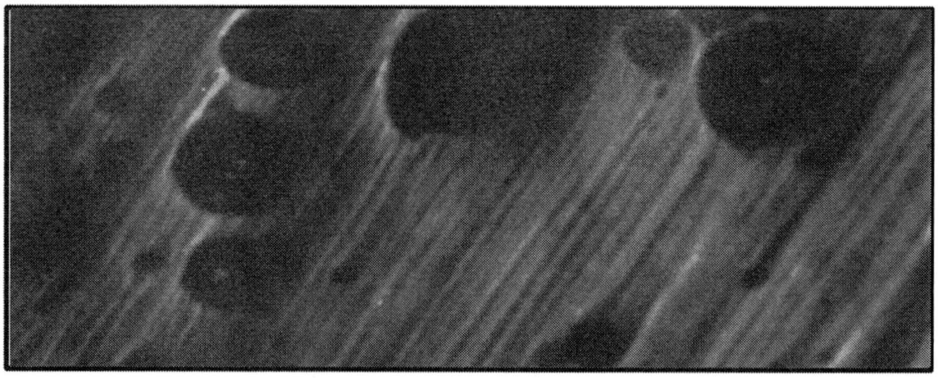

Figure 9. shows scanning electron microscope (SEM) of the laser crack created at speed (45 mm/sec) and laser power (60 watt).

4.2. The effect of laser crack on the tensile stress of acrylic properties

The heat of laser beam creates the crack's edge in acrylic specimens which has impeded the crack failure growth through the acrylic specimen which is caused by tensile stress. It has affected the specimen's resistance to failure compared to the original specimen (O-type) due to absence of laser beam effect. The crack laser beam is created according to the function which is consisting of an angle (41o) to redirection the crack failure path.

The laser's cracks are created according to polynomial's function (eq.3) contributed to prolonging the path of crack failure growth thus for increasing the tensile stress required for failure. The experimental results shown increased tensile stress on the tensile test specimens exposed to the laser beam compared to the original acrylic specimen (O-type). Figure 10 shows the experimental results comparison of tensile stress-strain of acrylic specimens (O, A1, A2, and A3) at constant laser speed (15 mm/sec) and different laser powers (40, 50 and 60) watt. The experimental results show increasing in tensile stress with increasing in average rate (12.25%).

The laser speed is increased to be equal (30 mm/sec) at different power levels of laser (40, 50 and 60) watt as shown in figure 11. The copmaring of the experimental results of specimens type (O and A4) shows improving in tensile stress at average rate (12.87%) and copmaring the experimental results of specimens type (O, A5 and A6) shows incresing in the tensile stress with average rate (14.088%).

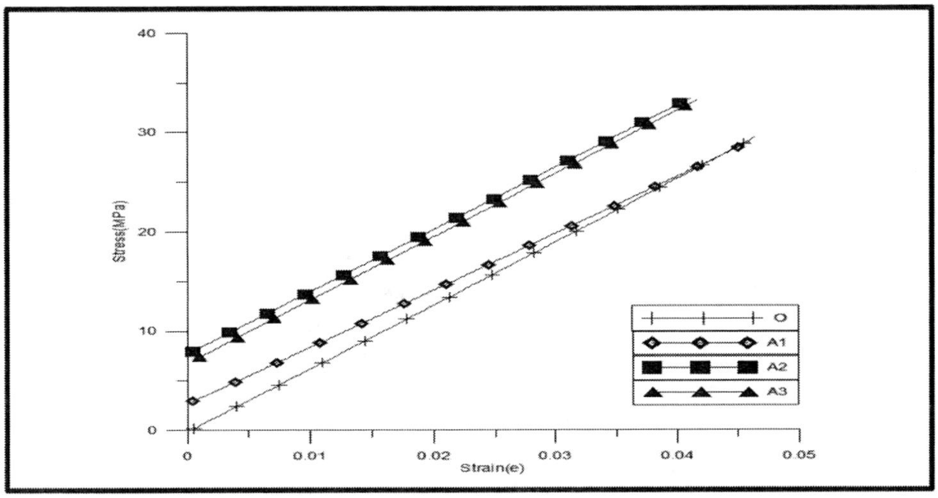

Figure 10. show the compare the tensile stress-strain results of acrylic specimens (O, A1, A2, and A3) (ASTM D412).

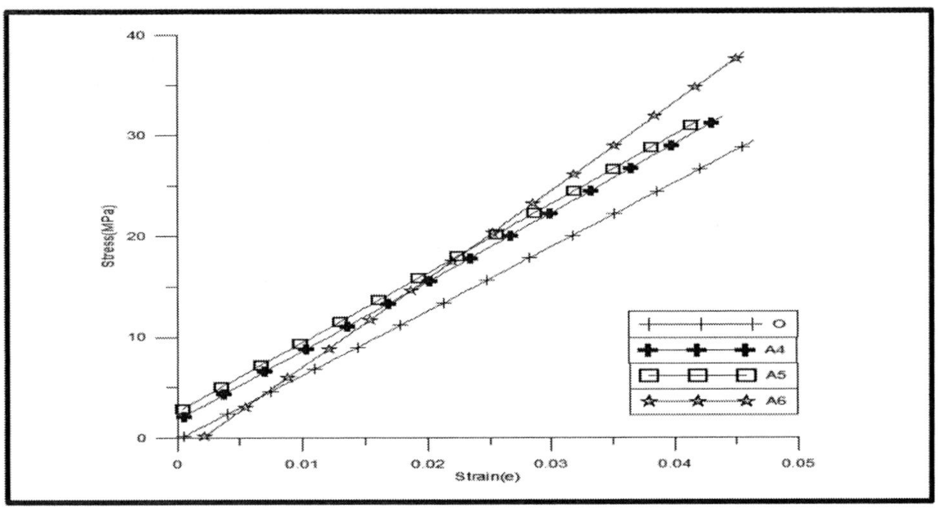

Figure 11. shows the compare the tensile stress-strain results of acrylic specimens (O, A4, A5, and A6)

Figure 12 shows the comparing of the experimental results of acrylic specimens type (O, A7, A8 and A9) at different laser beam powers (40, 50 and 60) watt and laser beam speed equals to (45 mm/sec). The experimental results show increasing in tensile stress at rates (6.85, 17.488 and 22.95%), respectively. This is caused as a result of increasing the laser beam power at constant speed (45 mm/sec) which increases the heating created by the laser beam on the surface of the acrylic specimens due to change the microstructure in it.

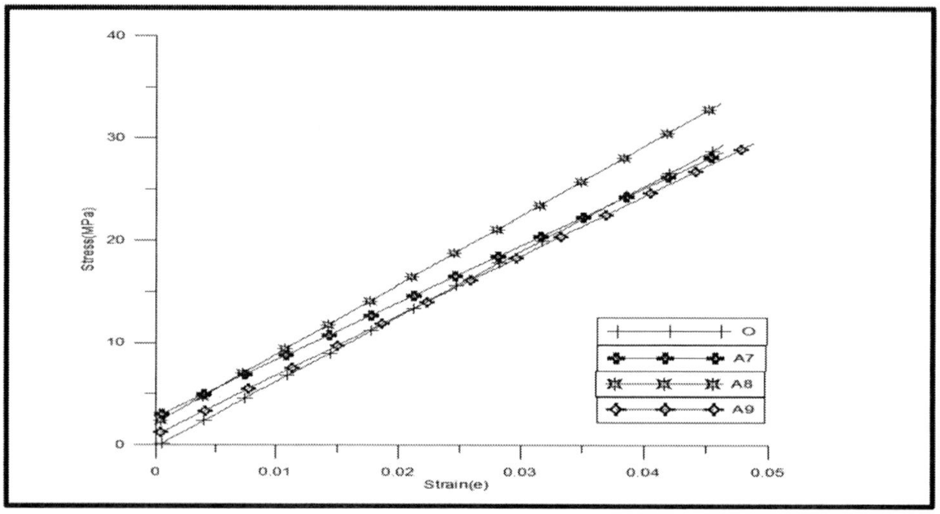

Figure 12. shows the compare the tensile stress-strain results of acrylic specimens (O, A7, A8, and A9)

5. The Conclusions:

The study concluded that the laser beam crack according to the function of the g (x) post increases the tensile stress of the acrylic specimens by (12.25%) at (15mm / sec) and different laser powers (40, 50, 60) watt and the stress increases by (14.088%) when comparing the results of the type of specimens (O and A5). The tensile stress of the type of acrylic specimens (A7, A8 and A9) at the laser beam powers (40, 50 and 60) watt and speed (45 mm / sec) when compared to the O-type specimen shows increases by (6.85, 17.488 and 22.95 %) respectively. The study creates three relationship functions f1,2,3(x,y) at different conditions of speed and laser beam power. The study explains the increase in tensile stress due to a change in the microstructure of the acrylic material on the edges of the laser beam caused by the heat of the laser beam that resists and redirects the progress of the crack failure to follow the longer path. In microstructure pictures creates the cavities on the internal surface of the laser beam crack due to dispersion of the stress concentration points which led to increase the failure resistance.

Acknowledgment

We would like to thank the University of Technology, especially the Petroleum Technology and center of training and factory for their supports in completing the research to develop the mechanical properties of acrylic panels using laser beam without any additives. I would also like to thank the President of the University Prof. Dr. Amin Daway Thamir and the Head of the Department of Petroleum Technology Dr. Aqeel Shakir Ghani Al-Adili and head of the Department.

References

[1] Hofmann K, Neubaur F, Holazer M, Mann V, Hugger F, Roth S and Schmidt M 2016 Effect of laser beam alloying strategies on the metallurgical and mechanical properties of hot forming tool steels *9th International Conference on Photonic Technology* (physics Procedia 83 pp. 246-276)

[2] Abouda E, Dal M, Aubry P, Tarfa T N, Demirci I, Gomy C and Malot T 2016 Effect of laser cladding parameter on the microstructure and properties of high chromium hard facing alloy *9th International Conference on Photonic Technology* (ELSEVIER, physics Procedia 83, pp. 684-696)

[3] Palmieri F L, Hopkins J, Wohl C J, Lin Y, Connell J W, Belcher M A and Blohowiak K Y 2016 Laser surface preparation of epoxy composites for secondary bonding: optimization of ablation depth *NASA Langley Research Center* (Hampton, VA 2368-2199)

[4] Akpinar Y Z, Kepceoglu A, Yavuz T, Aslan M A, Demirtag Z, Kılıc H S and Usumez A 2015, Effect of femtosecond laser beam angle on bond strength of zirconia-resin cement *Lasers Med Sci* DOI 10.1007/s10103-015-1762-1

[5] Wahab M S, Abd. Rahim E and Mustapa M S 2012 The effect of laser cutting parameters on the aerospace structure panel of CFRP composite material *Applied Mechanics and Materials* **225** pp 127-131

[6] Mandal R K, Bandyopadhyay A and Das S 2014 An experimental investigation on laser beam welding of acrylics *INDIAN WELDING JOURNAL* **47** 4 October.

[7] Noor M, Kadirgama K, Rahman M M, Zuki N M, Rejab M R, Muhamad K F and Mohamed J J 2009 Prediction modelling of surface roughness for laser beam cutting on acrylic sheets *Advance Material Research* ISSN:1022-6680

[8] Hosford W F 1992 *Overview of tensile testing, tensile testing* (P. Han Ed ASM International pp. 1–24)

[9] Gross D and Seelig T 2011 *Fracture Mechanics: with an introduction to micromechanics Mechanical Engineering Series* (Springer-Verlag Berlin Heidelberg)

[10] Khan I, Verma M and Mann S S 2012, Optimization of burr height in drilling of commercial acrylic sheet using taguchi method, *International Journal of Engineering Research and Applications* **2** 1 305-310

IOP Conf. Series: Materials Science and Engineering **579** (2019) 012043 doi:10.1088/1757-899X/579/1/012043

The influence of pile groups configuration on its stability in dry sand under lateral loads

Mohammed A. Al-Neami[#], Zeena W. Samueel, Marwah M. Al-Noori
Building and Construction Engineering Department, University of Technology, Baghdad, Iraq.

[#]Corresponding author's e-mail address: 40008@uotechnology.edu.iq

Abstract. This paper aims to investigate the effect of the arrangement of aluminum pipe piles closed from both ends on its response under lateral loads in terms of a lateral load and bending moment. Twenty - four model tests for laterally loaded aluminum pipe piles were investigated to study the effect of sand relative density, the ratio (S/D=3, 4 and 5) of spacing between pile in the group(s) / diameter of aluminum pipe pile (D), loading direction or piles configuration. It was found that the lateral resistance of the pipe piles loaded with a configuration of 2×1 is higher than that loaded in the direction of the configuration1×2. Relative sand density has a significant effect on the lateral capacity of the pile group. The increase of the relative density caused increase load resistance but the bending moment increased. The effect of (S/D) is very obviously on the behavior of bending moment and the load resistance of piles models, bending moment increased as S/d increased as well as for load resistance.

1. Introduction

Pile foundations are used to transmit a superstructure load to the deeper load bearing strata, and to resist lateral loads generated from many sources as well as earthquake, waves, and soil movement (Poulos and Davis, 1980). They are usually constructed from steel, concrete or timber. Piles are usually subjected to the axial or lateral loading or a combination of lateral loads, axial loads and moments. Piles subjected to high lateral loads are called laterally loaded piles. Understanding the behavior of laterally loaded piles is one of the major problems concerned with the soil structure interaction. The laterally loaded piles have a number of uses as foundations of structures, such as high-rise buildings subjected to wind and earthquake loadings and offshore structures (e.g., Platforms) (A1 Mandeel, 2000).

(Price and Wardle) (1979) conducted a series of tests on single piles in London clay to observe the deflection of the pile at different times of the year.

The deflection of the pile from an adjacent trench was measured by means of probes under static and cyclic loading applied to the pile head. They concluded that the deflected shape of the pile changes due to seasonal effects and this has an effect on the horizontal subgrade reaction when the piles are statically or cyclically loaded. Monitoring of the adjacent pile showed that the unloaded pile was affected by the movements of the adjacent laterally loaded pile.

Al-Taie (2004) studied the behavior of laterally loaded piles embedded in sandy soil with cavities. The pile model is made of solid steel metal with a circular cross section of 15 mm diameter. The tests were performed with loose sand with a dry unit weight of 15.6 kN/m^3; he considers the load corresponding to a horizontal displacement equals to 2 mm as working load. The results indicated that the number of cavities and their location have a combined effect on the behavior of laterally loaded pile model. The existence of a cavity is in front and in touch with a pile model that face causes decrease in the working load as long as it is located in the upper two-thirds of the pile model length. Whereas a cavity that exists in the back and in touch with the pile model face causes a decrease in working load as long as it is located in the lower one-third of pile model length.

In some works the ground excavation is to construct underground structures as well as reservoir, ground oil tanks and so on, a lateral displacement is occurred causing a lateral loading on the foundation near these projects therefore, this study aims to investigate the influence of the piles arrangement on the response of pipe piles closed from both ends under a lateral load.

2. Testing Equipment

To study the behavior of laterally loaded piles in sand, laboratory tests were conducted on a small scale model of aluminum pipe piles closed from both ends and have a hollow core, The pile model used has the outer diameter of 18 mm and the inner diameter of 15mm with a thickness equal to (1.5) mm.

The size of the piles model was chosen after reviewing the literature about a suitable pile size that could be considered representative. Although, Vesic (1977) stated that " *scale effects will be complex for model of piles smaller than (35 mm) in diameter*", many researchers used smaller diameters in their tests: Al-Mhaidib (1999) used steel piles with (30 mm) in diameter, Boominathan and Lakshmi (2000) used aluminum pile with a diameter of (19 mm) and Al-Mhaidib (2006) used steel piles with (25 mm) in diameter. The dimensions of the pile model used in this study were also selected to minimize the boundary effects of the soil container in the experimental setup.

The piles model used is made of smooth aluminum tube covered with a plastic sleeve to protect strain gauges. Pile-embedment ratio (depth to diameter) (L/d) used in testing of single and group piles was (30). Two different arrangements of the pile groups 1×2 and 2×1 with different spacing ratios (S/d) (3, 4 and 5) were used for the testing group of piles. The mechanical properties of the pile used are shown in Table 1.

Table1. Mechanical properties of aluminum pile used.

The Pile length (mm)	The Outer diameter (mm)	The Wall thickness (mm)	The Bending stiffness, E_PI_P
570	18	1.5	0.18×10^6

Along the pile shaft, (8) electrical- resistance-half bridge type strain gauges linked to the logger were pasted to measure the bending strain at distances of (0 L, l/4 L, l/2 L, and 3/4 L) from the top of the pile group embedded in dry sand as shown in Plate 1.

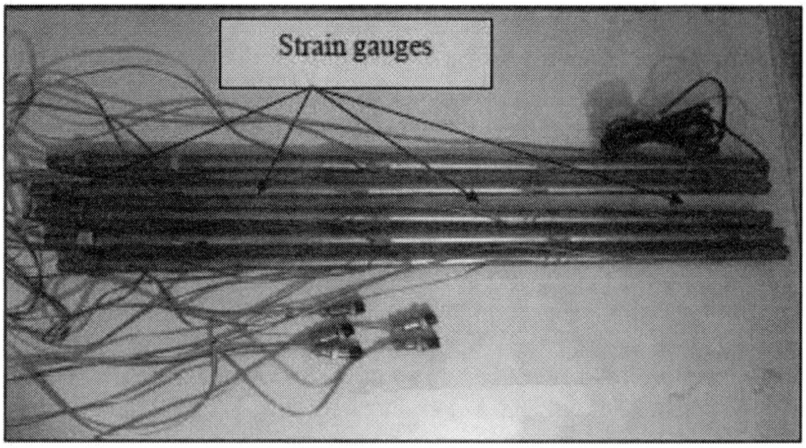

Plate 1. Pile model with strain gauge pasted on pile.

The pile cup used is made from steel with a square cross-section of 150mm length and 6mm thickness. Twenty four laboratory experimental models are conducted by taking the average of them; twelve models were used in result to study the behavior of laterally loaded aluminum pipe piles. The model piles were laterally loaded using a hydraulic jack which controlled by using AC-drive manufactured to control the loading of jack with a loading rate of 2.5 mm /min the jack fixed to a frame that welded with steel container of (750×750×750) mm in cross-section dimension.

Two-dial gauges with an accuracy of 0.01 mm were used to measure the pile's lateral displacements. The dial gauges were attached horizontally on the cap of the pile with the same loading level.

A steel screw shift was connected in between the hydraulic jack and load cell to measure the incremental load came with a digital weighing indicator to take load readings from it, a small steel cone was fixed to load cell for two purposes, first to have a point load and second to set a connection point between load cell and loading point at pile cap. A general layout of the equipment used in the present study is illustrated in Plate 2.

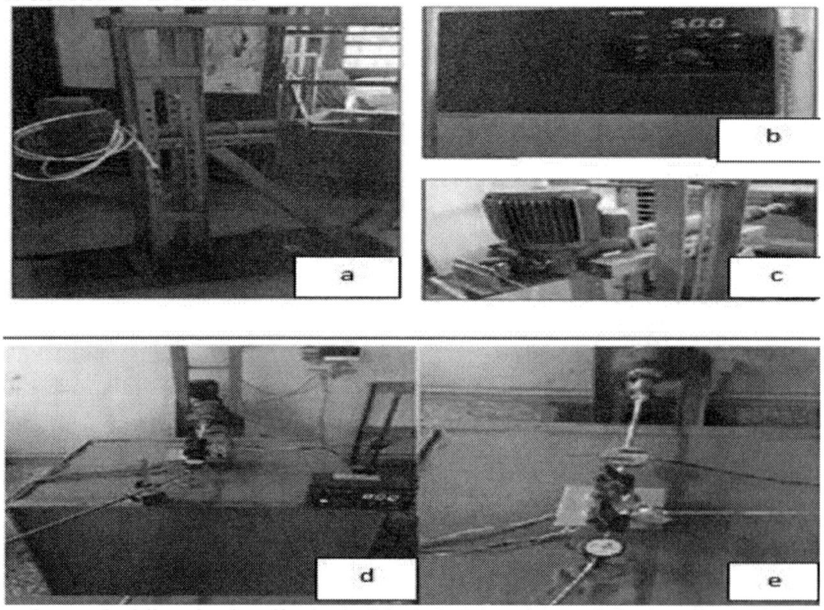

Plate2.(a) The frame holding hydraulic jack,(b) AC drive,(c) and hydraulic jack,(d, e) System of lateral loading.

The lateral displacement of the pile was considered as the average of the two dial gauges readings. Spirit level was used to ensure vertical and horizontal levels of the test setup.

The dimensions of the soil bin are big enough to overcome the effects of the boundary conditions on aluminum pipe piles response, whereas the dimensions of it are (750×750×750) mm as cubic shape. It is worth mentioning that the aluminum pipe piles length is 570 mm and the sand height is750 mm, and the loading level is the e=100mm distance between loading point and soil surface.

3. Soil Descriptions

Poorly graded sand is (SP) classified according to Unified Soil Classification System (USCS) which is brought up from the site in the middle of Baghdad at depth of (5-7) m.

Many standard tests are implemented on the soil samples to clarify the physical and mechanical properties of the sand used. The tests are carried out on sand samples with two different relative densities; medium sand (60%) and dense sand (75%). The specifications of tests followed in the experimental work are shown in Table 2 conformity to the ASTM specifications.

Table 2. Physical properties of the sand used.

Specification	Value		Index Property
ASTM D 422 and ASTM D 2487 (2007	0.14, 0.18, 0.22, 0.24		Effective sizes ,(mm) D_{10} , D_{30} , D_{50} , D_{60}
ASTM D 422 and ASTM D 2487 (2007)	1.7		Coefficient of uniformity (Cu)
ASTM D 422 and ASTM D 2487 (2007)	0.96		Coefficient of curvature (Cc)
ASTM D 854 (2006)	2.66		Specific gravity (Gs)
ASTM D 4253-(2000)	16.7		Maximum dry unit weight (kN/m^3)
ASTM D 4253-(2000)	13.8		Minimum dry unit weight (kN/m^3)
	0.93		Maximum void ratio
	0.59		Minimum void ratio
	75%	60%	Relative density (RD)%
	15.93	15.5	Dry unit weight (γ_d) kN/m^3
	0.67	0.72	Void ratio,(e)

4. Pile Group Configuration (Or Loading Direction)

The configuration of the piles in a group with two arrangements was designed to be loaded in two directions as 2×1(as a pile in line) and 1×2 (as a pile in raw) as shown in Figure 1.

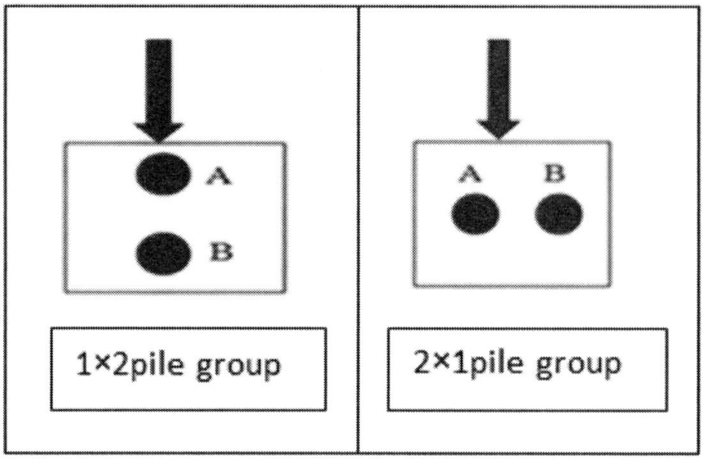

Figure 1. Pile group configuration according to the load direction.

5. Experimental Procedure

The piles in group vertically placed in the center of soil container according to the testing program and, the sand was formed in the soil container in layers with 50 mm thickness per each layer. To ensure the homogeneity of the sand formation, a designed weight of sand with an accuracy of 0.001 kN, was formed into a certain volume of the soil container by compaction to give the specified relative density of 60%, and75% according to a planned testing program. The compaction was carried out manually using a rammer weighing of 40.0 N and of (150 ×150×6) mm square hammer. The top surface of the formed sand was leveled using a sharpened straight steel plate. After the soil had been prepared and the piles installed, then after that a lateral loading system work, where the device is prepared and installed with the system for the test. The load cell connected to the shaft of a hydraulic jack and the strain gauge linked to the logger using the data wire connection. Two dial gauges attached to the pile cap, one of them is attached to the upper surface of a pile cap made to read the vertical displacement and the other dial gauges connected to the thin side of the pile cap made to read the lateral displacement, which face the load cell at the other side of pile cap as shown in Plate (2) previously. After that, the Ac drive operates to run the hydraulic jack and control the rate of movement according to the required work. After this point, the shaft starts to move subsequently to the jack movement to give the incremental load reading which recorded by digital weighing indicator that indicates what load cell had been sensed during jack moving. The loading is stopped when the reading of lateral dial gauge reach to 12mm displacement reading, the load and strain gauge logger data recorded and analyzed it to give the proper results.

6. Test Results and Discussion

A total of 24 tests were carried out on laterally loaded aluminum pipe piles with closed ends embedded in the dry sand. The response of laterally loaded aluminum pipe piles investigated under the effect of sand relative density, (S/D) ratio of spacing between pile in the group(s) / diameter of aluminum pipe pile (d), loading direction or piles configuration. The maximum lateral load reached of vertical piles was obtained from the load-displacement curves. The lateral displacement at pile cap (L.d) is expressed in a non-dimensional form in terms of outer pile diameter (D) as (L.d/ D) present as the aspect ratio. The maximum lateral capacity of the piles is obtained from the load-displacement curve as the pronounced peaks, after which the pile's lateral displacement reached to 12 mm displacement of the test end is equal to (0.67 mm/mm) aspect ratio.

The maximum loads for the pile in medium and dense sand for different studied parameters are measured and tabulated in Table 3. Variations of lateral force (P) with horizontal displacement, (L.d) of model piles that embedded in medium and dense sand for the ratios (S/D=3, 4 and 5). In this series, the sand relative density, (Rd equals 60% and 75%), Pile-embedment ratio (depth to diameter) (L/d) used in testing group piles was (30%), embedment ratio, and Load eccentricity (e) were kept a constant while (S/D) of spacing between pile in group(s)/diameter of pile (D) is (S/D) which have different values of 3, 4, and 5.

6.1 Lateral Load with Lateral Displacement:
Table 3 shows the summary of maximum lateral load 12 mm lateral displacement (equal to 0.67 mm/mm aspect ratio) at the end of the test.

Table 3. Maximum lateral load for laterally loaded pile group with different sand relative densities.

(S/D) ratio	Maximum load (2×1)		Maximum Load (1×2)	
	R.D		R.D	
	60%	75%	60%	75%
3	151.6	247	158.4	252.6
4	169.4	292.3	173.6	289.4
5	221.7	304.1	184.4	298.7

Figures (2a, b) and (3a, b) present the development of the lateral load with the aspect ratio of the pile group constructed in medium and dense sand at different aspect ratios for a lateral load of 2×1 and 1×2 configuration pile group. It can be seen that the lateral load increases as a relative density increased from 60% to 75%, and at the specific load intensity, the aspect ratio decreased as a relative density increased from 60% to 75%, this result is similar to El Wakil and Nazir (2013). In addition, the lateral load increased as S/D increased from 3 to 5 for both configurations of pile groups as observed in Figure (4 a & b).

The comparison results depicted in Figures (2a, b), (3a, b) and Figure (4 a & b), it can be observed that the lateral resistance of the pile group of 1×2 is higher than the pile group of 2×1 configuration at different S/D.

At the same load intensity, a pile group of 1×2 configuration has a lateral displacement less than 2×1 pile group and this behavior may be due to the interaction effect between piles in the group. It is worth to mention that as a closely spaced pile groups move laterally, the failure zones for front or trailing rows piles overlapped with leading row piles and decreases a lateral resistance.

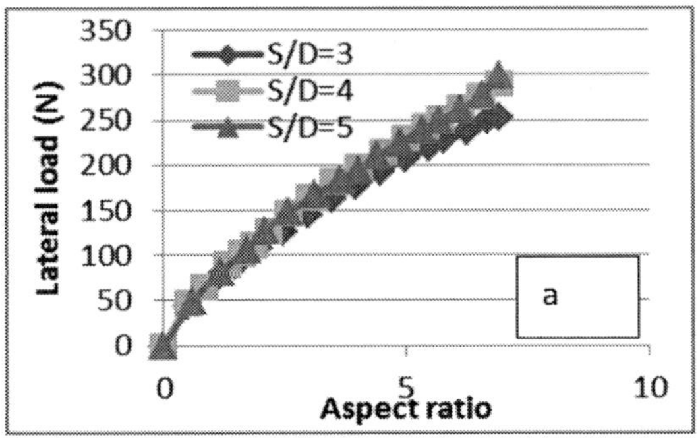

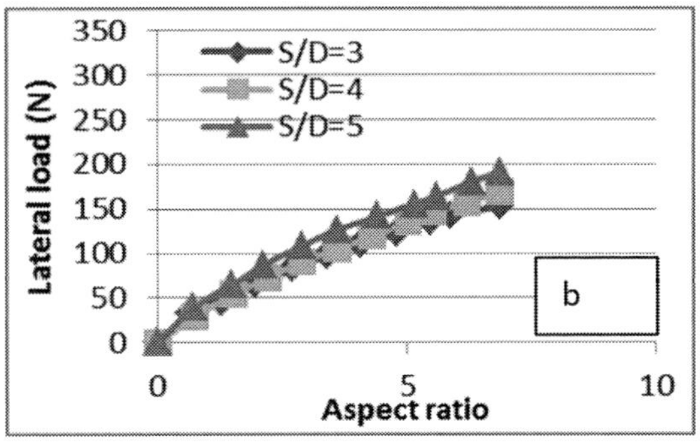

Figure 2. Variation of the lateral load with an aspect ratio of 2×1 pile group,a)Dense sand of R.D=60% b) Medium sand of R.D=75%.

IOP Conf. Series: Materials Science and Engineering **579** (2019) 012043 doi:10.1088/1757-899X/579/1/012043

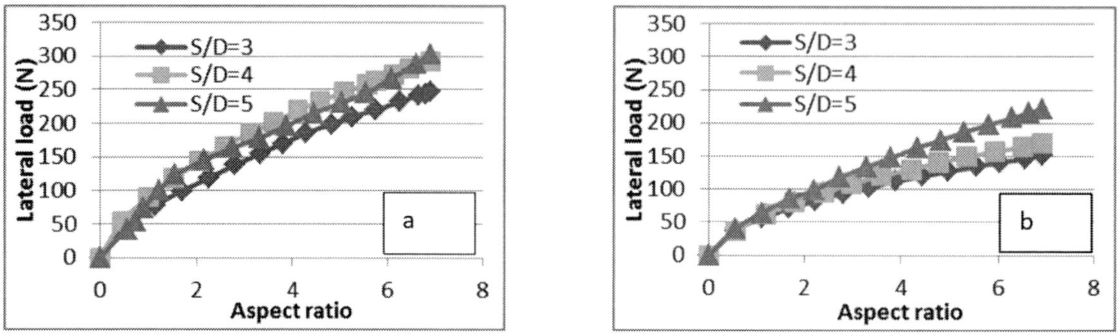

Figure 3. Variation of lateral load with an aspect ratio of 1×2 pile group.a) Dense sand of R.D=60% b) Medium sand of R.D=75%.

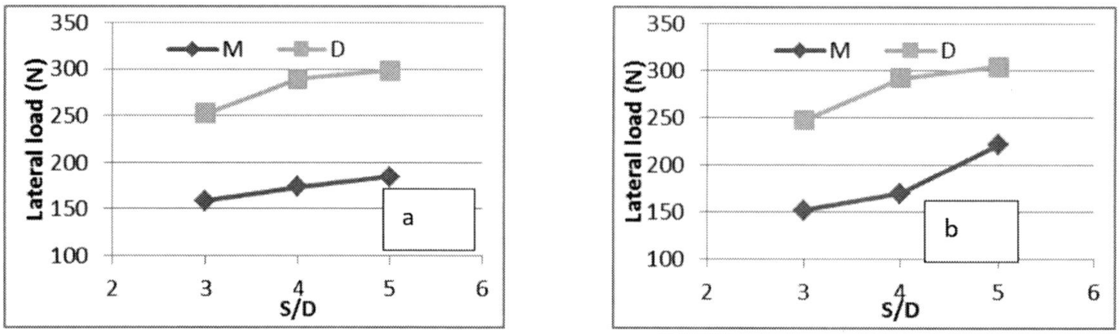

Figure 4 a & b. maximum lateral load with the ratio (S/D) for 2×1 and 1×2 pile group in the sand with different densities, respectively.

6.2 Result of Bending Moment with Pile Depth:

Tables (4) and (5) illustrate the summary of maximum negative and positive bending moments respectively at 12mm lateral displacement (equal to 0.67 mm/mm aspect ratio) at the end of the test. Both the negative and positive bending moment, occurred and the negative bending moment was shown at the upper part of pile depth. Figures (5a, b & c) and Figure (6 a, b, c, d, e & f) show the bending moment of 2×1and 1×2 pile group configurations with a respect to the pile depth constructed in medium and dense sand at S/D=3, 4 and 5.

Table 4 a, b. Negative maximum bending moment of laterally loaded piles group.

a) Maximum bending moment in line (1×2) .

(S/D) ratio	Maximum bending moment in line (1×2)			
	R.D=60%		R.D=75%	
	Front pile (A)	Rear pile (B)	Front pile (A)	Rear pile (B)
3	7.6	9.6	13	14
4	4.05	9.2	17	14
5	7.7	10	11	13

b) Maximum bending moment in raw (2×1).

(S/D) ratio	Maximum bending moment in raw (2×1) (N.m)	
	R.D=60%	R.D=75%
	Both pile facing load	Both pile facing load
3	6.5	10
4	9.4	12
5	8	13

IOP Conf. Series: Materials Science and Engineering **579** (2019) 012043 doi:10.1088/1757-899X/579/1/012043

Table 5 a, b. Positive maximum bending moment of laterally loaded piles group with different sand relative densities.

a) Maximum bending moment in line (1×2).

(S/D) ratio	Maximum bending moment in line (1×2) (N.m)			
	R.D=60%		R.D=75%	
	Front pile (A)	Rear pile (B)	Front pile (A)	Rear pile (B)
3	8.975	8.3	11.33	13.9
4	8.8	6.83	11.28	12.6
5	6.13	8	9.65	14

b) Maximum bending moment in raw (2×1).

(S/D) ratio	Maximum bending moment in raw (2×1) (N.m)	
	R.D=60%	R.D=75%
	Both pile facing load	Both pile facing load
3	8.4	11.5
4	9.53	11.6
5	7.23	12.13

In general, it can be concluded that the maximum bending moment increases as a relative density increased from 60% to 75% and this behavior is due to the increases of shear strength as the relative density increases. This behavior also supported with Albuhlale (2015) for single pile result. Also, it can be mentioned that the bending moment has no significant differences in magnitude as S/D increases from 3to5.

The negative bending moment for a font pile (A) is higher than for a rear pile while the positive bending moment of a front pile is higher than for a rear pile. This may happen because the lateral load subjected at pile cap causing tilting to the pile cap that making the front pile to uplift and the rear pile to compression this behavior make the bending moment decrease. This behavior is due to the increase of interaction effect of pile group where the front pile works as an active pile making the rear pile having the largest bending moment.

The comparison results between 2×1 and 1×2 pile groups showed that the bending moment of 2×1 pile group configuration has a bending moment higher than 1×2 pile group configuration at different S/D ratios.

The position of the maximum bending moment occurred approximately at the midpoint of pile shaft and generally, the bending moment of a single pile is higher than for the pile group. Also, the position of maximum bending moment increases with a depth as a result of the increase in the relative density of sandy soil due to increasing in the rigidity of pile-soil interaction zone which developed by confining zone available by a high relative density that agreed with Allami (2017).

IOP Conf. Series: Materials Science and Engineering **579** (2019) 012043 doi:10.1088/1757-899X/579/1/012043

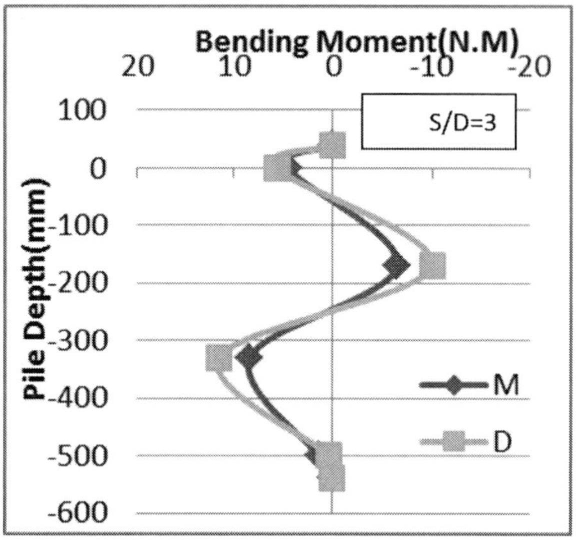

(a)

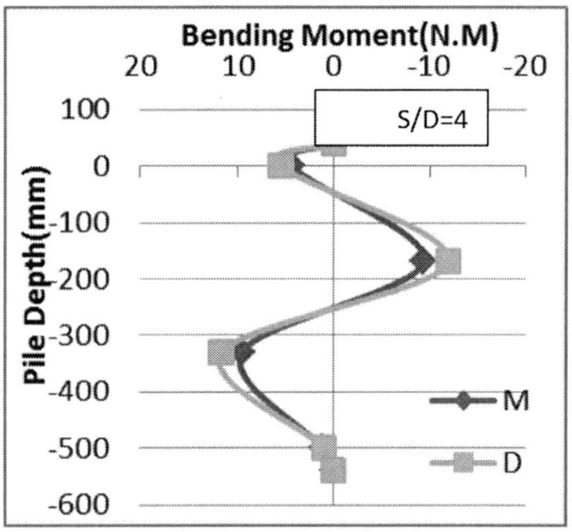

(b)

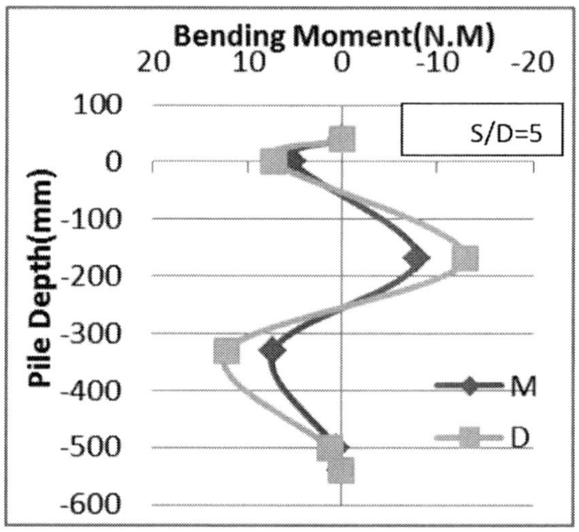

(c)

Figure 5. Bending moment of 2×1 pile group configuration with respect to pile depth constructed in medium and dense sand at a)S/D=3, b) S/D=4, c) S/D=5.

IOP Conf. Series: Materials Science and Engineering **579** (2019) 012043 doi:10.1088/1757-899X/579/1/012043

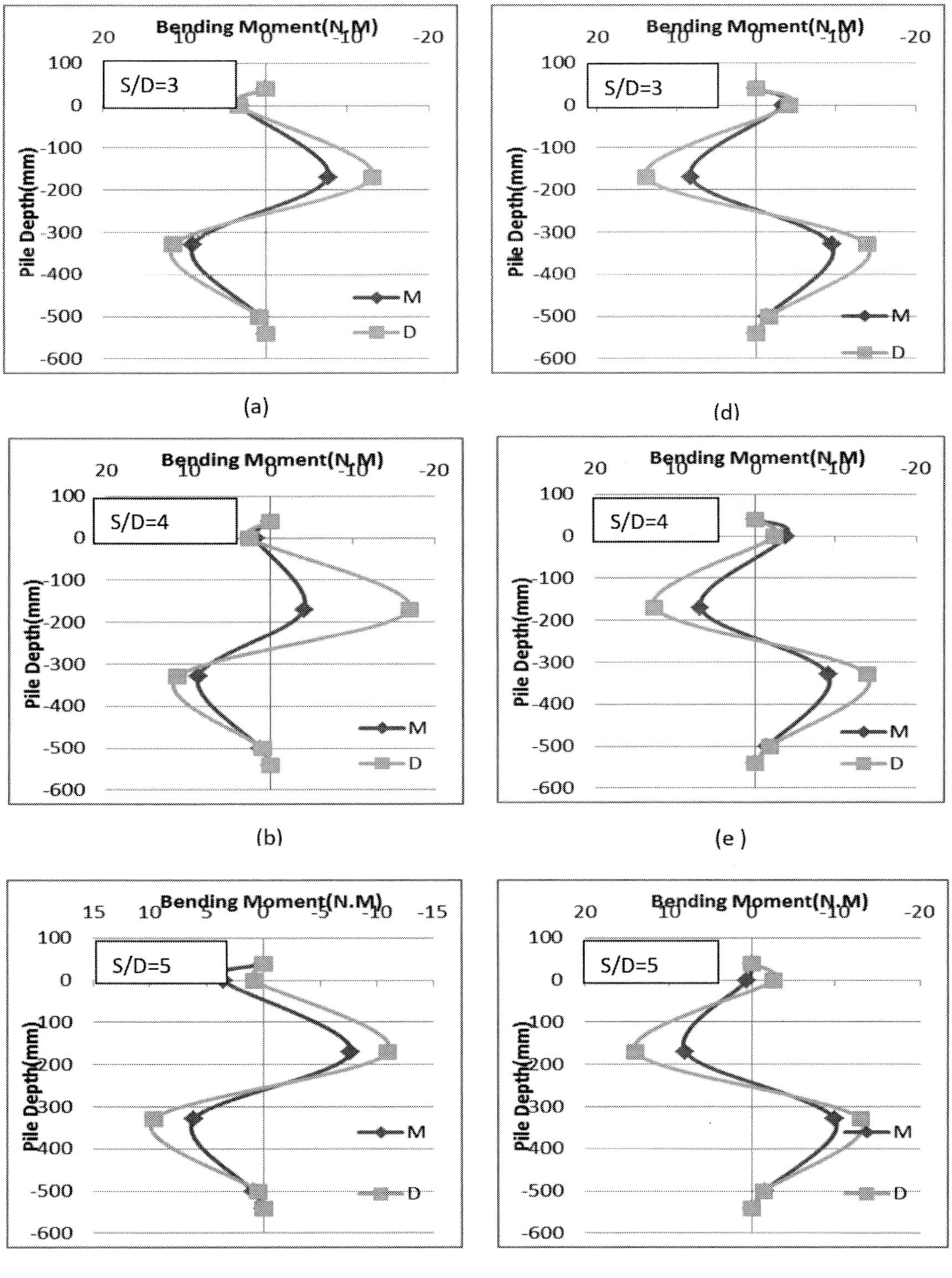

Figure 6. Bending moment of 1×2 pile group configuration with respect to pile depth constructed in medium and dense sand at (a, b, c) for front (A)pile and (d, e, f) of the rear pile (B).

7. Conclusions

From the analysis of experimental results, the following conclusions are obtained:

Sand relative density has a significant effect on the lateral load and bending moment of the pile group. The high sand relative density make high lateral load of the piles in the group. Bending moment increased when the relative density increased but S/D ratio increase has less effect on the bending moment.

The increases in aspect ratio cause a decreasing in the lateral displacement for both pile group configurations. At different S/D ratio, the bending moment of 2×1 pile group configuration has bending moment higher than 1×2 pile group configuration at different S/D ratio.

The lateral resistance of 1×2 pile group configuration is higher than in the pile group of 1×2 configurations. A Front pile in pile group with 1×2 configuration has lateral load more than rear pile but has a bending moment lower than rear pile.

At same load intensity pile group of 1×2 configuration has lateral displacement less than 2×1 pile group. The maximum bending moment approximately occurring at the midpoint of pile shaft.

References

[1] Albuhlale A. J. 2015 *Effect of Particle Size Distribution of Cohesionless Soils on the Behavior of Laterally Loaded Single Pile Model M.Sc. Thesis* (University Of Technology: Building and Construction Engineering Department, Iraq)

[2] Al Mandeel, S. J. 2000 *Behaviour of Laterally Loaded Pile Group Model in Sand M.Sc. Thesis* (Dahran, Saudia Arabia: King Fahd University of Petroleum & Minerals)

[3] Al-Mhaidib A. I. 1999 Bearing capacity of a model pile in sand under different loading rates *Proceedings of the (9th) International offshore and Polar Engineering Conference,* Vol. 1 (Brest, France) p 724-730

[4] Al-Mhaidib A. I. 2006 Experimental investigation of the behavior of pile groups in sand under different loading rates *Journal of Geotechnical and Geological Engineering* **24** 889-902

[5] Poulos, H. G., and Davis, E. H., (1980) Pile Foundation Analysis and Design *Series in Geotechnical Engineering* (John Wiley and Sons)

[6] Price, C. and Wardle, I. F. 1979 The Deformation of Vertical Piles in London Clay under Static and Cyclic Horizontal Working Loads *Proc. of the Conf. at IEE.* (London) p 87-94

[7] Vesic A. S. 1977 *Design of pile foundation", National Cooperative Highway Research Program, Synthesis of Highway Practice, No. 42, Transportation Research Board, Washington, D.C. Lateral Excitation of Pile Group Model in Sandy Soil Ph.D. Thesis* (University of Technology, Baghdad, Iraq)

[8] Al-Taie, Saheem M.G. 2004 *Performance of Laterally Loaded Piles Embedded in Sandy Soils with Cavities M.Sc. Thesis* (University of Baghdad)

[9] Boominathan A. and Lakshmi T. 2000 Dynamic characteristics of pile group under vertical vibration *the 12thworld conference on Earthquake Engineering* (Auckland, New Zealand) p 1-8

[10] ELEWI, 2017 *Response Of Single Pile And Pile Groups To Lateral Sandy Soil Movement Ph.D. Thesis*. (University of Technology, Baghdad, Iraq)

[11] El-Wakil, A. Z., And Nazir, A. K. 2013 Behaviour of Laterally Loaded Small Scale Barrettes in Sand Ain *Shams Engineering Journal* **4** 344-350

[12] Patra, N. R. and P. J. Pise. 2001 Ultimate Lateral Resistance of Pile Groups in Sand *Journal of Geotechnical and Geoenvironmental Engineering* **127** 481-487

The effect of coarse aggregate inclusion on the performance of reactive powder concrete exposed to oil products.

T S al-Attar[1], A S Ali[2] and B S Al-Numan[3]

[1] University of Technology, Civil Engineering Department, Baghdad, Iraq
[2] Al-Nahrain University, Civil Engineering Department, Baghdad, Iraq
[3] Ishik University, Civil Engineering Department, Erbil, Iraq

Corresponding author's e-mail address:40076@uotechnology.edu.iq

Abstract. Reactive powder concrete, RPC, is an ultra-high strength and high ductility cementitious composite with advanced mechanical and physical properties. RPC is composed of very fine powders (cement, sand, quartz powder and silica fume), steel fibers (optional) and superplasticizer. A basic principle, which was presented by the first developers of RPC, is to eliminate the coarse aggregate in mixture to enhance homogeneity. The main purpose of the present investigation is to modify RPC by including natural graded coarse aggregate in the mixture in different approaches. The investigation was devoted to monitor the strength development, total absorption and permeability of RPC after exposure to two types of oil products, kerosene and gas oil, up to 180 days. The current results showed that eliminating coarse aggregate as a principle could be gone-beyond. The better way for including coarse aggregate in RPC was to replace the fine sand and the most useful FA/A ratio is expected to be within the range of 0 - 25 % by weight. This range was found to create an optimum situation between strength and absorption for kerosene and gas oil exposures. It is not recommended to include the coarse aggregate in RPC mixtures as a replacement to the binder specially when dealing with oil products storage.

1. Introduction

Reactive Powder Concrete, RPC, was first developed in France in the early 1990s and the world's first Reactive Powder Concrete structure, the Sherbrook Bridge in Canada, was erected in July 1997. RPC is an ultra-high strength and high ductility cementitious composite with advanced mechanical and physical properties. It consists of a special concrete where the microstructure is optimized by precise gradation of all particles in the mix to yield maximum density. It uses extensively the pozzolanic properties of highly refined silica fume and optimization of the Portland cement chemistry to produce the highest strength hydrates [1].

RPC is composed of very fine powders (cement, sand, quartz powder and silica fume), steel fibers (optional) and superplasticizer. The superplasticizer, used at its optimal dosage, decreases the water to cement ratio (w/c) while improving the workability of the concrete. A very dense matrix is achieved by optimizing the granular packing of the dry fine powders. This compactness gives RPC ultra-high strength and durability. RPCs have compressive strength ranging from 200 MPa to 800 MPa [2].

The following basic principles were highly emphasized by the developers of the RPC [3]:

- Enhancement of homogeneity by elimination of coarse aggregates;
- Enhancement of compacted density by optimization of the granular mixture, and application of pressure before and during setting;
- Enhancement of the microstructure by post-set heat-treating;
- Enhancement of ductility by incorporating small-sized steel fibers;
- Maintaining mixing and casting procedures as close as possible to existing practice.

Collepardi et al. [4] stated that this prescribed mixture could not be considered as concrete due to the

absence of coarse aggregate. Moreover, the cement factor of the RPC is as high as 900-1000 kg/m^3 due to the use of very fine sand instead of ordinary aggregate. This unusual cement factor could increase drying shrinkage and creep strain of the RPC with respect to ordinary concrete with cement factor usually in the range of 300-500 kg/m^3. As a results of these claims, Collepardi et al. had modified the original RPC mixture to include some coarse aggregate and investigated the properties of the modified RPC mixtures in terms of required mixing water, compressive and flexural strength, shrinkage, swelling and creep. They concluded that the replacement of the fine ground quartz sand by an equal volume of well graded natural aggregate did not change the compressive strength of the RPC at the same water-cement ratio and that was in contradiction to Richard and Cheyrezy [3].
In the present work the same modification that proposed by Collepardi et al [4] has been adopted to study the effect of coarse aggregate inclusion on the performance of RPC when being exposed to kerosene and gas oil. In addition to that the high-reactivity Metakaolin was used as supplementary cementitious material instead of Silica fume and the butadiene styrene rubber was used as high-range water reducer. The investigation was devoted to monitor the strength development, total absorption and permeability of RPC.

2. Experimental Work

2.1 Materials

The binder in the current RPC mixes consists of 80 percent by weight of ordinary Portland cement (ASTM C150- Type I) [5] and 20 percent high reactivity Metakaolin (ASTM C618- Type N) [6]. Table 1 lists the chemical and physical properties of both materials.

Table 1. Chemical and physical properties of used binders.

No.	Property		OPC	HRM
1	Oxide Content, %.	CaO	60.5	0.28
		SiO_2	21.4	54.36
		Al_2O_3	5.98	32.76
		Fe_2O_3	3.9	1.32
		MgO	1.36	0.18
		SO_3	1.82	0.12
2	LOI		1.80	7.97
3	IR		1.04	----
4	LSF		0.92	----
5	Specific surface (Blaine method), cm^2/gm		3650	5741
6	Soundness, % (Autoclave method)		0.25	----
7	Setting time (vicat's apparatus): Initial, hrs: min. Final, hrs: min.		1:15 2:45	----
8	Compressive strength: 3days, N/mm^2 7days, N/mm^2		25 32.3	----
9	Specific gravity		3.15	3.11
10	Strength activity index, %		100	164

Very fine sand, with maximum size of 0.6 mm, was used as fine aggregate (BS 882:1992 - fine) [7], meanwhile natural gravel, with maximum size of 14 mm, was incorporated as coarse aggregate (BS 882:1992, 5 –14 mm) [7]. Table 2 shows the sieve analysis and some characteristics of the used aggregates. Straight steel fibers, with a length of 13 mm and diameter of 0.18 mm, were used to produce RPC. These fibers have 1800 MPa tensile strength and 210 GPa modulus of elasticity.

Styrene butadiene rubber, SBR, emulsion was employed successfully as high-range water reducer to control the workability of the RPC mixes [8]. This latex is typically included in concrete in the form of a colloidal suspension polymer in water.

Table 2. Properties of used aggregate.

No.	Test		FA	CA	BS 882:1992 Limitations [7]	
1	Sieve Analysis	Sieve Size, mm	% Cum. Passing		FA	CA
		14	---	100	---	90-100
		9.5	100	85	100	50-85
		4.75	100	10	100	0-10
		2.36	100	2	80-100	---
		1.18	100	---	70-100	---
		0.6	100	---	55-100	---
		0.3	45	---	5-70	---
		0.15	5	---	0-15	---
2	Sp. Grav., SDD		2.73	2.75	---	---
3	SO_3 Content, %		0.08	0.05	---	---
4	Absorption, %		0.70	0.63	---	---

The mix proportioning procedure, which was recommended by Richards and Cheyrezy [3], was adopted. To achieve 130 MPa compressive strength at 28 days with workability of 150 ± 5 mm, flow table, many trials were done. Finally the proportions shown in Table 3 were used for the reference RPC mix, M_0.

Two approaches were adopted to investigate the effect of coarse aggregate inclusion in the mix. The first is changing the fine aggregate to total aggregate by weight ratio, FA/A. The following ratios were investigated: 100, 52, 25 and 0 percent for mixes M_0, M_1, M_2 and M_3 respectively (series I).

In the second approach the effect of variation in the aggregate to binder by weight ratio, A/B, on the behavior of RPC was studied. Only the coarse aggregate content was varied in this part of the study. The investigation included four weight ratios: 0.91, 1.11, 1.23 and 1.35 for mixes M_0, M_4, M_5 and M_6 respectively (series II). Table 3 lists the details of the all investigated mixes.

Table 3. Details of the all investigated mixes.

Mix	Binder, kg/m³		Aggregate, kg/m³		FA/A ratio %	A/B ratio by weight	Water, kg/m³
	OPC	HRM	FA	CA			
M_0	920	230	1050	0	100	0.91	212
M_1			550	500	52		212
M_2			260	790	25		203
M_3			0	1050	0		194
M_4	835	210	1050	111	90	1.11	201
M_5	790	197		160	87	1.23	198
M_6	750	185		215	83	1.35	203

A constant volume fraction for steel fibers of 2 % was used for all mixes. The SBR content was kept constant also at 14 % by weight of cement.

2.2 Curing procedure

RPC mixes were cured in water at 90 °C for 48 hrs and after that in water at 20°C for 24 hrs and finally in air at 20°C till the age of 28 days. After the age of 28 days, tested specimens were grouped in three series. The first continued to be cured with water, meanwhile, the second cured with kerosene and the third with gas oil until the age of test. Kerosene and gas oil were used to simulate the exposure conditions in structures for storage or transportation of oil. Table 4 shows the properties of these products.

Table 4. Properties of used oil products.

Property	Kerosene	Gas oil
Moisture content, % by volume.	0	0
Sulfur content, % by weight.	0.31	0.80
pH	7.6	6.3
Specific gravity at:		
25C°	0.708	0.825
40C°	0.770	0.813
Viscosity (centipoises) at:		
25C°	1.092	3.960
40C°	0.855	2.943

2.3 Testing program

Table 5 lists the conducted tests, the adopted standards, types and dimensions of tested specimens and age of test for the hardened RPC mixes.

Table 5. Details of tests, adopted standards, specimen's types and testing age.

Test	Adopted standard	Specimen type	Dimensions, mm	Age of test, days
Compressive Strength	BS EN 12390-3 [9]	Cube	100*100*100	3, 7, 28, 30,90 and 180
Splitting Tensile Strength	ASTM C496 [10]	Cylinder	d=100 h= 200	3, 7, 28, 30,90 and 180
Absorption	ASTM C 267 [11]	Cube	100*100*100	7, 28, 60, 90 and180,

The permeability test, a non-standard test, was performed by the oil research and development center/ Ministry of Oil. The test was carried out on cylindrical specimens of 38 mm in diameter and 50 mm in thickness. The specimens were oven dried for one week, then cooled for 24 hrs and placed in the pressure vessel, to accelerate the saturation process. The sample was subjected to a pressure of 3.5 N/mm^2 for a period of one week under the effect of the oil products after which, the testing sample was inserted in the core holder. Compressed air was used to pressurize the specimens to 0.241 N/mm^2, equivalent of a head of 30.01 m gas oil and 31.01 m kerosene, hydrostatic pressure was maintained for 4 days.

Darcy's law [12] was used to determine the permeability coefficient from the following equation:

$$\frac{dq}{dt}\left(\frac{1}{A}\right)=K_o\left(\frac{dh}{L}\right) \quad (1)$$

Where:

dq/dt: rate of flow through the sample, cm^3/sec.
A: the cross sectional area of the specimens exposed to the fluid, cm^2.
dh: thickness of the specimen, cm.
K_o: coefficient of permeability, cm/sec.

3. Results and Discussion

3.1 Strength development

Irrespective of the exposure conditions, all mixes show a strength gain with aging and that is the result of continuous hydration of the high cement factor. Fig.1 displays the compressive strength development for all mixes which were cured in water after the age of 30 days. Mixes in series I: M_1, M_2 and M_3, have yielded higher strength values than the reference mix M_0 for the same age. For series I, the amount of fine sand was gradually replaced by natural graded coarse aggregate without changing the cement factor. The decrease in the FA/A ratio caused a reduction in the aggregate total surface area and consequently a reduction in the water requirements or in other words a decrease in the W/B ratio. Also increasing the coarse aggregate content inherited denser concrete mass and better bond. In series II, the graded coarse aggregate, had replaced part of the binder, Portland cement and HRM. Mixes M_4, M_5 and M_6 attained lower compressive strength values than mix M_0 for the same age. The probable cause for this reduction in strength is the increase in the W/B ratio or the reduction in the binder content.

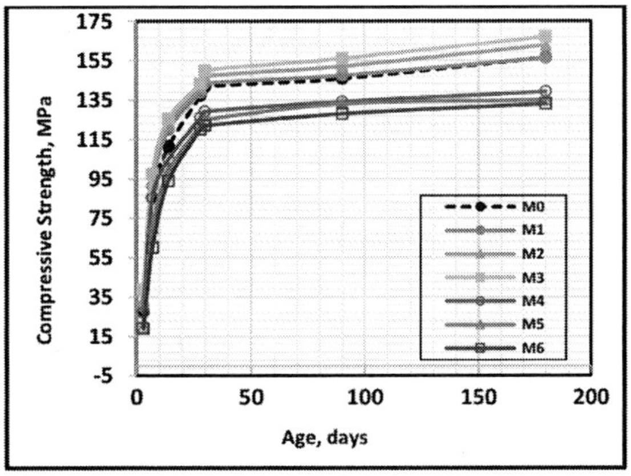

Fig. 1. Compressive strength development for RPC mixes.

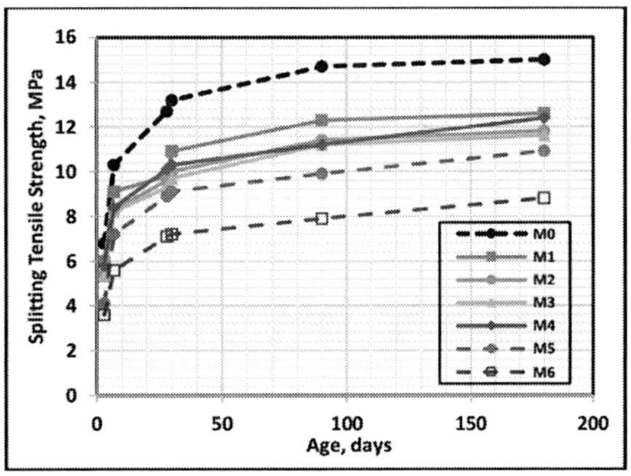

Fig. 2. Splitting tensile strength development for RPC mixes.

The effect of graded natural coarse aggregate inclusion on the splitting tensile strength development is shown in Fig.2. The mixes in the two investigated series I and II reached lower ultimate strength values with respect to the mix M_0 for the same age of test. This trend could be attributed to the effect of smooth surfaces of used aggregate on the bond strength between aggregate and mortar [13, 14] and many researchers [14, 15] had stated that tensile strength of high strength concrete is more sensitive to microcracking due to coarse aggregate inclusion than compressive strength.

In order to magnify the effect of exposure to kerosene and gas oil on the compressive strength development of RPC mixes, the value of strength for specimens soaked in oil products were compared to those attained by specimens cured in water for the same ages. Figs. 3 and 4 show the development of the ratios f_{cK}/f_{cW} and f_{cG}/f_{cW} with curing age. For mixes in series I, Figs. 3a and 4a, curing with kerosene and gas oil have positive effect on strength gain till the ages of 120 – 150 days. Afterward, the curing in water has overtook and yielded higher values. Such behavior could be attributed to the densification of concrete caused by the ingress of oil product which have lower viscosity than water [16, 17]. After the assigned period, 120 – 150 days of soaking, the oil products have a slight negative effect in decreasing strength. This trend is diagnosed in literature but higher rates were reported [18, 19]. The gas oil showed more harmful effect than kerosene and that may be due to its higher sulphate content.

Figs. 3b and 4b illustrate the development of the ratios f_{cK}/f_{cW} and f_{cG}/f_{cW} with curing age for mixes in series II. These mixes were produced with a varying A/B ratio. The degradation in strength was recorded in earlier ages than mixes in series I, as early as 60 days. Lowering the binder content could be the cause of such a behavior. Firstly, the hydration process will be reduced in rate and value. Secondly, reducing the content of finely divided binder, cement and HRM, would decrease the pore refinement physical effect in the microstructure of RPC [20].

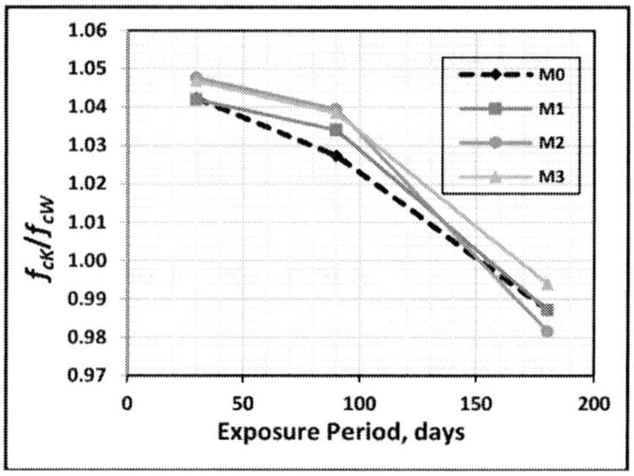

Fig. 3a. Comparison of compressive strength development between kerosene and water curing for series I mixes.

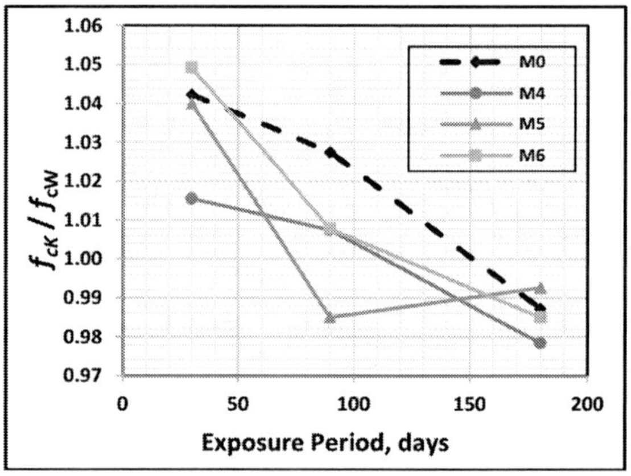

Fig. 3b. Comparison of compressive strength development between gas oil and water curing for series II mixes.

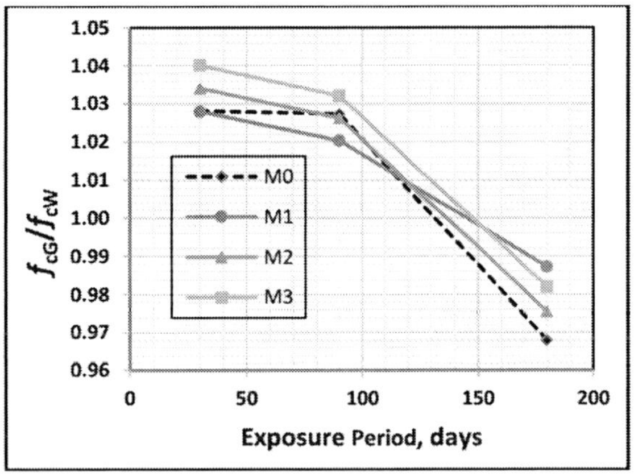

Fig. 4a. Comparison of compressive strength development between kerosene and water curing for series I mixes.

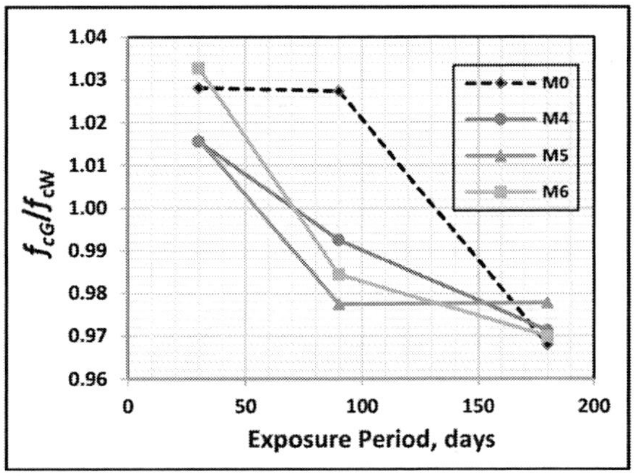

Fig. 4b. Comparison of compressive strength development between gas oil and water curing for series II mixes.

3.2 Total absorption and permeability

Figs. 5a, b and c show the results of total absorption for tested RPC mixes in water, kerosene and gas oil respectively for the assigned ages. Absorption values depend mainly on the viscosity of the penetrating fluid and the porosity of the absorbing media. The values of absorption were the highest in water and the lowest in gas oil. This descendant order is due to the variation in viscosity of these fluids. For all exposure conditions, there was a deceleration in the rate of absorption. The rate was higher for the period 7 - 90 days than that of the period 90 – 180 days. This trend could be resulted from the continuous hydration of the binder.

Figs. 6a, b and c display the relationship between the FA/A ratio and total absorption for the studied mixes, series I, when they are cured in water, kerosene and gas oil respectively. For the investigated ratios, 52, 25 and 0 %, mix 2, with the FA/A = 25%, has the lowest absorption values when the curing liquid is water or kerosene. For gas oil, mix 3, with no fine sand, showed the lowest absorption values. This behavior could be related to the viscosity of curing liquid. Based on these observations, it could be concluded that replacing fine sand with natural graded coarse aggregate could be most useful with FA/A = 25% although the strength was not the highest, mix 2. Absorption is a crucial property when dealing with oil products storage.

Furthermore, Figs. 7a, b and c show the relationship between the A/B ratio and total absorption for the studied mixes, series II, when they are cured in water, kerosene and gas oil respectively. The studied ratios were 1.11, 1.25 and 1.35 by weight of binder. This approach leads to the increase in total absorption for all mixes and for all ages. Therefore, it is not recommended to make the inclusion of coarse aggregate as a replacement of the binder.

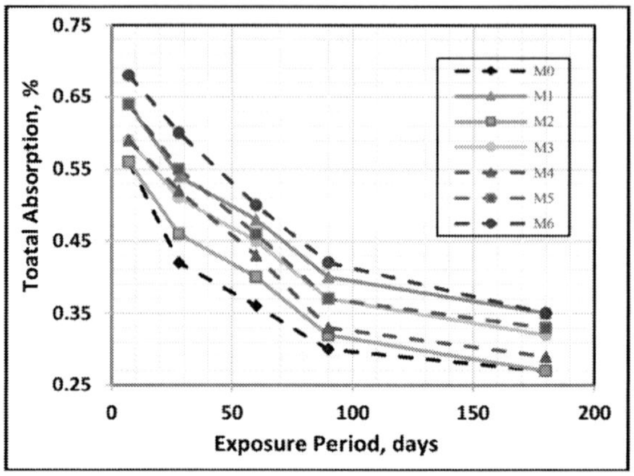

Fig. 5a. Total absorption of water-cured RPC mixes.

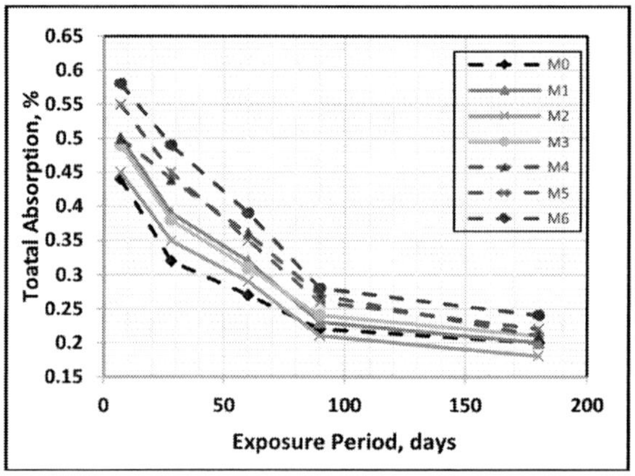

Fig. 5b. Total absorption of kerosene-cured RPC mixes.

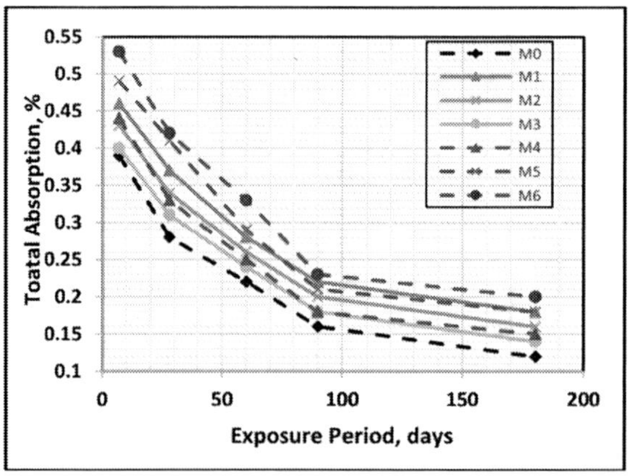

Fig. 5c. Total absorption of gas oil-cured RPC mixes.

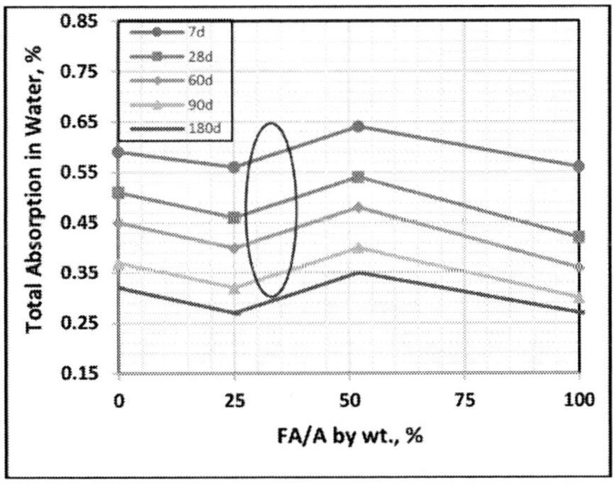

Fig. 6a. Relationship between absorption and FA/A for water-cured RPC mixes.

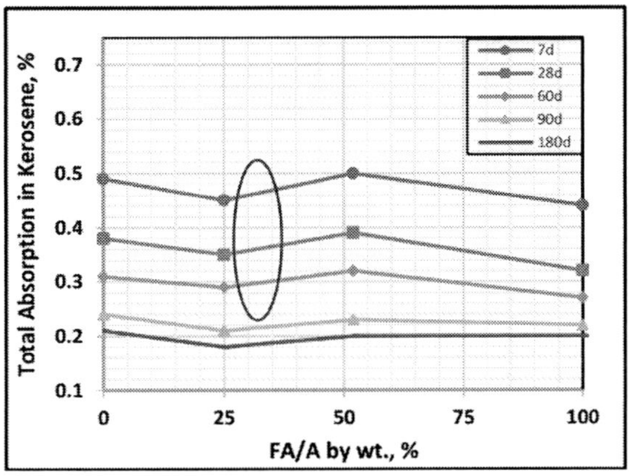

Fig. 6b. Relationship between absorption and FA/A for kerosene-cured RPC mixes.

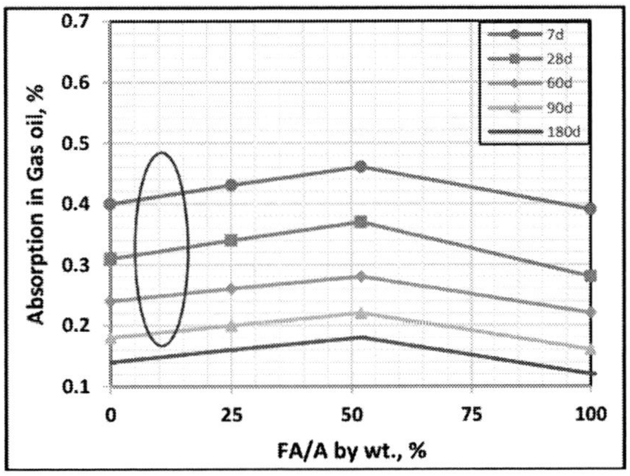

Fig. 6c. Relationship between absorption and FA/A for gas oil-cured RPC mixes.

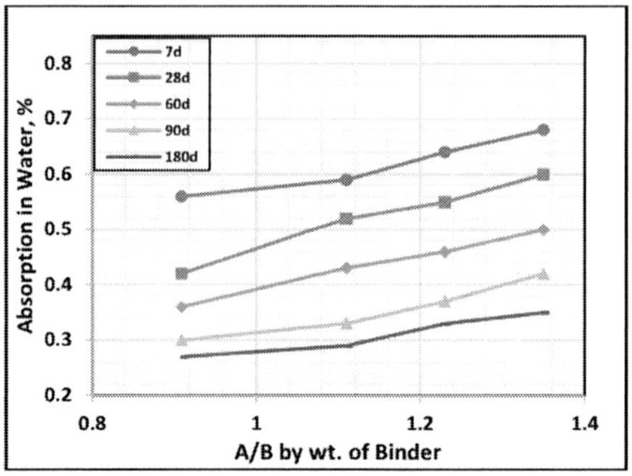

Fig. 7a. Relationship between absorption and A/B for water-cured RPC mixes.

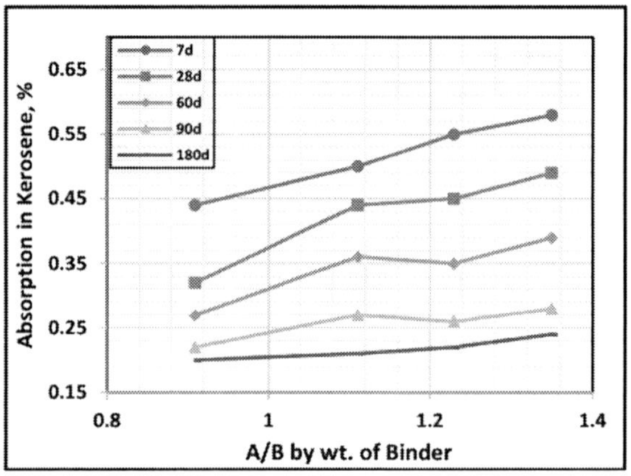

Fig. 7b. Relationship between absorption and A/B for kerosene-cured RPC mixes.

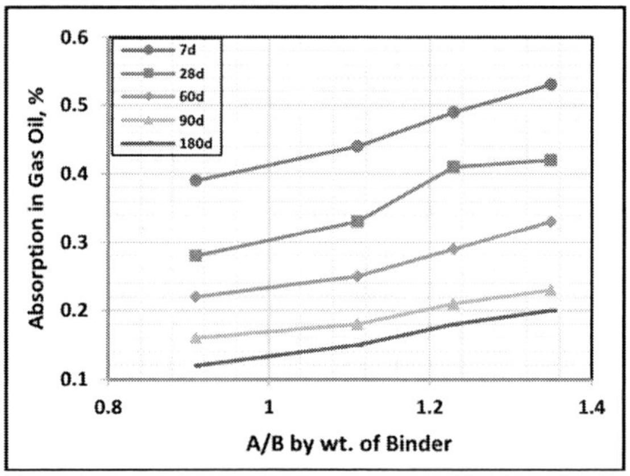

Fig. 7c. Relationship between absorption and A/B for gas oil-cured RPC mixes.

The results of the conducted permeability test for the RPC at ages of 60 and 90 days, cured in water or exposed to oil products, have no conclusive indication. All mixes have **zero** value for the coefficient of permeability. These values could not be considered as an indication for the impermeability of RPC mixes to water or oil products. In spite of the dense microstructure and low porosity of these mixes due to heat treatment and their high ability to be pressurized, but it seems that this test is not sensitive to the variation in microstructure between tested mixes.

IOP Conf. Series: Materials Science and Engineering **579** (2019) 012044 doi:10.1088/1757-899X/579/1/012044

4. Conclusions

The following findings are based on the results of the current work:

a. The basic principle of eliminating coarse aggregate in RPC to minimize the heterogeneity of the mixture could be gone-beyond.
b. The approach of replacing fine sand partially or totally by natural graded coarse aggregate without changing the binder content could be adopted. The most useful FA/A weight ratio is expected to be within the range of 0 - 25 %. This range was found to create an optimum situation between strength and absorption for kerosene and gas oil exposures.
c. It is not recommended to include the coarse aggregate in RPC mixtures as a replacement to the binder specially when dealing with oil products storage.

Abbreviations

A/B Aggregate to binder ratio by weight.
ASTM American society for testing and materials.
BS British standards.
CA Coarse aggregate.
FA/A Fine aggregate to total aggregate ratio by weight.
f_{cG} Compressive strength of gas oil-cured concrete, MPa.
f_{cK} Compressive strength of kerosene-cured concrete, MPa.
f_{cW} Compressive strength of water-cured concrete, MPa.
HRM High reactivity Metakaolin.
RPC Reactive powder concrete
W/B Water to binder ratio by weight.

References

[1] Lee N B and Chisholm D H 2006 *BRANZ* **146**
[2] Cheyrezy M, Maret V and Frouin L 1995 *Cem. Conc. Res.* **25** 7
[3] Richard P and Cheyrez M 1995 *Cem. Conc. Res.* **25** 7
[4] Collepardi S, Coppola L, Troli R and Collepardi M 1997 *ACI Pub. SP* **173**
[5] *ASTM* C150 2015
[6] *ASTM* C 618 2015
[7] *BS 882* 1992
[8] Al-Attar T S, Ali A S and Al-Numaan B S 2012 *Brit. Mat. Com.* **10**
[9] *BS EN* 12390-3 2002
[10] *ASTM* C496 2015
[11] *ASTM* C267 2015
[12] Neville A M, 2011 *Prop. of Conc.*, 5th ed
[13] Slate F O and Hover K C 1984 *Fract. Mech. Conc.*
[14] CEB-FIB 1984 Const. Mod H. St. H.P. Conc.
[15] Al-Kubaisy M A and Young A G 1975 *Mag. Conc. Res.* **27** 92
[16] *US Patent No.* 4 475952 1984
[17] Blaszczynski T Z 1996 *Statyba* **2** 6
[18] Lea F M 1974 *Chem. Cem. Conc.* 3rd ed
[19] Wilson S A, Langdon N J and Walden P J 2001 *Proc. ICE* 3
[20] Aitcin P C 1998 *Hi. Perf. Conc.*

A comparison of Al-Fe pillared Iraqi clays for Catalytic wet air oxidation

G Y Al Kindi* **and F H AL Ani**

Civil Engineering Department, University of Technology, Baghdad, Iraq

***Corresponding author's e-mail address:** *40126@uotechnology.edu.iq

Abstract. Five types of Al-Fe pillared Iraqi clays have been selected depending on a natural clays regions (Anbar- Mosul- Basra- Baghdad- and Erbil) have been used in preparation it, and tested for catalytic wet air oxidation (CWAO) using phenol as target compound at different parameters (temperature, pressure, phenol loading, catalyst loading). The Al-Fe pillared of Anbar- Mosul –and Erbil clay showed a higher activity than Baghdad and Basra for phenol oxidation in Batch reactor. After 8 days continues test for activating AL-Fe PC in PBR, Deactivation appears on catalyst of Anbar in different temperature, but in Erbil with increase temperature, Mosul catalyst has higher activity than other catalyst. Mosul catalyst has higher stability than Erbil and Anbar in blocking of surface area, Leaching Fe was negligible, Fairly different values of apparent activation energy were obtained for this reaction with these five types of catalysts, suggesting that the prevailing pathway in air decomposition is different. From all above discussion, Mosul catalyst in considered the best catalyst.

1. Introduction

Quantities of water is used in industrial processes, which discharged to surface water system before treatment [1], some Organic pollutant load in aqueous wastewaters are toxic [2], such as phenol, polyphenol, and BTEX (benzene, toluene, ethylbenzene, 1, 2-dichlorobenzene, o-, m-, p-xylene) [3], Phenol one of these pollutant is awful odor and taste to water, even at concentration in the parts per billion (ppb)[4], it is present in the effluent from major chemical and pharmaceutical industries, such as petrochemical industries, petroleum refineries, cool gasification operation, liquefaction process, resin manufacturing industries, dye synthesis units, paper mills [5], some level of phenol has an effect on human, organisms, and water, (10-24mg/l) the level of phenol has an effect on human, (150mg/100ml) that level has an effect on lethal blood [6], also effect on organism, (9-25mg/l) influence on fish [6], and as standard of Environment Product Agency (EPA) for drinking water 0.5μg/l [7], and for surface water less than 1 part per billion (ppb)[8].

Treatment for toxic effluents wastewater by biological process may not be suitable so that, continuous efforts is required to remove phenol from wastewater in order to reduce treatment cost [9], some of these treatments steam distillation [10] Separation by extraction [11] [6], Separation by adsorption [12], Separation by membrane [13], electrochemical oxidation [14], Biochemical abatement [15], Incineration, and advance treatment. (WAO) wet air oxidation is a method for oxide to organic compound with oxygen or air under high temperature and pressure, catalyst wet air oxidation process (CWAO), can speed the reaction temperature and pressure, the organic compound can be partly converted to less toxic intermediate or complete converted to CO_2, and H_2O (16). Two types of solids have been employed to catalyst for oxidation organic compound in wastewater, metal and noble metal oxide, the later have expansively used as active compounds (17), many advantages from combining two or more metal oxide catalysts, that improved selectivity , activity, and increasing rate of process, to increase the efficiency or process of wet oxidation, which has been modified or used suitable combination of process (18), Sugunan, 2002 [19], show that mixed Fe with Al in different ratio, the Fe $-Al_{0.5}$ that the best ratio to converted benzoyl chloride.

Also Bankovi, 2009 [20] mixed different contains of Fe with clay shows the composition texture and morphological properties different only in Fe contain according to the Fe amount of Fe added to clay. The CWAO for wastewater carried out in batch or fixed bed reactor to choose the suitable reactor, according to process of selectivity and stability (21). To understand the CWAO process through laws that describe the establishment of both reaction pathway during the oxidation reactions, and design of CWAO units, its important is to correct and safety design chemical reactors [22], [23], because in this reactor the reaction complete conversion to CO_2 and H_2O [26], apparent kinetic model usually favored for industrial researchers for scale-up of commercial reactor [24]. The oxidation of organic compound in three phase catalytic reactions, gas, liquid phase reactants are in contact with the solid phase [27], when the gas dissolved in liquid phase and diffusion in the bulk of the liquid, it is a physical state, where in the chemical state, both gaseous and liquid phase reactants diffuse into the porous structure of the catalyst and react with active sites, the products diffuse out of the particle in opposite direction [28].

The goal of this research is selecting the five types of Al-Fe pillared Iraqi clay according to Iraqi regions to study the wet air oxidation for phenol degradation from synthases wastewater and to evaluate the catalyst as activity and stability and to choose the best one.

2. Materials and Methods

2.1 Pillared clay characteristics

Five types of (extrudates 2*8 mm) AL-Fe pillared clays selected according to clays regions (Erbil, Baghdad Anbar- Basra- and Mosul) have been brought to study physicochemical properties, by (XRD) X-ray diffraction, (XRF) X-ray Florescent and BET (Blumenauer Emmer Teller) to determinate the surface area.

2.2 CWAO Experiments

The experimental work for phenol degradation by CWAO with different pillared clays were performed in a Basket stirred tank reactor (BSTR), the reactor consists of equipment shown in Figure (1), The experimental work procedure of (BSTR) is shown in Figure 2.

The filtered samples were analyzed by Germany High Performance Liquid Chromatography (HPLC) YL 9100 HPLC System, YL 9101 Vacuum Degasser. And YL 9110 Quaternary pump, YL 9120 UV/Vis Detector with C18 column. A mobile phase H_2SO_4 of variable composition was programmed at a 1.2 ml/min flow rate. The wave length between 450-550λ detected phenol and low molecular weight Intermediates compound. The (UV –VV-1100, UEBO9025) Spectrophotometer was used to detect phenol concentration. The total organic carbon (TOC) has been measured by Shimadzu TOC-LCSH/CSN Standalone Model TOC analyzer. Residual Fe concentration in the reaction media was also analyzed by Atomic Absorption.

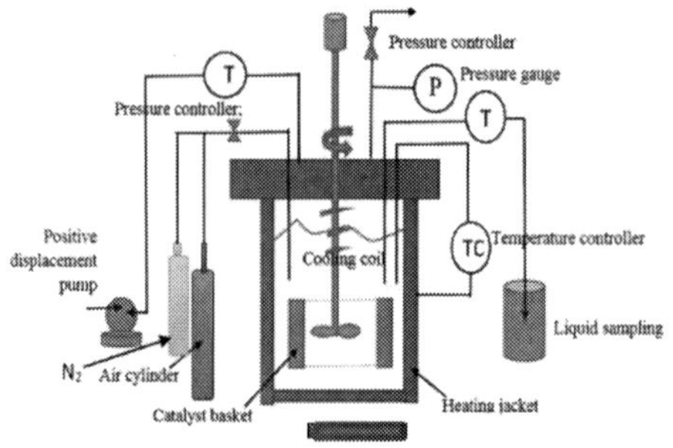

Figure 1. Diagram of the experimental setup of (BSTR)

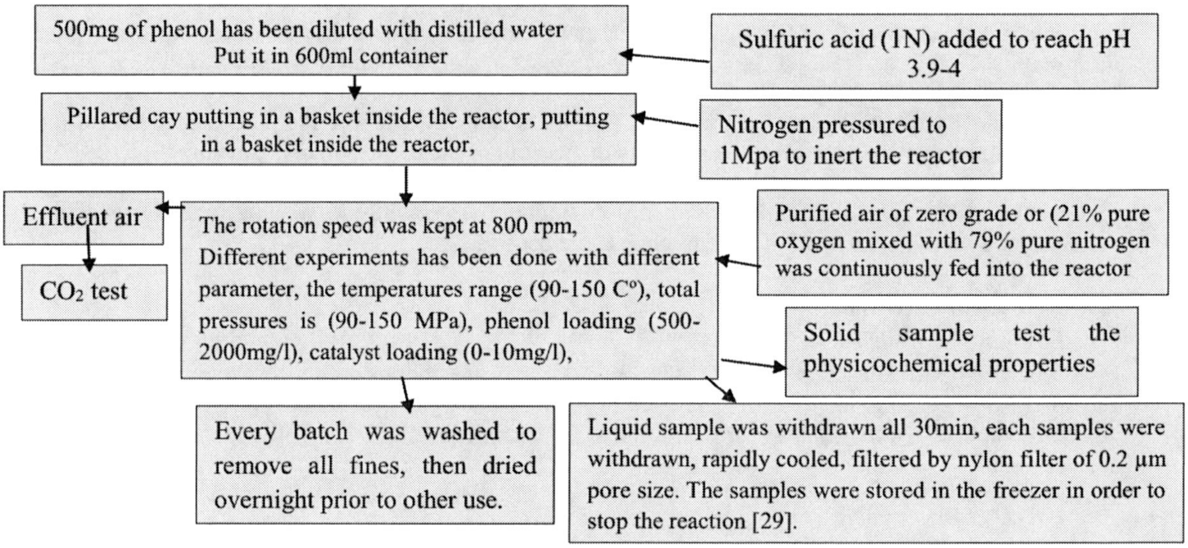

Figure 2: Diagram of the procedure of experimental work in (BSTR)

2.3 Stability experiments

The experiments of catalyst stability for CWAO of phenol in packed bed reactor with Al-Fe pillared clays for Anbar, Erbil, Mosul regions, the reactor system consist as in Figure (3),

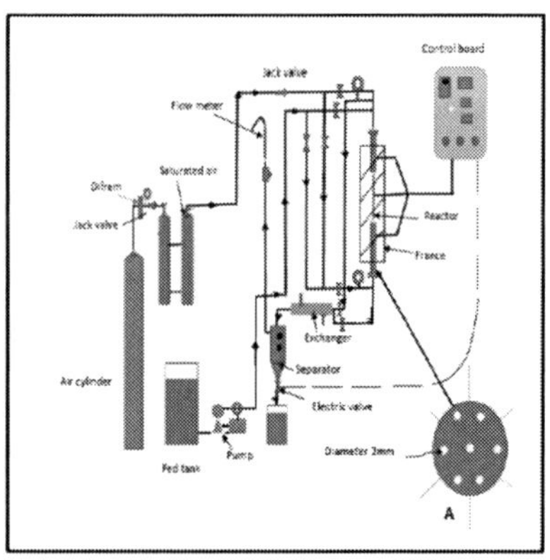

Figure 3. Diagram of the experimental setup of (PBR)

IOP Conf. Series: Materials Science and Engineering **579** (2019) 012045 doi:10.1088/1757-899X/579/1/012045

The experimental work was done with different gas superficial velocity between (0.2-0.8 cm/s) and liquid superficial velocity between (0.2-0.8 hr^{-1}) as stated by AL Dahhan (2005) [29] to keep the flow regime in bubble flow. The 500 gm of phenol has been diluted with distilled water to be used as synthesis's wastewater, as ideal wastewater, which recommend (500-1500mg/l) concentration of phenol in typical wastewaters from oil refineries of phenol mixed. The synthesis's wastewater pumped from fed tank to the reactor, in the limited velocity to wetted the catalyst, with Controlled the pressure to limit (3.2 Mpa), than the temperature raised to 150°C, the pressure and temperature should be controlled along the space time, when the air mixed with liquid after the needle valves, The liquid and gas mixed and preheated in the inert zone before entering the reactor from the bottom through a distributor, when arrived to steady state in 3.2Mpa, 150C° and in desired (gas , liquid superficial velocity), the liquid cooled in exchanger, and it will be collected in separator to turn off the sign and the liquid is drained after 8 hr till 8 day, samples of liquid collected filtered in micro nylon filter and stored in freezer to time of test, to study phenol and intermediate phenol. Some Instruments perversely used to measure the phenol, intermediate, and TOC.

3. Results and Discussion

Result of physicochemical characteristics of crystalline structure for Al-Fe pillared (Erbil, Baghdad Anbar- Basra- and Mosul) by XRD at 2θ between (0-70), show 3 peaks of 2θ appeared between 20 to 27, observed in Figure (4). From XRF the result of concentration iron, aluminum and calcium oxides has been shown in Table (1), Basra has the highest concentration of Fe_2O_3, while Baghdad have highest value of Al_2O_3.

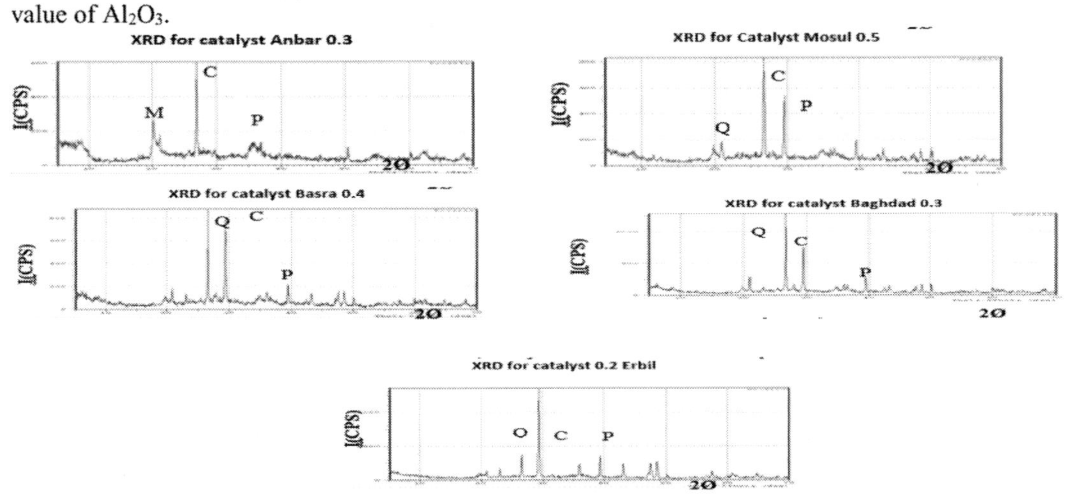

Figure 4. XRD result for Al -Fe pillared clay for Erbil, Mosul, Anbar, Baghdad, Basra

Table 1: main composition of five types of pillared clays

Ratio	Anbar catalyst	Mosul catalyst	Baghdad catalyst	Basra catalyst	Erbil catalyst
$\%SiO_2$	20.26	32.35	39.04	28.7	55.36
$Fe_2O_3\%$	7.98	9.82	7.8	14.81	9.6
$AL_2O_3\%$	15.37	14.35	29.49	21.73	23.4
CaO%	21.44	11.3	5.21	9.28	6.51

IOP Conf. Series: Materials Science and Engineering **579** (2019) 012045 doi:10.1088/1757-899X/579/1/012045

The Result of BET has been shown in Table (2), that is to say that the higher surface area for Erbil pillared clay than Anbar, Mosul, Basra and Baghdad respectively, as well as the pore volume and pore size, according to the International Union of Pure and Applied Chemistry (IUPAC) classification of pore size (Å) as macropore (≥ 500), mesopore (20 to 500) supermicropore (7-20) and ultramicropore (≤ 7) [30]. From the results, obtained Baghdad has the highest value of pore size width 150.4 Å, but Basra has the lowest value 75.1 Å, all regions are mesopore size [31].

Table 2: show result of BET surface area of five types of pillared clays

Samples Region	(BET) Surface area (m ²/g)	Average Pore volume m ³/g)(	Pore size width Å
Anbar Catalyst	97.815	0.0962	78.4
Erbil Catalyst	100.74	0.1221	124.2
Baghdad Catalyst	56.973	0.0907	150.4
Mosul Catalyst	93.605	0.1430	89.7
Basra Catalyst	61.371	0.1109	75.1

3.1 Effect of Parameters

3.1.1 Effect of Temperature

Figure (5) show the result of TOC conversion with different temperature for all catalyst types, the higher TOC conversion for Anbar catalyst reach to 88% than Erbil 79.28%, Mosul 78.2%, Baghdad 66.5%, and Basra 66.46%. The phenol conversion has been shown in Figures (6). The result shows phenol conversion increase with increase temperature with another conditions were constant, the results reveal that a temperature of 150°C is necessary to obtain an appreciable pheno conversion along the process. The reaction rate is increasing gradually with time and more influence on the reaction rate is at a temperature of 150, Erbil has high value of phenol degradation rate then Mosul, Anbar, Baghdad, and Basra respectively. The highest rate of reaction phenol, from the result phenol conversion is 96.2, 95.44, 93.7, 80, 76.3% for Erbil, Mosul, Anbar, Baghdad, Basra, respectively.

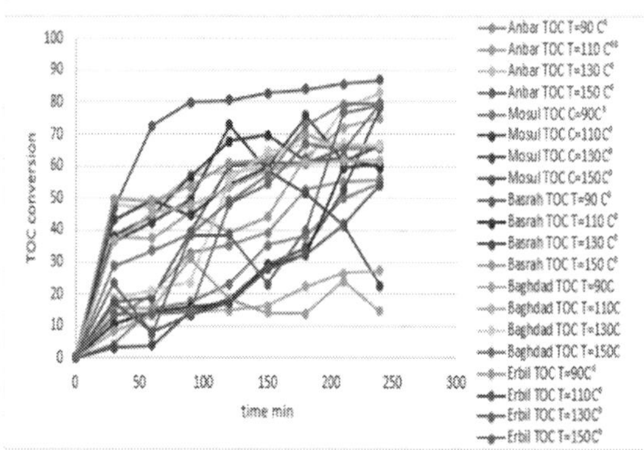

Figure 5. TOC conversion during 4hr in catalyst loading 7mg/l, pressure 2.5Mpa, 500mg/l phenol concentration, for all regions.

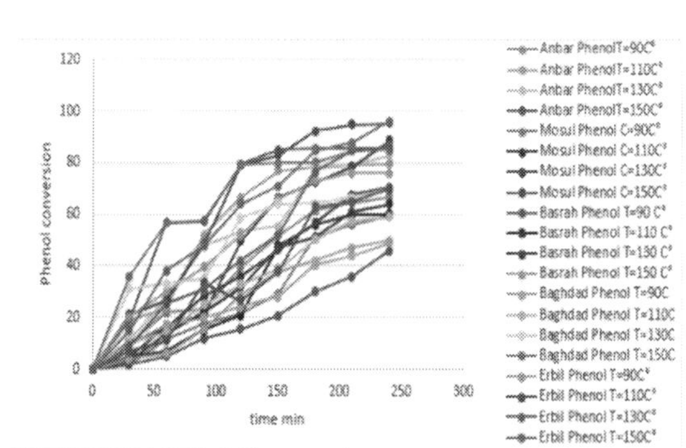

Figure 6. Phenol conversion during 4hr in catalyst loading 7mg/l, pressure 2.5Mpa, 500mg/l phenol concentration for all regions

3.1.2 Effect of pH

Figure (7) shows phenol conversion profiles for reaction tests carried out with an initial air concentration. Initial pH of 3.9 and (5.48) and temperature 110°C to 130°C, the later corresponds to the natural pH of the wastewater. Without pH adjustment, phenol conversion curves show a sigmoidal profile, which is typical of radical reactions. Also from the figure show an increase in the decomposition rate when the pH is 3.9, while this rate is less at pH of 5.48 for all regions. Because pH values between 3 and 4 led to the least decomposition of air and higher amounts of available hydroxyl radicals in the solution [32]. The degradation rate in Anbar is higher than that in the rest of the regions, followed by Mosul, Erbil, Baghdad, and Basra, respectively as a result of the nature clay (pore size and surface area) of the pillared clay (catalyst). Figure (8) shows TOC conversion profiles for reaction tests for all catalyst regions.

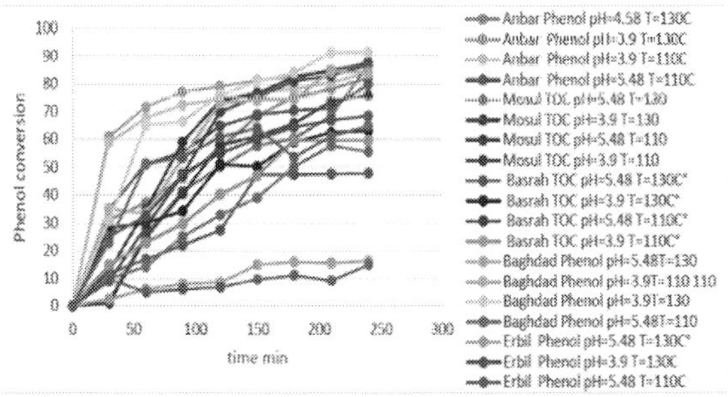

Figure 7. Phenol conversion during 4hr in different pH, catalyst loading 7 mg/l, pressure 1.5Mpa, 500mg/l phenol concentration, for all regions.

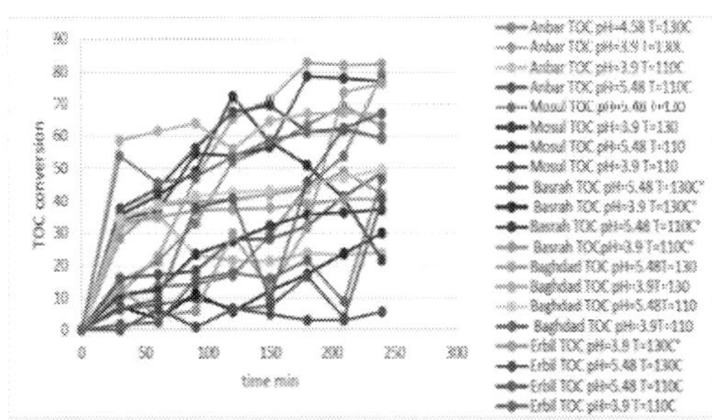

Figure 8. TOC conversion during 4hr in different pH, in catalyst loading 7 mg/l, pressure 1.5Mpa , 500mg/l phenol concentration,. for all regions

3.1.3 Effect of Phenol Concentration

The result of phenol conversion is presented in Figure (9) for five regions. It is known that during the CWO of phenol, polymeric compounds can be formed and promoted by acidic sites on the catalyst, which can adsorbed at the surface of the catalyst and decrease its activity [33]. Baghdad pillared clay has higher value of degradation, then Mosul, Anbar, Erbil, and Basra. The lower phenol conversion result shows that with high initial phenol concentration of 2000 mg/l, the pillared clay remained active in some region and complete mineralization after 4hr, while in another region uncompleted as Anbar catalyst. Deactivation appears on the surface of pillared clay during the CWO of phenol, the high conversion of TOC is shown in the Figure (10).

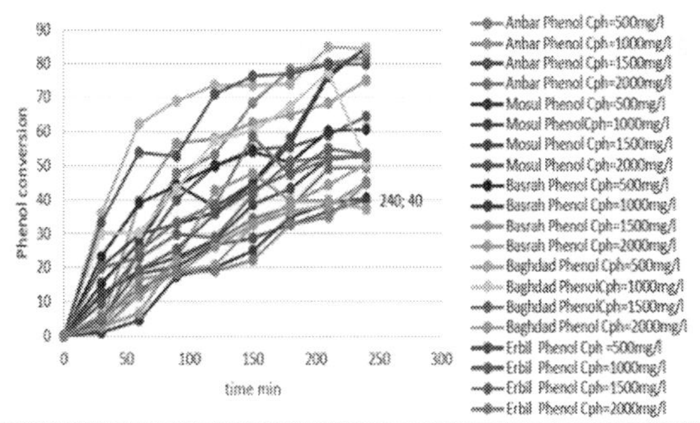

Figure 9. Phenol conversion during 4hr in different phenol concentration, catalyst loading 7mg/l, pressure 2. 5Mpa, for all regions.

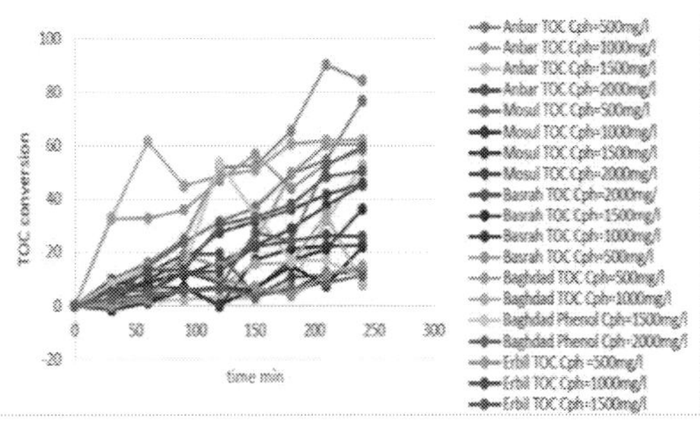

Figure 10. TOC conversion during 4hr in different phenol concentration, catalyst loading 7mg/l, pressure 2. 5Mpa, . for all regions

3.1.4 The Air Pressure Effect

Figure (11) shows the effect of pressure (0.8-2.0 -3.5Mpa) on phenol conversion at constant temperature 130°C for all catalyst regions. For all regions the high oxidation of phenol at 3.5 Mpa when equilibrium vapor pressure is 0.25Mpa in 130°C. From these figures, the highest phenol conversion was for AL Mosul, Anbar, Erbil, Basra, and Baghdad respectively, while Figure (12) shows the TOC conversion for all catalyst region. As a result, the highest conversion TOC was in AL Anbar, then Mosul, Erbil, Basra, and Baghdad because these catalyst different in pore size and surface area, so that the result of Anbar, Erbil, and Mosul convergent. The applied air pressure exceeded the vapor pressure of the liquid phase at a given temperature, so that the wet air oxidation was kept in the liquid phase. Corresponding to the temperature range of 90-150oC, the value of equilibrium vapor pressure is from 0.1 to 0.45 MPa [34]. The oxygen is capable to oxide the phenol and substitute oxygen atom an aromatic ring to be quinine or dihydric phenol, and also capable to attract t phenol double bond of Carbone to form carboxylic acid, that represents the very low concentration in aqueous phase.

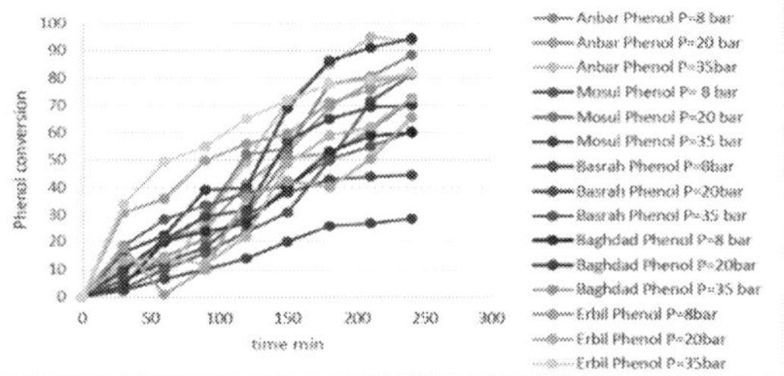

Figure 11. Phenol conversion during 4hr at different air pressure, in catalyst loading 7mg/, 130 °C, 500 mg/l phenol concentration, for all catalyst region.

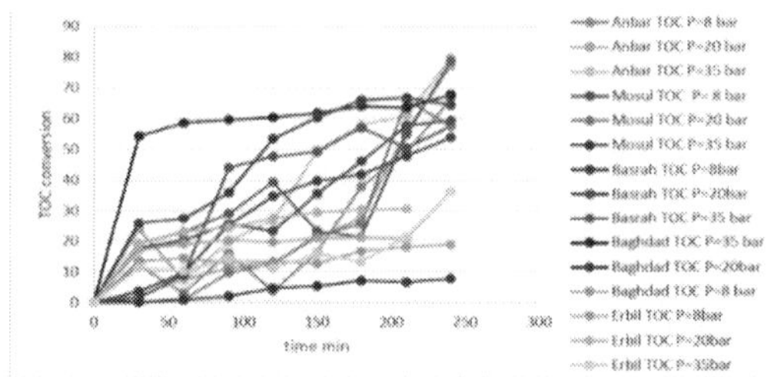

Figure 12. TOC conversion during 4hr at different air pressure,in catalyst loading 7mg/, 130 °C, 500 mg/l phenol concentration for all catalyst region.

3.1.5 Effect of Catalyst Loading

The different catalyst loading (0-5-8-10mg/l) investigated in this work for oxide phenol at 110°C and 1.5 Mpa, are shown in Figure (13) as conversion of phenol degradation during 4hr of reaction, From these results, it can be noted that unanalyzed lead (zero catalyst) is gradually increased in phenol concentration removal during 4hr, but considerably less than other value of catalyst loading. As a result, the air or oxygen is not significant on homogenous reaction but has an effect on heterogeneous reaction to oxidation of phenol. During the heterogeneous reaction, the phenol oxidation on active site of catalyst increases the acidity of solution. But for all types of catalysts the deactivation is not significant in these experiments. As a result, it has been noted that the best catalyst load were found in Anbar and Basra (8mg/l) then Mosul (10mg/l). Figures (14) shows the TOC conversion.

The results show that Anbar- Erbil- and Mosul have higher phenol conversion > 90% than other catalyst regions, so that these catalyst used in experimental of packed bed reactor determine the flow regime and to study the stability.

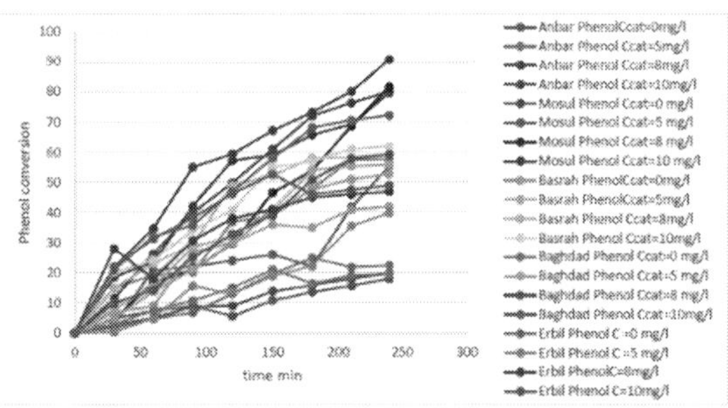

Figure 13. Phenol conversion during 4hr in different catalyst loading, temperature 130°C, 500mg/l phenol concentration for region.all

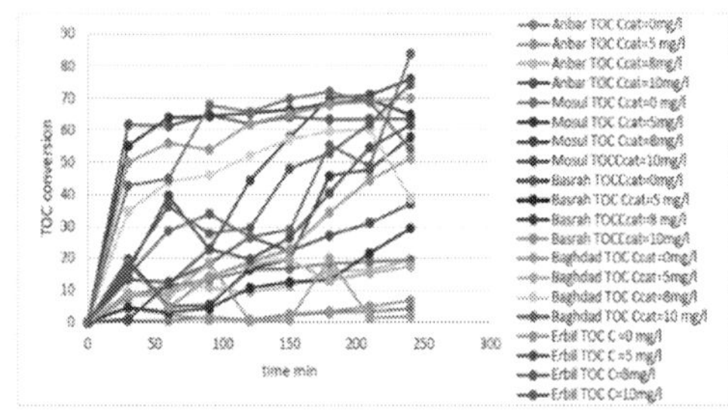

Figure 14. TOC conversion during 4hr in different catalyst loading, temperature 130°C, 500mg/l phenol concentration for all region

The aromatic intermediates evolve to carboxylic acids, the Hydroquinone, Malonic acid, Formic acid,

IOP Conf. Series: Materials Science and Engineering **579** (2019) 012045 doi:10.1088/1757-899X/579/1/012045

Malic acid, Oxalic acid, Acetic acid are intermediate of phenol for Anbar catalyst, and the Hydroquinone is higher concentration of phenol intermediate, while Malic acid is lowest concentration. The three intermediate phenol (Hydroquinone, Formic acid, and Oxalic acid) are appeared for Mosul catalyst, hydroquinone has higher concentration, while formic acid is lowest concentration. The intermediate for Erbil, Baghdad and Basra the two intermediate phenol formic acid and oxalic acid are appeared, the oxalic acid has higher concentration than formic acid.

The result of experimental hydrodynamic was done in packed bed reactor with different gas-liquid superficial velocity shows in the Figure (15) for different gas superficial velocity, but Figure (16) for different gas superficial velocity, from the result the suitable condition ($0.2hr^{-1}$) for liquid superficial velocity, (0.6 cm/s) for gas superficial velocity, these conditions keep the flow in bubble regime, in that regime the efficient wet is 100%, that refer to the reaction in liquid phase.

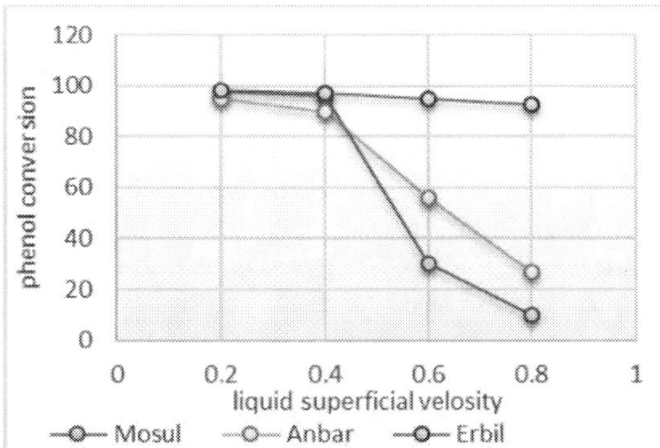

Figure 15. phenol conversion at different LHSV (Pressure= 3.2 MPa, phenol=500 ppm, and u_G= 0.28 cm/s).

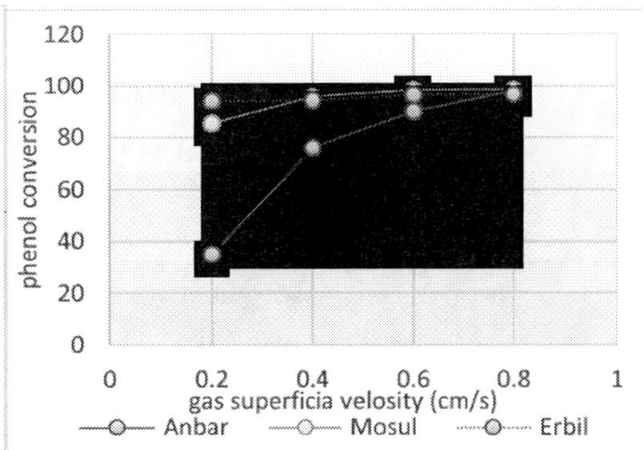

Figure 16. Effect of gas velocity on phenol conversion at LHSV=$0.6hr^{-1}$.

From the Leaching Fe test in batch reactor, the result was negligible along the reaction time, which required for practically complete phenol conversion, all result in part per billion. Deactivation appears on samples of Anbar in different temperature as shown in Figure (17), while Erbil Deactivation appear in high temperature as shown in Figure (18), Figure (19) shows result of leaching Fe for Mosul catalyst.

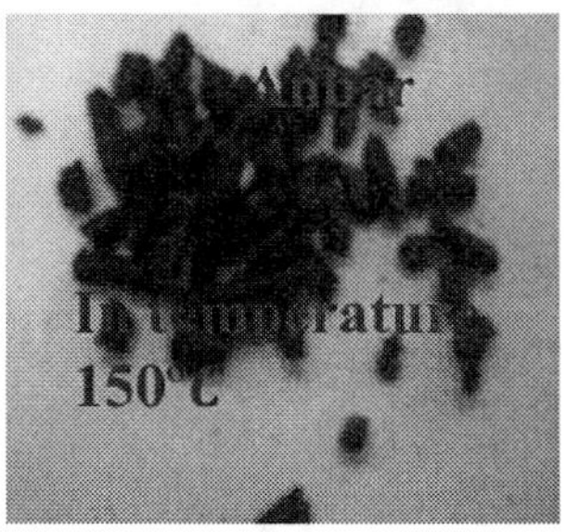

Figure 17. deactivation in Anbar catalyst during phenol oxidation in Batch reactor

Figure 18. deactivation in Erbil catalyst during phenol oxidation in Batch reactor

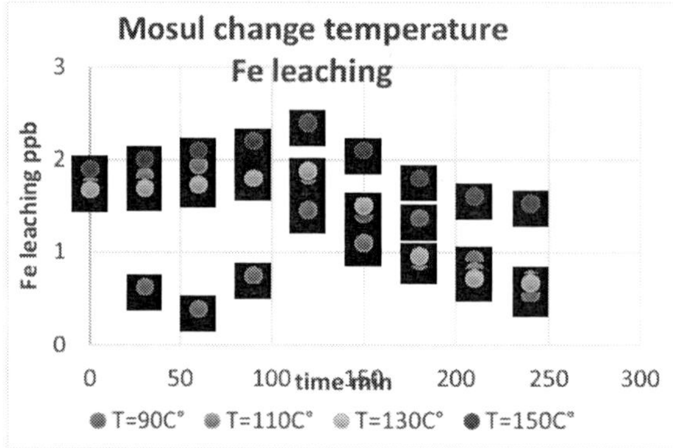

Figure 19. Fe concentration with different temperature in batch reactor

The above result shows that Anbar, Mosul and Erbil catalyst have the highest surface area, pore volume in mesoporous and have higher removal of phenol than Baghdad and Basra catalyst.

The three active catalyst (Anbar, Mosul, and Erbil) used to study activity and stability in PBR, from result the Mosul catalyst has highest conversion of phenol reach to (98.7%) at stable condition, LHSV equal to 0.6hr^{-1} and gas superficial velocity equal to 0.2 cm/s. in the same condition the higher conversion of phenol for Anbar catalyst reach to (98%), while Erbil catalyst shows the higher conversion of phenol reach to (96.7%), that indicate two these catalysts have high activity.

The physical blocking on active sites of catalyst during operation time (192 hr) was observed through the data collection in different operation condition. The physical blocking on active sites not consider in PBR operation of Erbil and Mosul catalyst for phenol removal, due to low leaching Fe value so that these catalyst can be continuously operated, but for Anbar catalyst the blocking appeared catalyst was crashed after 100hr as shown in Figure (20).

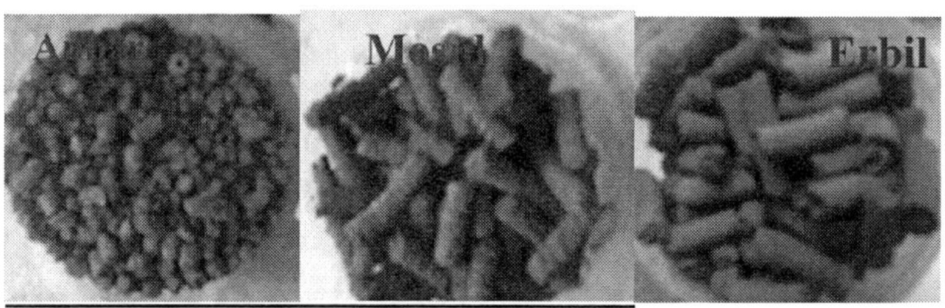

Figure 20. deactivation in Anbar, Mosul, and Erbil catalyst during phenol oxidation in PBR

The catalyst blocking after PBR also measured by changing the surface area (BET), after testing the surface area of Anbar catalyst is 79.043 m ^{2}/g, so that the blogging reach to (19.2%), while for Erbil catalyst blocking equal to (63%) when surface area after test equal to 37.26 m ^{2}/g, and surface area for AL-Mosul catalyst equal to 80.17 m ^{2}/g, so that the blogging equal to (14.35%), from all above discussion AL-Mosul catalyst is considered the best catalyst stability.

3.2 The apparent kinetic models

The apparent kinetic models are generally expressed in terms of either the simple power law or more complex equation based on adsorption–desorption mechanism, i.e. Langmuir- Hinshelwood- Hougen-Watson (LHHW) model, for the heterogeneous catalytic oxidation of aqueous organic pollutants, these models shows in Table (3) [35] [36].This fact is accounted for in the kinetic model by expressing the catalyst concentration in terms of an empirical correlation [37], thus, the actual phenol disappearance rate could be expressed as equation (1).

$$\frac{dA}{dt} = rH\ Ccat^{n} \tag{1}$$

Where $[A] = [A]_0$ that represent the initial conditions: Power law expressions has been typically pointed out. These equations are of the general form of Eq. 2.

$$r_{org} = k_0{}^{\beta} exp(-E_a/RT)\ [C_{O2}]^{a}\ [C_{org}] \tag{2}$$

In this study MATLAB is used to solve the kinetic model equations. Figure (21) shows the flow diagram that is used for parameters determinations.

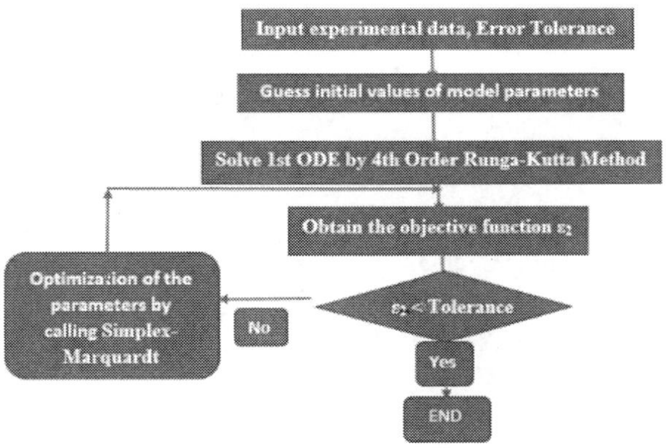

Figure 21. Schematic diagram for model parameter identification.

Table 3. Kinetic models proposed for air CWO heterogeneous reaction

Types	Models	Mechanical	
Power law	M1 $r_H = k_1 [A]^{P} [O_2]^{q}$	(empirical approach)	(3)
LHHW	M2 $r_H = \frac{k1\ kA\ ko2\ [A][O2]}{(1-KA\ [A]+KO2[O2])^{2}}$	single site O_{2+}	(4)
LHHW	M3 $r_H = \frac{k1\ kA\ ko2\ [A]^{0.5}[O2]^{0.5}}{(1+KA[A]^{0.5}+KO2[O2]^{0.5})^{2}}$	single site O_{2+}	(5)
LHHW	M4 $r_H = \frac{K1\ KA\ KO2\ [A]^{0.5}[O2]^{0.5}}{(1+KA[A])(1+KO2[O2]^{0.5})^{0.5}}$	Dual site. O2 +2	(6)

The experimental data is fitted with the proposed models M1-M4.The determination scheme was an algorithm coupling the 4th order Runga- Kutta numerical method [38] with the nonlinear Simplex-Marquardt algorithm [39]. Further optimization techniques are implemented to minimize the error between theoretical (computed) results and experimental data by a least-square (equation 7) so that the minimum of residual sum of squares (SQR) could be identified eventually for models M1-M4, respectively.

$$SQR=\sum_{i=1}^{n}\left(\frac{[A]_{sim}-[A]_{Exp}}{[A]_{Exp}}\right)^2 \tag{7}$$

IOP Conf. Series: Materials Science and Engineering **579** (2019) 012045 doi:10.1088/1757-899X/579/1/012045

Where the subscripts *Exp* and *Sim* represent the experimental and simulated values. the previously mentioned models have been tested for r_H with the kinetic parameter values, in Tables (4) were given the parameter for the best model for all regions Anbar, Mosul, Basra, Baghdad, and Erbil respectively.The actual phenol disappearance rate could be expressed as equation (8) as the average absolute relative errors errors (AAAR).

$$AAAR= \sum_{i=1}^{N} \left| (y_{sim, I} - y_{Exp, I}) / y_{Exp, I} \right| /N. \quad (8)$$

The models for all regions can be checked by comparing the simulation data with experimental data. Results of perfect models are shown in Figures (22), (23), (24), (25), (26) for Anbar, Mosul, Basra, Baghdad, and Erbil respectively, exhibits parity plots that compare the calculated with those experimental observations from the suitable models for all regions. From the result of the kinetic models M1, M2, M3, and M4 for all regions.

Table 4. Rate parameters for CWAO heterogeneous reaction for Anbar

Catalyst region	Stabile Model r_H	Parameter k_1		q	p	n	SQR	AAAR
Anbar	**M2**	**K**	**1.056E3e(-34213.05/RT)**	----	---	**0.87**	**1.23**	**0.0811**
		Kc_{ph}	**0.5E3e(-32958.38/RT)**					
		K_{co2}	**3E3e(-30216.8/RT)**					
Mosul	**M4**	**K**	**0.7625E3e(-32151.1/RT)**	---	---	**0.9**	**1.74**	**0.075**
		Kc_{ph}	**1.0325E3e(-4638.9/RT)**					
		K_{co2}	**3.857E3e(-36810.7/RT)**					
Basra	**M2**	**K**	**9E5e(-42176.16/RT)**	---	---	**0.89**	**1.12**	**0.102**
		Kc_{ph}	**0.28E5e(-41662.03/RT)**					
		K_{co2}	**5E5e(-32081.45/RT)**					
Baghdad	**M3**	**K**	**1.07E2e(-32346.6/RT)**	---	---- -	**0.65**	**1.3**	**0.082**
		Kc_{ph}	**1.193E2e(-35177.4/RT)**					
		K_{co2}	**1.475E2e(-40088.4/RT)**					
Erbil	**M2**	**K**	**0.523E4e(-25111.11/RT)**	----	---- --	**0.802**	**3.721**	**0.145**
		Kc_{ph}	**1.15E4e(-21232.4/RT)**					
		K_{co2}	**2.564E4e(-20022.9/RT**					

* Apparent activation energy in J/mol

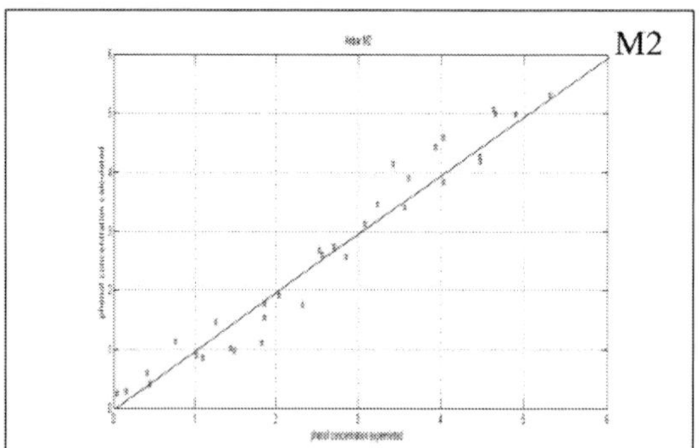

Figure 22. Parity plot for phenol concentration, (mol/m^3): for Anbar employing model M1; (b)employing model M2; (c) employing modelM3; (d) employing model M4.

IOP Conf. Series: Materials Science and Engineering **579** (2019) 012045 doi:10.1088/1757-899X/579/1/012045

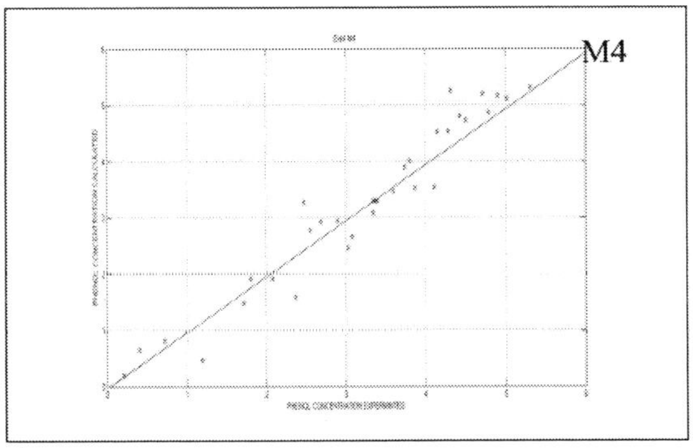

Figure 23. Parity plot for phenol concentration, (mol/m^3): for Erbil employing model M1; (b)employing model M2; (c) employing modelM3; (d) employing model M4.

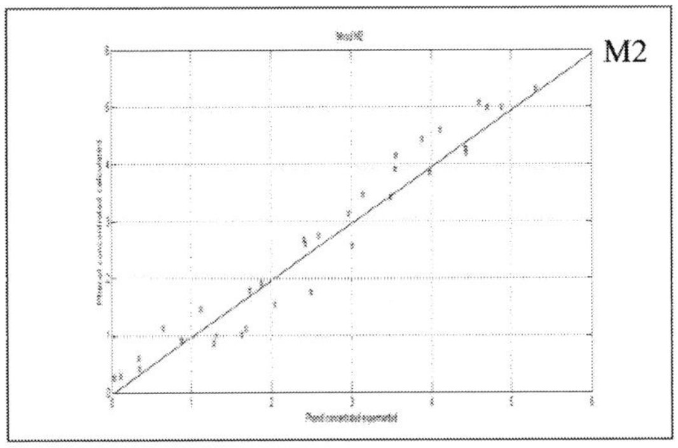

Figure 24. Parity plot for phenol concentration, (mol/m^3): for Mosul employing model M1; (b)employing model M2; (c) employing modelM3; (d)

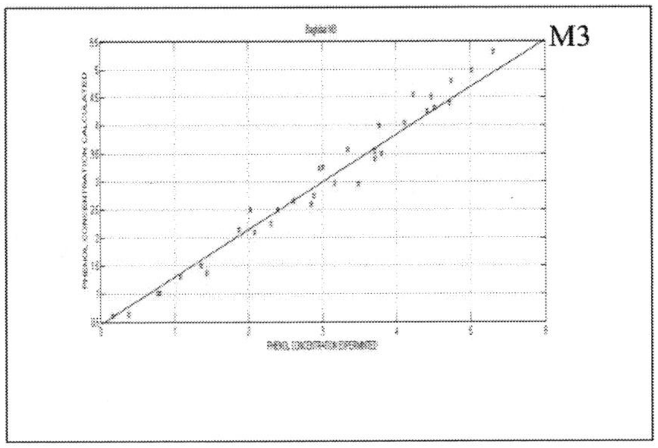

Figure 25. Parity plot for phenol concentration, (mol/m^3): for Baghdad employing model M1; (b)employing model M2; (c) employing modelM3; (d) employing model M4.

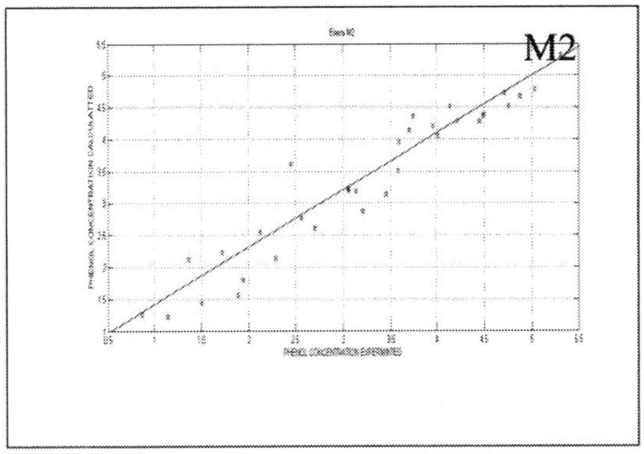

Figure 26. Parity plot for phenol concentration, (mol/m^3): for Basra employing model M1; (b)employing model M2; (c) employing modelM3; (d) employing

The reproductions of some representative experimental data are shown in Figures (27), (28), (29), (30), and (31) for Anbar, Mosul, Basra, Baghdad, and Erbil respectively, The best models with good agreement have been shown in Anbar then Baghdad, Erbil, Mosul, and Basra models respectively. M2 is slightly better in terms of the statistical standard for Anbar, Mosul and Basra. M3 Mode is better in Baghdad models. M4 Model is better in Erbil. Table (5) shows the best models equations obtained for all catalysts in this study

IOP Conf. Series: Materials Science and Engineering **579** (2019) 012045 doi:10.1088/1757-899X/579/1/012045

Table 5. Final kinetic model equation results for five catalyst regions

Region	Models	Model equations
Anbar	M2	$-\frac{dA}{dt}=\frac{k1\ kA\ ko2\ [A][O2]}{(1-KA\ [A]+KO2[O2])^{2}}*C_{cat}^{0.87}$
Erbil	M4	$-\frac{dA}{dt}=\frac{K1\ KA\ KO2\ [A]^{0.5}[O2]^{0.5}}{(1+KA[A])(1+KO2[O2]^{0.5})^{0.5}}*C_{cat}^{0.82}$
Mosul	M2	$-\frac{dA}{dt}=\frac{k1\ kA\ ko2\ [A][O2]}{(1-KA\ [A]+KO2[O2])^{2}}*C_{cat}^{0.75}$
Baghdad	M3	$-\frac{dA}{dt}=\frac{k1\ kA\ ko2\ [A]^{0.5}[O2]^{0.5}}{(1+KA[A]^{0.5}+KO2[O2]^{0.5})^{2}}*C_{cat}^{0.65}$
Basra	M2	$-\frac{dA}{dt}=\frac{k1\ kA\ ko2\ [A][O2]}{(1-KA\ [A]+KO2[O2])^{2}}*C_{cat}^{0.89}$

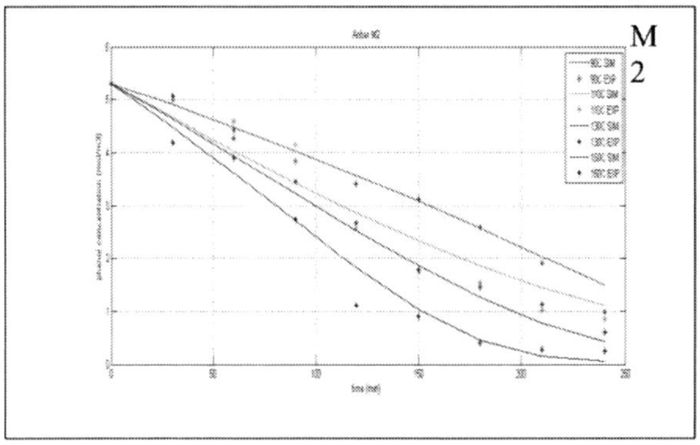

Figure 27. Measured and predicted time profiles of phenol removal at different Temperature, employing the best model M2 for Anbar.

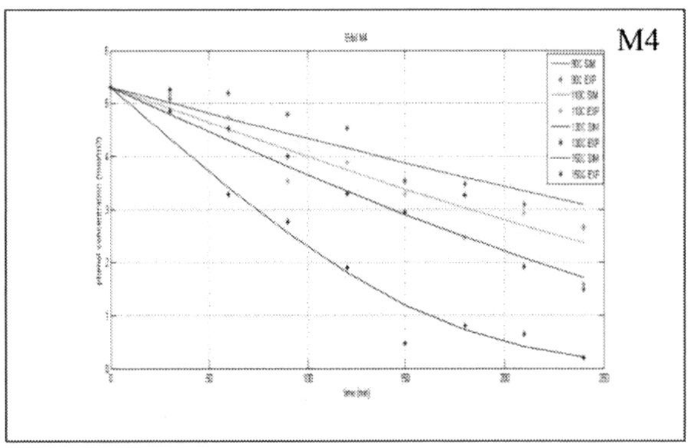

Figure 28. Measured and predicted time profiles of phenol removal at different Temperature, employing the best model M4; for Erbil.

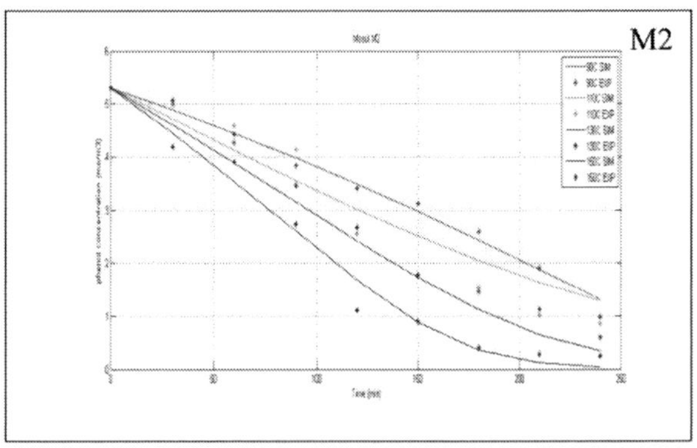

Figure 29. Measured and predicted time profiles of phenol removal at different temperature: (a) employing model M2; for Mosul.

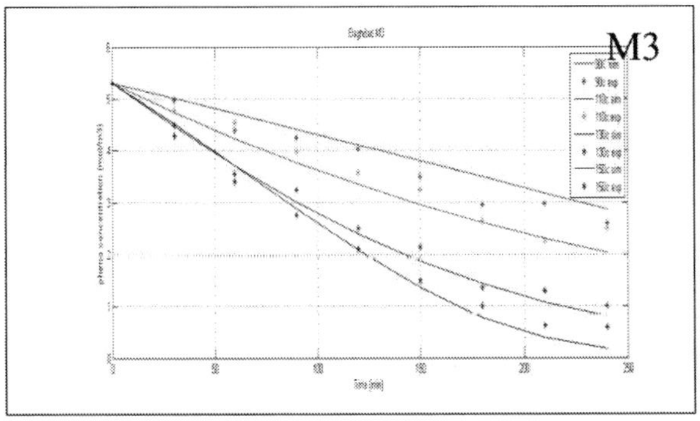

Figure 30. Measured and predicted time profiles of phenol removal at different temperature, employing the best model M3 for Baghdad.

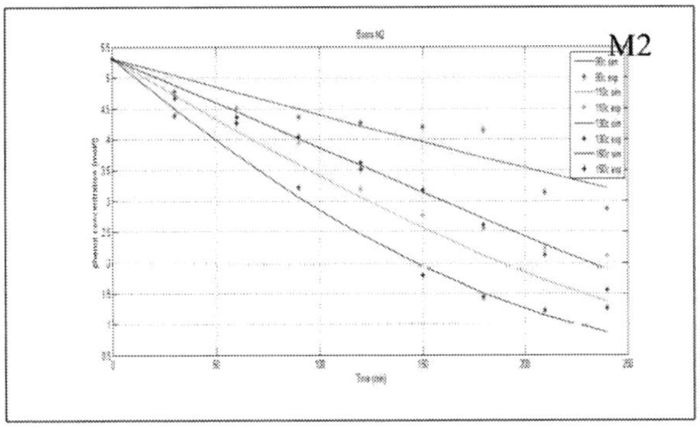

Figure 31. Measured and predicted time profiles of phenol removal at different Temperature, employing the best model M3 for Basra.

4. Conclusions

Five types of Catalysts based on clays regions (Anbar- Mosul- Basra- Baghdad- and Erbil) with Al-Fe have been tested for phenol oxidation with air in different condition of temperature , pressure, phenol loading, and catalyst loading, using phenol as target compound. The catalyst of Anbar, Mosul and Erbil showed a higher activity for phenol conversion and TOC removal for batch reactor test, promotes mainly the generation of OH radicals, in the other cases a substantial percentage of air must be giving rise to O2, with a low oxidation capacity at these mild operating conditions used, Mosul catalyst has higher activity than other catalyst. Mosul catalyst higher stability than Erbil and Anbar in surface area blocking, Leaching Fe was negligible within the reaction time required for a practically complete phenol conversion, all result in part per billion. The best model obtained was for each region as M2 model for Anbar, Mosul, and Basra, M3 model for Baghdad, and M4 model for Erbil, The activation energies of each reactions model with different catalysts are equal to (37114.014 j/mol) for Anbar (37795, 48783.9, 36628, 40785 j/mol) for Erbil, Mosul, Baghdad and Basra respectively. So that Mosul catalyst is considered the best catalyst.

References

[1] Tensel B 2008 *Chem. Eng.* 1 **17** 26.
[2] Luck F 1996 *Catal. Today* **27** 195-202
[3] Rubalcaba A M E, Suarez –Ojeda F, Stuber A, Fortuny C, Bengoa I, Metcalf J, Font J, Carrera and A . Fabregat 2007 *Water Science & Tecnology*, 55 **12** 221-227
[4] Kulkarni U S and Dixit S G 1991 *Ind Eng Chem Res* 30 1916-1920
[5] Matatov-Meytal Y I and Sheintuch M 1998 *Ind. Eng. Chem. Res.* 37 309-326
[6] Sunil J, Kulkarui, Jayant and P.Kaware 2013 *Journal Scientific & Reseach Oublication* **3** 4
[7] Rodriguez I and Liompart M P 2000 *Chramatogr* 885 291-304
[8] Busca G, Berardinelli S, Resini C, and Arrighi L 2008 *Journal of Hazardous Materials* 160 265-288
[9] Rubalcaba A, Suarez –Ojeda F, Stuber A, Fortuny C, Bengoa I, Metcalf J, Font J, Carrera and Fabregat A 2007 *Journal Water Science & Tecnology* 55 **12** 221-227
[10] Janda V and Krijt K 1984 *J. Chromatography* **283** 309
[11] Greminger D C, Burns G P, Lynn S, Hanson D N and King C J 1982 *Ind. Eng. Chem. Process Des. Dev.* 21 **1** 51–54
[12] Dabrowski A P, Podkoscielny Z, Hubicki and Barczak M 2005 *Chemosphere* 58 **8** 1049-70

[13] Xiao M, Zhou J, Tan Y, Zhang A, Xia Y and Ji L 2006 *Desalination* **195** 281-293
[14] Chen G 2004 *Purif. Techhnol* **38** 11- 41
[15] Saravanan P K, Pakshirajan and Saha P 2008 *Bioresour Technol.* 99 **1** 205-209
[16] Hu X, Lei L, and Chen G 2001 *J. Water Res* 35 8 2078-2080
[17] Joglekar H, Samant S D, and Joshi J B 1991 *J. Water Res* 25 **2** 135-145
[18] Suranjana C, and Krishna G, Bhattacharyy 2006 *Journal of Chemical Technology* **13** 499-504
[19] Sugunan S and Renuka N K 2002 *Ind. J. Chem.* 41A 304
[20] Bankovic P, Milutionovic A, Nikolic A, Rosic N, Jovic – Jovani and Jovanovic 2009 *Journal of Physical Chemistry* 83 **9** 1485-1489
[21] Stüber F, Polaert I, Delmas H, Font J, Fortuny A and Fabregat A 2001 *J. Chem.Technol. Biotechnol.* **76** 743–751
[22] Rob J, Berger E, Hugh Stitt, Guy B, Marin Freek Kapteijn, Jacob A and Moulijn 2001 *EUOKEN* **5** 1 USA
[23] Eftaxias Athanasios 2002 Catalytic Wet Air Oxidation of Phenol in a Trickle Bed Reactor: Kinetics and Reactor Modelling Doctoral thesis in Chemical Engineering of the Rovirai Virgili University
[24] Bos A N R, Lefferts L, Marin G B and Steijns M H G M 1997 *Appl. Catal. A*: General, 160 185
[25] Santos A, Yustos P, Durban B and Garcia Ochoa F 2001 *Ind Eng Chem Res* **40** 2773–2781
[26] Mishra V S V V, Mahajani J B, and Joshi 1995 *Ind. Eng. Chem. Res.* 34 **1** 2-48
[27] Kapteijn F, Johan J H, Xander T A N, and Moulijn J A 1999 *Baltzer Science* **3** 24-38
[28] Trunfio G 2008 "Catalyst development for the catalytic wet air oxidation (CWAO) of phenol PhD Thesis, University of Messina
[29] AL-Dahhan Guo 2005 *J. Chem. Eng. Sci*, **60** 735-746
[30] Nicoleta Platon, Ana-maria Rosu, Vasilica Alisa Arua, Denisa Ileana nistor, and Ilie Siminiceanu 2013 *Journal of Engineering Studies and Research* **19** 4 52-58
[31] Lastoskie Christian, Keith E G ubbins, and Nicholas Quirke 1993 *journal phiscal and chemical* **97** 4786-4796
[32] Bishop D F, Stern G, Fleischman M and Marshall L S 1968 *Ind. Eng. Chem. Process Des. Dev.* **7** 110
[33] Pintar A, Levec J 1992 *Chem. Eng. Sci.* **47** 9-11 2395-2400
[34] Lin S H, Ho S J, and Wu C L 1996 *Catal. Today* **29** 317-322
[35] Aurora S, Pedro Y, Beatriz D, and Felix G O 2001 *Ind. Eng. Chem. Res.* **40** 2773
[36] Guo J and Al-Dahhan M H 2004 *Chem. Eng. Sci* **59** 2023–2037
[37] Rubalcaba A, Suarez –Ojeda M E, Stuber F, Fortuny A, Bengoa C, Metcalf I, Font J, Carrera J and Fabregat A 2007 *Journal water Science & Tecnology* **55** 12 221-227
[38] Press H W, Teukolsly S A, Vetterling W T and Flannery B P 1992 *Numerical Recipes, the Art of Scientific Computing*, Cambridge University Press: Cambridge, U.K
[39] Duggleby R G 1984 *Comput. Biol. Med.* **14** 447

IOP Conf. Series: Materials Science and Engineering **579** (2019) 012046 doi:10.1088/1757-899X/579/1/012046

Impact of Mixing process on the physical and Rheological Characteristics of Asphalt Binder Modified with Nano- Silica Powder

Hussein H Zghair[1], Hasan H Joni[1] and Maan S Hassan[1]

[1]Civil Engineering Department, University of Technology, 10066, Baghdad, Iraq

*E-mail: 11539@uotechnology.edu.iq (Hussein H Zghair)

Abstract. In Iraq pavements materials are exposed to various distresses due to the adverse environmental conditions and heavy traffic. The improvement of performance of flexible pavements is necessary for decreasing these distresses. The objective of this research is to improve the mixing efficiency of nano-silica particles with asphalt binder. Two mixer type (mechanical and high shear), rotational speeds (2000, 4000 and 6000 rpm), and mixing period (30 and 60 minutes) were assessed. Three nano-silica per cent 2%, 4%, and 6% by weight of asphalt were added and the asphalt binder mixtures were prepared by heating at mixing temperature 140 °C. Furthermore, comparisons were made to assess the modified asphalt binder characteristics such as penetration value, softening point, temperature sensitivity (penetration index), ductility and rotational viscosity. Results showed that the use of nano silica as an additive was useful in improving the asphalt binder characteristics. Generally, has been found that the penetration value decreased and softening point increased, as well as the viscosity, were increased and the temperature sensitivity was developed. Additionally, high shear mixing was more useful especially to obtain a good dispersion of nano silica powder in asphalt binder and got a homogenous composite material.

1. Introduction

Pavements of a highway are exposed to various distresses in Iraq, due to the adverse environmental conditions and heavy traffic. The improvement of performances of flexible pavements is necessary for decreasing these distresses. In order to improve the performances properties and the long-time behaviour, a new group of asphalt mixture has been developed through the combination of various types of modifiers [1–6]. During the last decades, there have been a variety of asphalt binder additives such as styrene-butadiene-styrene (SBS) [7–10], lime [11, 12], crumb rubber [13] and electronic waste powders [14-16] used to develop physical and rheological characteristics of asphalt binder.

In recent years, many researchers are interested in using the nanomaterial in improving asphalt binder properties and performance [17]. Nanomaterials are particles that have at least one dimension less than 100 nm [18]. There are several types of nano materials that are commonly used in asphalt modification such as nano silica, nano-clay, carbon nanotube, nano titanium oxide, and nano-hydrated lime(19-24).Many researchers concluded that nano silica powder has been used to reinforce polymers (25), Portland cement concrete mixtures (26, 27). Addition of nano silica with polymer modified asphalt to enhance the rheological properties and increased rutting resistance (28). Asphalt binder was modified with nano-silica and seen that nano silica increased penetration value, reduced softening point and increased the rotational viscosity (29).Nano-silica has several characteristics such as high chemical purity, good dispersal ability, strong adsorption, large specific surface area, excellent and stability. Using nano silica powder by 2-4% of binder weight has been presented to decrease the rut depth by about half (30).

The major challenge in developing nano composite is the nanoparticles dispersion or chemical compatibility with composite materials. As a result, researchers are using proper procedures means to make uniform nano materials and stably disperse these in the asphalt binder. They found out that the issue of improving the nano silica (nS)/matrix interfacial adhesion and whole distribution must be resolved before completing the full potential of nano-reinforced composite materials. Dispersion of nanoparticles has been one of the main challenges due to the aggregation of the nanoparticles [31].

The object of study is to assess the mixing efficiency of a conventional mechanical mixer and the high shear mixer on a dispersion of nano silica into the asphalt binder by three contents 2%, 4% and 6% by weight. Also, evaluating the influence of mixing time on the physical and rheological properties of the modified asphalt binder.

2. Material and Methodology

2.1. Asphalt cement

An original asphalt 60/70 penetration grade obtained from the Dora refinery at the middle region of Iraq was used in this study. The basic characteristics of the asphalt binder were tested in the laboratory and showed in Table 1.

Table 1. Properties of original asphalt

Test	Test Conditions	Standard	Results	Standard Limits of Iraqi Specification (SCRB/ R9, 2003)[37]
Penetration	(0.1mm)	ASTM D5[32]	66	60-70
Ductility	cm	ASTM D113 [33]	130	>100
Flash point Fire point	…..	ASTM D92 [34]	Flash 302°C Fire 310°C	> 232 °C ---
Thin – film oven test	163 °C, 50gm, 5 hr.	0.222<1 Mass loss	Pen. 89.9 Duct. 120	>52 >50
Rotational Viscosity	Pa. sec	ASTM D4402 [35]	@ 135 °C 430 @ 165 °C 128	---
Softening point	…..	ASTM D36 [36]	49.5	---
Penetration index	…..	ASTM D36 [36]	-0.665	---

2.2. Nano size of Sio_2 powder (nS)

Nano silica is a white colour and hydrophobic powdered, a synthetic amorphous silica purchased from Aerosol Company in North America and shown in plate1 ,was used as the modifier of the binder. The main characteristics of nano silica, according to the manufacturer's sheets are listed in Tables 2& 3.

Plate 1. Nano Size of SiO2 Powder

Table 2. Properties of Nano SiO2 Powder According to the Manufacturer Characteristics.

Property	Result
Physical form	Powder
Color	White
Size	11- 12nm
Bulk Density (gr/cm^3)	0.1<
True Density (gr/cm^3)	2.4
Specific surface area, m2/g	200

Table 3. Chemical Properties of Nano SiO_2Powder According to the Manufacturer Characteristics.

Property	Result
SiO_2 content based on ignited material , wt.%	99.8
PH	3.7-4.5

2.3. Mixer Type

Conventional mechanical mixer and high shear mixer are apparatuses of the laboratory used for mixing modifiers such as polymers and nanomaterials with asphalt binder. Therefore, each mixer has its special qualifications. For example, due to the constant rotational speed of a mechanical mixer, its mixing time is supposed as a variable. And the rotational speed of the mixer is considered as variables for a high shear mixer.

2.3.1 Mechanical Mixer (M.M) is mixing technique as shown in the plate 2, the high-speed rotary motion of the mixer produces a high centrifugal force that helps to mix asphalt binder with additives. The rotational speed of the motor is constant through the time of mixing. The rotational speed based on the additive type and the operator's opinion. The mixer is used by assuming its mixing time parameter variable, while the rotational speed of the motor is constant in each minute, the composite material must be mixed as extended to obtain a homogenous material.

IOP Conf. Series: Materials Science and Engineering **579** (2019) 012046 doi:10.1088/1757-899X/579/1/012046

2.3.2 Shear Mixer (S.M.) is a special design of the tip that makes the high shear speed most proper mixers the previously mention mixers are used to mix asphalt binder with nanomaterial as shown in the plate 2. High shear mixer joins between the high shear speed degree and low pumping and circulations level [38]. High revolution rate pulls the materials from the bottom of the container to the top then a centrifugal force drives materials quickly out through the stator screen into lower speed area of the mix. Speed difference creates a hydraulic shearing force which produces homogenous composite materials [39].

Plate 2. Mixers Used to Mix Nano SiO_2 Powder with Asphalt Binder, Mechanical Mixer at Left Side and High Shear Mixer at Right Side.

2.4 Mixing Nano SiO_2 Powder with Asphalt Binder

Two mixers were used to mix nano SiO2 powder with asphalt binder. The mechanical mixer can be set at a constant rotational speed (2000) rpm and high shear mixer can be set at angular velocities (2000, 4000 and 6000) rpm. Mixing time parameter is considered as a variable for mixers ranges (30 to 60) minutes to produce a homogeneous composite material. The asphalt binder was preheated to 140°C, based on the viscosity of base asphalt binder. Then, nano SiO_2 powder was gradually added in three percent (2%, 4% and 6 %) to the heated asphalt binder as shown the plate 3; and then tested to assess the physical and rheological properties of the modified asphalt binder.

Plate 3. Different Mixing Process of Nano SiO2 Powder with Asphalt Binder.

2.5 Laboratory Testing
Nano SiO2 powder was mixed with a 60-70 penetration grade asphalt binder at 140 °C using two various mixers. The characteristics of the control and modified asphalts were assessed by using penetration grade, softening point temperature, temperature sensitivity, Brookfield rotational viscosity and ductility property.

3. Results and Discussions

Figures 1 and 2 illustrated the relationship of penetration grade and softening point temperature for two mixing processes of non- modified and modified asphalt binder. It can be observed that the penetration grade is declined and the softening point going up by adding the nano-silica concentration for two mixing processes. This development in the stiffness of modified asphalt binder may consequence from adsorption and diffusion of nano-silica in asphalt binder. Therefore, indications to loss and absorption oily materials in the maltene phase by nano-silica and alter to the resin in the asphalting phase of modified asphalt. As well, the stiffness of nano silica particles is harder than the stiffness of base asphalt binder [40]. Furthermore, the great decrease in penetration value and growth of softening point seen within the mechanical mixing process, due to the fact that the non-homogenous dispersion of nano silica powder and leads to the agglomeration.

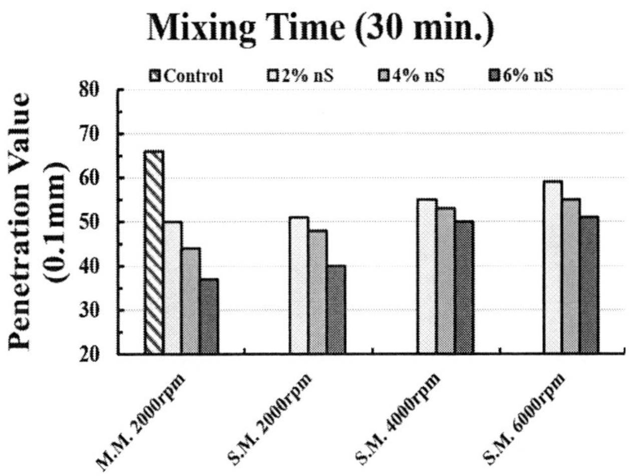

Figure 1. Penetration Value with Different Contents of Nano-Silica for Two Types of Mixer at Mixing Period 30 Minute.

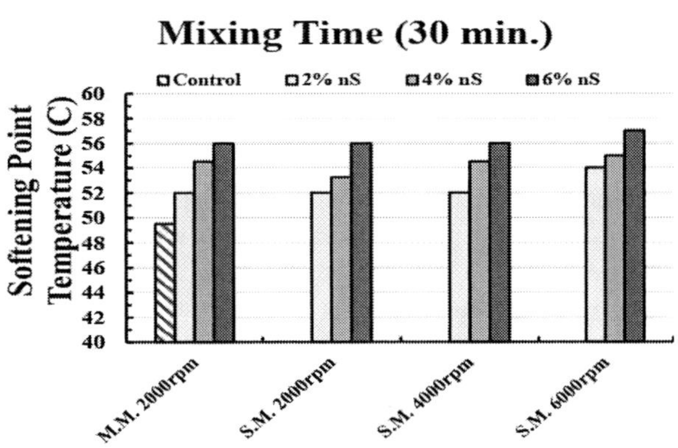

Figure 2. Softening Point with Different Contents of Nano-Silica for Two Types of Mixer at Mixing Period 30 Minute. .

Figures 3 and 4, display the prolonged mixing time improving the physical and rheological characteristics up to 60 minutes for the high shear mixing process. Despite the fact, the mechanical mixing process leads to produce the hard asphalt binder attributed to increasing nano silica agglomeration.

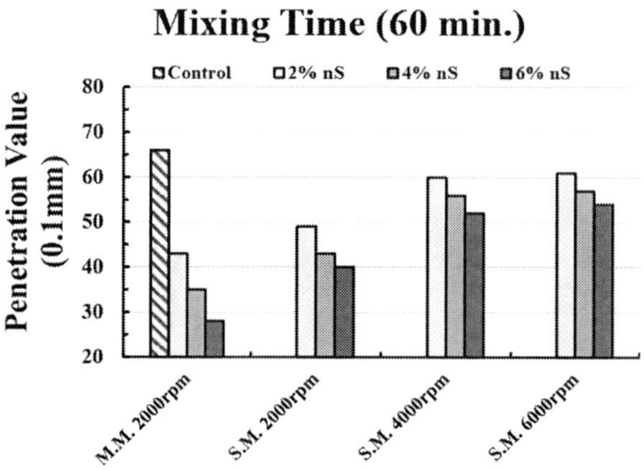

Figure 3. Penetration Value with Different Contents of Nano-Silica for Two Types of Mixer at Mixing Period 60 Minute.

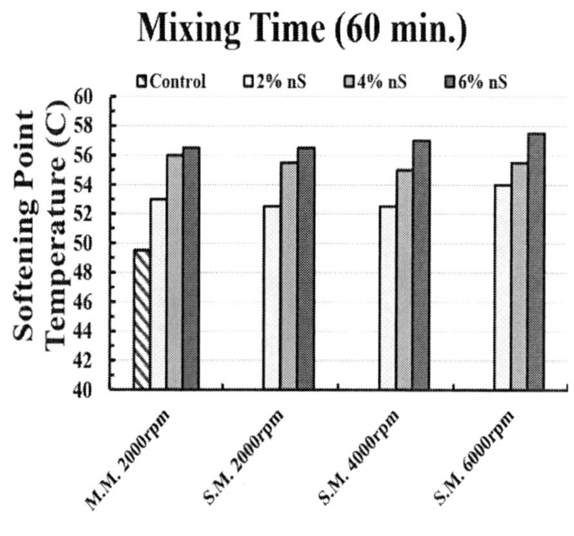

Figure 4. Softening Point with Different Contents of Nano-Silica for Two Types of Mixer at Mixing Period 60 Minute.

Figure 5 showed penetration index (PI) values against nano silica content. It can be observed that the PI value is reduced by adding the nano-silica for two mixing processes, which was rotational speed set at 2000 rpm. While the PI values for the high shear mixing process, which was rotational speed set at (4000 to 6000) rpm are increased by adding the nano-silica content. The incompatible influence by adding the nano-silica concentration for PI values, due to that the non-homogenous dispersion of nano silica powder and leads to the agglomeration [41].

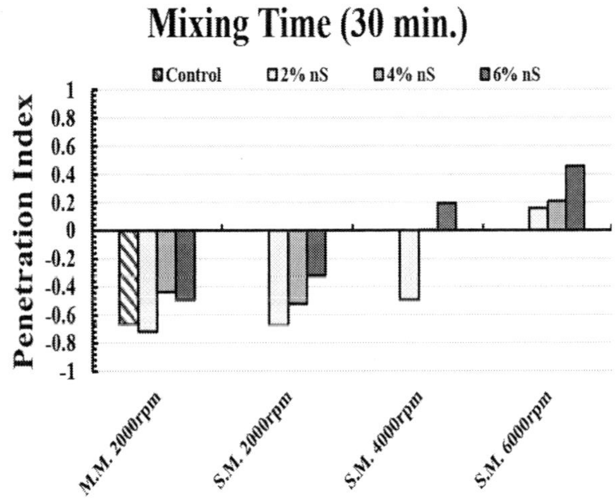

Figure 5. Penetration Index with Different Contents of Nano-Silica for Two Types of Mixer at Mixing Period 30 Minute.

Figure 6, exhibited the extended mixing time leads to great growth in PI value within high shear mixing process indicates that the dispersion of nano silica was more homogenous than the mechanical mixing process [42]. Representing that the rutting resistance increases with an addition of nano silica concentration attributed to form an exfoliated structure [43]. On the other hand the PI values still within the limits of specifications (+2.0 to -2.0), that can be used in pavement materials in reference [44].

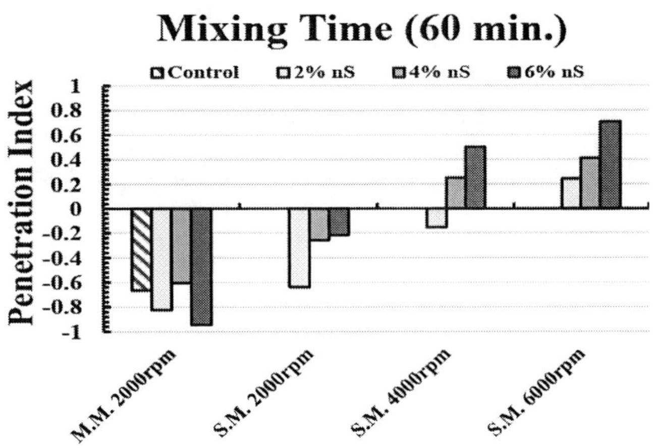

Figure 6. Penetration Index with Different Contents of Nano-Silica for Two Types of Mixer at Mixing Period 60 Minute.

Figure 7, Presented the ductility values against nano silica content. It can be observed that the ductility is declined by adding the nano-silica. An indication of loss and absorption oily materials in the maltene phase by nano silica particles. Furthermore, a great decline in ductility value seen within the mechanical mixing process, due to the fact that the non-homogenous dispersion of nano silica powder and leads to the agglomeration, as shown in plate 4.

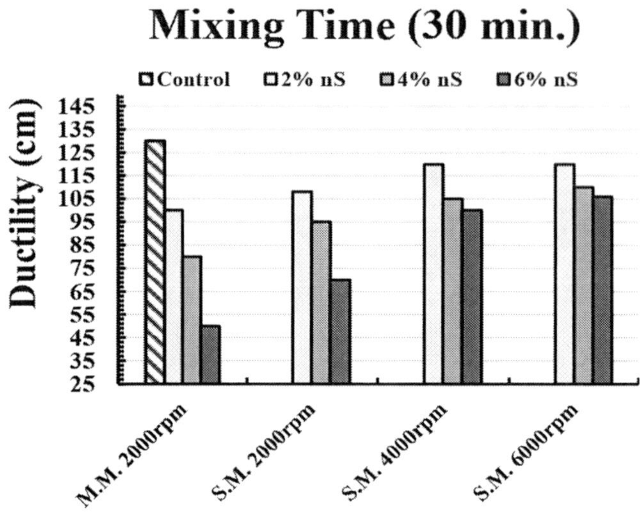

Figure7. Ductility value with Different Contents of Nano-Silica for Two Types of Mixer at Mixing Period 30 Minute.

Plate 4. Dispersion of Nano-Silica particles in Asphalt Binder for Different Mixer (Mechanical Mixer at Left Side and High Shear Mixer at Right Side).

Figure 8, shows that the extended mixing time enhancing the ductility property up to 60 minutes for the high shear mixing process. Although the huge decline of ductility value is seen within the mechanical mixing process of modified asphalt up to 60 minutes, indicating that the agglomeration of nano silica particles [42].

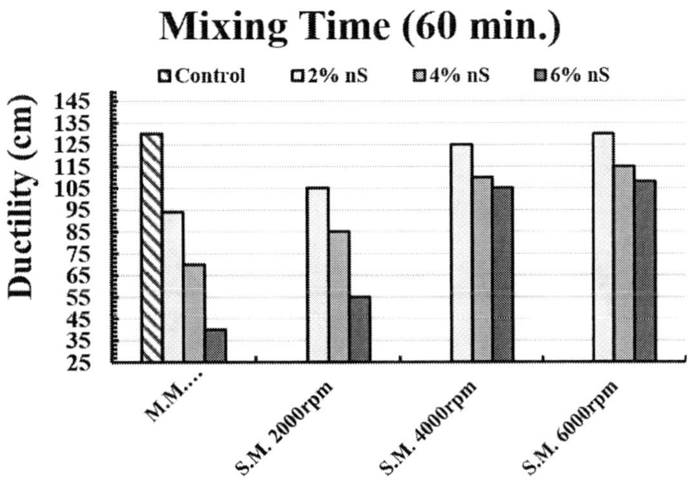

Figure 8. Ductility value with Different Contents of Nano-Silica for Two Types of Mixer at Mixing Period 60 Minute.

Figures 9 - 12, illustrate the relationship of rotational viscosity values and various nano silica content at a temperature of 135 and 165°C. As can be observed that the viscosity is improved by the addition of nano silica content and it can consequence from adsorption and diffusion of nano-silica in asphalt binder and growth the stiffness of binder. Also, the great increase of viscosity value is seen within mechanical mixing process up to 60 minutes, indicating that the non-homogenous dispersion of nano silica powder and leads to the agglomeration, as shown in plate 4. While the slight a reduction of viscosity value is seen within high shear mixing of modified asphalt binder up to 60 minutes as shown in figures, as agreed with references [40&41].

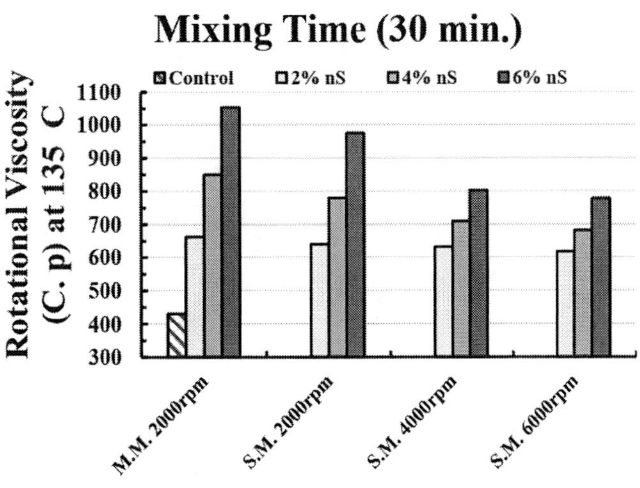

Figure 9. Viscosity value with Different Contents of Nano-Silica for Two Types of Mixer at Mixing Period 30 Minute.

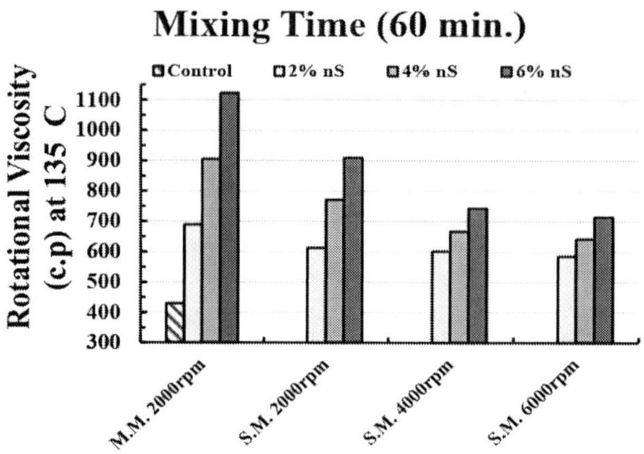

Figure 10. Viscosity value with Different Contents of Nano-Silica for Two Types of Mixer at Mixing Period 60 Minute.

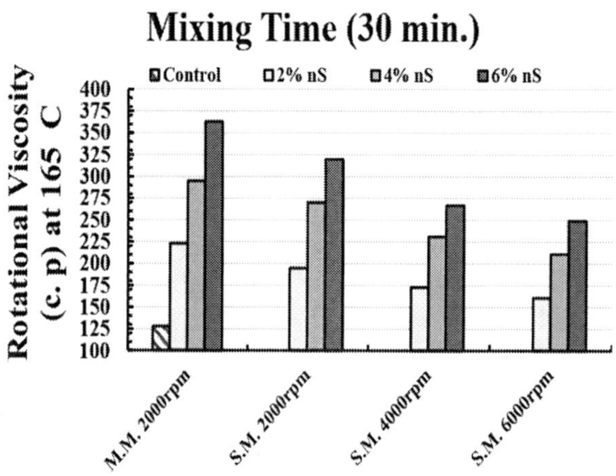

Figure 11. Viscosity value with Different Contents of Nano-Silica for Two Types of Mixer at Mixing Period 30 Minute.

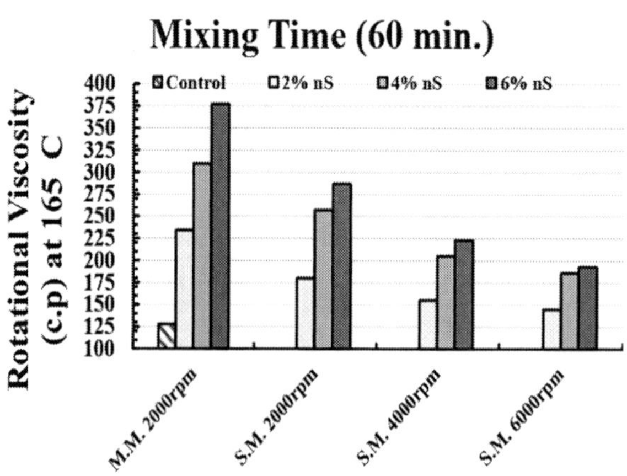

Figure 12. Viscosity value with Different Contents of Nano-Silica for Two Types of Mixer at Mixing Period 60 Minute.

IOP Conf. Series: Materials Science and Engineering **579** (2019) 012046 doi:10.1088/1757-899X/579/1/012046

4. Conclusions

With a view to enhance the mixing efficiency of the mechanical mixer and the high shear mixer on a dispersion of nano silica into the asphalt binder. Moreover, the following conclusions were made according to the results obtained:

1. Using nano silica powder as an asphalt modifier reduces the penetration value and increases the softening point temperature.
2. The great decrease in penetration value and growth of softening point is seen within the mechanical mixing process, due to that the non-homogenous dispersion of nano silica powder and leads to the agglomeration.
3. A sensitivity of temperature (PI) is reduced by adding the nano-silica for two mixing processes, which was rotational speed set at 2000 rpm. While the PI values for the high shear mixing process, which was rotational speed set at (4000 to 6000) rpm are increased by adding the nano-silica content.
4. The ductility value declined by adding the nano-silica per cent. Also, a great decline in ductility value is seen within the mechanical mixing process, due to the non-homogenous dispersion of nano silica powder and leads to the agglomeration.
5. Long mixing time is enhancing the ductility property up to 60 minutes for the high shear mixing process. Although the huge declination of ductility value is seen within the mechanical mixing process of modified asphalt up to 60 minutes, indicating that the agglomeration of nano silica particles.
6. The viscosity is improved by the addition of nano silica content .It can consequence growth the stiffness of binder. The great increase of viscosity value is seen within mechanical mixing process up to 60 minutes. Although the slight declination of viscosity value is seen within high shear mixing of modified asphalt up to 60 minutes. This effect depends on the dispersion of nano silica powder and the agglomeration process.
7. High shear mixing set at (4000 to 6000) rpm for mixing time reached (30) minutes at 140°C mixing temperature was more useful especially to obtain a good dispersion of nano silica powder in asphalt binder and got a homogenous composite material. The prolonged mixing time enhances the physical and rheological characteristics of asphalt binder up to 60 minutes, but it is considered more consumption of energy and more costly.
8. Based on the characteristics of modified asphalt binder, the 4% of nano silica content was satisfactory to enhance the physical characteristics. Additionally, it become more appropriate for construction the pavements of highway in hot climate conditions.

References

[1] Collins J H, Boudin M G, Gelles R and Berker A. 1991 Improved Performance Of Paving Asphalts By Polymer Modification *Proc Assoc Asphalt PavTechnol* 43–79

[2] Saeed Ghaffarpour Jahromi and Ali Khodaii 2009 Effects of Nano clay on Rheological Properties of Bitumen Binder *Construction and Building Materials* **23**, 2894–2904

[3] Zhanping You and Julian Mills-Beale 2011 Nano clay-modified asphalt materials: Preparation and characterization *Construction and Building Materials* **25** 1072–1078

[4] Saeed Sadeghpour Galooyak and Bahram Dabir 2010 Rheological Properties and Storage Stability of Bitumen/SBS/Montmorillonite Composites *Construction and Building Materials* 24 300–307

[5] Sengoz B and Isikyakar G 2007 Evaluation of the Properties and Microstructure of SBS and EVA Polymer Modified Bitumen *Construction and Building Materials* **22** 1897–905

[6] Lu X and Isacsson U 2001 Modification of Road Bitumen with Thermoplastic Polymers *PolymTest* **20** 77–86

[7] Ouyang C, Wang S, Zhang Y and Zhang Y 2006 Improving the Ageing Resistance of Styrene-Butadiene-Styrene Tri-Block Copolymer Modified Asphalt by Addition of Antioxidants *Polymer Degradation and Stability* **91** 795–804

[8] Kok B V and Yilmaz M 2009 The Effects of Using Lime and Styrene Butadiene- Styrene on Moisture Sensitivity Resistance of Hot Mix Asphalt *Construction and Building Materials* **23** 1999–2006

[9] Khattak MJ, Baladi GY and Drzal LT 2007 Low-Temperature Binder-Aggregate Adhesion and Mechanistic Characteristics of Polymer Modified Asphalt Mixtures *Journal of Materials in Civil Engineering* **19** 5 411–422

[10] Hasan H Joni and Ethar K Shaker 2017 Determination of the Acceptable Range of Mixing and Compaction Temperatures for Modified Asphalt Mixture with Styrene Butadiene Styrene (SBS) *International Journal of Current Engineering and Technology*

[11] Huang SC, Robertson RE, Branthaver JF and Petersen JC 2005 Impact of Lime Modification of Asphalt and Freeze-Thaw Cycling on the Asphalt-Aggregate Interaction and Moisture Resistance to Moisture Damage *Journal of Materials in Civil Engineering* **17** 6 711–718

[12] Lesueur D, Petit J and Ritter HJ 2013 The Mechanisms of Hydrated Lime Modification of Asphalt Mixtures: A State-of-The-Art Review *Road Materials and Pavement Design* **14** 1 1–16

[13] Lo Presti D 2013 Recycled Tire Rubber Modified Bitumen's for Road Asphalt Mixtures: A Literature Review *Construction and Building Materials* **49** 863–881

[14] Colbert BW and You Z 2012 Properties of Modified Asphalt Binders Blended with Electronic Waste Powders *Journal of Materials in Civil Engineering* **24** 10 1261–1267

[15] Joni H H and Zghair H H 2016 Effect of Adding Used-Foundry Sand on Hot Asphalt Mixtures Performance *Eng. &Tech .Journal* **34** (A) 6

[16] Zghair HH, Joni HH and Hassan MS 2018 Influence of Micro- Size Silica Powder on Physical and Rheological Characteristics of Asphalt binder *International Journal of Engineering & Technology* **7** 4.20 180-184

[17] Hasan Z, Kamran R, Mohammad F, Ahmad G. and Hosein F 2001 Evaluation of Different Conditions on the Mixing Bitumen and Carbon Nano-tubes *International Journal of Civil & Environmental Engineering* **12** 6 12-53

[18] Faramarzi M, Arabani M, Haghi A and Mottaghitalab V 2015 Carbon Nanotubes-Modified Asphalt Binder: Preparation and Characterization *International Journal of Pavement Research and Technology* **8** 1 29-37

[19] Fang C, Yu R, Liu S and Li Y 2013 Nanomaterials Applied in Asphalt Modification: A Review *Journal of Materials Science & Technology* **29** 7 589-594

[20] Shafabakhsh G H, Mirabdolazimi S M and Sadeghinejad M 2015 Evaluation the Effect of Nano-Tio2 on the Rutting and Fatigue Behavior of Asphalt Mixtures *Journal of Construction and Building Materials* **54** 1 566-572

[21] Shafabahsh G H and Jafari Ani O 2015 Experimental Investigation of Effect of Nano TiO2/SiO2 Modified Bitumen on The Rutting and Fatigue Performance of Asphalt Mixtures Containing Steel Slag Aggregates *Journal of Construction and Building Materials* **85** 1 136-143

[22] Yusoff N I M, Breem A A S, Alattug H N M, Hamim A and Ahmad J 2014 The Effects of Moisture Susceptibility and Ageing Conditions on Nano-Silica/Polymer-Modified Asphalt Mixtures *Journal of Construction and Building Materials* **72** 1 139-147

[23] You Z, Mills-Beale J, Foley J M, Roy S, Odegard G M, Dai Q and Goh S W 2011 Nano Clay Modified Asphalt Materials: Preparation and Characterization *Construction and Building Materials* **25** 2 1072-1078

[24] Khattak M J, Khattab A, Rizvi HR and Zhang P 2012 The Impact of Carbon Nano-Fiber Modification on Asphalt Binder Rheology *Construction and Building Materials* **30** 5 257-264

[25] Chrissafis K, Paraskevopoulos KM, Papageorgiou GZ, Bikiaris DN 2008 Thermal and Dynamic Mechanical Behavior of Bio-Nanocomposites; Fumed Silica Nanoparticles Dispersed in Poly (Vinyl Pyrrolidone), Chitosan, and Poly (vinyl alcohol) *J. Appl. Polym. Sci.* 110 1739-1749

[26] Jaber A, Gorgis I and Hassan M 2017 Relationship between Splitting Tensile and Compressive Strengths for Self-Compacting Concrete Containing Nano-and Micro Silica *MATEC Web of Conferences* **162** 02013

[27] Quercia G and Brouwers HJH 2010 Application of Nano-Silica (Ns) in Concrete Mixtures In 8th Fib International PhD *Symposium in Civil Engineering* Lyngby **20** 431-436

[28] Alhamali DI, Wu J, Liu Q, Hassan NA, Yusoff NIM and Ali SIA 2015 Physical and Rheological Characteristics of Polymer Modified Bitumen with Nano silica Particles *Arab Journal of Sci. Eng.* **15**

[29] Ezzat H, El Badawy S, Gabr A, Zaki EI and Breakah J 2016 Evaluation of Asphalt Binders Modified with Nano clay and Nanosilica. *Advances in Transportation engineering*: The Third International Conference on Transportation Geotechnology **143** 1260-1267

[30] Yao H, You Z, Li L, Lee CH, Wingard D, Yap YK, Shi X, Goh SW 2012 Rheological Properties and Chemical Bonding of Asphalt Modified with Nano Silica *J. Mater. Civ. Eng.* **25** 11 1619-1630

[31] Hussain F et al. 2006 Review Article: Polymer–Matrix Nanocomposites, Processing, Manufacturing, and Application: An Overview *J Compos Mater* **40** 17 1511–75

[32] ASTM D5 2015 Standard Test Method for Penetration of Bituminous Materials *Annual Book of Standards American Society for Testing and materials* **04.03**

[33] ASTM-D113 2015 Standard Test Method for Ductility of Bituminous Materials. *Annual Book of Standards American Society for Testing and materials* **04.03**.

[34] ASTM D92 2015 Standard Test Method for Flash and Fire Points by Cleveland Open Cup Tester *Annual Book of Standards American Society for Testing and materials* **04.03**

[35] ASTM D4402 2015 Standard Test Method for Viscosity Determination of Asphalt at Elevated Temperatures Using a Rotational Viscometer *Annual Book of Standards American Society for Testing and materials* **04.03**

[36] ASTM D36 2015 Standard Test Method for Softening Point of Bitumen (Ring-and-Ball Apparatus) *Annual Book of Standards American Society for Testing and materials* **04.03**

[37] State Organization for Roads and Bridges 2003 Hot Asphaltic Concrete Pavement, Iraqi Standard Specification *Ministry of Housing and Construction, Department of Design and Study* Section R9

[38] Ames G 2014 Mixing, Blending and Size Reduction Handbook. Available from: www.admix.com/pdfs/resourcelibrary-tech-mixhandbook.pdf

[39] Silverson 2016 Mixing with a High Shear Laboratory Mixer Available fromhttp://www.silverson.com/us/products/laboratory-mixers/how-it-works

[40] Taherkhani H and Afroozi S 2016 The Properties of Nano Silica-Modified Asphalt Cement. *Petroleum Science and Technology* 1381-1386

[41] Mahmoud Enieb and Aboelkasim Diab 2017 Characteristics of Asphalt Binder and Mixture Containing Nano silica *International Journal of Pavement Research and Technology* **10** 148-157

[42] Taherkhani H and Afroozi S 2017 Investigating the Performance Characteristics of Asphaltic Concrete Containing Nano-Silica. *Civil Engineering Infrastructures Journal* **50** 1 75 – 93

[43] Nura Bala, Ibrahim Kamaruddin, Madzlan Napiah and Nasiru Danlami 2017 Rheological and Rutting Evaluation of Composite Nano Silica/ Polyethene Modified Bitumen *Materials Science and Engineering* 201

[44] White Oak D and Read J 2015 *Shell Bitumen Handbook* Shell Bitumen London UK

Properties of self compacting concrete exposed to wetting and drying cycles in oil products

S A Salih[1], W I Khalil[2#] and A A Ali[3]

[1, 2] Professor in the Building and Construction Engineering Department, University of Technology, Baghdad, Iraq
[3]An Engineer in Iraqi ministry of oil, Baghdad, Iraq

[#]Corresponding author's e-mail address: 100108@uotechnology.edu.iq

Abstract. Although the early deterioration of concrete exposed to petroleum products has being observed yet, the behavior of Self Compacting Concrete (SCC) being exposure to oil products is still unknown. In the present investigation, the mechanical and physical properties of hardened SCC cyclically being exposed to oil products (crude oil, gas oil, motor oil and fuel oil) have tested and discussed. The numbers of exposure cycles are 2, 3, 4, 5 and 6. Each exposure cycle consists of twenty days immersion in water or different oil products followed by air drying at ambient temperature for a period of ten days. The results have been compared to other specimens cured in water or continuously exposed to oil products for the same period. The results show that the effect of continuous exposure to oil products on the mechanical properties is more severe than that of cyclic exposure of SCC specimens. The compressive strength, splitting tensile strength, modulus of rupture, and static modulus of elasticity of SCC specimens have slightly deteriorated until the fifth cycle of exposure for all types of oil products comparing to reference specimens (cured in water). Eventually at the sixth cycle all properties have decreased.

1. Introduction

Oil has become one of the most vital energy resources from the beginning of the pervious century for its unique economic and operative characteristics. This has enabled it to exceed the other available power resources, and its importance has increased rapidly with its wide spread use and the discovery of huge oil reservoir in different parts of the world [1]. There are limited information and studies about the behavior of concrete which have been exposed to petroleum products especially the new types of concrete such as self-compacted concrete (SCC). In spite of the advantages of concrete structures such as shock and fire resistant, low cost of maintenance, yet the use of concrete structures as oil storage still limited due to restrictions. The unknown behavior of concrete exposed to oil products, the penetration of oil through concrete or concrete cracks, and the difficulty to modify or repair [2]. Due to the dominant improvements in the properties of concrete containing different types of admixtures, it is very important to study the effect of admixtures on the behavior of concrete exposed to oil products to improve their properties [3]. The behavior of SCC after exposure to oil products is still a fogy area and requires more investigations.

2. Research Significance

The main aim of this research is to investigate the effect of different exposure cycles (wetting and drying) to water or different oil products such as crude oil, gas oil, motor oil, and fuel oil on SCC properties and comparing these properties with those of SCC continuously exposed to water or the same different oil products for corresponding periods of exposure.

3. Experimental Works

3.1 Materials

3.1.1 Cement

Sulphate resistance Portland cement type (V) is used in this research. Its chemical and physical properties conform to ASTM specifications C150-04[4].

3.1.2 Coarse Aggregate

Iraqi natural crashed gravel is used. Its grading conforms to the requirement of ASTM C33-03[5] size No7. The specific gravity, sulfate content and absorption of the coarse aggregate are 2.64, 0.096%, and 0.7% respectively.

3.1.3 Fine Aggregate

Iraqi natural sand is used as fine aggregate. Its gradation was within the requirements of ASTM C33-03[5]. The specific gravity, sulfate content and absorption of the fine aggregate are 2.6, 0.19%, and 0.75% respectively.

3.1.4 Superplasticizer

The Superplasticizer that has been used throughout this study is commercially known as "structure 520" [6]. It is based on a unique carboxylic ether polymer with long lateral chains. It is suitable for the production of SCC. It is free from chlorides and complies with ASTM-C494 type F [7]. Its specific gravity is 1.1 with PH value 6.5 and alkali contents less than 1.5 gm of Na_2O equivalent per liter of admixture.

3.1.5 Water

Potable water is used throughout this experiment for mixing the SCC and curing of the hardened SCC specimens.

3.1.6 Silica Fume

Condensed silica fume is used to produce SCC with reliable fresh concrete properties. Its accelerated pozzolanic strength activity index with Portland cement at 7 days is 106%. The chemical composition and physical requirements shown in Tables (1) and (2) respectively indicate that the silica fume conforms to the chemical and physical requirements of ASTM C1240 specifications [8].

3.1.7 Oil Products

Oil products from Iraqi ministry of oil are used in this investigation. Table (3) shows the chemical analysis of the oil products used in this investigation to expose different SCC specimens for specific period.

3.2 Concrete Mixes

Self-compacted concrete mix has been designed according to European guidelines for testing fresh SCC EFNARC [5] to obtain a minimum cubic compressive strength of 65 MPa at 28 days. The mix proportions are (1:1.72:1.97) by weight with 10% silica fume as addition by weight of cement and a w/c ratio of 0.38. Several trail mixes are carried out in order to select the optimum dosage of Superplasticizer (SP) (2.8 liter per 100 kg of cementitious materials) which is determined by using all workability testes of SCC (slump flow test, V- funnel test, L-Box test, and J-Ring test).

3.3 Preparation of SCC Specimens

3.3.1 Mixing of SCC

SCC has been mixed in a rotary mixer with capacity of 0.1 m^3. Fine aggregate has added to the mixer with 1/3 quantity of water and being mixing lasts for 1.5 minutes. Cement and silica fume are added with the end 1/3 of water and mixing continued for 3 minutes. Then half the quantity of coarse aggregate is added with the remaining quantity of water and 1/3 the dosage of superplasticizer (2.8 liter per 100 kg cementitious materials) and mixed for another 1.5 minute. The mixture rest for about 0.5 minute, the remaining quantity of coarse aggregate and superplasticizer added and remixed for 1.5 minutes. After that the fresh properties test were performed to assure the self-compatibility requirements.

3.3.2 Casting and Curing of SCC Specimens

Standard moulds have been prepared for SCC casting. They include 100mm cubes for compressive strength according to BS 1881: part 116 [9], 100*200mm cylinders for splitting tensile strength according to ASTM C496 [10], 150*300mm cylinders for elasticity modulus according to ASTM C469 [11], and 100*100*400 mm prisms for rupture modulus according to ASTM C78 [12]. SCC mixture does not require any compacting procedure, so the mix is poured into the tight steel moulds until completely filled and has leveled easily. The moulds have covered with polyethylene sheet for about 24 hours, and then the specimens are demoulded. To develop the strength of specimens, they are cured in water tanks for 27 days. After 28 days, age the specimens have divided into five groups, according to the goal of the study. These groups are cyclically (six cycles) exposed to the following fluid:-
Group 1: exposed to water.
Group 2: exposed to crude oil.
Group 3: exposed to gas oil.
Group 4: exposed to motor oil.
Group 5: exposed to fuel oil.
Each cycle exposure consist of twenty days of immersion in water or the various oil products followed by air drying at ambient temperature for a period of ten days.
To examine the effect of such condition, the mechanical and physical properties of the SCC specimens have assessed throughout experimental tests as detailed in section 3.3.2.

IOP Conf. Series: Materials Science and Engineering **579** (2019) 012047 doi:10.1088/1757-899X/579/1/012047

Table 1. Chemical analysis of silica fume*

Oxide Composition	Oxide content (%)	ASTM C1240-03
SiO_2	93.94	Min. 85%
Al_2O_3	0.7	<1%
Fe_2O_3	0.45	<2.5%
CaO	0.88	<1%
SO_3	0.93	< 1%
K_2O+Na_2O	1.37	< 3%
L.O.I	3.96	Max. 6%
Cl	0.17	< 0.2%
C(free)	3.1	< 4%

* Chemical analysis was conducted by National Center for Construction Laboratories and Researches

Table 2. Physical properties of silica fume*

Property	Result	ASTM C1240-03
Strength activity index	106%	≥ 105
Specific gravity, kg/m^3	2.2	-
Physical form	Powder	-
Color	Grey	-
Size	0.15	0.15 micron
Density	0.5	0.5± 0.1kg/liter (dry bulk)
Moisture	0%	< 2%
Specific surface, m^2/g	16	≥ 15

* Physical properties were conducted by National Center for Construction Laboratories and Researches

Table 3. Properties of different oil products*

Oil inspection data	Fuel oil results	Gas oil results	Motor oil results	Crude oil results	Method
Carbon residue wt%	8.7	0.09	-	-	ASTM-D524
Sulfur content % by weight	4.24	1.2%	-	0.597	ASTM-4294
Sulfur content % by weight	-	-	1.45		ASTM-D648
Cetane No.	-	57.8	-	-	IP-218
Specific gravity (gm/cm^3) at 15.6 C° 25.0 C° 30.0 C°	0.9625	0.8383 0.8186 0.8151	0.898	0.8067	ASTM-D1298
Specific gravity at 15.6 °C	-	-	0.898	-	ASTM-D4052
Viscosity (centistokes) at 10 °C 20 °C 30 °C 37.8 °C 40 °C 50 °C 100 °C	- - - - 234 -	- - - 3.4 - -	- 478.7 249.3 - 141.49 86.27 15.24	5.8 3.71 - 2.24 1.6 -	ASTM-D445
Moisture content % by volume	-	-	Nil	-	ASTM-D95
Flash point (pm) **°C**	90.6	88.6	236 Min	-	ASTM-D93
Pour point	+3	-	-9 Max.	-9 Max	ASTM-D97
Water &sediment % vol.	Trace	-	-	-	ASTM-D1796
Water &sediment % vol.	-	-	-	0.05	IP-75
Diesel index	-	62.1	-	-	IP -21
Ash wt.%	-	Nil	-	-	ASTM-874
Ash wt.%	-	-	0.579	-	ASTM-D482
Sulfate Ash % by weight	-	-	0.7231	-	ASTM-D874
API GRAVITY @15.6 °C	-	-	-	43.9	ASTM-D1298
water content vol %	-	-	-	traces	IP-74
Wax content,%	-	-	-	3.49%	-
Density @15 °C	-	-	-	0.8064	-
Salt content % by weight	-	-	-	0.0006	IP-77
KUOP Characterization Factor	-	-	-	12.1	UOP method 375

All oil products test were carried out by the laboratory of Ministry of Oil

4. Results and Discussion
4.1 Fresh Properties of SCC

Fresh properties of SCC have been tested according to the procedures European guidelines of testing fresh SCC EFNARC [13]. Three characteristics have been achieved by conducting three tests which are, flow ability, passing ability and segregation resistance. After specifying the weight mix proportions of 1:1.72:1.97 (cement: sand: Gravel), several mixtures were carried out as trial to select the optimum dosage of superplasticizer which produce SCC. It is 2.8% liter per100kg, 10% silica fume as addition by weight of cement is used with w/c ratio 0.38. The ingredients of the mixture to produce SCC with acceptable criteria are shown in Table (4).

Table 4. Mix proportions and fresh properties for the selected SCC mix

<table>
<tr><th colspan="2">Mix proportions for the selected SCC mix</th><th colspan="2">Type of test</th><th>Mix Designation</th><th>Limits of EFNARC 2005 [13]</th></tr>
<tr><td>Cement (kg/m^3)</td><td>450</td><td colspan="2">Slump Flow D(mm)</td><td>750</td><td>650 - 800</td></tr>
<tr><td>Sand (kg/m^3)</td><td>778</td><td colspan="2">T50 (sec)</td><td>4</td><td>2 - 5</td></tr>
<tr><td>Gravel (kg/m^3)</td><td>890</td><td colspan="2">V-Funnel (sec) (Tv)</td><td>10</td><td>6 - 12</td></tr>
<tr><td>Silica fume (kg/m^3)</td><td>45</td><td colspan="2">T5 min</td><td>3</td><td>0 - 3</td></tr>
<tr><td>Water (l/m^3)</td><td>188.1</td><td rowspan="3">L-box</td><td>T20 (sec)</td><td>6</td><td>-</td></tr>
<tr><td>SP% liter per 100kg</td><td>2.8</td><td>T40 (sec)</td><td>8.5</td><td>-</td></tr>
<tr><td>w/cm</td><td>0.38</td><td>BR% (h2/h1)</td><td>0.95</td><td>0.8 - 1</td></tr>
<tr><th colspan="2">Compressive Strength (MPa)</th><td rowspan="4">J-ring</td><td>TJ500 sec</td><td>4.5</td><td>-</td></tr>
<tr><td>7 days</td><td>28 day</td><td>DJ mm</td><td>7°0</td><td>-</td></tr>
<tr><td rowspan="2">45</td><td rowspan="2">73.9</td><td>BJ mm</td><td>0</td><td>-</td></tr>
<tr><td>Deviation D-DJ mm</td><td>0</td><td>0 - 10</td></tr>
</table>

4.2 Hardened Properties of SCC

4.2.1 Compressive Strength

The results summarized in Table (5) and Figure (1) represent compressive strength values of SCC exposed to various types of immersion and drying cycles in water or oil products. The compressive strength of SCC specimens increases as the number of wetting and drying cycles in water has increased. This may be due to the continuous hydration process of cement paste which forms new hydration product within the SCC mass. This leads to improve the bonds between cement paste and aggregate [14, 15]. The presence of silica fume in SCC improves the mechanical properties and durability of concrete [16]. Generally, the results indicate that the compressive strength of SCC specimens being exposed to cycles of immersion and drying in oil products (all types) decreases comparing to the reference specimens (exposed to the same number of wetting and drying cycles in water). The reduction of compressive strength may probably be due to the extension of gel pores and spreading solid hydration components caused by oil products penetration into the microstructure of SCC leading to adhesion and cohesion forces in the cement paste matrix reduction. In addition to the effects of oil products on SCC surface interactions which have been confirmed by other investigations [17, 18, 19, 20]. There is a slight decline in compressive strength of SCC specimens exposed to immersion and drying cycles in oil products till the exposure period of 150 days (5 cycles) relative to specimens exposed to same cycles in water. While specimens exposed to six cycles show higher reduction in compressive strength. This could be attributed to the pores of SCC which is still partially filled with water and leads to further hydration that delay the deterioration of SCC [21, 22]. The presence of silica fume leads to a modification of the microstructure of concrete, especially at later ages that significantly reduces the permeability of concrete. This will delay the deterioration of concrete subjected to oil products [23].

It is observed that specimens exposed to six cycles of immersion in crude and gas oil show higher reduction in compressive strength than those exposed to motor and fuel oil. This is due to the difference in the viscosity of these products which has great effect in concrete properties [24]. The higher the viscosity of the oil, the less dangerous it is on concrete [22]. Table (2) shows that crude and gas oils have less viscosity than motor and fuel oils.

The results of the present study have been compared with the results of a study conducted by [25]. Figure (2) investigates the properties of SCC continuously exposed to oil products. The figure shows that the compressive strength of SCC specimens continuously immersed in water is higher than that exposed to wetting and drying cycles in water for the same of period. This can be attributed to continuous hydration of cement for specimens continuously immersed in water which improves compressive strength. Figure (3) shows the decline of compressive strength of SCC specimens exposed continuously to oil products and specimens immersed in oil products for different cycles relative to reference specimens (exposed to water for the same period). It can be concluded that the continuous exposure to oil products is more severe than the exposure to wetting and drying cycles. This may be due to the deterioration of specimen's surface being exposed continuously to oil product [17]. Moreover the compressive strength of high density concrete depends to a large extent on cement hydration. When coating a concrete surface or un hydrated cement grains in the concrete, the oil film blocks moisture from the coated surfaces and concrete grains, arresting their further hydration. Therefore, increasing concrete strength has slowed down or stopped altogether [26].

4.2.2 *Splitting Tensile Strength*

Splitting tensile strength results of SCC specimens exposed to wetting and drying cycles in water or different oil products are shown in Table (5) and Figure (4). Generally, the test results indicate that, the pattern of behavior of the splitting tensile strength of SCC is similar to that of compressive strength. The test results for SCC specimens have being subjected to wetting and drying cycles in water show continuous increase in splitting tensile strength as the number of cycles have increased.

It is clear that increasing compressive strength will lead to an increase in splitting tensile strength [27]. The results show that the splitting tensile strength of SCC specimens exposed to immersion and drying cycles in different oil products slightly decrease for immersion and drying 2,3,4 and 5 cycles as compared to reference specimen (exposed to water for the same number of cycles). The decrease is only 0.36%, 0.18%, 0.17% and 0.86% for crude oil and 0.36%, 1.4%, 0.34% and 0.86% for gas oil, while the decrease is about 2.14%, 1.58%, 0.52% and 0.17% for motor oil and 2.14%, 1.93%, 0.52% and 0.69% for fuel oil, respectively. The logical explanation to this phenomenon is that the pores in concrete are still partially filled with water, when causes further hydration in cement that increases the strength of concrete. The low penetration of oil products cannot permit to loss the bond due to low permeability of SCC produced in this investigation (containing silica fume) that greatly enhances the durability of concrete subjected to oil products. The splitting tensile strength for SCC specimens exposed to crude oil and gas oil has significantly decreases as the number of immersion and drying cycles in oil products increased to six cycles compared to the reference. The decrease ratio is about 24.1% and 14.85% respectively. Specimens being exposed to motor oil and fuel oil show slight decrease in splitting tensile strength in comparison with the reference (2.56% and 2.39% respectively). This is due to the low viscosity of crude and gas oil. As for compressive strength, splitting tensile strength of SCC specimens continuously immersed in water is higher than that being exposed to wetting and drying cycles in water as shown in Figure (5). The continuous exposure to oil products is more severe than immersion and drying cycles of exposure on splitting tensile strength of SCC, as shown in Figure (6).

Table 5. Properties of SCC specimens exposed to oil products

Exposure Type	Properties	No. of Cycles of Exposure (Age of Specimens, days)					
		1 (28)	2 (60)	3 (90)	4 (120)	5 (150)	6 (180)
Water	Compressive strength (MPa)	73.89	75.2	76.95	82.8	84.4	86.1
	Splitting tensile strength (MPa)	5.5	5.62	5.71	5.82	5.84	5.86
	Modulus of rupture (MPa)	7.28	7.36	7.38	7.48	7.51	7.78
	Static modulus of elasticity (GPa)	54.58	59.78	62.72	63.7	64	64.93
	Total absorption (%)	3.16	3.07	2.7	2.5	2.46	1.93
Crude oil	Compressive strength (MPa)		74.23	76.44	82	80	63.6
	Splitting tensile strength (MPa)		5.6	5.7	5.81	5.79	4.45
	Modulus of rupture (MPa)		7.35	7.37	7.45	7.4	6.6
	Static modulus of elasticity (GPa)		56.27	60	63.5	63	47
	Total absorption (%)		1.39	1.25	1.16	0.84	0.79
Gas oil	Compressive strength (MPa)		74.8	76.72	81	80.4	70
	Splitting tensile strength (MPa)		5.6	5.63	5.8	5.79	4.99
	Modulus of rupture (MPa)		7.35	7.37	7.42	7.41	7.2
	Static modulus of elasticity (GPa)		57.18	61.1	63.4	63.29	52.2
	Total absorption (%)		2.39	1.44	1.51	1.13	1.24
Motor oil	Compressive strength (MPa)		73.95	75.2	80.1	80.9	77.7
	Splitting tensile strength (MPa)		5.5	5.62	5.79	5.83	5.71
	Modulus of rupture (MPa)		7.29	7.36	7.41	7.5	7.39
	Static modulus of elasticity (GPa)		54.92	59.78	63.4	64	62.8
	Total absorption (%)		2.42	2.02	1.21	1.12	0.8
Fuel oil	Compressive strength (MPa)		73.95	74.8	80.1	80.9	77.7
	Splitting tensile strength (MPa)		5.5	5.6	5.79	5.8	5.72
	Modulus of rupture (MPa)		7.29	7.36	7.41	7.5	7.39
	Static modulus of elasticity (GPa)		55.02	57.18	63.9	64.5	62.9
	Total absorption (%)		2.36	1.73	1.57	1.5	0.97

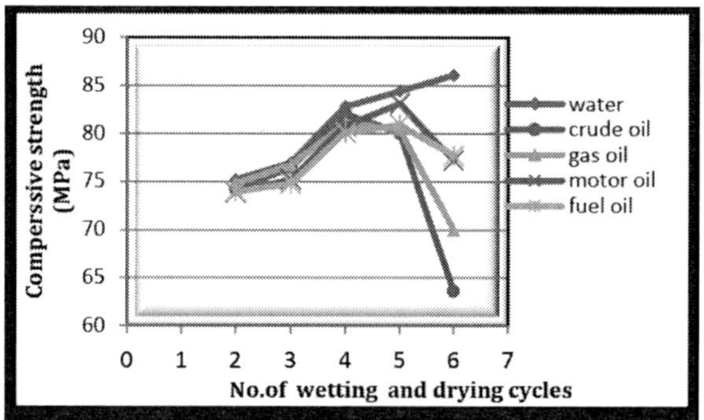

Fig.1. Compressive strength for SCC specimens exposed to wetting and drying cycles in water or

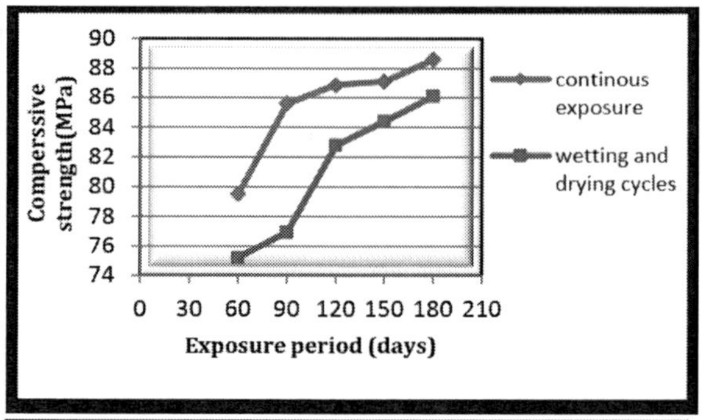

Fig.2. Effect of water exposure type on compressive strength of SCC (The continuous

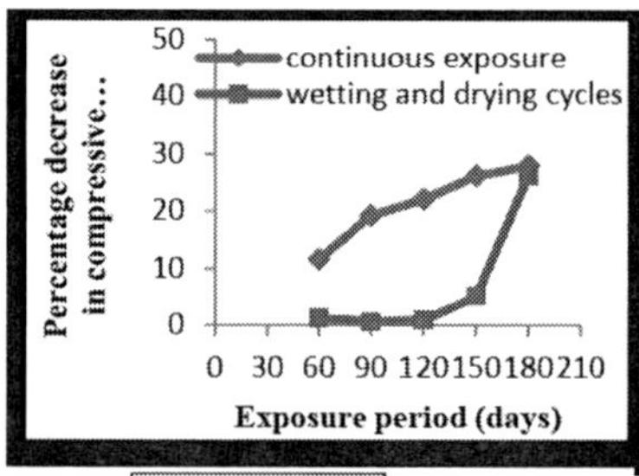

A- Crude

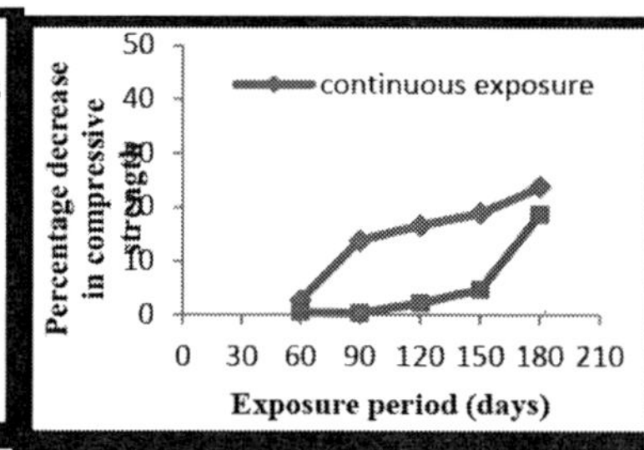

B- Gas oil

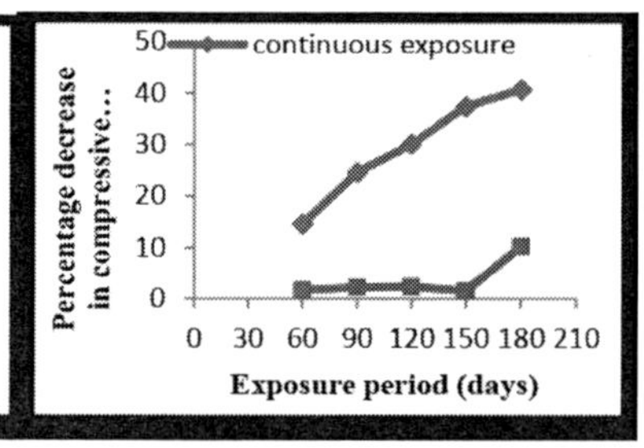

C- Motor oil

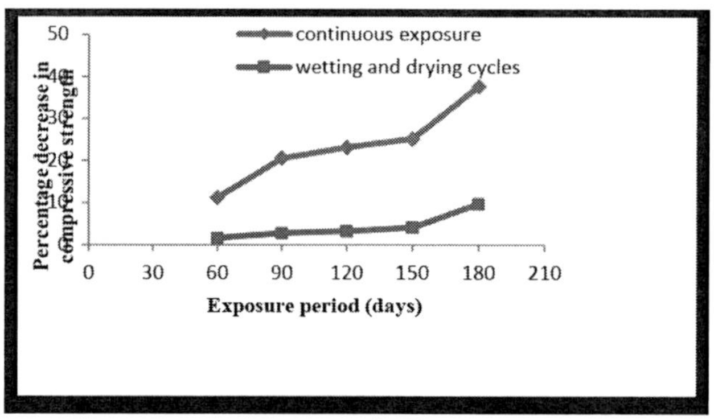

D – Fuel oil

Fig.3. Effect of type of exposure to oil products on compressive strength of SCC (The continuous exposure results are from reference 15)

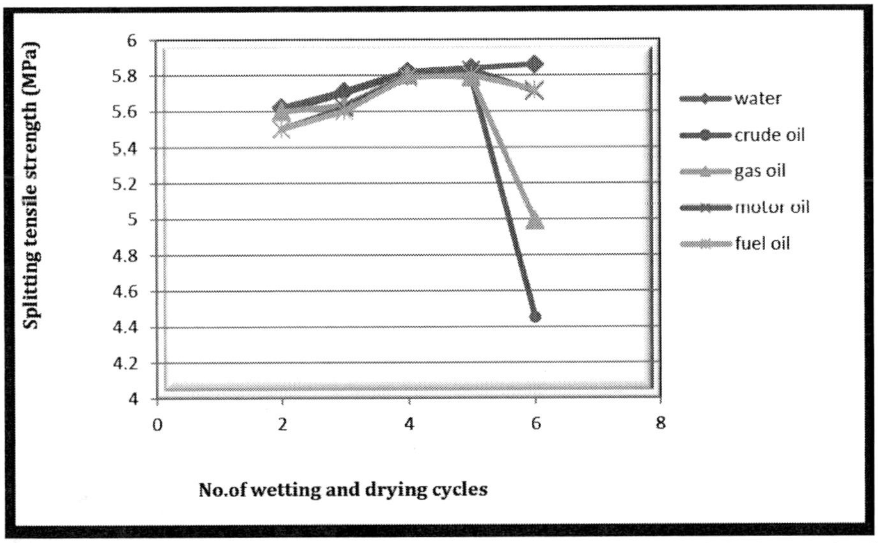

Fig.4. Splitting tensile strength of SCC specimens exposed to wetting and drying cycles in water or different oil products

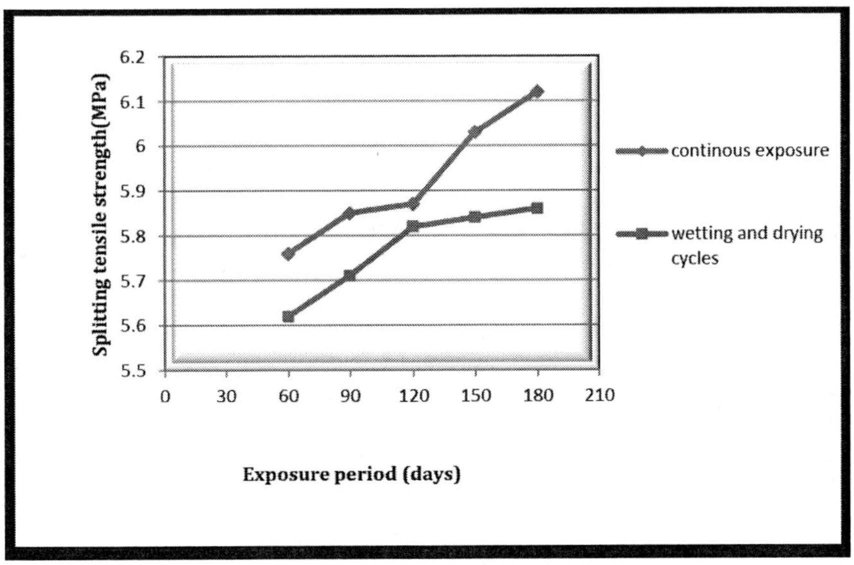

Fig.5. Effect of water exposure type on splitting tensile strength of SCC (The continuous exposure results are from reference 15)

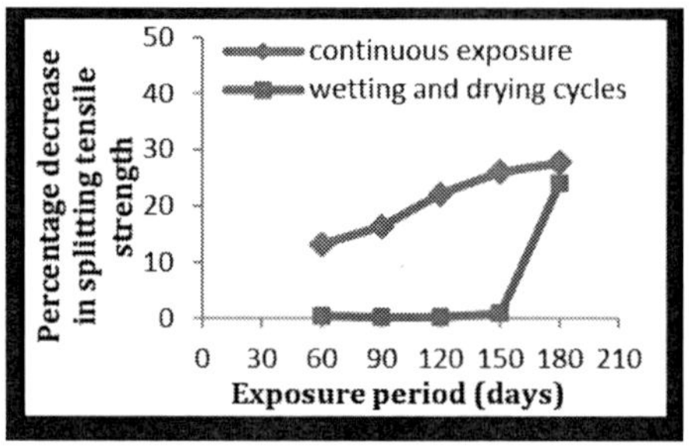

A- Crude oil

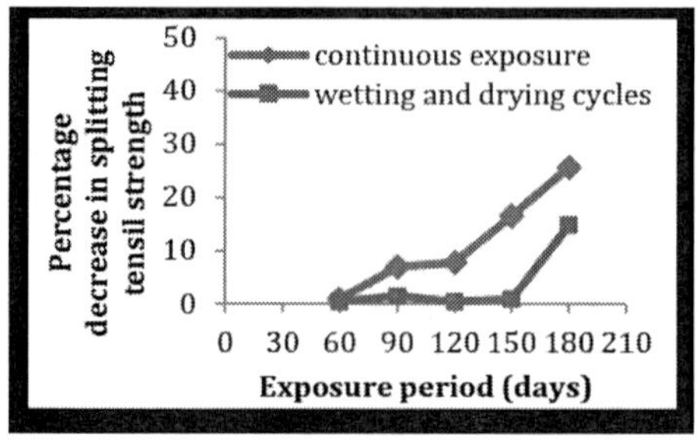

B- Gas oil

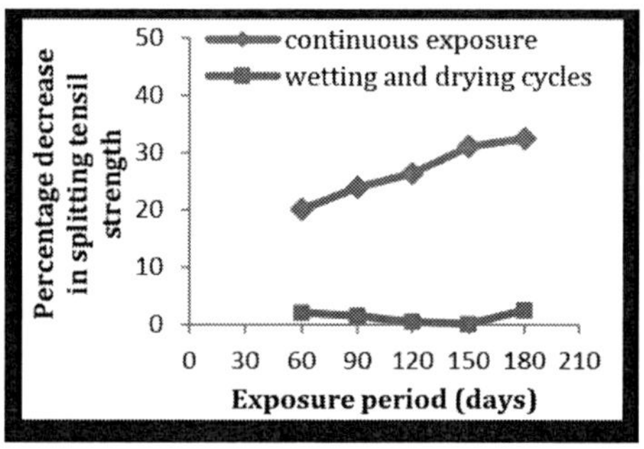

C- Motor oil

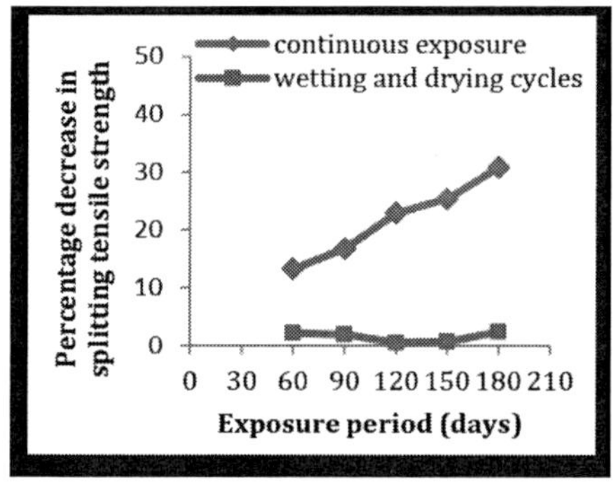

D – Fuel oil

Fig.6. Effect of exposure type to oil products on splitting tensile strength of SCC (The continuous exposure results are from reference 15)

4.2.3 Modulus of Rupture

Modulus of rupture of SCC exposed to wetting and drying cycles in water or different oil products are summarized in Table (5) and Figure (7). The results indicate that there is a continuous increase in modulus of rupture as the number of wetting and drying cycles in water increased. The percentage has increased about 1.09%, 1.37%, 2.74%, 3.15% and 6.86% being exposure in 2, 3, 4, 5 and 6 cycles respectively comparing to the reference (not exposed to wetting and drying cycles). The modulus of rupture of SCC specimens exposed to wetting and drying cycles in different oil products has slightly decreased being exposure in 2, 3, 4and 5 cycles in comparison with reference specimens (exposed to water for the same number of cycles). This is due to the closing and autogenously healing of crack and flaws in SCC due to possible volume change by effect of products [28]. The modulus of rupture of SCC specimens exposed to different oil products began to decrease as the number of wetting and drying cycles to oil products increased to six cycles relative to the reference. The decrease ratio in modulus of rupture of SCC specimens exposure to six cycles in crude oil, gas oil, motor oil and fuel oil is 15.17%, 7.46%, 5% and 5% respectively relative to the reference. Figure (8) shows that modulus of rupture of SCC specimens immersed in water for a period is more than that which exposed to wetting and drying cycles for all exposure periods. This may be attributed to the continuous hydration of cement due to continuous supply of water which can lead to increase in modulus of rupture. The comparison between the decrease in modulus of rupture of SCC specimens continuously exposed to oil products and specimens exposed to wetting and drying cycles in oil products relative to the reference specimens is shown in Figure (9). It can be concluded that the continuous exposure to oil products is more severe than wetting and drying exposure. This might be attributed to the continuous losses of some oil by evaporation during the drying cycles and hence reduces its negative affect [22]. While for continuous exposure, oil film coated the unhydrated cement grains blocks moisture from the coated surfaces and concrete grains, ceasing further hydration. Therefore, the strength of concrete has slowed or stopped [26].

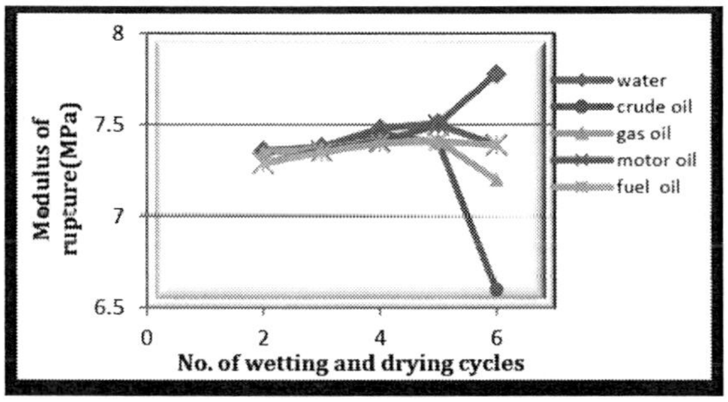

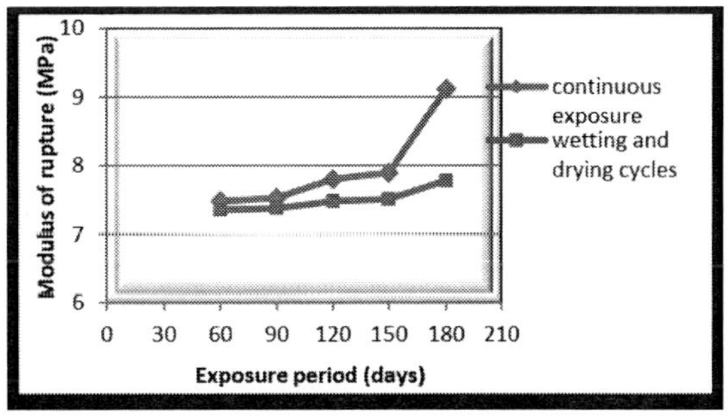

Fig.7. Modulus of rupture of SCC exposed to wetting and drying cycles in water or oil products

Fig.8. Effect of water exposure type on modulus of rupture of SCC (The continuous exposure results are from reference 15)

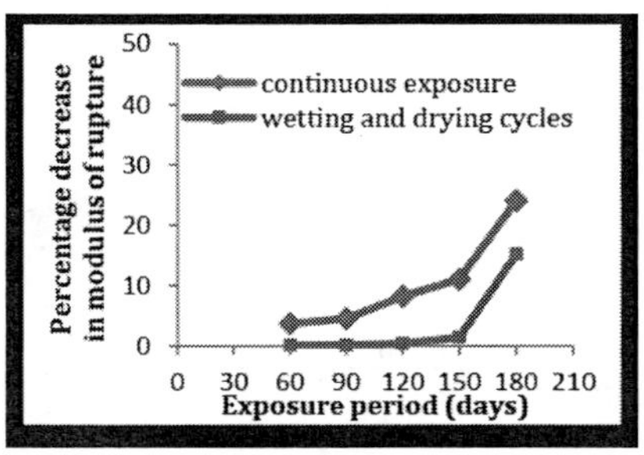

A- Crude oil

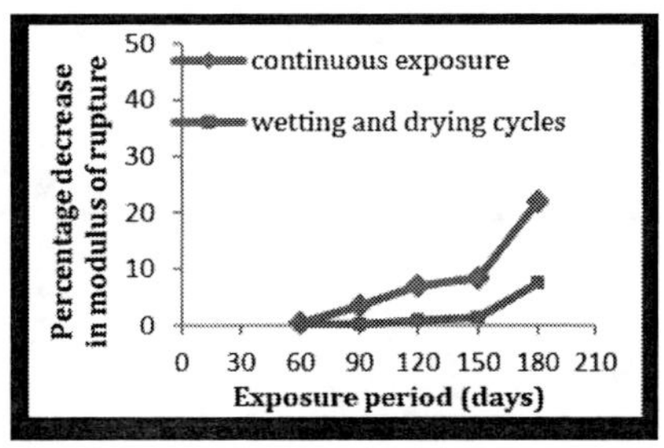

B- Gas oil

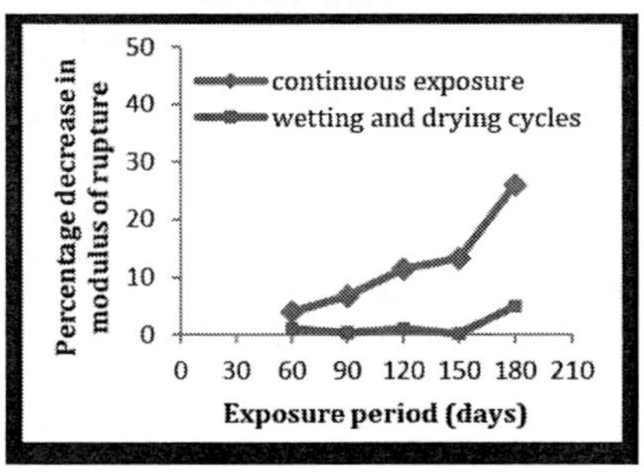

C- Motor oil

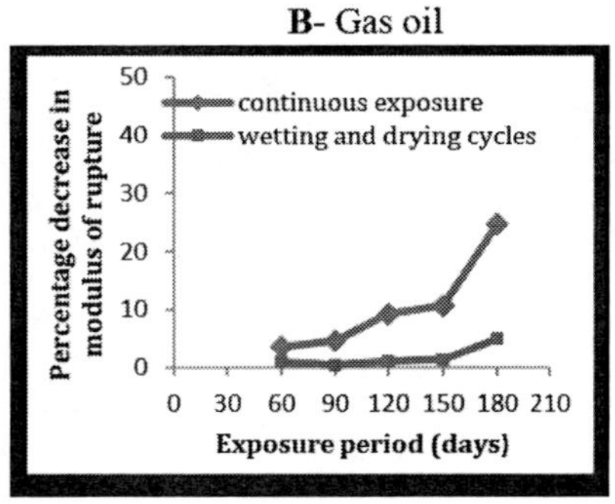

D – Fuel oil

Fig.9. Effect of exposure type to oil products on modulus of rupture of SCC (The continuous exposure results are from reference 15)

4.2. 4 Static Modulus of Elasticity

The results of static modulus of elasticity of SCC specimens exposed to wetting and drying cycles in water and oil products are summarized in Table (5) and Figure (10). The results illustrate that there is a continuous increase in static modulus of elasticity as the number of wetting and drying cycles in water increased. Static modulus of elasticity for SCC specimens exposed to immersion and drying cycles in different oil products slightly decreases for 2, 3, 4 and 5 cycles relative to reference specimens (exposed to water for the same number of cycles). The pattern of decrease in compressive strength leads to a decrease in modulus of elasticity for concrete. SCC specimens exposed to crude oil and gas oil show the higher reduction in static modulus of elasticity when it exposed to six cycles in comparison to the reference. The decrease ratio is about 27.6% and 19.6% respectively. This is because crude and gas oils have lower viscosity than motor and fuel oils, so the aggressivity of crude and gas oils is higher than that of motor and fuel oil. This reduction in modulus of elasticity attributed to:

- a- The presence of the oil might cause a weak effect on the cement paste components due to the dilution of the cement paste and the reduction of the adhesive forces between the gel particles.
- b- The presence of micro cracks at the aggregate-matrix interfaces due to the volume changes during setting and hardening [22].

Figure (11) shows that static modulus of elasticity for SCC specimens continuously immersed in water is more than that exposed to wetting and drying cycles in water for all exposure periods. From Figure (12), it can be concluded that the continuous exposure of SCC to different oil products is more severe than wetting and drying exposure. This may be due to the deterioration of specimen's surface continuously exposed to oil products [17].

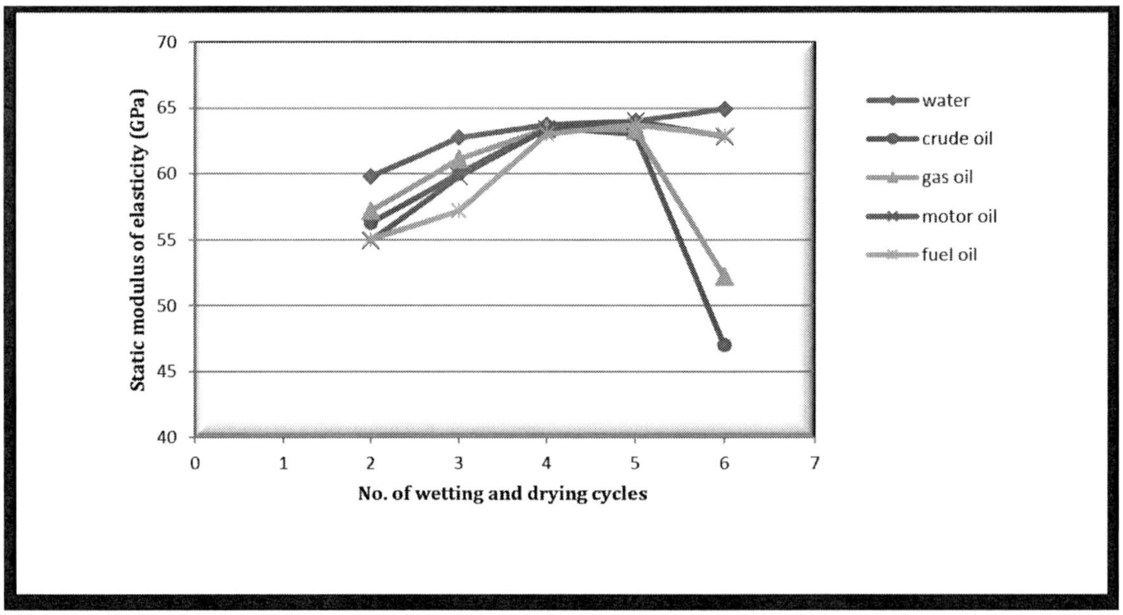

Fig.10. Relationship between static modulus of elasticity for SCC specimens and wetting and drying cycles in water or oil products

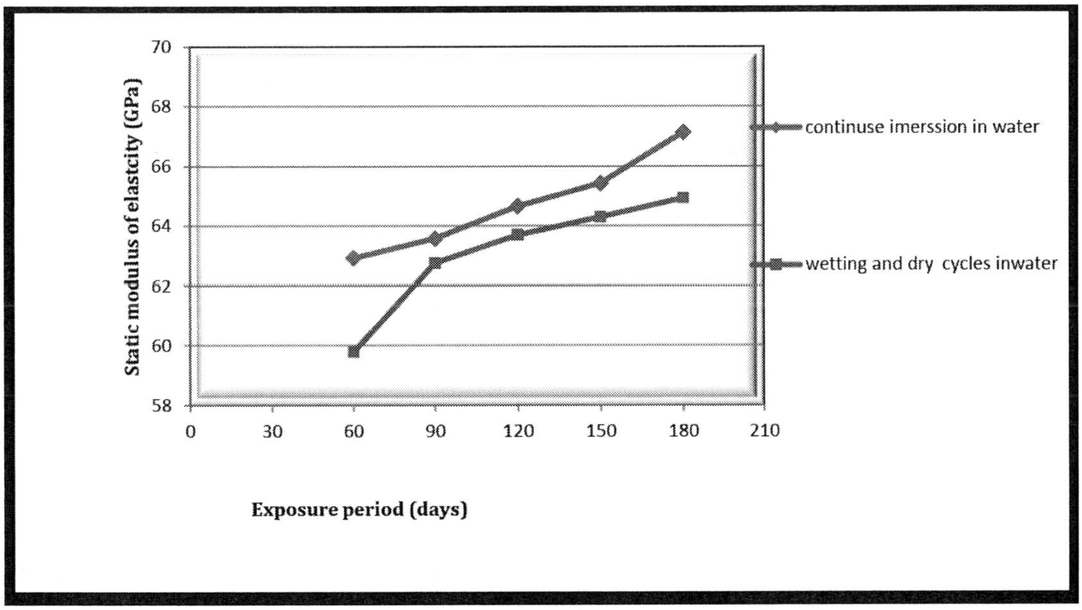

Fig.11. Static modulus of elasticity of SCC specimens exposed to water (continues exposure and wetting and drying cycle exposure)

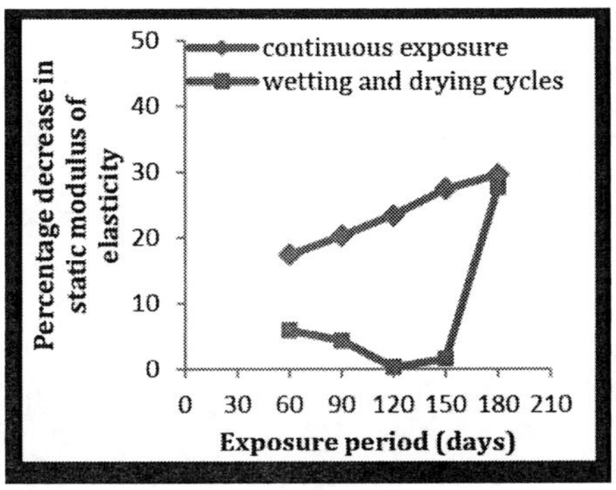

A- Crude oil

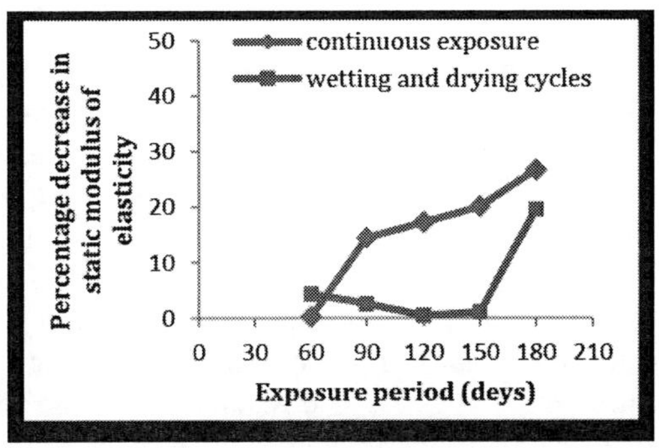

B- Gas oil

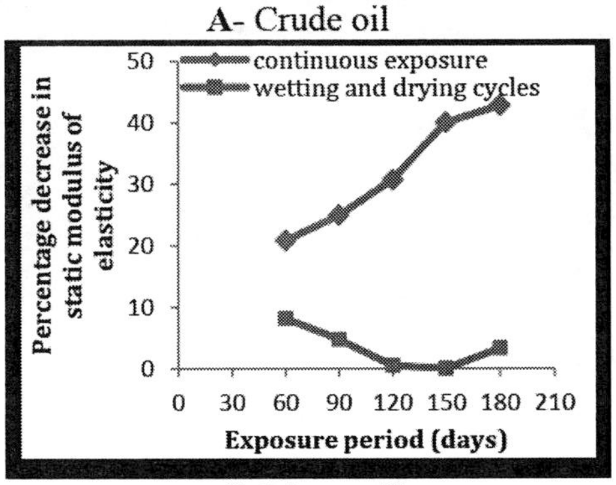

C- Motor oil

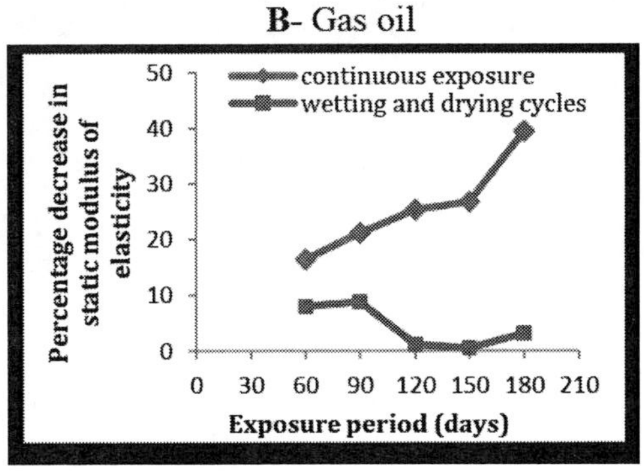

D – Fuel oil

Fig.12. Effect of exposure type to different oil products on static modulus of elasticity for SCC specimens (The continuous exposure results are from reference 15)

4.2.5 Total Absorption

The results show that the total absorption of all SCC specimens exposed to water or different oil products are below 10 % by weight as shown in Table (5). This gives an indication of concrete with low permeability [29]. The total absorption for SCC specimens exposed to wetting and drying cycles in water decreases as the number of wetting and drying cycles increased. This is attributed to the continuous hydration of cement which decreases the absorption of SCC. The use of low water/cement ratio and silica fume which modifies the microstructure of concrete and reduces the capillary porosity leading to better packing and increase the density then reduces the absorption ratio [30]. The results also show that the total absorption of SCC specimens exposed to wetting and dry cycles in crude oil, motor oil and fuel oil products has decreased as the number of cycles increased. Specimens have been exposed to wetting and drying cycles in water show higher total absorption than those exposed to wetting and drying cycles in different oil products for all exposure cycles. This may be related to the large molecular size of oil products and its viscosity compared to water. Also, the SCC which has been produced in this investigation has small pores in its microstructure since silica fume leads to pores size refinement and this will need more time for oil products to penetrate in comparison with water.
Generally, total absorption of SCC exposed to wetting and drying cycles in crude oil is lower than that of SCC exposed to cycles in gas oil, motor oil and fuel oil. This can be attributed to the deposits and the wax that are found in the chemical composition of crude oil as shown in Table (3) which can block some pores.

4.2.6 Density

The results of density of SCC specimens exposed to wetting and drying cycles in water and oil products are shown in Figure (13). The results indicate that there is a slight increase in the density as the number of wetting and drying cycles in water have increased. This can be attribute to the continuous hydration process of cement paste which forms a new hydration product within the SCC mass, then increases the bond between cement paste and aggregate [29, 14] .
From the results, it can be concluded that there is no significant effect on the density of SCC exposed to 2, 3, 4 and 5 wetting and drying cycles in different oil products, but as the exposure cycles increased to six cycles, the density slightly decreases relatively to reference the specimen (exposed to water for the same number of cycles). The results of SCC specimens which have been exposed to six wetting and drying cycles in crude oil and gas oil show the higher reduction in density (1.42% and 1.1%) compared to the reference. This may be due to the slight deterioration of specimen's surface exposed to oil product [17].

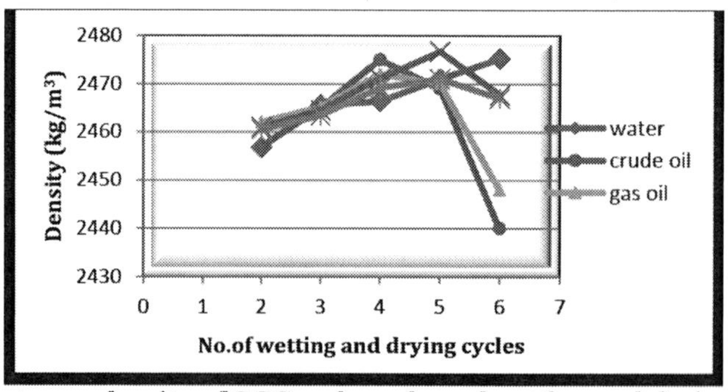

Fig.13. Relationship between density of SCC and number of wetting and drying cycles in water or oil products

IOP Conf. Series: Materials Science and Engineering **579** (2019) 012047 doi:10.1088/1757-899X/579/1/012047

5. Conclusions

This research presents an experimental study of the properties of SCC exposed oil products for different cycles. The following conclusions are drawn from the test results:

1. The compressive strength, splitting tensile strength, modulus of rupture, and static modulus of elasticity, of SCC specimens exposed to wetting and drying cycles in water have increased as the number of cycles increased. The percentage of increase is between 1.77-16.52 % for compressive strength, 2.18-6.54% for splitting tensile strength 1.09 -6.86% for modulus of rupture, and 9.50 - 18.96% for static modulus of elasticity for 2-6 cycles respectively.
2. The compressive strength, splitting tensile strength, modulus of rupture, and static modulus of elasticity of SCC specimens exposed to wetting and drying cycles in all oil products generally have decreased in comparison with the reference specimen (exposed to water for the same number of cycles). Specimens exposed to crude oil and gas oil show a significant decrease as the number of wetting and drying cycles has increased to six cycles. The decrease ratio is about 26.13%, 18.7% for compressive strength, 24%, 14.85% for splitting tensile strength, 15.2%, 7.46% for modulus of rupture, and 27.6%, 19.6% for static modulus of elasticity respectively relative to the reference (exposed to water for the same number of cycles).
3. Continuous exposure to oil product is more severe than wetting and drying exposure cycles on compressive strength, splitting tensile strength, modulus of rupture, and static modulus of elasticity of SCC specimens.
4. Total absorption for SCC specimens exposed to wetting and drying cycles in water decreases as the number of cycles increased. Water absorption is about 3.1%, 2.7%, 2.5%, 2.46% and 1.9% for wetting and drying cycles of 2,3,4,5 and 6 respectively. While the total absorption for SCC specimens exposed to wetting and drying cycles in oil products is lower than that exposed to wetting and drying cycles in water.
5. The density of SCC is slight increase as the number of wetting and drying cycles in water increased, while there is no significant effect on the density of SCC exposed to 2, 3, 4 and 5 wetting and drying cycles in different oil products, but the density slightly decreases as the exposure cycles increased to six cycles relatively to SCC exposed to water for the same number of cycles.

References

[1] Ra'ed KHA *Behavior of High Performance Concrete in Direct Contact with an Oil Products* MSc Thesis University of Technology Baghdad (2002)
[2] Rashed L *Behavior of Fiber Reinforced Concrete Exposed to Oil* MSc Thesis University of Technology Baghdad (1998)
[3] Al hamadani ZK *Improvement of the Performance of Concrete Against Oil Products* MSc Thesis University of Technology Baghdad (1997)
[4] ASTM C150 2015 Portland *Cement Annual Book of Standards American Society for Testing and Materials* **04.01**
[5] ASTM C33 2015 *Concrete Aggregate Annual Book of Standards American Society for Testing and Materials* **04.02**
[6] Fosroc constructive solution *Structuro 520* fosam company limited FO SA/STRUCTURO 520/07/09
[7] ASTM C494 2015 *Standard Specification for Chemical Admixtures for Concrete, Annual Book of Standards, American Society for Testing and Materials* **0402**
[8] ASTM C1240 2015 *Standard Specification for the use of Silica Fume as a Mineral Admixture in Hydraulic Cement Concrete, Mortar, and Grout, Annual Book of Standards, American Society for Testing and Materials* **04.02**

[9] BS 1881 part 116 1989 *Method of Determination of Compressive Strength of Concrete Cubes, British Standards Institution*

[10] ASTM C496 2015 *Standard Test Method for Splitting Tensile Strength of Cylindrical Concrete Specimens, Annual Book of Standards, American Society for Testing and Materials* **0402**

[11] ASTM C469 2015 *Standard Test Method for Static Modulus of Elasticity and Poisson's Ratio of Concrete in Compression, Annual Book of Standards, American Society for Testing and Materials* **04.02**

[12] ASTM C78 2015 *Standard Test Method for Flexural Strength of Cement, Annual Book of Standards, American Society for Testing and Materials* **04.02**

[13] EFNARC *The European Guidelines for Self-Compacting Concrete Specification, Production and Use* (May 2005)

[14] Shetty M S 2000 *Concrete Technology India*

[15] Neville AM and Brooks JJ **2010** *Properties of Concrete* (Second Edition Prentic Hall England)

[16] Siddique R 2011 Utilization of Silica Fume in Concrete: Review of Hardened Properties, *Resources, Conservation and Recycling* **55** 923-932

[17] Onabolu OA 1989 Some Properties of Crude Oil Soaked Concrete Exposure at Ambient Temperatures *ACI Material Journa*l **26** 2 150-158

[18] Onabolu OA 1989 Some Properties of Crude Oil Soaked Concrete Exposure at Elevated Temperatures *ACI Material Journal* **86** 3 205-213

[19] Ejeh SP and Uche OAU 2009 Effect of Crude Oil Spill on Compressive Strength of Concrete Materials *Journal of Applied Sciences Research* **5** 10 1756-1761

[20] Ogbonna AC, Okoro EE and Daraha T 2013 Laboratory Investigation of the Effect of Crude Oil Spill on Engineering Properties of Concrete *IOSR Journal of Mechanical and Civil Engineering* **10** 2 11-17

[21] AL- Harby MG 1989 Effect of Oil Products on Structures *Nationalist Center of Constructional Laboratories Baghdad*

[22] Abdul- Moghni HSM 1999 *Durability of High Performance Concrete Exposed to Oil Products* PhD Thesis University of Technology Baghdad Iraq

[23] Mehta KP and Monteiro PJM 2006 *Concrete Microstructure, Properties and Material* Third Edition Mc Graw-Hill, London

[24] Biczok I 1967 *Concrete Corrosion and Concrete Protection* Publishing House of the Hungarian Academy of Science Budapest

[25] Al Mashhadany SA, Khalil WI and Ali AA 2014 Properties of Self Compacting Concret Continuously Exposed to Oil Products *Engineering and Technology Journal* **32** Part (A) 8

[26] Diab H 2012 Compressive Performance of Low- and High- Strength Concrete Soaked in Mineral Oil *Construction and Building Materials* **33** 25-31

[27] Khaleel O R 2007 *The Effect of Coarse Aggregate Properties on the Behavior of Self-Compacting Concrete* MSc Thesis University of Technology

[28] Matti M 1976 *Some Properties and Permeability of Concrete in Direct Contact with Crude Oil* PhD Thesis University of Sheffield

[29] Neville AM 2005 *Properties of Concrete* Pearson Education Limited Fourth and Final Edition

[30] *Guide for the Use of Silica Fume in Concrete* ACI 234R (2006)

Response of Soft Soil Mixing with Recycled Gypsum (Plasterboard) as Stabilized Agent for Soil underneath oil tank as a case study.

A Al-Adili[#] N M Salim and K. Y. Al-Soudany

[#]University of Technology, Baghdad, Iraq

Corresponding author's e-mail address: aqeeladili@hotmail.com

Abstract. Soil stability plays a vital rule in projects' infrastructure such as embankment, retaining walls dams and highway. Structure built on soft soil may experience uncontrollable settlement and critical bearing capacity. This may not meet the design requirements for the geotechnical engineers. Soil stabilization is a process to change such undesirable properties in order to meet the requirements. The main objective of this study is to evaluate the use of recycled gypsum, which is derived from gypsum waste plasterboard, as a stabilized agent for soft clay. Twenty eight experimental tests have been conducted to improve a soil brought from a site in Basra (Garma Ali /south of Iraq) using four different recycled gypsum percentage varying from 0 to 15%. The properties which have been studied are grain size distribution, Atterberg limits, unconfined compressive strength, and compressibility. The results indicate that as the gypsum contents increase, the liquid limit decreases up to gypsum content of 3% and then increases. The plastic limit has been decrease up to 7 % of the gypsum content and then increases. Furthermore, the maximum dry density decreases while the optimum moisture content is increased as the percentage of gypsum content has increased. The compression index (C_c) has increased as the gypsum content increases while the swelling index has increased up to 5% then it has decreased. The unconfined compressive strength has increased by adding recycled gypsum up to 5% while it is reduced as the percentages of gypsum has increased beyond 5%. Adding 5% of recycled gypsum raise the bearing capacity to approximately 167% compared with the bearing capacity of untreated one.

1. Introduction

Stabilization is the method of merging different types of soil or adding additives to soil to improve characteristic, such as strength, gradation and plasticity which make the soil more stable. The main purpose of stabilization can be listed as follows:

1- Decrease volume expansion
2- Increase strength
3- Decrease cohesiveness

There are different methods of soil stabilization such as:

a) - Mechanical stabilization.
b) - Chemical stabilization.
c) - Both mechanical and chemical stabilization

Chemical stabilization depends mainly on the chemical reaction between the additive material (cementation) and soil minerals (pozzolanic material) to achieve desired effect. Cement, lime, fly ash bitumen and combination of these can be used in this type of stabilization. Usually stabilized soils have a higher strength and lower compressibility than the original soils, that's why many researchers have used these types of stabilization. [1,7].

Nowadays, the researchers around the world have directed their research on improving soils by utilizing the millions of tons of local waste material, such as using waste plasterboard. A large quantity of gypsum plasterboard is produced during the three stages of production, construction and demolition. So, to avoid the increase of gypsum amount waste plasterboard that is disposed in landfill, it is necessary to find alternative ways to reduce the dispose waste which causes an increase in the cost of disposal and decease the available ground space that are specified for landfills sites. Furthermore, gypsum waste plasterboard in landfills under certain circumstances the emission cause of hydrogen sulphide gases, which are potentially harmful and lethal. In general, recycled gypsum (Bassanite) which is derived from gypsum plasterboard has the potential to be used as a cementation material because gypsum is the main key of cementation materials.

Ahmed et al., [8], demonstrated that the results obtained from laboratory investigations for the utilization of recycled gypsum in ground improvement are not adequate to evaluate the use of recycled gypsum as a stabilized material due to the difficulty in stabilizing soft clay in the field and this attributed to the following:

1-The in-situ behavior of clay soil is mostly very complex due to various geological patterns of the soil in the same site

2- The water contained in the clay soil changes from time to time during the construction period due to weathering factors such as rainfall, humidity, temperature and season changes.

3- It is difficult to mix the stabilized material with clay in the field because clay is very sticky and it is difficult to bring homogeneous and isotropic properties to the mixture as compared to laboratory.

4- Information about the use of recycled gypsum in ground improvement project has not been clarified yet and the data for this issue is limited.

2. Previous studies

Different materials have been used for soil stabilization; the most common those which have been used for chemical are or mechanical soil stabilization such as lime and Portland cement. In addition there are several studies which have investigated the use different types of waste materials as stabilizing agents to improve the mechanical properties of soil. Example of using waste material to improve the strength of Iraqi soil, using nylon bags by product with different percentages corresponding to 1%, 3% and 5% of dried weight. The main conclusions drawn that, as the percentage of additive has been increased, the liquid limit has decreased s while the plastic limit has increased causes, the plasticity index to be decreased. The maximum dry density decreases while the optimum moisture content increases as the Nylon fibre percentage has been raised. The compression index decline as the Nylon fibre has inclined and reaches a maximum value of 43% by adding 5% of nylon fibre. Finally, the undrained shear strength has increased as nylon fibre increased as well [9].

The use of recycled gypsum, which is derived from gypsum waste plasterboard, in ground improvement is limited around the world are very limited researches available in this area. Therefore, this section deals with previous studies using this type of additive.

Kamei et al., [10]. used different contents of basanite, which derived from gypsum waste plasterboard to improve sand and clay soils. They concluded that using the stabilization agent has improved the stress strain behaviour of the used soil., [11].The study conducted a series of unconfined compression, splitting tensile and capillary rise tests to study the effect of using sandy soil reinforced with gypsum waste plasterboard. Plaster waste was used as a cementation agent to improve the compressive strength while strips of waste plastic trays were used to enhance the tensile strength of the improved soil with gypsum wastes. The results reveal an increase in both compressive and splitting strengths, which happened with the presences of recycled gypsum content.

3. Soil Properties

In this part, the experimental work has been carried out using standard procedures. The untreated soils have been tested for their classification and index properties, as well as their consistency properties. In addition, testing program has conducted on the clayey soil stabilized with different percentage of recycled gypsum, including Atterberg's limits, static compaction, consolidation test. The grain size distribution (sieve analysis and hydrometer test) was performed in accordance with ASTM D421-58 as shown in Fig. 1. All the physical and mechanical properties of tested soil are listed in Table 1. The soil is classified as clay with low plasticity (CL) according to the unified soil classification system (USCS).

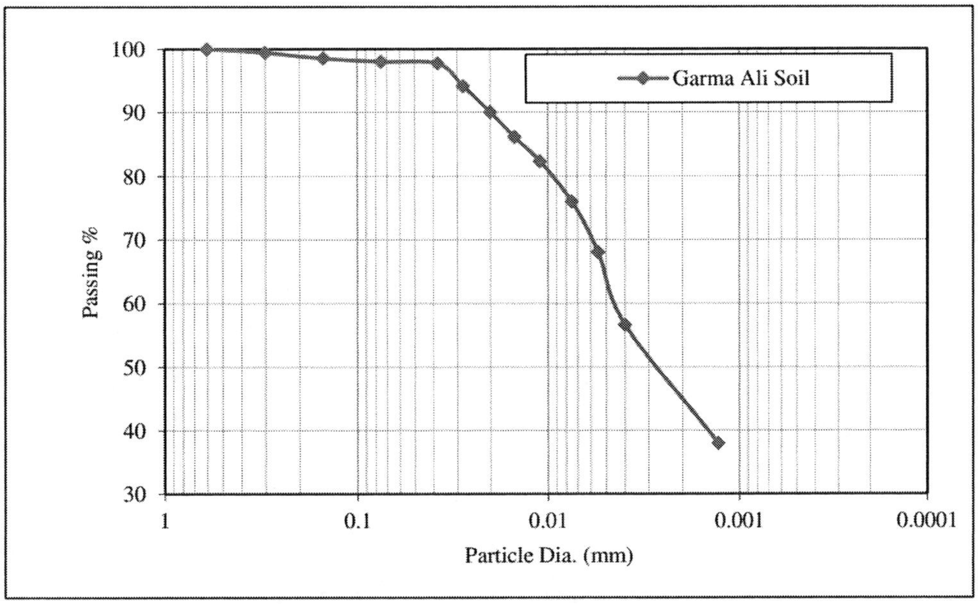

Fig. 1 Grain size distribution curve for soil used

Table 1. Physical and mechanical properties of tested soil

Engineering property	value
Liquid Limit, LL, %	46
Plastic Limit, PL, %	20
Plasticity index, PI, %	26
Shrinkage Limit, SL, %	17
Activity of Clay, AT, *	0.6
Specific gravity, G_s	2.69
Clay content ($<$ 0.005 mm), C %	64
Silt (0.005 - 0.075 mm), M	34

%	
Sand (0.075 to 4.75 mm), S	2
%	
Classification (USCS)	CL
Organic Matter (O.M,%	< 0.01
SO_3 Content, %	0.38
Total Solved Salt (TSS),%	6.89
pH value, %	8.69
MDD (kN/m^3)	18
OMC, %	15.71
Initial water content, (%)	3.1

4. Material and Methods

4.1. Soil Samples

Gray slant to green clayey soil is brought from a site in Bsara (Garma Ali, South of Iraq). The natural moisture content is determined immediately in the soil mechanic laboratory, University of Technology. After that soil samples are spread out for two weeks for air-drying at room temperature. Then the soil has sieved to obtain the final soil sample for the test. To investigate the effect of recycled gypsum material on the geotechnical properties of the soft soil, soil-gypsum mixture have been prepared using three different percentages of gypsum material (5, 10 and 15%) passing sieve No (0.425 mm).

4.2. Recycled Gypsum

The recycled gypsum which is derived from gypsum waste plasterboard, used in this project has been collected from different construction sites, crushed by crushing machine and then screened to remove any solid waste such as paper, fibres, wood. The crushed gypsum $(CaSo_4.2H_2O)$ is heated at temperature 45^0C to produce
Bassanite $CaSo_4$.

4.3. Experimental Tests

In this study, laboratory experiments are achieved to assess the effect of recycled gypsum on the physical properties, compaction, undrained shear strength and consolidation parameters of the untreated (unstabilized) and treated (stabilized) soft clayey soil samples.

As aforementioned, all the tests have been conducted after choosing the worst case, i,e., by taking the lowest density and maximum water content. Then, static compaction is applied by using the modified proctor mold to study the effect of different gypsum percentages on clay by checking the unconfined compressive strength, and consolidation tests.

5. Effect of recycled gypsum on Soil Properties

For the natural and stabilized samples (after adding 5, 10 and 15 % gypsum), the above-mentioned same tests and procedures are as followed.

5.1. Specific gravity

The specific gravity of soil is determined according to ASTM D 854-02-Standard test for specific gravity of soil solids by water pycnometer. Fig. 2 shows the specific gravity variation with different gypsum contents.

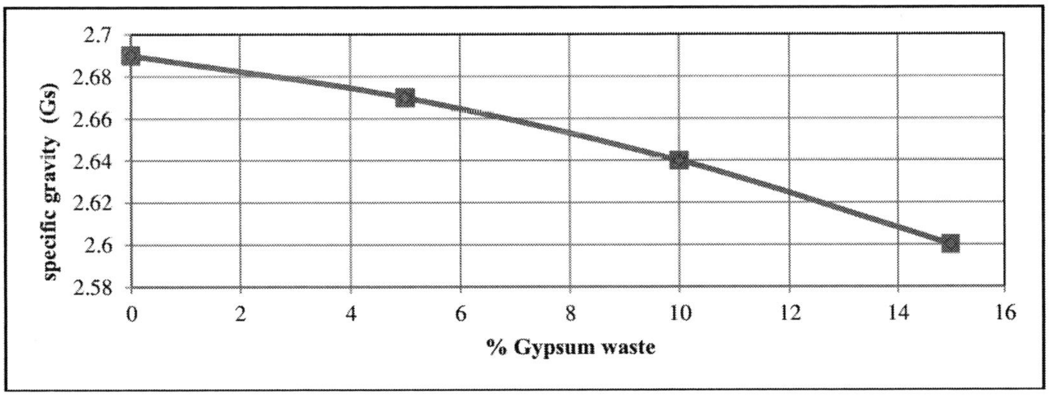

Fig. (2) Specific gravity variation with different gypsum contents.

5.2. Compaction

This test is performed according to ASTM D 1557to define the relation between the dry unit weight and the moisture content of the soil. Fig. 3 shows the optimum moisture content (OMC) obtained from compaction test for untreated (0 % gypsum) and treated (5, 10 and 15% gypsum) soils.

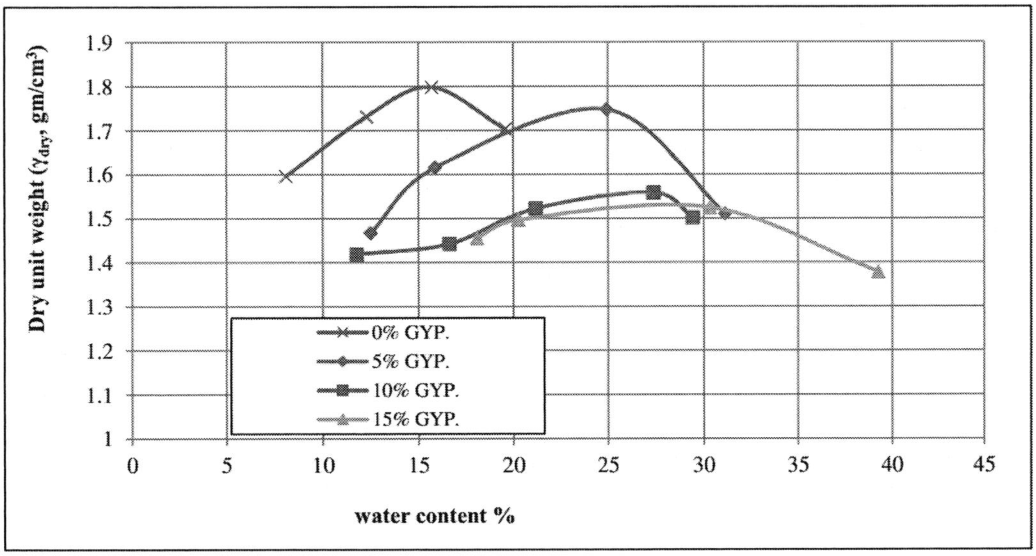

Fig. (3) Optimum moisture content (OMC) obtained from the compaction test for untreated and treated soils.

From compaction test, the maximum dry density variation and the optimum moisture content (O.M.C) with different gypsum contents are presented in Fig. 4 and 5.

The effect of adding recycled gypsum beyond 5% causes an increase in optimum moisture reached to a maximum value 25% by adding 15% recycled gypsum, then there is a sharp drop in dry unit weight beyond 5% recycled gypsum additive

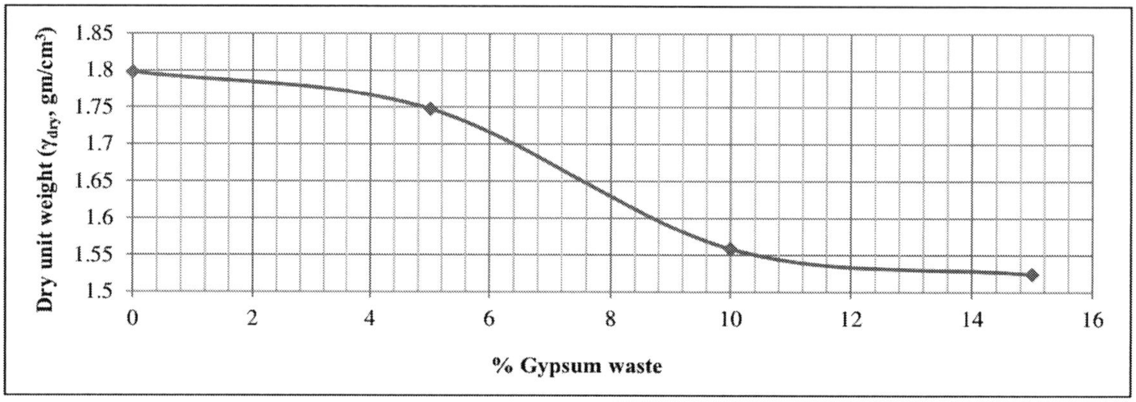

Fig. (4) Maximum dry unit weight variation with different gypsum contents.

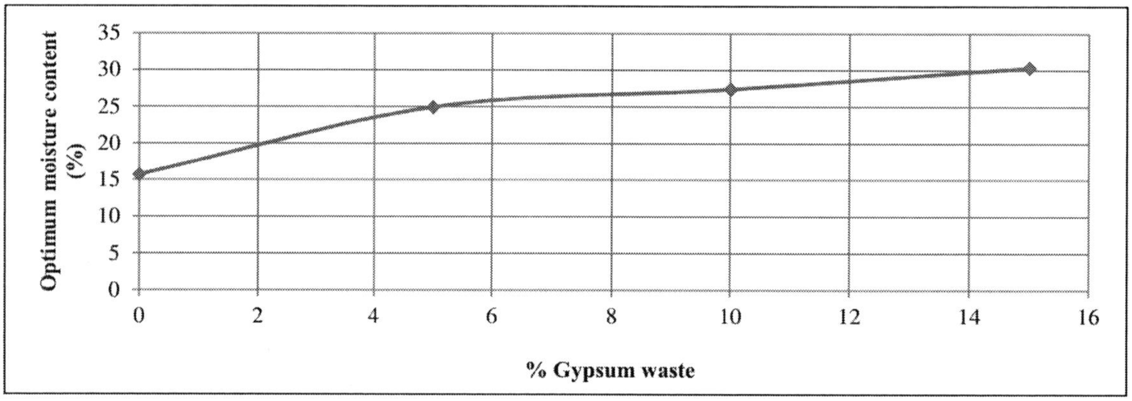

Fig. (5) Optimum moisture content variation with different polymer contents.

5.3. Atterberg's limits

Wide varieties of soil engineering properties have been correlated with consistency limits (Atterberg's limits).Liquid Limit (L.L) and plastic limit (P.L) are carried out according to the ASTM D 4318-00. Figs 6-8 illustrates the variations of these limits with gypsum content. Adding recycled gypsum up to 5% causes a reduction in liquid limit, this behavior is due to the bonds between the soil particles and recycled gypsum which causes the change in chemical composition of bassanite ($CaSo_4$) to hydrate calcium sulphate ($CaSo_4.2H_2O$) followed by water being absorbed from the clay soil In addition the chemical composition of the hydrate calcium sulphate as oxide contains a significant amount of calcium lime (Cao). The increase of silica content has caused the increase of pozzolanic reactions which produces increased quantity of the water holding gelatinous products. This causes an increase in liquid limit.

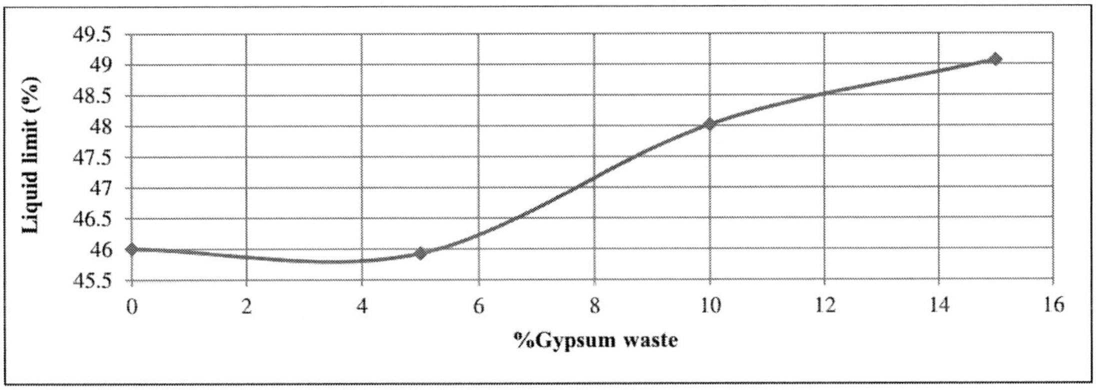

Fig. (6) Liquid limits variation with gypsum content.

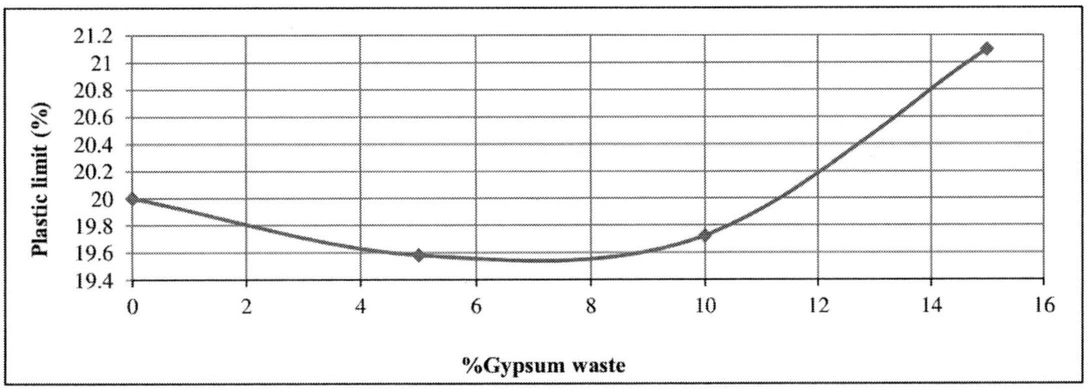

Fig. (7) Plastic limits variation with gypsum content.

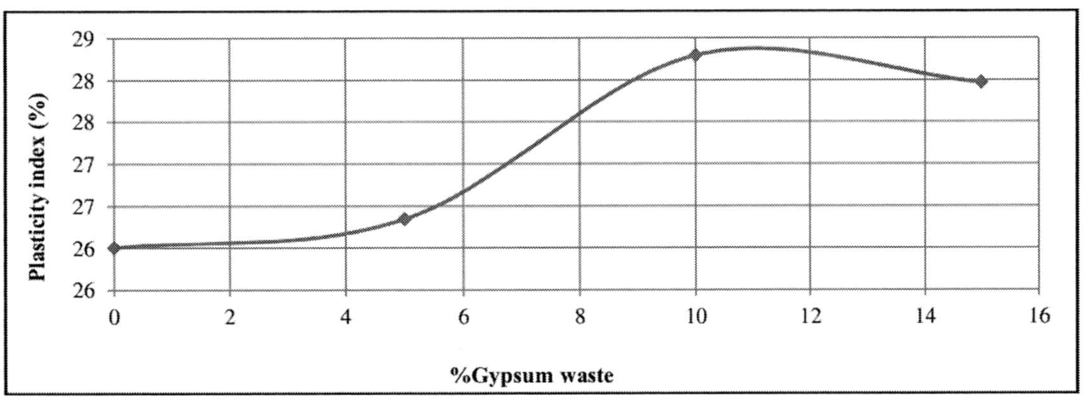

Fig. (8) Plasticity index variation with gypsum content.

5.4. Unconfined compressive strength

Unconfined compressive test is conducted to determine the unconfined compressive strength (UCS,

q_u) following the procedure of the ASTM D 2166-00 Standard Test Method for a cohesive soil. Then, the undrained shear strength (or undrained cohesion, s_u) of a cohesive soil is calculated which is equal to one-half of the unconfined compressive strength (q_u). Figure 9 illustrates the stress-strain relationship from the unconfined compression test for the treated and untreated soils. The results of undrained shear strength values of UCS tests on natural and stabilized soils with different gypsum contents are shown in Figs. 9 and 10. It is clear that the maximum increase in compressive strength with 5% of increase of recycled gypsum, this can be attributed to the hardening of the soil particles in the clay soil after adding the recycled gypsum resulting in an increase in the cohesion strength between soil particles. The hardening between the soil particles and recycled gypsum is duc to thc change in chemical composition of bassanite ($CaSo_4.\frac{1}{2}\ H_2o$) to hydrate calcium sulphate ($CaSo_4.2H_2O$) followed by water being absorbed from the clay soil, which reduced the voids between the soil particles and increase the soil strength. The chemical composition of the hydrate calcium sulphate as oxide contains a significant amount of calcium lime (Cao) which has been act like cohesive agent in soils because of its ability to increase the strength of cohesive (Bell, 1996). Finally, Gypsum contains a significant amount of calcium sulphate, when is mixed with soil, dissolve slowly and dissociates into Ca^{++} and $So4$—ions. Calcium positively charged and becomes attracted to clay particles, which having negative charges. This attraction between the soil particles and calcium, improves the bond between the soil particles and strengthens the soil.

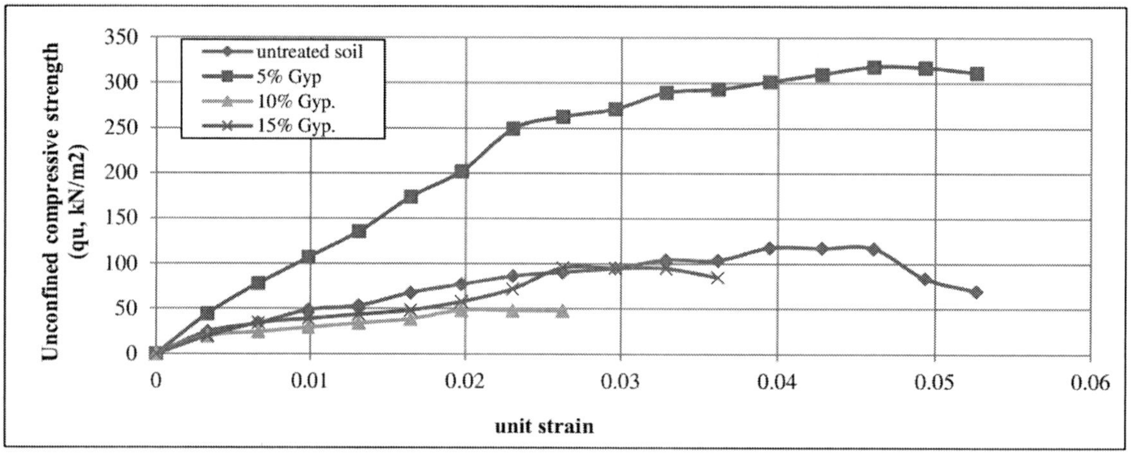

Fig. (9) Effect of gypsum content on the unconfined compressive strength.

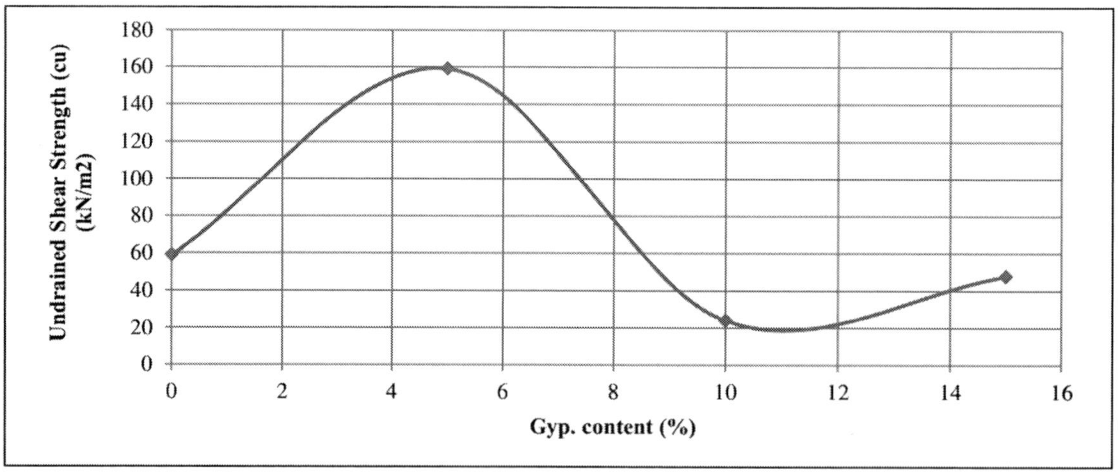

Fig. (10) Effect of gypsum fibers content on undrained shear strength of soil.

5.5. Consolidation test

This test is performed according to ASTM D 2435-02 Standard Test Method for One-Dimensional Consolidation Properties of Soils. Loading pressures are 25, 50, 100, 200, 400 and 800 kPa and reloading pressures are 800, 200and 50 kPa. From e-log p relation, the compression index (C_c) and recompression index or swelling index (C_r) during unloading are determined at different gypsum contents. Fig. 11 shows the consolidation test results by relating e-log p with gypsum addition.
Fig. 12 shows The compression index (Cc) increase sharply up to 5% of recycled gypsum additive. This increment is reduced beyond 5%.

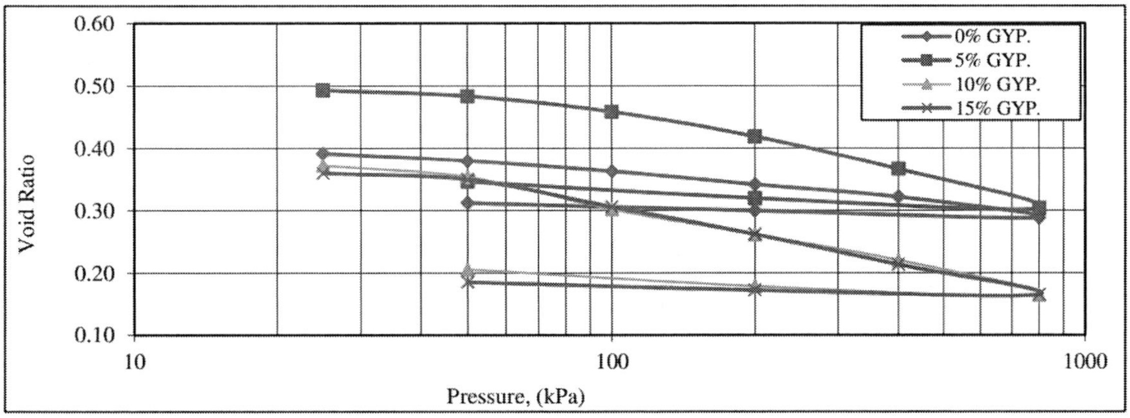

Figure (11) Relation of e-log p for untreated and treated soils with gypsum addition.

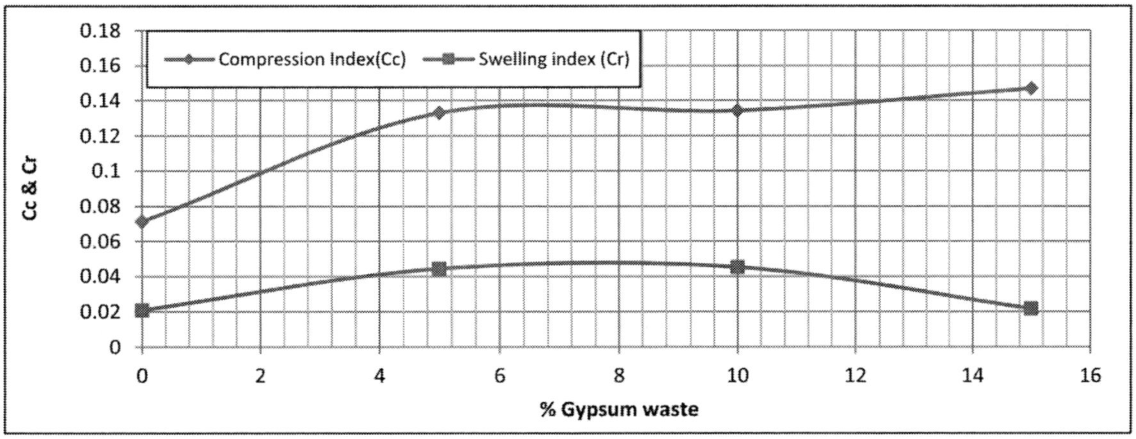

Fig. (12): The relationship between compression Index (Cc) and rebound (swelling) Index (Cr) with different Gypsum Content.

6. Effect of Recycled Gypsum additive on the Bearing Capacity

Construction on soft clay is a great challenge. Soft clays are available in the middle and southern parts of Iraq. Al-Basra governorate is one of the most regions in the ancient and modern Iraq. Al Basra is the most Socio-economic hub of the southern of Iraq and becomes an important commercial and industrial centre, which demands the construction of large structures such as electrical power stations, port platforms, and a huge oil tanks. Since the undrained shear, strength of base soil is low which restricts the construction of such type of structures. These problems can possibly be solved soil improvement, using recycled gypsum. The ultimate bearing capacity of soft clay can be compared when with and without additive by using the following equation[12].

$$q_{ult} = CNc \qquad (1)$$

$$q_{ult} = 5.7\ Cu \qquad (2)$$

Where Nc; is the bearing capacity factor, and
Cu; is the undrained shear strength of soil
For oil tank, the processes of filling and empty occur rapidly, so this causes to use the undrained shear strength, which means that there is not enough time for excess pore water pressure to dissipate.
Case 1: The ultimate bearing capacity without soil improvement:
The maximum undrained shear strength ($Cu = 60\ kN/m^2$) so the

$$q_{ult} = 5.7\ x\ 60 = 342\ kPa$$

Case 2: with 5% recycled Gypsum the undrained shear strength ($Cu = 160\ kN/m^2$) so the

$$q_{ult} = 5.7\ x\ 160 = 912\ kPa$$

Percentage of increase in bearing capacity = $\frac{912-342}{342} = 167\ \%$

To find $q_{all} = \frac{q_{ult}}{2.5} = 137$ kPa (without soil improvement)

$q_{all} = \frac{q_{ult}}{2.5} = 365$ kPa (with soil improvement)

If we compared these values with average oil tanks load
Average tank diameter = 21-46 m
The weight of tank =1800 kN -3600 kN
The height of oil in Tank = 16 m- 29 m
Unit weight of the oil =11.1 kN/m3
Total weight = 1980 kN- 35600 kN
Pressure excretes under the tank vary from 5.7 kPa -214 kPa
From these results, it is clearly shown that tanker with 46 m diameter cannot stand over soft clay without improvement with recycled Gypsum board because pressure due to the oil tank (214) and this is greater than the allowable load of soft clay (137kPa).

7. Conclusions

The main purpose of this work is to investigate and confirm the potential use of recycled gypsum as a stabilizer material for soft clay. Reducing the cost of soil improvement and in addition to improve the sustainability of the environment. Based on the results of the tests on strength, compressibility and durability, the following conclusions have been drowned:

- The effect of adding recycled gypsum beyond 5% causes an increase in the optimum moisture content and reached a maximum value by adding 15% recycled gypsum while these is a sharp drops in dry unit weight is beyond 5% recycled gypsum additive.
- By adding recycled gypsum up to 5%, there is a reduction in liquid limit while beyond this value; there is a great increase in liquid limit.
- The compression index (Cc) increase sharps up to 5% of recycled gypsum additive. This increment reduced beyond the 5% gypsum additive.
- The Percentage of increase in bearing capacity approaches 167% by using 5% Gypsum board as stabilized material compared with untreated one.

References

[1] Al-Adili A, Yali K and Shakir A 2017 Investigation of crude oil pollution effect on stiffness characteristics of sandy and gypsums soil *Mechanics and Foundation Engineering journal* **54,**

4 276 -282

[2] Al-Adili A and Sabs M 2019 Effect of salty soil on subsurface concrete strength *International Journal of Civil Engineering and Technology (IJCIET)* **10** 1 2556–2565

[3] Abd El-Aziz M A, Abo-Hashema M A and El-Shourbagy M 2004 *Engineering Conference, Faculty of Engineering Man sour University* 6 Sharm pp. 1-018

[4] Geiman C M 2005 Stabilization of soft Clay Sub grade in Virginia: Phase I Laboratory Study *M.Sc thesis Virginia Polytechnic Institute and State University, at Blacksburg, VA.*

[5] Harichane K, Ghrici M, Kenai S, Grine K 2011b **29** 5 759

[6] Kumar N, Swain S and Sahoo U 2012 *Journal of the Geotechnical and Geological Engineering* **24** 2 DOI 10, 1007/s 10706-012-9532-3

[7] Al-Adili A, Abd Al-Ameer O and Raheem E 2015 Investigation of incorporation of two waste admixtures effect on some properties of concrete *Energy Procedia Elsevier* **74** 652 – 662

[8] Ahmed A, Ugai K and Kamei T 2011 *Building and Construction Materials Journal* **25** 1 208-217

[9] Al-Adili A and Al-Soudany K 2018 Effect of random inclusion of date palm leaf fibers on some soil properties *International Journal of Earth Sciences and Engineering* **11** 1 28-34

[10] Kamei T, Kato T and Shuku T 2007 *Geotechnical Society Electronic Journals*, **2** 3 245– 252

[11] Ahmad A, Ugai K and Kamei T 2010 *Proceeding of International Geoshanghai Conference, Ground Improvement and Geosynthetics* Geotechnical Special Publication ASCE 207 165-173

[12] Terzaghi P 1943 *Theoretical Soil Mechanics* (John Wiley 2nd ed, New York)

IOP Conf. Series: Materials Science and Engineering **579** (2019) 012049 doi:10.1088/1757-899X/579/1/012049

Evaluation of Soil Structure Interaction Effect for Tanks under Earthquake with Different Foundations Soils

Mohammed Elwi and Marawan Mohammed

Building and Construction Engineering Department, University of Technology, Baghdad.

Abstract. Though the structures are supported on soil, most of the designers do not consider the soil structure interaction and its subsequent effect on the structure during an earthquake. Different soil properties can affect seismic waves as they pass through a soil layer. When a structure is subjected to an earthquake excitation, it interacts with the foundation and soil, and thus changes the motion of the ground. It means that the movement of the whole ground structure system is influenced by type of soil as well as by the type of an engineering structure. Tanks are supposed to be of construction in sense that they might have been analyzed and designed to meet the provision of relevant codes of practice and building bye-laws. IS 1893: 2002 "Criteria for Earthquake Resistant Design of Structures" gives response spectrum for different types of soil such as hard, medium and soft. An attempt has been made in this paper to study the effect of Soil-structure interaction on multi storeyed buildings with various foundation systems as well as to study the response of buildings subjected to seismic forces with Rigid and Flexible foundations. Tanks with fixed and flexible support subjected to seismic forces were analysed under different soil conditions like hard, medium and soft. The buildings were analysed by Response spectrum method using software by ABAQUS program . A structural model using the generalized single degree of freedom (SDF) system is proposed for seismic design of concrete rectangular Liquid Containing Structures (LCS). The proposed model considers the effect of flexibility of tank wall on hydrodynamic pressures and uses the consistent mass approach. The proposed model is compared with the results obtained using the current practice as well as the finite element method. It is concluded that the current approach in design codes and standards does not truly represent the behaviour of LCS. The proposed model using the generalized SDF system can be simply used in seismic design of LCS.

Keyword: response spectrum, ABAQUS program, earthquake, soil structure interaction

1 Introduction

Dynamic forces are present in every engineering problem although they are only considered to be significant in few specific scenarios. The highest profile situation is when an engineering structure is subjected to earthquake induced loading. Finite element is the most commonly accepted analysis tool for a solution of engineering problems. Effective pre and post processing capabilities make modelling and interpretation of results simple. It is relatively easy to incorporate changes, if any, and repeat the analysis without much loss of time. Viewing of animated mode shapes and dynamic response makes understanding of the dynamic behaviour of the machine foundation system relatively simpler [1].

Auditing of vivified mode shapes and powerful response makes viewing the direct component of the machine foundation structure reasonably less complex [2].

As waves from a tremor land at A structure, they plan movements in the structure. These movements depend on the structure's vibrational viewpoints and the plan of structure. As those structure will react with the movement, it necessities with succeed its very own lethargy constrain, which realizes a joint effort amid the centre of those structure and the dirt. Those degrees with which the basic response changes those viewpoints about seismic tremor movements viewed toward this system level depends on the relative impostor and immovability thus properties of the soil and the structure. Thusly, this physical property of the system medium is a foremost computed in this shake response from asserting structures supported on it. This approach for which the response of the dirt effects this development of the structure and the development of the structure impacts the response of the earth will be named Similarly as soil-structure affiliation (SSI) [3].

The Investigation of the component collaboration about soils and structures changes fundamentally Concerning delineation of the specific physical issue changes. There are no under four evolving soil-structure interactional issues (Christian Furthermore Hall, 1982):. 1. Vibrating mechanical assembly for the most part is decently little appeared differently in relation to this level of the setting up soil, and the basic states would Typically the people Throughout oversaw activity of the provisions. Hence, a persevering state result to A structure found at then again shut the outside of a half-space is commonly satisfactory. The device attempts to compelled What's increasingly known frequencies.

2. The following influence effect following on the other hand affects security incorporates a transient issue for amazingly few, regularly one, beats about overpressure. Unflinching state response and the pointless contemplations and direct free to move around at will times are not especially fundamental.

3. Seismic tremors incorporate long data signs that have rich repeat substance Also huge segments cycles about load reversal. An immediate outcome the waves proliferate In a good separation, the showing for considerable Furthermore frustrated geometries is a basic issue. The enter development is normally not extraordinary known Also might be routinely determined amid a darken then again severely masterminded region [4].

4. Immense toward the ocean structures information precisely transient reaction, yet the persisting state light of wave stacking is by far most enormous for exhaustion split issues.

The earth direct strategy under component stacking depends on some number elements, including. 1. The method for the dirt (penetrability, relative thickness, texture, etc.

2. Nature's area of the earth (static pressure state and water weight).

3. The methods for the dynamic stacking (strain greatness, strain rate, Also number for cycles for stacking).

Liquid holding structures (LCS) Concerning representation and just normal building workplaces are essentially used for water sewage prescription plants. Regularly, they require help developed from guaranteeing strengthened. Concrete in the structure of rectangular on the other hand equipment arrangements. At present is there a couple of codes Furthermore estimates open for seismic arrangement from asserting LCS secured close by North America. On Just about the larger part for codes Also gauges, those Housner's model (Housner, 1963) require to component Investigation of LCS. This model approximates the effect of hydrodynamic load to a twofold-symmetric-liquid compartment oppressed on level quickening as demonstrated over figure 1(a). Those hydrodynamic loads that provoked Eventually Tom's examining tremors would be separated under two sections of indiscreet convective fragments which would approximate eventually Tom's scrutinizing those lumped included massenet. The included impostor As far as hasty weight might be acknowledged rigidly regarding this tank divider and the included impostor. As far as convective weight might be acknowledged related with the tank divider using versatile springs to reproduce the effect about sloshing development. In this model, the limit condition in the check about hydrodynamic loads might be dealt with Concerning representation inflexible [4].

Despite the way that the Housner's model needs to be associated in the seismic arrangement from asserting LCS before, later examinations show that on account of the doubt of the lumped included impostor and the inflexible tank divider, this technique prompts exorbitantly preservationist impacts. Chen Also Kianoush (2005) framed a framework implied on Similarly as those back to back system to enrolling hydrodynamic loads subordinate upon a two-dimensional model to rectangular tanks On which the effect about flexibility of tank divider may have been contemplated. Ghaemian et al. (2005) associated the stunned procedure on deal with the coupled liquid limit tank issues in. Three-dimensional space. Appeared differently in relation to those Housner's model, these Outcomes show that by and vast the lumped impostor strategy overestimates base shear and base moment basically. Chen and Kinaoush (2007) recommended A summed up single dimension of adaptability (SDF) structure for dynamic examination for LCS. The unfaltering impostor philosophy and the effect for versatility about tank divider around hydrodynamic loads were seen as the embraced vibration state limits addressing such mode shapes for those cantilever divider limit conditions that were affirmed. In this paper, this prescribed auxiliary model using the summed up SDF system will be looked at for the Housner's model grasped in the present blueprint codes for more standards. Those framework diagrams for the included impostor of liquid in view of imprudent hydrodynamic weight and the relating convincing stature would be displayed. The responsibility from asserting higher methods of the changing response about LCS might be incorporated into the prescribed model. Base of entire of money about square (SRSS) technique is used for mixing the underlying two modes. The outcomes are contrasted and the people got using those Housner's model and moreover those restricted part strategy. It might be recommended that present setup philosophy necessity should make changes this prescribed auxiliary model using the summed up SDF system could be seen as clear model on beat those present lacks for layout for LCS [5].

2. The comparison of the investigation model by claiming movement.

The Houser's model (1963) is shown in figure 1(a). Figure 1(b) shows An cantilever tank divider with the scattered impostor m(y) and solidness EI(y) for every unit stature laid open of the tremor ground reviving üg(t). This divider shows a limitless amount of degrees about chance to flexural system for response. Due to the possibility that there would be a percentage for destined shapes to estimate the vibration of the framework, during that perspective the development of the skeleton camwood makes depicted Eventually Tom's perusing a single variable, or summed dependent upon organize over which just An solitary DOF exists. The skeleton appreciated in that capacity may be alluded with Concerning illustration the summed up SDF frameworks. In this examination, the summed up SDF skeleton may be associated with disentangle dynamic response of liquid stockpiling tanks uncovered will seismic tremors. Those condition for development of the summed up SDF skeleton is that [7]:.
M-. U+ c-. What's to come for U +k-u = p-(1).
The place m-,u, c-,k-, p-are portrayed Concerning illustration those summed up course of action from claiming mass, damping, solidness What's more force independently.

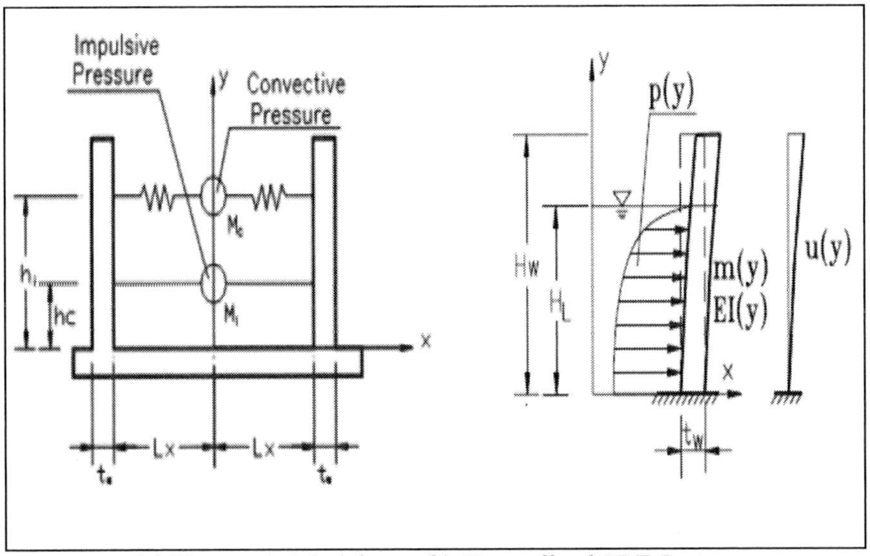

(a) Housner's Model (b) Generalized SDF System

Figure 1 Analysis Model[3]

For straightforwardness, the endorsed vibration shape work SF1 speaking to the primary mode shape for the cantilever divider limit condition can be utilized in powerful investigation as follows:

SF1(y)=Ψ(y)=(3y2 – y3)/2H2w (2)

The legitimacy of the shape work SF1 was checked and examined in the past examination (Chen and Kianoush, 2007).

The immediate coupling strategy is utilized in the dynamic investigation. The collaboration among fluid and tank divider is fathomed straightforwardly in the condition of movement utilizing the additional mass strategy [6].

$$p = \sum_{i=1}^{\infty} \frac{2\cdot\rho_l\cdot\tanh(\lambda_i\cdot L_x)}{\lambda_i\cdot H_L}\cdot\cos(\lambda_i\cdot y)\cdot\int_0^{H_L}\cos(\lambda_i\cdot y)\cdot\ddot{u}(t)dy \quad (3)$$

The place λi = (2i-1)π/2HL. For illustrating the plan in the over state mix quick, the polar introductory for be three money house under the course of action would be used to practical provisions.

When using the summed up SDF skeleton in the changing examination for LCS, the hydrodynamic weight may be combined under those coupling examination through the extra impostor. The summed is dependent upon the great incorporated impostor of.
Liquid due to hasty hydrodynamic weight, 1 m ~ What's more 1 m , might a chance to be controlled using Eqns. 2. 4 and 2. 5 independently.

2.1. Included impostor of fluid.

The hydrodynamic weight camwood gives a chance to be handled using the separation for Components method which fulfills those limit states. The hydrodynamic weight scattering on the versatile divider state might make communication follows:.

$$\tilde{m}_L = \sum_{i=1}^{\infty} \frac{2 \cdot \rho_l}{\lambda_{i,n} \cdot H_L} \tanh(\lambda_{i,n} L_x) [\int_0^{H_L} \cos(\lambda_{i,n} y) \cdot \psi(y) dy]^2$$

$$m_L = \sum_{i=1}^{\infty} \frac{2 \cdot (-1)^{i+1} \rho_l}{\lambda_{i,n}^2 \cdot H_L} \tanh(\lambda_{i,n} L_x) \int_0^{H_L} \cos(\lambda_{i,n} y) \cdot \psi(y) dy \quad (4)$$

In perspective of those Housner's model, the extent of the fruitful included impostor from asserting liquid due to hasty hydrodynamic weight I m of the absolute impostor for liquid in the control l m is imparted as:.

Mi/ML =tanh[0. 866(Lx/HL)]/0. 866(Lx/HL) (5).

Relatively to the summed up SDF framework, the extent of summed up and fruitful included impostor for liquid on account of hydrodynamic load to the embraced mode condition of the half impostor for liquid to LCS, I. E. L m~/M and l m/M , could be figured. It will be important that appeared differently in relation to the out and out impostor for liquid in the Housner's model, only An expansive part of the impostor about liquid will be seen as in the summed up SDF system In light of those two-overlay symmetric fluid basic model.

Figures 2(a) and 2(b) show the included impostor of liquid in light of impulsive load regarding to the Hounser's model and the summed up SDF structure using state work ψ(y) =1 which require help for both contrasting and an unyielding tank divider, and the extents from guaranteeing l m~/M and l m/M Concerning delineation limits of the extent about width from asserting tank will be significance of liquid Lx/HL. The shape limit SF1 might be used for the essential mode recognizing the flexibility about tank divider which is done by unique analyzation. It is important that figure 2 will be principle for the full tank condition, I. E. Hl = HW[9].

Figure 2 exhibits that the example for bends to the Housner's model and state work SF1 for the essential mode state using the summed up SDF structure is equivalent. In any case, the results got using by Housner's model need help of multiple occasions from asserting the people got using ψ(y) =1 to the summed up SDF system perceiving those overlay symmetric-liquid auxiliary model. The reason behind the Contrast as a result of the resistance is a direct result of the particular schedules used inside the calculation from guaranteeing hydrodynamic weight. In this investigation, the hydrodynamic weight might be registered using those speed probability systems. It will be acknowledged that the liquid will be perfect, which might be incompressible What's more inviscid. Nonetheless, A direct newtonian gooey shear model might be used inside the Housner show which may furthermore realizes firm response.

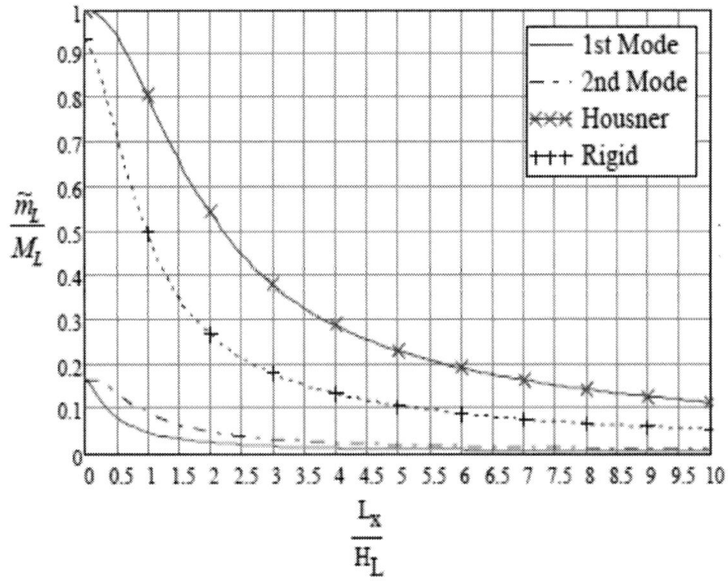

(a) Generalized Added Mass

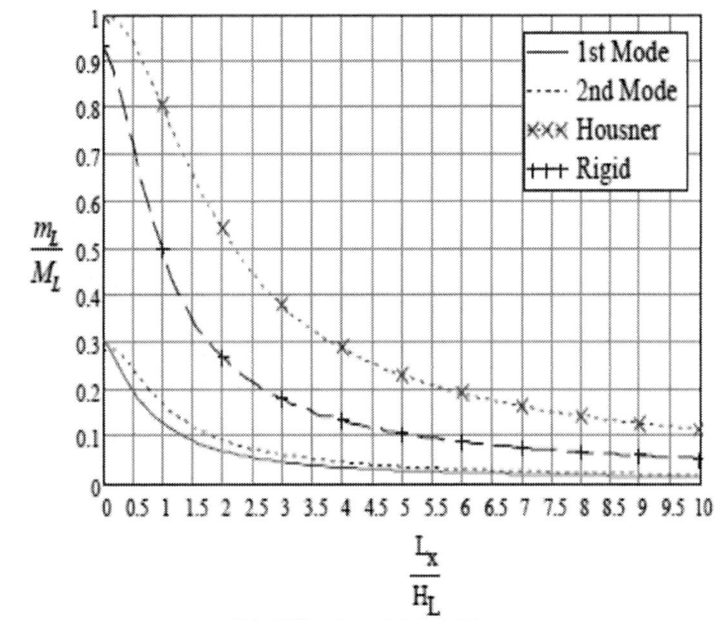

(b) Effective Added Mass

Figure 2 shows the proportion from the claiming included impostor for fluid due to incautious hydrodynamic weight versus Lx/HL proportion (HL = HW[4].

2. 2 powerful statures.

In the exhibit arrangement practice, the inertial mass of robust divider and the extra impostor about liquid due to hydrodynamic weight would be lumped during the described capable statures. The inertial mass about robust tank divider is lumped at the point of convergence of gravity of the tank divider. In the off chance that the tank divider will be uniform [10], the the greater part of crazy

IOP Conf. Series: Materials Science and Engineering **579** (2019) 012049 doi:10.1088/1757-899X/579/1/012049

inertial mass for tank divider is lumped at those mid-stature of the divider. Those extra impostors about liquid due to hasty hydrodynamic weight may be lumped at the centroid of the rash parallel energy. This stature might a chance to be controlled using Eqns. 2. 7 Furthermore 2. 8 as pursues (ACI 350. 3, 2006):.

For tanks with Lx/HL < 1. 333 , hello there/HL=0. 5-0. 09375 (Lx/HL) (6).

To tanks with Lx/HL ≥ 1. 333 ,.

Greetings there/HL=0. 375 (7).

In the summed dependent upon SDF framework, the fruitful statures during which the feasible included impostor from claiming liquid due to hydrodynamic weight will be connected, the camwood be controlled Likewise pursues:

(8).

Figure 3 shows the institutionalized great tallness in which the hydrodynamic weight will be associated as ability of the extent about A large portion width of tank on liquid profundity Lx/HL to the full tank state for instance hl = HW. Those figure shows the starting two modes, the unyielding divider limit condition ψ(y) =1 and the Housner's model. It has a tendency to be uncovered that those urging statures greetings obtained from the Hounser's model and the firm divider cutoff state ψ(y) =1 are similar.

For liquid holding structures, the feasible tallness in which the finish capable sidelong force may joined camwood a chance to be dictated using Eqn. 2. 10. This explanation incorporates both those effects from claiming inertial mass about tank divider and the extra impostor about liquid due to hydrodynamic weight [11].

$$h = \frac{m_w \cdot h_w + m_L \cdot h_i}{m_w + m_L} \qquad (9).$$

It will be significant to consider the versatility about tank divider, the forcing tallness h at which the all level interesting control is associated is higher than that procured from the unyielding divider condition.

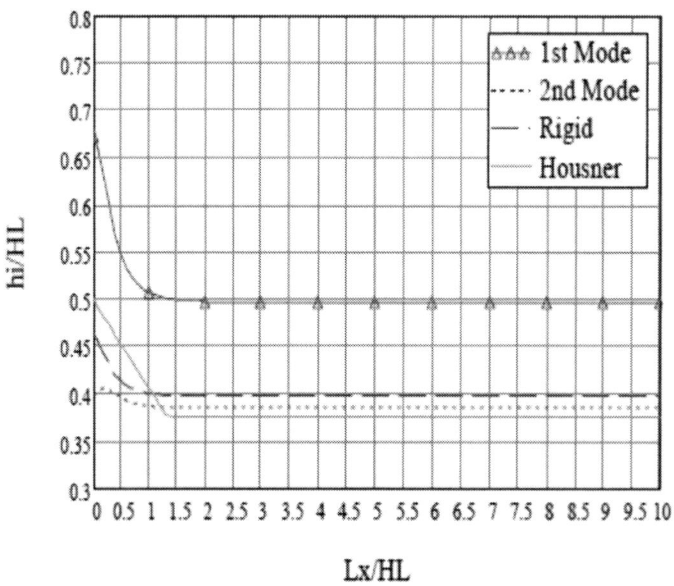

Figure 3 shows thae compelling tallness Components for incautious hydrodynamic weight versus Lx/HL proportion (HL = HW) [6].

2.3. impact about higher Modes.

the similar technique is used on exceptional examination of LCS to principal mode camwood a chance to be joined on the element examination to higher modes. Those square base from claiming aggregate from claiming square (SRSS) methodology camwood a chance to be used to the blend for higher modes. Toward and large, those thought of the introductory two modes ought with provide for enough exact conclusions to setup purposes [12].

3. Finite-Element Implementation for Dynamic-Analysis of LCS

In this investigation, another FEM procedure In light of the sureness that hydrodynamic weight spread is enacted by wave scientific proclamation secured close by liquid Web-area will be utilized to affirm occurs. Tolerating that water will be incompressible Furthermore ignoring its consistency, those little adequacy silly development of water will be enacted Toward those two-dimensional wave condition:

∇2P(x, y, t) =0 (10).

In the coupling system for liquid – structure the loads require help associated of the structure surface outline the heaps on the compartment dividers. The numerical entire explanation about fluid – structure may have an opportunity to be created in the going with shape:

$$[M]\{\ddot{U}\}+[C]\{\dot{U}\}+[K]\{U\}=\{f_1\}-[M]\{\ddot{U}_g\}+[Q]\{P\}=\{F_1\}+[Q]\{P\}$$
$$[G]\{\ddot{P}\}+[C']\{\dot{P}\}+[H]\{P\}=\{F\}-\rho[Q]^T(\{\ddot{U}\}+\{\ddot{U}_g\})=\{F_2\}-\rho[Q]^T\{\ddot{U}\} \quad (11)$$

To which, [M], [C] Also [K] would mass, damping What's more solidness frameworks from claiming structure same time [H] Also [G] need aid talking should robustness Furthermore impostor to liquid range. The haul [C'] is the damping grid for liquid which may be subject of the thickness about liquid and wave digestion over liquid space What's more cutoff points. The system [Q] exchanges the liquid weight of the structure in the same way that assistant reviving of the liquid space [13].

A 8-hub isoparametric part with two interpretations level from claiming good fortune for each centre may be used to show the tank dividers and station. The liquid space is showed by using four-hub isoparametric fluid parts with weight level from claiming chance done each centre. Those restricted part presentation is used on investigate the behaviour of a tall tank Toward ABAQ|US system.

Displaying.
In this examination, a tall tank acknowledged officially will be for the illustration an arrangement model. the estimations and properties of the tank would as for every those be as follows:
Lx=9. 8m, Hw=10m, HL=10m, tw =1. 2m, Ec=20. 776x103MPa, ρw= 2300 kg/m3, ρl =1000 kg/m3, ν= 0. 17.
Those want response reach subject to ASCE 7-05 will be used to get those responses ghostly expanding pace. Those parameters for the structure response extend would be[13]:.
(1) brief time the vast majority amazing extraordinary response expanding speed: Ss=1. 25 (2) 1-second best ghostly response quickening: S1=0. 60 (3) site class b. The figuring using those set part technobabble (FEM) and the ACI 350. 3 code would be presented in this examination. Those set part method recommended in the secret word examination (Chen Furthermore Kianoush, 2005) to model 4 is used to the affirmation. The predictable impostor for both tank divider and the included impostor of liquid due to rash hydrodynamic weight are recognized in the FEM. Likewise, those conclusions using those suggested FEM systems are contrasted with the individuals related for incorporated impostor FEM method and ACI code. Those Indeed going a piece recorded for 1940 El-Centro is used as

excitation of the schema. Those considerably section might have been scaled Along these lines that peak ground speeding up accomplishes 0. 4g. The model configuration will be depicted done figure 4. ACI 350. 3 (2006) code recognizes the effect for adaptability through those.
(2) reaction progress component r. It may be huge that the response transform component r and the hugeness variable I, need aid not viewed as in this examination (for sample r and i need aid acknowledged Concerning illustration solidarity). In this way, the correspondence between the suggested model and ACI 350. 3 code may be dependent upon versant examination.

A substitute sorts for segments were used with presentation the issue something like that as to get the inside anxieties Also part instance in the calculated strip offset Also worries in the foundation dirt. Those bar part might have been used will show those steel backing. A essential bilinear weight strain twist will be used On steel fortress will exhibit the yield stress On weight and weight which depended upon those sort of the used steel bars [10, 14].

Plain strain isoperimetric four-hub quadrilateral segments were used clinched alongside two situations. The principal might have been used the show the strip offset attaching under possibility the non-linearity about bond [9, 10]. The material model identifies with parts of bond done biaxial weight states and provides for the softening and pulverizing cases of bond. The basic key for performing non-direct examination about bond is a straight, versant and powerless material over pressure, also elasto-plastic done weight. The robust need a confined breaking point Previously, restricting strain, furthermore is hence allowed should break The point when the standard anxieties surpass those sensible pliable weight (st). The second sort of part might have been used to present the soil media, contemplating the nonlinearity for dirt Eventually Tom's perusing using Duncan-Mohr-Coulomb altered model [9, 11,12]. In last, the breaking points of the soil networking were shown by left What's more straight two-hub unbounded components, which portray the soil rationality [9, 13]. Inference of the key numerical states analyzings to diverse parts might have been as of late acquainted Toward [9-14]. Hence worth of effort from claiming such segments On reenacting those parity dirt issue could indicate certified issues [14]. A pc project might have been made uncommonly for this examination Previously, which the thought something like straight What's more non-direct constrained What's more wearisome parts of the model were executed. The proposed instance size of the building 30m X 30m, sort of structure : RCC Multi story outlines, seismic zone v , reaction decline component - 3, significance component – 1 , stature of the building-40m, no about storey– 10, stature for floor – 4m, forced load – 4 kN/m2, Materials – M25(beam), M30(column), Fe415, profundity of the area – 150mm, unit load for RCC – 25kN/m3, sort from claiming dirt – difficult , medium and Soft, reaction spectra – may be 1893(Part I) 2002 , Damping 5% , profundity of stronghold - 1. 5m, divider thickness-230mm. Table 1 exhibits those estimation from claiming dirt strength done level (x and z) What's more verthandi (y) direction[8].

Table1. Soil Stiffness Values For Buildings With Flexible Base[10]

type of soil	Soil Stiffness (kN/m)		
	Kx	Ky	Kz
hard	8000	10000 0	8000
Medium	4000	50000	4000
soft	1500	25000	1500

Figure 4 Finite Element Mesh Configuration

The estimation results are condensed in Table 1 for the tall tank. The correlation of the outcomes got utilizing both FEM methods and the proposed model shows great understanding. In any case, the base shear got utilizing ACI 350.3 Code is about 1.85 occasions higher than that acquired utilizing the proposed summed up SDF framework. The base minute for ACI 350.3 Code when contrasted with the proposed summed up SDF framework is about 1.36 occasions higher. It is presumed that the structure utilizing the Hounser's model received in the present plan measures and codes is excessively traditionalist. Table (2 and 3) and Figs. (5 and 6) demonstrate Lateral redirection of tanks with settled base. Figs. (5 and 6) show horizontal diversion with adaptable base.

Table (2) Lateral deflection of tanks with fixed base (mm)

Height	Soft soil	Medium soil	Hard soil
1st	7	6	5
2nd	20	18	17
3rd	38	30	20
4th	42	38	30
5th	58	50	38
6th	70	58	42
7th	80	62	50
8th	82	77	58
9th	90	78	59
10th	99	80	60

Table (3) Lateral deflection of tanks with flexible base (mm)

Height	Soft soil	Medium soil	Hard soil
1st	148	80	48
2nd	152	90	50
3rd	162	95	56
4th	170	100	60
5th	182	105	63
6th	190	112	72
7th	200	113	80
8th	202	120	83
9th	206	135	85
10th	211	141	91

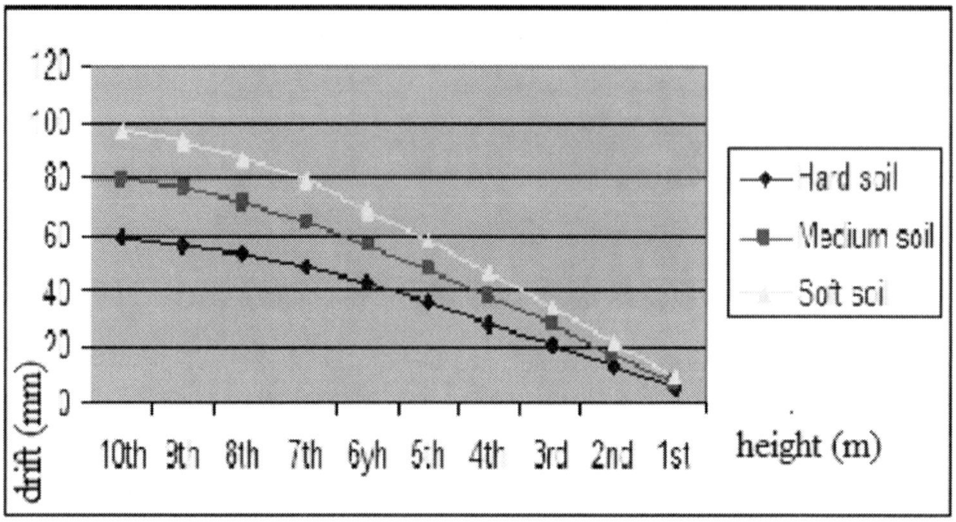

Fig. (5) Lateral deflection of tanks with fixed base

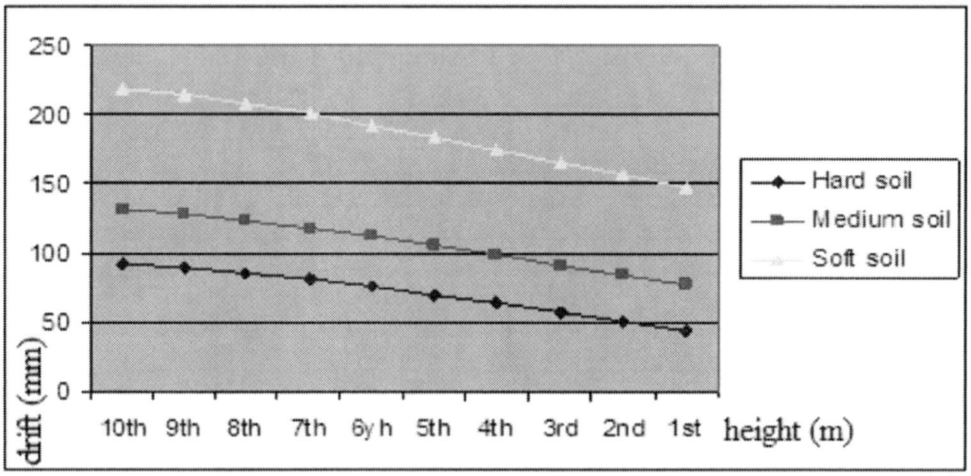

Fig. (6) Lateral deflection of tanks with flexible base

4. Conclusion

Due to lateral deflection, Base shear values increase when the type of soil changes from hard to medium and medium to soft for fixed and flexible base buildings. Because of lateral deflection, Base shear and Moment values of fixed base building was found to be lower as compared to flexible base tanks. The change type of foundation for building that will change all behaviour of tanks for the hard, Medium, Soft soil since the lateral deflection increase about (54.6% , 62.4% , 127.6%). Hence suitable foundation system considering the effect of Soil stiffness has to be adopted while designing tanks for seismic forces.

REFERENCES

[1] Anand N and Mightraj C 2010 Seismic behavior of RCC shear wall under different soil conditions In *Proceedings of the Indian Geotechnical Conference on GEO trendz*

[2] Pandey A D, Prabhat Kumar and Sharad Sharma 2011Seismic soil-structure interaction of buildings on hill slopes *Journal of Civil and Structural Engineering*

[3] Amin M R and Hasan P 2011 Effect of storey on multistoried reinforced concrete building frame In *Proceedings of 4th Annual Paper Meet and 1st Civil Engineering Congress*

[4] Jenifer Priyanka R M and Anand N 2012 Effect of lateral force on tall buildings with different type of irregularities *In Proceedings of the INCACMA Conference on Advances in Construction Manufacturing and Automation Research*

[5] IS 1893 (Part 1)-2002 *Indian Standard Criteria for Earthquake Resistant Design of Structures* Part 1–General Provisions and Buildings (Fifth Revision, Bureau of Indian Standards New Delhi)

[6] IS 456 – 2000 *Indian standard Criteria for Plain and Reinforce Structure* New Delhi

[7] *International Journal of Emerging Technology and Advanced Engineering Website*: www.ijetae.com (ISSN 2250-2459, ISO 9001:2008 Certified Journal **2** 12 December 2012

[8] American Concrete Institute ACI 350.3 2006 *Seismic design of Liquid Containing Concrete structures* Farmington Hills MI US

[9] American Society of Civil Engineers (ASCE) 2005 *Minimum Design Loads for Buildings and Other Structures* ASCE 7 ASCE Standard SEI/ASCE 7-02 Reston VA US

[10] Chen J Z and Kianoush M R 2007 Generalized SDF System for Analysis of Concrete Rectangular Liquid Storage Tanks *Proceedings of the 9th Canadian Conference on Earthquake Engineering* Ottawa Ontario

[11] Chen J Z and Kianoush M R 2005 Seismic Response of Concrete Rectangular Tanks for Liquid Containing Structures *Canadian Journal of Civil Engineering* 32 739-752

[12] Ghaemian M, Kianoush M R and Mirzabozorg H 2005 Time Domain Dynamic Analysis of Rectangular Liquid Containers in Three-Dimensional Space *Journal of European Earthquake Engineering* XIX 2 3-9

[13] Housner G W 1963 The Dynamic Behavior of Water Tanks *Bulletin of the Seismological Society of American* **53** 2

[14] Christian J T and Hall J R 1982 Soil-Structure Interaction Problems *Proceedings of the Fourth International Conference on Numerical Methods in Geotnechanics* Edmonton, Canada May 19

Linear and Non-Linear Stress Analysis for the Prediction of Fracture Toughness for Brittle and Ductile Material using ASTM E399 and ASTM E1290 By ANSYS Program package.

A A Nassar[1#] and E G Fayyad[2]

[1]Basra university /Mechanical Engineering Department/Basra/Iraq
[2]Basra Oil Company / South Rumelia/design & project Department/ Basra/Iraq

[#]Corresponding author's e-mail address: en.ebtesam@gmail.com

Abstract. Prediction of fracture toughness value of most the metallic materials is a very important issue to be consider in designing of engineering structural components like pipes, flow tanks and pressure vessel where it can be used in the evaluation of critical crack length that is consider very important factor in non-destructive testing because during inspections of these components, the critical crack length is being compared with the minimum allowable flow size. Fracture toughness value can be predicted by using two different types of experimental tests such as ASTM-E399 (which is used to predict fracture toughness under plane strain condition for brittle materials which undergo to small plastic deformation) and ASTM-E1290 (which is used to predict fracture toughness for ductile materials which undergo to large plastic deformation). These experimental tests utilize specimens such as compact tension specimen CT and three-point bend specimen SENB which must be pre-cracked. However, manufacturing of pre-cracked in metallic structural components before testing is expensive and time-consuming process, thus in the current study, fracture toughness will be predicted by using economized method in both time and cost that is the finite elements method by using ANSYS PROGRAM Where, a three-dimensional model has been designed by using two different types of elements (plane-82) and (solid-95).Firstly, two-dimensional model mesh will be idealized by using element type (plane-82) due to ANSYSY PROGRAM can pick–up singularity for two dimensional model only, but in order to obtain more accurate results, a three dimensional model will be designed by utilizing Sweep Model with restriction with ANSYSY abilities in the treatment of fracture mechanics problems for three dimensional model analysis where the element length must be ranging from 1to4 in all directions. In elastic region, the fracture toughness is been predicted directly from ANSAYS PROGRAM using the specification in ASTM-E-399 for compact tension specimen for practical design problems. But in elastic-plastic region, the fracture toughness is been predicted by using the crack tip opening displacement model (CTOD-Model) using the specification in ASTM-E-1290 for three-point bend specimen for practical design problem. The critical value for crack tip opening displacement will be calculated after extracting load-displacement data from ANSAYS PROGRAM then using these data in (CTOD-Model) which is separated into two components, elastic and plastic, the elastic component has been estimated using Dug Dual-Model, while the plastic component will be estimated using Plastic Hinge-Model. However, the fracture toughness value that has been predicted by finite element method from non –linear elastic analyses and from non –linear elastic-plastic analyses gives excellent results which are very close to the experimental results with error ratio ranging between (10% to 14%).

IOP Conf. Series: Materials Science and Engineering **579** (2019) 012050 doi:10.1088/1757-899X/579/1/012050

1.Introduction

Every structure contains defects such as small flaws or cracks of sizes and distribution are dependent on the material and its manufacturing processing. These may vary from nonmetallic inclusions and micro-voids to weld defects, grinding cracks, quench cracks, surface laps etc. The Presence of these defects in the components creates problems that can adversely affect the integrity of the structural components and shortens their service life. Therefore, Fracture Mechanics methodology has been developed to figure out if these small flaws might grow into larger cracks causing the catastrophic failure of these structural components. The analysis of Fracture Mechanics is become very important for all engineers are researching about the reasons of failure of structural components in different scientific fields. The determination of the stress distribution near cracks especially at crack tip region can be used for proper predictions regardıng the consequences of a crack which could lead to rapid fracturing [1, 2].

The basic principal of Liner Elastic Fracture Mechanic theory is "the fracture occurs when the stress intensity factor is equal or greater than plane strain fracture toughness or the fracture occurs when the stress intensity factor is equal or greater than plane stress fracture toughness" i.e. crack propagation occurs when KI >KIC or KI>KC. Consequently, using (LEFM) theory which is limited with (plane strain condition & Small-Scale Yielding) engineers can design a structure with the stress intensity factors corresponding to different fracture modes. while, the stress intensity factor does not represent the true fracture mode beyond the elastic region ;therefore, after this region the Non-Linear Elastic Plastic Fracture Mechanics theory must be used and as shown in (Figure 1)where a is a crack length and F is an applied force. There are two major branches in Elastic Plastic Fracture Mechanics, the CTOD-Model and the J-Integral Model which have been developed to understand and solve the non-linear behavior of cracked structures and characterizing crack tip toughness [3].

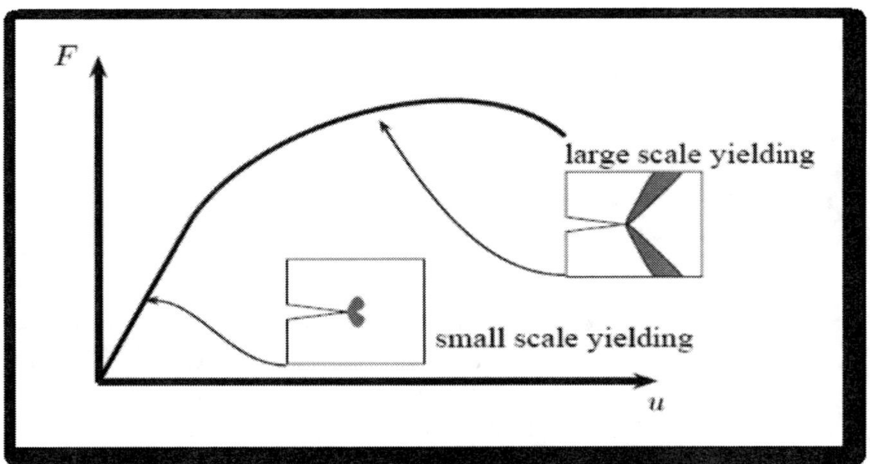

Figure 1. The EC experimental setup.

2. Fracture Toughness Concept

In physical science, Fracture toughness is an indication of the resistance of a material to physical separation by a process of unstable macro-crack propagation. Conceptually, it is an intrinsic material parameter that should not vary with changes in specimen size, speed of loading, temperature, etc. In material science, Fracture toughness is an empirical material property that is determined by one or more of several standard fracture toughness test methods and is one of the most important parameters to evaluate the mechanical properties of any materials for virtually all design applications [4,5].

3. Fracture Toughness Parameters

There are two fracture toughness parameters commonly obtained from testing [6]: -

3-1 Stress Intensity Factor (KI)

The stress intensity factor can be considered as a stress based-estimate of fracture toughness. Depending on this parameter, plane strain fracture toughness under plane strain and small-scale yielding condition for most metallic materials can be predicted according to (ASTME-399) specification. Experimental international data for practical design problems will be used to predict plane strain fracture toughness while keeping the validity conditions according to recommendation in (ASTM E-399) being very important in choosing of the suitable thickness of the design structural components.

3-2 Crack Tip Opening Displacement (CTOD)

Linear elastic analysis is used to predict the value of plane strain fracture toughness of high strength materials according to (ASTM E-399) that is constrained with small plastic deformation zones that being involved with the creation of large plastic deformation zones and of the non-linear elastic-plastic behavior for most low-to-medium strength materials .Therefore, another standard and model have been utilized to predict fracture toughness value ;that is the (ASTM E-1290) (Standard Test Method for Crack-Tip Opening Displacement CTOD-Fracture Toughness Measurement).

The crack-tip opening displacement CTOD can be considered as a strain-based estimate of the fracture toughness where it is one of a family of fracture mechanics tests that measures material resistance to growing crack; Similar tests (KIC and JIC) can be used to determine fracture resistance of a material. CTOD is used to characterize the crack tip field at extensive plastic deformation as in ductile materials such as steel structures which is used in the constructions of pipes, flow tank, pressure vessel and ships. The critical value of crack-tip opening displacement CTOD can be separated into elastic and plastic components. The elastic component of CTOD can be estimated using Dug Dual Model while the plastic component is only obtained from Plastic hinge model by converting the crack mouth opening displacement CMOD into crack tip opening displacement CTOD using the relationships of similar triangles. These two models will be used in the current study to predict fracture toughness value.

The specimens which have been used to predict and calculate of the fracture toughness values through testing in the laboratory the structural components of metallic materials should be pre-cracked as shown in (Figs. 2 and 3). The presence of sharp pre-crack in notched specimen is a major requirement in fracture toughness testing procedure that has been executed by employing servo-hydraulic machine, the generation of a pre-crack specimen is an expensive and time-consuming process [7].

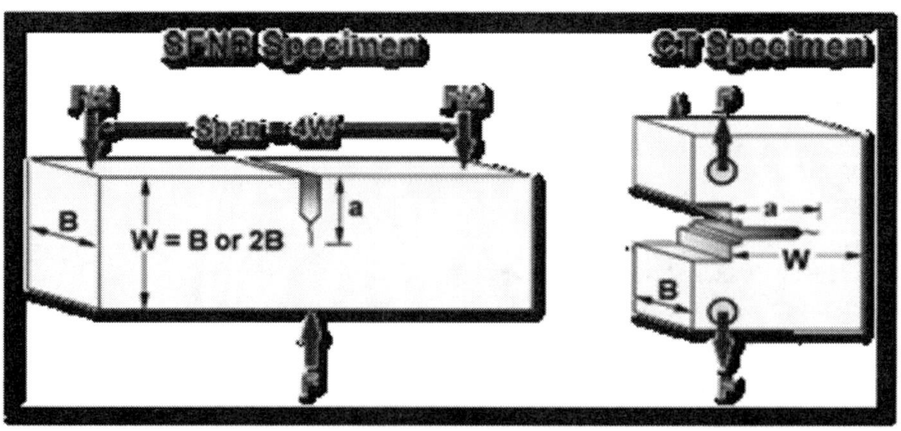

Figure 2. Fracture Toughness Test Specimens Types

IOP Conf. Series: Materials Science and Engineering **579** (2019) 012050 doi:10.1088/1757-899X/579/1/012050

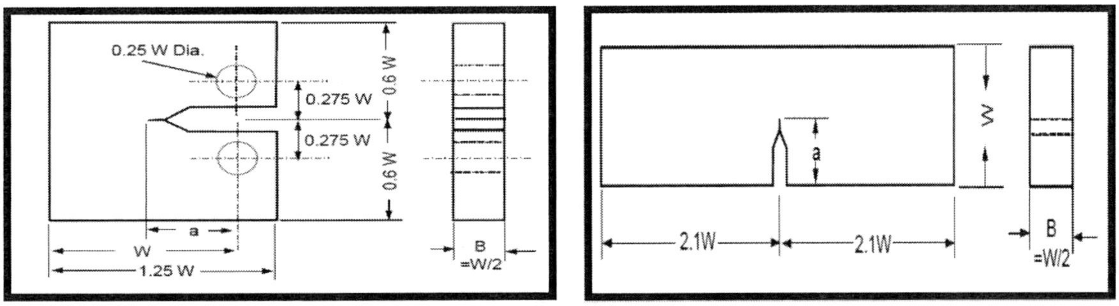

Figure 3. Pre-Crack in Fracture Toughness test Specimens

4. Modeling and Meshing of Test Specimens

A three-dimensional finite element model of the CT and SENB fracture test specimen will be utilized to predict the value of stress intensity factor at critical load i.e. fracture toughness near the crack tip in two cases. The first case is small scale yielding where the dimension of model will be prepared according to ASTM E-399 and the second case is large scale yielding where the dimension of model will be prepared according to ASTM E-1290. Firstly, two-dimensional continuum finite element model will be created by using element type (plane-82) under plane strain condition. This element is defined by eight nodes, one at each corner and each mid-side. Then, generation of three-dimensional crack elements by defining element type (Solid-95) through Sweep/Extrusion Techniques as shown in fig. 4a and fig. 4b.

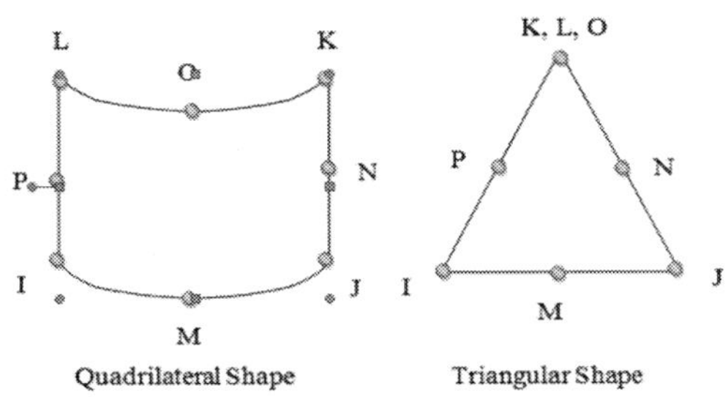

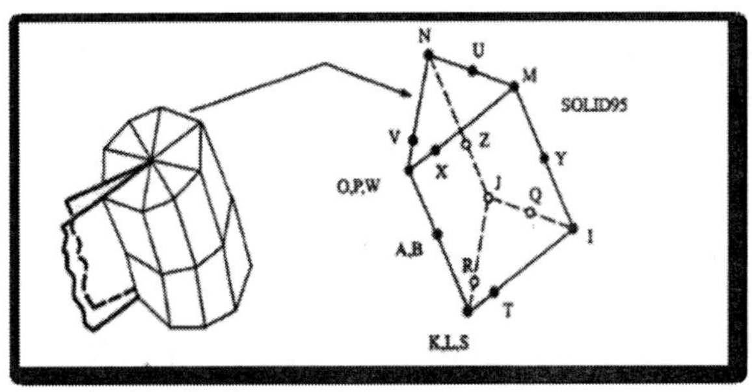

Figure 4.a (plane-8) 8-nodes structural solid **Figure 4.b** (Solid-95) 20-nodes structural solid

5. Results and Discussion

The fracture toughness will be predicted for brittle and ductile materials by performing the non-linear three-dimensional finite elements model with using the specification in ASTM E-399 and ASTM E-1290. According to ATM E-399 that has used to predict fracture toughness under plane strain condition for brittle materials, the non-linear three-dimensional finite elements model will be designed for CT specimen and according to ASTM E-1290 that has used to predict fracture toughness for ductile materials, another non-linear three-dimensional finite element model will be designed for SENB specimen for practical design problems.

6. Numerical Model Analysis for the Prediction of Fracture Toughness According To ASTME-399 Specification

The practical design problems have been selected to predict fracture toughness numerically. The dimensions of CT specimen are listed in (Table 1), elastic and plastic material properties are listed in (Table 2), Experimental load and fracture toughness data are listed in (Table 3) for design problems A:

Table 1. Specimen Dimensions for Design Problem A

Design problem	Crack size	Thickness	width
A	38.1mm	46mm	101.6mm

Table 2. Material properties for Design Problem A

Design problem	Yield Strength MP	Tangent Modulus Gp	Poisons ratio	Elasticity Modulus GP
A	1172.15	200	0.3	210

Table 3. Experimental load and KIC data

Design problem	Experimental maximum load	Experimental critical load	Experimental Fracture Toughness
A	245.351KN	231.296 KN	109.9MPa√m

7. Numerical Analysis Procedure

The half model as shown in (Figure 5) has used during this analysis,2-D finite elements mesh idealization for half model of this specimen is shown in (Figure 6) which has performed using element type (plane-82) where the maximum elements numbers were 1621 and the maximum nodes numbers were 5001, and the element length is 2.5. Then, by defining element type (solid-95) and giving two divisions through thickness, the Sweep Model will be generating for this problem .3-D finite elements mesh for half model of this specimen will became as shown in (Figure 7) where the maximum elements numbers are 4863 and the maximum nodes number are 18383.The boundary condition has been applied as shown in (Figure 8).

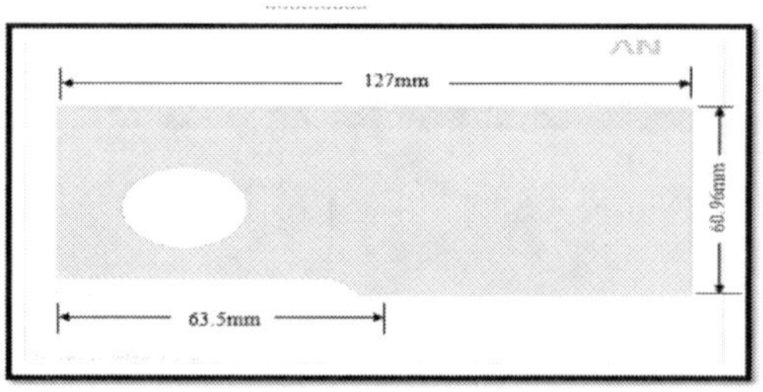

Figure 5. half model idealization for CT

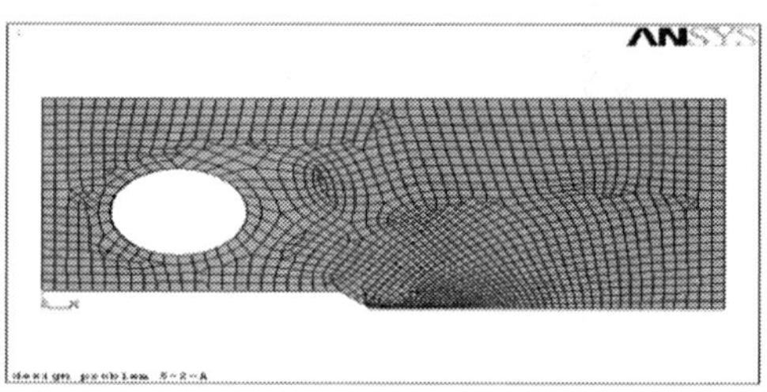

Figure 6. 2-D finite elements mesh idealization

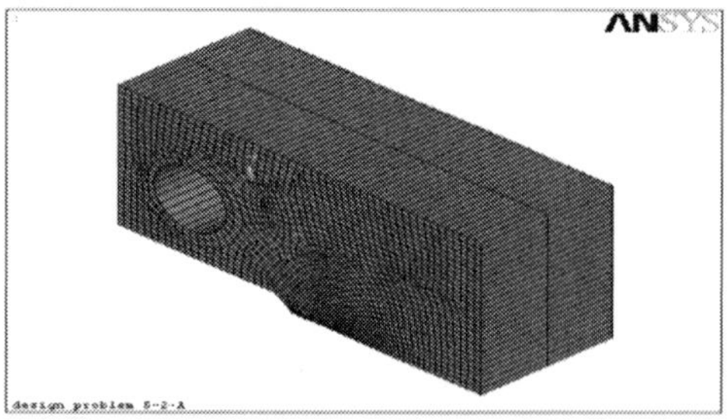

Figure 7. 3-D finite elements mesh idealization

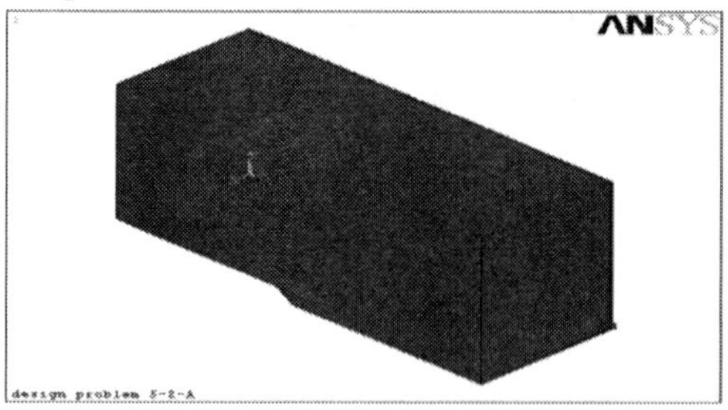

Figure 8. Application of B.C.

After running the program, the deform shape for this specimen is shown in(Figure 9) which appears that the maximum displacement which occur in the y-direction is about 0.49423mm, the nodal solution is shown in (Figs. 10 and 11) which appears that the maximum displacement occurring in the y-direction at node No.5323 is about 0.447412mm.The element solution shows that the maximum stress occurring in the y-direction is about 11046MPA as shown in (Figure 12).Thus, the maximum failure load can be concluded from reaction solution in y-direction FY which is about 230.930KN at last step and about 217.345KN at step number 16 which represents the critical load. Consequently, fracture toughness can be extracted directly from ANSYS which is about 96.39MPa.m⅕case.

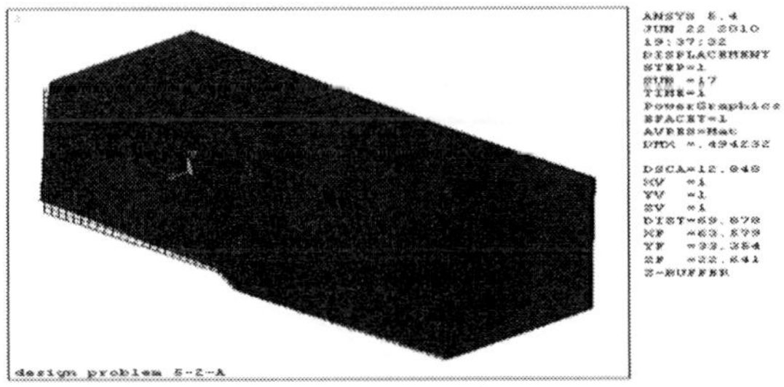

Figure 9. Deforms shape of the CT specimen

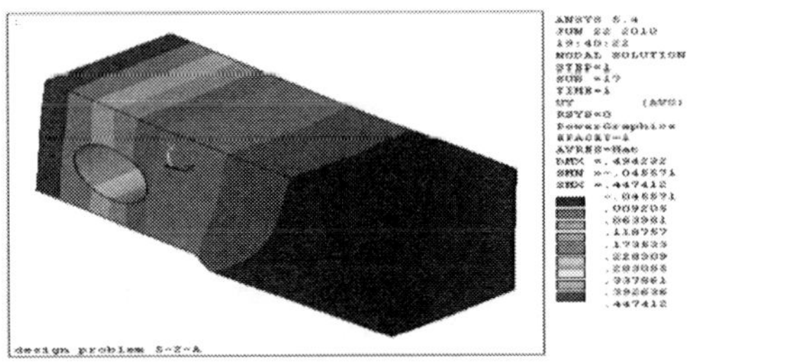

Figure 10. Maximum displacement in y-direction

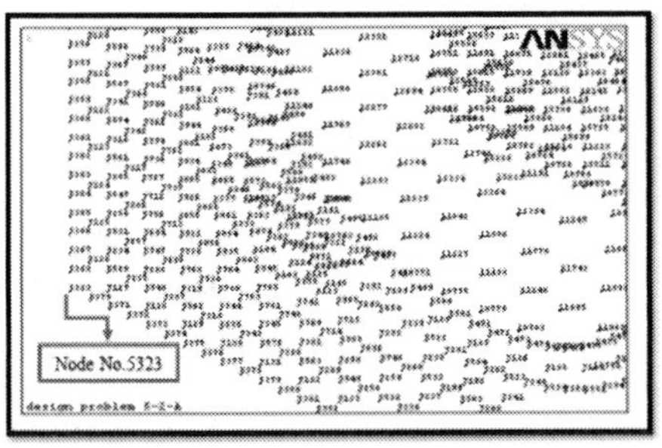

Figure 11. Node have max. Displacement

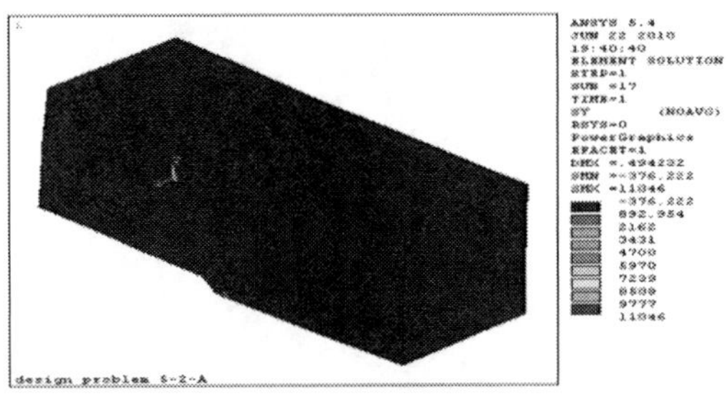

Figure 12. Maximum stress in the y-direction

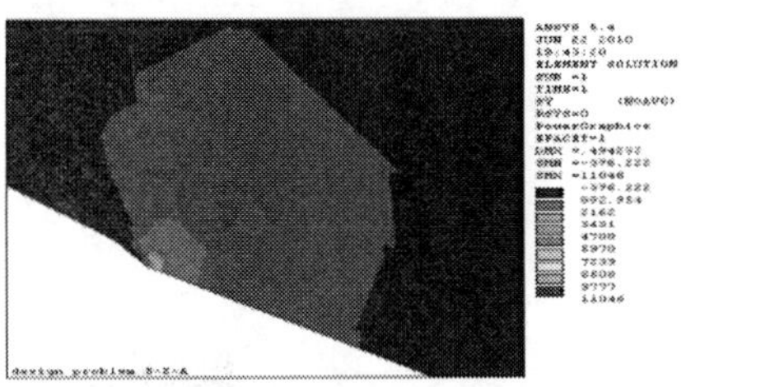

Figure 13. maximum stress at crack tip region

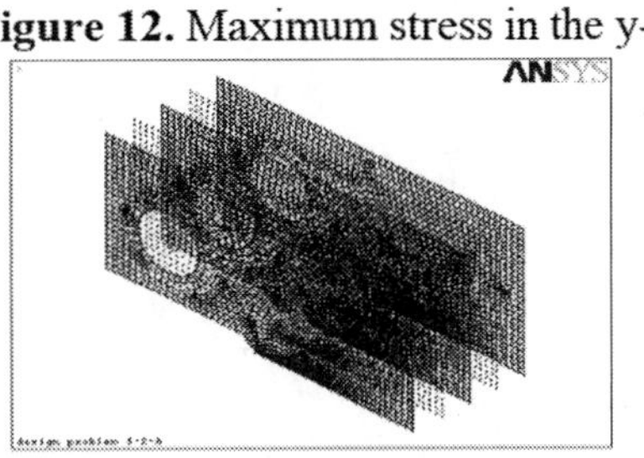

Figure 14. Drawing of two division through thickness

8. Numerical Analysis of prediction Fracture Toughness Using CTOD

A three-point bend specimen with S = 250 mm, W = 60 mm, a = 30 mm, and B = 30 mm is used to determine the critical crack tip opening displacement of steel plate. The load versus crack mouth displacement (P-V) of the test shows that the Maximum load is 31.600 KN. The elastic material property which will be used as an elastic input data for ANSYS to estimate the elasticity response of the material, also the plastic material properties (tangent modulus and effective yield strength) will be used as plastic input data to estimate the plasticity response of the material.

The half model of geometry of three-point bend specimen is shown in (Figure 15), the 2-D finite elements mesh idealization for half model is shown in (Figure 16) which has been also designed using element type plane-82 where the maximum elements numbers were 740 and the maximum nodes number were 2285. By defining element type solid-95 and giving two divisions through thickness, the sweep model will be generated, as shown in (Figure 17) for full model and the 3-D finite elements mesh idealization for half model of this specimen will become as shown in (Figure 18) where the maximum elements numbers were 2220 and the maximum nodes number were 8401.

IOP Conf. Series: Materials Science and Engineering **579** (2019) 012050 doi:10.1088/1757-899X/579/1/012050

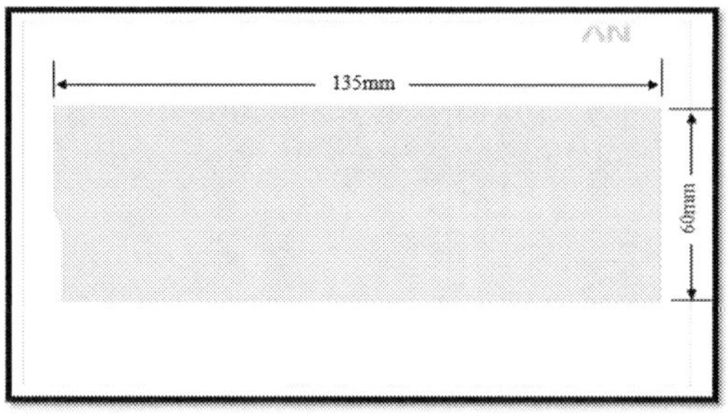

Figure 15. half model idealization

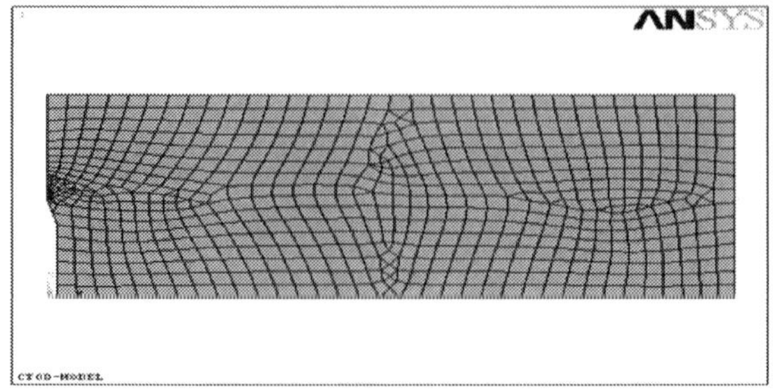

Figure 16. 2-D finite elements mesh idealization

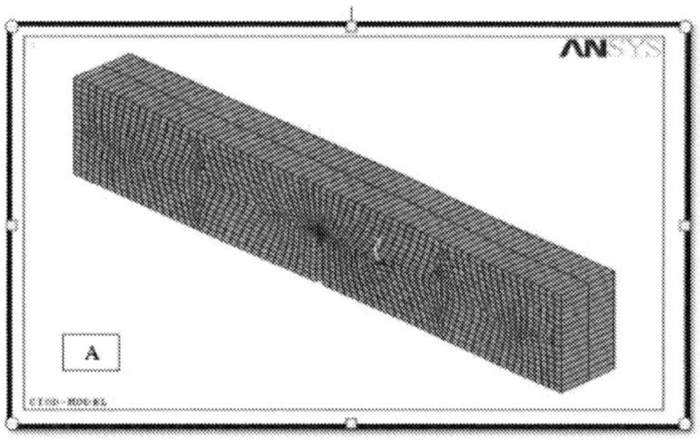

Figure 17. 3-D Finite elements mesh idealization

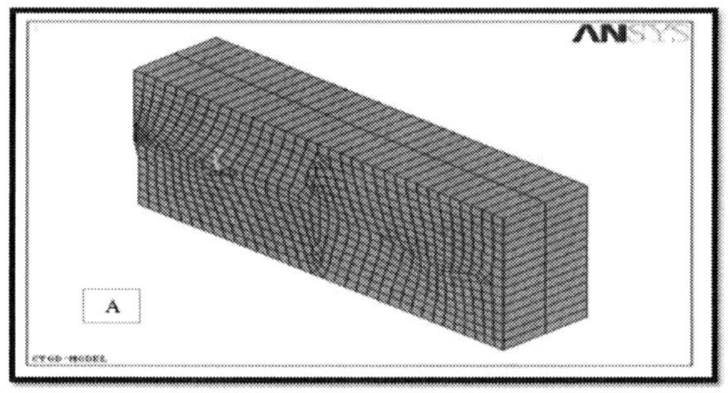

Figure 18. 3-D Finite elements mesh idealization

Then, the boundary condition will be applied as shown in (Figure 19), After running the program and obtaining solution, the deform shape for this specimen is shown in (Figure 20) which shows that the maximum displacement that occurring in the y-direction is about 2.364 mm, Then, the maximum failure load can be concluded from reaction solution in y-direction FY which is about 34.589 KN at last step.

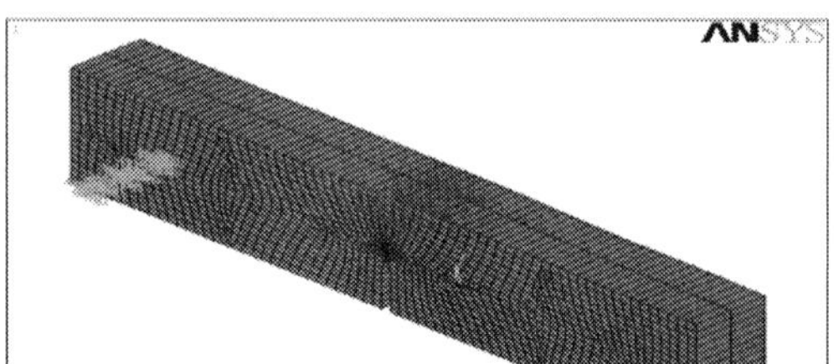

Figure 19. Application of B.C.

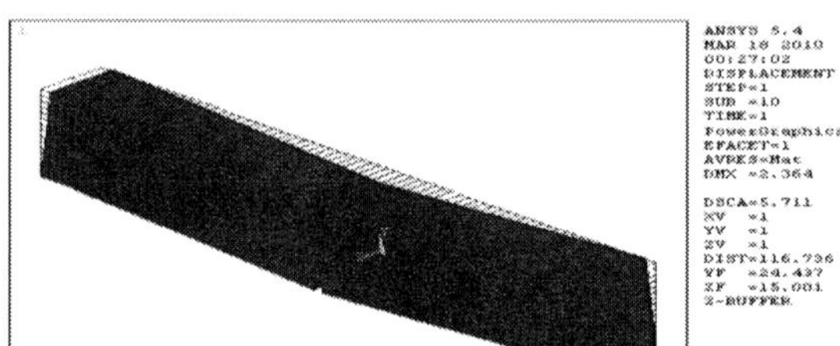

Figure 20. Deform shape for bending specimen

9. Numerical Results Discussion of compact specimen

The non-linear three-dimensional finite elements model according to ASTM E-399 specification provides excellent results predicting of the fracture toughness directly from ANSYS as shown in (Table 4), it also provides excellent results for maximum loading from ANSYS as shown in (Table 5). Adding to that, it provides excellent results for critical loading from ANSYS as shown in (Table 6).

IOP Conf. Series: Materials Science and Engineering **579** (2019) 012050 doi:10.1088/1757-899X/579/1/012050

Table 4. Differences between experiment and finite elements result for prediction of fracture toughness

Design problem	Predictive of Fracture Toughness by experiment	Predictive of Fracture Toughness by FEA	Differences between experiment and FEA
A	109.9MPa√m	96.39MPa√m	14 %

Table 5. Differences between experiment and finite elements result for maximum loading

Design problem	Maximum load by experiment	Maximum load by FEA	Differences between experiment and FEA
A	245.351KN	230.930KN	6.24%

Table 6. Differences between experiment and FE results for Critical loading

Design problem	Critical load by experiment	Critical load by FEA	Differences between experiment and FEA
A	231.296KN	217.345KN	6.41%

Difference between FEA and experiment results for compact specimen as shown in (Figure 21): -

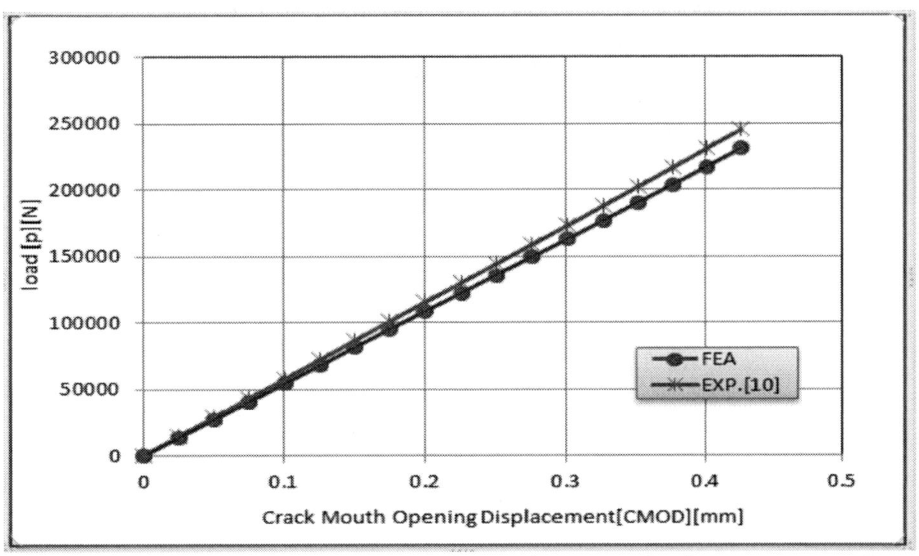

Figure 21. Difference between FEA and Experiment results for compact specimen

10. Numerical Results Discussion of bending specimen

The differences between experimental and finite element results of three-point bend specimen, for maximum loading are shown in (Table 7) and in (Figure 22), clip gage displacement (V_p) can be concluded after drawing a straight line which is equal to linear elastic portion which was about (0.85mm).The differences between the experimental and finite element value for (V_p) is shown in (Table 8), the plastic part of crack tip opening displacement $(\delta_t)_{pl.}$ can be estimated from plastic hinge model which was about (0.242mm). The differences between the experimental and finite element value for $(\delta_t)_{pl.}$ are shown in (Table 9), the elastic part of crack tip opening displacement $(\delta_t)_{el.}$ can be estimated from dug dual model which was about (0.007mm).The differences between the experimental and finite element value for $(\delta_t)_{el.}$ as shown in (Table 10),critical value of crack tip opening displacement can be concluded by adding the elastic part to the plastic part which was about (0.249mm) then fracture toughness value is about (311.636MPa√m).

Table 7. Differences between experiment and finite elements result for maximum loading

Design problem	Maximum load by experiment	Maximum load by FEA	Differences between experiment and FEA
SENB	31.6 KN	34.589KN	9.45%

Table 8. Differences between experiment and finite elements result for clip gage displacement

Design problem	(V_p) by experiment	(V_p) by FEA	Differences between experiment and FEA
SENB	1mm	0.85mm	17.64%

Table 9. Differences between experiment and finite elements result for plastic part of CTOD

Design problem	$(\delta_t)_{pl.}$ by experiment	$(\delta_t)_{pl.}$by FEA	Differences between experiment and FEA
SENB	0.286mm	0.242mm	18.18%

Table 10. Differences between experiment and finite elements result for elastic part of CTOD

Design Problem	$(\delta_t)_{el.}$ by experiment	$(\delta_t)_{el.}$by FEA	Differences between experiment and FEA
SENB	0.006mm	0.007mm	16.66%

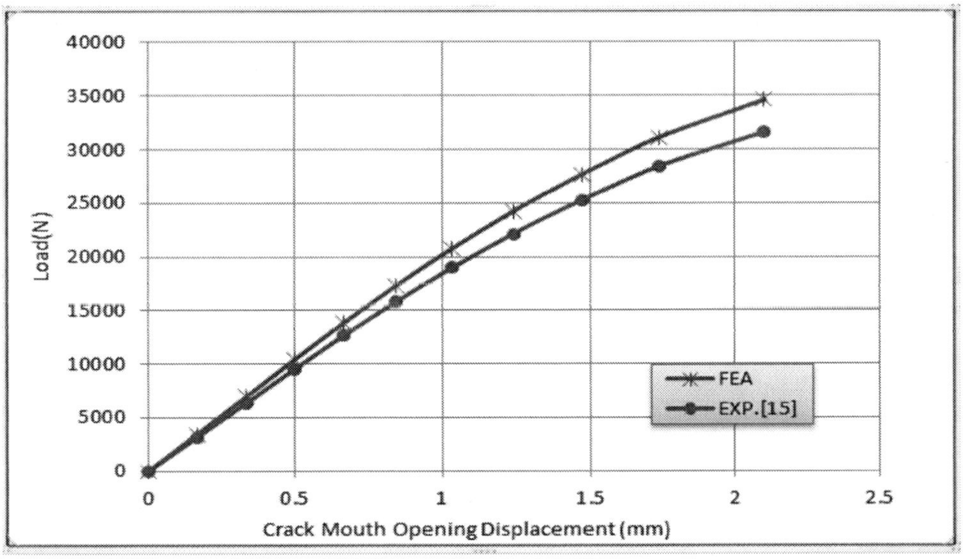

Figure 22. Differences between experimental and finite elements results for bending specimen

11. Conclusions

The following conclusions from the present study can be drawn: -

- The analysis of the non-linear three-dimensional finite elements model according to ASTM E-399, which has been used to predict fracture toughness directly from ANSYS for compact tension specimens, it provides excellent results for output load-displacement data from ANSYS where it has high accuracy. Thus, the fracture toughness value according to this type of analysis is of high accuracy.
- The analysis of non-linear three-dimensional finite elements model according to ASTM E-1290, is used to predict fracture toughness by using CTOD-Model for three-point bend specimen, it provides excellent results for the output load-displacement data from ANSYS where it also has high accuracy. The critical value, the elastic and the plastic part for crack tip opening displacement and clip gage displacement according to this type of analysis are accurate.
- Although, the non-linear analyses have been used for the prediction of fracture toughness according toASTME-399, but the relation between load and displacement is linear; this means that no or little plastic deformation occurs in the elements which have been utilized in this program because the behavior of fracture being from the type of brittle mode.

Acknowledgment

First and foremost, all sincere praises and thanks are due to our God. Special appreciation is extended to Assist. Prof. Dr. Abdul-Kareem and Assist. Prof. Dr. Ouday for the sincere and untiring efforts by them in preparing the program and set-ups utilized in this study.Finally, the words are not enough to dedicate thanks to My Family and My Faithful Friends and Colleagues who have never forgotten me.

IOP Conf. Series: Materials Science and Engineering **579** (2019) 012050 doi:10.1088/1757-899X/579/1/012050

References

[1] Hedia H S, Shabara MA, Fattah A A and Helal M M 2005 Effect of Crack Configuration and Pre-Crack Length on Stress Intensity Factors *Journal of Material prüfung* **47** 10 2–7
[2] Hedia H S, Shabara M A , Fattah A A and Helal M M 2005 Effect of Notch Configuration and Pre-Crack Length on Stress Intensity Factors *Journal of Material prüfung* **47** 10 2–8
[3] Perez N 2004 *Fracture Mechanics* 1st Ed. (Boston: Kluwer Academic Publishers) pp 47–96
[4] Sih G C 1976 *Fracture Toughness Concept* America Society For Testing & Materials Special Technical Publication-605 **ASTM STP-605** pp 3–15
[5] Standard Test Method for Linear-Elastic Plane-Strain Fracture Toughness K_{IC} of Metallic Materials 2008 **ASTM E–399**
[6] Tornblom S 2007 *Under Matching Butt Weldes in High Strength Steel* **MSc thesis** Department of Civil & Environomental Engineering Lulea University of Tecnology
[7] Hedia HS and Shabara MA 2002 Effect of Notch Tip Sharpness and Pre-Crack on the Stress Intensity Factor *MP* **44**

IOP Conf. Series: Materials Science and Engineering **579** (2019) 012051 doi:10.1088/1757-899X/579/1/012051

New coumain derivatives as corrosion inhibitor

Y K Al-Majedy[1], H H Ibraheem[1#], M S Falih[2] and A A Al-Amiery[1]

[1]Department of Applied Science, University of Technology, Baghdad, Iraq.,
[2]Department of Chemistry, College of Science, Al- Mustansiriyah University.

[#]Corresponding e-mail address: yasmin.chem79@gmail.com

Abstract. Electrocoagulation can be considered as a simple and an efficient method for treating both wastewater and water. In this work, a synthetic bentonite suspension water was treated using this technique to reduce turbidity caused by solid particles pollutants. This process was carried out in a batch glass reactor of 1000 ml operated with aluminium electrodes with and without agitation. The removal efficiency as a turbidity removal procedure has been examined using bentonite clay as a turbidity source. The effects of some parameters such as contact time, and electrolyte concentration on the removal efficiency were investigated and compared. The mixing effect over the range (0-800 rpm) on the removal efficiency was also evaluated. The process was found to achieve excellent turbidity removal with mild mixing (550 rpm). In addition, the effect of using Ultrasonic as a mixing source has also been studied. The lowest residual turbidity was found to be 1.0 NTU for the samples with initial turbidities of 500 NTU. Similar final turbidity of 1.0 NTU was obtained after 25 min with initial turbidity of 200 NTU. This was obtained at a current of 750 mA, a contact time of 60 min, and a calculated dissolved aluminium concentration of 10 mg/l.

1. Introduction

Temperature greatly affects the rate of metal electrochemical corrosion. If there should arise an occurrence of corrosion in a nonpartisan arrangement, the expansion of temperature favorably affects the over capability of oxygen depolarization and the rate of oxygen dissemination, however it prompts a decline of oxygen solvency. If there should be an occurrence of corrosion in a corrosive medium, the consumption rate increments with temperature increment in light of the fact that the hydrogen advancement over potential abatements. Temperatures impacts on acidic erosion, regularly in sulphuric and hydrochloric acids, have been the protest of an extensive number of examinations [1-10].

Corrosion is continuous annihilation of materials typically metals, by concoction response with its condition which implies electrochemical oxidation of metals in response with an oxidation, for example, oxygen.Consumption happens in metals as well as in non-metals like plastics and glass [11]. Corrosion inhibitor is a synthetic aggravates that diminishes the corrosion rate of a material when added to a fluid or a gas [12].

The normal system for hindering corrosion includes development passivation, covering which forestalls access of the substance to the metal. Inhibitors are added to numerous frameworks, for example, cooling frameworks, concoction, oil and gas creation unit, boilers and refinery unit to forestall corrosion. By and large, the compelling inhibitors utilized contain a heteroatom, for example, oxygen, nitrogen, sulfur, and numerous bonds in their structure through which they are adsorbed on the surface of the metal [13]. That adsorption depends exclusively on certain physiochemical properties of the inhibitor gatherings, for example, useful gatherings, electronic skeleton of the particles and electron thickness at the contributor iota [14]. Recent studies have demonstrated that natural compound containing polar useful gatherings are very productive in limiting the impact of consumption notwithstanding the heterocyclic compound containing π-electrons and polar gatherings. The design of coumarin compounds has consisted of NH_2, carbonyl, and π-electron, which feasibly add to the hindrance of mellow steel consumption in acidic media. The proposed structure of the novel synthesized as corrosion inhibitor is appeared in scheme 1.

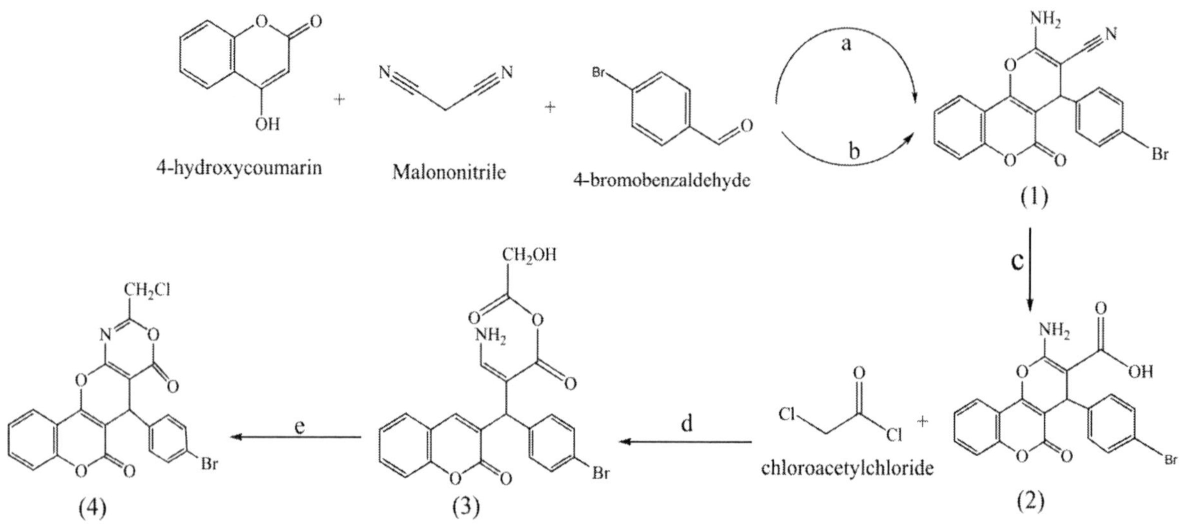

Reagents and Conditions: a = Tetrabutylammoniumbromide (TBAB) / reflux; b = diammonium hydrogen phosphate (DAHP) / reflux; c = hydrolysis; d = benzene/ reflux; e = H_2SO_4 / reflux.

Scheme 1. Reaction of the new coumarin compounds

2. Experimental Section

2.1. General Information

The compounds and solvents utilized were provided by Sigma/Aldrich. The IR spectra of prepared molecules have been done by Nicolet 6700 spectrophotometer and the qualities are communicated in cm^{-1}. NMR spectra were recorded utilizing an AVANCE III 600 MHz spectrometer (Bruker, Billerica, MA, USA), utilizing DMSO-d6 as an interior standard. Elemental analysis was done on an Elementary Vario El III Carlo Erba 1108 basic analyzer (Italy).

2.2. Chemistry

2.2.1- Synthesis of 2-amino-4-(4-bromophenyl)-5-oxo-4, 5-dihydropyrano (3, 2-c) chromene-3-carbonitrile (1)

a. Synthesis of *COMP.(1)* by using Tetrabutylammonium Bromide (TBAB) as catalyst

4-hydroxy-2H-chromen-2-one (0.162 g; 10 mmol) was refluxed with 4-bromobenzaldehyde (5 mmol), malononitrile (15 mmol) and tetrabutylammonium bromide (10 mol %). Then, cooled the mixture at 25°C. Then filtered solid mass , dried and recrystallized from ethanol [15]. Then the compound was characterized by spectroscopic and physical data, yields 75%.M.P (222-224°C).

b.Synthesis of compound (1) by using Diammonium Hydrogen Phosphate (DAHP) as catalyst

In 50 ml of aqueous ethanol (50% ethanol: 50% water), with 4-hydroxy-2H-chromen-2-one (0.162 g; 10 mmol), 4-bromobenzaldehyde (10 mmol), malononitrile (12 mmol) , (DAHP) (26.4 mg, 10 mol %), were mixture (4 hr). at 25°C. when the time was finished, the product was filtered then washed with ethanol and water [16]. Then the compound was characterized by spectroscopic and physical data. , yields 70%. M.P (221-223°C); ^{1}H-NMR: δ 4.26(s, 1H, CH),δ 7.84 (dd, 2H), δ 7.45 (dd, 2H), δ 7.41-7.83 (m, 1H, C-H benzen ring), δ8.5 (s, NH_2); FT-IR: 3381.5 cm^{-1} (NH_2), 3183.0 cm^{-1} (C-H aromatic), 1705.4 cm^{-1} (C=O, lactone), 2211cm^{-1} (C≡N), 1635.0 cm^{-1} (C=C aromatic); Analysis: The theoretical calculation for $C_{19}H_{11}BrN_2O_3$: C 57.7 % H 2.81 % N 7.09 %, while the experimental calculation shows C 57.4% H 2.5% and N 7.01%.

2.2.2-Synthesis of 2-amino-4-(4-bromophenyl)-5-oxo-4, 5-dihydropyrano (3, 2-c) chromene-3-carboxylic acid (2)

Aqueous solution of (NaOH) (1 M, 0.3 ml) was heated with compound (1) for 5 min at 140°C. Then we added HCl(1 M, 0.3 ml) to neutralized the reaction and we got compound (2) at 19–20 min. After fulfillment of the reaction, the product was filtered and washed with ethanol and water, [17], yields 45%. M.P (235-237°C); ^{1}H-NMR: δ 3.83 (s, 1H, CH), δ 7.14 (dd, 2H), δ 7.77 (dd, 2H), δ 7.10-7.60(m, 1H, C-H aromatic ring), δ8.3 (s, NH_2) , δ 11.0 (s, OH); FT-IR: 3220(NH_2), 3309 cm^{-1} (OH); 3085.4 cm^{-1} (C-H aromatic), 1652cm^{-1} (C=O, lactone), 1710 cm^{-1} (C=O, carbooxylic), 1625.9 cm^{-1} (C=C aromatic); Analysis: The theoretical calculation for $C_{19}H_{11}BrNO_5$: C 55.09 % H 2.92 % N 3.38 %, while the experimental calculation shows C 55.01% H 2.81% and N3. 29%.

2.2.3-Synthesis of 3-amino-2-((4-bromophenyl)(2-oxo-2H-chromen-3-yl)methyl)acrylic 2-chloroacetic anhydride (3)

In 100 ml RBF, (0.02 mol) of compound (2) in 4ml ethanol was mixed on the attractive stirrer for 10 min. Four equivalent of chloroacetylchloride (0. 02 mol) was included drop shrewd for 1 hr. Response blend was kept for mixing for 24 squirm response blend was cooled, filled super cold water (50 ml) containing a drop of pyridine and mixed until the point when the oil sets. Unrefined item was sifted, washed with icy water and dried. The product was recrystallized from ethanol [18] , yields 40%. M.P (277-279°C); ^{1}H-NMR: δ 3.55 (s, OH) δ 4.39 (s, 1H, CH), δ 6.54(s, 1H, CH)δ 7.01(s, 1H, CH), δ 7.11 (dd, 2H), δ 7.81 (dd, 2H), δ 7.4-7.79 (m, 1H, C-H benzen ring), δ8.3 (s, NH_2) ; FT-IR: 3431 cm^{-1} (NH_2), 3050.1cm^{-1} (C-H aromatic), 1654cm^{-1} (C=O) 1611.5 cm^{-1} (C=O, lactone); Analysis: The theoretical calculation for $C_{21}H_{16}BrNO_6$: C 55.04% H 3.52 % N 3.06 %, while the experimental calculation shows C 54.9% H 3.11% and N 2.9%.

2.2.4 -Synthesis of 7-(4-bromophenyl)-10-(chloromethyl) -6H-chromeno [3', 4': 5,6] pyrano [2,3-d][1,3] oxazine-6,8(7H)-dione (4)

To a solution of 2 mmol of compound (3) in 50 mL of anhydrous methanol was added concentrated sulphuric acid (80 mmol). The mixture was refluxed on a water bath for 3-4h. Subsequent to cooling, the response content was filled ice-water blend. The precipitated solid was collected by filtration, washed with water and recrystallized from methanol to obtain compound (4) [19] , yields 25%. M.P (215-217°C); ^{1}H-NMR: δ 3.29 (s,2H,CH_2Cl), δ 4.39 (s, 1H, CH), δ 7.11 (dd, 2H), δ 7.79(dd, 2H), δ 7.3-7.79 (m, 1H, C-H aromatic ring) ; FT-IR: 3180.5cm^{-1} (C-H aromatic), 1670.2,1613.2 cm^{-1} (C=O, lactone), 1571.3 cm^{-1} (C=C aromatic); Analysis: The theoretical calculation for C21H11BrClNO5: C 53.36 % H 2.35 % N 2.96 %, while the experimental calculation shows C 52.91% H 2.1% and N 2.88%.

2.3 Gravimetric Experiments

2.3.1 Mild steel specimens.

All through this investigation with the dynamic surface zone of 4.5 cm^2 the Mild steel examples got from the Metal Samples Company were utilized as the working anodes. The piece (wt%) of the mellow steel was as per the following: Fe, 99.21; C, 0.21; Si, 0.38; P, 0.09; S, 0.05; Mn, 0.05; and Al, 0.01. The estimations were coordinated in flowed air through, non-blended 1.0 M HCl arrangements containing distinctive convergences of coumarin compound.

2.3.2 Weight Loss Method

The specimens in duplicate were suspended in the test solution (200 mL), with the four compounds and without the four compounds. Corrosion Inhibitor of different concentration ($0x10^{-2}$ mM, $5x10^{-2}$ mM, $10x10^{-2}$ mM, $15x10^{-2}$ mM, $20x10^{-2}$ mM, $25x10^{-2}$ mM and $50x10^{-2}$ mM),for 1, 2, 3, 4, 5, 10, 24, 48 and 72 hr. at room temperature, [20] The weight of the specimens before and after immersion was determined. Inhibition efficiency of the mild steel was determined using the average of the duplicate value using the equation:

Inhibition Efficiency (IE %) = $(1-W_2/W_1)$ X100

Where, $W_1$and $W_2$are weight loss of the mild steel without and with the four compounds Inhibitor respectively.

3. Results and Discussion

3.1 Weight Loss Method

Table (1-4) shows the inhibitor efficiency for various concentrations of coumarin compounds (1-4) in 1M HCl at various time interim. It demonstrates that as the inhibitor fixation expands, level of inhibitor proficiency additionally builds which implies inhibitor effectiveness is specifically relative to inhibitor focus at a specific time. It likewise uncovered that at a predefined fixation, rate inhibitor proficiency is conversely corresponding to time that is as time expands, rate inhibitor productivity diminishes. The ideal inhibitor productivity was around 91.2% which relates to $50x10^{-2}$ mM grouping of compound (2) in 1.0M HCl at 72 hours [21].

Table 1: Inhibitor efficiency %IE of compound (1) in 1.0M HCl

Concentration of compound (1)	**IE at %**				
	1hr.	3 hr.	24 hr.	48 hr.	**72 hr.**
$0x10^{-2}$ mM	0.00	0.00	0.00	0.00	**0.00**
$10x10^{-2}$ mM	10	15	26	33	**38**
$15x10^{-2}$ mM	13	16	28	34	**39**
$20x10^{-2}$ mM	15	18	29	40	**41**
$25x10^{-2}$ mM	16	20	31	42	**45**
$50x10^{-2}$ mM	17	31	36	44	**50**

Table 2: Inhibitor efficiency %IE of compound (2) in 1.0M HCl

Concentration of compound (2)	IE at %				
	1hr.	3 hr.	24 hr.	48 hr.	**72 hr.**
$0x10^{-2}$ mM	0.00	0.00	0.00	0.00	**0.00**
$10x10^{-2}$ mM	20	26	34	41	**52**
$15x10^{-2}$ mM	23	28	36	44	**54**
$20x10^{-2}$ mM	26	29	40	49	**57**
$25x10^{-2}$ mM	29	33	45	51	**60**
$50x10^{-2}$ mM	54	66	76	85	**91.2**

Table 3: Inhibitor efficiency %IE of compound (3) in 1.0M HCl

Concentration of compound (3)	IE at %				
	1hr.	3 hr.	24 hr.	48 hr.	**72 hr.**
$0x10^{-2}$ mM	0.00	0.00	0.00	0.00	**0.00**
$10x10^{-2}$ mM	22	25	34	41	**50**
$15x10^{-2}$ mM	24	29	35	47	**51**
$20x10^{-2}$ mM	25	30	41	48	**52**
$25x10^{-2}$ mM	30	32	46	55	**61**
$50x10^{-2}$ mM	53	62	74	81	**90.3**

Table 4: Inhibitor efficiency %IE of compound (4) in 1.0M HCl

Concentration of compound (4)	IE at %				
	1hr.	3 hr.	24 hr.	48 hr.	**72 hr.**
$0x10^{-2}$ mM	0.00	0.00	0.00	0.00	**0.00**
$10x10^{-2}$ mM	9	13	20	30	**38**
$15x10^{-2}$ mM	11	14	25	34	**39**
$20x10^{-2}$ mM	13	18	27	39	**40**
$25x10^{-2}$ mM	14	20	30	40	**40.5**
$50x10^{-2}$ mM	15	23	33	40.3	**41**

Figure 1,2,3,4 demonstrates the chart of rate inhibitor proficiency against inhibitor focus for different time for compound 1,2,3,4 respectively, The results of the experiment have shown that compounds 2and 3 decreased the corrosion rate as high as around 91.2 %, 90.3% for this specific investigation. The inhibitor proficiency is enabling and this can be used in the ventures as a substitute for the imported synthetic inhibitors. Abdul Amir et.al new coumarin derivative was synthesized and the IE was investigated on the corrosion of MS in 1.0 M HCl but the yield of these compounds is very low and the cost of synthesis is very high compared with our prepared compounds [22].

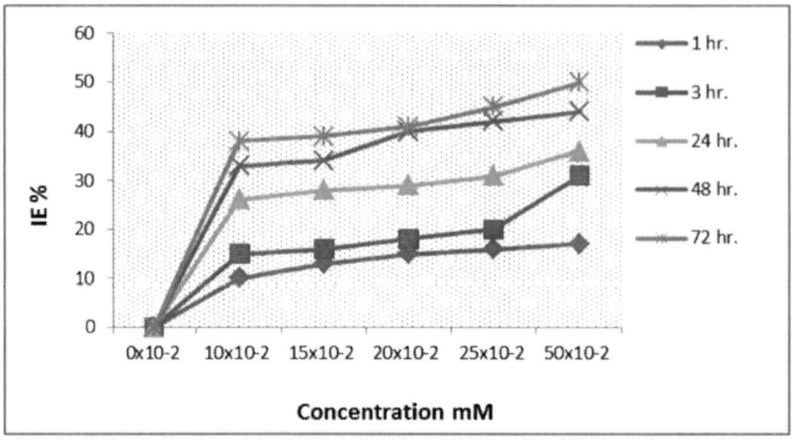

Figure 1: Chart of % IE for compound (1) against concentration at different time

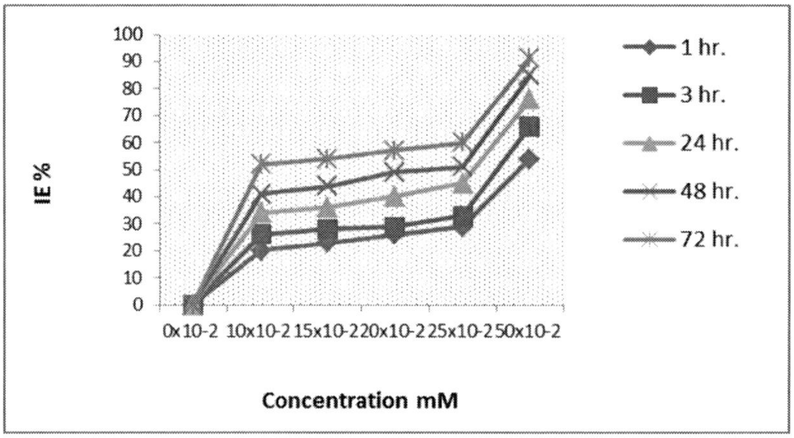

Figure 2: Chart of % IE for compound (2) against concentration at different time

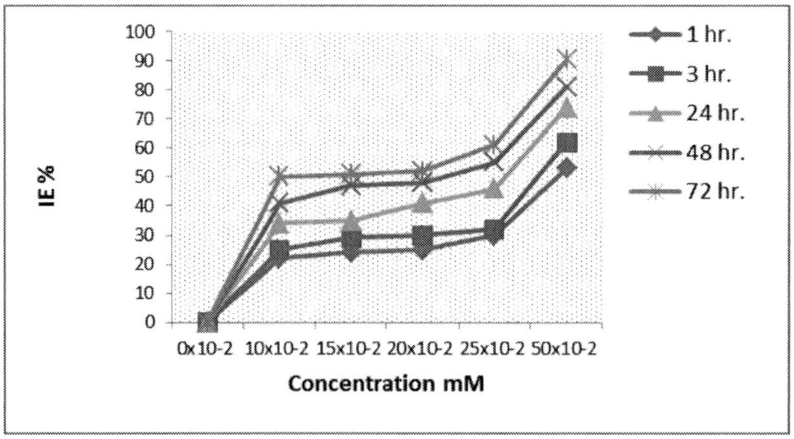

Figure 3: Chart of % IE for compound (3) against concentration at different time

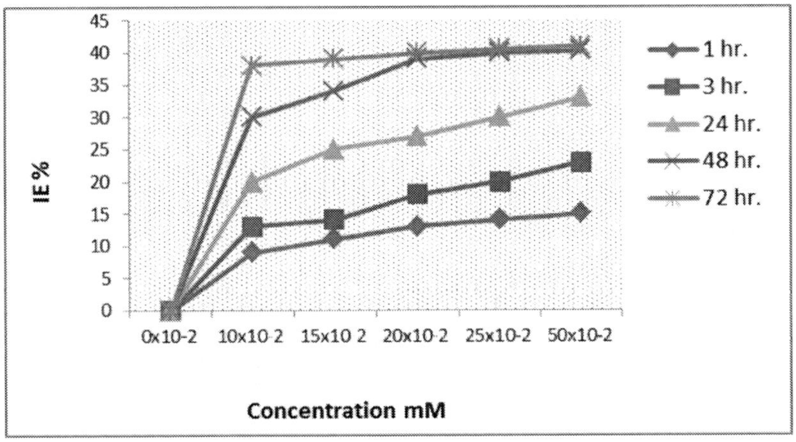

Figure 4: Graph of % IE for compound (4) against concentration at different time

3.2 Scanning Electron Microscopy (SEM) Analysis

The SEM test was directed at the Electron Microscopy Unit of applied science department. Figure 5 shows the mild steel without corrosion inhibitor and Figure 6 shows that the corrosion inhibitor (compound 2) given security to the mellow steel from the erosion assault brought about by HCl.

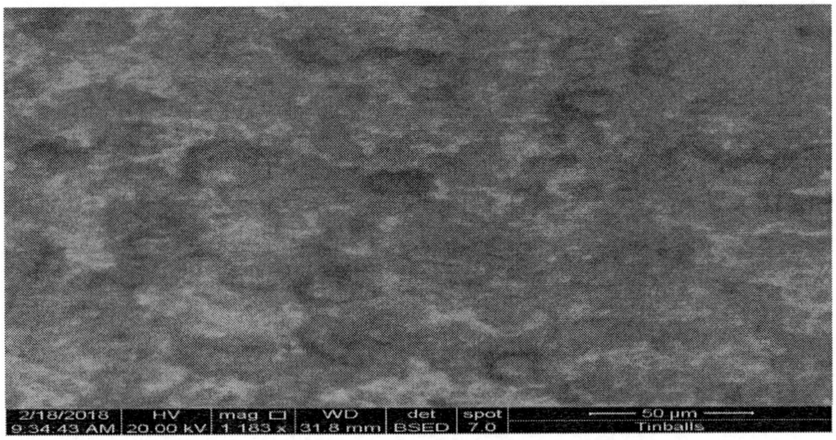

Figure 5.The SEM micrographs 5000X, for MS in 1.0 M HCl

Figure 6. The SEM micrographs 5000X, for MS in 1.0 M HCl with 0.5 mM of the compound (2)

3.3 Proposed Mechanisms Actions of Synthesized Compound as Corrosion Inhibitor

Chemically the inhibitor is adsorbed on the metal surface and structures a defensive thin film or synthetic bonds frame by response between the inhibitor and metal. The adsorption segment of regular inhibitors can proceed by methods for one of these courses. first, the cooperation between unpaired electrons and the metal surface. second, charged atoms and metal pull in electrostatic-ally. third, cooperation between π-electrons and the metal surface [23-27]. The restraint effectiveness of our erosion inhibitor against the consumption of mellow steel in 1M of (HCl) can be uncovered by the amount of charge thickness, adsorption locales, sub-atomic size, method of association with the metal surface and capacity of arrangement of metallic insoluble complex. The π electrons for the twofold bonds and free electrons on the oxygen and nitrogen molecules shape synthetic bonds with the metal surface as showed up Figure 7 and Figure 8.

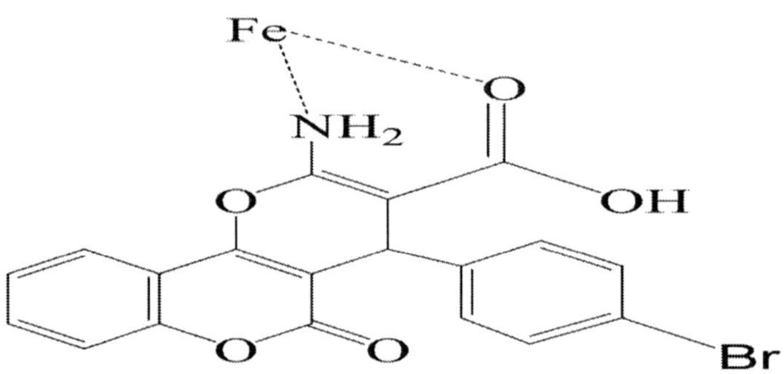

Figure7. The proposed mechanism of compound(2) as inhibitor

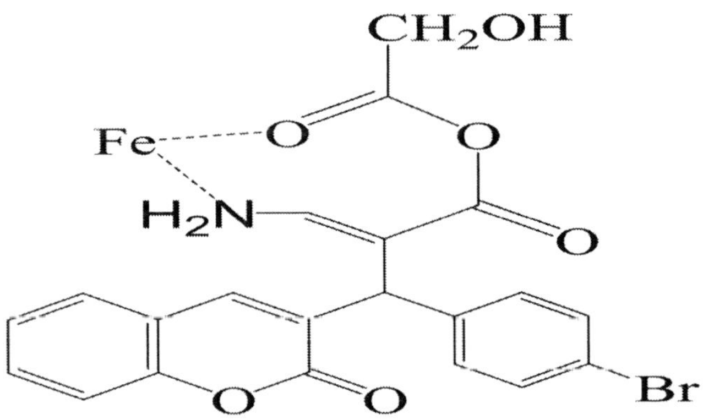

Figure 8.The proposed mechanism of compound(3) as inhibitor

4 Conclusions

This study has shown that compound (1-4) can be utilized as consumption inhibitor. As the convergence of inhibitor delivered expands, the consumption rate diminishes and the inhibitor has an ideal effectiveness of about 91.2% and 90.3 % for mixes 2, 3 individually which showed that its utilization in the process ventures will diminish unquestionably the erosion rate.

Acknowledgment

we are grateful to the staff membered at Branch of Chemistry, Department of Applied Science, University of Technology (UOT), for kind help & support.

References

[1] Zhuk N P 1976 *Course on Corrosion and Metal Protection* Metallurgy Moscow
[2] Perboni G and Rocchini G 1988 *Proceedings of the 10th ICMC* Madras India 2763
[3] Omar I H Trabanelli G and Zucchi F 1988 *Proceedings of the 10th ICMC* Madras, India, p. 2723
[4] Mathur R P and Vasudevan T 1982 *Corrosion* **38** 171
[5] Zucchi F and Trabanelli G 1990 *Proceedings of the 7th European Symposium on Corrosion Inhibitors* Ferrara 339
[6] McCafferty E and Hackerman N 1972 *J. Electrochem. Soc.* 119 - 999
[7] Hariharaputhran R, Subramanian A, Antony A A, Sankar P M, Gopalan A, Vasudevan T and Vencatakrishna I S 1998 *Br. Corros. J.* 33 214
[8] Ivanov E S 1986 *Inhibitors for Metal Corrosion in Acid Media, Metallurgy* Moscow
[9] Foroulis Z A 1985 *Proceedings of the 6th European Symposium on Corrosion Inhibitors* Ferrara 48
[10] Slaiman Q J and Al-Saaty H M 1990 *Proceedings of the 7th European Symposium on Corrosion Inhibitors* Ferrara p149
Sadeddin K, Naser A and Firas A 2011 Removal of turbidity and suspended solids by electro-coagulation to improve feed water quality of reverse osmosis plant *Desalination* **268** 204–207
[11] El Ouali I, Hammouti B, Aouniti A, Ramli Y, Azougagh M, Essassi E M and Bouachrine M 2010 *J. Mater. Envir. Sci.* **1** 1-8
[12] Gerhardus H K, Michiel P H, Bronger N G, Thompson Y, Paul V and Payer J H 2002 Corrosion Cost and Preventive Strategies in the United States *Supplement to Materials Performance.* Report Number FHWA RD-01-156 Federal Highway Administration Mclean
[13] Yuhazri M Y, Jeefferie A R, Haeryip S and Siti S 2011 Effect of Coating Controlled in Tin Coating on the Mild Steel Substrate *International Journal of Engineering Science and*

Technology **3** 2113-2117

[14] Yuhazri MY, Jeefferie A R, Sihambing H, Nooririnah O. and Warikh A R 2011 Coating effect Condition on the Corrosion Properties of Mild Steel Substrate *International Journal of Applied Science and Technology* **1** 45-49

[15] Singh A, Ebenso E E and Quraishi M A 2012 Corrosion Inhibition of Carbon Steel in Hydrochloric Acid Solution by Some Plant Extract *International Journal of Corrosion* 1-20

[16] Khurana, J M and Kumar S 2009 Tetrabutylammonium bromide (TBAB): a neutral and efficient catalyst for the synthesis of biscoumarin and 3, 4-dihydropyrano [c] chromene derivatives in water and solvent-free conditions *Tetrahedron Letters* **50** 28 4125-4127

[17] Abdolmohammadi S and Balalaie S 2007 Novel and efficient catalysts for the one-pot synthesis of 3, 4-dihydropyrano [c] chromene derivatives in aqueous media *Tetrahedron Letters* **48** 18 3299-3

[18] Reid A E et al 2011 Radiosynthesis of C- 11 labeled auxin (3- indolyl [1- 11C] acetic acid) and its derivatives from gramine. Journal of Labelled Compounds and Radiopharmaceuticals **54** 8 433-437

[19] Patel H, Desai H and Mistry H 2004 Synthesis and antimicrobial activity of some new piperazine derivaties containing aryl sulfonyloxy group *Journal of Chemistry* **1** 2 93-98

[20] Serban G 2015 5-Arylamino-1, 3, 4-thiadiazol-2-yl acetic acid esters as intermediates for the synthesis of new bisheterocyclic compounds *Farmacia* 63 146-149

[21] Salami L et al. 2012 A study of the corrosion inhibitor of mild steel in sulphuric acid using Musa sapientum peels extract *Global Engineers and Technologists Review* **2** 12 1-6

[22] Tao Z et al. 2012 A study of differential polarization curves and thermodynamic properties for mild steel in acidic solution with nitrophenyltriazole derivative *Corrosion science* 60 205-213

[23] Kadhum A H et al. 2014 Inhibition of Mild Steel Corrosion in Hydrochloric Acid Solution by New Coumarin *Materials* 7 4335-4348

[24] Sokolova V et al. 2011 Characterisation of exosomes derived from human cells by nanoparticle tracking analysis and scanning electron microscopy *Colloids and Surfaces B: Biointerfaces 87* 1 146-150

[25] Al‐ Amiery A, Kadhum A, Alobaidy A, Mohamad A and Hoon P 2014 Novel corrosion inhibitor for mild steel in HCl **Materials 7** 2 662–672

[26] Obayes H, Alwan G, Alobaidy A, Al‐ Amiery A, Kadhum A and Mohamad A 2014 Quantum chemical assessment of benzimidazole derivatives as corrosion inhibitors *Chem Cent J* 8 21-27

[27] Junaedi S, Kadhum A A H, Al‐ Amiery A A, Mohamad A B and Takriff M S 2012 Synthesis and characterization of novel corrosion inhibitor derived from oleic acid: 2‐ Amino 5‐ Oleyl‐ 1,3,4 Thiadiazol (AOT) *Int J Electrochem Sci* 7 3543–3554

New derivative of 1,2-Dihydropyridine-3-Carbonitrile as Corrosion Inhibitor

Hiba H. Ibraheem
Chemistry division, Department of applied science, University of Technology (UOT)/Baghdad, Iraq

Corresponding Author e mail address: dr.hiba1982@gmail.com

Abstract
New 1,2-dihydropyridine-3-carbonitrile derivative compound as corrosion inhibitor namely, 3,3'-(3,3'-(1E,1'E)-1,4-phenylenebis(diazene-2,1-diyl)bis(4-(dimethylamino)-3,1-phenylene))bis(2-oxo-6-(2,4,6-trihydroxyphenyl)-1,2-dihydropyridine-3-carbonitrile) (PDPTD) was synthesized by cyclization of compound (1) with 1-(2,4,6-trihydroxyphenyl) ethenone in presence of ethyl cyano acetate and ammonium acetate. The new synthesized compound has been characterized using Melting point, FT-IR spectroscopy and ^{1}H-NMR. The new synthesized compound (PDPTD) has been tested on mild steel (MS) in 1 M HCl media by a weight loss method and identification of the surface technique employed by scanning electron microscope (SEM) studies. The weight loss investigation demonstrated that inhibition performances have been improved via concentration raising of inhibitor. To support our results scanning electron microscopy has employed about clarify the surface from the MS with and without (PDPTD) in 1.0 M HCl media. The inhibition performance rate was 83% at the maximum utilized concentration of (PDPTD).

Keywords: benzene-1,4-diamine; 4-(dimethylamino)benzaldehyde; SEM.

1. Introduction
Pyridine derivatives are the majority important heterocycles which found in medicinal chemistry such they have been an excellent biological activity for a wide range from applications, including antimicrobial [1], antiinflammatory [2], anti-HIV [3], antiplasmodial [4], antitubercular [5], antibacterial [6], anticonvulsant [7], COX inhibitor [8], anticancer effects [9], blood platelet aggregation inhibitors [10], antioxidant [11], antitumor [12] and antifungal activities [13]. The Corrosion inhibitor is a

substance which has retarded that inhibitors function at one either more ways about control corrosion: through adsorption from a thin film at the surface from a corroding material, through inducing the formation from a thick corrosion creator, or through changing the characteristic from the environment resulting at reduced offensive [14]. The Corrosion is a natural process, which changes a refined metal about a more chemically-stable shape, like its oxide,hydroxide, or sulfide. It has been the gradual demolition from materials (customarily metals) through chemical and/either electrochemical reaction for their medium. The Corrosion engineering is the field appropriated about controlling and pausing corrosion. The use from chemical inhibitors about diminish the rate from corrosion processes has been quite varied. At the oil extraction and making industries, inhibitors have constantly considered about the first line from protect against corrosion. A large number from scientific studies have devoted about the subject from the corrosion inhibitors [15]. The Corrosion inhibitors have been one from the practical and have cost efficient methods from controlling metallic corrosion. They have been inorganic either organic substances. Inorganic substances appropriate such metal corrosion inhibitor must simply oxidize the metal about form an inefficient layer which prevents immediate ions-metal interaction and hence prevent the rate from metal dissolution at the medium. The organic accordance, on the another hand, must own features containing presence from heteroatoms and/either double bonds, big surface area, efficient centre, etc. which at adsorption to the metal surface will be blanket a big area from the metal and so isolate it of the offensive ions present at the medium[16].

The aim of this work is to about study the inhibiting effect from 3,3'-(3,3'-(1E,1'E)-1,4-phenylenebis (diazene-2,1-diyl)bis(4-(dimethylamino)-3,1-phenylene))bis(2-oxo-6-(2,4,6-trihydroxyphenyl)-1,2-dihydropyridine-3-carbonitrile) (PDPTD) at mild steel at one normal hydrochloric acid solution. The results were supplemented by SEM, IR and NMRstudies. Suggested structure for combined consumption inhibitor is appeared as shown in Figure (1).

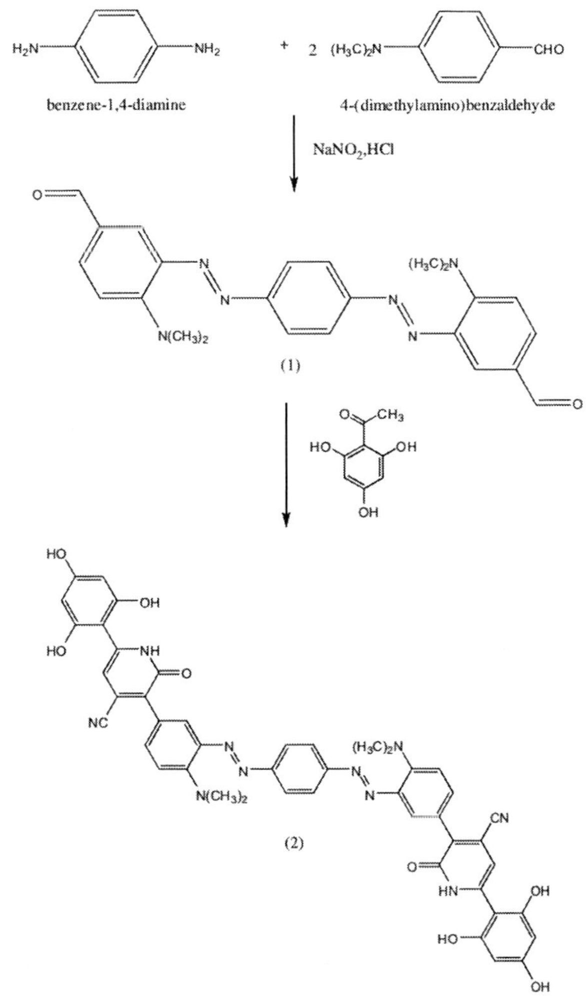

Figure 1. Structure that suggested of the prepared corrosion inhibitor (PDPTD)(2).

2.Experimental

2.1 Synthesis of corrosion Inhibitor:

2.1.1 Synthesis of Azo Derivative (1)

Benzene-1,4-diamine (0.01 mole) was added to a mixture of water (4 ml) and concentrated hydrochloric acid (4.5 ml). The resulting solution was stirred for 10 min., before being cold to (0-5) ^{0}C. A solution of sodium nitrite (1.38g.,0.02 mole) in water (2.5 ml) is added dropwise [17], after being stirred for 10 min. The resulting solution of diazonium salt was added dropwise to a mixture of 4-(dimethylamino)benzaldehyde (2.98g., 0.02 mole) in ethanol and 10% NaOH (20ml.)at (0-5)^{0}C at pH=5.5 and the disappearance of reactants was figured according to Thin-layer chromatography (TLC) technique. After the addition was completed, the mixture was stirred for further 20 min., then it left for 1 hr. The resulting solid was filtered off, dried and recrystallized from ethanol to get the compound (1), red precipitate ,yield 84%, M.P (80-82°C).

2.1.2 Synthesis of PDPTD (2)

A solution of compound (1) (0.001 mole) in ethanol (10 ml.), 1-(2,4,6-trihydroxyphenyl) ethanone (0.002 mole), ethyl cyano acetate (0.226g.,0.002 mole), ammonium acetate (1.2g.,0.016 mole) were added [18]. The reaction mixture was refluxed for 2hrs and the disappearance of reactants was figured according to TLC technique. The consequent formed has filtered off then washed by water, dried and recrystallized of ethanol to get the target compound (2), yellow precipitate ,yield 68%, M.P (112-115°C).

2.2 Corrosion technique:

2.2.1 MS specimens

MS specimens got by the Metal Samples Company were employed all through this investigation. The formation (wt %) from the mild steel had been as follows: iron, 99.21; carbon, 0.21; silicon, 0.38; phosphorous, 0.09; sulfur, 0.05; manganese, 0.05; and aluminum, 0.01. The specimen was cleaned in accordance with procedure G1-03 that mentioned in reference [19]. The measurements were conducted in aerated [20]. The components were analyzed in Malaysia by UKM University with X-Ray Fluorescence, without stirring 1.0 M HCl solutions including different concentrations from (PDPTD) compound.

2.2.2 Technique of Weight Loss

Copouns in duplicate were suspended in 200 mL of the examine media, with & without the PDPTD (2) corrosion Inhibitor of different concentration (0 mM, 0.1 mM, 0.15 mM, 0.2 mM, 0.25 mM, 0.5 mM), for 1, 2, 24, 48 and 72 hrs. at room temperature [21], Specimens weights before/after immersed were computed. The inhibition efficiency of the mild steel was determined by using the average of the duplicated value using the equation [22]:

Inhibition Efficiency (IE %) = (1-W2/W1) × 100

Where, W1and W2 are damage weight from Mild Steel (MS) without and with (PDPTD) (2).

3. Results and Discussion

3.1 Identification of PDPTD (2)

The synthesis of an azo dye needs two organic compounds a diazonium salt and a coupling component. The diazonium salt like an electrophile interacts with an electron-rich coupling component, as a phenol derivative over an electrophilic aromatic substitution reaction. The FTIR spectrum of compound (1) has showed an absorption band at (2906 cm^{-1}) because of stretching vibration for C-H arom., band at (2820,2804) for C-H aliph.,,band at (1737 cm^{-1}) because of stretching vibration at C=O of aldehyde, band at (1541 cm-1) indicating the (N=N).

The inhibitor 3,3'-(3,3'-(1E,1'E)-1,4-phenylenebis (diazene-2,1-diyl)bis (4 (dimethylamino)-3,1-phenylene))bis (2-oxo-6-(2,4,6-trihydroxyphenyl)-1,2 dihydropyridine-3-carbonitrile) (PDPTD) was prepared in good yield by reflection reaction of compound (1) and 1-(2,4,6-trihydroxyphenyl)ethanone. The compound (2) was identified by FTIR and ^{1}HNMR spectroscopy. The FTIR spectrum of the compound (2) shows bands at the frequency of (1705 cm^{-1}) owing to the carbonyl group from amide appears, the NH near (3150) cm^{-1}. And the appearance from absorption band during (2208.5) cm^{-1} for CN group. The ^{1}HNMR spectrum of compound (2) shows the following data: 3.09 (s,12H,$N(CH_3)_2$), 6 (s,2H, C=CH for pyridinone ring), 7.1- 7.8 (m,14H,Ar-H), 8 (s,2H, NH), 10 (s, 2H, -OH), and 11.9 (s, 4H, -OH).

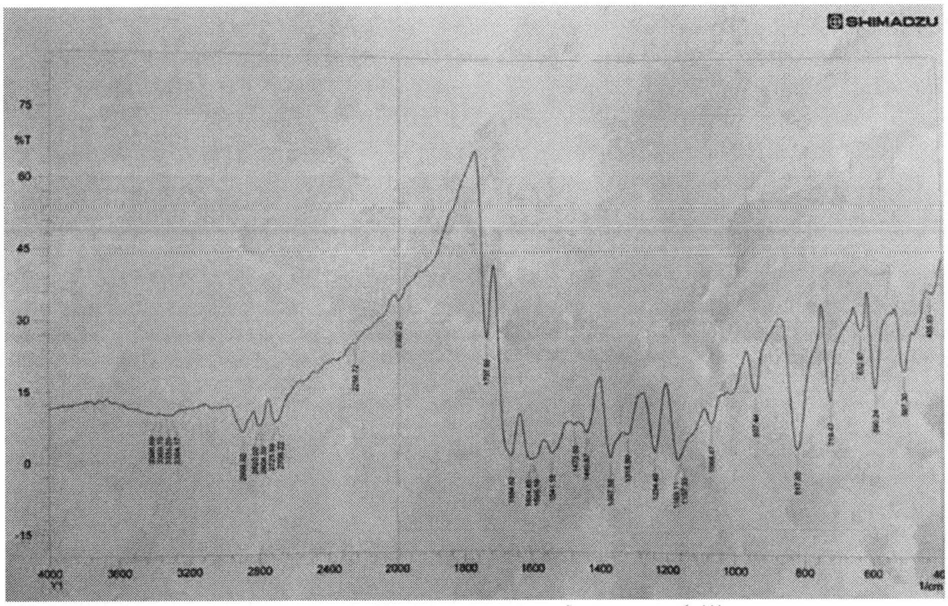

Figure 2. The FTIR spectrum of compound (1).

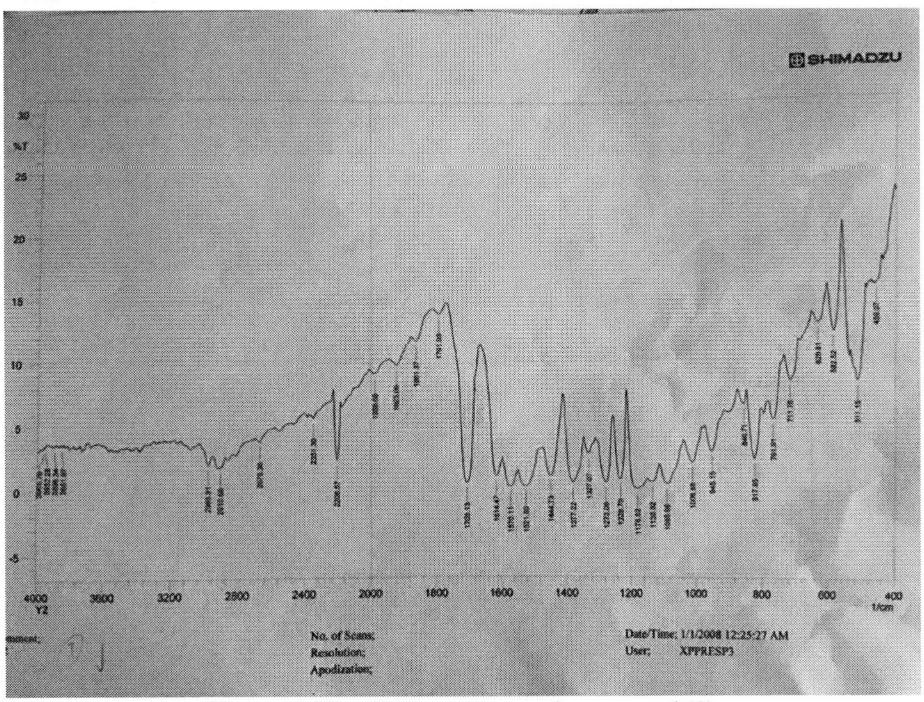

Figure 3. The FTIR spectrum of compound (2).

3.2 Technique of Weight Loss (Concentration Effect)

The inhibition efficiency (IE) esteems computed from weight reduction estimations for MS in the corrosive mixture that contain different concentrations of PDPTD (2) at encompassing temperature are appeared in Figure (4), that the PDPTD diminished the corrosion from MS in corrosive especially. The IE% expanded with a rise in concentration of PDPTD (2) and attained upper IE at 0.5 mM concentration from PDPTD. The raising in IE% for raise at concentrations were prompsed from the raise at the extent from performance of PDPTD preservation. The inhibition efficiency from inhibitor into the corrosion from steel has obtained by using the following equation [23]:

Inhibition Efficiency (IE %) = (1-Wcorr/W°corr) × 100

Wcorr and W°corr are the corrosion rate from steel samples with and without the inhibitor.

The best of inhibitor efficiency has been given 83% which agree to 0.5 mM concentration of PDPTD in 1.0M HCl at 72 hours.

Table 1. Inhibitor efficiency into various concentrations from PDPTD (2) in 1.0M HCl.

Concentration of PDPTD	%IE at				
	1hr.	3 hr.	24 hr.	48 hr.	72 hr.
0 mM	0.00	0.00	0.00	0.00	0.00
0.1 mM	0.00	0.1	0.47	2.3	3.46
0.15 mM	0.01	0.14	1.43	7	12
0.2 mM	0.03	0.35	3	15	32
0.25 mM	0.04	0.4	4.5	21	51
0.5 mM	0.05	1.2	12	42	83

Figure (4) shows the graph of percentage inhibitor efficiency against inhibitor concentration into several time for PDPTD (2) respectively, the results of the experiment have shown that PDPTD (2) reduced the corrosion rate to 83 % into this study. The inhibitor efficiency has been encouraging and this can be used at the industries such a substitute into the significant chemical inhibitors.

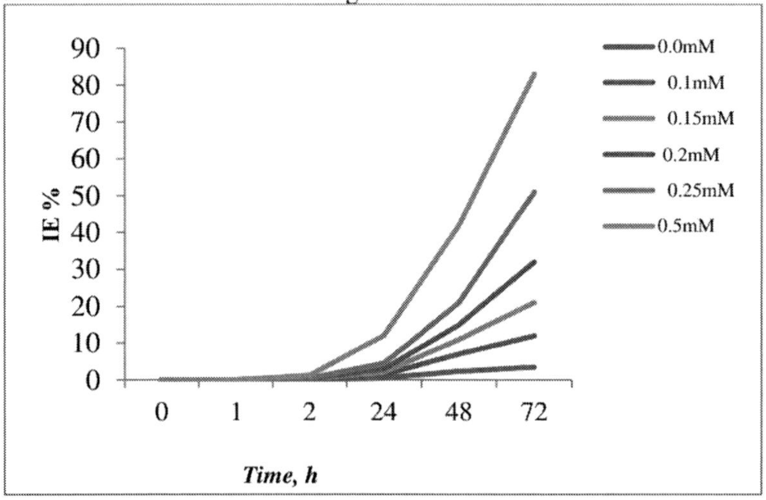

Figure 4. Different concentrations of PDPTD vis different immersion periods time for Inhibition Efficiency (IE %) of mild steel.

3.3. Scanning Electron Microscopy (SEM) Analysis

SEM analysis had performed about discuss the surface morphology from the mild steel after inundation at 1.0 M HCl at the absence and the presence from PDPTD, Figure (5). Destroyed surface had observed at the absence from PDPTD because of high dissolution rate from iron by such pH, yet a thin and uniform layer during the metal surface had observed at the presence from PDPTD (2), the cracks at the film have been because from the dehydration from surface ago the surface had dried previously the SEM portrait. This was proof that PDPTD had absorbed by the mild steel surface and insulates the surface of the acidic intermediate [24].

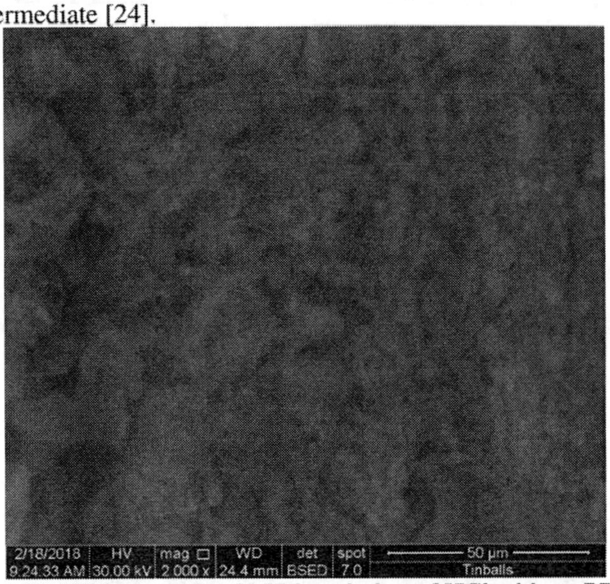

Figure 5. The SEM pictures, of MS in solution of HCl without PDPTD (2).

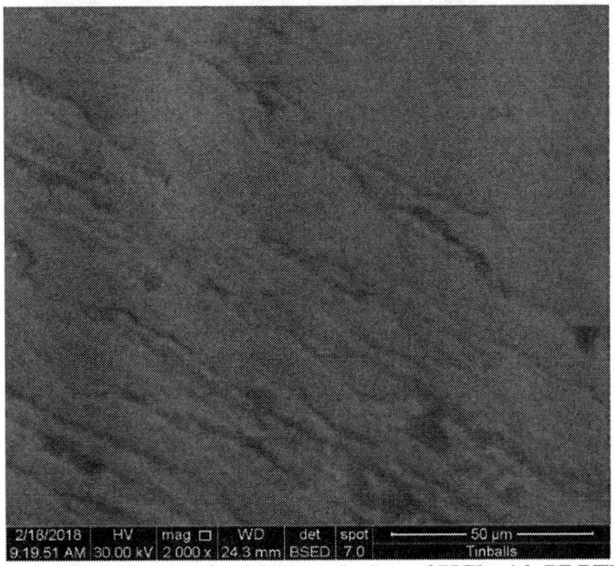

Figure 6. SEM picture, for MS in solution of HCl with PDPTD (2).

3.4. PDPTD (2) as Inhibitor- Action Mechanisms

Mostly inhibitors have adsorbed at the mild steel and structures a paint film neither a synthetic bonds structure through restraint from inhibitor and metal and of inorganic complex (free electrons from inhibitor-ligand for the unoccupied d-orbital into the metal). The action mechanism from PDPTD (2) such inhibitor can move ahead through means from interaction. Interaction of PDPTD (2), metal electrostatically neither interaction from unpaired electrons from hetro-atoms from PDPTD (2) for metal. IE from PDPTD (2) has been demonstrated about the hetro-atoms, charge, size, stereo-hindrance and ability from consistence from metallic insoluble complex. PDPTD (2) work like bi-dentate ligand unpaired electrons from oxygen and nitrogen about compose coordinated bonds for metal Figure (7).

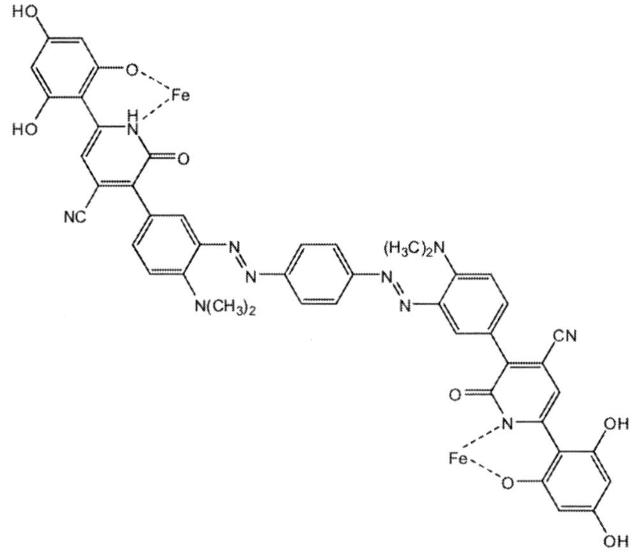

Figure 7. The suggested mechanism of PDPTD (2) as corrosion inhibitor

4. Conclusions

3,3'-(3,3'-(1E,1'E)-1,4-phenylenebis(diazene-2,1-diyl)bis(4(dimethylamino)-3,1-phenylene))bis(2-oxo-6-(2,4,6-trihydroxyphenyl)-1,2 dihydropyridine-3-carbonitrile) as new inhibitor for surface of mild steel was prepared compound (1) and the structure of it molecule was identified based on NMR and FT-IR spectroscopy. The results of this investigations revealed that the prepared 1,2-Dihydropyridine-3-Carbonitrile derivative namely PDPTD worked as an excellent corrosion inhibitor in corrosive media for surface MS in solution of HCl with concentration subordinate mode. The inhibition performance of PDPTD as an inhibitor most extreme hindrance proficiency was up to 83% at 0.5 mM PDPTD concentration. PDPTD (2) shows the productive natural inhibitor having great inhibitive characteristics because of quality of atoms namely nitrogen with oxygen. SEM estimations bolstered development of a defensive layer through PDPTD (2) on the MS surface. The anti-corrosion investigation for PDPTD (2) showed the protection of surface of MS in corrosive solution.

Acknowledgements

This study was supported by the University of Technology, Department of Applied Science at Branch of Chemistry, Baghdad, Iraq.

References

[1] Gaonkar S L, Rai K M L and Prabhuswamy B 2007 *Med. Chem. Res.* **15** 407–417
[2] Hosni H M and Abdulla M M 2008 *Acta Pharm.* **58** 175–186
[3] Ali M A, Yar M S, Siddiqui A A, Sriram D, Yogeeswari P and de Clercq E 2007 *Acta Pol. Pharm.* **64** 423–428
[4] Kumar S, Das S K, Dey S, Maity P, Guha M, Choubey V, Panda G and Bandyopadhyay U 2008 *Agents Chemother.* **52** 705 715
[5] Lourenço M C S, de Souza M V N, Pinheiro A C , de Lima Ferreira M, Goncalves R S B, Nogueira T C M and Peralta M A 2007 *ARKIVOC* p 181–191
[6] Sharma P C and Jain S 2008 *Acta Pol. Pharm.* **65** 551–556
[7] Shafiee A, Rastkari N and Sharifzadeh M 2004 *Daru* **12** 81–86
[8] Chintakunta V K, Akella V, Vedula M S, Mamnoor P K, Mishra P, Casturi S R, Vangoori A and Rajagopalan R 2002 *Eur. J. Med. Chem.* **37** 339–347
[9] Thapa P. Karki R . Thapa U. Jahng Y. Jung M -J. Nam J M. Na Y Kwon Y and Lee E-S 2010 *Bioorg. Med. Chem.* **18** 377–386
[10] Sotelo E Fraiz N Yanez M Terrades V Laguna R Cano E and Ravina E 2002 *Bioorg. Med. Chem.* **10** 2873–2882
[11] Caliskan E B, Sukuroglu M, Coban T, Banoglu E and Suzen S 2008 *J. Enz. Inhib. Med. Chem.* **23** 225–229
[12] Zhang S-X Bastow K F Tachibana Y Kuo S –C Hamel E Mauger A Narayanan V L and Lee K –H 1999 *J. Med. Chem.* **42** 4081–4087
[13] Wu J, Kang S, Luo L, Shi Q Ma J Yin J Song B Hu D and Yang S 2013 *Chem. Cent. J.* **7** 64
[14] Khaled K F 2003 *Electrochimica Acta* **48** 2493-2503
[15] Jones L W 1988 Corrosion and Water Technology for Petroleum Producers, Tulsa, Okla, Oil and Gas Consultants International
[16] Lebe A, Israel O and Emeka E 2014 *International Journal of Materials Engineering* **4** 171-179
[17] Gupta P, Gupta JK and Halve AK 2014 *Int J Res Pharm Sci* **4** 13-20
[18] Beheshtia Y, Khorshidi M, Heravi M and Baghernejad B 2010 *Eur. J.Chem.* **1** 232-235
[19] ASTM G1-3 2003 Standard Practice for Preparing, Cleaning, and Evaluating Corrosion Test Specimens. *ASTM International:* (West Conshohocken, PA USA)
[20] Al-Amiery AA, Al-Majedy YK, Kadhum AAH, Mohamad AB 2016 *Res Chem Intermed* **42** 3905-3918
[21] Salami L OP Akinyemi RJ Patinvoh 2012 *Global Engineers and Technologists Review* **2** 1-6
[22] Aytac A Uand Ozmen M 2005 *Mater. Chem. Phys.* **89** 176-181
[23] Touir R Dkhireche N Ebn Touhami M Sfaira M Senhaji O Robin J J Boutevin B Cherkaoui M 2010 *Mat. Chem. Phys.* **122** 1-9
[24] Sokolova V and Anna-Kristin 2011 *Biointerfaces* **87** 146-150

IOP Conf. Series: Materials Science and Engineering **579** (2019) 012053 doi:10.1088/1757-899X/579/1/012053

An Experimental Study for the effect of Water in Crude Oil on its Rheological Properties and Flow Characteristics

Farhan Lafta Rashid, Samah Zaki Naji[#]
Petroleum Engineering Department, University of Kerbala, Iraq

[#]Corresponding author's e-mail address: samah.alrashid@uokerbala.edu.iq

Abstract. An experimental study for the effect of water in Iraqi crude oil on rheological and flow characterization was achieved by the experimental rig. The crude oil used was from Al-Najaf refinery with a density of 916.6 kg/m^3, dynamic viscosity of 24.94 cp at a temperature of 25 °C. A centrifugal pump was used with a straight impeller, one suction and one discharge line. The results show that the density of the crude oil increases as the water/oil ratio increases. While the viscosity of the crude oil will be reduced. Shear stress is increased by adding water to the crude oil, the maximum value of shear stress is observed at 12.5 %. The two-phase flow rate increases as water/oil ratio increased, for the flow of water-oil, the pump head increases as water/oil ratio increases reaching to 0.8 water/oil ratio, which is the point of inversion, and then reduces. In addition, the volumetric flow rate of the two-phase flow increases as water/oil ratio is increasing and the increase in water-oil flow velocity causes rising in the pump head.

1. Introduction

The term multiphase flow refers to any flow of fluid for more than one component or phase. These components or phases are perfectly mixed above the molecular level. Thus, the flows have some level of components or phase separation at a well scaled above the molecular level. This leaves a large spectrum of various multiphase flows. They can be classified according to the state of phases or components and may refer to solids/gas flows, or solids/liquid flows or gas/particle flows [1, 2, 3]. A modified model for the evaluation of pump head for the two-phase flow was adopted by Poullikkas [4]. The model deals with the effect of phase separation and compressibility and gas phase condensation.

Most of the experimental researches focus on the mixture flow of oil-water in pipelines were concentrated to investigate different aspects of multiphase flows in pipe for oil and water that involve flow patterns identification such as the dispersed, intermittent, stratified, and the combination of them [5-7]. Fluid flow characteristics and hydrodynamic behaviour on horizontal and vertical flows in pipe for different volume fraction were investigated experimentally, so the effect of the density and viscosity on the fluid flow behaviour is presented [5-8-11]. Also, the highest viscosity of crude and heavy oil was used in experiments [8,12,13]. Experimental work of the continuous flow phenomena for the case of oil-water mixture flow for horizontal pipelines was achieved in [9, 13, 14, 15]. The inversion flow phenomena in a vertical pipe for the dispersed oil/water flows were experimentally studied by Hu & Angeli [8]. Pressure was observed to be reduced in both cases when water was added into the oil and when oil was added into the water. Shoulong and Aifen studied the effect of temperature and water content on the rheological properties of heavy oil. It has been shown that when the water content is in the range of 0 to 30%, the viscosity of heavy oil first changes little, then increases, and finally decreases with the increase of water content [16].

The current work deals with the effect of water in crude oil on rheological characterization and on pump performance, where a centrifugal pump of straight impeller type was used.

2. The System of Oil-Water Flow

The flow system of water/oil includes a mixing tank of water/oil, a stirrer for mixing oil and water in the mixing tank, centrifugal pump (single suction-single discharge type, straight impeller type, maximum head, flow rate, power) are 40 m, 40 l/min, and 1.25 hp, respectively. Valves used to control the flowrate, and a storage tank. The instruments used are: flowmeter, two-gauge pressures (one for suction side, P1, and the other for the discharge side, P2) and pvc tubes of 1 inch. Figure (1) shows the schematic diagram of the oil-water flow system.

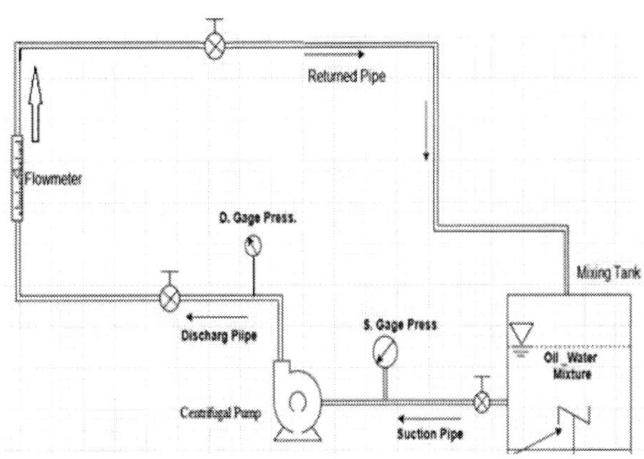

Figure 1. Schematic diagram of water-oil flow system

3. Experimental Method and Procedure

The pump performance in the case of two-phase flow depends on the gas-liquid interaction behaviour in the pup.

The two-phase flow mixture density is given as;

$$\bar{\rho} = \frac{\iiint_{V_g V_f} \rho dV}{(V_g+V_f)} = \alpha\rho_g + (1-\alpha)\rho_f \quad (1)$$

Where $\alpha = \frac{V_g}{V_g+V_f}$, $(1-\alpha) = \frac{V_f}{V_g+V_f}$

The term of void fraction was replaced by volume fraction, (volume ratio of phase 1 to the total volume of the flow) when the two phase was liquid/liquid (water/oil) as in this work. Then Equation (1) can be modified to represent the water/oil mixture density:

$$\bar{\rho} = V_{oil}\rho_{oil} + V_{water}\rho_{water} \quad (2)$$

The two-phase flow pump head was estimated as the ratio of the ressure difference through the pump to the two-phase mixture specific weight;

$$H_P = \frac{\Delta P}{\gamma_{water-oil}} = \frac{P_2-P_1}{\bar{\rho}.g} \quad (3)$$

For the water-oil flow system, the following procedure was used:

1. Filling the mixing tank with water (40 liters) and pumping it till it reached a steady state, during the flow, measuring the suction pressure (P1), discharge pressure (P2), and the flow rate.
2. Adding (1 liter of oil) to the water in the mixing tank (means 97.57% water and 2.43 % oil) and repeating the measurement of suction, discharge pressure and flow rate.
3. Repeating step 2 by adding (2, 3,4… liter of oil), and for each addition process, measuring suction, discharge pressure and flow rate.
4. From Eq. (2), calculating the two-phase (water-oil) density ($\bar{\rho}$), from eq. (3) calculate the pump head (H_P), for each reading.

Density of the crude oil is measured by using Density Meter (v(B)3.0.0.207[DDM2910]), viscosity is measured by using Viscometer (minimum sample volume 100μL, viscosity range 1-3500cSt, operating range 18-40 °C, output range 20-100 °C). Shear stress and shear rate are measured by using ARES Rheometer (temperature -10 to 600 °C, shear rate 10^{-4} to 10^{4} (1/s), viscosity 10^{-2} to 10^{5} Pa.s).

4. Results and Discussion

4.1 Rheological Characterization

Iraqi crude oil is tested in the laboratory of Material Engineering College/ Babylon University when it is treated with different water/oil ratio ranged (0, 5, 7.5, 10, and 12.5%) to investigate the effect of water in oil on rheological characterization. Fig. (2) shows that the crude oil density increases as the water/oil ratio increases. On the other hand, the viscosity of the crude oil will be reduced as shown in figure (3).

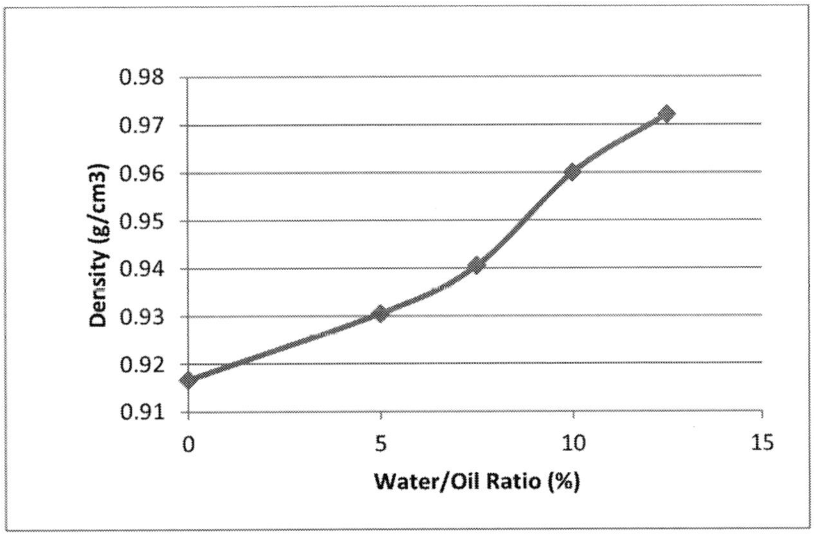

Figure 2. Variation of density with water/oil ratio

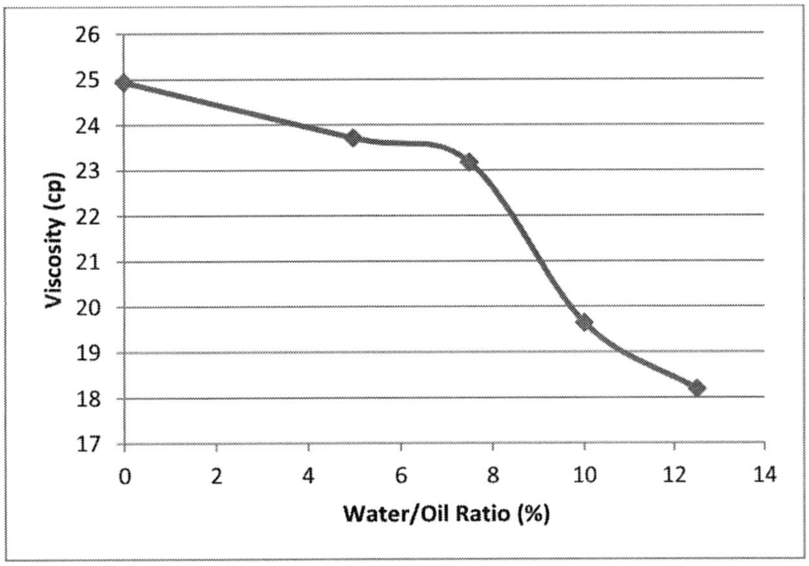

Figure 3. Variation of viscosity with water/oil ratio

Figure (4) shows the behaviour of shear stress and shear rate of the crude oil treated with different water/oil ratio (0, 5, 7.5, 10, and 12.5 %). It is observed that at water/oil ratio of 0 ppm (pure crude oil), almost a linear increase in the shear stress is shown with the shear rate. This ratio exhibits the highest values of the shear stress at all values of the shear rate compared with other ratios. In addition, a considerable increase in the shear stress is obtained by adding water to the crude oil. This increment varies as the water/oil ratio varies and the highest increment (highest shear stress) is observed at 12.5 %. It is also observed that the shear stress is increased as the water/oil ratio is increased, which follow non-Newtonian shear thinning behaviour.

Figure (5) shows the viscosity–shear rate relationship for the crude oil treated with different water/oil ratio (0, 5, 7.5, 10, and 12.5 %) at a constant temperature of 25 °C. From this figure, it has been shown that adding water reduces significantly the viscosity of the crude oil and the reduction depends on water/oil ratio. Consequently, the lowest and highest reduction is obtained at 5 and 12.5 %, respectively. This means that the influence of water in oil increases the pipeline pressure drop which does not refer to its effect on viscosity but it is attributed mainly to its effect in increasing the turbulence energy degree inside the pipeline network, thereby increasing gradually the shear stress which is clearly presented in figure (6).

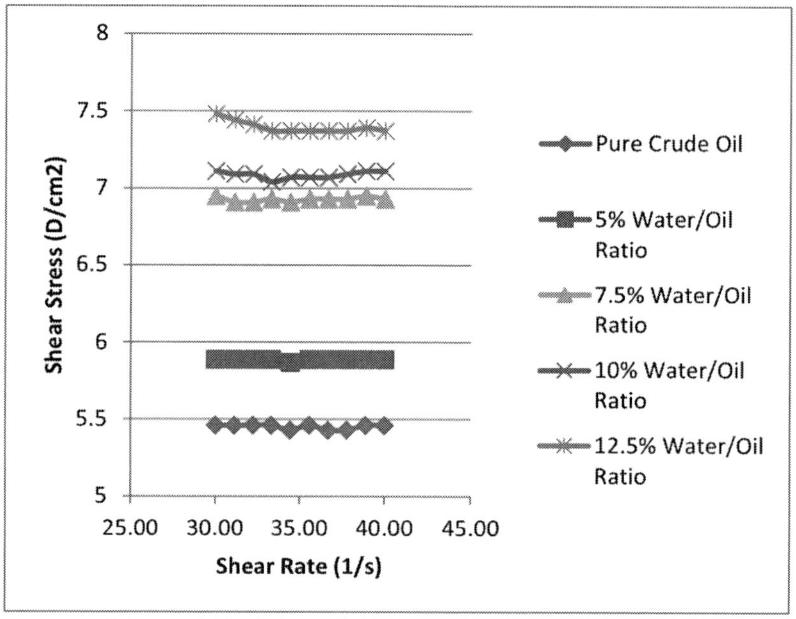

Figure 4. Variation of shear stress with shear rate

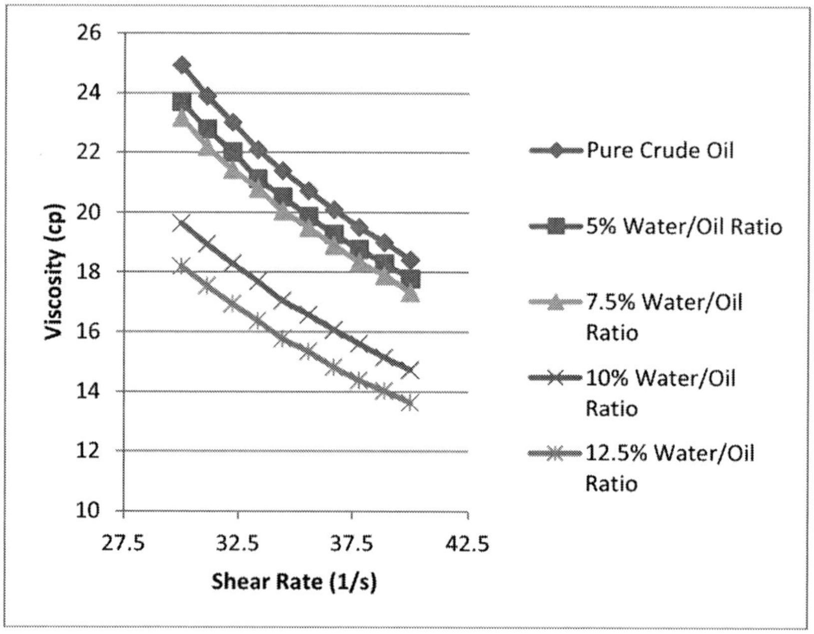

Figure 5. Variation of viscosity with shear rate

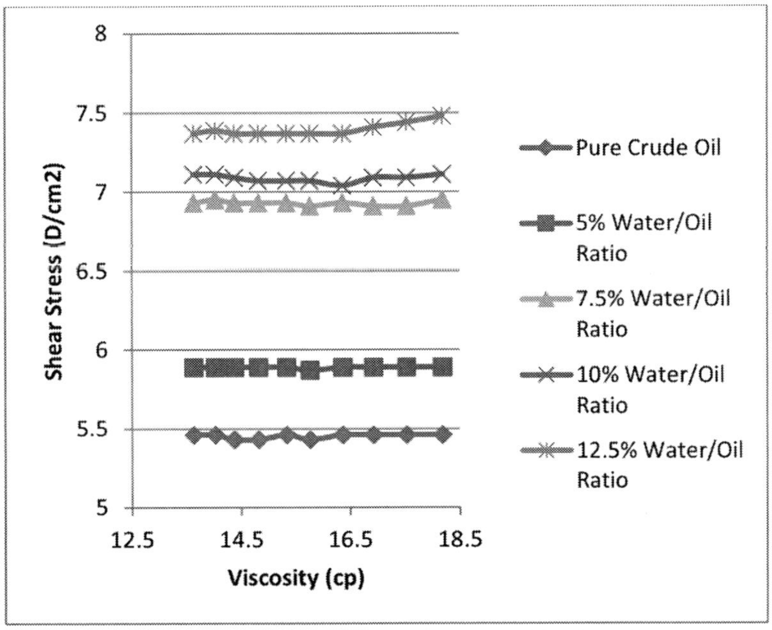

Figure 6. Variation of shear stress with viscosity

4.2 Water-Oil Flow Characterization

Figure (7) shows variation of pressure drop with water/oil ratio. It has been shown that the pressure drop increases with increasing the water/oil ratio. This is due to the fact that the addition of water to oil reduces the viscosity and therefore the Reynolds number will increase which leads to increase velocity, thereby the pressure drop increases. The profile of the present work shows a concave down at water-oil ratio about 0.8, which is the point of inversion.

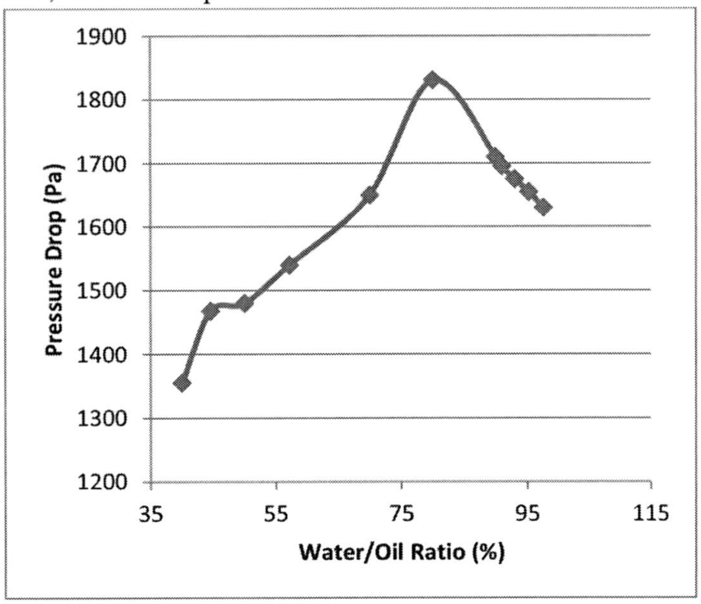

Figure 7. Variation of pressure drop with water/oil ratio for flow rate of (34 l/min.)

Figure (8) indicates the relationship between pump head and water/oil ratio. This curve shows that the pump head increases as water/oil ratio increases, and this is because of decreasing in the density of the two-phase as water/oil ratio increased.

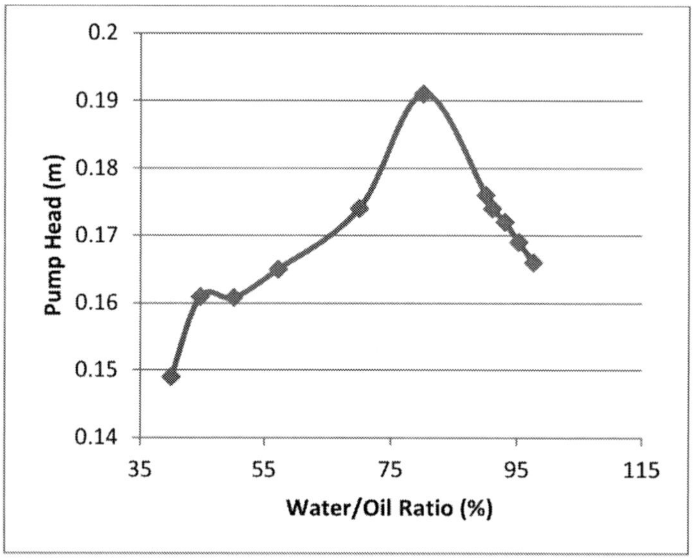

Figure 8. Variation of pump head with water/oil ratio for flow rate of (34 l/min.)

Figure (9) shows the relationship between the volumetric flow rate of the two-phase flow and water/oil ratio. In the previously mentioned figure, the volumetric flow rate of the two-phase flow increases as water/oil ratio increased (i.e. decreasing oil/water ratio). The cause belongs to the decreasing in the viscosity of the mixture as water/oil ratio increased, which leads to increase in the velocity of the two-phase flow.

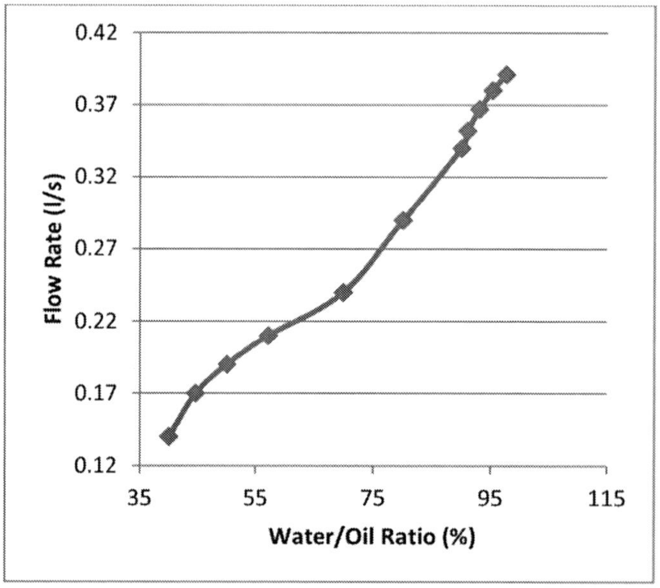

Figure 9. Variation of flow rate with water/oil ratio for flow rate of (34 l/min.)

Figure (10) shows variation of pressure drop with water/oil ratio for different flow rate (35, 25, and 15 l/min.). The increase in two-phase flow velocity leads to increase in pressure drop.

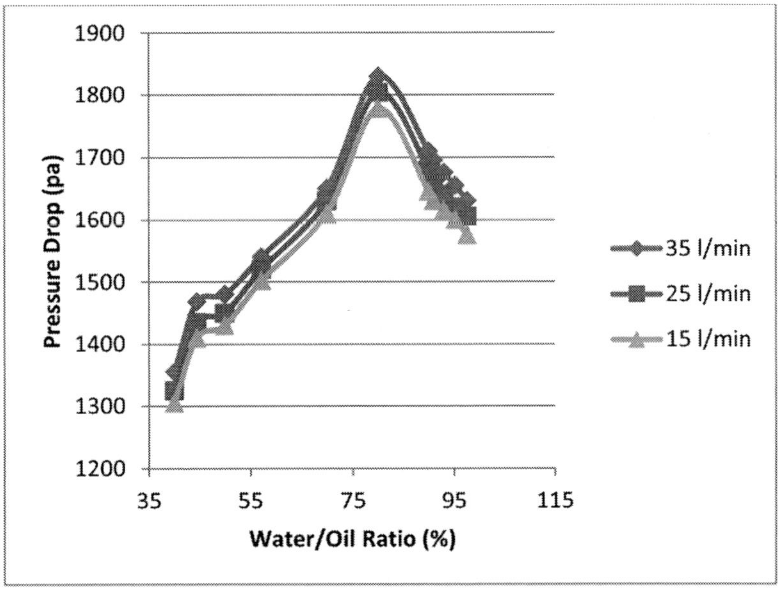

Figure 10. Variation of pressure drop with water/oil ratio

Figure (11) indicates the relationship between pump head and water/oil ratio for different flow rate (35, 25, and 15 l/min.). This curve shows that the pump head increases as water/oil ratio increases, and the increase in two-phase flow velocity cause an increase in pressure drop.

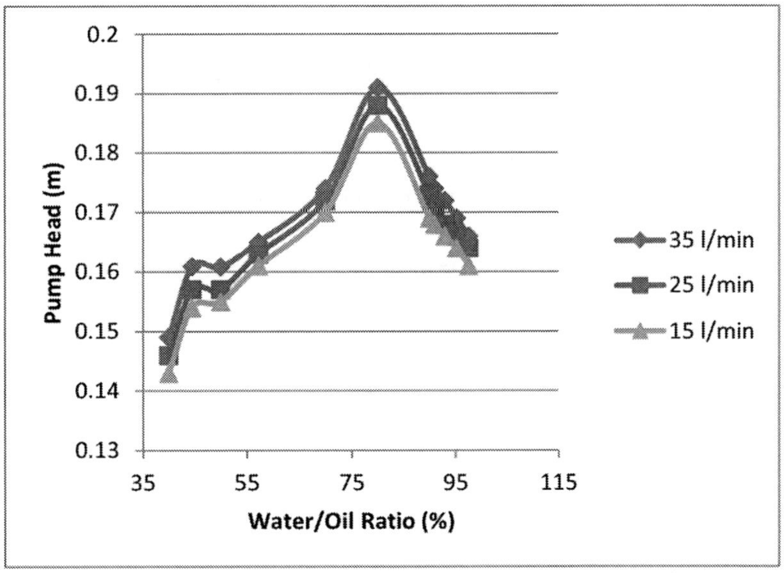

Figure 11. Variation of pump head with water/oil ratio for different flow rate

5. Conclusions

Experiments were done for measuring the rheological and flow characterization of water-crude oil two-phase flow by using a centrifugal pump, including viscosity, shear stress, shear rate, pump head, flow rate, etc.. The following conclusions were obtained:

1.The density of the crude oil increases as the water/oil ratio increases. While the viscosity of the crude oil will be reduced.

2. Adding water to the crude oil causes an increase in the shear stress. This increment varies as the water/oil ratio varies and the highest increment (highest shear stress) is observed at 12.5 %. The basis of concentration selection is 2.5% increment.

3. The two-phase flow rate increases as water/oil ratio increased.

4. For the flow of water-oil, the pump head increases as water/oil ratio increases reaching to 0.8 water/oil ratio, which is the point of inversion, and then reduces.

5. The volumetric flow rate of the two-phase flow increases as water/oil ratio increases.

6. The increase in water-oil flow velocity causes an increase in pump head.

Acknowledgements

We would like to express our deep thanks and respect to all members of (Faculty of Materials Engineering, Polymers and Petrochemical Industries Department/ Babylon University) for their support as well as to Al-Najaf refinery for providing crude oil with its properties.

References

[1] Adam, S. and Schnerr, G.H. 1997 Instabilities and bifurcation of non-equilibrium two-phase flows *J. Fluid Mech.* **348** 1-28

[2] Atkinson, C.M. and Kyt¨omaa, H.K. 1992 Acoustic wave speed and attenuationin suspensions *Int. J. Multiphase Flow* **18** 577-592.

[3] Biesheuvel, A. and Gorissen, W.C.M. 1990 Void fraction disturbances in a uniform bubbly fluid *Int. J. Multiphase Flow* **16** 211-231

[4] Poullikkas, A. 1999 Compressibility and condensation effects when pumping gas- liquid mixtures *J. of Fluid Dynamics Research* **25** 57-62

[5] ME Charles, GW Govier, GW Hodgson 1961 The horizontal pipeline flow of equal density oil-water mixtures *The Canadian Journal of Chemical Engineering* **39** 27-36

[6] XX Xu 2007 Study of oil-water two phase flow in horizontal pipelines *Journal of Petroleum Science and Engineering* **59** 43-58

[7] W Wang, J Gong, P Angeli 2011 Investigation of heavy crude-water two phase flow and related flow characteristics *International Journal of Multiphase Flow* **37** 1156-1164

[8] B Hu, P Angeli 2006 Phase inversion and associated phenomena in oil-water vertical pipeline *The Canadian Journal of Chemical Engineering* **84** 1-14

[9] JY L Lum, T Al Wahaibi, P Angeli 2006 Upward and downward inclination oil-water flows *International Journal of Multiphase Flow* **32** 413-435

[10] K Piela, R Delfos, G Ooms, J Westerweel, RVA Oliemans 2006 Experimental investigation of phase inversion in an oil-water flow through a horizontal pipe loop *International Journal of Multiphase Flow* **32** 1087-1099

[11] K Piela, R Delfos, G Ooms, J Westerweel, RVA Oliemans 2008 On the phase inversion process in an oil-water pipe flow *International Journal of Multiphase Flow* **34** 665-677

[12] AC Bannwart, OMH Rodriguez, CHM de Carvalho, IS Wang, RMO Vara 2004 Flow patterns in heavy crude oil-water flow *Trans ASME* **126** 184-189

[13] J Lovick, P Angeli 2004 Experimental studies on the dual continuous flow pattern in oil-water flows *International Journal of Multiphase Flow* **30** 139-157

[14] F Greifzu, C Kratzsch, T Forgber, F Lindner, R Schwarze 2016 Assessment of particle-tracking models for dispersed particle laden flows implemented in OpenFOAM and ANSYS FLUENT *Engineering Applications of Computational Fluid Mechanics* **10** 30-43
[15] H Pouraria, JK Seo, JK Paik 2016 Numerical modelling of two-phase oil-water patterns in a subsea pipeline *Ocean Engineering* **115** 135-148
[16] Shoulong W. and Aifen L. 2018 Effect of water content and temperature on the rheological behavior of Caoqiao heavy oil *Journal of Petroleum Science and Technology* **36** 1-5

IOP Conf. Series: Materials Science and Engineering **579** (2019) 012054 doi:10.1088/1757-899X/579/1/012054

An Overview on the Recent Techniques for Improving the Flowability of Crude Oil in Pipelines

Raheek I Ibrahim, Manal K Odah, and Amna Al-Mufti

Electromechanical Engineering Department, University of Technology, Baghdad, Iraq

Corresponding author's e-mail address: doctorraheek@yahoo.com

Abstract. Crude oil is considered as a major source of energy all over the world. Midstream refers to the various methods of transportation. Important issues that oil transportation methods face are spills and inadvertent emissions. Therefore, most of the technology developments in transportation methods are aimed at reducing emissions, increasing efficiency, or preventing spills and leaks. Piping systems are the safest and most efficient and cost-effective way for crude oil transportation. The process suffers from serious problems such as: asphaltene and paraffin structures interactions, pipes pressure drop, and high pumping energy consumption. One of the economical important challenges in oil pipeline transportation is to keep the flowability and reduce the pressure drop along the pipe. The current review highlights the recent and most development techniques which have been used to enhance the fluidity of crude oil in pipelines.

1. Introduction

The economic development and the dramatic population growth recorded in recent decades, cause a rising demand for fossil fuels, that result in a gradual decline in the conventional oil reserves, including both light and medium crude oil reserves, which have become scarce and insufficient to meet the continuous increase in energy demand. Unconventional oil reserves, including heavy oil, extra heavy oil, oil sand, tar sands, oil shale and bitumen reserves represent an alternative resource for fossil fuels. However, in contrast with conventional oil, unconventional oil cannot be recovered and subsequently transported in their natural state by the typical production and transportation methods, it generally requires an additional effort to guarantee its flowability in an acceptable flow rate. Therefore, the development of new technologies becomes crucial for the economical production of unconventional oils. Heavy oils are expected to become a great alternative to conventional oil in the future.

2. Fossil Fuel

Fossil fuels are hydrocarbons created from dead animals and plants decomposition, that were sunk to the bottom of the ocean, covered by deposits of sedimentary rock, subjected to intense heat and pressure over millions of years, and then transformed into oil and gas. Hydrocarbons migrate through permeable sedimentary source rock toward the earth surface, trapped beneath impermeable sedimentary cap rock, and then formed oil and gas reservoirs. The origin of oil is shown in (Figure 1). A reservoir is a porous sedimentary rock that has tiny spaces which can hold large quantities of water, gas, and oil. Three different types of fossil fuels were formed, coal, gas, and oil, that depend on the organic material combination, burial depth and time length, and pressure and temperature conditions. About 82% of global energy is supplied by this type of fuel, also it considered as non-renewable energy source, as it has taken millions of years to form [2, 3].

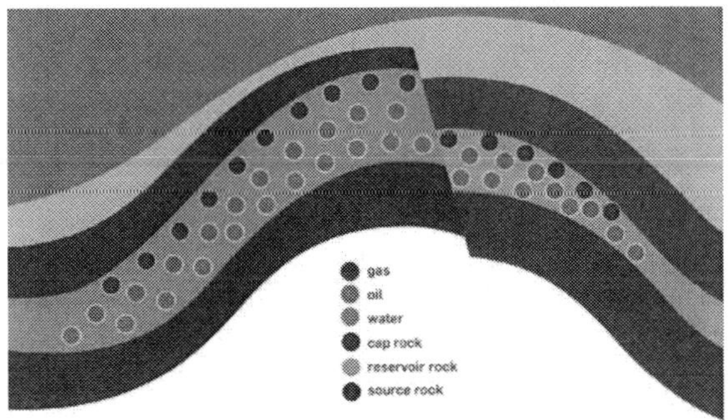

Figure 1. Origin of oil [1].

3. Crude Oil

Crude oil is extracted from a reservoir by drilling a well, then pumping the oil up the well. As soon as the oil is recovered, it is transported to a refinery where it undergoes a complex refining process that produces different petroleum products. Crude oil can be classified into different categories.

Light and heavy crude oil deals with oil density.

Sweet and sour crude oil deals with the sulfur content, sour crude oil has sulfur greater than 0.5%.

Conventional and unconventional crude oil deals with different extraction methods, conventional crude oil is extracted by traditional extraction methods and requires minimal processing techniques after being produced, while for unconventional crude oil, it is extracted by advanced extraction methods and requires additional processing techniques after being produced in order to be transported and refined, since unconventional oil cannot be recovered and transported in its natural state [2-3].

3.1 Heavy Oil

Heavy oil is one form of unconventional oil, that is characterized by high viscosity, high density, high specific gravity, low H/C ratio, as well as high contents of asphaltenes, resins, sulfur, and heavy metals. API (American Petroleum Institute) is a measure of oil lightness and heaviness as:

$$API=\frac{141.5}{SG}-131.5$$

Where specific gravity is the ratio of oil density to density of water at standard conditions [2-3]. The classification of crude oil with respect to viscosity and API is shown in Figure 2.

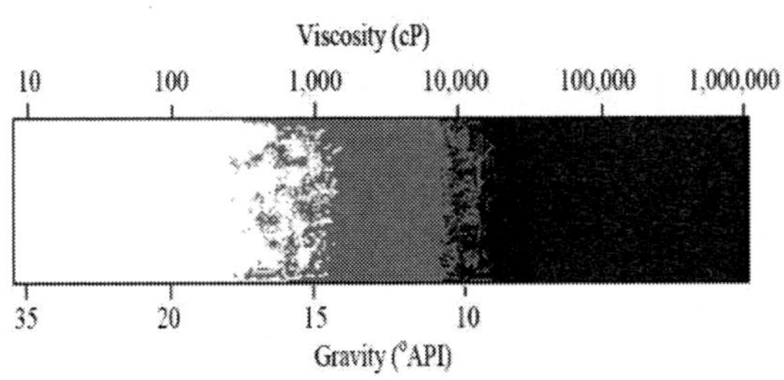

Figure 2: Classifications of crude oil [4].

4. Oil Transport through Pipelines

Midstream refers to the various transportation methods used to move oil from one location to another, that includes tank trucks, rail cars, tanker ships, and pipelines. Crude oil is extracted in remote sites away from where it is stored, processed, or consumed. Therefore, transportation networks have been built to transport crude oil from extraction site to storage facilities or refineries where it is stored or processed, and to transport refined products from refineries to distribution facilities where it is consumed. In the near future, it is expected that these four transportation methods will continue to be used, unless a new transportation method is found. Therefore, most of the technology developments in transportation methods are aimed at increasing efficiency, reducing emissions, or preventing spills and leaks [5-8]. Transportation methods comparison is summarized in Table 1.

4.1 Pipelines

About 70% of all petroleum products are transported by pipelines. Pipelines have several competitive advantages over other forms of oil transportation. They are considered as an expensive technique that has high capital and maintenance costs, and takes a significant time to set up. Moreover, pipelines require a series of pumping stations, usually referred to as booster pumps, situated at various intervals along each route, thus further adding to transportation costs. However, they transport more crude oil and refined products at less cost and require less human capital capacity. Pipelines are eco-friendly and considered as the most cost efficient and the most energy efficient form of oil transportation, that uses less fuel and has less rates of carbon footprint. They perform the best related to spill risks, human fatalities, environmental impacts, and property destruction. Spill risks are rare in relation to the massive volume of oil being transported. Spill risks caused by pipelines are 4.5 times less compared to rails [5-8]. Kirkuk - Ceyhan pipeline of 970 km used to deliver crude oil to Botas terminal in Ceyhan, Turkey, is shown in Figure 3.

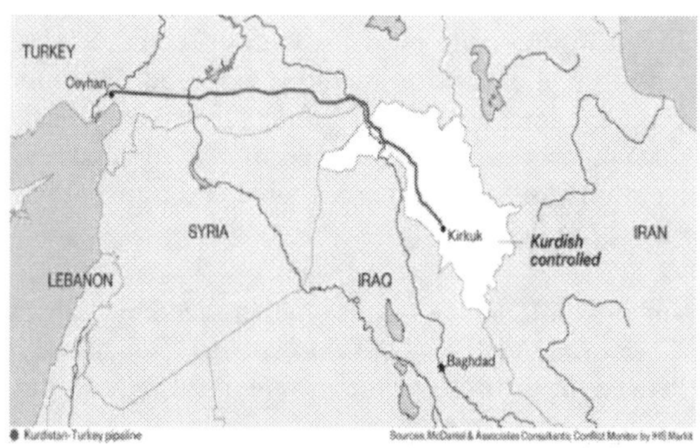

Figure 3: Kirkuk - Ceyhan pipeline [9].

5. Flow Improving Techniques

There are several techniques used to enhance the crude oil transportation via pipelines, the most popular and common techniques are: dilution, heating, core annular flow, drag reduction additives, and electric or magnetic field. The following paragraphs review briefly the flow improving techniques.

5.1 Dilution

Dilution is generally the oldest and most popular techniques to reduce the high viscosity of heavy oil since the 1930s. The basis of this technique is that the addition of less viscous oil to high viscous crude oil decreases the viscosity of the later to an acceptable level for pumping, since the mixture of two oils with different flow properties results in a mixed oil with a flow property between those of the initial oils. Dilution is an effective technique to improve the crude oil flow properties, and to facilitate certain operations such as dehydration and desalting. Dilution requires two pipelines, one for the oil and the other for the diluent. A ratio of 20% to 30% of solvent is sufficient to enhance the crude oil transportation via pipelines. According to [10], dilution requires large amount of diluent to reduce the heavy oil viscosity to an acceptable limit for transportation, that might reach 30%, that causes an increase in the transportation volume, which requires large pipeline capacity. Therefore, it requires substantial investments in pipelines, that results in a cost increase. As explained by [11], dilution causes asphaltenes precipitation, that results in an instability during transportation and storage, which leads to a subsequent pipeline blockage, and therefore requires compatibility testing between the oils. Moreover, the oil mixture has a lower selling price than the diluent, due to the lower quality of the heavy oil. Based on the ideas of [12], dilution requires predetermine the mixing ratio and reliable measurement of mixture viscosity and compatibility, since simple mixing rules do not directly apply, and any change in oil composition may affect the required mixing ratio.

5.1.1 Light crude oil as diluent

On the other hand [13] suggested a recycling of diluents for re-use, due to the increasing scarcity of light oil and the declining reserves of conventional light crude oil. However, it requires installing a diluent recovery unit, that separates the diluents and subsequently returns it to the oil production site, which requires substantial investments to operate an additional pipeline system, and therefore results in a cost increase. Dilution with light oil is less efficient than with condensate in the dilution of heavy oil. According to [14], dilution with light oil has much better results than with alcohol. In an experiment with heavy oil, a dilution with 10% of light oil that has a viscosity of 300 cP, causes a decrease in heavy oil viscosity from 10000 cP to 1200 cP. In an experiment with heavy oil, a dilution with 20% of light oil that has a viscosity of 300 cP, causes a decrease in heavy oil viscosity from 10000 cP to 350 cP. As explained by [15], dilution with light oils is an expensive technique that has low efficiency in the dilution of heavy oil, causes asphaltenes precipitation and requires further heating. In an experiment with heavy oil, a dilution with 30% of light oil, and an increase in the applied temperature from 20 °C to 30 °C, causes a decrease in heavy oil viscosity from 15000 cP to 1000 cP. In an experiment with heavy oil, a dilution with 30% of light oil, and an increase in the applied temperature from 20 °C to 50 °C, causes a decrease in heavy oil viscosity from 15000 cP to 300 cP.

5.1.2 Alcohol as diluent

Based on the ideas of [14], dilution with light oil has much better results than with alcohol. In an experiment with heavy oil, a dilution with 10% of alcohol, causes a decrease in heavy oil viscosity from 10000 cP to 2000 cP. In an experiment with heavy oil, a dilution with 20% of alcohol, causes a decrease in heavy oil viscosity from 10000 cP to 940 cP. According to [16], dilution with alcohol is doubly effective than with kerosene in the dilution of heavy oil.

5.1.3 Naphtha as diluent

As explained by [16], naphtha is a liquid hydrocarbons mixture separated from the crude oil, that has low density, low viscosity, and high API gravity. Dilution with naphtha is an alternative to condensate, that has good compatibility with asphaltenes and high efficiency in heavy oil dilution. A mixture of naphtha and organic solvent causes a decrease in the amount of diluent required to reduce the heavy oil viscosity to an acceptable limit for transportation. Based on the ideas of [17], in an experiment with crude oil, a dilution with naphtha, causes a decrease in crude oil viscosity from 250 cP to 40 cP.

5.1.4 Condensate as diluent

According to [18], condensate is a liquid hydrocarbons mixture separated from natural gas, that has low density, low viscosity, and high API gravity. Dilution with condensate causes asphaltenes precipitation. Condensate availability depends on natural gas. Therefore it is insufficient to meet the increasing heavy oil production needs. As explained by [17], in an experiment with crude oil, a dilution with condensate causes a decrease in crude oil viscosity from 250 cP to 50 cP.

5.1.5 Kerosene as diluent
Based on the ideas of [13], dilution with kerosene is an expensive technique, that requires large amount of diluent to reduce the heavy oil viscosity to an acceptable limit for transportation, ranges from 0% to 20% for heavy oil, and from 25% to 50% for bitumen.

5.2 Heating
Heating is generally the most commonly used technique to reduce the high viscosity of heavy oil carried via pipelines. The basis of this technique is that viscosity decreases with increasing temperature, making it less difficult to pump oil. Heating is an attractive technique to facilitate crude oil flow properties mobility in the pipeline, and to decrease the resistance to flow. According to [19], preheating the heavy oil then subsequent heating the pipeline is required to reduce heavy oil viscosity and to improve heavy oil flow ability. However, the increase in temperature affects the heavy oil macromolecular structures, keeping monomer units scattered, that results in an instable flow. Moreover, heating losses occur along the pipeline to the surroundings during crude oil transportation. Therefore heating requires number of heating stations, that causes an increase in implementation costs. Crude oils with viscosity less than 500 cP can be economically pumped. As explained by [13], pipeline insulation is required to reduce heat losses and to maintain the high temperature at which the oil is extracted at the wellhead. For short distances, pipelines are insulated and crude oils are moved as quickly as possible to minimize temperature losses. However, for longer lengths of pipelines, insulation becomes ineffective. Moreover, temperature variation causes pipeline expansion. Sudden expansion and contraction causes challenging problems. Therefore, for above ground pipelines, expansion loops are used to absorb the expansion, while for buried pipelines, thicker steel is required. Based on the ideas of [10], reheating the heavy oil at pumping stations through direct fired heaters is required to minimize temperature losses and to compensate gradual cooling in the pipeline. A direct-fired heater is generally used to raise the temperature of the oil. Heaters can be natural gas or fuel oil fired. Plugging generally occurs when the pipelines are cooled down to ambient temperature. Therefore displacement oil must be used during start-up and shutdown operations to secure future start-up without difficulty. According to [14], electrically heated subsea pipeline is a difficult technique because of the cooling effect exerted by the seawater, the complexity of the pipeline design, the difficulty of installing, maintaining, and power supplying submerged subsea pumping and heating stations. As explained by [20], in an experiment with medium crude oil that has a density of 0.929 g mL^{-1}, an increase in the applied temperature from 10 °C to 30 °C, causes a decrease in viscosity from 700 cP to 300 cP. Based on the ideas of [11], heated pipeline is an expensive technique that has high capital and operational costs and involves technical considerations regarding pipeline expansion, pipeline internal corrosion, heat loss estimations, number of pumping and heating stations, and the fact that it is necessary to heat the whole pipeline length. On the other hand [21] suggested an electrically heated subsea pipeline of a concentric pipe-in-pipe configuration for offshore pipeline transportation of heavy oil at cold temperature of subsea environment, where the electrical current flows along an inner electrically insulated pipe, which is surrounded concentrically by an outer pipe.

5.3 Core Annular Flow
Core annular flow is a water-oil two-phase flow technique. The basis of this technique is to surround the crude oil and to lubricate the pipeline wall with water, which acts as a lubricant that absorbs the shear stress existing between the oil and the pipeline wall, since the less viscous phase migrates to the high shear region near the pipeline wall. In this regard, the water flows as the annulus while the crude oil flows as the core. A ratio of 10% to 30% of water is sufficient to enhance the crude oil transportation via pipelines. According to [22], oil tends to adhere to the pipeline walls. The gradual adherence of oil leads to a subsequent pipeline blockage, that can be aggravated during a shutdown

operation, where stratification of the oil and water occurs. However, increasing the pressure restores the system. As explained by [11], large density difference between the fluids causes a buoyancy force, that pushes the oil core up to the pipeline wall, which leads to an instable core annular flow. The stable and instable core annular flow is shown in Figure 4.

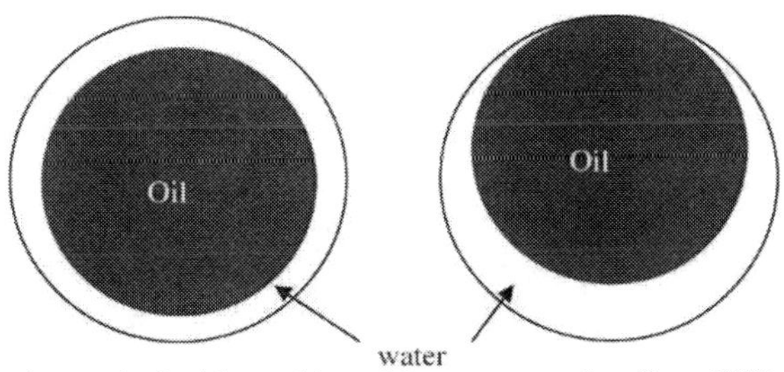

Figure 4: Stable and instable core annular flow [19].

Moreover, velocity difference between the fluids creates waves at the oil-water interface, that result in a wavy core annular flow. However, increasing the velocity enhances the core stability. Based on the ideas of [23], core annular flow causes a pressure drop reduction by 90%, compared to that without water lubrication. According to [24], perfect core annular flow is very rare and may only exist for fluids of possibly similar densities. There have been only two industrial core annular flow examples since 1970, and it seems that no project concerning this method is planned today.

5.4 Drag Reduction Additives

Drag force is defined as the force that must be overcome to drive the fluid through the pipeline during transportation. The basis of this technique is that the drag reduction additive molecules can be stretched out to long chains due to their elastic properties, and form films or layers inside crude oil matrix, that act as shock absorbers, that have the ability to damp and suppress smaller eddies, which are responsible for the drag in turbulent flow, and cause a drag reduction or a decrease in pipeline turbulence, and therefore an increase in flow rate. Drag reduction additives can be classified into three main categories: surfactants, fibers, and polymers. Surfactants reduce friction near the pipeline walls, while fibers and polymers suppress the growth of turbulent eddies within the turbulent fluid core. It was first discovered in 1948. As explained by [25], drag reduction is difficult with heavy oils, due to their high viscosity. The increase in viscosity causes an increase in frictional loss, that results in an inefficient flow and wasted energy, which requires an additional dilution or heating technique. Moreover, the transportation over long distances causes an increase in pressure drop, that requires an application of more pressure to maintain a constant flow rate. However, specifications of pipeline design may limit the amount of pressure required, and therefore causes a decrease in investment costs. Based on the ideas of [26], high molecular weight polymers are by far the most efficient drag reducers, that cause a drag reduction with heavy oils by 28% to 36%, compared to commercial drag reduction additives, that cause no drag reduction with heavy oils. According to [27], a mixture of polymer and surfactant causes more efficient drag reduction than that of additives alone. When the shear stress increases and reaches a critical shear stress value, the drag reduction caused by the surfactant solution alone disappears, while that of polymer remains. As explained by [28], polymers should have high molecular weight, shear degradation resistance, quick solubility in the fluid, and stability against heat and chemical agent. Current generation of drag reduction additives have ultra-high molecular weight polymers, that have long chain of hydrocarbons, which act as an intermediate layer between the fluid and the inner wall of the pipe to reduce energy loss caused by turbulence. However, commercial drag reduction polymers do not perform well with heavy oils having low API gravities and high asphaltene content. Based on the ideas of [29], surfactants cause less drag reduction than polymers. However, after the critical shear stress is decreased to a certain level, surfactants have the ability to recover its microstructures and re-assume its own drag reduction. According to [30], a dosage of 10 ppm of polyisobutylene was required to keep a constant pressure drop. However, determining the dosage required to maintain constant pressure drop is challenging. Moreover, smaller diameter and rougher surface pipes cause an increase in Reynolds number, that cause an increase in drag reduction. As explained by [31], drag reduction additives of polymeric suspensions have a difficulty when dissolved in heavy oils, and a tendency to separate when stored, and degrade when passed through a pump or severe constrictions, that limit their use and reduce its effectiveness dramatically.

5.5 Electric and Magnetic Field

This electrical field technique does not change the crude oil temperature, it temporary aggregates paraffin and asphaltene suspended particles inside the crude oil into short chains along the field direction. The particle aggregation changes the crude oil rheological property and leads to viscosity reduction. The mechanism of electrical field technique is shown by Figure 5.

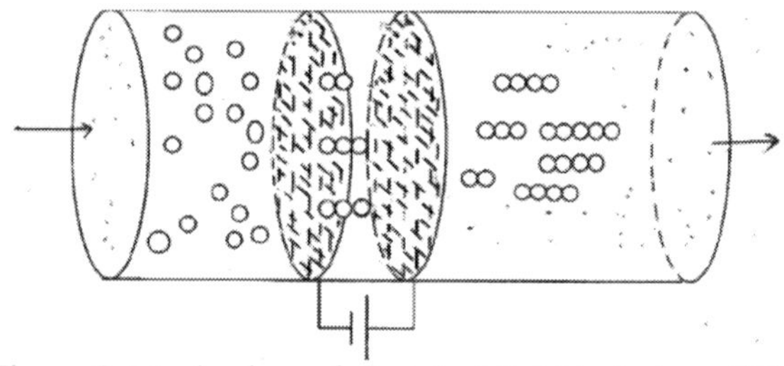

Figure 5: Mechanism of electrical field technique [36].

However, this viscosity reduction is temporary and not permanent, it lasts for hours. Therefore, reapplication of the electric field along the way is needed. It is well-known that there is always some water content inside asphalt-base crude oil, and accidental discharge sparks or dielectric breakdown may occur. However, it does not cause any fire due to the lack of oxygen, since the electrodes are fully immersed inside the crude oil, For the magnetic field technique, it is considered safer than the electric field technique because it does not have that accidental discharge spark issue. Based on the ideas of [32], bunker diesel is the cheapest and heaviest fuel that has high amounts of sulfur and serious pollutants issue, that is obtained after gasoline and diesel distilled away from crude oil. Bunker diesel is any type of fuel oil used aboard ships, that cannot be used on cars and trucks, because it needs to be heated in order to make it flow. In an experiment with bunker diesel fuel, an application of 1.5 kV mm^{-1} of electric field for 18.8 s, causes a decrease in viscosity by 50%, that results in an increase in flow rate by 30%. According to [33], biodiesel is an alternative diesel engines fuel that is produced by a reaction of a vegetable oil or animal fat with an alcohol which is not derived from petroleum, has higher viscosity and less rates of carbon footprint compared to petroleum products. In an experiment with biodiesel, an application of 1.667 kV mm^{-1} of electric field for 10.86 s, causes a decrease in viscosity from 3.65 cP to 2.91 cP, by 25.5%, that results in an increase in flow rate from 1.4 mg s^{-1} to 1.77 mg s^{-1}, by 20%, which leads to a decrease in droplets size by 41%, and therefore causes an increase in engine combustion by 20%. As explained by [34], in the experiment with asphaltene base crude oil, an application of 1.612 kV mm^{-1} of electric field at 23 °C, causes a decrease in viscosity from 715.8 cP to 383.4 cP, an application of 1.704 kV mm^{-1} of electric field at 16 °C, causes a decrease in viscosity from 1821.2 cP to 1022.0 cP, an application of 1.704 kV mm^{-1} of electric field at 12 °C, causes a decrease in viscosity from 2750.5 cP to 1559.2 cP, and an application of 1.612 kV mm^{-1} of electric field at 6 °C, causes a decrease in viscosity from 6003.7 cP to 3594.0 cP. On the other hand [35] used different values of applied voltage, electrodes distance, treatment time, and nanosilica concentration, that ranges between 140-220 V, 2010 cm, 0-60 s, and 0-700 mg L^{-1} respectively. In the experiment with 31.2 °API crude oil, an application of 188 V of voltage for 32 s, at 10 °C, with a distance of 6.11 cm between the electrodes, causes a decrease in viscosity without nanosilica addition from 32.5 cSt to 20.479 cSt, by 37%, that results in an increase in flow rate by 45.6%, and causes a decrease in viscosity with 100 mg L^{-1} of nanosilica addition from 32.5 cSt to 12.8 cSt, by 60.6%, that results in an increase in flow rate by 77.8%. The electrical capacitor design is shown in Figure 6.

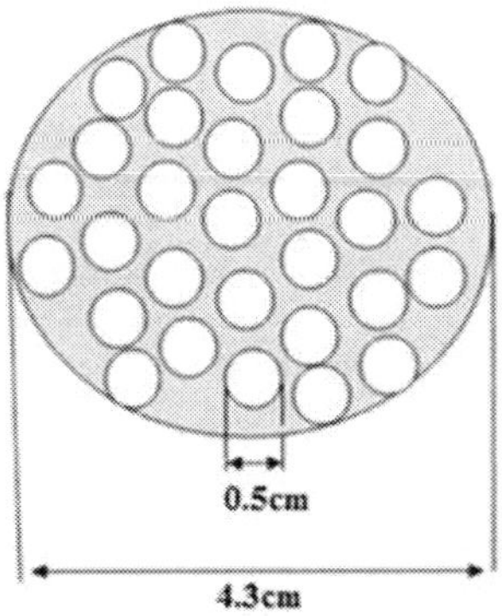

Figure 6: Electrical capacitor designed by [35].

Based on the ideas of [36], in the experiment with 34 °API crude oil, an application of 1.6 kV mm^{-1} of electric field at 1.5 °C, causes a decrease in viscosity from 105.4 cP to 53.9 cP, by 48.9%, that results in an increase in flow rate from 4.95 mg s^{-1} to 10.1 mg s^{-1}, by 104%, an application of 1.6 kV mm^{-1} of electric field at -1.4 °C, causes a decrease in viscosity from 149.9 cP to 56.7 cP, by 62.2%, that results in an increase in flow rate from 3.633 mg s^{-1} to 9.71 mg s^{-1}, by 167%, and an application of 1.6 kV mm^{-1} of electric field at -3.1 °C, causes a decrease in viscosity measured by a capillary tube viscometer from 261.3 cP to 46.7 cP, by 82.1%, that results in an increase in flow rate from 2.08 mg s^{-1} to 11.66 mg s^{-1}, by 460.6%, and causes a decrease in viscosity measured by a Brookfield rotational viscometer from 261.3 cP to 121.1 cP, by 53.7%, that increased to 151.2 cP, by 42.1% after 12 h, that further increased to 172.4 cP, by 34% after 24 h. According to [37], sulfur particles have high dielectric constant. Ultra low sulfur diesel viscosity reduction is much less than that of high sulfur diesel, therefore it needs much stronger electric field. In the experiment with high-sulfur diesel, an application of 1 kV mm^{-1} of electric field for 2 s, causes a decrease in viscosity from 4.6 cP to 4.18 cP, by 10%. However, its viscosity takes more than 60 min to return to its original value. In the experiment with ultra-low-sulfur diesel, an application of 2 kV mm^{-1} of electric field for 5 s, causes a decrease in viscosity by 23%, that results in an increase in flow rate by 30%, which leads to a decrease in droplet size by 41%. However, its viscosity reduction does not last long, its viscosity takes about 4 min to return to its original value. As explained by [38], a decrease in the fuel viscosity, causes a decrease in the fuel droplets size, that results in an increase in the fuel droplets surface area, which leads to a decrease in the incomplete combustion and carbon emissions, and therefore causes a cleaner and more efficient combustion. Diesel fuel viscosity is much higher than that of gasoline, therefore diesel fuel has much bigger droplets size. However, its viscosity takes quite a while to return to its original value. In an experiment with diesel fuel, an application of 1 kV mm^{-1} of electric field for 2 s, causes a decrease in viscosity from 4.6 cP to 4.18 cP, by 9%, that results in an increase in the droplets number from 5.3% to 15.3%. In the experiment with gasoline, an application of 1.2 kV mm^{-1} of electric field, causes an increase in the droplets number from 17.6% to 20.7%, by 18%.

Acknowledgements
This study was carried out on Electromechanical Engineering Departmentequipment of the University of Technology, Baghdad- Iraq.

IOP Conf. Series: Materials Science and Engineering **579** (2019) 012054 doi:10.1088/1757-899X/579/1/012054

References

[1] https://krisenergy.com/company/about-oil-and-gas/origin.html
[2] Hyne N J 2012 *Nontechnical Guide to Petroleum Geology, Exploration, Drilling, and Production* (Tulsa, Okla: PennWell Books) ed. 3
[3] Ala M 2017 An Introduction to Petroleum Geoscience. London: World Scientific Publishing.
[4] Aske N, Kallevik H, Johnsen E E and Sjoblom J 2002 *Energy & Fuels* **16** 1287
[5] Rodrigue J-P, Comtois C, and Slack B 2017 *The Geography of Transport Systems* (New York: Routledge) ed. 4
[6] Hénaut I, Forestiere A, Heraud J P and Argillier J F 2007 *US Patent Application 20070295642*
[7] Argillier J F, Henaut I and Gateau P 2006 *US Patent Application 20060118467*
[8] Van den Bosch P J W M and Schrijvers F A M 2006 *US Patent Application 20060144754*
[9] http://www.digirev.us/kirkuk-map.html
[10] Martínez-Palou R, Mosqueira M L, Zapata-Rendón B, Mar-Juárez E, Bernal-Huicochea C, Clavel-López J C and Aburto J 2011 *Journal of Petroleum Science & Engineering* **75** 274
[11] Sanière A, Henaut I and Argillier J F 2004 *Oil & Gas Science & Technology* **59** 455
[12] Zaman M, Bjorndalen N and Islam M R 2004 *Petroleum Science & Technology* **22** 1119
[13] Santos R G, Loh W, Bannwart A C and Trevisan O V 2014 *Brazilian Journal of Chemical Engineering* **31** 571
[14] Hasan S W, Ghannam M T and Esmail N 2010 *Fuel* **89** 1095
[15] Yaghi B M and Al-Bemani A 2002 *Energy Sources* **24** 93
[16] Gateau P, Henaut I, Barre L and Argillier J F 2004 *Oil & Gas Science & Technology* **59** 503
[17] Myers R D, MacLeod J B, Ghosh M and Chakrabarty T 2000 *US Patent Application 6096192*
[18] Shigemoto N, Al-Maamari R S, Jibril B Y and Hirayama A 2006 *Energy & Fuels* **20** 2504
[19] Hart A 2014 *Journal of Petroleum Exploration & Production Technology* **4** 327
[20] Ghannam M T and Esmail N 2006 *Petroleum Science & Technology* **24** 985
[21] Langner C G and Bass R M 2001 *US Patent Application 6264401*
[22] Wylde J J, Leinweber D, Low D, Botthof G, Oliveira A P, Royle C and Kayser C 2012 Heavy Oil Transportation: Advances in Water Continuous Emulsion Methods. In: Proceedings of the world heavy oil congress, Aberdeen.
[23] Bensakhria A, Peysson Y and Antonini G 2004 *Oil & Gas Science & Technology* **59** 523
[24] Bannwart A C 2001 *Journal of Petroleum Science & Engineering* **32** 127
[25] Johnston R, Lauzon P and Pierce J 2008 *Hydrocarbon Engineering* **13** 89
[26] Milligan S N, Johnston R L, Burden T L, Dreher W R, Smith K W and Harris W F 2008 *US Patent Application 20080149530*
[27] Matras Z, Malcher T and Gzyl-Malcher B 2008 *Thin Solid Films* **516** 8848
[28] Peysson Y, Bensakhria A, Antonini G and Argillier J F 2007 *SPE Production & Operations* **22** 135
[29] Zhou T, Leong K C and Yeo K H 2006 *International Journal of Heat & Mass Transfer* **49** 1462
[30] Mowla D and Naderi A 2006 *Chemical Engineering Science* **61** 1549
[31] Harris W F, Smith K W, Milligan S N, Johnston R L and Anderson V 2006 *US Patent Application 20060148928*
[32] Du E, Xu X, Huang K, Tang H and Tao R 2018 *International Journal of Modern Physics B* **32**
[33] Du E, Cai L, Huang K, Tang H, Xu X and Tao R 2018 *Fuel* **211** 194
[34] Du E, Zhao Q, Xiao Y, Cai L and Tao R 2018 *Fuel* **220** 358
[35] Ibrahim R I, Oudah M K and Hassan A F 2017 *Journal of Petroleum Science & Engineering* **156** 356
[36] Tao R and Tang H 2014 *Fuel* **118** 69
[37] Du E, Tang H, Huang K and Tao R 2011 *Journal of Intelligent Material Systems & Structures* **22** 1713
[38] Tao R, Huang K, Tang H and Bell D 2008 *Energy & Fuels* **22**, 3785

Institute of Physics
Dirac House, Temple Back
Bristol BS1 6BE UK

ISSN: 1757-8981
ISBN 978-1-5108-9450-1